Classics in Mathematics

Robert M. Switzer

Algebraic Topology–Homotopy and Homology

Springer
Berlin
Heidelberg
New York
Barcelona
Hong Kong
London
Milan
Paris
Tokyo

Robert M. Switzer

Algebraic Topology – Homotopy and Homology

Reprint of the 1975 Edition

Robert M. Switzer
Georg-August-Universität
Mathematisches Institut
Bunsenstrasse 3–5
37073 Göttingen
Germany
e-mail: switzer@math.uni-goettingen.de

Originally published as Vol. 212 of the
Grundlehren der mathematischen Wissenschaften

Cataloging-in-Publication Data applied for

Die Deutsche Bibliothek - CIP-Einheitsaufnahme
Switzer, Robert M.:
Algebraic topology : homotopy and homology / Robert M. Switzer. - Reprint of the 1975 ed.. - Berlin;
Heidelberg; New York; Barcelon; Hong Kong; London; Milan; Paris; Tokyo: Springer, 2002
(Classics in mathematics)
ISBN 3-540-42750-3

Mathematics Subject Classification (2000): Primary 55-02;
Secondary 57A55, 57A65, 57D75

ISSN 1431-0821
ISBN 3-540-42750-3 Springer-Verlag Berlin Heidelberg New York

Springer-Verlag Berlin Heidelberg New York
a member of BertelsmannSpringer Science+Business Media GmbH

http://www.springer.de

Printed in Germany

Printed on acid-free paper SPIN 10855114 41/3142ck-5 4 3 2 1 0

Robert M. Switzer

Algebraic Topology— Homotopy and Homology

Springer-Verlag
Berlin Heidelberg New York 1975

Robert M. Switzer
Wissenschaftlicher Rat and Professor, Mathematisches Institut der Universität Göttingen

AMS Subject Classification (1970): Primary 55-02;
Secondary 57A55, 57A65, 57D75

ISBN 3-540-06758-2 Springer-Verlag Berlin Heidelberg New York
ISBN 0-387-06758-2 Springer-Verlag New York Heidelberg Berlin

Library of Congress Cataloging in Publication Data
Switzer, Robert M. 1940–. Algebraic topology—homotopy and homology.
(Die Grundlehren der mathematischen Wissenschaften in Einzeldarstellungen mit besonderer Berücksichtigung der Anwendungsgebiete; Bd. 212)
"This book is the result of lecture courses... given by the author at the University of Manchester in 1967–1970, at Cornell University in 1970–1971, and at the Georg August University, Göttingen, in 1971–1972" Bibliography: p. Includes index.
1. Algebraic topology. 2. Homotopy theory. 3. Homology theory. I. Title. II. Series: Die Grundlehren der mathematischen Wissenschaften in Einzeldarstellungen; Bd. 212.
QA612.S9. 513'.23. 74-22378.

Typesetting: William Clowes & Sons Ltd., Colchester, Essex/GB. Printing and bookbinding: Konrad Triltsch, Würzburg.

To Karen, Waltraut and Elisabeth

Introduction

This book is the result of lecture courses on algebraic topology given by the author at the University of Manchester in 1967–1970, at Cornell University in 1970–1971 and at the Georg August University, Göttingen, in 1971–1972. The level of the material is more advanced than that of a first-year graduate course in algebraic topology; it is assumed that the student has already had a course on basic algebraic topology which included singular homology, the fundamental group and covering spaces. Moreover, a student who has never encountered differentiable manifolds will probably have difficulty with Chapter 12. On the other hand no knowledge of homotopy theory beyond the fundamental group is assumed.

The last few years have seen the publication of several excellent text-books on basic algebraic topology, most notably the book by Spanier [80], which I suggest as a companion volume to this one. There is a certain over-lap between Spanier's book and this text—particularly in Chapters 0–6, 14 and 15—but the present book goes considerably further and has as its goal that the reader should be brought to a point from which he could begin research in certain areas of algebraic topology: stable homotopy theory, *K*-theory, cobordism theories.

Despite the title "Algebraic Topology" this book does not (and could not) pretend to achieve the same very advanced level in all areas of this subject. The choice of topics to be emphasized is, of course, heavily influenced by the research interests of the author. Thus, for example, unstable homotopy theory is only developed to the point at which it really begins to be interesting and is then dropped in favor of stable homotopy theory. The reader who finds that his appetite for unstable homotopy theory has been whetted is advised to follow the signposts set up by Adams in [10]. Another important branch of algebraic topology which is omitted is obstruction theory—partly due to lack of space, partly because one could scarcely give a better introduction than Thomas' *Seminar on fibre spaces* [86].

The following basic idea occurs repeatedly as a leitmotiv in this text: the majority of the problems which have been solved by means of algebraic topology have first been reduced to the question of the existence or non-existence of a continuous function $f: X \to Y$ between two given topological spaces X, Y. One tries to prove the non-existence, for example, by finding an appropriate functor F from the category of topological spaces to some algebraic category—that is for every space X we are given $F(X)$, a group, ring, module, ..., and for every continuous function $f: X \to Y$ we have $F(f): F(X) \to F(Y)$, a homomorphism preserving the given algebraic structure. Then one seeks to demonstrate that the algebraic map $F(f): F(X) \to F(Y)$ cannot possibly exist. (Proofs of existence are in general more difficult to handle). From this point of view the richer the natural algebraic structure on $F(X)$ the better: if $F(X)$, $F(Y)$ have a very complex algebraic structure, then there will not be many homomorphisms $\phi: F(X) \to F(Y)$ preserving this structure, and thus the chances of showing $F(f)$ cannot exist are good. At several points, then, (cf. Chapters 2, 13, 17, 18) we shall strive to enrich the natural algebraic structure available on our functors. In Chapter 19 we make the happy discovery that we have a sufficiently complex natural algebraic structure on $F(X)$ that we can (under favorable circumstances) say precisely which algebraic maps $\phi: F(X) \to F(Y)$ are of the form $\phi = F(f)$ for some continuous function $f: X \to Y$. At this point existence proofs become possible.

Chapters 0 and 1 contain respectively certain results from set-theoretic topology which are repeatedly used in the text and the basic definitions of category theory; both chapters should be in the nature of a review for the reader. Chapter 2 takes up the sets $[X, Y]$ of homotopy classes of maps $f: X \to Y$ and deals with such questions as: under what conditions on X or Y is $[X, Y]$ a group, when is a sequence

$$[X, W] \xleftarrow{f^*} [Y, W] \xleftarrow{g^*} [Z, W]$$

exact, etc. (enrichment of structure!). In Chapter 3 we then specialize to $X = S^n$ and consider $\pi_n(Y, y_0) = [S^n, s_0; Y, y_0]$, which is always a group of $n \geqslant 1$—the *nth homotopy group* of Y. The more elementary properties of these groups are demonstrated in this chapter. In Chapter 4 we define the notions of *fibration* and *weak fibration* and show that for a weak fibration $p: E \to B$ with fibre $F = p^{-1}(b_0)$ there is an exact sequence

$$\cdots \longrightarrow \pi_n(F, e_0) \xrightarrow{i_*} \pi_n(E, e_0) \xrightarrow{p_*} \pi_n(B, b_0) \xrightarrow{\partial} \pi_{n-1}(F, e_0) \longrightarrow \cdots.$$

We define the geometrically very important notion of a *fibre bundle* and show that every fibre bundle is a weak fibration. There follow some important examples of fibre bundles:

$$O(n) \to O(n)/O(n-k), \quad O(n)/O(n-k) \to O(n)/O(n-k) \times O(k)$$

and others. The chapter is concluded by remarking that a covering $p:X' \to X$ is a fibre bundle with discrete fibre and by using this remark to compute the homotopy groups of S^1 and $T^n = S^1 \times S^1 \times \cdots \times S^1$.

Proofs of deeper results for arbitrary topological spaces X, Y are difficult; it is not easy to demonstrate, for example, the existence of continuous functions $f:X \to Y$. We turn therefore to the smaller category of CW-complexes. Since CW-complexes are built up by glueing cells D^n together, it is possible to construct maps and homotopies cell by cell. This property permits strong statements about $[X, Y]$ when X or Y is a CW-complex. In Chapter 5 we define CW-complexes and prove some straightforward properties. Chapter 6 contains some deeper homotopy results, such as: $\pi_n(X, x_0)$ depends only on the cells of dimension at most $n+1$; the suspension homomorphism

$$\Sigma: \pi_q(X, x_0) \to \pi_{q+1}(SX, *) \qquad ([f] \mapsto [1 \wedge f])$$

is an isomorphism for $q < 2n+1$ if X is an n-connected CW-complex; a map $f:X \to Y$ between CW-complexes induces an isomorphism $f_*: \pi_q(X, x_0) \to \pi_q(Y, y_0)$, $q \geqslant 0$, if and only if f is a homotopy equivalence.

At this point we turn from homotopy theory to homology and cohomology theories. A *generalized homology theory* is a family of functors $\{h_n : n \in \mathbb{Z}\}$ from the category of topological spaces to the category of abelian groups which satisfies the first six of the seven Eilenberg-Steenrod axioms (i.e. the "Dimension Axiom" is not necessarily satisfied). Chapter 7 is an investigation of the properties of such theories which follow directly from the axioms; this amounts to carrying out such parts of the program of Eilenberg and Steenrod [40] as still go through without the seventh axiom.

Chapter 8 contains a construction of Boardman's stable category of spectra and demonstrations of some of its most important properties—in particular that $\Sigma: [E, F] \to [E \wedge S^1, F \wedge S^1]$ is a bijection for all spectra E, F. We show how to construct a homology theory E_* and a cohomology theory E^* for every spectrum E. Then in Chapter 9 we prove that we have already constructed all possible cohomology theories in Chapter 8 (Brown's representation theorem). Then come the three most important known examples of homology and cohomology theories: ordinary homology (Ch. 10), K-theory (Ch. 11) and bordism (Ch. 12). In Chapter 10 we also show how to compute the singular homology groups of an arbitrary CW-complex and prove the Hurewicz isomorphism theorem. Chapter 11 contains the computation of the homotopy groups of the stable groups O, U, Sp.

Next comes a chapter on products in homology and cohomology. Chapter 13 begins with the universal coefficient theorems (how does one express $H_*(X; G)$ and $H^*(X; G)$ in terms of $H_*(X; \mathbb{Z})$ and G?) and the

Künneth theorem (how does one express $H_*(X \times Y)$ in terms of $H_*(X)$, $H_*(Y)$?). Next the $\times$-, $\cup$- and $\cap$-products for singular homology and cohomology are briefly discussed. Then we make a digression in order to construct the smash product of two spectra, whereupon we can describe products in the generalized homology and cohomology theories E_*, E^* associated to a *ring spectrum* E. We describe explicitly the products for ordinary homology, K-theory and bordism.

Chapter 14 then applies what we have learned about products to investigations of duality (Alexander, Lefschetz, Poincaré for manifolds; Spanier-Whitehead for finite spectra) and to questions of orientability of manifolds with respect to generalized cohomology theories. Spanier-Whitehead duality also permits a proof of a representation theorem for homology theories similar to the one for cohomology theories in Chapter 9.

In Chapter 15 the level of difficulty increases with the introduction of spectral sequences. Everyone finds spectral sequences baffling at the first encounter. Experience, however, shows that spectral sequences are among the topologist's most effective tools, so that the effort required to master their use is well worth while. We develop the Atiyah–Hirzebruch–Whitehead and Leray–Serre spectral sequences and make some important applications of them: Gysin, Wang and Serre exact sequences, Leray–Hirsch theorem, Thom isomorphism theorem.

Chapter 16 is concerned with the calculation of the homology and cohomology rings $E_*(BG)$, $E^*(BG)$ of the stable classifying spaces BG, $G = O$, U and Sp. Here the Atiyah–Hirzebruch–Whitehead spectral sequence proves very useful. In the process we construct certain classes $c_i \in E^*(BG)$ with whose help we can form invariants $c_i(\xi) \in E^*(X)$ of isomorphism classes of G-vector bundles $\xi \to X$—the so-called *characteristic* classes. The chapter ends with a proof of the Bott periodicity theorem $BU \simeq \Omega_0^2 BU$.

Chapter 17 represents a large step in our natural-algebraic-structure-enrichment program: we show how to make $E^*(X)$ into a module over an algebra $A^*(E) = E^*(E)$ and $E_*(X)$ into a comodule over a Hopf algebra $A_*(E) = E_*(E)$ in a natural way (under favorable conditions on E). We compute the Hopf algebras $A_*(E)$ for $E = MU$, MSp, K and KO and find they have satisfyingly complex algebraic structures (recall our leitmotiv). As an application we find bounds on the image of the Hurewicz homomorphisms

$$h_K : \pi_q(MU) \to K_q(MU)$$
$$h_{KO} : \pi_q(MSp) \to KO_q(MSp)$$

for small values of q.

Chapter 18 contains a determination of the algebras $A^*(H(\mathbb{Z}_2))$ and $A_*(H(\mathbb{Z}_2))$ including the construction of certain elements $Sq^i \in A^*(H(\mathbb{Z}_2))$. The analogous results for $A_*(H(\mathbb{Z}_p))$, $A^*(H(\mathbb{Z}_p))$ are stated without proof.

In Chapter 19 we enjoy the triumph of our natural-algebraic-structure-enrichment program: using the comodule structure of $E_*(X)$ over $E_*(E)$ constructed in Chapter 17 we can build a spectral sequence (the *Adams spectral sequence*) $\{E_r^{st}, d_r\}$ whose E_2-term is the purely algebraic construct

$$\mathrm{Ext}_{E_*(E)}^{s,t}(\tilde{E}_*(S^0), E_*(X))$$

($\mathrm{Ext}_C^{**}(-,-)$ is a functor derived from $\mathrm{Hom}_C^*(-,-)$, C a coalgebra, and thus has to do with algebraic maps) and which converges (for connected E) to $\pi_*^S(X)/D_*$, the stable homotopy of X modulo a certain subgroup D_* which depends on E. For $E = MU$ or MSp we show that $D_* = 0$; for $E = H(\mathbb{Z}_p)$ D_* is the subgroup of elements of finite order prime to p. Thus in cases where the spectral sequence proves manageable one can start from a knowledge of $E_*(X)$ as $E_*(E)$-comodule and compute $\pi_*^S(X)/D_*$. We show how the Adams spectral sequence has permitted the determination of $\pi_q^S(S^0)$ for small q. We then turn to a consideration of the spectral sequence for $E = K$, KO. Here E is not connected, so the spectral sequence may not converge, but it still provides homomorphisms

$$e_C : \ker k_K \to \mathrm{Ext}_{K_*(K)}^{1,q+1}(\tilde{K}_*(S^0), K_*(X))$$
$$e_R : \ker h_{KO} \to \mathrm{Ext}_{KO_*(KO)}^{1,q+1}(\widetilde{KO}_*(S^0), KO_*(X)).$$

The Ext-groups for $X = S^0$ are computed and the result is used to localize a non-trivial direct summand in $\pi_*^S(S^0)$ whose order is related to the Bernoulli numbers.

Chapter 20 then represents an extended application of the Adams spectral sequence. The cases $E = H(\mathbb{Z}_p)$ and $X = MG$, $G = O$, SO, or U are, fortunately, of the sort in which the Adams spectral sequence is manageable and permits a complete determination of $\pi_*(X) = \pi_*(MG) \cong \Omega_*^G$, the G-cobordism ring. We also prove the theorem of Hattori and Stong, which describes the image of

$$h_K : \pi_*(MU) \to K_*(MU).$$

After reading Chapter 20 the student should be able to understand without undue difficulty the papers [13, 14] of Anderson, Brown and Peterson in which Ω_*^{SU} and Ω_*^{Spin} are determined.

This summary undoubtedly makes clear that the level of mathematical expertise demanded of the reader rises rather markedly from Chapter 2, say, to Chapter 20. The student who begins with the minimal prerequisites described in the first paragraph will not acquire the facility needed for understanding the later chapters merely by reading straight through. He must try to master the material in each chapter to such an extent that he can apply it to the solution of problems other than those worked out in the text. In some cases he will find it valuable to seek further applications in the books and articles listed at the end of each chapter.

The bibliography included here does not attempt to be comprehensive. Steenrod's valuable compendium of all mathematical reviews having to do with topology makes such a comprehensive bibliography unnecessary. Instead this bibliography has two goals: (1) to suggest to the student where he might begin to pursue a given topic further and (2) to acknowledge the sources from which much of the material in this text is drawn.

In addition, however, I wish to acknowledge quite explicitly and with gratitude my debt to Frank Adams. It is no exaggeration to say that most of what I know about algebraic topology I learned from him. Anyone familiar with his work will recognize the influence of his way of looking at algebraic topology on the presentation of the subject given here. Moreover, on certain topics (e.g. Chapter 8, Chapter 9 in the case of finite CW-complexes, the construction of $E \wedge F$) I have largely reproduced his presentation of the topic with only small alterations (which may not have been for the better).

I further wish to express my deep gratitude to Egbert Brieskorn for his encouragement and helpful suggestions. My thanks also go to Fräulein Ingrid Sochaczewsky and Frau Christiane Preywisch for help in typing the manuscript.

The manuscript of this book was submitted to the Georg August University, Göttingen, in January 1973 as Habilitationsschrift.

Göttingen, November, 1973 ROBERT M. SWITZER

Table of Contents

Chapter 0

Some Facts from General Topology

It is assumed that the reader is familiar with the elements of general topology—e.g. the most important properties of continuous functions, compact sets, connected sets, etc. Nevertheless, certain general topological results which will be used repeatedly in this book are assembled here for the reader's convenience.

0.1. Let X be a topological space and $A_1, A_2, \dots, A_n$ be closed subspaces such that $X = \bigcup_{i=1}^{n} A_i$.

Suppose $f_i: A_i \to Y$ is a function, $1 \leqslant i \leqslant n$; there is a function $f: X \to Y$ such that $f|A_i = f_i$, $1 \leqslant i \leqslant n$, if and only if $f_i|A_i \cap A_j = f_j|A_i \cap A_j$, $1 \leqslant i \leqslant n$, $1 \leqslant j \leqslant n$. In this case f is continuous if and only if each f_i is.

We shall often set about defining a continuous function $f: X \to Y$ by cutting up X into closed subsets A_i and defining f on each A_i separately in such a way that $f|A_i$ is obviously continuous; we then have only to check that the different definitions agree on the "overlaps" $A_i \cap A_j$.

0.2. *The universal property of the cartesian product:* let $p_X: X \times Y \to X$, $p_Y: X \times Y \to Y$ be the projections onto the first and second factors respectively. Given any pair of functions $f: Z \to X$ and $g: Z \to Y$ there is a unique function $h: Z \to X \times Y$ such that $p_X \circ h = f$, $p_Y \circ h = g$. h is continuous if and only if both f and g are. This property characterizes $X \times Y$ up to homeomorphism. The unique h will often be denoted by (f, g).

In particular, to check that a given function $h: Z \to X \times Y$ is continuous it will suffice to check that $p_X \circ h$ and $p_Y \circ h$ are continuous.

0.3. *The universal property of the quotient:* let α be an equivalence relation on a topological space X, let X/α denote the space of equivalence classes and $p_\alpha: X \to X/\alpha$ the natural projection. Given a function $f: X \to Y$, there is a function $f': X/\alpha \to Y$ with $f' \circ p_\alpha = f$ if and only if $x\alpha x'$ implies $f(x) = f(x')$. In this case f' is continuous if and only if f is. This property characterizes X/α up to homeomorphism.

An important example of a quotient which we frequently encounter is that of a space X with a closed subspace A collapsed to a point. Explicitly, if $A \subset X$ is a closed non-empty subspace, we take the relation

$$\alpha = A \times A \cup \{(x,x): x \in X\} \subset X \times X$$

and let $X/A = X/\alpha$. For $A = \varnothing$ we adopt the convention that $X/A = X/\varnothing = X^+ = X \cup \{\text{a disjoint point}\}$. This convention will be seen to be justified by the fact that every theorem we state about X/A will hold equally well for $X/\varnothing$.

0.4. *Product and quotient combined:* if α is an equivalence relation on a topological space X and β is an equivalence relation on Y, then there is an obvious equivalence relation $\alpha \times \beta$ on $X \times Y$:

$$(x,y)\ \alpha \times \beta\ (x',y') \Leftrightarrow x\,\alpha\,x' \quad \text{and} \quad y\,\beta\,y'.$$

There is a unique continuous function

$$\phi: \frac{X \times Y}{\alpha \times \beta} \to (X/\alpha) \times (Y/\beta),$$

such that $\phi \circ p_{\alpha\times\beta} = p_\alpha \times p_\beta$ and which is even bijective—but not necessarily a homeomorphism. One important case in which ϕ is a homeomorphism is that where β is the identity relation 1 and Y is locally compact.

Proof: We must prove that if $T \subset \dfrac{X \times Y}{\alpha \times 1}$ is open then $\phi(T)$ is open in $(X/\alpha) \times Y$. T open means $T' = p_{\alpha\times 1}^{-1}(T)$ is open in $X \times Y$. Suppose $(x_0, y_0) \in \phi(T)$ and choose $x_0 \in X$ with $p_\alpha(x_0') = x_0$. By the definitions of ϕ and T', we have $(x_0', y_0) \in T'$; and since T' is open, there exist neighborhoods U of x_0' and K of y_0 in X, Y respectively such that $U \times K \subset T'$. We may assume K is compact.

Let $J' = \{x' \in X : x' \times K \subset T'\}$. We first show that J' is open. For every $x' \in J'$ we can, because K is compact, find an open neighborhood $U_{x'}$ of x' such that $U_{x'} \times K \subset T'$, which proves J' is open. If $J = \{x \in X/\alpha : x \times K \subset \phi(T)\}$, then clearly $p_\alpha^{-1}(J) = J'$, so J is open in X/α. Also $(x_0, y_0) \in J \times K \subset \phi(T)$, so $\phi(T)$ is open. ▯

Example. The unit interval $I = [0,1]$ is compact, so $\dfrac{X \times I}{\alpha \times 1} \cong (X/\alpha) \times I$.

0.5. A *homotopy* from X to Y is a continuous function $F: X \times I \to Y$. For each $t \in I$ one has $F_t: X \to Y$ defined by $F_t(x) = F(x,t)$ for all $x \in X$. The functions F_t are called the "stages" of the homotopy. If $f,g: X \to Y$ are two maps, we say f is *homotopic* to g (and write $f \simeq g$) if there is a homotopy $F: X \times I \to Y$ such that $F_0 = f$ and $F_1 = g$. In other words, f can be continuously deformed into g through the stages F_t. If $A \subset X$ is a subspace, then F is a homotopy *relative to A* if $F(a,t) = F(a,0)$, all $a \in A$, $t \in I$.

0.6. *The relation $\simeq$ is an equivalence relation.*

Proof: $f \simeq f$ is obvious; take $F(x,t) = f(x)$, all $x \in X$, $t \in I$. If $f \simeq g$ and F is a homotopy from f to g, then $G: X \times I \to Y$ defined by $G(x,t) = F(x, 1-t)$, $x \in X$, $t \in I$, is a homotopy from g to f—i.e. $g \simeq f$. If $f \simeq g$ with homotopy F and $g \simeq h$ with homotopy G, then $f \simeq h$ with homotopy H defined by

$$H(x,t) = \begin{cases} F(x,2t) & 0 \leqslant t \leqslant 1/2 \\ G(x,2t-1) & 1/2 \leqslant t \leqslant 1. \end{cases}$$

Note that here we use 0.1 to show H is continuous. ▯

0.7. Thus the set of all continuous functions $f: X \to Y$ is partitioned into equivalence classes under the relation $\simeq$. The equivalence classes are called *homotopy classes*, and the set of all homotopy classes is denoted by $[X; Y]$. If $f: X \to Y$, then the homotopy class of f is denoted by $[f]$.

As an application of 0.4 we get the following proposition.

0.8. Proposition. *If α is an equivalence relation on a topological space X and $F: X \times I \to Y$ is a homotopy such that each stage F_t factors through X/α—i.e. $x\alpha x' \Rightarrow F_t(x) = F_t(x')$—then F induces a homotopy $F': (X/\alpha) \times I \to Y$ such that $F' \circ (p_\alpha \times 1) = F$.*

Proof: The hypothesis on F is precisely that F factors through $\dfrac{X \times I}{\alpha \times 1}$—i.e. there exists a continuous function $F'': \dfrac{X \times I}{\alpha \times 1} \to Y$ such that $F'' \circ p_{\alpha \times 1} = F$. By 0.4 $\phi: \dfrac{X \times I}{\alpha \times 1} \to (X/\alpha) \times I$ is a homeomorphism. Then $F' = F'' \circ \phi^{-1}$: $(X/\alpha) \times I \to Y$ is the required homotopy. ▯

Example. If A is a closed subspace of X and $F: X \times I \to Y$ is a homotopy such that $F(a,t) = F(a',t)$ for all $a, a' \in A$, $t \in I$, then F induces a homotopy $F': (X/A) \times I \to Y$.

0.9. *Function spaces:* if X and Y are topological spaces, we let Y^X denote the set of all continuous functions $f: X \to Y$. We give this set a topology, called the *compact-open topology*, by taking as a subbase for the topology all sets of the form $N_{K,U} = \{f: f(K) \subset U\}$, $K \subset X$ compact, $U \subset Y$ open.

0.10. The *evaluation function* $e: Y^X \times X \to Y$ defined by $e(f,x) = f(x)$ is continuous if X is locally compact. In all the cases we shall consider X will be $I = [0,1]$.

0.11. *The exponential law:* if X, Z are Hausdorff spaces and Z is locally compact, then the natural function

$$E: Y^{Z \times X} \to (Y^Z)^X$$

defined by $(Ef(x))(z) = f(z,x)$ is a homeomorphism.

0.12. *Base points:* in what follows we shall often have to consider not just a topological space X but rather a space X together with a distinguished point $x_0 \in X$ called the *base point*. The pair (X, x_0) is called a *pointed space* (one also speaks of pointed sets). When we are concerned with pointed spaces (X, x_0), (Y, y_0), etc. we always require that all functions $f: X \to Y$ shall preserve base points—i.e. $f(x_0) = y_0$—and that all homotopies $F: X \times I \to Y$ be relative to the base point—i.e. $F(x_0, t) = y_0$, all $t \in I$—unless an explicit disclaimer to the contrary is made. We shall use the notation $[X, x_0;\ Y, y_0]$ to denote the homotopy classes of base point-preserving functions—where homotopies are rel x_0, of course. $[X, x_0;\ Y, y_0]$ is a pointed set with base point f_0, the constant function: $f_0(x) = y_0$, all $x \in X$.

Let us use the notation $(Y, B)^{(X, A)}$ to denote the subspace of Y^X consisting of those functions such that $f(A) \subset B$, where $A \subset X$, $B \subset Y$ are subspaces. There are obvious generalizations of this notation, such as $(Y, B, B')^{(X, A, A')}$, etc. In particular, if (X, x_0), (Y, y_0) are pointed spaces, then we have the space $(Y, y_0)^{(X, x_0)}$ of base point-preserving functions. It has as base point f_0, where $f_0(x) = y_0$, all x.

If (X, x_0), (Y, y_0) and (Z, z_0) are three pointed spaces, we can form

$$((Y, y_0)^{(Z, z_0)}, f_0)^{(X, x_0)}, \quad f_0(z) = y_0, \quad \text{all } z \in Z,$$

which is a subspace of $(Y^Z)^X$. Thus we may ask what subspace of $Y^{Z \times X}$ corresponds to it under the exponential function of 0.11. One readily sees that the answer is

$$(Y, y_0)^{(Z \times X,\ Z \times \{x_0\} \cup \{z_0\} \times X)}.$$

We use the notation $Z \vee X$ for the subspace $Z \times \{x_0\} \cup \{z_0\} \times X$ of $Z \times X$. It can be thought of as the result of taking the disjoint union $Z \cup X$ and identifying z_0 with x_0. $Z \vee X$ is again a pointed space with base point (z_0, x_0). Then we have

0.13. Proposition. *The exponential function $E: Y^{Z \times X} \to (Y^Z)^X$ induces a continuous function*

$$E: (Y, y_0)^{(Z \times X,\ Z \vee X)} \to ((Y, y_0)^{(Z, z_0)}, f_0)^{(X,\ x_0)}$$

which is a homeomorphism if Z and X are Hausdorff and Z is locally compact.

Remark. The subspace $Z \times \{x_0\} \cup \{z_0\} \times X \subset Z \times X$ is called the *wedge sum* of Z and X and is characterized by the property that for any continuous functions $f: (Z, z_0) \to (W, w_0)$, $g: (X, x_0) \to (W, w_0)$ there is a unique continuous function $h: (Z \vee X, *) \to (W, w_0)$ such that $h|Z = f$, $h|X = g$. (Compare this with 0.2.) The unique function h will be denoted by

$(Z \vee X, *) \xrightarrow{(f,g)} (W, w_0)$. Of course, given continuous functions $f:(X,x_0) \to (X',x_0')$ and $g:(Y,y_0) \to (Y',y_0')$ there is a continuous function $f \vee g:(X \vee Y, (x_0,y_0)) \to (X' \vee Y', (x_0',y_0'))$ which is f on X, g on Y. We use $*$ to denote the base point $(x_0,y_0) \in X \vee Y$.

Exercise. Use 0.11 to give an independent proof of 0.4 in the case where X and Y are Hausdorff.

Chapter 1

Categories, Functors and Natural Transformations

In modern mathematics whenever one defines a new class of mathematical objects one proceeds almost in the next breath to say what kinds of functions between objects will be considered; thus, for example, topological spaces and continuous functions, groups and homomorphisms, rings and ring homomorphisms. If we formalize this observation, we are led to the notion of a category.

1.1. Definition. A *category* is

a) a class of *objects* (e.g. spaces, groups, etc.);

b) for every ordered pair (X, Y) of objects a set $\hom(X, Y)$ of *morphisms* with *domain* X and *range* Y; for $f \in \hom(X, Y)$ we write $f: X \to Y$ or $X \xrightarrow{f} Y$;

c) for every ordered triple (X, Y, Z) a function $\hom(Y,Z) \times \hom(X, Y) \to \hom(X,Z)$ called *composition*. If $f \in \hom(X, Y)$ and $g \in \hom(Y,Z)$ then the image of (g,f) in $\hom(X,Z)$ will be denoted by $g \circ f$.

These objects and morphisms are required to satisfy two axioms:

C1) If $f \in \hom(X, Y)$, $g \in \hom(Y,Z)$, $h \in \hom(Z, W)$, then $h \circ (g \circ f) = (h \circ g) \circ f$ in $\hom(X, W)$.

C2) For every object Y there is a $1_Y \in \hom(Y, Y)$ such that $1_Y \circ g = g$ for every $g \in \hom(X, Y)$ and $h \circ 1_Y = h$ for every $h \in \hom(Y,Z)$, all X,Z.

One can show that 1_Y is unique.

1.2. Definition. Two objects X, Y are called *equivalent* if there are morphisms $f \in \hom(X, Y)$ and $g \in \hom(Y, X)$ such that $g \circ f = 1_X$ and $f \circ g = 1_Y$. f and g are called *equivalences*.

1.3. Examples. i) The category $\mathcal{S}$ of all sets and all functions.

ii) The category $\mathcal{T}$ of all topological spaces and all continuous functions.

iii) The category $\mathcal{PS}$ of pointed sets and functions preserving base point.

iv) The category $\mathcal{PT}$ of pointed topological spaces and continuous functions preserving base point.

v) The category $\mathcal{G}$ of groups and homomorphisms.

vi) The category $\mathcal{A}$ of abelian groups and homomorphisms.

vii) The category $\mathcal{M}_R$ of left R-modules (R some fixed ring) and R-homomorphisms.

viii) The category $\mathcal{T}'$ in which the objects are topological spaces but $\hom(X, Y) = [X, Y]$. Given $[f] \in \hom(X, Y)$, $[g] \in \hom(Y, Z)$, we define $[g] \circ [f]$ to be $[g \circ f]$. One readily checks that $[g] \circ [f]$ is well defined and that C1), C2) are satisfied. In like fashion we have $\mathcal{PT}'$.

Note that X and Y are equivalent in $\mathcal{T}$ if and only if they are homeomorphic, whereas in $\mathcal{T}'$ they are equivalent if and only if they are *homotopy equivalent*.

In algebraic topology one attempts to assign to every topological space X some algebraic object $F(X)$ in such a way that to every continuous function $f: X \to Y$ there is assigned a homomorphism $F(f): F(X) \to F(Y)$. One advantage of this procedure is, for example, that if one is trying to prove the non-existence of a continuous function $f: X \to Y$ with certain properties, one may find it relatively easy to prove the non-existence of the corresponding algebraic function $F(f)$ and hence deduce that f could not exist. In other words, F is to be a "homomorphism" from one category (e.g. $\mathcal{T}$) to another (e.g. $\mathcal{G}$ or $\mathcal{A}$). If we formalize this notion, we are led to define a functor.

1.4. Definition. A *functor* from a category $\mathcal{C}$ to a category $\mathcal{D}$ is a function which

a) to each object $X \in \mathcal{C}$ assigns an object $F(X) \in \mathcal{D}$;

b) to each $f \in \hom_{\mathcal{C}}(X, Y)$ assigns a morphism

$$F(f) \in \hom_{\mathcal{D}}(F(X), F(Y)).$$

F is required to satisfy the two axioms:

F1) For each object $X \in \mathcal{C}$ we have $F(1_X) = 1_{F(X)}$.

F2) For $f \in \hom_{\mathcal{C}}(X, Y)$, $g \in \hom_{\mathcal{C}}(Y, Z)$ we have

$$F(g \circ f) = F(g) \circ F(f) \in \hom_{\mathcal{D}}(F(X), F(Z)).$$

In the arrow notation we have that if $f: X \to Y$ then $F(f): F(X) \to F(Y)$. We also have the notion of cofunctor; cofunctors "reverse the arrow".

1.5. Definition. A *cofunctor* F^* from the category $\mathcal{C}$ to the category $\mathcal{D}$ is a function which

a) to each object $X \in \mathcal{C}$ assigns an object $F^*(X) \in \mathcal{D}$,

b) to each $f \in \hom_{\mathscr{C}}(X, Y)$ assigns a morphism

$$F^*(f) \in \hom_{\mathscr{D}}(F^*(Y), F^*(X))$$

satisfying the two axioms:
CF1) For each object $X \in \mathscr{C}$ we have $F^*(1_X) = 1_{F^*(X)}$.
CF2) For each $f \in \hom_{\mathscr{C}}(X, Y)$ and $g \in \hom_{\mathscr{C}}(Y, Z)$ we have

$$F^*(g \circ f) = F^*(f) \circ F^*(g) \in \hom_{\mathscr{D}}(F^*(Z), F^*(X)).$$

Remark. In the literature functors are often referred to as covariant functors and cofunctors as contravariant functors.

1.6. Examples. i) Define $F: \mathscr{T} \to \mathscr{S}$ as follows: if $X \in \mathscr{T}$, let $F(X)$ be the underlying set (forget the topology), and if $f: X \to Y$ is a continuous function, let $F(f)$ be the underlying function (forget continuity). For obvious reasons F is called a "forgetful functor". One can think of many examples of forgetful functors.

ii) Given a fixed ring R and fixed (left) R-module K, we can define a functor $F_K: \mathscr{M}_R \to \mathscr{A}$; we take $F_K(M) = \operatorname{Hom}_R(K, M)$, $M \in \mathscr{M}_R$, and for any homomorphism $\phi: M \to M'$, we take

$$F_K(\phi) = \operatorname{Hom}_R(1_K, \phi): \operatorname{Hom}_R(K, M) \to \operatorname{Hom}_R(K, M').$$

Similarly we get a cofunctor $F_K^*: \mathscr{M}_R \to \mathscr{A}$ by taking

$$F_K^*(M) = \operatorname{Hom}_R(M, K), \quad F_K^*(\phi) = \operatorname{Hom}_R(\phi, 1_K).$$

iii) In a similar vein, given a fixed pointed space $(K, k_0) \in \mathscr{P}\mathscr{T}$, we define a functor

$$F_K: \mathscr{P}\mathscr{T} \to \mathscr{P}\mathscr{S}$$

as follows: for each $(X, x_0) \in \mathscr{P}\mathscr{T}$ we take $F_K(X, x_0) = [K, k_0;\ X, x_0]$. Given $f: (X, x_0) \to (Y, y_0)$ in $\hom((X, x_0), (Y, y_0))$ we define $F_K(f)$ by

$$F_K(f)[g] = [f \circ g] \in [K, k_0;\ Y, y_0]$$

for every $[g] \in [K, k_0;\ X, x_0]$.

Similarly we obtain a cofunctor F_K^* by taking $F_K^*(X, x_0) = [X, x_0;\ K, k_0]$ and for $f: (X, x_0) \to (Y, y_0)$ in $\hom((X, x_0), (Y, y_0))$

$$F_K(f)[g] = [g \circ f] \in [X, x_0; K, k_0]$$

for every $[g] \in [Y, y_0;\ K, k_0]$.

Observe that $f \simeq f'$ rel x_0 implies $F_K(f) = F_K(f')$ and likewise $F_K^*(f) = F_K^*(f')$. Therefore F_K and F_K^* can equally well be regarded as defining a functor $\mathscr{P}\mathscr{T}' \to \mathscr{P}\mathscr{S}$, respectively a cofunctor.

iv) Define a functor $C: \mathscr{T} \to \mathscr{T}$ by taking $C(X) = \hat{X} =$ one point compactification of X for every $X \in \mathscr{T}$. If $f: X \to Y$ is a continuous function, then it defines a unique continuous function $\hat{f}:\hat{X} \to \hat{Y}$ such that $\hat{f}|X = f$. Take $C(f) = \hat{f}$ for each $f \in \hom(X, Y)$.

v) Let $\mathscr{I}$ be the category of all integral domains and ring monomorphisms, $\mathscr{F}$ the category of all fields and ring homomorphisms. Define $Q:\mathscr{I} \to \mathscr{F}$ by taking $Q(A)$ to be the field of fractions of A for every integral domain A. Given a monomorphism $\phi:A \to B$ of integral domains there is a unique homomorphism $Q(\phi):Q(A) \to Q(B)$ which restricts to ϕ on $A \subset Q(A)$. Q is then a functor.

vi) Let $\mathscr{B}$ be the category of Banach spaces over $\mathbb{R}$ and bounded linear maps. Define $D:\mathscr{B} \to \mathscr{B}$ by taking $D(X) = X^*$ = Banach space of bounded linear functionals on X and $D(f) = f^*$ for $f:X \to Y$ a bounded linear map. Then D is a cofunctor.

In order to compare two functors $F, G:\mathscr{C} \to \mathscr{D}$ we need the notion of natural transformation.

1.7. Definition. Let $\mathscr{C}, \mathscr{D}$ be two categories and $F, G:\mathscr{C} \to \mathscr{D}$ two functors from $\mathscr{C}$ to $\mathscr{D}$. A *natural transformation* T from F to G is a function which

i) to each $X \in \mathscr{C}$ assigns a morphism $T(X) \in \hom_{\mathscr{D}}(F(X), G(X))$, i.e. $T(X):F(X) \to G(X)$;

ii) for each morphism $f \in \hom_{\mathscr{C}}(X, Y)$ satisfies

$$T(Y) \circ F(f) = G(f) \circ T(X).$$

$$\begin{array}{ccc} F(X) & \xrightarrow{F(f)} & F(Y) \\ \downarrow T(X) & & \downarrow T(Y) \\ G(X) & \xrightarrow{G(f)} & G(Y) \end{array}$$

There is a corresponding notion for a natural transformation between two cofunctors.

1.8. If $T(X)$ is an equivalence in the category $\mathscr{D}$ for every $X \in \mathscr{C}$, then we call T a *natural equivalence* and say that F and G are *naturally equivalent.*

1.9. Examples. i) Suppose $\phi:(K, k_0) \to (L, l_0)$ is a morphism in $\mathscr{PT}$. Then we define $T_\phi:F_L \to F_K$ as follows:

$$T_\phi(X, x_0)[g] = [g \circ \phi] \in [K, k_0; X, x_0]$$

for every (X, x_0) and $[g] \in F_L(X, x_0) = [L, l_0; X, x_0]$. $T_\phi(X, x_0)$ sends constant map to constant map, so $T_\phi(X, x_0) \in \hom_{\mathscr{PS}}(F_L(X, x_0), F_K(X, x_0))$.

Moreover, for every $f:(X,x_0) \to (Y,y_0)$ the square

$$\begin{array}{ccc} [L,l_0;\,X,x_0] & \xrightarrow{F_L(f)} & [L,l_0;\,Y,y_0] \\ \Big\downarrow T_\phi(X,x_0) & & \Big\downarrow T_\phi(Y,y_0) \\ [K,k_0;\,X,x_0] & \xrightarrow{F_K(f)} & [K,k_0;\,Y,y_0] \end{array}$$

does commute, for if $[g] \in [L,l_0;\,X,x_0]$, then

$$\begin{aligned} T_\phi(Y,y_0) \circ F_L(f)[g] = T_\phi(Y,y_0)[f \circ g] = [f \circ g \circ \phi] &= F_K(f)[g \circ \phi] \\ &= F_K(f) \circ T_\phi(X,x_0)[g]. \end{aligned}$$

Exercise. Show that $\phi:(K,k_0) \to (L,l_0)$ is a homotopy equivalence if and only if T_ϕ is a natural equivalence.

ii) Consider the duality functor $D:\mathscr{B} \to \mathscr{B}$ of 1.6.vi). $D^2 = D \circ D$ is also a functor. We also have the identity functor $1:\mathscr{B} \to \mathscr{B}$. Define $T:1 \quad D \circ D$ as follows: for every $X \in \mathscr{B}$ let $T(X):X \to D^2X = X^{**}$ be the natural inclusion—that is, for $x \in X$ we have $[T(X)(x)](f) = f(x)$ for every $f \in X^*$. T is a natural transformation. On the subcategory of finite-dimensional Banach spaces T is even a natural equivalence. The largest subcategory of $\mathscr{B}$ on which T is a natural equivalence is called the category of reflexive Banach spaces.

Chapter 2

Homotopy Sets and Groups

The functors F_K and cofunctors F_K^* introduced in 1.6.iii) are especially important in algebraic topology. In this chapter we give conditions on (K,k_0) which guarantee that $F_K(X,x_0)$ and $F_K^*(X,x_0)$ are groups for all pointed spaces (X,x_0)—i.e. conditions on (K,k_0) which make F_K (resp. F_K^*) a functor (resp. cofunctor) from $\mathscr{PT}$ (or $\mathscr{PT}'$) to $\mathscr{G}$. In particular, if $S^n = \{x \in \mathbb{R}^{n+1} : \|x\| = 1\}$ is the n-sphere and $s_0 = (1,0,0,\ldots,0)$, then (S^n, s_0) is a pointed space such that $F_{S^n} : \mathscr{PT}' \to \mathscr{G}$, and we investigate some of the properties of $\pi_n(X,x_0) = F_{S^n}(X,x_0)$, $n \geqslant 0$.

In this chapter and from now on we shall use the word "map" to mean a continuous function between two topological spaces.

2.1. A *path* in a topological space X is a map $w: I \to X$. w is said to be a path from $w(0)$ to $w(1)$. Thus X^I is the space of all paths in X with the compact-open topology. We introduce a relation $\sim$ on X by saying $x \sim y$ if and only if there is a path $w: I \to X$ from x to y. $\sim$ is clearly an equivalence relation, and the set of equivalence classes is denoted by $\pi_0(X)$. The elements of $\pi_0(X)$ are called the *path components* or 0-*components* of X. If $\pi_0(X)$ contains just one element, then X is called *path connected* or 0-*connected*.

Exercise. Show that 0-connected $\Rightarrow$ connected and give a counter-example for the converse statement.

If (X,x_0) is a pointed space, then we may regard $\pi_0(X)$ as a pointed set with the 0-component of x_0 as base point. We use the notation $\pi_0(X,x_0)$ to denote $\pi_0(X)$ thought of as a pointed set. If $f: X \to Y$ is a map then f sends 0-components of X into 0-components of Y and hence defines a function $\pi_0(f): \pi_0(X) \to \pi_0(Y)$. Similarly a base point-preserving map $f: (X,x_0) \to (Y,y_0)$ induces a morphism of pointed sets $\pi_0(f): \pi_0(X,x_0) \to \pi_0(Y,y_0)$. It is immediate that $\pi_0(1_X) = 1_{\pi_0(X)}$ and $\pi_0(f \circ g) = \pi_0(f) \circ \pi_0(g)$. Thus π_0 is a functor $\mathscr{T} \to \mathscr{S}$ or $\mathscr{PT} \to \mathscr{PS}$.

We can give another description of π_0. Let $*$ denote any space with one point. Let S^0 denote the 0-sphere $\{+1,-1\} \subset \mathbb{R}^1$; we give S^0 the base point $s_0 = +1$.

2.2. Proposition. *There is a natural equivalence* $[*;\ X] \leftrightarrow \pi_0(X)$ *given by* $[f] \mapsto$ {0-*component of* $f(*)$}.

Proof: There is obviously a one–one correspondence between maps $f: * \to X$ and points of X given by $f \mapsto f(*)$. f and g are homotopic if and only if $f(*)$ and $g(*)$ lie in the same path component. ▯

The corresponding result in the category $\mathscr{PT}$ is the following.

2.3. Proposition. *There is a natural equivalence* $[S^0, s_0;\ X, x_0] \leftrightarrow \pi_0(X, x_0)$ *given by* $[f] \mapsto$ {0-*component of* $f(-1)$}.

2.4. If $(X,x_0), (Y,y_0) \in \mathscr{PT}$, we define the *smash product* $(X \wedge Y, *) \in \mathscr{PT}$ to be the quotient

$$X \wedge Y = X \times Y / X \vee Y$$

with the point $* = p(X \vee Y)$ as base point. Here $p: X \times Y \to X \wedge Y$ is the projection. For any $(x,y) \in X \times Y$ we denote $p(x,y) \in X \wedge Y$ by $[x,y]$. Given maps $f:(X,x_0) \to (X',x_0')$ and $g:(Y,y_0) \to (Y',y_0')$, the map $f \times g: X \times Y \to X' \times Y'$ maps $X \vee Y$ into $X' \vee Y'$ and so induces a map $f \wedge g: X \wedge Y \to X' \wedge Y'$.

Remark. In the notation $X \wedge Y$ the base points are suppressed. Strictly speaking we should write $(X,x_0) \wedge (Y,y_0)$, but for simplicity's sake we take x_0 and y_0 to be understood. This is common procedure; for example, the topologies on X and Y are also suppressed in the notation throughout.

Exercise. Show that $X \wedge (Y \wedge Z) \cong (X \wedge Y) \wedge Z$.

2.5. Theorem. *If* $(X,x_0), (Y,y_0), (Z,z_0) \in \mathscr{PT}$, X, Z *Hausdorff and* Z *locally compact, then there is a natural equivalence*

$$A: [Z \wedge X, *;\ Y, y_0] \to [X, x_0;\ (Y,y_0)^{(Z,z_0)}, f_0]$$

defined by $A[f] = [\hat{f}]$, *where if* $f: Z \wedge X \to Y$ *is a map then* $\hat{f}: X \to Y^Z$ *is given by* $(\hat{f}(x))(z) = f[z,x]$.

Proof: i) A is surjective: given any map $f': (X,x_0) \to ((Y,y_0)^{(Z,z_0)}, f_0)$ we know from 0.13 that the function $\bar{f}: (Z \times X, Z \vee X) \to (Y,y_0)$ defined by $\bar{f}(z,x) = (f'(x))(z)$ is continuous ($\bar{f} = E^{-1} f'$). By 0.3 $\bar{f}$ defines a map $f: (Z \wedge X, *) \to (Y,y_0)$ such that $f[z,x] = \bar{f}(z,x) = (f'(x))(z)$. Thus $\hat{f} = f'$— i.e. $A[f] = [f']$.

ii) A is injective: suppose $f,g: (Z \wedge X, *) \to (Y,y_0)$ are two maps such that $A[f] = A[g]$—i.e. $\hat{f} \simeq \hat{g}$. Let $H': X \times I \to (Y,y_0)^{(Z,z_0)}$ be the homotopy rel x_0. By 0.13 the function $\bar{H}: Z \times X \times I \to Y$ defined by $\bar{H}(z,x,t) = (H'(x,t))(z)$ is continuous ($\bar{H} = E^{-1} H'$). For each $t \in I$ we have $\bar{H}((Z \vee X) \times \{t\}) = y_0$, so by 0.8 there is a homotopy $H: (Z \wedge X) \times I \to Y$ rel $*$ such that $H([z,x],t) = \bar{H}(z,x,t) = (H'(x,t))(z)$. Thus $H_0([z,x]) = (H_0'(x))(z) = (\hat{f}(x))(z) = f[z,x]$ and similarly $H_1 = g$. Thus $[f] = [g]$. □

A particularly important case of 2.5 is that where Z is the 1-sphere $S^1 = \{x \in \mathbb{R}^2 : \|x\| = 1\}$ and $z_0 = s_0 = (1,0)$.

2.6. If $(Y, y_0) \in \mathscr{PT}$, we define the *loop space* $(\Omega Y, \omega_0) \in \mathscr{PT}$ of Y to be the function space

$$\Omega Y = (Y, y_0)^{(S^1, s_0)}$$

with the constant loop ω_0 ($\omega_0(s) = y_0$, all $s \in S^1$) as base point.

2.7. If $(X, x_0) \in \mathscr{PT}$, we define the *suspension* $(SX, *) \in \mathscr{PT}$ of X to be the smash product $(S^1 \wedge X, *)$ of X with the 1-sphere.

2.8. Corollary. *If (X, x_0), $(Y, y_0) \in \mathscr{PT}$ and X is Hausdorff, then there is a natural equivalence*

$$A : [SX, *; Y, y_0] \to [X, x_0; \Omega Y, \omega_0].$$

Remarks. i) We actually have two new functors here which are worth taking note of; they both go from $\mathscr{PT}$ to $\mathscr{PT}$. The *suspension functor* S takes (X, x_0) to $(SX, *)$. Given $f : (X, x_0) \to (X', x_0')$, we let Sf be the map

$$1 \wedge f : (S^1 \wedge X, *) \to (S^1 \wedge X', *).$$

The *loop functor* Ω takes (Y, y_0) to $(\Omega Y, \omega_0)$. Given $g : (Y, y_0) \to (Y', y_0')$, we let $\Omega g : (\Omega Y, \omega_0) \to (\Omega Y', \omega_0')$ be the map defined by $\Omega g(\omega)(s) = g \circ \omega(s)$, $s \in S^1$. That Ωg so defined is continuous follows from 0.10 and 0.13. Because of 2.8 S and Ω are said to be *adjoint* to one another in $\mathscr{PT}'$.

ii) The correspondence A of 2.8 is natural in the following sense: if $f : (X', x_0') \to (X, x_0)$ and $g : (Y, y_0) \to (Y', y_0')$ are maps then the squares

$$\begin{array}{ccc} [SX, *; Y, y_0] & \xrightarrow{A} & [X, x_0; \Omega Y, \omega_0] \\ \downarrow Sf^* & & \downarrow f^* \\ [SX', *; Y, y_0] & \xrightarrow{A} & [X', x_0'; \Omega Y, \omega_0] \end{array}$$

$$\begin{array}{ccc} [SX, *; Y, y_0] & \xrightarrow{A} & [X, x_0; \Omega Y, \omega_0] \\ \downarrow g_* & & \downarrow \Omega g_* \\ [SX, *; Y', y_0] & \xrightarrow{A} & [X, x_0; \Omega Y', \omega_0'] \end{array}$$

commute. Here $f^* = F^*_{\Omega Y}(f)$, $g_* = F_{SX}(g)$, etc.

Now we turn to the question of properties on spaces (K, k_0) which make F_K into a group-valued functor or F_K^* into a group-valued cofunctor. We take the following description of a group as our guide.

Definition. A *group* is a pointed set (G,e) with a *multiplication* $\mu : G \times G \to G$ and an *inverse* $\nu : G \to G$ such that the following diagrams commute:

i)
$$\begin{array}{ccccc} G & \xrightarrow{(e,1)} & G \times G & \xleftarrow{(1,e)} & G \\ & {\scriptstyle 1}\searrow & \downarrow{\scriptstyle \mu} & \swarrow{\scriptstyle 1} & \\ & & G & & \end{array}$$
(e is a two-sided identity),

ii)
$$\begin{array}{ccc} G \times G \times G & \xrightarrow{\mu \times 1} & G \times G \\ \downarrow{\scriptstyle 1 \times \mu} & & \downarrow{\scriptstyle \mu} \\ G \times G & \xrightarrow[\mu]{} & G \end{array}$$
(associativity),

iii)
$$\begin{array}{ccccc} G & \xrightarrow{(\nu,1)} & G \times G & \xleftarrow{(1,\nu)} & G \\ & {\scriptstyle e}\searrow & \downarrow{\scriptstyle \mu} & \swarrow{\scriptstyle e} & \\ & & G & & \end{array}$$
(inverse).

Here $e : G \to G$ is the constant map $e(g) = e$, all $g \in G$. $(e,1)$ means the map such that $(e,1)(g) = (e,g)$, etc. G is called *commutative* or *abelian* if in addition the following diagram commutes

$$\begin{array}{ccc} G \times G & \xrightarrow{T} & G \times G \\ & {\scriptstyle \mu}\searrow \quad \swarrow{\scriptstyle \mu} & \\ & G & \end{array}$$

where $T : G \times G \to G \times G$ is the *switch* map $T(g_1,g_2) = (g_2,g_1)$, all $(g_1,g_2) \in G \times G$.

Motivated by this definition of a group we make the following definitions.

2.9. Definition. An *H-space* is a pointed space (K,k_0) with a multiplication map $\mu : K \times K \to K$ such that k_0 is a *homotopy identity*—that is, the diagram

2.10.
$$\begin{array}{ccccc} K & \xrightarrow{(k_0,1)} & K \times K & \xleftarrow{(1,k_0)} & K \\ & {\scriptstyle 1}\searrow & \downarrow{\scriptstyle \mu} & \swarrow{\scriptstyle 1} & \\ & & K & & \end{array}$$

commutes up to homotopy: $\mu \circ (1,k_0) \simeq 1 \simeq \mu \circ (k_0,1)$. We say μ is *homotopy associative* if the diagram

2.11.

$$\begin{array}{ccc} K \times K \times K & \xrightarrow{\mu \times 1} & K \times K \\ \downarrow{\scriptstyle 1 \times \mu} & & \downarrow{\scriptstyle \mu} \\ K \times K & \xrightarrow{\mu} & K \end{array}$$

commutes up to homotopy: $\mu \circ (\mu \times 1) \simeq \mu \circ (1 \times \mu)$. A map $v:(K,k_0) \to (K,k_0)$ is called a *homotopy inverse* if the diagram

2.12.

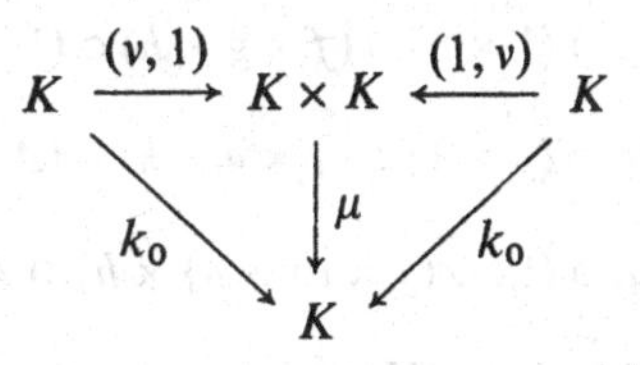

commutes up to homotopy: $\mu \circ (v,1) \simeq k_0 \simeq \mu \circ (1,v)$. We say μ is *homotopy commutative* if the diagram

2.13.

$$\begin{array}{ccc} K \times K & \xrightarrow{T} & K \times K \\ & {\scriptstyle \mu}\searrow \quad \swarrow{\scriptstyle \mu} & \\ & K & \end{array}$$

commutes up to homotopy: $\mu \circ T \simeq \mu$. An *H-group* is an H-space (K,k_0) with homotopy associative multiplication μ and homotopy inverse v. Note: All maps and homotopies should be relative to the base point. $K \times K$ has base point (k_0,k_0).

The reason for this definition is the following proposition.

2.14. Proposition. *If (K,k_0) is an H-group with multiplication μ and homotopy inverse v, then for every $(X,x_0) \in \mathscr{PT}$ the set*

$$[X,x_0;\ K,k_0]$$

can be given the structure of a group if we define the product $[f]\cdot[g]$ to be the homotopy class of the composition

$$X \xrightarrow{\Delta} X \times X \xrightarrow{f \times g} K \times K \xrightarrow{\mu} K.$$

Here Δ is the diagonal map given by $\Delta(x) = (x,x)$. The identity of the group is the class $[k_0]$ of the constant map, and the inverse is given by $[f]^{-1} = [v \circ f]$. If μ is homotopy commutative, then $[X,x_0;\ K,k_0]$ is abelian. Every map $f:(X,x_0) \to (Y,y_0)$ induces a homomorphism $f^*:[Y,y_0;\ K,k_0] \to [X,x_0;\ K,k_0]$.

Proof: First we show $[f]\cdot[g]$ is well defined. Suppose $H: X \times I \to K$ is a homotopy between f and f', $G: X \times I \to K$ a homotopy between g and g'.

Define a homotopy $M: X \times I \to K$ by

$$M(x,t) = \mu(H(x,t), G(x,t)).$$

Then $M_0 = \mu \circ (f \times g) \circ \Delta$ and $M_1 = \mu \circ (f' \times g') \circ \Delta$.

That the resulting multiplication defines a group is virtually obvious from 2.10–2.13. We prove associativity to illustrate.

$$\begin{aligned}
[f]\cdot([g]\cdot[h]) &= [\mu \circ (f \times \{\mu \circ (g \times h) \circ \Delta\}) \circ \Delta] \\
&= [\mu \circ (1 \times \mu) \circ (f \times g \times h) \circ (1 \times \Delta) \circ \Delta] \\
&= [\mu \circ (\mu \times 1) \circ (f \times g \times h) \circ (\Delta \times 1) \circ \Delta] \quad \text{by 2.11} \\
&= [\mu \circ (\{\mu \circ (f \times g) \circ \Delta\} \times h) \circ \Delta] \\
&= ([f]\cdot[g])\cdot[h]. \quad \square
\end{aligned}$$

Thus if (K, k_0) is an H-group then F_K^* is a cofunctor from $\mathscr{PT}'$ to $\mathscr{G}$.

2.15. Examples. i) Clearly every topological group is an H-group.

ii) The most important example of an H-group is the loop space $(\Omega Y, \omega_0)$ of any $(Y, y_0) \in \mathscr{PT}$. Before defining

$$\mu: \Omega Y \times \Omega Y \to \Omega Y$$

we make the following remark: there is an obvious homeomorphism

$$(I/\{0,1\}, *) \to (S^1, *),$$

and in fact we have a homeomorphism

$$(Y, y_0)^{(S^1, *)} \cong (Y, y_0)^{(I, \{0,1\})}.$$

For many purposes it will be more convenient to regard ΩY as being $(Y, y_0)^{(I, \{0,1\})}$—i.e. as the space of paths beginning and ending at y_0. In these terms we define μ by

$$\mu(\omega, \omega')(t) = \begin{cases} \omega(2t) & 0 \leqslant t \leqslant 1/2, \\ \omega'(2t-1) & 1/2 \leqslant t \leqslant 1. \end{cases}$$

What we have done, in fact, is define a map

$$\bar{\mu}: \Omega Y \times \Omega Y \times I \to Y$$

which is continuous because of 0.1, 0.2 and 0.10. Then the exponential law 0.11 gives us our

$$\mu: \Omega Y \times \Omega Y \to \Omega Y.$$

We shall use the notation $\omega * \omega'$ for the loop $\mu(\omega, \omega')$. Schematically we could represent $\omega * \omega'$ by the diagram

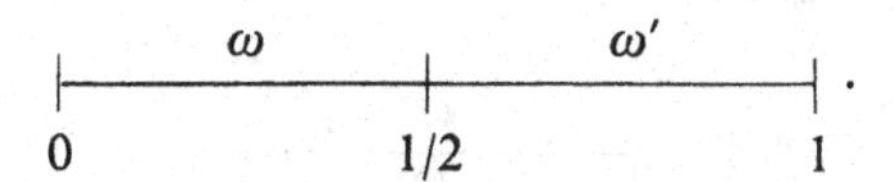

We must show that the constant loop ω_0 is a homotopy identity. In other words we want a homotopy $H: \Omega Y \times I \to \Omega Y$ from $\mu \circ (1 \times \omega_0)$ to 1. For a fixed $\omega \in \Omega Y$ our H will look as follows schematically

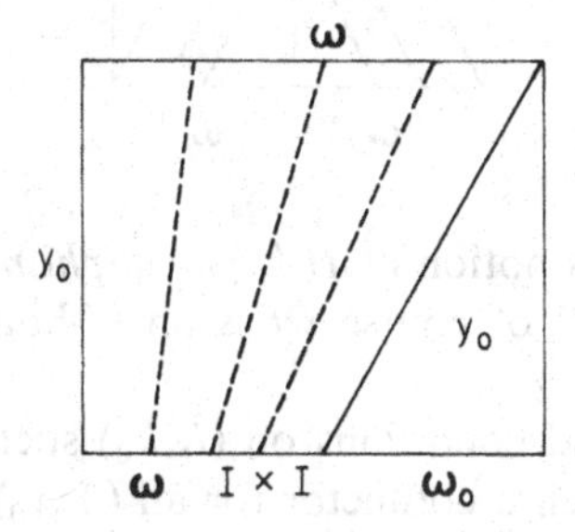

where the dashed lines denote "isoclines". Explicitly, we define an $\bar{H}$: $\Omega Y \times I \times I \to Y$ by

$$\bar{H}(\omega, t, s) = \begin{cases} \omega\left(\dfrac{2s}{t+1}\right) & 0 \leqslant s \leqslant \dfrac{t+1}{2}, \\ y_0 & \dfrac{t+1}{2} \leqslant s \leqslant 1. \end{cases}$$

Again H is continuous because of 0.1 and 0.10 and hence defines a homotopy $H: \Omega Y \times I \to \Omega Y$. It is immediate that $H_0 = \mu \circ (1 \times \omega_0)$, $H_1 = 1$. That $\mu \circ (\omega_0 \times 1) \simeq 1$ is proved similarly.

For the homotopy $\mu \circ (1 \times \mu) \simeq \mu \circ (\mu \times 1)$ we give only the schematic diagram. The reader should try to write down the explicit homotopy.

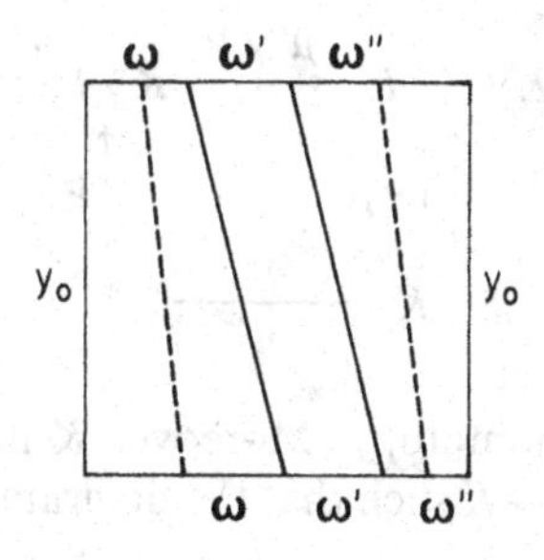

We define a homotopy inverse $\nu:\Omega Y \to \Omega Y$ by $\nu(\omega) = \omega^{-1}$, where $\omega^{-1}(t) = \omega(1-t)$, $t \in I$. The schematic diagram of a homotopy between $\mu \circ (1, \nu)$ and ω_0 is given by

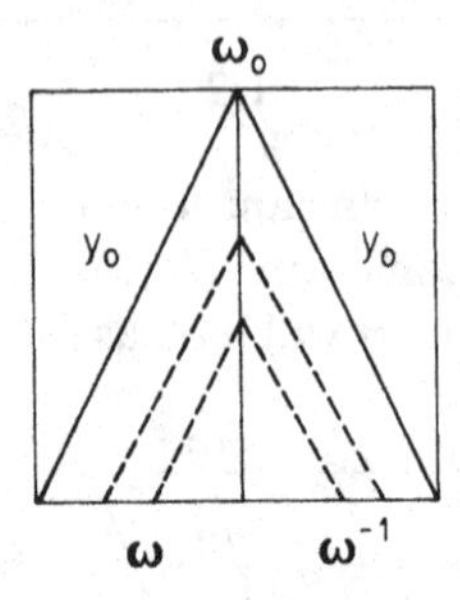

There is an obvious notion of *H-homomorphism* $\phi:(K,k_0) \to (K',k_0')$ between H-groups, and of course Ωf is an H-homomorphism for any $f:(Y,y_0) \to (Y',y_0')$.

Thus we have found conditions on (K,k_0) such that $F_K^*:\mathscr{PT}' \to \mathscr{G}$. In particular $F_{\Omega Y}^*$ is such a cofunctor for all $(Y,y_0) \in \mathscr{PT}$. We now ask when F_K is group-valued and are led to the dual notion of H-cogroups.

2.16. Definition. An *H-cogroup* is a pointed space (K,k_0) together with a continuous *comultiplication* $\mu':K \to K \vee K$ such that k_0 is a *homotopy identity*, i.e. the diagram

2.17.

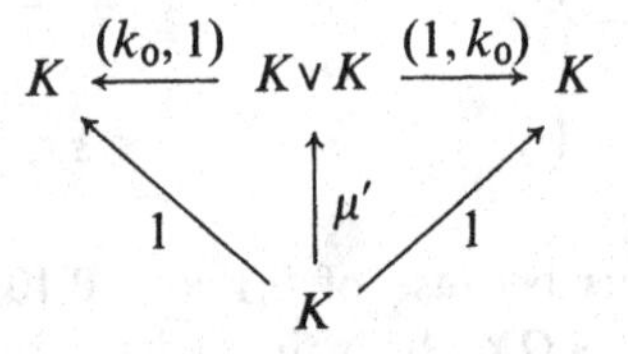

commutes up to homotopy (here $(k_0, 1)$ denotes the map such that $(k_0,1)(k,k_0) = k_0$ and $(k_0,1)(k_0,k) = k$ for all $k \in K$). μ' is further required to be *homotopy associative*—i.e. the diagram

2.18.

$$\begin{array}{ccc} K \vee K \vee K & \xleftarrow{\mu' \vee 1} & K \vee K \\ \uparrow{\scriptstyle 1 \vee \mu'} & & \uparrow{\scriptstyle \mu'} \\ K \vee K & \xleftarrow[\mu']{} & K \end{array}$$

must commute up to homotopy. Moreover K must have a continuous *homotopy inverse* $\nu':K \to K$ such that the diagram

2.19.

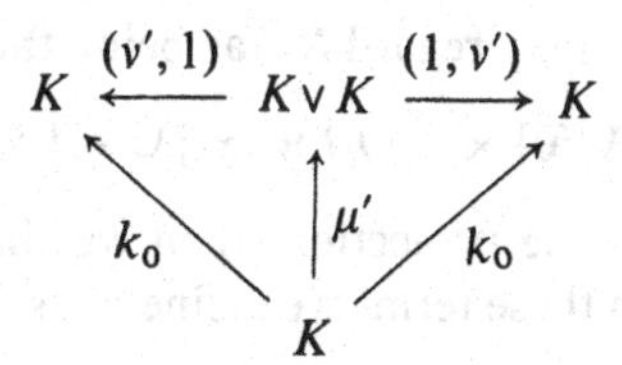

commutes up to homotopy. Finally K is called a *homotopy commutative H-cogroup* if in addition the diagram

2.20.

$$\begin{array}{ccc} K \vee K & \xrightarrow{T} & K \vee K \\ & {}_{\mu'}\nwarrow \quad \nearrow_{\mu'} & \\ & K & \end{array}$$

commutes up to homotopy.

Metamathematical remark: The observant reader may have deduced that the way to "dualize" in this subject is to replace each construct by its "dual" (e.g. $K \times K$ by $K \vee K$) and reverse all arrows. This procedure will occur again in other contexts.

We now get the proposition dual to 2.14.

2.21. Proposition. *If (K, k_0) is an H-cogroup with comultiplication μ' and homotopy inverse ν', then for every $(X, x_0) \in \mathscr{PT}$ the set*

$$[K, k_0;\ X, x_0]$$

can be given the structure of a group if we define the product $[f] \cdot [g]$ to be the homotopy class of the composition

$$K \xrightarrow{\mu'} K \vee K \xrightarrow{f \vee g} X \vee X \xrightarrow{\Delta'} X.$$

Here Δ' is the folding map given by $\Delta'(x, x_0) = x = \Delta'(x_0, x)$. The identity of the group is the class $[x_0]$ of the constant map, and the inverse is given by $[f]^{-1} = [f \circ \nu']$. If μ' is homotopy commutative, then $[K, k_0;\ X, x_0]$ is abelian. Every map $f: (X, x_0) \to (Y, y_0)$ induces a homomorphism

$$f_*: [K, k_0;\ X, x_0] \to [K, k_0;\ Y, y_0].$$

The proof is again quite obvious.

2.22. If the principal example of an H-group is ΩY, any Y, then in the light of 2.8 it should be no surprise that the principal example of an H-cogroup is SX, any X. Before defining

$$\mu': SX \to SX \vee SX$$

we make the following remark: because of the homeomorphism

$(S^1, s_0) \cong (I/\{0,1\}, *)$ we may regard SX as being the quotient

$$I \times X/\{0\} \times X \cup I \times \{x_0\} \cup \{1\} \times X.$$

If $p: I \times X \to SX$ is the projection, then we shall denote $p(t,x) \in SX$ by $[t,x]$, $t \in I$, $x \in X$. In these terms we define μ' as follows:

$$\mu'[t,x] = \begin{cases} ([2t,x], x_0) & 0 \leqslant t \leqslant 1/2, \\ (x_0, [2t-1,x]) & 1/2 \leqslant t \leqslant 1. \end{cases}$$

If we represent SX by the picture

X × {1}
X × {1/2} = X
X × {0}

then μ' is represented by the picture

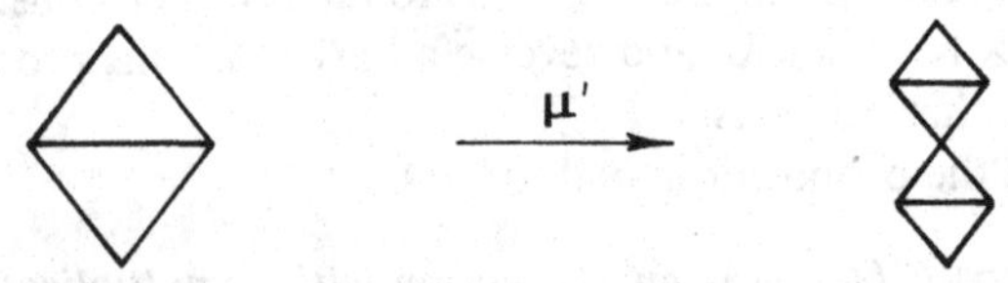

In other words μ' "pinches SX at its middle". The homotopy inverse $\nu': SX \to SX$ "turns SX upside down"; explicitly:

$$\nu'[t,x] = [1-t,x], \quad t \in I, \quad x \in X.$$

The proof that $(SX, *)$ equipped with μ', ν' is an H-cogroup is quite straightforward. The homotopies required have exactly the same schematic diagrams as those for ΩY; only the interpretations are a little different.

2.23. Proposition. *The adjoint correspondence*

$$A: [SX, *; Y, y_0] \to [X, x_0; \Omega Y, \omega_0]$$

is an isomorphism of groups.

Proof: Given $f, g: (SX, *) \to (Y, y_0)$ let $f \circ g$ denote the composite

$$SX \xrightarrow{\mu'} SX \vee SX \xrightarrow{f \vee g} Y \vee Y \xrightarrow{\Delta'} Y.$$

Given $f', g': (X, x_0) \to (\Omega Y, \omega_0)$ let $f' * g'$ denote the composite

$$X \xrightarrow{\Delta} X \times X \xrightarrow{f' \times g'} \Omega Y \times \Omega Y \xrightarrow{\mu} \Omega Y.$$

We shall prove that $\widehat{(f \circ g)} = \hat{f} * \hat{g}$. For all $x \in X$, $t \in I$ we have

$$((\widehat{f \circ g})(x))(t) = (f \circ g)[t,x] = \begin{cases} f([2t,x]) & 0 \leqslant t \leqslant 1/2, \\ g([2t-1,x]) & 1/2 \leqslant t \leqslant 1. \end{cases}$$

$$(\hat{f} * \hat{g}(x))(t) = \begin{cases} (\hat{f}(x))(2t) \\ (\hat{g}(x))(2t-1) \end{cases} = \begin{cases} f([2t,x]) & 0 \leqslant t \leqslant 1/2, \\ g([2t-1,x]) & 1/2 \leqslant t \leqslant 1. \end{cases} \quad \square$$

Now suppose that (K,k_0) is an H-cogroup and (L,l_0) is an H-group. Then we have two possible ways of defining a group structure on $[K,k_0;\ L,l_0]$; are they the same? The answer is "yes", and the reason applies to a more general setting.

2.24. Proposition. *Let X be a set with two multiplications $\circ$ and $*$ satisfying*

i) *there is a mutual two-sided identity e; that is, for all $x \in X$ $x \circ e = x * e = x = e * x = e \circ x$;*

ii) *$\circ$ and $*$ are mutually distributive; that is, for all $x, x', y, y' \in X$ $(x \circ x') * (y \circ y') = (x * y) \circ (x' * y')$. Then $\circ$ and $*$ are equal and both are associative and commutative.*

Proof: For x, y, $z \in X$ we have

a) $x \circ y = (x * e) \circ (e * y) = (x \circ e) * (e \circ y) = x * y$. Thus $\circ$ and $*$ agree.

b) $x \circ y = (e * x) \circ (y * e) = (e \circ y) * (x \circ e) = y * x = y \circ x$. Thus $\circ$ (and hence $*$) is commutative.

c) $x \circ (y \circ z) = (x * e) \circ (y * z) = (x \circ y) * (e \circ z) = (x \circ y) \circ z$. Thus $\circ$ (and hence $*$) is associative. $\square$

2.25. Proposition. *If (K,k_0) is an H-cogroup and (L,l_0) is an H-group, then the two multiplications on $[K,k_0;\ L,l_0]$ agree and are both commutative.*

Proof: By 2.24 it suffices to show the two multiplications are mutually distributive, since we already know the class $[l_0]$ of the constant map is a mutual two-sided identity. Let μ' be the comultiplication of K and μ the multiplication of L. We continue to use the notation of the proof of 2.23. We shall show that for any $f, f', g, g' : (K,k_0) \to (L,l_0)$

$$(f \circ f') * (g \circ g') = (f * g) \circ (f' * g').$$

In the diagram below the composite along the top is $(f \circ f') * (g \circ g')$, that along the bottom $(f * g) \circ (f' * g')$.

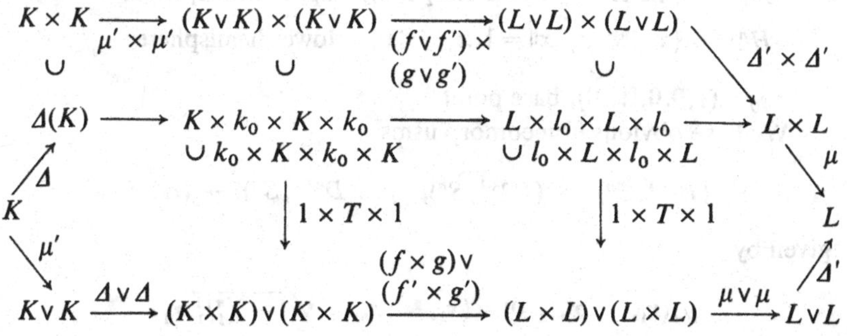

$$\begin{array}{ccccccc}
K \times K & \xrightarrow{\mu' \times \mu'} & (K \vee K) \times (K \vee K) & \xrightarrow{(f \vee f') \times (g \vee g')} & (L \vee L) \times (L \vee L) & \searrow^{\Delta' \times \Delta'} & \\
\cup & & \cup & & \cup & & \\
\Delta(K) & \longrightarrow & K \times k_0 \times K \times k_0 \cup k_0 \times K \times k_0 \times K & \longrightarrow & L \times l_0 \times L \times l_0 \cup l_0 \times L \times l_0 \times L & \longrightarrow & L \times L \\
\nearrow_{\Delta} & & \downarrow 1 \times T \times 1 & & \downarrow 1 \times T \times 1 & & \searrow^{\mu} \\
K & & & & & & L \\
\searrow^{\mu'} & & & & & & \nearrow_{\Delta'} \\
K \vee K & \xrightarrow{\Delta \vee \Delta} & (K \times K) \vee (K \times K) & \xrightarrow{(f \times g) \vee (f' \times g')} & (L \times L) \vee (L \times L) & \xrightarrow{\mu \vee \mu} & L \vee L
\end{array}$$

That the middle square commutes is obvious. That the two pentagons at the ends commute is readily checked. □

That $[SX,*;\ \Omega Y,\omega_0]$ turns out to be an abelian group may seem to be an unexpected bonus.

We can form iterated loop spaces $\Omega^n Y$ inductively: $\Omega^n Y = \Omega(\Omega^{n-1} Y)$, $n \geqslant 1$, $\Omega^0 Y = Y$. Similarly we get iterated suspensions: $S^n X = S(S^{n-1} X)$, $n \geqslant 1$, $S^0 X = X$. Then $[SX,*;\ \Omega Y,\omega_0]$ is isomorphic to $[X,x_0;\ \Omega^2 Y,\omega_0]$ and also to $[S^2 X,*;\ Y,y_0]$.

2.26. Proposition. *$\Omega^n Y$ is a homotopy commutative H-group for $n \geqslant 2$. Similarly $S^n X$ is a homotopy commutative H-cogroup for $n \geqslant 2$.*

Proof: For all $(X,x_0) \in \mathscr{PT}$ we know that

$$[X,x_0;\ \Omega^n Y,\omega_0] \cong [SX,*;\ \Omega^{n-1} Y,\omega_0] = [SX,*;\ \Omega(\Omega^{n-2} Y),\omega_0]$$

is abelian. Thus if $[f]$, $[g] \in [X,x_0;\ \Omega^n Y,\omega_0]$, then we have $[f]\cdot[g] = [g]\cdot[f]$—that is $\mu \circ (f \times g) \circ \Delta \simeq \mu \circ (g \times f) \circ \Delta$. Let us take $X = \Omega^n Y \times \Omega^n Y$, $f = p_1$ (projection on first factor) and $g = p_2$. Then for $(x,y) \in \Omega^n Y \times \Omega^n Y$ we have $(f \times g) \circ \Delta(x,y) = (p_1 \times p_2)((x,y),\ (x,y)) = (x,y)$—i.e. $(f \times g) \circ \Delta = 1_X$. On the other hand $(g \times f) \circ \Delta(x,y) = (p_2 \times p_1)((x,y),(x,y)) = (y,x) = T(x,y)$—i.e. $(g \times f) \circ \Delta = T$. Thus we have proved $\mu \simeq \mu \circ T$—i.e. μ is homotopy commutative. The proof that $S^n X$ is homotopy commutative is similar. Naturally one can also give a direct proof. □

2.27. Lemma. *For all $n \geqslant 0$ $S^1 \wedge S^n$ is homeomorphic to S^{n+1}.*

Proof: We regard S^n and S^{n+1} as sitting in $\mathbb{R}^{n+2}$. In fact, we use the following subsets of $\mathbb{R}^{n+2}$:

$S^{n+1} = \{x \in \mathbb{R}^{n+2} : \|x\| = 1\}$, unit sphere;

$S^n = \{x \in \mathbb{R}^{n+2} : \|x\| = 1, x_{n+2} = 0\}$, equator;

$D^{n+1} = \{x \in \mathbb{R}^{n+2} : \|x\| \leqslant 1, x_{n+2} = 0\}$, disk in equatorial plane;

$H_+^{n+1} = \{x \in \mathbb{R}^{n+2} : \|x\| = 1, x_{n+2} \geqslant 0\}$, upper hemisphere;

$H_-^{n+1} = \{x \in \mathbb{R}^{n+2} : \|x\| = 1, x_{n+2} \leqslant 0\}$, lower hemisphere;

$s_0 = (1,0,0,\ldots,0)$, base point.

We have obvious homeomorphisms

$$p_+ : (D^{n+1}, S^n) \to (H_+^{n+1}, S^n), \quad p_- : (D^{n+1}, S^n) \to (H_-^{n+1}, S^n)$$

given by

$$p_+(x_1,\ldots,x_{n+1},0) = (x_1,\ldots,x_{n+1}, \sqrt{1 - \textstyle\sum_{i=1}^{n+1} x_i^2})$$

and

$$p_-(x_1,\ldots,x_{n+1},0) = (x_1,\ldots,x_{n+1}, -\sqrt{1-\textstyle\sum_{i=1}^{n+1} x_i^2}).$$

If $x \in S^n$, then for every $t \in I$ we have $tx + (1-t)s_0 \in D^{n+1}$.

We define $h: S^1 \wedge S^n \to S^{n+1}$ by

$$h[t,x] = \begin{cases} p_-(2tx + (1-2t)s_0) & 0 \leqslant t \leqslant 1/2 \\ p_+(2(1-t)x + (2t-1)s_0) & 1/2 \leqslant t \leqslant 1 \end{cases}, \quad x \in S^n.$$

Here we are again regarding $S^1 \wedge S^n$ as

$$I \times S^n/\{0\} \times S^n \cup I \times \{s_0\} \cup \{1\} \times S^n.$$

It is easy to see that this function is well defined, bijective and continuous. Since both $S^1 \wedge S^n$ and S^{n+1} are compact, it follows h is a homeomorphism. ▯

2.28. Corollary. *For every $n \geqslant 0$ $S^n X$ is homeomorphic to $S^n \wedge X$.*

Proof: $S^0 = \{-1,1\}$ and so $S^0 \wedge X = \{-1,1\} \times X/\{1\} \times X \cup \{-1,1\} \times \{x_0\} \cong \{-1\} \times X \cong X = S^0 X$. If we have proved that $S^n X \cong S^n \wedge X$, then $S^{n+1} X = S(S^n X) \cong S(S^n \wedge X) = S^1 \wedge (S^n \wedge X) \cong (S^1 \wedge S^n) \wedge X \cong S^{n+1} \wedge X$. Therefore by induction the statement is true for all $n \geqslant 0$. ▯

Once an algebraic topologist has groups his next thought will be to seek exact sequences connecting these groups. We now set about constructing exact sequences for homotopy sets and groups.

2.29. If (A,a_0), (B,b_0), $(C,c_0) \in \mathscr{PS}$ and $f:(A,a_0) \to (B,b_0)$, $g:(B,b_0) \to (C,c_0)$ are two functions, then the sequence

$$(A,a_0) \xrightarrow{\ f\ } (B,b_0) \xrightarrow{\ g\ } (C,c_0)$$

is called *exact* if $\operatorname{im} f = g^{-1} c_0$. If A, B, C are groups with identities a_0, b_0, c_0 respectively and if f, g are homomorphisms, then the sequence is exact in the usual sense for groups (i.e. $\operatorname{im} f = \ker g$) if and only if it is exact as a sequence of sets in the sense just defined. A long sequence

$$(A_1,a_1) \xrightarrow{\ f_1\ } (A_2,a_2) \xrightarrow{\ f_2\ } \cdots \xrightarrow{\ f_r\ } (A_{r+1},a_{r+1})$$

is called *exact* if each short sequence

$$(A_{k-1},a_{k-1}) \xrightarrow{\ f_{k-1}\ } (A_k,a_k) \xrightarrow{\ f_k\ } (A_{k+1},a_{k+1})$$

is exact for all k, $2 \leqslant k \leqslant r$.

2.30. Definition. A sequence $(X,x_0) \to (Y,y_0) \to (Z,z_0)$ in $\mathscr{PT}$ is called *coexact* if for every $(W,w_0) \in \mathscr{PT}$ the sequence of pointed sets

$$[X,x_0;W,w_0] \xleftarrow{f^*} [Y,y_0;W,w_0] \xleftarrow{g^*} [Z,z_0;W,w_0]$$

is an exact sequence of pointed sets as in 2.29. (Here we have abbreviated $F_W^*(f)$ as f^*; we shall usually do this in what follows.)

What we wish to do now is to show that any map $f:(X,x_0) \to (Y,y_0)$ in $\mathscr{PT}$ can be embedded in a coexact sequence

$$(X,x_0) \xrightarrow{f} (Y,y_0) \xrightarrow{g} (Z,z_0)$$

for a suitable choice of (Z,z_0) and g.

2.31. Definition. For any $(X,x_0) \in \mathscr{PT}$ we define the *cone* $(CX,*) \in \mathscr{PT}$ to be the smash product

$$(CX,*) = (I \wedge X,*),$$

where the base point of I is assumed to be 0. Explicitly CX is the quotient

$$CX = I \times X/\{0\} \times X \cup I \times \{x_0\}.$$

As with the suspension we use the notation $[t,x]$ to denote the image of $(t,x) \in I \times X$ in CX. The map $i: X \to CX$ defined by $i(x) = [1,x], x \in X$, is a homeomorphism from X to $\operatorname{im} i$, so we identify X with $\operatorname{im} i$ and think of X as a subspace of CX. Note that $CX/X \cong SX$.

2.32. Proposition. *A map $f:(X,x_0) \to (Y,y_0)$ is homotopic* rel x_0 *to the constant map (null homotopic) if and only if f has an extension $g:(CX,*) \to (Y,y_0)$ to CX.*

Proof: Suppose f is homotopic rel x_0 to the constant map y_0. Let $H: X \times I \to Y$ be the homotopy from y_0 to f. If we define $h: I \times X \to Y$ by $h(t,x) = H(x,t)$, then $h(\{0\} \times X \cup I \times \{x_0\}) = y_0$, so h induces a map $g: CX \to Y$ which satisfies $g[1,x] = h(1,x) = H_1(x) = f(x)$, all $x \in X$. That is $g|X = f$.

Conversely, give $g:(CX,*) \to (Y,y_0)$ extending f, let $H: X \times I \to Y$ be the composite

$$X \times I \xrightarrow{T} I \times X \xrightarrow{q} CX \xrightarrow{g} Y,$$

where q is the projection. Then $H_0(x) = g([0,x]) = y_0$, all $x \in X$, $H_1(x) = g[1,x] = f(x)$, all $x \in X$ and $H(x_0,t) = g[t,x_0] = y_0$, all $t \in I$. Thus H is a homotopy rel x_0 from y_0 to f. $\square$

2.33. Given a map $f:(X,x_0)\to(Y,y_0)$ in $\mathscr{PT}$ we construct the *mapping cone* $Y\cup_f CX$ as follows: $Y\cup_f CX$ is obtained from $Y\vee CX$ by identifying $[1,x]\in CX$ with $f(x)\in Y$ for all $x\in X$. Intuitively we "glue the base of the cone CX to Y by means of f". The projection $q:Y\vee CX\to Y\cup_f CX$ clearly defines a homeomorphism from Y to $q(Y)$, so we regard Y as a subspace of $Y\cup_f CX$.

2.34. Proposition. *For any map $g:(Y,y_0)\to(Z,z_0)$ we have $g\circ f\simeq z_0$ if and only if g has an extension $h:(Y\cup_f CX,*)\to(Z,z_0)$ to $Y\cup_f CX$.*

Proof: The proof is an obvious generalization of that for 2.32. ▯

2.35. Proposition. *For every $f:(X,x_0)\to(Y,y_0)$ in $\mathscr{PT}$ the sequence*

$$(X,x_0)\xrightarrow{f}(Y,y_0)\xrightarrow{j}(Y\cup_f CX,*)$$

is coexact, where j is the inclusion.

Proof: We must show that for every $(W,w_0)\in\mathscr{PT}$ the sequence of sets

$$[X,x_0;W,w_0]\xleftarrow{f^*}[Y,y_0;W,w_0]\xleftarrow{j^*}[Y\cup_f CX,*;W,w_0]$$

is exact. First observe that since $j:Y\to Y\cup_f CX$ has an extension to $Y\cup_f CX$, namely $1_{Y\cup_f CX}$, $j\circ f\simeq *$. Thus $f^*\circ j^*=(j\circ f)^*=*$, so $\operatorname{im} j^*\subset f^{*-1}(*)$. Now suppose $[h]\in[Y,y_0;\ W,w_0]$ lies in $f^{*-1}(*)$; i.e. $h\circ f\simeq *$. By 2.34 this implies h has an extension h' to $Y\cup_f CX$. But then $j^*[h']=[h'\circ j]=[h]$, so $[h]\in\operatorname{im} j^*$. ▯

Now we can iterate the procedure of taking mapping cones.

2.36. Lemma. *For any map $f:(X,x_0)\to(Y,y_0)$ in $\mathscr{PT}$ the sequence*

$$(X,x_0)\xrightarrow{f}(Y,y_0)\xrightarrow{j}(Y\cup_f CX,*)\xrightarrow{k}$$

$$((Y\cup_f CX)\cup_j CY,*)\xrightarrow{l}(((Y\cup_f CX)\cup_j CY)\cup_k C(Y\cup_f CX),*)$$

is coexact, where j, k, l are the inclusions.

Proof: We must check coexactness for each pair (f,j), (j,k) and (k,l). But each of these is of the type

$$(U,u_0)\xrightarrow{\phi}(V,v_0)\xrightarrow{j}(V\cup_\phi CU,*)$$

for appropriate U, V, ϕ. Thus coexactness follows from 2.35. ▯

Now the spaces $(Y\cup_f CX)\cup_j CY$ and

$$((Y\cup_f CX)\cup_j CY)\cup_k C(Y\cup_f CX)$$

are rather unwieldy, but we shall see that we can replace them with homotopy equivalent spaces which are much handier. Notice that if we collapse CY to a point in $(Y \cup_f CX) \cup_j CY$ the resulting space $(Y \cup_f CX) \cup_j CY/CY$ is homeomorphic to SX. Similarly

$$((Y \cup_f CX) \cup_j CY) \cup_k C(Y \cup_f CX)/C(Y \cup_f CX) \cong SY.$$

We would like to know that this collapsing process does not change the homotopy type of the spaces involved. We therefore set about proving that this is so.

2.37. Lemma. *For any map $f:(X,x_0) \to (Y,y_0)$ the projection*

$$q:(Y \cup_f CX) \cup_j CY \to (Y \cup_f CX) \cup_j CY/CY$$

is a homotopy equivalence.

Proof: Let $Z = (Y \cup_f CX) \cup_j CY$; observe that Z is obtained from $CX \vee CY$ by identifying $[1,x] \in CX$ with $[1,fx] \in CY$ for each $x \in X$. Define $H: Z \times I \to Z$ as follows:

$$H([s,y],t) = [(1-t)s,y], \quad s,t \in I, \quad y \in Y,$$

$$H([s,x],t) = \begin{cases} [(1+t)s,x] & 0 \leqslant s \leqslant \dfrac{1}{1+t} \\ [2-(1+t)s,fx] & \dfrac{1}{1+t} \leqslant s \leqslant 1 \end{cases}, \quad t \in I, \quad x \in X.$$

Then H is continuous and $H_0 = 1_Z$, $H_1(CY) = *$. Thus H_1 induces a map $r: Z/CY \to Z$ such that $r \circ q = H_1$. Clearly $r \circ q \simeq 1_Z \operatorname{rel} *$. Also note that $H_t(CY) \subset CY$ for every $t \in I$, so $q \circ H_t(CY) = *$ in Z/CY for every t. By 0.8 $q \circ H$ induces $\bar{H}: Z/CY \times I \to Z/CY$ such that $\bar{H} \circ (q \times 1) = q \circ H$. Then for every $z \in Z$ we have $\bar{H}_0(q(z)) = q \circ H_0(z) = q(z)$. Since q is surjective, it follows $\bar{H}_0 = 1_{Z/CY}$. Similarly $\bar{H}_1(q(z)) = q \circ H_1(z) = q \circ r \circ q(z)$. Thus $\bar{H}$ is a homotopy from 1 to $q \circ r$ rel $*$. In other words, r is a homotopy inverse of q, so q is a homotopy equivalence. ∎

Naturally 2.37 also implies that

$$q':((Y \cup_f CX) \cup_j CY) \cup_k C(Y \cup_f CX) \to ((Y \cup_f CX) \cup_j CY) \cup_k C(Y \cup_f CX)/C(Y \cup_f CX)$$

is a homotopy equivalence.

2.38. Proposition. *If $A \subset X$ is a subspace then $X \cup_i CA/CA$ is homeomorphic to X/A. Here $i: A \to X$ is the inclusion.*

Proof: Let $q: X \cup_i CA \to X \cup_i CA/CA$ be the projection and $j: X \to X \cup_i CA$ the inclusion. Then $q \circ j(A) = *$, so $q \circ j$ induces a map $\phi: X/A \to$

$X \cup_\iota CA/CA$ such that $\phi \circ p = q \circ j$ ($p: X \to X/A$ the projection). On the other hand, we can define a map $k: X \vee CA \to X/A$ from the wedge sum to X/A by $k|X = p$, $k|CA = *$. k factors through $X \cup_\iota CA$ to give a map $\bar{k}: X \cup_\iota CA \to X/A$ which clearly satisfies $\bar{k}(CA) = *$ and thus in turn induces $\psi: X \cup_\iota CA/CA \to X/A$. It is easy to see that $\phi \circ \psi = 1$, $\psi \circ \phi = 1$. □

In particular $(Y \cup_f CX) \cup_j CY/CY \cong Y \cup_f CX/Y \cong CX/X \cong SX$ and

$$((Y \cup_f CX) \cup_j CY) \cup_k C(Y \cup_f CX)/C(Y \cup_f CX) \cong (Y \cup_f CX) \cup_j CY/Y \cup_f CX \cong CY/Y \cong SY.$$

Combining these homeomorphisms with the homotopy equivalences provided by 2.37, we get homotopy equivalences $\bar{q}: (Y \cup_f CX) \cup_j CY \to SX$ and $\bar{q}': ((Y \cup_f CX) \cup_j CY) \cup_k C(Y \cup_f CX) \to SY$. In fact, we find that the following diagram commutes up to homotopy

2.39. $X \xrightarrow{f} Y \xrightarrow{j}$

$$\begin{array}{ccccc}
Y \cup_f CX & \xrightarrow{k} & (Y \cup_f CX) \cup_j CY & \xrightarrow{l} & ((Y \cup_f CX) \cup_j CY) \cup_k C(Y \cup_f CX) \\
& \searrow^{v' \circ k'} & \downarrow{\scriptstyle v' \circ \bar{q}} & & \downarrow{\scriptstyle \bar{q}'} \\
& & SX & \xrightarrow{Sf} & SY
\end{array},$$

where $k' = \bar{q} \circ k$ is also the composite $Y \cup_f CX \to Y \cup_f CX/Y \cong SX$ and $v': SX \to SX$ is the homotopy inverse of the H-cogroup SX as in 2.22. Recalling that the top line of 2.39 is coexact we obtain the following.

2.40. Lemma. *For any map $f: (X, x_0) \to (Y, y_0)$ the sequence*

$$(X, x_0) \xrightarrow{f} (Y, y_0) \xrightarrow{j} (Y \cup_f CX, *) \xrightarrow{k'} (SX, *) \xrightarrow{Sf} (SY, *)$$

is coexact.

Proof: For any $(W, w_0) \in PT$ the maps $\bar{q}^*, \bar{q}'^*, v'^*$ are bijections. □

2.41. Theorem. *For any map $f: (X, x_0) \to (Y, y_0)$ the sequence*

$$(X, x_0) \xrightarrow{f} (Y, y_0) \xrightarrow{j} (Y \cup_f CX, *) \xrightarrow{k'} (SX, *) \xrightarrow{Sf}$$
$$(SY, *) \xrightarrow{Sj} \cdots \longrightarrow (S^n X, *) \xrightarrow{S^n f} (S^n Y, *) \xrightarrow{S^n j}$$
$$(S^n(Y \cup_f CX), *) \xrightarrow{S^n k'} \cdots$$

is coexact.

Proof: Let $(W, w_0) \in \mathscr{PT}$; we must check the exactness of the long sequence of homotopy sets. Every possible consecutive pair of morphisms will occur in a sequence like

$$[S^n X, *; W, w_0] \xleftarrow{S^n f^*} [S^n Y, *; W, w_0] \xleftarrow{S^n j^*} [S^n(Y \cup_f CX), *; W, w_0]$$
$$\xleftarrow{S^n k'^*} [S^{n+1} X, *; W, w_0] \xleftarrow{S^{n+1} f^*} [S^{n+1} Y, *; W, w_0]$$

for some $n \geqslant 0$. Leaving out base points for brevity, we get a commutative diagram

$$\begin{array}{ccccc}
[S^n X; W] & \xleftarrow{S^n f^*} & [S^n Y; W] & \xleftarrow{S^n j^*} & [S^n(Y \cup_f CX); W] \\
\cong \downarrow A^n & & \cong \downarrow A^n & & \cong \downarrow A^n \\
[X; \Omega^n W] & \xleftarrow{f^*} & [Y; \Omega^n W] & \xleftarrow{j^*} & [Y \cup_f CX; \Omega^n W]
\end{array}$$

$$\begin{array}{ccccc}
\xleftarrow{S^n k'^*} & [S^{n+1} X; W] & \xleftarrow{S^{n+1} f^*} & [S^{n+1} Y; W] \\
& \cong \downarrow A^n & & \cong \downarrow A^n \\
\xleftarrow{k'^*} & [SX; \Omega^n W] & \xleftarrow{Sf^*} & [SY; \Omega^n W].
\end{array}$$

The lower sequence is exact, and hence the upper one must be as well. $\square$

Naturally the sequence

$$[X, x_0; W, w_0] \underset{f^*}{\longleftarrow} [Y, y_0; W, w_0] \underset{j^*}{\longleftarrow} [Y \cup_f CX, *; W, w_0]$$
$$\underset{k'^*}{\longleftarrow} [SX, *; W, w_0] \underset{Sf^*}{\longleftarrow} [SY, *; W, w_0] \longleftarrow \cdots$$

is exact in the sense of groups from $[SX, *; W, w_0]$ on to the right. To the left of $[SX, *; W, w_0]$ it is exact in the sense of sets, but in fact at $[Y \cup_f CX, *; W, w_0]$ we can make a slightly stronger exactness statement: there is an action of the group $[SX, *; W, w_0]$ on the set $[Y \cup_f CX, *; W, w_0]$ such that the sets $j_*^{-1} x$, $x \in [Y, y_0; W, w_0]$ are precisely the orbits.

2.42. Definition. A group G *acts* (on the left) on a set A if there is a function $\alpha: G \times A \to A$ such that the following diagrams commute:

$$\text{i)} \quad \begin{array}{ccc}
A & \xrightarrow{(e,1)} & G \times A, \\
& {\scriptstyle 1} \searrow & \downarrow \alpha \\
& & A
\end{array}
\qquad
\text{ii)} \quad \begin{array}{ccc}
G \times G \times A & \xrightarrow{1 \times \alpha} & G \times A \\
\downarrow \mu \times 1 & & \downarrow \alpha \\
G \times A & \xrightarrow{\alpha} & A,
\end{array}$$

where $\mu: G \times G \to G$ is the multiplication map of G and $(e, 1)(x) = (e, x)$

for all $x \in A$, e being the identity of G. The *orbits* of the action are the sets $Gx = \{gx : g \in G\}$ for $x \in A$.

2.43. Definition. An H-cogroup (K, k_0, μ') *coacts* on a space (X, x_0) if there is a *coaction map* $\alpha' : X \to K \vee X$ such that the following diagrams commute up to homotopy:

$$\text{i)}\quad \begin{array}{ccc} X & \xleftarrow{(x_0,\,1)} & K \vee X, \\ & {}_{1}\nwarrow & \uparrow{\scriptstyle \alpha'} \\ & & X \end{array} \qquad\qquad \text{ii)}\quad \begin{array}{ccc} K \vee K \vee X & \xleftarrow{1 \vee \alpha'} & K \vee X \\ \uparrow{\scriptstyle \mu' \vee 1} & & \uparrow{\scriptstyle \alpha'} \\ K \vee X & \xleftarrow{\alpha'} & X \end{array}$$

The following is then obvious.

2.44. Proposition. *If the H-cogroup (K, k_0, μ') coacts on the space (X, x_0) with coaction map α', then for every space (W, w_0) there is a natural action of the group $[K, k_0;\ W, w_0]$ on the set $[X, x_0;\ W, w_0]$ given by*

$$\alpha([f], [g]) = [\Delta' \circ (f \vee g) \circ \alpha']$$

for $f : (K, k_0) \to (W, w_0)$, $g : (X, x_0) \to (W, w_0)$.

2.45. As indicated above our principal example will be the map

$$\alpha' : Y \cup_f CX \;\to\; SX \vee (Y \cup_f CX)$$

defined as follows:

$$\alpha'(y) = (*, y)$$

$$\alpha'[t, x] = \left.\begin{cases} ([2t, x], *) & 0 \leqslant t \leqslant 1/2 \\ (*, [2t-1, x]) & 1/2 \leqslant t \leqslant 1 \end{cases}\right\} \quad x \in X.$$

The schematic picture looks like this

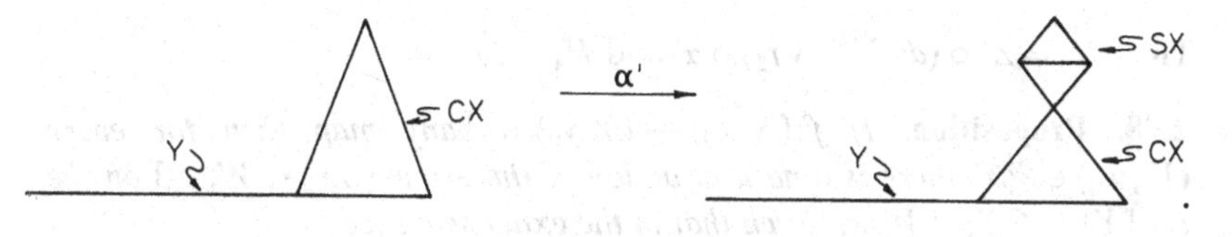

2.46. Lemma. α' *is a coaction map and in addition has the following properties:*

i) $\alpha' \circ j = (*, j)$,

ii) $\mu' \circ k' = (1 \vee k') \circ \alpha'$.

The proof is almost immediate.

Given two maps $f_1\colon Y\cup_f CX\to W$, $f_2\colon Y\cup_f CX\to W$ such that $f_1|Y=f_2|Y$ we can form

$$d(f_1,f_2)\colon SX\to W$$

defined by

$$d(f_1,f_2)[t,x]=\begin{cases} f_1[2t,x] & 0\leqslant t\leqslant 1/2\\ f_2[2-2t,x] & 1/2\leqslant t\leqslant 1.\end{cases}$$

It is clear that if $f_1\simeq f_1'$ rel Y, $f_2\simeq f_2'$ rel Y, then $d(f_1,f_2)\simeq d(f_1',f_2')$.

2.47. Lemma. *For any two maps $f_1,f_2\colon Y\cup_f CX\to W$ such that $f_1|Y=f_2|Y$ we have*

$$\Delta'\circ(d(f_1,f_2)\vee f_2)\circ\alpha'\simeq f_1\text{—i.e. } [d(f_1,f_2)]\cdot[f_2]=[f_1].$$

Proof:

$$\Delta'\circ(d(f_1,f_2)\vee f_2)\circ\alpha'(y)=f_2(y)=f_1(y),\quad y\in Y,$$

$$\Delta'\circ(d(f_1,f_2)\vee f_2)\circ\alpha'[t,x]=\begin{cases} f_1[4t,x] & 0\leqslant t\leqslant 1/4\\ f_2[2-4t,x] & 1/4\leqslant t\leqslant 1/2\\ f_2[2t-1,x] & 1/2\leqslant t\leqslant 1.\end{cases}$$

We define $H\colon(Y\cup_f CX)\times I\to W$ by

$$H(y,t)=f_1(y)$$

$$H([s,x],t)=\begin{cases} f_1[(4-3t)s,x] & 0\leqslant s\leqslant \dfrac{1}{4-3t}\\[2ex] f_2\left[\dfrac{4-t-(2-t)(4-3t)s}{2},x\right] & \dfrac{1}{4-3t}\leqslant s\leqslant\dfrac{1}{2-t}\\[2ex] f_2[(2-t)s+t-1,x] & \dfrac{1}{2-t}\leqslant s\leqslant 1.\end{cases}$$

Then $H_0=\Delta'\circ(d(f_1,f_2)\vee f_2)\circ\alpha'$ and $H_1=f_1$. □

2.48. Proposition. *If $f\colon(X,x_0)\to(Y,y_0)$ is any map then for every $(W,w_0)\in\mathcal{PT}$ there is a natural action of the group $[SX,*;\,W,w_0]$ on the set $[Y\cup_f CX,*;\,W,w_0]$ such that in the exact sequence*

$$[X;W]\xleftarrow{f^*}[Y;W]\xleftarrow{j^*}[Y\cup_f CX;W]\xleftarrow{k'^*}[SX;W]\xleftarrow{Sf^*}[SY;W]$$

the following hold:

i) *for* $x_1, x_2 \in [Y \cup_f CX; W]$ *we have* $j^*x_1 = j^*x_2 \Leftrightarrow$ *there is a* $\theta \in [SX; W]$ *with* $\theta x_2 = x_1$;

ii) *for* $y_1, y_2 \in [SX; W]$ *we have* $k'^*(y_1 + y_2) = y_1 \cdot k'^*(y_2)$;

iii) *for* $y_1, y_2 \in [SX; W]$ *we have* $k'^* y_1 = k'^* y_2 \Leftrightarrow$ *there is a* $\gamma \in [SY; W]$ *with* $y_2 = y_1 + Sf^*(\gamma)$;

iv) $k'^*(y) = y \cdot [w_0]$ *for all* $y \in [SX; W]$.

Proof: i) Suppose x_1, x_2 are represented by $f_1, f_2: Y \cup_f CX \to W$ and $j^*x_1 = j^*x_2$. Then $f_2|Y \simeq f_1|Y$ with homotopy H, say; extend H to a homotopy $\bar{H}:(Y \cup_f CX) \times I \to W$ with $\bar{H}_0 = f_2$ as follows. Let $r: I \times I \to I \times \{0\} \cup \{0, 1\} \times I$ be a retraction, say the projection from $(1/2, 2)$ as in

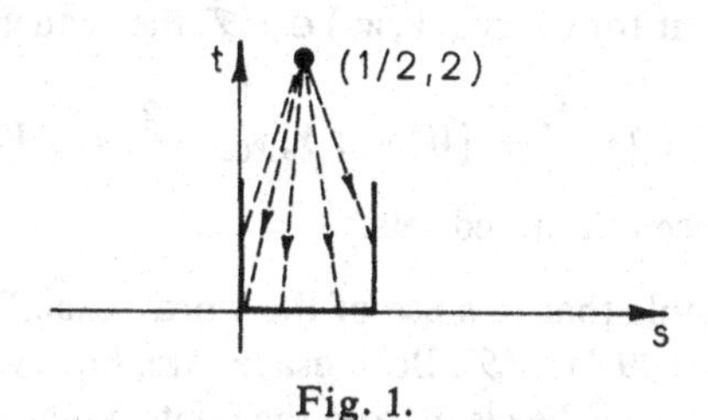

Fig. 1.

Fig. 1. We define $F: Y \times I \cup (Y \cup_f CX) \times \{0\} \to W$ by $F|Y \times I = H$ and $F|(Y \cup_f CX) \times \{0\} = f_2$. Then $\bar{H}:(Y \cup_f CX) \times I \to W$ defined by

$$\bar{H}(y, t) = F(y, t)$$
$$\bar{H}([s, x], t) = F([\pi_1 r(s, t), x], \pi_2 r(s, t)),$$

where $\pi_i: I \times I \to I$, $i = 1, 2$, are the projections, is the desired homotopy with $\bar{H}_0 = f_2$. Thus $f_2' = \bar{H}_1$ is another representative for x_2 and $f_2'|Y = f_1|Y$. Then we have

$$x_1 = [d(f_1, f_2')] \cdot [f_2'] = \theta x_2 \quad \text{for} \quad \theta = [d(f_1, f_2')] \in [SX; W].$$

On the other hand, if $[g] \in [SX; W]$, then

$$j^*([g]x_1) = j^*[\Delta' \circ (g \vee f_1) \circ \alpha'] = [\Delta' \circ (g \vee f_1) \circ \alpha' \circ j]$$
$$= [\Delta' \circ (g \vee f_1) \circ (*, j)] = [f_1 \circ j] = j^* x_1.$$

ii) Suppose y_1, $y_2 \in [SX; W]$ are represented by $f_1: SX \to W$, $f_2: SX \to W$ respectively. Then $k'^*(y_1 + y_2)$ is represented by

$$[\Delta' \circ (f_1 \vee f_2) \circ \mu' \circ k'] = [\Delta' \circ (f_1 \vee f_2) \circ (1 \vee k') \circ \alpha'] = y_1 \cdot k'^*(y_2).$$

iii) Therefore if $k'^*(y_1) = k'^*(y_2)$, we have

$$[w_0] = k'^*(-y_1 + y_1) = (-y_1) \cdot k'^*(y_1) = (-y_1) \cdot k'^*(y_2) = k'^*(-y_1 + y_2).$$

This shows that there is a $\gamma \in [SY;\ W]$ such that $Sf*(\gamma) = -y_1 + y_2$—i.e. $y_2 = y_1 + Sf^*(\gamma)$. On the other hand, for any γ

$$k'^*(y_1 + Sf^*(\gamma)) = y_1 \cdot k'^* Sf^*(\gamma) = y_1 \cdot [w_0] = k'^*(y_1 + 0) = k'^*(y_1).$$

iv) For $y \in [SX;\ W]$ we have $k'^*(y) = k'^*(y + 0) = y \cdot k'^*(0) = y \cdot [w_0]$. □

At this point the reader should be very surprised and disappointed if there were no results dual to 2.30 through 2.48. Of course there are such results.

2.49. A sequence $(X, x_0) \xrightarrow{f} (Y, y_0) \xrightarrow{g} (Z, z_0)$ in $\mathscr{PT}$ is called *exact* (in the topological sense) if for every $(W, w_0) \in \mathscr{PT}$ the sequence of sets

$$[W, w_0;\ X, x_0] \xrightarrow{f_*} [W, w_0;\ Y, y_0] \xrightarrow{g_*} [W, w_0; Z, z_0]$$

is an exact sequence of pointed sets.

Note. It is regrettable that this use of the word "exact" for $\mathscr{PT}$ conflicts with that given in 2.29 for $\mathscr{PS}$. Both usages are, however, common in the literature. It will always be clear from the context which sense of "exact" is meant.

2.50. We use the notation PY for the space $(Y, y_0)^{(I,0)}$ of paths in Y starting at y_0. One readily sees that the exponential law gives a one–one correspondence

$$[CX, *;\ Y, y_0] \leftrightarrow [X, x_0; PY, \omega_0].$$

The map $p: PY \to Y$ defined by $p(w) = w(1)$ is continuous. Note that $p^{-1}(y_0) = \Omega Y$. The dual of 2.32 is the following.

2.51. Proposition. *A map $f: (X, x_0) \to (Y, y_0)$ is null homotopic if and only if f has a lifting $g: (X, x_0) \to (PY, \omega_0)$ such that $p \circ g = f$.*

2.52. Given a map $f: (X, x_0) \to (Y, y_0)$ in $\mathscr{PT}$ we define P_f to be the following: $P_f = \{(x, w) \in X \times PY: f(x) = p(w) = w(1)\} \subset X \times PY$. The map $\pi: P_f \to X$ given by $\pi(x, w) = x$ is clearly continuous.

Dual to 2.34 is the following.

2.53. Proposition. *For any map $g: (Z, z_0) \to (X, x_0)$ we have $f \circ g \simeq y_0 \Leftrightarrow g$ has a lifting $h: (Z, z_0) \to (P_f, *)$—i.e. such that $\pi \circ h = g$.*

2.54. Proposition. *For every $f: (X, x_0) \to (Y, y_0)$ in $\mathscr{PT}$ the sequence*

$$(P_f, *) \xrightarrow{\pi} (X, x_0) \xrightarrow{f} (Y, y_0)$$

is exact.

2.55. Lemma. *For any map $f:(X,x_0) \to (Y,y_0)$ in $\mathcal{PT}$ the sequence*

$$(P_\rho,\omega_0) \xrightarrow{\sigma} (P_\pi,\omega_0) \xrightarrow{\rho} (P_f,*) \xrightarrow{\pi} (X,x_0) \xrightarrow{f} (Y,y_0)$$

is exact, where π, ρ, σ are the natural projections.

Again we are able to identify P_π and P_ρ with more familiar spaces—namely ΩY and ΩX, respectively.

2.56. Lemma. *For any map $f:(X,x_0) \to (Y,y_0)$ the map $q:\Omega Y \to P_\pi$ defined by $q(\omega) = (x_0,\omega,\omega_0)$ is a homotopy equivalence.*

Proof: Note that $P_\pi = \{(x,w,w') \in X \times PY \times PX : f(x) = w(1) \text{ and } x = w'(1)\}$. Thus $q(\omega)$ does lie in P_π for all $\omega \in \Omega Y$. We define three other maps

$$r:P_\pi \to \Omega Y$$
$$H:\Omega Y \times I \to \Omega Y$$
$$K:P_\pi \times I \to P_\pi$$

as follows:

$$r(x,w,w') = w * (f \circ w')^{-1}, \quad (x,w,w') \in P_\pi$$

$$H(\omega,t)(s) = \begin{cases} \omega\left(\dfrac{2s}{t+1}\right) & 0 \leqslant s \leqslant \dfrac{t+1}{2}, \\ y_0 & \dfrac{t+1}{2} \leqslant s \leqslant 1, \end{cases}$$

$$K(x,w,w',t) = (w'(t), \alpha(w,w',t), w'_t),$$

where $w'_t(s) = w'(st)$, $s, t \in I$, $w' \in PX$ and

$$\alpha(w,w',t)(s) = \begin{cases} w(s(2-t)) & 0 \leqslant s \leqslant \dfrac{1}{2-t}, \quad t \in I, \quad w \in PY \\ f \circ w'(s(t-2)+2) & \dfrac{1}{2-t} \leqslant s \leqslant 1, \quad w' \in PX. \end{cases}$$

r, H, K are readily seen to be continuous. It is also immediate that $H_0 = r \circ q$, $H_1 = 1_{\Omega Y}$, $K_0 = q \circ r$ and $K_1 = 1_{P_\pi}$. Thus r is a homotopy inverse of q. ▯

From 2.39 it of course follows that $q':\Omega X \to P_\rho$ defined by $q'(\omega) = (x_0,\omega_0,\omega,\omega_0)$ is a homotopy equivalence. Also the following diagram commutes up to homotopy

2.57.

$$\begin{array}{ccccccccc}
P_\rho & \xrightarrow{\sigma} & P_\pi & \xrightarrow{\rho} & P_f & \xrightarrow{\pi} & X & \xrightarrow{f} & Y \\
{\scriptstyle q'}\uparrow & & \uparrow{\scriptstyle q\circ\nu} & \nearrow{\scriptstyle \rho'\circ\nu} & & & & & \\
\Omega X & \xrightarrow[\Omega f]{} & \Omega Y & & & & & &
\end{array},$$

where $\rho' = \rho \circ q$ is the map $\rho'(\omega) = (x_0, \omega)$ and $\nu : \Omega X \to \Omega X$ is the homotopy inverse of the H-group ΩX as in 2.15. Since the top line of 2.57 is exact we obtain the following.

2.58. Lemma. *For any map* $f:(X,x_0) \to (Y,y_0)$ *the sequence*

$$(\Omega X, \omega_0) \xrightarrow{\Omega f} (\Omega Y, \omega_0) \xrightarrow{\rho'} (P_f, *) \xrightarrow{\pi} (X, x_0) \xrightarrow{f} (Y, y_0)$$

is exact.

2.59. Theorem. *For any map* $f:(X,x_0) \to (Y,y_0)$ *the sequence*

$$\cdots \longrightarrow (\Omega^n P_f, \omega_0) \xrightarrow{\Omega^n \pi} (\Omega^n X, \omega_0) \xrightarrow{\Omega^n f} (\Omega^n Y, \omega_0) \longrightarrow \cdots$$

$$(\Omega X, \omega_0) \xrightarrow{\Omega f} (\Omega Y, \omega_0) \xrightarrow{\rho'} (P_f, *) \xrightarrow{\pi} (X, x_0) \xrightarrow{f} (Y, y_0)$$

is exact.

2.43 through 2.48 have appropriate duals also. Finally, we turn to a particularly important set of homotopy groups.

2.60. Definition. For any $(X,x_0) \in \mathscr{PT}$ we have already defined $\pi_0(X,x_0)$, a pointed set. For $n \geqslant 1$ we define the *nth homotopy group* $\pi_n(X,x_0)$ of (X,x_0) by

$$\pi_n(X, x_0) = \pi_0(\Omega^n X, \omega_0).$$

Now by 2.3 $\pi_0(Y,y_0)$ can be regarded as the set $[S^0,+1;\ Y,y_0]$, so we obtain the following isomorphisms

$$\begin{aligned}
\pi_n(X, x_0) = \pi_0(\Omega^n X, \omega_0) &= [S^0, +1; \Omega^n X, \omega_0] \\
&\cong [S^1, s_0; \Omega^{n-1} X, \omega_0] \\
&\cong [S^2, s_0; \Omega^{n-2} X, \omega_0] \\
&\cong \cdots \cong [S^n, s_0; X, x_0].
\end{aligned}$$

Thus we might equally well have defined $\pi_n(X,x_0)$ to be the set of homotopy classes of maps $(S^n,s_0) \to (X,x_0)$. In this formulation it is clear that $\pi_n(X,x_0)$ is a group for $n \geqslant 1$ and even abelian for $n \geqslant 2$. In Chapters 3, 4 and 6 we shall investigate the properties of these groups.

Comments

The homotopy sets $[X, Y]$ or $[X,x_0;\ Y,y_0]$ were first systematically studied by M. G. Barratt in [20]. He introduced in that paper the exact sequence

$$[X, W] \xleftarrow{f^*} [Y, W] \longleftarrow [Y \cup_f CX; W] \longleftarrow [SX, W] \longleftarrow \cdots.$$

An exhaustive study of this sequence and its properties was made by Puppe in [71] (hence the name "Puppe sequence"). The duality which we have described in this chapter (for example between CX and PY, between $Y \cup_f CX$ and P_f, etc.) was first systematically investigated by Eckmann and Hilton (see e.g. C. R. Acad. Sci. Paris 246 (1958), 2444–2447). It is often called Eckmann–Hilton duality.

References

1. M. G. Barratt [20]
2. D. Puppe [71]

Chapter 3

Properties of the Homotopy Groups

In this chapter we shall demonstrate some of the elementary properties of the homotopy groups. Then we shall construct the relative homotopy groups $\pi_n(X, A, x_0)$ of a pointed topological pair $x_0 \in A \subset X$ and show how they fit into a long exact sequence with the groups $\pi_n(A, x_0)$ and $\pi_n(X, x_0)$. Finally we shall discuss the relation between the groups $\pi_n(X, x_0)$ and $\pi_n(X, x_1)$ for different base points.

3.1. π_n is a functor from $\mathscr{PT}'$ to $\mathscr{G}$, $n \geqslant 1$ (or even to $\mathscr{A}$ for $n \geqslant 2$): given $f:(X, x_0) \to (Y, y_0)$ we have

$$\pi_0(\Omega^n f): \pi_0(\Omega^n X, \omega_0) \to \pi_0(\Omega^n Y, \omega_0)$$

defining $\pi_n(f): \pi_n(X, x_0) \to \pi_n(Y, y_0)$. In fact, Ω^n is a functor $\mathscr{PT}' \to \mathscr{PT}'$ ($f \simeq f' \Rightarrow \Omega^n f \simeq \Omega^n f'$) and π_0 is a functor $\mathscr{PT}' \to \mathscr{PS}$, which on $\operatorname{im} \Omega^n$ happens to take group values. The composite of two functors is a functor, so $\pi_n = \pi_0 \circ \Omega^n$ is a functor. It is clear that the isomorphism $\pi_n(X, x_0) \cong F_{S^n}(X, x_0)$ demonstrated in 2.60 is in fact a natural equivalence of functors.

We shall always write f_* for $\pi_n(f)$ unless some confusion could arise.

3.2. As with all functors if f is an equivalence in $\mathscr{PT}'$ then $f_* = \pi_n(f)$ is an equivalence in $\mathscr{G}$. $[f] \in [X, x_0;\ Y, y_0]$ is an equivalence in $\mathscr{PT}'$ if and only if there is a $[g] \in [Y, y_0;\ X, x_0]$ such that $[g] \circ [f] = [1_X]$, $[f] \circ [g] = [1_Y]$—i.e. $g \circ f \simeq 1_X \operatorname{rel} x_0$, $f \circ g \simeq 1_Y \operatorname{rel} y_0$. We have called such an f (or g) a homotopy equivalence and (X, x_0), (Y, y_0) are said to have the same *homotopy type*. Thus if f is a homotopy equivalence, then $f_*: \pi_n(X, x_0) \cong \pi_n(Y, y_0)$, all $n \geqslant 0$, and if (X, x_0) and (Y, y_0) have the same homotopy type, then $\pi_n(X, x_0) \cong \pi_n(Y, y_0)$, all $n \geqslant 0$.

3.3. Definition. A pointed space (X, x_0) is called *contractible* if the inclusion $i:(\{x_0\}, x_0) \to (X, x_0)$ is a homotopy equivalence. There is, of course, a unique map $c:(X, x_0) \to (\{x_0\}, x_0)$, and clearly $c \circ i = 1_{\{x_0\}}$. Thus (X, x_0) is contractible if and only if there is a homotopy $H: X \times I \to X \operatorname{rel} x_0$ with $H_0 = 1_X$, $H_1 = i \circ c$—i.e. $H_1(X) = x_0$. H is called a *contracting homotopy*.

By 3.2 $\pi_n(X,x_0) \cong \pi_n(\{x_0\},x_0)$, all $n \geqslant 0$. But for each $n \geqslant 0$ there is but one map $(S^n,s_0) \to (\{x_0\},x_0)$, namely the "constant" map. Thus $\pi_n(\{x_0\},x_0) = 0 =$ group (set) with one element. Thus we have the following.

3.4. Proposition. *If (X,x_0) is contractible then $\pi_n(X,x_0) = 0$ for all $n \geqslant 0$.*

3.5. Examples. The n-disk (D^n,x) is contractible for all $n \geqslant 0$ (D^0 is taken to be the single point 0) and $x \in D^n$ any point. Therefore $\pi_r(D^n,x) = 0$, all $r \geqslant 0$, $n \geqslant 0$. Similarly

$$\pi_r(\mathbb{R}^n,x) = 0, \quad \text{all } r \geqslant 0, \quad n \geqslant 0, \quad x \in \mathbb{R}^n.$$

Another simple space is the n-sphere S^n and one might hope to be able to compute $\pi_r(S^n,s_0)$. We shall show in Chapter 6 that $\pi_r(S^n,s_0) = 0$ for $r < n$ and $\pi_n(S^n,s_0) \cong \mathbb{Z}$, $n \geqslant 1$. The groups $\pi_r(S^n,s_0)$, $r > n$, are not necessarily zero and are not known in general. The computation of these groups is a major area of research in algebraic topology at the moment.

3.6. Definition. If $A \subset X$ is a subspace and $i: A \to X$ is the inclusion map, then A is called a *retract* of X if there is a map $r: X \to A$ with $r \circ i = 1_A$. Applying the functor π_n, $n \geqslant 2$, we get $r_* \circ i_* = (r \circ i)_* = (1_A)_* = 1_{\pi_n(A,x_0)}$, $x_0 \in A$ any base point. That is, r_* is a left inverse of i_*. Thus $i_*: \pi_n(A,x_0) \to \pi_n(X,x_0)$ is a monomorphism. More is true; we even have

$$\begin{aligned}\pi_n(X,x_0) &\cong \operatorname{im} i_* \oplus \ker r_* \\ &\cong \pi_n(A,x_0) \oplus \ker r_*.\end{aligned}$$

For in fact $\operatorname{im} i_* \cap \ker r_* = \{0\}$ and every $\alpha \in \pi_n(X,x_0)$ can be written

$$\alpha = i_*(r_*(\alpha)) + (\alpha - i_*(r_*(\alpha)))$$

where $i_*(r_*(\alpha)) \in \operatorname{im} i_*$, $\alpha - i_*(r_*(\alpha)) \in \ker r_*$.

We shall later see that $\ker r_*$ can be identified with the relative group $\pi_n(X,A,x_0)$.

For $n = 1$ we still find that $i_*: \pi_1(A,x_0) \to \pi_1(X,x_0)$ is a monomorphism, $\ker r_*$ is a normal subgroup of $\pi_1(X,x_0)$ and every $\alpha \in \pi_1(X,x_0)$ can be written uniquely as $\alpha' \cdot \alpha''$ with $\alpha' \in \operatorname{im} i_*$, $\alpha'' \in \ker r_*$. $\pi_1(X,x_0)$ is the so-called semi-direct product of $\operatorname{im} i_*$ with $\ker r_*$.

3.7. Definition. If $A \subset X$ is a retract (with retraction r) and if in addition $i \circ r \simeq 1_X$, then A is called a *deformation retract* of X. If the homotopy $i \circ r \simeq 1_X$ is rel A, then A is called a *strong deformation retract* of X. The homotopy $H: X \times I \to X$ such that $H_0 = 1_X$, $H_1(X) \subset A$ is called the *deformation* of X into A. If $x_0 \in A$ and H is a homotopy rel x_0 (which will certainly be true for a strong deformation retract) then the inclusion $i: (A,x_0) \to (X,x_0)$ is a homotopy equivalence and hence $i_*: \pi_n(A,x_0) \to \pi_n(X,x_0)$ is an isomorphism for all $n \geqslant 0$. As a special case observe that (X,x_0) is contractible if and only if $\{x_0\}$ is a strong deformation retract of X.

3.8. Definition. Let $\mathcal{T}^2$ be the category of topological pairs (X,A), $A \subset X$ a subspace, and maps $f:(X,A) \to (Y,B)$. Similarly $\mathcal{PT}^2$ is the category of pointed pairs (X,A,x_0), $x_0 \in A \subset X$, and maps $f:(X,A,x_0) \to (Y,B,y_0)$. We also have homotopy categories $\mathcal{T}^{2'}$, $\mathcal{PT}^{2'}$ where the morphisms are homotopy classes of maps of pairs.

Given $(X,A,x_0) \in \mathcal{PT}^2$ let

$$P(X;x_0,A) = (X,x_0,A)^{(I,0,1)},$$

the space of paths in X starting at x_0 and ending in A. We have a continuous map

$$\pi: P(X;x_0,A) \to A$$

defined by $\pi(w) = w(1)$. $P(X;x_0,A)$ is in fact an old friend, namely P_i for the inclusion $i:A \to X$. Or rather, we have a homeomorphism

$$h: P(X;x_0,A) \to P_i$$

defined by $h(w) = (w(1),w)$. $h^{-1}: P_i \to P(X;x_0,A)$ is given by $h^{-1}(a,w) = w$.

We now define the *nth relative homotopy set* $\pi_n(X,A,x_0)$ of $(X,A,x_0) \in \mathcal{PT}^2$, $n \geqslant 1$, by

$$\begin{aligned}\pi_n(X,A,x_0) &= \pi_{n-1}(P(X;x_0,A),\omega_0)\\ &= \pi_0(\Omega^{n-1}P(X;x_0,A),\omega_0).\end{aligned}$$

Then $\pi_n(X,A,x_0)$ is a group for $n \geqslant 2$ and abelian for $n \geqslant 3$. π_n is a functor $\mathcal{PT}^{2'} \to \mathcal{PS}$, $\mathcal{G}$ or $\mathcal{A}$. If we define $\partial: \pi_n(X,A,x_0) \to \pi_{n-1}(A,x_0)$ to be $\pi_{n-1}(\pi): \pi_{n-1}(P(X;x_0,A),\omega_0) \to \pi_{n-1}(A,x_0)$, then ∂ is a natural transformation from the functor $\pi_n: \mathcal{PT}^{2'} \to \mathcal{PS}$ to the functor $\pi_{n-1} \circ R$, where $R: \mathcal{PT}^{2'} \to \mathcal{PT}'$ is given by $R(X,A,x_0) = (A,x_0)$. We define $j_*: \pi_n(X,x_0) \to \pi_n(X,A,x_0)$ to be $\pi_{n-1}(\rho'): \pi_{n-1}(\Omega X,\omega_0) \to \pi_{n-1}(P(X;x_0,A),\omega_0)$.

3.9. Proposition. *For every pair* $(X,A,x_0) \in \mathcal{PT}^2$ *the sequence*

$$\cdots \longrightarrow \pi_{n+1}(X,A,x_0) \xrightarrow{\partial} \pi_n(A,x_0) \xrightarrow{i_*} \pi_n(X,x_0) \xrightarrow{j_*}$$

$$\pi_n(X,A,x_0) \xrightarrow{\partial} \cdots \pi_1(X,A,x_0) \xrightarrow{\partial} \pi_0(A,x_0) \xrightarrow{i_*} \pi_0(X,x_0)$$

is exact.

Proof: From 2.52 we know the sequence

$$\cdots \longrightarrow \Omega^n A \xrightarrow{\Omega^n i} \Omega^n X \xrightarrow{\Omega^{n-1}\rho'} \Omega^{n-1}P(X;x_0,A) \xrightarrow{\Omega^{n-1}\pi}$$

$$\Omega^{n-1}A \longrightarrow \cdots \xrightarrow{\rho'} P(X;x_0,A) \xrightarrow{\pi} A \xrightarrow{i} X$$

is exact. Thus we get an exact sequence of sets and groups if we apply the functor $[S^0,+1;-]$ to this sequence. But by 2.3 $[S^0,+1;-]=\pi_0(-)$, so this sequence is precisely our sequence 3.9. $\square$

The exact sequence 3.9 is called the *exact homotopy sequence* of the pair (X,A,x_0).

3.10. Just as we showed that $\pi_n(X,x_0)$ could be identified with homotopy classes of maps of the n-sphere into X, so we can show $\pi_n(X,A,x_0)$ is homotopy classes of maps $(D^n,S^{n-1},*)\to(X,A,x_0)$. As in 2.43 the exponential law gives us a one–one correspondence

$$[CX,X,*;\ Y,B,y_0]\leftrightarrow[X,x_0;P(Y;y_0,B),\omega_0].$$

Thus we get

$$\begin{aligned}\pi_n(X,A,x_0)&=\pi_{n-1}(P(X;x_0,A),\omega_0)\\&\cong[S^{n-1},s_0;P(X;x_0,A),\omega_0]\\&\cong[CS^{n-1},S^{n-1},*;X,A,x_0].\end{aligned}$$

But we have a homeomorphism $h:CS^{n-1}\to D^n$ defined by $h([t,x])=tx+(1-t)s_0$, $t\in I$, $x\in S^{n-1}$. Therefore we also have $\pi_n(X,A,x_0)\cong[D^n,S^{n-1},s_0;\ X,A,x_0]$. In these terms $\partial:\pi_n(X,A,x_0)\to\pi_{n-1}(A,x_0)$ is given by $\partial[f]=[f|S^{n-1}]$.

3.11. If we consider the exact homotopy sequence of the pointed pair $(X,\{x_0\},x_0)$ for $(X,x_0)\in\mathscr{PT}$, we see that since $\pi_r(\{x_0\},x_0)=0$ for $r\geqslant 0$ we have $j_*:\pi_r(X,x_0)\to\pi_r(X,\{x_0\},x_0)$ is an isomorphism. Thus we may regard absolute groups as being special cases of relative groups whenever that is convenient. Moreover, if we identify $\pi_r(X,x_0)$ with $\pi_r(X,\{x_0\},x_0)$, then for any pair (X,A,x_0) the morphism $j_*:\pi_r(X,x_0)\to\pi_r(X,A,x_0)$ may be regarded as induced by the inclusion

$$j:(X,\{x_0\},x_0)\ \to\ (X,A,x_0).$$

3.12. Definition. A pair $(X,A)\in\mathscr{T}^2$ is called *0-connected* if every path component of X meets A. (X,A) is called *n-connected* if and only if (X,A) is 0-connected and $\pi_r(X,A,a)=0$ for $1\leqslant r\leqslant n$ and for all $a\in A$. A space X is called *n-connected* if and only if $\pi_k(X,x)=0$ for $0\leqslant k\leqslant n$ and all $x\in X$.

From the exactness of the sequence of the pair (X,A,x_0) it is clear that we have the following.

3.13. Proposition. *For any pair $(X,A)\in\mathscr{T}^2$ (X,A) is n-connected, $n\geqslant 0$, if and only if*

$$i_*:\pi_r(A,x_0)\ \to\ \pi_r(X,x_0)$$

is a bijection for $r<n$ and a surjection for $r=n$, all $x_0\in A$.

In particular (X,X) is n-connected, all $n\geqslant 0$.

The following condition will often be useful for characterizing those maps $f:(D^n, S^{n-1}, s_0) \to (X, A, x_0)$ which define the zero element in $\pi_n(X, A, x_0)$.

3.14. Proposition. *A map $f:(D^n, S^{n-1}, s_0) \to (X, A, x_0)$ defines the zero class in $\pi_n(X, A, x_0)$ if and only if f is homotopic rel S^{n-1} to a map f' with $f'(D^n) \subset A$.*

Proof: Suppose $f \simeq f'$ rel S^{n-1} and $f'(D^n) \subset A$. Let H be the homotopy. Define $G:(D^n \times I, S^{n-1} \times I, s_0 \times I) \to (X, A, x_0)$ by

$$G(x,t) = \begin{cases} H(x, 2t) & 0 \leqslant t \leqslant 1/2 \\ f'((2-2t)x + (2t-1)s_0) & 1/2 \leqslant t \leqslant 1. \end{cases}$$

Then $G_0 = f$, $G_1 = x_0$.

Conversely, suppose $G:(D^n \times I, S^{n-1} \times I, s_0 \times I) \to (X, A, x_0)$ is a null homotopy of f. Define $H: D^n \times I \to X$ by

$$H(x,t) = \left\{ \begin{array}{ll} G\left(\dfrac{2x}{2-t}, t\right) & 0 \leqslant \|x\| \leqslant 1 - t/2 \\ G\left(\dfrac{x}{\|x\|}, 2 - 2\|x\|\right) & 1 - t/2 \leqslant \|x\| \leqslant 1 \end{array} \right\} t \in I.$$

Then $H_0 = G_0 = f$; for all $x \in S^{n-1}$, $t \in I$, $H(x,t) = G(x,0) = H(x,0)$ and $H_1(D^n) \subset A$. □

Thus (X, A) is n-connected if and only if every map $f:(D^k, S^{k-1}) \to (X, A)$ is homotopic rel S^{k-1} to a map into A, $0 \leqslant k \leqslant n$ (where $D^0 =$ point, $S^{-1} = \varnothing$).

3.15. Given any map $f:(X, x_0) \to (Y, y_0)$ it is possible to fit $f_*:\pi_r(X, x_0) \to \pi_r(Y, y_0)$ into a long exact sequence. We employ a standard trick to "convert f into an inclusion". The *reduced mapping cylinder* $\tilde{M}_f$ of f is the space obtained from $(I \times X/I \times \{x_0\}) \vee Y$ by identifying $[1, x] \in I \times X/I \times \{x_0\}$ with $f(x) \in Y$ for every $x \in X$. If $q:(I \times X/I \times \{x_0\}) \vee Y \to \tilde{M}_f$ is the projection, then we denote by $[t, x]$ the image $q([t, x])$ for $[t, x] \in (I \times X/I \times \{x_0\})$ and by $[y]$ the image $q(y)$ for $y \in Y$. There are the following maps

$$\begin{aligned} &i:(X, x_0) \to (\tilde{M}_f, *), \quad i(x) = [0, x]; \\ &j:(Y, y_0) \to (\tilde{M}_f, *), \quad j(y) = [y]; \\ &r:(\tilde{M}_f, *) \to (Y, y_0), \quad r([y]) = y, \quad r[s, x] = f(x); \\ &H:\tilde{M}_f \times I \to \tilde{M}_f, \quad H([y], t) = [y], \quad H([s, x], t) = [t + s - st, x]. \end{aligned}$$

These satisfy the following relations:

i) $r \circ i = f$,
ii) $r \circ j = 1_Y$,
iii) $H_0 = 1_{\tilde{M}_f}$,
iv) $H_1 = j \circ r$.

Thus r, j are homotopy equivalences. i is clearly a homeomorphism of X onto $i(X) \subset \tilde{M}_f$, so we shall think of X as a subspace of $\tilde{M}_f$.

Note 1. $\tilde{M}_f / X \cong Y \cup_f CX$.

Note 2. If we are working in $\mathscr{T}$ rather than $\mathscr{PT}$, then for $f: X \to Y$ we take $M_f = (I \times X) \cup Y$ with $[1, x] \sim f(x)$, all $x \in X$, the *unreduced mapping cylinder*.

Now consider the exact homotopy sequence of the pair $(\tilde{M}_f, X)$:

$$\cdots \longrightarrow \pi_{r+1}(\tilde{M}_f, X, *) \xrightarrow{\partial}$$

$$\pi_r(X, x_0) \xrightarrow{i^*} \pi_r(\tilde{M}_f, *) \xrightarrow{k^*} \pi_r(\tilde{M}_f, X, x_0) \xrightarrow{\partial} \cdots$$

$$f_*: \pi_r(X, x_0) \searrow \pi_r(Y, y_0), \qquad r_*: \pi_r(\tilde{M}_f, *) \underset{\cong}{\rightleftarrows} \pi_r(Y, y_0): j_*$$

The diagram commutes, and r_*, j_* are isomorphisms, since r, j are homotopy equivalences. Thus we get a long exact sequence

3.16.

$$\cdots \longrightarrow \pi_{r+1}(\tilde{M}_f, X, *) \xrightarrow{\partial} \pi_r(X, x_0) \xrightarrow{f_*} \pi_r(Y, y_0) \xrightarrow{k_* \circ j_*}$$

$$\pi_r(\tilde{M}_f, X, *) \xrightarrow{\partial} \cdots$$

called the *exact homotopy sequence of f*. With this sequence statements about when f_* is an isomorphism can be converted into statements about the connectivity of the pair $(\tilde{M}_f, X, *)$.

3.17. Definition. A map $f: X \to Y$ is called an *n-equivalence*, $n \geqslant 0$, if for all $x_0 \in X$ we have $f_*: \pi_r(X, x_0) \to \pi_r(Y, y_0)$ is a bijection for $r < n$ and a surjection for $r = n$. f is called a *weak homotopy equivalence* if it is an n-equivalence for all $n \geqslant 0$. Note that every homotopy equivalence (in $\mathscr{T}$) is a weak homotopy equivalence.

Exercise. Give an example to show the following definition is not useful: "$f: (X, x_0) \to (Y, y_0)$ is a weak homotopy equivalence if $f_*: \pi_r(X, x_0) \to \pi_k(Y, y_0)$ is a bijection for $k \geqslant 0$".

From the exact sequence 3.16 we obtain the following.

3.18. Proposition. *A map $f: X \to Y$ is an n-equivalence (respectively,* *weak homotopy equivalence) if and only if for all* $x_0 \in X$ $(\tilde{M}_f, X)$ *is 0-connected and* $\pi_r(\tilde{M}_f, X, *) = 0$ *for* $1 \leqslant r \leqslant n$ *(respectively for* $1 \leqslant r$*).*

3.19. A *triple* (X, A, B, x_0) is a topological space X and two subspaces $B \subset A \subset X$ with base point $x_0 \in B$. A triple generates a remarkable plethora of inclusion relations:

$$i_1 : A \to X, \quad i_2 : B \to A, \quad i_3 : B \to X,$$

$$j_1 : (X, \{x_0\}) \to (X, A), \quad j_2 : (A, \{x_0\}) \to (A, B), \quad j_3 : (X, \{x_0\}) \to (X, B)$$

$$I : (A, B) \to (X, B), \quad J : (X, B) \to (X, A).$$

We can fit all the induced maps of homotopy groups and the appropriate boundary morphisms ∂_1, ∂_2, ∂_3 into the following elaborate diagram

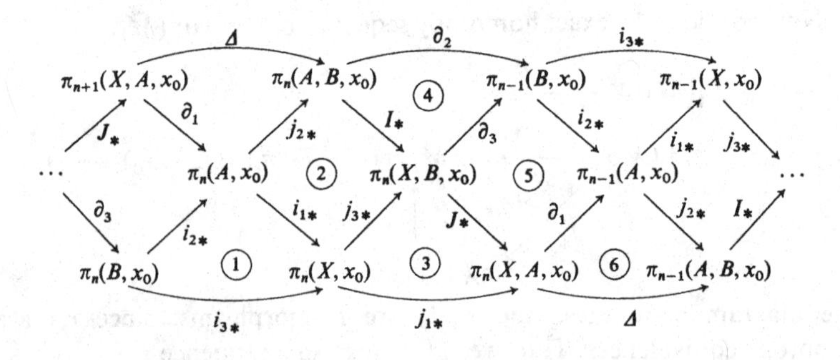

where Δ is defined to be $j_{2*} \circ \partial_1$. The diagram is commutative: triangles ① and ③ and square ② involve only morphisms induced by inclusions. Hence they commute because π_n is a functor. Triangle ④ and square ⑤ commute by naturality of the boundary ∂. ⑥ commutes by definition of Δ.

3.20. Theorem. *For any triple* (X, A, B, x_0) *the sequence*

$$\cdots \longrightarrow \pi_{n+1}(X, A, x_0) \xrightarrow{\Delta} \pi_n(A, B, x_0) \xrightarrow{I_*} \pi_n(X, B, x_0) \xrightarrow{J_*}$$
$$\pi_n(X, A, x_0) \xrightarrow{\Delta} \cdots \longrightarrow \pi_1(A, B, x_0) \xrightarrow{I_*}$$
$$\pi_1(X, B, x_0) \xrightarrow{J_*} \pi_1(X, A, x_0)$$

is exact.

Proof: The reader should check that the following proofs can be made to work for $n = 1$. We must check six different things:

i) $I_* \circ \Delta = 0 : I_* \circ \Delta = I_* \circ j_{2*} \circ \partial_1 = j_{3*} \circ i_{1*} \circ \partial_1 = 0$, since $i_{1*} \circ \partial_1 = 0$.

ii) $\Delta \circ J_* = 0 : \Delta \circ J_* = j_{2*} \circ \partial_1 \circ J_* = j_{2*} \circ i_{2*} \circ \partial_3 = 0$, since $j_{2*} \circ i_{2*} = 0$.

iii) $J_* \circ I_* = 0$: $J \circ I$ is equal to the following composition of inclusions $(A,B) \to (A,A) \to (X,A)$, and since $\pi_n(A,A,x_0) = 0$, it follows $J_* \circ I_* = 0$.

iv) $\ker I_* \subset \operatorname{im} \Delta$: suppose $x \in \pi_n(A,B,x_0)$ and $I_* x = 0$. Then $\partial_2 x = \partial_3 I_* x = 0$, so there is a $y \in \pi_n(A,x_0)$ with $j_{2*} y = x$. Then $j_{3*} i_{1*} y = I_* j_{2*} y = I_* x = 0$, so there is a $z \in \pi_n(B,x_0)$ with $i_{3*} z = i_{1*} y$. Then $i_{1*}(y - i_{2*} z) = i_{1*} y - i_{1*} i_{2*} z = i_{1*} y - i_{3*} z = 0$, so there is a

$$w \in \pi_{n+1}(X,A,x_0)$$

with $\partial_1 w = y - i_{2*} z$. Then $\Delta w = j_{2*} \partial_1 w = j_{2*}(y - i_{2*} z) = j_{2*} y - j_{2*} i_{2*} z = x$. Hence $x \in \operatorname{im} \Delta$.

v) $\ker \Delta \subset \operatorname{im} J_*$: suppose $x \in \pi_n(X,A,x_0)$ and $\Delta x = 0$. Then $j_{2*} \partial_1 x = 0$, so there is a $y \in \pi_{n-1}(B,x_0)$ with $i_{2*} y = \partial_1 x$. $i_{3*} y = i_{1*} i_{2*} y = i_{1*} \partial_1 x = 0$, so there is a $z \in \pi_n(X,B,x_0)$ with $\partial_3 z = y$. Then $\partial_1(x - J_* z) = \partial_1 x - \partial_1 J_* z = \partial_1 x - i_{2*} \partial_3 z = \partial_1 x - i_{2*} y = 0$, so there is a $w \in \pi_n(X,x_0)$ with $j_{1*} w = x - J_* z$. Then $J_*(j_{3*} w + z) = j_{1*} w + J_* z = x$. Hence $x \in \operatorname{im} J_*$.

vi) $\ker J_* \subset \operatorname{im} I_*$: suppose $x \in \pi_n(X,B,x_0)$ and $J_* x = 0$. Then $i_{2*} \partial_3 x = \partial_1 J_* x = 0$, so there is a $y \in \pi_n(A,B,x_0)$ with $\partial_2 y = \partial_3 x$. Thus $\partial_3(x - I_* y) = \partial_3 x - \partial_3 I_* y = \partial_3 x - \partial_2 y = 0$, so there is a $z \in \pi_n(X,x_0)$ with $j_{3*} z = x - I_* y$. Then $j_{1*} z = J_* j_{3*}(z) = J_*(x - I_* y) = 0$, so there is a $w \in \pi_n(A,x_0)$ with $i_{1*} w = z$. Then $I_* j_{2*} w = j_{3*} i_{1*} w = j_{3*} z = x - I_* y$, so $x = I_*(j_{2*} w + y)$. Hence $x \in \operatorname{im} I_*$. □

It is clear that the sequence 3.20 is natural in the sense that if $f:(X,A,B,x_0) \to (X',A',B',x_0')$ is a map of triples, then the following diagram commutes:

$$\begin{array}{ccccccc}
\cdots \longrightarrow & \pi_{n+1}(X,A,x_0) & \xrightarrow{\Delta} & \pi_n(A,B,x_0) & \xrightarrow{I_*} \\
& \downarrow f_* & & \downarrow (f|A)_* & \\
\cdots \longrightarrow & \pi_{n+1}(X',A',x_0') & \xrightarrow{\Delta} & \pi_n(A',B',x_0') & \xrightarrow{I_*}
\end{array}$$

$$\begin{array}{ccccc}
\pi_n(X,B,x_0) & \xrightarrow{J_*} & \pi_n(X,A,x_0) & \xrightarrow{\Delta} & \cdots \\
\downarrow f_* & & \downarrow f_* & & \\
\pi_n(X',B',x_0') & \xrightarrow{J_*} & \pi_n(X',A',x_0') & \xrightarrow{\Delta} & \cdots .
\end{array}$$

Of course if we take the triple $(X,A,\{x_0\},x_0)$ then the exact sequence 3.20 reduces to the exact sequence 3.9 of the pair (X,A,x_0) after we make the identifications $\pi_n(A,\{x_0\},x_0) = \pi_n(A,x_0)$, etc.

3.21. Let us return to the case of a retraction $r: X \to A$ of X onto a subset as in 3.6. Since $r_* \circ i_* = (r \circ i)_* = 1_*$ and hence i_* is a monomorphic, we see that in the exact sequence of the pair (X, A, x_0)

$$\cdots \xrightarrow{\partial} \pi_n(A, x_0) \xrightarrow{i_*} \pi_n(X, x_0) \xrightarrow{j_*} \pi_n(X, A, x_0) \xrightarrow{\partial} \cdots$$

in fact ∂ must always be zero, so that we obtain a collection of short sequences

$$0 \longrightarrow \pi_n(A, x_0) \underset{r_*}{\overset{i_*}{\rightleftarrows}} \pi_n(X, x_0) \xrightarrow{j_*} \pi_n(X, A, x_0) \longrightarrow 0$$

in which $r_* \circ i_* = 1$. In this situation it always follows that the middle group $(\pi_n(X, x_0))$ is the direct sum of the two end groups

$$(\pi_n(A, x_0) \oplus \pi_n(X, A, x_0)),$$

provided all are abelian groups.

3.22. Lemma. *For any short exact sequence*

$$0 \longrightarrow A \xrightarrow{i} B \xrightarrow{j} C \longrightarrow 0$$

of abelian groups the following are equivalent.

i) *There is a morphism $r: B \to A$ such that $r \circ i = 1_A$.*
ii) *There is a morphism $q: C \to B$ such that $j \circ q = 1_C$.*

In either case we have an isomorphism $B \cong A \oplus C$.

Proof: i) Suppose given $r: B \to A$ with $r \circ i = 1_A$. Then we can define $\phi: B \to A \oplus C$ by $\phi(b) = (rb, jb)$. ϕ is a homomorphism. If $\phi b = 0$, then $jb = 0$, so there is an $a \in A$ with $ia = b$. But $a = ria = rb = 0$, so $b = ia = 0$. Thus ϕ is a monomorphism. Given any $(a, c) \in A \oplus C$, choose a $b' \in B$ such that $jb' = c$. Let $b = ia + b' - irb'$; then $rb = ria + rb' - rirb' = a + rb' - rb' = a$ and $jb = jia + jb' - jirb' = jb' = c$. Hence $\phi(b) = (a, c)$, so ϕ is an epimorphism. Finally we define $q: C \to B$ by $q(c) = \phi^{-1}(0, c)$. Then q is a homomorphism and $j \circ q(c) = j \circ \phi^{-1}(0, c) = c = 1_C(c)$ for all $c \in C$—i.e. $j \circ q = 1_C$.

ii) Suppose given $q: C \to B$ with $j \circ q = 1_C$. For every $b \in B, j(b - qjb) = jb - jqjb = jb - jb = 0$; thus since i is a monomorphism there is a unique $a_b \in A$ such that $i(a_b) = b - qjb$. Let $r(b) = a_b$. Since $i(a_b + a_{b'}) = (b - qjb) + (b' - qjb') = (b + b') - qj(b + b') = i(a_{b+b'})$, it follows $r(b + b') = rb + rb'$—i.e. r is a homomorphism. Moreover, for any $a \in A$, $ir(ia) = ia - qj(ia) = ia$; since i is monomorphic, we have $ri(a) = a$ for all a—i.e. $r \circ i = 1_A$. Then by i) above it follows $B \cong A \oplus C$. $\square$

If either i) or ii) (and hence both) holds, we say the short exact sequence is *split*. If the groups are not abelian, then ii) holds if and only if B is isomorphic to a semi-direct product or "split extension" of C by A. i) and ii) are no longer equivalent.

We end this chapter with a consideration of the question: if we choose a different base point, do we get isomorphic homotopy groups? That the answer is "no" in general is easily seen from the following example. In the next chapter we shall show that $\pi_1(S^1, s_0) \cong \mathbb{Z}$. Let X be the disjoint union $S^1 \cup \{x_0\}$, where x_0 is a single point. Since S^1 is 0-connected, it follows that for every continuous map $f:(S^1, s_0) \to (X, y)$, $y \in X$, $f(S^1)$ must lie in the 0-component of y. Therefore $\pi_1(X, s_0) \cong \pi_1(S^1, s_0) \cong \mathbb{Z}$ but $\pi_1(X, x_0) \cong \pi_1(\{x_0\}, x_0) = 0$. Thus one is led to see that if X is not 0-connected then we cannot expect $\pi_n(X, x_0) \cong \pi_n(X, x_1)$ in general. 0-connectedness, however, is sufficient, as we shall show.

3.23. Definition. Let G, A be two groups with multiplication functions $\mu: A \times A \to A$, $\mu': G \times G \to G$. An action $\alpha: G \times A \to A$ of G on the set A as in 2.42 is *compatible* with the multiplication μ on A if and only if the following diagram commutes

$$\begin{array}{ccc}
G \times A \times A & \xrightarrow{1 \times \mu} & G \times A \\
\downarrow{\scriptstyle \Delta \times 1} & & \\
G \times G \times A \times A & & \\
\downarrow{\scriptstyle 1 \times T \times 1} & & \downarrow{\scriptstyle \alpha} \\
G \times A \times G \times A & & \\
\downarrow{\scriptstyle \alpha \times \alpha} & & \\
A \times A & \xrightarrow{\mu} & A.
\end{array}$$

In other words $g(a_1 a_2) = (ga_1)(ga_2)$ for $g \in G$, $a_1, a_2 \in A$.

Remarks. i) A group always acts compatibly on itself on the left by conjugation: $\alpha: G \times G \to G$ given by $\alpha(g, a) = gag^{-1}$.

ii) If A is abelian, then giving a compatible action of G on A is equivalent to giving A the structure of a module over the integral group ring $\mathbb{Z}[G]$ of G. For example, given an action $\alpha: G \times A \to A$, we define α': $\mathbb{Z}[G] \otimes A \to A$ by $\alpha'(\sum n_i g_i \otimes a_i) = \sum n_i \alpha(g_i, a_i)$, and given a module action $\alpha': \mathbb{Z}[G] \otimes A \to A$ we define $\alpha: G \times A \to A$ by $\alpha(g, a) = \alpha'(g \otimes a)$.

Just as we translated the notion of group into the notion of H-group so we can translate the notion of one group acting compatibly on another into that of one H-group acting compatibly on another.

3.24. Definition. A *compatible* action of an H-group (K,k_0,μ_K) on another (L,l_0,μ_L) is an action $\alpha: K\times L\to L$ such that the following diagram commutes up to homotopy:

$$\begin{array}{ccc}
K\times L\times L & \xrightarrow{1\times\mu_L} & K\times L \\
\downarrow{\scriptstyle \Delta\times 1} & & \\
K\times K\times L\times L & & \\
\downarrow{\scriptstyle 1\times T\times 1} & & \downarrow{\scriptstyle \alpha} \\
K\times L\times K\times L & & \\
\downarrow{\scriptstyle \alpha\times\alpha} & & \\
L\times L & \xrightarrow{\mu_L} & L.
\end{array}$$

Just at this point we are more interested in the dual notion of an H-cogroup coacting compatibly on another.

3.25. Definition. If (K,k_0,μ_K'), (L,l_0,μ_L') are H-cogroups, then a *compatible* coaction of K on L is a coaction $\alpha': L\to K\vee L$ such that the dual (turn arrows around, replace $\times$ by $\vee$) of 3.24 commutes up to homotopy. The following propositions are evident.

3.26. Proposition. *If (K,k_0,μ_K), (L,l_0,μ_L) are H-groups and $\alpha: K\times L\to L$ is a compatible action of K on L, then for every $(W,w_0)\in\mathscr{PT}$ the natural action of the group $[W,w_0;K,k_0]$ on the group $[W,w_0;L,l_0]$ defined by*

$$[f]\cdot[g]=[\alpha\circ(f\times g)\circ\Delta]$$

for $f:(W,w_0)\to(K,k_0)$, $g:(W,w_0)\to(L,l_0)$ is compatible.

3.27. Proposition. *If (K,k_0,μ_K'), (L,l_0,μ_L') are H-cogroups and $\alpha': L\to K\vee L$ is a compatible coaction of K on L, then for every $(W,w_0)\in\mathscr{PT}$ the natural action of the group $[K,k_0;W,w_0]$ on the group $[L,l_0;W,w_0]$ defined by*

$$[f]\cdot[g]=[\Delta'\circ(f\vee g)\circ\alpha']$$

for $f:(K,k_0)\to(W,w_0)$, $g:(L,l_0)\to(W,w_0)$ is compatible.

3.28. Example. We are going to construct a compatible coaction of (S^1,s_0) on (S^n,s_0) for every $n\geqslant 1$. In fact, if $\mu': S^n\to S^n\vee S^n$ is the comultiplication on S^n defined in 2.22, then $C\mu': CS^n \to C(S^n \vee S^n) \simeq CS^n \vee CS^n$ is a comultiplication on $CS^n = D^{n+1}$ restricting to μ' on S^n and making CS^n into an H-cogroup. We shall define a compatible coaction $\alpha': CS^n \to S^1 \vee CS^n$ such that $\alpha'|S^n$ is the coaction of S^1 on S^n mentioned above.

We regard S^n as $S^{n-1} \times I$ with $S^{n-1} \times \{0\}$ collapsed to a point (the base point) and $S^{n-1} \times \{1\}$ collapsed to another point (this is the so-called "unreduced suspension of S^{n-1}") for $n \geqslant 1$. We consider I as having base point 1. Then we define $\beta': CS^n \to I \vee CS^n$ by

$$\beta'([s,x,t]) = \begin{cases} (2st, *) & 2st \leqslant 1 \\ \left(1, \left[\dfrac{2st-1}{2t-1}, x, 2t-1\right]\right) & 2st > 1 \end{cases} \quad s,t \in I,\ x \in S^{n-1}.$$

Checking continuity of this function is not immediate but not difficult. The restriction $\beta'|S^n$ defines a map $\beta: S^n \to I \vee S^n$ in each case. If we break with the usual convention by taking the base points of $I \vee S^n$ and $I \vee CS^n$ to be $(0, *)$ in both cases, then β and β' will be base point-preserving maps. The map β' can be represented schematically as follows:

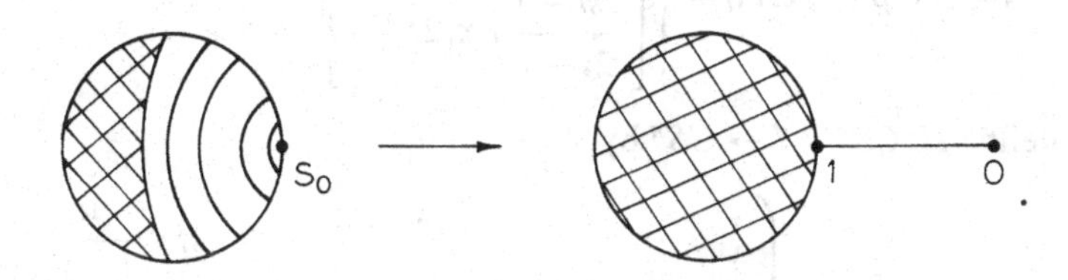

s_0 goes to 0, the part with curved lines goes to I, the curves drawn in are isoclines, and the cross-hatched part is spread out to cover D^n.

Let $I * I = (I,1) \vee (I,0)$ with base point $(1,1)$ and define $\hat{\mu}: I \to I * I$ by

$$\hat{\mu}(t) = \begin{cases} (2t, 0) & 0 \leqslant t \leqslant 1/2, \\ (1, 2t-1) & 1/2 \leqslant t \leqslant 1. \end{cases}$$

Then for any paths w, w' in X with $w(1) = w'(0)$ we see that $w * w'$ is the path

$$I \xrightarrow{\hat{\mu}} I * I \xrightarrow{w \vee w'} X \vee X \xrightarrow{\Delta'} X,$$

where X must be given the base point $w(1) = w'(0)$.

The properties of β' are summarized as follows.

3.29. Lemma. *The following diagrams commute up to homotopy:*

$$\text{i)}\quad \begin{array}{ccc} CS^n & \xleftarrow{(*,1)} & I \vee CS^n \\ & {\scriptstyle 1}\nwarrow & \uparrow{\scriptstyle \beta'} \\ & & CS^n \end{array} \qquad \text{ii)}\quad \begin{array}{ccc} (I * I) \vee CS^n \cong I \vee (I \vee CS^n) & \xrightarrow{1 \vee \beta'} & I \vee CS^n \\ \uparrow{\scriptstyle \hat{\mu} \vee 1} & & \uparrow{\scriptstyle \beta'} \\ I \vee CS^n & \xleftarrow{\beta'} & CS^n \end{array}$$

$$\text{iii)}\quad \begin{array}{ccc} I \vee CS^n \vee CS^n & \xleftarrow{1 \vee C\mu'} & I \vee CS^n \\ \uparrow{\scriptstyle \bar{\Delta}} & & \Big\downarrow{\scriptstyle \beta'} \\ (I \vee CS^n) \vee (I \vee CS^n) & & \\ \uparrow{\scriptstyle \beta' \vee \beta'} & & \\ CS^n \vee CS^n & \xleftarrow{C\mu'} & CS^n \end{array}$$

where $\bar{\Delta} = (\bar{\Delta}_1, \bar{\Delta}_2)$ is defined by $\bar{\Delta}_1(t,x) = (t,x,*)$ and $\bar{\Delta}_2(t,x) = (t,*,x)$.

Proof:

$$\text{i)}\quad (*,1) \circ \beta'([s,x,t]) = \begin{cases} * & 2st \leqslant 1, \\ \left[\dfrac{2st-1}{2t-1}, x, 2t-1\right] & 2st > 1. \end{cases}$$

We define $H: CS^n \times I \to CS^n$ by

$$H([s,x,t],u) = \begin{cases} * & st \leqslant \dfrac{u}{1+u}, \\ \left[\dfrac{st(u+1)-u}{t(u+1)-u}, x, t(u+1)-u\right] & st > \dfrac{u}{1+u}. \end{cases}$$

Then H is a homotopy and $H_0 = 1$, $H_1 = (*,1) \circ \beta'$.

ii)

$$(1 \vee \beta') \circ \beta'([s,x,t]) = \begin{cases} (2st, 0, *) & 0 \leqslant st \leqslant 1/2, \\ (1, 4st-2, *) & 1/2 \leqslant st \leqslant 3/4, \\ \left(1, 1, \left[\dfrac{4st-3}{4t-3}, x, 4t-3\right]\right) & 3/4 < st \leqslant 1. \end{cases}$$

$$(\hat{\mu} \vee 1) \circ \beta'([s,x,t]) = \begin{cases} (4st, 0, *) & 0 \leqslant st \leqslant 1/4, \\ (1, 4st-1, *) & 1/4 \leqslant st \leqslant 1/2, \\ \left(1, 1, \left[\dfrac{2st-1}{2t-1}, x, 2t-1\right]\right) & 1/2 < st \leqslant 1. \end{cases}$$

We define $H: CS^n \times I \to I \vee (I \vee CS^n)$ by

$$H([s,x,t],u) = \begin{cases} \left(\dfrac{4st}{u+1}, 0, *\right) & 0 \leqslant st \leqslant \dfrac{u+1}{4}, \\ (1, 4st-u-1, *) & \dfrac{u+1}{4} \leqslant st \leqslant \dfrac{u+2}{4}, \\ \left(1, 1, \left[\dfrac{4st-u-2}{4t-u-2}, x, \dfrac{4t-u-2}{2-u}\right]\right) & \dfrac{u+2}{4} < st \leqslant 1. \end{cases}$$

Then H is a homotopy and $H_0 = (\mu' \vee 1) \circ \beta'$, $H_1 = (1 \vee \beta') \circ \beta'$.

iii) If $n \geqslant 2$ then diagram iii) even commutes strictly because β' and $C\mu'$ involve distinct suspension coordinates. For the case $n = 1$ the proof is tedious, though elementary. It is analogous to proving that

$$\omega * w * w' * \omega^{-1} \simeq \omega * w * \omega^{-1} * \omega * w' * \omega^{-1}. \quad \square$$

Now let $q: I \to S^1$ be the projection ($q(t) = e^{2\pi i t}$). Then we define $\alpha': CS^n \to S^1 \vee CS^n$ and $\alpha: S^n \to S^1 \vee S^n$ by $\alpha' = (q \vee 1) \circ \beta'$, $\alpha = (q \vee 1) \circ \beta$. Then $\alpha' | S^n = \alpha$, and inserting factors of $q \vee 1$ at appropriate places in i), ii) and iii) above we see that α', α are coactions.

3.30. Proposition. *For every pair $(X, A, x_0) \in \mathscr{PT}^2$ there is a compatible action of $\pi_1(A, x_0)$ on $\pi_n(X, A, x_0)$, $n \geqslant 2$, and of $\pi_1(X, x_0)$ on $\pi_n(X, x_0)$, $n \geqslant 1$, satisfying:*

i) *the actions are natural; i.e. for every $f: (X, A, x_0) \to (Y, B, y_0)$ there are commutative diagrams*

$$\begin{array}{ccc} \pi_1(A,x_0) \times \pi_n(X,A,x_0) & \to & \pi_n(X,A,x_0) \\ \downarrow (f|A)_* \times f_* & & \downarrow f_* \\ \pi_1(B,y_0) \times \pi_n(Y,B,y_0) & \to & \pi_n(Y,B,y_0) \end{array} \qquad \begin{array}{ccc} \pi_1(X,x_0) \times \pi_n(X,x_0) & \to & \pi_n(X,x_0) \\ \downarrow f_* \times f_* & & \downarrow f_* \\ \pi_1(Y,y_0) \times \pi_n(Y,y_0) & \to & \pi_n(Y,y_0); \end{array}$$

ii) *$\partial: \pi_n(X, A, x_0) \to \pi_{n-1}(A, x_0)$ is a homomorphism of groups with $\pi_1(A, x_0)$ action for $n \geqslant 2$; i.e. the following square commutes*

$$\begin{array}{ccc} \pi_1(A,x_0) \times \pi_n(X,A,x_0) & \to & \pi_n(X,A,x_0) \\ \downarrow 1 \times \partial & & \downarrow \partial \\ \pi_1(A,x_0) \times \pi_{n-1}(A,x_0) & \to & \pi_{n-1}(A,x_0); \end{array}$$

iii) *the action of $\pi_1(X, x_0)$ on itself is given by conjugation; i.e. for $\gamma, \alpha \in \pi_1(X, x_0)$ we have $\gamma \cdot \alpha = \gamma\alpha\gamma^{-1}$.*

Proof: The action $\pi_1(X,x_0) = [S^1,s_0;\ X,x_0]$ on $\pi_n(X,x_0) = [S^n,s_0;\ X,x_0]$ is given by α and 3.27. The action of $\pi_1(A,x_0)$ on $\pi_n(X,A,x_0) = [CS^{n-1},S^{n-1},*;\ X,A,x_0]$ is defined in a similar manner using α'. The naturality is immediate and ii) follows from the facts that $\alpha'|S^{n-1} = \alpha$ and $\partial[f] = [f|S^{n-1}]$ for $f:(CS^{n-1},S^{n-1},*) \to (X,A,x_0)$. To prove iii) we observe that $\alpha: S^1 \to S^1 \vee S^1$ is given by

$$\alpha([t]) = \begin{cases} ([4t], *) & 0 \leqslant t \leqslant 1/4 \\ (*, [2t - 1/2]) & 1/4 \leqslant t \leqslant 3/4 \\ ([4(1-t)], *) & 3/4 \leqslant t \leqslant 1. \end{cases}$$

Hence if $\gamma = [f]$, $\alpha = [g]$, then $\gamma\alpha$ is represented by $h: I \to S^1$, where

$$h(t) = \begin{cases} f(4t) & 0 \leqslant t \leqslant 1/4, \\ g(2t - 1/2) & 1/4 \leqslant t \leqslant 3/4, \\ f(4(1-t)) & 3/4 \leqslant t \leqslant 1. \end{cases}$$

On the other hand $(\gamma\alpha)\gamma^{-1}$ is represented by h', where

$$h'(t) = \begin{cases} f(4t) & 0 \leqslant t \leqslant 1/4, \\ g(4t - 1) & 1/4 \leqslant t \leqslant 1/2, \\ f(2(1-t)) & 1/2 \leqslant t \leqslant 1. \end{cases}$$

It is clear that $h \simeq h' \operatorname{rel}\{0,1\}$. ☐

We now turn to homotopy groups based at different points. Let $PX[x_0,x_1]$ be the set of homotopy classes of paths from x_0 to x_1. For any pair (X,A) and any x_0, $x_1 \in A$ we define

$$\theta: PA[x_0,x_1] \times \pi_n(X,A,x_1) \to \pi_n(X,A,x_0)$$

by $\theta([w],[f]) =$ the homotopy class of

$$(CS^{n-1}, S^{n-1}, s_0) \xrightarrow{\beta'} (I \vee CS^{n-1}, I \vee S^{n-1},(0,*)) \xrightarrow{w \vee f}$$

$$(A \vee X, A \vee A, (x_0, x_1)) \xrightarrow{\Delta'} (X,A,x_0),$$

where in the wedge $A \vee X$ we give A and X both base point x_1. From 3.29 we immediately obtain:

i) for all $[f] \in \pi_n(X,A,x_0)$ we have $\theta([\omega_0],[f]) = [f]$;

ii) for all $[w] \in PA[x_0,x_1]$, $[w'] \in PA[x_1,x_2]$ and $[f] \in \pi_n(X,A,x_2)$ we have $\theta([w * w'],[f]) = \theta([w],\theta([w'],[f]))$;

iii) for all $[w] \in PX[x_0,x_1]$, $[f]$, $[f'] \in \pi_n(X,A,x_1)$ we have $\theta([w],[f]+[f']) = \theta([w],[f]) + \theta([w],[f'])$.

For fixed $[w]$, $\theta([w],-): \pi_n(X,A,x_1) \to \pi_n(X,A,x_0)$ is a homomorphism because of iii). From i) and ii) we see that

$$\theta([w^{-1}],\theta([w],[f])) = \theta([w^{-1} * w],[f]) = \theta([\omega_0],[f]) = [f].$$

Thus $\theta([w^{-1}],-)$ is the inverse of $\theta([w],-)$. Hence for every $[w] \in PX[x_0,x_1]$ we get an isomorphism $\theta([w],-):\pi_n(X,A,x_1) \cong \pi_n(X,A,x_0)$. Thus we have proved the following.

3.31. Proposition. *For any pair (X,A) if A is 0-connected, then $\pi_n(X,A,x_0) \cong \pi_n(X,A,x_1)$ for all $x_0,x_1 \in A$, $n \geqslant 1$. Similarly if X is 0-connected then $\pi_n(X,x_0) \cong \pi_n(X,x_1)$ for all $x_0,x_1 \in X, n \geqslant 0$.*

Remark. From the definitions it is clear that for $\gamma \in \pi_1(A,x_0)$, $\alpha \in \pi_n(X,A,x_0)$ we have $\gamma\cdot\alpha = \theta(\gamma,\alpha)$. One can show that we get a *unique* isomorphism $\pi_n(X,A,x_1) \cong \pi_n(X,A,x_0)$ if and only if the action of $\pi_1(A,x_0)$ on $\pi_n(X,A,x_0)$ is trivial—i.e. $\gamma\cdot\alpha = \alpha$ for all $\gamma \in \pi_1(A,x_0)$, $\alpha \in \pi_n(X,A,x_0)$.

Remark. Since the action of $\pi_1(X,x_0)$ on $\pi_n(X,x_0)$, $n \geqslant 1$, is natural, it is clear that (X,x_0) can have the same homotopy type as (Y,y_0) only if $\pi_1(X,x_0) \cong \pi_1(Y,y_0)$ and $\pi_n(X,x_0)$, $\pi_n(Y,y_0)$ are isomorphic as modules over $Z[\pi_1(X,x_0)]$ for $n \geqslant 2$. This is a stronger condition than merely requiring that $\pi_n(X,x_0)$, $\pi_n(Y,y_0)$ be isomorphic as groups for all $n \geqslant 0$.

3.32. Exercise. Show that for the unreduced mapping cylinder M_f of a map $f:X \to Y$ the retraction $r:M_f \to Y$ induces an isomorphism $r_*:\pi_q(M_f,x_0) \to \pi_q(Y,y_0)$, $q \geqslant 0$. Thus in the exact sequence 3.16 we may replace $\tilde{M}_f$ by M_f.

3.33. Exercise. Show that if $f:X \to Y$ is a homotopy equivalence (in $\mathscr{T}$) then $f_*:\pi_n(X,x) \to \pi_n(Y,f(x))$ is an isomorphism for all $n \geqslant 0$, all $x \in X$.

Comments

The fundamental group was defined by Poincaré. There was comparatively little interest in the higher homotopy groups until Hopf discovered in 1931 that $\pi_3(S^2,s_0) \neq 0$ [47]. Since that time their importance in topology and geometry has steadily grown. In Chapter 12, for example, we shall see how a classification of smooth manifolds can be reduced to the computation of certain homotopy groups.

References

1. P. J. Hilton [42]
2. H. Hopf [47]
3. S.-T. Hu [48]
4. E. H. Spanier [80]

Chapter 4

Fibrations

The reader familiar with homology will have noticed that in Chapter 3 we proved the analogs for $\pi_n(X,A,x_0)$ of the first five Eilenberg–Steenrod axioms and in place of the seventh axiom ("dimension axiom") we have $\pi_n(\{x_0\},x_0)=0$ for all $n \geqslant 0$. It is the sixth axiom, the "excision axiom", however, which makes homology computable for such a wide class of spaces (including the spheres S^n), and we shall see in Chapter 6 that excision holds for homotopy only under very restricted circumstances. Homotopy does have this redeeming feature, though: it behaves well with respect to fibrations. This chapter is devoted to a proof of this fact and a brief investigation of its consequences. We begin with a simple case.

4.1. Theorem. *For any* (X,x_0), $(Y,y_0) \in \mathscr{PT}$ *we have an isomorphism*

$$(p_{X*}, p_{Y*}): \pi_n(X \times Y, (x_0, y_0)) \to \pi_n(X, x_0) \times \pi_n(Y, y_0) \quad \textit{for } n \geqslant 0.$$

Proof: The proof consists of two applications of the universal property of products 0.2. Suppose we have $([f],[g]) \in \pi_n(X,x_0) \times \pi_n(Y,y_0)$. Then there is a map $(f,g):(S^n,s_0) \to (X \times Y,(x_0,y_0))$ such that $p_X \circ (f,g)=f$, $p_Y \circ (f,g)=g$. Thus $(p_{X*},p_{Y*})[(f,g)]=([f],[g])$, so (p_{X*},p_{Y*}) is surjective.

Suppose $h:(S^n,s_0) \to (X \times Y,(x_0,y_0))$ is such that $p_X \circ h \simeq x_0$, $p_Y \circ h \simeq y_0$. Let $H:S^n \times I \to X, G:S^n \times I \to Y$ be the null homotopies. Then $K=(H,G):S^n \times I \to X \times Y$ is a homotopy. $p_X \circ K_0=H_0=p_X \circ h$, $p_Y \circ K_0=G_0=p_Y \circ h$. By the uniqueness of (H_0,G_0), it follows $K_0=h$. Clearly $K_1=(x_0,y_0)$. Thus $h \simeq (x_0,y_0)$, so (p_{X*},p_{Y*}) is injective. ▯

The map $p_X: X \times Y \to X$ is a particular example of a fibration.

4.2. Definition. A map $p:E \to B$ is said to have the *homotopy lifting property* (HLP) with respect to a space X if for every map $f:X \to E$ and homotopy $G:X \times I \to B$ of $p \circ f$ there is a homotopy $F:X \times I \to E$ with $f=F_0$ and $p \circ F=G$. (F is said to be a *lifting* of G.) The diagram looks as follows,

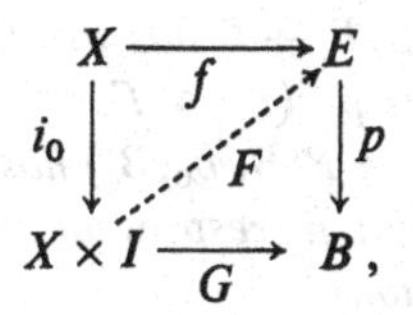

where $i_0(x) = (x,0)$, $x \in X$. p is called a *fibration* if it has the HLP for all spaces X and a *weak fibration* if it has the HLP for all disks D^n, $n \geqslant 0$. If $b_0 \in B$ is the base point, then the space $F = p^{-1}(b_0)$ is called the *fibre* of p.

The projection $p_B: B \times F \to B$ is clearly a fibration and is called the *trivial* fibration over B with fibre F. We have already encountered several nontrivial fibrations.

4.3. Proposition. *The map $\pi: PX \to X$ is a fibration with fibre ΩX.*

Proof: Suppose we are given maps $f: Y \to PX$ and $G: Y \times I \to X$ with $G_0 = \pi \circ f$. We define $F': Y \times I \times I \to X$ by

$$F'(y,t,s) = \begin{cases} (f(y))(s(t+1)) & 0 \leqslant s \leqslant \dfrac{1}{t+1} \\ G(y, s(t+1)-1) & \dfrac{1}{t+1} \leqslant s \leqslant 1 \end{cases} \quad \begin{matrix} t \in I, \\ \\ y \in Y. \end{matrix}$$

F' is continuous and defines a map $F: Y \times I \to X^I$ which clearly satisfies $F(y,t)(0) = f(y)(0) = x_0$, all y, t, i.e. $F(y,t) \in PX$, $F(y,0)(s) = f(y)(s)$, all y, s, i.e. $F_0 = f$, and $\pi \circ F(y,t) = G(y,t)$, all y, t. Thus F is the required lifting of G.

Clearly $\pi^{-1}(x_0) = \{w \in PX: w(1) = x_0\} = \Omega X$. ▯

A fact about PX which we shall need shortly is the following.

4.4. Proposition. *(PX, ω_0) is contractible for every (X, x_0).*

Proof: We define $H': PX \times I \times I \to X$ by

$$H'(w,t,s) = w(s(1-t)).$$

H' is continuous and hence defines a map $H: PX \times I \to PX$.

$H_0(w)(s) = w(s)$, so $H_0 = 1$.

$H_1(w)(s) = w(0) = y_0 = \omega_0(s)$, so $H_1 = \omega_0$.

$H(\omega_0, t)(s) = \omega_0(s(1-t)) = y_0 = \omega_0(s)$, so H is a homotopy rel ω_0. ▯

The following proposition gives the reason why homotopy groups behave well on fibrations.

4.5. Proposition. *Suppose $p:E \to B$ has the HLP with respect to a space $I \times X$, $b_0 \in B' \subset B$ and $E' = p^{-1}(B') \subset E$.*

Then $P(p):P(E; e_0, E') \to P(B; b_0, B')$ has the HLP with respect to X. In particular, if p is a fibration (resp. weak fibration) then $P(p)$ is also a fibration (resp. weak fibration).

Proof: Suppose we are given $f:X \to P(E; e_0, E')$ and $G:X \times I \to P(B; b_0, B')$ such that $G_0 = P(p) \circ f$.

$$\begin{array}{ccc} X & \xrightarrow{f} & P(E; e_0, E') \\ {\scriptstyle i_0}\downarrow & \overset{H}{\nearrow} & \downarrow{\scriptstyle P(p)} \\ X \times I & \xrightarrow[G]{} & P(B; b_0, B') \end{array} \qquad \begin{array}{ccc} I \times X \times \{0\} \cup \{0\} \times X \times I & \xrightarrow{f''} & E \\ {\scriptstyle i'}\downarrow & \overset{H'}{\nearrow} & \downarrow{\scriptstyle p} \\ I \times X \times I & \xrightarrow[G']{} & B \end{array}$$

The exponential law then gives maps $f':I \times X \to E$ and $G':I \times X \times I \to B$ such that $G'_0 = p \circ f':I \times X \to B$. We define $f'':I \times X \times \{0\} \cup \{0\} \times X \times I \to E$ by $f''|I \times X \times \{0\} = f'$, $f''|\{0\} \times X \times I = e_0$. Then $G'|(I \times X \times \{0\} \cup \{0\} \times X \times I) = p \circ f''$. Moreover, we have an obvious homeomorphism $h':I^2 \to I^2$ taking $I \times \{0\} \cup \{0\} \times I$ onto $I \times \{0\}$.

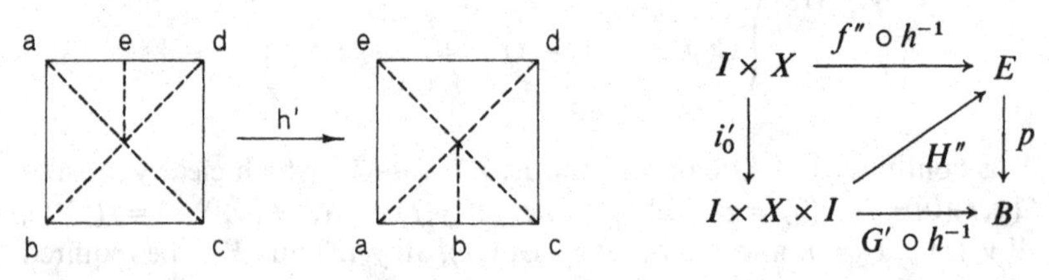

h' induces a homeomorphism $h:I \times X \times I \to I \times X \times I$ carrying $I \times X \times \{0\} \cup \{0\} \times X \times I$ onto $I \times X \times \{0\}$. Applying the HLP to $f'' \circ h^{-1}$, $G' \circ h^{-1}$, we get a lifting $H'':I \times X \times I \to E$ of $G' \circ h^{-1}$ with $H''_0 = f'' \circ h^{-1}$. Then $H' = H'' \circ h$ is a lifting of G' with $H'_0 = f'$ and $H'(0,x,t) = e_0$, all $x \in X, t \in I$. Since $p \circ H'(1,x,t) = G'(1,x,t) \in B'$, it follows

$$H'(1,x,t) \in p^{-1}(B') = E',$$

so the dual $H: X \times I \to E^I$ is in fact a map of $X \times I$ into $P(E; e_0, E')$ such that $P(p) \circ H = G$ and $H_0 = f$. ▯

4.6. Theorem. *If $p:E \to B$ is a weak fibration, $b_0 \in B' \subset B$ and $E' = p^{-1}(B')$, then $p_*:\pi_n(E, E', e_0) \to \pi_n(B, B', b_0)$ is a bijection for every $n \geqslant 1$.*

Proof: The proof will be by induction on n, beginning with $n = 1$. Given any path $\omega:(I,0,1) \to (B, b_0, B')$ we can find a lift $\omega':I \to E$ with $\omega'(0) = e_0$ (here we use the HLP for $D^0 = 0$). Since $p \circ \omega'(1) = \omega(1) \in B'$, we see that $\omega'(1) \in E'$. Hence $[\omega'] \in \pi_1(E, E', e_0)$ and clearly $p_*[\omega'] = [\omega]$. Thus p_* is surjective. Now suppose $w_1, w_2: I \to E$ are two paths with $w_\varepsilon(0) = e_0$, $w_\varepsilon(1) \in E'$, $\varepsilon = 0, 1$, such that $p \circ w_1 \simeq p \circ w_2 \,\mathrm{rel}\, 0$ in B. Let $G: I \times I \to B$ be the homotopy. We have $f: I \times \{0\} \cup \{0\} \times I \cup I \times \{1\} \to E$ defined by $f|I \times \{0\} = w_1$, $f|I \times \{1\} = w_2$, $f|\{0\} \times I = e_0$. Then $p \circ f = G|(I \times \{0\} \cup \{0\} \times I \cup I \times \{1\})$. Since there is a homeomorphism $I^2 \to I^2$ carrying $I \times \{0\} \cup \{0\} \times I \cup I \times \{1\}$ onto $I \times \{0\}$, we may repeat the trick of 4.5 and find an $H: I \times I \to E$ extending f and lifting $G(I \cong D^1)$. But then H is a homotopy from w_1 to w_2. Thus p_* is injective.

Now suppose the theorem proved for all weak fibrations and all $1 \leqslant n < m$. For any fibration $p: E \to B$ we have a commutative diagram

$$\begin{array}{ccc}
\pi_m(E, E', e_0) & \xrightarrow{p_*} & \pi_m(B, B', b_0) \\
\| & & \| \\
\pi_{m-1}(P(E;\, e_0, E'), \omega_0) & \xrightarrow{P(p)_*} & \pi_{m-1}(P(B;\, b_0, B'), \omega_0) \\
\cong \downarrow j_* & & j_* \downarrow \cong \\
\pi_{m-1}(P(E;\, e_0, E'), PE', \omega_0) & \xrightarrow{P(p)_*} & \pi_{m-1}(P(B;\, b_0, B'), PB', \omega_0).
\end{array}$$

j_* is an isomorphism because PB', PE' are contractible. But by 4.5 $P(p): P(E;\, e_0, E') \to P(B;\, b_0, B')$ is also a weak fibration, so by the inductive hypothesis $P(p)_*$ is a bijection. Therefore p_* is an isomorphism, completing the induction. ▯

Combining the exact sequence of the pair (E, F, e_0) with the isomorphism of 4.6 we get a commutative diagram

$$\begin{array}{ccccc}
\cdots \longrightarrow & \pi_{n+1}(E, F, e_0) & \xrightarrow{\partial} & \pi_n(F, e_0) & \xrightarrow{i_*} \\
{\scriptstyle p_*} \searrow & \cong \downarrow p_* & & & \\
 & \pi_{n+1}(B, \{b_0\}, b_0) \cong \pi_{n+1}(B, b_0) & & &
\end{array}$$

$$\begin{array}{ccccc}
\pi_n(E, e_0) & \xrightarrow{j_*} & \pi_n(E, F, e_0) & \xrightarrow{\partial} & \cdots \\
 & {\scriptstyle p_*} \searrow & \cong \downarrow p_* & & \\
 & & \pi_n(B, b_0) & &
\end{array}$$

This gives rise to the *exact homotopy sequence of the weak fibration* $p: E \to B$

4.7.

$$\cdots \xrightarrow{p_*} \pi_{n+1}(B,b_0) \xrightarrow{\partial'} \pi_n(F,e_0) \xrightarrow{i_*} \pi_n(E,e_0) \xrightarrow{p_*} \pi_n(B,b_0)$$
$$\cdots \xrightarrow{\partial'} \pi_0(F,e_0) \xrightarrow{i_*} \pi_0(E,e_0) \xrightarrow{p_*} \pi_0(B,b_0).$$

4.8. Corollary. *If $p:E \to B$ is a weak fibration with E contractible, then $\partial':\pi_n(B,b_0) \to \pi_{n-1}(F,e_0)$ is an isomorphism for $n \geqslant 1$.*

Example. In 4.3 we proved that $PX \xrightarrow{\pi} X$ is a fibration for any X and in 4.4 that PX is contractible. The fibre is ΩX, so 4.8 gives an isomorphism $\partial':\pi_n(X,x_0) \cong \pi_{n-1}(\Omega X,\omega_0)$—hardly a surprise!

We are now going to consider a class of fibrations which turns up most frequently. The definition will be stated in such a way that it will not be at all obvious that the result is a fibration, and we shall content ourselves with showing only that it is a weak fibration.

4.9. Definition. A *fibre bundle* is a quadruple (B,p,E,F) where p is a map $p:E \to B$ such that B has an open covering $\{U_\alpha\}_{\alpha \in A}$, and for each $\alpha \in A$ there is a homeomorphism

$$\phi_\alpha: U_\alpha \times F \to p^{-1}U_\alpha$$

such that $p \circ \phi_\alpha = p_{U_\alpha}: U_\alpha \times F \to U_\alpha$. In other words, locally $p:E \to B$ looks like a trivial fibration.

If B is paracompact, one can show that $p:E \to B$ is a fibration (see e.g. [34]).

4.10. Proposition. *If (B,p,E,F) is a fibre bundle then $p:E \to B$ is a weak fibration.*

Proof: Suppose given maps $f:D^n \to E$, $G:D^n \times I \to B$ such that $G_0 = p \circ f$. Choose an open cover $\{U_\alpha\}_{\alpha \in A}$ of B and homeomorphisms $\phi_\alpha: U_\alpha \times F \to p^{-1}U_\alpha$ as in 4.9. The open sets $\{G^{-1}U_\alpha\}_{\alpha \in A}$ cover $D^n \times I$. Since $D^n \times I$ is a compact metric space, there is a number $\lambda > 0$ (the Lebesgue number of the covering) such that every subset $S \subset D^n \times I$ of diameter $<\lambda$ is contained in some $G^{-1}U_\alpha$. Triangulate D^n so finely that every simplex has diameter $<\lambda/2$ and then subdivide I by $0 = t_0 < t_1 < \cdots < t_k = 1$ so that each set $\sigma \times [t_i,t_{i+1}]$ has diameter $<\lambda, \sigma$ simplex of D^n, $0 \leqslant i < k$.

Suppose H has already been constructed on $D^n \times [t_0,t_i]$($H|D^n \times \{0\} = f$). We shall construct H on $D^n \times [t_i,t_{i+1}]$ simplex by simplex by induction over $\dim\sigma$. If $\dim\sigma = 0$, we choose an $\alpha \in A$ so that $G(\sigma \times [t_i,t_{i+1}]) \subset U_\alpha$. Since $p \circ H(\sigma,t_i) = G(\sigma,t_i)$, we have $H(\sigma,t_i) \subset p^{-1}U_\alpha$. Define $H(\sigma,t) = \phi_\alpha(G(\sigma,t), \pi_F\phi_\alpha^{-1}(H(\sigma,t_i)))$, all $t \in [t_i,t_{i+1}]$. Then H is continuous on $\sigma \times [t_i,t_{i+1}]$ and agrees with the previous definition on $\sigma \times \{t_i\}$. Suppose H has been constructed on $\sigma' \times [t_i,t_{i+1}]$ for all σ' with

$\dim \sigma' < n$ and suppose $\dim \sigma = n$. Again choose an $\alpha \in A$ such that $G(\sigma \times [t_i, t_{i+1}]) \subset U_\alpha$. H is already defined on $\sigma \times \{t_i\} \cup \dot{\sigma} \times [t_i, t_{i+1}]$ (where $\dot{\sigma}$ means boundary of σ). We now use once again the trick we used in 4.5 and 4.6: there is a homeomorphism of $\sigma \times [t_i, t_{i+1}]$ with itself carrying $\sigma \times \{t_i\} \cup \dot{\sigma} \times [t_i, t_{i+1}]$ onto $\sigma \times \{t_i\}$. Since $p_{U_\alpha}: U_\alpha \times F \to U_\alpha$ is a fibration, we can find a lifting of $G|\sigma \times [t_i, t_{i+1}]$ to $U_\alpha \times F$ extending $\phi_\alpha^{-1} \circ H|(\sigma \times \{t_i\} \cup \dot{\sigma} \times [t_i, t_{i+1}])$. We follow this lifting by ϕ_α, obtaining a lifting $H|\sigma \times [t_i, t_{i+1}]$. This completes the induction, and hence we can extend H over $D^n \times [0, t_{i+1}]$. By induction over i we then construct H on all of $D^n \times I$, lifting G and with $H_0 = f$. ▯

4.11. One important way in which fibre bundles arise is the following. Suppose G is a topological group and $H \subset G$ is a closed subgroup. We can consider the space G/H of left cosets of H and the projection $p: G \to G/H$; of course G/H is a group only if H is a normal subgroup of G.

4.12. Definition. A map $p: E \to B$ has a *local cross-section* at a point $x \in B$ if there is a neighborhood U of x in B and a map $\lambda: U \to E$ with $p \circ \lambda = 1_U$.

Remark. In order that $p: G \to G/H$ shall have a local cross-section at every point of G/H it suffices that p should have a local cross-section at the coset H. For G acts transitively on the space G/H, the action being given by $g \cdot (g'H) = (gg')H$. Suppose (U, λ) is a local cross-section for p at H. For any other point gH of G/H the set $g \cdot U$ is a neighborhood of gH and the function $\lambda_g: g \cdot U \to G$ given by $\lambda_g(g'H) = g \cdot \lambda(g^{-1}g'H)$, $g'H \in g \cdot U$, is continuous and satisfies $p \circ \lambda_g = 1_{gU}$.

The spaces G/H are called homogeneous spaces and the above remark illustrates the aptitude of the name. In general the local structure at any point of G/H is the same as the local structure at H.

4.13. Theorem. *If $p: G \to G/H$ has a local cross-section at H, then for any closed subgroup $K \subset H$ the natural projection $p': G/K \to G/H$ is a fibre bundle with fibre H/K.*

Proof: According to the above remark we may assume that p has a local cross-section at every point of G/H. Suppose $x \in G/H$ and (U, λ) is a local cross-section of p at x. We define $\phi: U \times H/K \to G/K$ by $\phi(y, hK) = \lambda(y) \cdot hK$ for $y \in U$, $h \in H$. Then ϕ is continuous and $p' \circ \phi(y, hK) = p'(\lambda(y) \cdot hK) = \lambda(y) \cdot h \cdot H = \lambda(y) \cdot H = p(\lambda(y)) = y$ for $y \in U$, $h \in H$. We define $\psi: (p')^{-1}(U) \to U \times H/K$ by $\psi(gK) = (gH, \lambda(gH)^{-1} \cdot gK)$ for $gK \in (p')^{-1}(U)$. Then ψ is continuous and $\psi \circ \phi = 1_{U \times H/K}$, $\phi \circ \psi = 1_{(p')^{-1}(U)}$. ▯

Remark. The assumption that a local cross-section exists is necessary since every fibre bundle clearly has local cross-sections.

4.14. Examples. 1) Let $O(n)$ denote the group of orthogonal $n \times n$ matrices. We have an inclusion $O(k) \subset O(n)$ for $k \leqslant n$ if we regard $O(k)$ as the set of matrices of the form

$$\left[\begin{array}{c|c} A & 0 \\ \hline 0 & I_{n-k} \end{array}\right]$$

with A an orthogonal $k \times k$ matrix and I_{n-k} the $(n-k) \times (n-k)$ identity matrix.

$O(n)$ is regarded as a subspace of $\mathbb{R}^{n^2}$; with the relative topology $O(n)$ is compact and $O(k)$ is a closed subgroup.

Let $V_{k,n} = V_{k,n}(\mathbb{R}) = \{(v_1,\ldots,v_k): v_i \in \mathbb{R}^n, \langle v_i, v_j \rangle = \delta_{ij}, 1 \leqslant i,j \leqslant k\}$. An element of $V_{k,n}$ is called a *k-frame* in $\mathbb{R}^n$. We give $V_{k,n}$ a topology by regarding it as a subspace of $\mathbb{R}^{kn}$; it is then compact. $V_{k,n}$ is called the *Stiefel manifold* of k-frames in $\mathbb{R}^n$.

W define $\phi_{k,n}: O(n)/O(n-k) \to V_{k,n}$ by setting $\phi_{k,n}(AO(n-k)) = \{Ae_{n-k+1}, \ldots, Ae_n\}$ for $A \in O(n)$. Clearly $\phi_{k,n}(AO(n-k))$ depends only on the coset $AO(n-k)$. It is easy to see that $\phi_{k,n}$ is continuous and bijective. Since $O(n)/O(n-k)$ and $V_{k,n}$ are both Hausdorff and compact, it follows $\phi_{k,n}$ is a homeomorphism.

There is an obvious homeomorphism $V_{1,n} \cong S^{n-1}$ and hence $O(n)/O(n-1) \cong S^{n-1}$. Also clearly the diagram

$$\begin{array}{ccc} O(n)/O(n-l) & \xrightarrow{\phi_{l,n}} & V_{l,n} \\ {\scriptstyle p'}\downarrow & & \downarrow{\scriptstyle q} \\ O(n)/O(n-k) & \xrightarrow{\phi_{k,n}} & V_{k,n} \end{array}$$

commutes for $k \leqslant l$, where $q(v_1,\ldots,v_l) = (v_{l-k+1},\ldots,v_l)$.

We wish to show that $p: O(n) \to O(n)/O(n-k)$ has a local cross-section at $O(n-k)$. We do this by constructing a local cross-section of $q: V_{n,n} \to V_{k,n}$ at $(e_{n-k+1},\ldots,e_n)$. One easily sees that there is an open neighborhood U of $(e_{n-k+1},\ldots,e_n)$ in $V_{k,n}$ so that $(v_1,\ldots,v_k) \in U$ if and only if

$$(e_1,\ldots,e_{n-k}, v_1,\ldots,v_k)$$

is linearly independent. We define $\lambda: U \to V_{n,n}$ as follows. The Gram–Schmidt process gives a canonical and continuous procedure for taking a set $(u_1,\ldots,u_n)$ of linearly independent vectors in $\mathbb{R}^n$ and constructing an orthonormal set $(u_1',\ldots,u_n')$ with

$$u_n' = \frac{u_n}{\|u_n\|}, \quad u_{n-1}' = \frac{u_{n-1} - \langle u_{n-1}, u_n' \rangle u_n'}{\|u_{n-1} - \langle u_{n-1}, u_n' \rangle u_n'\|}, \quad \text{etc.}$$

For $(v_1,\ldots,v_k) \in U$ we let $\lambda(v_1,\ldots,v_k) = (e'_1,\ldots,e'_{n-k},v'_1,\ldots,v'_k)$. Since $v'_i = v_i$, $1 \leqslant i \leqslant k$, it follows $q \circ \lambda = 1_U$. Thus $p': O(n)/O(n-l) \to O(n)/O(n-k)$, $k \leqslant l$, is a fibre bundle with fibre $O(n-k)/O(n-l)$, and $q: V_{l,n} \to V_{k,n}$ is a fibre bundle with fibre $V_{l-k,n-k}$. In particular

$$p: O(n)/O(n-k-1) \to O(n)/O(n-k)$$

has fibre $O(n-k)/O(n-k-1) \cong S^{n-k-1}$.

2) Let us denote by $O(n-k) \times O(k)$ the subgroup of $O(n)$ consisting of matrices of the form

$$\left[\begin{array}{c|c} A & 0 \\ \hline 0 & B \end{array}\right]$$

with A an $(n-k) \times (n-k)$ orthogonal matrix, B a $(k \times k)$ orthogonal matrix. Let $G_{k,n} = G_{k,n}(\mathbb{R})$ denote the set of all k-dimensional linear subspaces of $\mathbb{R}^n$. There is an obvious surjection $\pi: V_{k,n} \to G_{k,n}$ given by $\pi(v_1,\ldots,v_k) = \{\text{subspace spanned by } v_1,\ldots,v_k\}$. We give $G_{k,n}$ the quotient topology induced by π. We write $\langle v_1,\ldots,v_k\rangle$ for $\pi(v_1,\ldots,v_k)$.

We can define $\psi_{k,n}: O(n)/O(n-k) \times O(k) \to G_{k,n}$ by $\psi_{k,n}(A(O(n-k) \times O(k)) = \langle Ae_{n-k+1},\ldots,Ae_n\rangle$. ψ clearly depends only on the coset $A(O(n-k) \times O(k))$ and the diagram

$$\begin{array}{ccc} O(n)/O(n-k) & \xrightarrow{\phi_{k,n}} & V_{k,n} \\ {\scriptstyle p'}\downarrow & & \downarrow{\scriptstyle \pi} \\ O(n)/O(n-k) \times O(k) & \xrightarrow{\psi_{k,n}} & G_{k,n} \end{array}$$

commutes. Hence $\psi_{k,n}$ is continuous. Since $\psi_{k,n}$ is bijective, it follows $\psi_{k,n}$ is a homeomorphism. In particular $O(n)/O(n-1) \times O(1) \cong G_{1,n} \cong RP^{n-1} = (n-1)$-dimensional projective space.

One can construct a local cross-section for $\pi: V_{k,n} \to G_{k,n}$ at $\langle e_{n-k+1},\ldots,e_n\rangle$ as follows. As neighborhood U' we take all k-dimensional subspaces W of $\mathbb{R}^n$ with $\langle e_1,\ldots,e_{n-k}\rangle \cap W = \{0\}$. For $W \in U'$ the orthogonal projections $\bar{e}_i$ of the e_i onto $W, n-k+1 \leqslant i \leqslant n$, are linearly independent. We apply the Gram–Schmidt process, obtaining a k-frame $(\bar{e}'_{n-k+1},\ldots,\bar{e}'_n)$, and we set $\lambda'(W) = (\bar{e}'_{n-k+1},\ldots,\bar{e}'_n) \in V_{k,n}$. Clearly $\pi \circ \lambda' = 1_{U'}$. If (U,λ) is the local cross-section for $q: V_{n,n} \to V_{k,n}$ constructed in 1) above, then $U \supset \lambda'(U')$ and $(U', \lambda \circ \lambda')$ is a local cross-section of $\pi \circ q$. It follows that $p': O(n)/O(n-k) \to O(n)/O(n-k) \times O(k)$ and $\pi: V_{k,n} \to G_{k,n}$ are fibre bundles with fibres $O(k)$ and $V_{k,k}$ respectively.

The $G_{k,n}(\mathbb{R})$ are called the *Grassmann manifolds* of k-planes in $\mathbb{R}^n$.

3) In $O(n)$ we have the subspace $SO(n)$ of orthogonal matrices with determinant 1. In fact there is an exact sequence

$$\{I\} \to SO(n) \to O(n) \xrightarrow{\det} \mathbb{Z}_2 \to \{1\},$$

where $\mathbb{Z}_2$ denotes the multiplicative subgroup $\{-1,+1\}$ of $\mathbb{R}^* = \mathbb{R} - \{0\}$. A cross-section for det is given by

$$\lambda(\pm 1) = \begin{pmatrix} \pm 1 & & & 0 \\ & 1 & & \\ & & \ddots & \\ 0 & & & 1 \end{pmatrix}.$$

4) Let $U(n)$ denote the group of all $n \times n$ unitary matrices regarded as a subspace of $\mathbb{C}^{n^2}$. We have inclusions $U(1) \subset U(2) \subset \cdots \subset U(n) \subset \cdots$, and $U(n)/U(n-k) \cong V_{k,n}(\mathbb{C})$ = space of all k-frames in $\mathbb{C}^n$. Local cross-sections are constructed as in 1).

5) $U(n)/U(n-k) \times U(k) \cong G_{k,n}(\mathbb{C})$ = space of all complex k-dimensional linear subspaces of $\mathbb{C}^n$.

6) We have the subgroup $SU(n) \subset U(n)$ of all unitary matrices with determinant 1. In this case we get an exact sequence

$$\{I\} \to SU(n) \to U(n) \xrightarrow{\det} S^1 \to \{1\},$$

where S^1 denotes the multiplicative group of all complex numbers of norm 1. A cross-section for det is given by

$$\lambda(z) = \begin{pmatrix} z & & & \\ & 1 & 0 & \\ & & \ddots & \\ 0 & & & 1 \end{pmatrix}, \qquad z \in S^1.$$

7) In a similar fashion we have fibrations $Sp(n) \to V_{k,n}(\mathbb{H})$ with fibre $Sp(n-k)$ and $V_{k,n}(\mathbb{H}) \to G_{k,n}(\mathbb{H})$ with fibre $V_{k,k}(\mathbb{H})$, where $Sp(n)$ is the group of $n \times n$ symplectic matrices. ($\mathbb{H}$ is the skew field of quaternions.)

Note that $V_{1,n}(\mathbb{C}) \cong S^{2n-1}$, $V_{1,n}(\mathbb{H}) \cong S^{4n-1}$, $G_{1,n}(\mathbb{C}) \cong CP^{n-1}$ and $G_{1,n}(\mathbb{H}) \cong HP^{n-1}$.

4.15. Proposition. *The inclusion $i: SO(n) \to O(n)$ induces an isomorphism $i_*: \pi_q(SO(n), 1) \to \pi_q(O(n), 1)$, $q \geqslant 1$. $\pi_q(SO(1), 1) = 0$, $q \geqslant 0$, $\pi_q(O(1), 1) = 0$, $q \geqslant 1$, $\pi_0(O(1), 1) = \mathbb{Z}_2$ and $\pi_q(SO(2), 1) \cong \pi_q(S^1, s_0)$, $q \geqslant 0$.*

Proof: From 4.14.3) we have the exact homotopy sequence

$$\cdots \to \pi_{q+1}(\mathbb{Z}_2, 1) \to \pi_q(SO(n), 1) \xrightarrow{i_*} \pi_q(O(n), 1) \to \pi_q(\mathbb{Z}_2, 1)$$

and $\pi_q(\mathbb{Z}_2,1)=0$ for $q \geqslant 1$. Thus the first assertion follows. The others follow from the facts $SO(1)=\{1\}$, $O(1)=\{-1,1\}$ and $SO(2)\cong S^1$. $\square$

4.16. Definition. A fibre bundle (B,p,E,F) with F discrete is called a *covering* of B. p is called a *covering projection* and E a *covering space* over B.

Since F is discrete, we have $\pi_n(F,e_0)=\pi_n(\{e_0\},e_0)=0$ for $n \geqslant 1$. Thus the next proposition follows.

4.17. Proposition. *If $p:\tilde{X}\to X$ is a covering of X, then $p_*:\pi_n(\tilde{X},\tilde{x}_0)\to \pi_n(X,x_0)$ is an isomorphism for all $n>1$ and a monomorphism for $n=1$. If $\tilde{X}$ is 0-connected, then the points of F are in 1–1 correspondence with the cosets of $p_*(\pi_1(\tilde{X},\tilde{x}_0))$ in $\pi_1(X,x_0)$.*

4.18. Examples. i) Regard S^1 as the circle in $\mathbb{C}:S^1=\{z\in\mathbb{C}:|z|=1\}$. Define $p:\mathbb{R}\to S^1$ by $p(t)=e^{2\pi it}$, $t\in\mathbb{R}$. We shall show that p is a covering projection. Let $U_1=S^1-\{1\}$ and $U_2=S^1-\{-1\}$. Then $\{U_1,U_2\}$ is an open cover of S^1. Define $\phi_1:U_1\times\mathbb{Z}\to p^{-1}U_1$ ($\mathbb{Z}$ = integers) by $\phi_1(z,n)=n+(1/2\pi i)\log z$, where $\log z$ denotes the principal branch of the log function on $\mathbb{C}-\{r\in\mathbb{R}:r\geqslant 0\}$. Similarly we define ϕ_2 by $\phi_2(z,n)=n+(1/2\pi i)\log z$, where here $\log z$ denotes the principal branch of the log function on $\mathbb{C}-\{r\in\mathbb{R}:r\leqslant 0\}$. In each case we have $p\circ\phi_i=p_{U_i}$. Moreover ϕ_1^{-1} is given by $\phi_1^{-1}(r)=(e^{2\pi ir},[r])$, where $[r]$ = the greatest integer $<r$ for $r\in\mathbb{R}-\mathbb{Z}=p_1^{-1}U_1$ and ϕ_2^{-1} is given by $\phi_2^{-1}(r)=(e^{2\pi ir},[r+\frac{1}{2}])$, $r\in p_2^{-1}U_2$. Thus ϕ_1,ϕ_2 are homeomorphisms as required.

Therefore by 4.17 we have $\pi_n(S^1,s_0)\cong\pi_n(\mathbb{R},0)$ for $n>1$. Since $\mathbb{R}$ is contractible, this implies that $\pi_n(S^1,s_0)=0$ for $n>1$. From the sequence 4.7 we also have $\pi_1(S^1,s_0)\cong\pi_0(F,0)=\pi_0(\mathbb{Z},0)=\mathbb{Z}$ as sets. In fact the class $[1_{S^1}]\in\pi_1(S^1,s_0)$ corresponds to $1\in\mathbb{Z}$, and one easily sees that the class of $[z\mapsto z^n]$ corresponds to $n\in\mathbb{Z}$. Also $[z\mapsto z^n]*[z\mapsto z^m]=[z\mapsto z^{n+m}]$, so $\pi_1(S^1,s_0)\cong\mathbb{Z}$ as a group.

ii) The n-dimensional torus T^n is the n-fold cartesian product $S^1\times S^1\times\cdots\times S^1$. The map $p:\mathbb{R}^n\to T^n$ defined by $p(r_1,r_2,\ldots,r_n)=(e^{2\pi ir_1},e^{2\pi ir_2},\ldots,e^{2\pi ir_n})$ is a covering projection. The fibre is the set of integer lattice points in $\mathbb{R}^n$. Since $\mathbb{R}^n$ is contractible it follows $\pi_k(T^n,*)=0$ for $k\geqslant 2$. This follows also from 4.1: $\pi_k(T^n,*)=\prod_{i=1}^n\pi_k(S^1,s_0)=0$, $k\geqslant 2$, $\pi_1(T^n,*)\cong\prod_{i=1}^n\mathbb{Z}\cong\mathbb{Z}\oplus\cdots\oplus\mathbb{Z}$.

The principal source of coverings is the following.

4.19. A topological group G *acts* on a topological space X if there is a (continuous) action map $\alpha:G\times X\to X$ such that 2.42i) and ii) with A replaced by X commute strictly—not just up to homotopy. The sets $Gx\subset X$ for fixed $x\in X$ are called the *orbits*. If we define $\sim$ on X by $x\sim x'$ if and only if there is a $g\in G$ with $gx=x'$, then the quotient $X/\!\sim$ is the space of orbits, denoted by X/G. A discrete group G with identity e is said to act *properly discontinuously* if a) for every $x \in X$ there is a neighborhood U_x such that $gU_x \cap U_x \neq \varnothing \Rightarrow g = e$ and b) for every y, $x \in X$, $y \notin Gx$ there are neighborhoods V_x, V_y of x,y such that $gV_x \cap V_y = \varnothing$, all $g \in G$. b) implies that X/G is Hausdorff. We show now that a) implies the projection $q: X \to X/G$ is a covering.

4.20. Proposition. *If the discrete group G acts on a space X properly discontinuously, then $(X/G, q, X, G)$ is a covering.*

Proof: For any $z \in X/G$ choose an $x \in X$ with $q(x) = z$ and a neighborhood U_x as in a). Let $W_z = q(U_x)$; then $q^{-1} W_z = \bigcup_{g \in G} gU_x$, so W_z is open and $z \in W_z$. Moreover $q|U_x$ is a homeomorphism of U_x onto W_z. We define $\phi_z: W_z \times G \to q^{-1} W_z$ by $\phi_z(y,g) = g \cdot (q|U_x)^{-1} y$, $g \in G$, $y \in W_z$. Then ϕ_z^{-1} is given by $\phi_z^{-1}(gy') = (q(y'), g)$ for $gy' \in gU_x$. Thus ϕ_z is a homeomorphism, and clearly $q \circ \phi_z = p_{W_z}$. The open sets W_z cover X/G, so $(X/G, q, X, G)$ is a covering. ▯

4.21. *Remarks.* i) Assume X is 0-connected. One can show that

$$\partial: \pi_1(X/G, *) \to \pi_0(F, x_0) = G$$

is a homomorphism, and of course $0 \to \pi_1(X, x_0) \xrightarrow{q_*} \pi_1(X/G, *) \xrightarrow{\partial} G \to 0$ is exact. The homeomorphisms $T_g: X \to X$ defined by $T_g(x) = gx$ define an action of G on $\pi_n(X, x_0)$, $n \geqslant 1$, by $g\gamma = \theta([w_g], T_{g*}(\gamma))$ for $\gamma \in \pi_n(X, x_0)$, w_g some fixed path from x_0 to $T_g(x_0)$. If $\pi_1(X, x_0) = 0$, then $\pi_1(X/G, *) \cong G$, and we have an action of G on $\pi_n(X/G, *)$ as in 3.30. One can show that $q_*: \pi_n(X, x_0) \to \pi_n(X/G, *)$ is a homomorphism of groups with G-action: $q_*(g\gamma) = g \cdot q_*(\gamma)$, $g \in G$, $\gamma \in \pi_n(X, x_0)$.

ii) A covering $p: \tilde{X} \to X$ is called *regular* if $p_*(\pi_1(\tilde{X}, \tilde{x}_0)) \subset \pi_1(X, x_0)$ is normal. Thus the coverings $q: X \to X/G$ of 4.20 are always regular. One can show that every regular covering arises as in 4.20.

iii) If G is a finite discrete group and acts on X without fixed points (i.e. $gx = x \Rightarrow g = e$) and X is Haudorff, then it is easy to see that G acts properly discontinuously.

iv) $\mathbb{Z}$ acts properly discontinuously on $\mathbb{R}: n \cdot r = r + n$.

There is a well-developed theory of covering spaces; the interested reader is referred to [80], for example. For an extensive discussion of fibre bundles see [81].

4.22. Exercise. i) Let $p: E \to B$ be a map and consider

$$Z = \{(e, w) \in E \times B^I : p(e) = w(0)\} \subset E \times B^I.$$

A *path lifting function* for p is a map $\lambda: Z \to E^I$ with $\lambda(e, w)(0) = e$ and $p \circ \lambda(e, w) = w$. Show that p is a fibration if and only if there is a path lifting function λ for p.

ii) Let $p: E \to B$ be a fibration with fibre F. Let $\lambda: Z \to E^I$ be a path lifting function for p and define maps

$$g: F \to P_p \quad (P_p \text{ as in 2.52})$$
$$h: P_p \to F$$

by

$$g(f) = (f, \omega_0) \in P_p, f \in F \quad (\omega_0 = \text{constant path in } B)$$
$$h(e, w) = [\lambda(e, w^{-1})](1).$$

Show that g, h are homotopy inverses of one another.

It follows that there is an exact sequence

$$\cdots \longrightarrow \Omega^2 B \longrightarrow \Omega F \xrightarrow{\Omega i} \Omega E \xrightarrow{\Omega p} \Omega B \xrightarrow{\bar{p}} F \xrightarrow{i} E \xrightarrow{p} B$$

where $\bar{p}(\omega) = [\lambda(e_0, \omega^{-1})](1)$, $\omega \in \Omega B$. Application of $[S^0, +1; -]$ to this sequence gives the exact homotopy sequence of the fibration $p: E \to B$, of course.

Comments

Fibre bundles play an important rôle in geometry quite apart from their nice homotopy properties—perhaps most of all in the theory of differentiable manifolds. We shall have more to say about them in Chapters 11, 12.

References

1. A. Dold [3]
2. D. Husemoller [47]
3. N. Steenrod [81]

Chapter 5

CW-Complexes

Computing homotopy groups is not easy in general, and one very quickly exhausts the supply of useful theorems about the homotopy groups of general topological spaces. One of the difficulties is that given two arbitrary topological spaces X, Y it is very difficult to construct any map $f: X \to Y$. If we restricted our attention to a class of spaces built up step by step out of simple building blocks (think of simplicial complexes, for example), then we might hope to construct maps step by step, extending them over the building blocks one at a time. In this chapter we describe a useful category of such spaces (CW-complexes) and display some of their elementary properties. In the next chapter we shall prove some much deeper homotopy properties of CW-complexes.

In this chapter all spaces are assumed to be Hausdorff.

5.1. Definition. A *cell complex* K on a space X is a collection $K = \{e_\alpha^n : n = 0, 1, 2, \ldots, \alpha \in J_n\}$ of subsets of X indexed by non-negative integers n and for each n by α running through some index set J_n. The set e_α^n is called a *cell of dimension n*. The cells must satisfy conditions which we list below.

Let $K^n = \{e_\alpha^r : r \leqslant n, \alpha \in J_r\}$, $n \geqslant 0$. If we are working in $\mathscr{T}$, then we take $K^{-1} = \varnothing$. If we are considering $\mathscr{PT}$, then we regard the base point x_0 as a cell of dimension $-\infty$, and in that case we take $K^n = \{\{x_0\}\}$, $n < 0$. K^n is called the *n-skeleton* of K. We let $|K^n| = \bigcup_{\substack{r \leqslant n \\ \alpha \in J_r}} e_\alpha^r$. Note that $|K^n|$ is a subspace of X, while K^n is a collection of cells. For each cell e_α^n let

$$\dot{e}_\alpha^n = e_\alpha^n \cap |K^{n-1}| = \textit{boundary} \text{ of } e_\alpha^n$$
$$\mathring{e}_\alpha^n = e_\alpha^n - \dot{e}_\alpha^n = \textit{interior} \text{ of } e_\alpha^n.$$

We require K to satisfy

i) $X = \bigcup_{n,\alpha} e_\alpha^n = |K|$;

ii) $\mathring{e}_\alpha^n \cap \mathring{e}_\beta^m \neq \varnothing$ implies $n = m$, $\alpha = \beta$;

iii) for each cell e_α^n there is a map

$$f_\alpha^n : (D^n, S^{n-1}) \to (e_\alpha^n, \dot{e}_\alpha^n)$$

which is surjective and maps $\mathring{D}^n$ homeomorphically onto $\mathring{e}_\alpha^n$.

The maps f_α^n are called the *characteristic maps* of the cells e_α^n. Authors differ as to whether the characteristic maps belong to the cell-structure; we choose to say they do.

iii) implies that each e_α^n is a compact subset of X and hence closed, since X is assumed to be Hausdorff. i) and ii) imply that X is the disjoint union of the interiors $\mathring{e}_\alpha^n$.

For any given n there may be no n-cells, i.e. $J_n = \varnothing$, but note that $J_0 = \varnothing \Rightarrow X = \varnothing$ in $\mathscr{T}$.

5.2. Definition. Let $\dim K = \sup\{n : J_n \neq \varnothing\}$. $\dim K$ may be ∞.

A cell e_β^m is called an *immediate face* of e_α^n if $\mathring{e}_\beta^m \cap e_\alpha^n \neq \varnothing$. Thus e_α^n is an immediate face of itself and if e_β^m is any other immediate face, then $m < n$. e_β^m is called a *face* of e_α^n if there exists a finite sequence

$$e_\beta^m = e_{\beta_0}^{m_0}, \quad e_{\beta_1}^{m_1}, \ldots, e_{\beta_s}^{m_s} = e_\alpha^n$$

such that $e_{\beta_i}^{m_i}$ is an immediate face of $e_{\beta_{i+1}}^{m_{i+1}}$ for $0 \leqslant i < s$. A cell is called *principal* if it is not a face of any other cell. For example, if $\dim K = n < \infty$, then all n-cells are principal.

N.B. It is not necessarily true that $\mathring{e}_\alpha^n$ is an open subset of X. Even the interiors of principal cells may not be open in a general cell complex. In fact infinite cell complexes can have very badly behaved topologies. Therefore we consider a smaller category of complexes.

5.3. Definition. A *CW-complex* K on a space X is a cell complex K on X satisfying

C) K is *closure-finite*—that is, each cell has only a finite number of (immediate) faces;

W) X has the *weak topology* induced by K—that is, a subset $S \subset X$ is closed if and only if $S \cap e_\alpha^n$ is closed in e_α^n for each n, α.

Since f_α^n induces a homeomorphism between e_α^n and a quotient space of D^n ($D^n/\sim$ where $x \sim y$ if and only if $f_\alpha^n x = f_\alpha^n y$), W) implies $S \subset X$ is closed if and only if $(f_\alpha^n)^{-1}(S)$ is closed in D^n for each n, α. We also see that interiors of principal cells are open in a CW-complex.

We shall often say that a space X "is a CW-complex" when some particular complex K on X is understood. In general, though, if X admits a CW-complex structure at all, then it admits many.

5.4. Examples. 1) Any simplicial complex is a CW-complex on its underlying polyhedron.

2) We give the n-sphere S^n a CW-complex structure in two different ways, both of which have their uses.

a) We take a single 0-cell s_0 and a single n-cell $e^n = S^n$. The characteristic map $f:(D^n, S^{n-1}) \to (e^n, \dot{e}^n) = (S^n, s_0)$ is the obvious map. For example, if we regard D^n as CS^{n-1}, then we can take f to be

$$CS^{n-1} \to CS^{n-1}/S^{n-1} \cong SS^{n-1} \cong S^n.$$

b) We take two cells of each dimension from 0 to n. Inductively, if we have already chosen the cells of S^{n-1}, then we regard $S^{n-1} \subset S^n$ as the equator, take $|K^{n-1}| = S^{n-1}$ and $e^n_+ = H^n_+$, $e^n_- = H^n_-$ – northern and southern hemispheres. The characteristic maps $f^n_\pm:(D^n, S^{n-1}) \to (e^n_\pm, \dot{e}^n_\pm) = (H^n_\pm, S^{n-1})$ are taken to be the maps $p_\pm$ defined in the proof of 2.27. In this case $f^n_\pm | S^{n-1}$ is also a homeomorphism—in fact it is $1_{S^{n-1}}$.

3) If we let $\mathbb{R}^\infty$ denote the set of all sequences $(x_1, x_2, \ldots)$ which are zero for all but a finite number of indices, then we have

$$\mathbb{R}^0 \subset \mathbb{R}^1 \subset \cdots \subset \mathbb{R}^n \subset \mathbb{R}^{n+1} \subset \cdots \subset \mathbb{R}^\infty.$$

We give $\mathbb{R}^\infty$ the weak topology—i.e. $S \subset \mathbb{R}^\infty$ is closed if and only if $S \cap \mathbb{R}^n$ is closed in $\mathbb{R}^n$ for each n. Then we have $S^\infty \subset \mathbb{R}^\infty$ and a sequence

$$S^0 \subset S^1 \subset \cdots \subset S^n \subset S^{n+1} \subset \cdots \subset S^\infty.$$

In fact $S^\infty = \bigcup_{n \geqslant 0} S^n$, and we give S^∞ a *CW*-complex structure with two cells in every dimension so that $|K^n| = S^n$.

4) Real projective n-space RP^n is the space of all lines through 0 in $\mathbb{R}^{n+1}$. We give RP^n a topology by regarding it as a quotient $\mathbb{R}^{n+1} - \{0\}/\sim$ where $x \sim x'$ in $\mathbb{R}^{n+1} - \{0\}$ if there is an $r \in \mathbb{R} - \{0\}$ with $x' = rx$. Let $q':\mathbb{R}^{n+1} - \{0\} \to RP^n$ be the projection. Then we denote the point $q'(x_1, x_2, \ldots, x_{n+1}) \in RP^n$ by $[x_1, x_2, \ldots, x_{n+1}]$. These are the so called "homogeneous coordinates" in RP^n. Let $q:S^n \to RP^n$ be the restriction $q'|S^n$. For any $x \in RP^n$, $q^{-1}x$ consists of two antipodal points in S^n, so RP^n can also be regarded as the quotient of S^n obtained by identifying antipodal points. This formulation makes it easy to see that RP^n is Hausdorff and compact.

Just as $\mathbb{R}^n$ is regarded as a subspace of $\mathbb{R}^{n+1}$, so we regard RP^{n-1} as a subspace of RP^n: $RP^{n-1} = \{[x_1, \ldots, x_{n+1}]: x_{n+1} = 0\}$. We give RP^n a *CW*-structure with one cell in each dimension from 0 to n. Inductively, if RP^{n-1} already has a *CW*-structure, then in RP^n we take $|K^{n-1}| = RP^{n-1}$ and $e^n = RP^n$. The characteristic map

$$f^n:(D^n, S^{n-1}) \to (e^n, \dot{e}^n) = (RP^n, RP^{n-1})$$

is to be the map

$$(D^n, S^{n-1}) \xrightarrow[p_+]{} (H^n_+, S^{n-1}) \subset (S^n, S^{n-1}) \xrightarrow[q]{} (RP^n, RP^{n-1}).$$

Thus f^n is given by

$$f^n(x_1,\ldots,x_n) = [x_1, x_2, \ldots, x_n, \sqrt{1 - \sum_{i=1}^n x_i^2}].$$

On $RP^n - RP^{n-1} = \mathring{e}^n$ we can define the inverse g^n of $f^n|\mathring{D}^n$ by

$$g^n([x_1, x_2, \ldots, x_{n+1}]) = \left(\frac{x_1 x_{n+1}}{\sqrt{\sum_{i=1}^{n+1}(x_i x_{n+1})^2}}, \ldots, \frac{x_n x_{n+1}}{\sqrt{\sum_{i=1}^{n+1}(x_i x_{n+1})^2}}\right).$$

Thus $f^n|\mathring{D}^n$ is indeed a homeomorphism of $\mathring{D}^n$ onto $\mathring{e}^n$.

As with S^∞ we can define RP^∞ to be the union $\bigcup_{n\geq 0} RP^n$ with the weak topology. Then RP^∞ has one cell in every dimension.

5) Complex projective n-space CP^n is the space of all complex lines through 0 in $\mathbb{C}^{n+1}$. We give CP^n a topology by regarding it as a quotient $\mathbb{C}^{n+1} - \{0\}/\sim$ where $x \sim x'$ in $\mathbb{C}^{n+1} - \{0\}$ if there is a $z \in \mathbb{C} - \{0\}$ with $x' = zx$. We let $q': \mathbb{C}^{n+1} - \{0\} \to CP^n$ be the projection and take homogeneous coordinates in CP^n: $[x_1, \ldots, x_{n+1}]$ is the point $q'(x_1, \ldots, x_{n+1})$ if $(x_1, \ldots, x_{n+1}) \in \mathbb{C}^{n+1} - \{0\}$. The restriction of q' defines a map $\eta: S^{2n+1} \to CP^n$ which also defines CP^n as a quotient of S^{2n+1}. The inverse image of a point in CP^n is a circle: $x' \in S^{2n+1}$ has the same image as $x \in S^{2n+1}$ if and only if $x' = zx$ for some $z \in S^1 = \{w \in \mathbb{C}: |w| = 1\}$.

Again the inclusions $\mathbb{C}^0 \subset \mathbb{C}^1 \subset \cdots \subset \mathbb{C}^n \subset \mathbb{C}^{n+1} \subset \cdots$ induce inclusions $CP^0 \subset CP^1 \subset \cdots \subset CP^{n-1} \subset CP^n \subset \cdots$. We give CP^n a CW-structure with one cell in every even dimension from 0 to $2n$; inductively, if CP^{n-1} already has a cell structure, then in CP^n we take $|K^{2n-1}| = CP^{n-1}$ and $e^{2n} = CP^n$. The characteristic map $f^{2n}: (D^{2n}, S^{2n-1}) \to (e^{2n}, \dot{e}^{2n}) = (CP^n, CP^{n-1})$ is to be the map given by

$$f^{2n}(x_1, x_2, \ldots, x_{2n-1}, x_{2n}) = [x_1 + ix_2, \ldots, x_{2n-1} + ix_{2n}, \sqrt{1 - \sum_{i=1}^{2n} x_i^2}].$$

On $CP^n - CP^{n-1}$ we can define the inverse g^{2n} of $f^{2n}|\mathring{D}^{2n}$ by

$$g^{2n}([z_1, z_2, \ldots, z_{n+1}]) = \left(\frac{\mathrm{Re}(z_1 \bar{z}_{n+1})}{|z_{n+1}|\sqrt{\sum_{i=1}^{n+1}|z_i|^2}}, \frac{\mathrm{Im}(z_1 \bar{z}_{n+1})}{|z_{n+1}|\sqrt{\sum_{i=1}^{n+1}|z_i|^2}}, \ldots, \right.$$

$$\left. \frac{\mathrm{Re}(z_n \bar{z}_{n+1})}{|z_{n+1}|\sqrt{\sum_{i=1}^{n+1}|z_i|^2}}, \frac{\mathrm{Im}(z_n \bar{z}_{n+1})}{|z_{n+1}|\sqrt{\sum_{i=1}^{n+1}|z_i|^2}}\right).$$

We can take $CP^\infty = \bigcup_{n\geq 0} CP^n$ with the weak topology. Then CP^∞ has one cell in every even dimension.

6) Quaternionic projective n-space HP^n is the space of all quaternionic lines through 0 in $\mathbb{H}^{n+1}$ (we regard $\mathbb{H}^{n+1}$ as a right module over $\mathbb{H}$). We

give HP^n a topology by regarding it as $\mathbb{H}^{n+1} - \{0\}/\sim$ where $x \sim x'$ in $\mathbb{H}^{n+1} - \{0\}$ if there is $q \in \mathbb{H} - \{0\}$ with $x' = xq$. Again we have $q' : \mathbb{H}^{n+1} - \{0\} \to HP^n$ and homogeneous coordinates $[x_1, x_2, \ldots, x_{n+1}]$ in HP^n. The restriction of q' defines a map $\nu : S^{4n+3} \to HP^n$ such that the inverse image of any point in HP^n is homeomorphic to S^3.

We give HP^n a CW-structure with one cell in every fourth dimension from 0 to $4n$ such that $|K^{4n-1}| = HP^{n-1} \subset HP^n$ and $e^{4n} = \dot{HP}^n$. The characteristic map $f^{4n} : (D^{4n}, S^{4n-1}) \to (HP^n, HP^{n-1})$ is given by

$$f^{4n}(x_1, \ldots, x_{4n}) =$$

$$[x_1 + ix_2 + jx_3 + kx_4, \ldots, x_{4n-3} + ix_{4n-2} + jx_{4n-1} + kx_{4n}, \sqrt{1 - \textstyle\sum_{i=1}^{4n} x_i^2}]$$

We let $HP^\infty = \bigcup_{n \geqslant 0} HP^n$ with the weak topology; HP^∞ has one cell in every fourth dimension.

We said at the beginning of this chapter that we wanted a category of spaces on which it was easy to construct maps step by step. The next proposition shows that CW-complexes do indeed have this property.

5.5. Proposition. *Let X be a CW-complex Y a topological space and $f : X \to Y$ a function. Then f is continuous if and only if $f|e^n_\alpha$ is continuous for each n, α, or equivalently if and only if $f \circ f^n_\alpha$ is continuous for each n, α.*

Proof: Suppose $A \subset Y$ is closed. Since X has the weak topology, $f^{-1}A$ is closed if and only if $(f^{-1}A) \cap \bar{e}^n_\alpha$ is closed in $\bar{e}^n_\alpha$ for each n, α. But this latter is true if and only if $f|\bar{e}^n_\alpha$ is continuous for each n, α.

Alternatively, we can take a copy D^n_α of the n-disk for each $\alpha \in J_n$, each $n \geqslant 0$, and define the "big characteristic map" $\chi : \bigcup_{n,\alpha} D^n_\alpha \to X$ from the disjoint union of all D^n_α to X such that $\chi|D^n_\alpha = f^n_\alpha$. Then X is a quotient space of $\bigcup_{n,\alpha} D^n_\alpha$, and hence f is continuous if and only if $f \circ \chi$ is continuous. But on the disjoint union $f \circ \chi$ is continuous if and only if $(f \circ \chi)|D^n_\alpha = f \circ f^n_\alpha$ is continuous for each n, α. □

We shall very often construct maps $X \to Y$ skeleton by skeleton. If we have constructed f on $|K^{n-1}|$, then in order to construct f on $|K^n|$, we have only to extend $f|\dot{e}^n_\alpha$ over $\bar{e}^n_\alpha$ continuously for each $\alpha \in J_n$. The result will be a continuous map on $|K^n|$. When we have done this for all n, the result will be a continuous function $f : X \to Y$. Expressed slightly differently, if $f\,||K^{n-1}|$ is defined, then for each $\alpha \in J_n$ we wish to extend $f \circ (f^n_\alpha|S^{n-1})$ over all of D^n continuously. If the result is $g : D^n \to Y$, then we can define $f|\bar{e}^n_\alpha$ by $f(f^n(x)) = g(x)$, $x \in D^n$. The result is well defined and continuous.

Homotopies can also be constructed cell by cell.

5.6. Proposition. *A function $F: X \times I \to Y$ is continuous if and only if $F|e_\alpha^n \times I$ is continuous for all n, α, or equivalently if and only if $F \circ (f_\alpha^n \times 1)$ is continuous for each n, α.*

Proof: Let $\chi: \bigcup_{n,\alpha} D_\alpha^n \to X$ be the "big characteristic map". By 0.8 F is continuous if and only if $F \circ (\chi \times 1)$ is continuous. But $(\bigcup_{n,\alpha} D_\alpha^n) \times I = \bigcup_{n,\alpha} (D_\alpha^n \times I)$ is a disjoint union, so $F \circ (\chi \times 1)$ is continuous if and only if $F \circ (f_\alpha^n \times 1)$ is continuous for each n, α. Again by 0.8 $F \circ (f_\alpha^n \times 1)$ is continuous if and only if $F|(e_\alpha^n \times I)$ is continuous. ▯

The following fact about CW-complexes is also useful time and again.

5.7. Proposition. *If K is a CW-complex on X, then any compact subset $S \subset X$ meets only a finite number of interiors of cells.*

Proof: For each n, α such that $S \cap \mathring{e}_\alpha^n \neq \varnothing$ choose a point $x_\alpha^n \in S \cap \mathring{e}_\alpha^n$. Let T be the set of all x_α^n; we wish to show that T is finite. Since K is closure-finite, each cell is contained in a finite union of interiors. Therefore any subset $T' \subset T$ meets each cell in a finite set. But in a Hausdorff space a finite set is closed, so $T' \cap e_\alpha^n$ is closed for each n, α. Thus T' is closed. This holds for every subset T' of T, so T is discrete. Also T itself is closed; but a closed subspace of a compact space is compact. Therefore T is discrete and compact, hence finite. ▯

5.8. Definition. If K is a cell complex on X and $L \subset K$, then L is called a *subcomplex* of K if and only if $e_\alpha^n \in L$ implies every face of e_α^n is in L. If L is a subcomplex, then L is a cell complex on $|L|$, and if K is a CW-complex on X, then L is a CW-complex on $|L|$. Each of the skeleta K^n is a subcomplex.

In 5.7 we proved that every compact subset S of X is contained in a finite subcomplex L: let L be the finite collection of cells whose interiors meet S together with all their faces.

5.9. Proposition. *If K is a CW-complex on X and L is a subcomplex, then $|L|$ is a closed subspace of X.*

Proof: $e_\alpha^n \cap |L|$ meets only finitely many interiors of cells in L, say $\mathring{e}_1^{n_1}, \ldots, \mathring{e}_k^{n_k}$. Thus $e_\alpha^n \cap |L| = \bigcup_{i=1}^k e_\alpha^n \cap e_i^{n_i}$ is a finite union of closed sets, hence closed. This is true for all n, α; hence $|L|$ is closed. ▯

So far we have taken the point of view that the space X is given and K is a prescription for cutting X up into cells. Frequently, however, we shall want to build X up cell by cell by successively glueing on new cells. We show now how this can be done. Here we shall work in $\mathscr{PT}$, although the methods can be adapted equally well for $\mathscr{T}$.

5.10. Definition. Let X be a space and $g: S^{n-1} \to X$ a map. Then we can form the mapping cone $X \cup_g CS^{n-1}$ as in Chapter 2. The result is called *X with an n-cell attached.* The map g is called the *attaching map* of the cell. The projection $q: X \vee CS^{n-1} \to X \cup_g CS^{n-1}$ restricts to give a map $f: CS^{n-1} \to X \cup_g CS^{n-1}$ which on the interior of CS^{n-1} is a homeomorphism. f is called the *characteristic map* of the cell. Since $CS^{n-1} \cong D^n$, we may regard f as a map

$$f:(D^n, S^{n-1}) \to (X \cup_g CS^{n-1}, X).$$

Note that $f|S^{n-1} = g$.

More generally if we have a map

$$g: \vee_\alpha S_\alpha^{n-1} \to X$$

of many $n-1$ spheres into X, then the mapping cone $X \cup_g C(\vee_\alpha S_\alpha^{n-1}) = X \cup_g \vee_\alpha (CS_\alpha^{n-1})$ is called *X with n-cells attached.* The subset $q(CS_\alpha^{n-1})$ is the n-cell e_α^n and its attaching map is $g|S_\alpha^{n-1}$. The characteristic map f_α^n of e_α^n is $q|CS_\alpha^{n-1}$. Attaching a 0-cell will mean adding a disjoint point.

5.11. Lemma. *Let K be an $(n-1)$-dimensional CW-complex on X. Attach n-cells e_α^n, $\alpha \in J_n$, by attaching maps $g_\alpha: S^{n-1} \to X$ and let $K' = K \cup \{e_\alpha^n : \alpha \in J_n\}$. Then K' is a CW-complex on $Y = X \cup_{\{g_\alpha\}} \vee_\alpha (CS_\alpha^{n-1})$.*

Proof: We must first show that the result Y is a Hausdorff space. Suppose $x, y \in Y$ are distinct points. If either lies in the interior of some n-cell e_α^n, then it is simple to find disjoint open neighborhoods of x and y in Y. Therefore let us suppose $x, y \in X$. In X we can find disjoint open neighborhoods U, V of x, y respectively. U, V need not be open in Y, so we proceed as follows. For each α the sets $g_\alpha^{-1} U$, $g_\alpha^{-1} V$ are open and disjoint in S^{n-1}. We can clearly find open disjoint sets U_α, V_α in D^n such that $U_\alpha \cap S^{n-1} = g_\alpha^{-1} U$, $V_\alpha \cap S^{n-1} = g_\alpha^{-1} V$. In $X \vee (\vee_\alpha D_\alpha^n)$ the open sets $U' = U \vee (\vee_\alpha U_\alpha)$, $V' = V \vee (\vee_\alpha V_\alpha)$ have the property that $q^{-1} \circ q(U') = U'$, $q^{-1} \circ q(V') = V'$, so that qU', qV' are open in Y. Also clearly $qU' \cap qV' = \varnothing$ and $x \in qU'$, $y \in qV'$.

Now the other properties of a CW-complex are simple. From the construction it is clear the sets $\mathring{e}_\alpha^n$ are disjoint from one another and also from X, hence from the $\mathring{e}_\beta^m$, $m < n$. It is also clear that $Y = X \cup \bigcup_\alpha e_\alpha^n = (\bigcup_K e_\beta^m) \cup \bigcup_\alpha e_\alpha^n$. The cells e_α^n have characteristic maps $f_\alpha^n = q|CS_\alpha^{n-1}$ which behave as required, and the cells of K come already equipped with characteristic maps. That each $e_\alpha^m \in K$ has only finitely many faces is a consequence of the closure-finiteness of K. Since each $\dot{e}_\alpha^n = g_\alpha(S^{n-1})$ is compact, it meets only finitely many interiors of cells of K, and hence each e_α^n has only finitely many faces. Finally, a subset $S \subset Y$ is closed if and only if $q^{-1} S \subset X \vee (\vee_\alpha CS_\alpha^{n-1})$ is closed; which is true if and only if

$S \cap X$ is closed in X and $S \cap e_\alpha^n$ is closed in e_α^n for each n, α. But $S \cap X$ is closed in X if and only if $S \cap e_\beta^m$ is closed in e_β^m for each $e_\beta^m \in K$. Thus Y has the weak topology. $\square$

5.12. Proposition. *Suppose we have a sequence* $\{x_0\} = X^{-1} \subset X^0 \subset X^1 \subset X^2 \subset \cdots \subset X^n \subset X^{n+1} \subset \cdots$ *of spaces such that* X^n *is obtained from* X^{n-1} *by attaching* n*-cells,* $n \geqslant 0$*. If we give* $X = \bigcup_{n \geqslant -1} X^n$ *the weak topology (i.e.* $S \subset X$ *closed if and only if* $S \cap X^n$ *closed in* X^n *for each* $n \geqslant -1$*), then* $K = \{$*all cells*$\}$ *is a* CW*-complex on* X*.*

Proof: Again the hard part is to show that X is Hausdorff. If $x, y \in X$ are distinct points, choose an n so that $x, y \in X^n$. By 5.11 and induction each X^n is a CW-complex, hence Hausdorff. Choose open disjoint neighborhoods U, V of x, y in X^n. Then we inductively construct disjoint open neighborhoods U_m, V_m of x, y in X^m, $m \geqslant n$, such that $U_n = U$, $V_n = V$. If U_{m-1}, V_{m-1} have already been constructed, then we "fatten" them up into open disjoint sets in X^m by the process described in the proof of 5.11. Finally $U' = \bigcup_{m \geqslant n} U_m$, $V' = \bigcup_{m \geqslant n} V_m$ are disjoint sets in X and since $U' \cap X^m = U_m$, $V' \cap X^m = V_m$ are open for all $m \geqslant n$, it follows that U', V' are open in X. Clearly $x \in U'$, $y \in V'$.

All the other properties required of K are immediate from the fact that X^n is a CW-complex for each n. $\square$

Thus we have given ourselves the opportunity of building CW-complexes to our own specifications.

5.13. Definition. We can generalize slightly the notion of CW-complex to that of relative CW-complex. A *relative CW-structure* on a pair (X, A) is a sequence

$$A = (X, A)^{-1} \subset (X, A)^0 \subset \cdots \subset (X, A)^n \subset (X, A)^{n+1} \subset \cdots \subset X$$

such that $(X, A)^n$ is obtained from $(X, A)^{n-1}$ by attaching n-cells, $n \geqslant 0$, $X = \bigcup_{n \geqslant -1} (X, A)^n$ and X has the weak topology: $S \subset X$ closed if and only if $S \cap (X, A)^n$ closed in $(X, A)^n$ for all $n \geqslant -1$.

We say $\dim(X, A) = n$ if $(X, A)^n = X$ and $(X, A)^{n-1} \neq X$.

Note that if $A = \{x_0\}$, then X is a CW-complex. Conversely, if X is a CW-complex and $A \subset X$ is any subcomplex, then (X, A) is a relative CW-complex.

Remark. If K is a cell complex on X and L is a cell complex on Y, then the set

$$K \times L = \{e_\alpha^n \times e_\beta^m : e_\alpha^n \in K, e_\beta^m \in L\}$$

is a cell complex on $X \times Y$. The characteristic map of $e_\alpha^n \times e_\beta^m$ is the map

$$D^{n+m} \cong D^n \times D^m \xrightarrow{f_\alpha^n \times f_\beta^m} e_\alpha^n \times e_\beta^m.$$

However, if K, L are CW-complexes $K \times L$ need not be a CW-complex on $X \times Y$. The difficulty is that the product topology on $X \times Y$ may not agree with the weak topology given by $K \times L$. In some situations the way out of this dilemma is simply to let $X \times Y$ have the weak topology. Otherwise some restriction must be imposed. For example, if either K or L is locally finite (i.e. every point has a neighborhood meeting only finitely many cells) then the two topologies agree. Thus, for example, $X \times I$ can always be given a CW-structure if X can.

5.14. Exercise. Show that if (X, A) is a relative CW-complex, then X/A is a CW-complex.

Hence under appropriate conditions $X \wedge Y$ is a CW-complex if X and Y are.

5.15. Exercise. Suppose (X, A) is a relative CW-complex and $p: E \to B$ is a weak fibration. Show that for any map $f: X \to E$ and homotopies $F: X \times I \to B$, $H: A \times I \to E$ with $F_0 = p \circ f$, $H_0 = f|A$ and $p \circ H = F|A \times I$ there is a homotopy $G: X \times I \to E$ lifting F with $G/A \times I = H$, $G_0 = f$ and $p \circ G = F$.

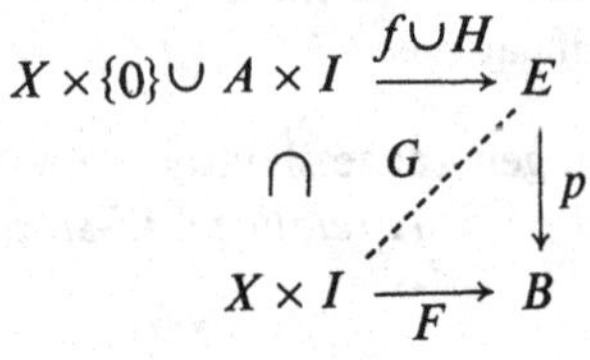

$$\begin{array}{ccc} X \times \{0\} \cup A \times I & \xrightarrow{f \cup H} & E \\ \cap & \overset{G}{\nearrow} & \downarrow p \\ X \times I & \xrightarrow[F]{} & B \end{array}$$

As a special case of Exercise 5.14 we have the following.

5.16. Corollary. If X is obtained from A by attaching n-cells $\{e_\beta^n : \beta \in B\}$, then $X/A \cong \vee_{\beta \in B} S_\beta^n$. The homeomorphism can be chosen so that the diagram

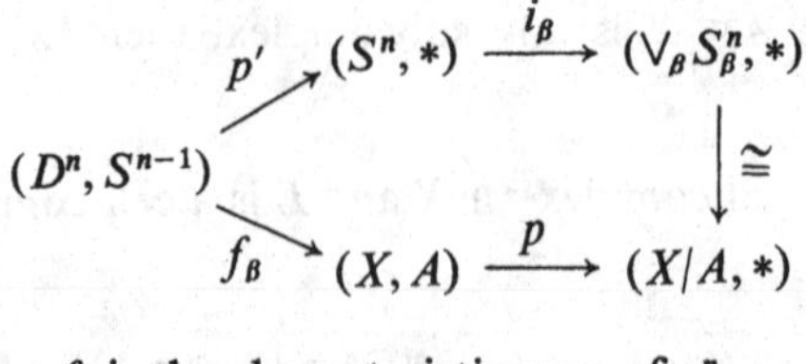

$$\begin{array}{ccccc} & & (S^n, *) & \xrightarrow{i_\beta} & (\vee_\beta S_\beta^n, *) \\ & \overset{p'}{\nearrow} & & & \downarrow \cong \\ (D^n, S^{n-1}) & & & & \\ & \underset{f_\beta}{\searrow} & & & \\ & & (X, A) & \xrightarrow{p} & (X/A, *) \end{array}$$

commutes, where f_β is the characteristic map of e_β^n.

5.17. Corollary. *For any relative CW-complex (X, A) with n-cells $\{e^n_\beta : \beta \in B\}$ we have*

$$(X, A)^n/(X, A)^{n-1} = \bigvee_{\beta \in B} S^n_\beta,$$

and the diagram

$$\begin{array}{ccccc} & \overset{p'}{\nearrow} & (S^n, *) & \xrightarrow{\quad i_\beta \quad} & (\bigvee_\beta S^n_\beta, *) \\ (D^n, S^{n-1}) & & & & \downarrow \\ & \underset{f^n_\beta}{\searrow} & ((X, A)^n, (X, A)^{n-1}) & \rightarrow & ((X, A)^n/(X, A)^{n-1}, *) \end{array}$$

commutes.

5.18. If we work in $\mathscr{T}$ instead of $\mathscr{PT}$, then we regard the attaching maps $g : \bigcup_\alpha S^n_\alpha \to X^n$ as being defined on the disjoint union and we take $X^{-1} = \varnothing$.

Comments

CW-complexes were first introduced by J. H. C. Whitehead in [95] and have gradually established themselves since as being the "right" sort of spaces with which to work for many purposes. In [63] Milnor investigated spaces having the homotopy type of a CW-complex, and in [91] Wall investigates such questions as "When does a space have the homotopy type of a finite CW-complex?" In Chapter 6 we shall begin to see more clearly just how pleasant CW-complexes are.

References

1. J. Milnor [63]
2. E. H. Spanier [80]
3. C. T. C. Wall [91]
4. J. H. C. Whitehead [95]

Chapter 6

Homotopy Properties of CW-Complexes

In this chapter we shall prove some quite deep results about the homotopy groups of CW-complexes and relative CW-complexes. Most of these hard theorems will be consequences of the simplicial approximation theorem. In addition, we shall show that if $f:X \to Y$ is a map between CW-complexes such that $f_*:\pi_n(X,x_0) \to \pi_n(Y,y_0)$ is an isomorphism for all $n \geqslant 0$, then f is a homotopy equivalence.

We begin with the notion dual to that of a fibration. Recall the diagram 6.1 used in the definition of a fibration $p:E \to B$.

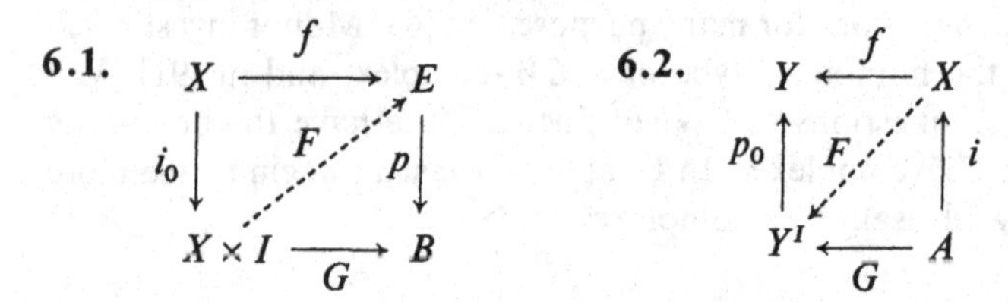

We obtain the notion of a cofibration $i:A \to X$ if we turn all the arrows around and replace $X \times I$ by its dual Y^I and i_0 by p_0 where $p_0(w) = w(0)$. Remembering that G can be regarded as a map $A \times I \to Y$ and likewise F, we are led to the following definition.

6.3. Definition. An inclusion $i:A \to X$ is said to have the *homotopy extension property* (HEP) with respect to a space Y if for every map $f:X \to Y$ and homotopy $G:A \times I \to Y$ of $f|A$ there is a homotopy $F:X \times I \to Y$ of f extending G. i is called a *cofibration* if it has the HEP with respect to all spaces Y.

Remark. It can be shown that the assumption that $i:A \to X$ be an inclusion is unnecessary. The other conditions on a cofibration imply that i is a homeomorphism of A onto a closed subspace of X.

6.4. Lemma. *For any map $g:(X,x_0) \to (Y,y_0)$ the inclusion $j:Y \to Y \cup_g CX$ is a cofibration.*

Proof: Let $r: I \times I \to I \times \{0\} \cup \dot{I} \times I$ be a retraction—for example regard $I \times I$ as sitting in $\mathbb{R}^2$ and project linearly from $(1/2, 2)$.

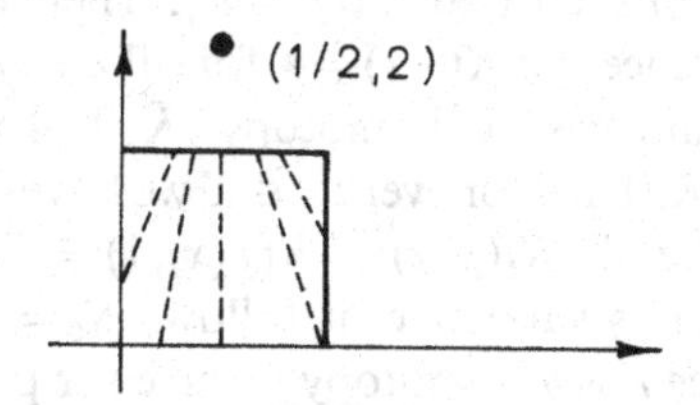

Now suppose given maps $f: Y \cup_g CX \to W$ and $G: Y \times I \to W$ with $G_0 = f | Y$. Define

$$G': CX \times \{0\} \cup X \times I \to W$$

by $G'|CX \times \{0\} = f|CX$ and $G'|X \times I = G \circ (g \times 1)$. Now we define $H: (Y \cup_g CX) \times I \to W$ by taking $H|Y \times I = G$ and $H([s,x],t) = G'([p_1 \circ r(s,t), x], p_2 \circ r(s,t))$ for $[s,x] \in CX$, $t \in I$, (p_1, p_2 are the restrictions to $I \times \{0\} \cup \dot{I} \times I$ of the projections $p_1, p_2: I \times I \to I$). Then H is well defined, continuous and $H_0 = f$. ▯

This proof was given earlier in proving 2.48.

In particular if X is obtained from A by attaching n-cells, then $X = A \cup_g C(\vee_\alpha S_\alpha^{n-1})$ so the inclusion $A \to X$ is a cofibration.

6.5. Proposition. *If (X, A) is a relative CW-complex then the inclusion $i: A \to X$ is a cofibration.*

Proof: For each n the space $(X, A)^n$ is obtained from $(X, A)^{n-1}$ by attaching n-cells. Thus $(X, A)^{n-1} \subset (X, A)^n$ is a cofibration. Given $f: X \to Y$ and $G: A \times I \to Y$ a homotopy of $f|A$, we can use 6.4 and induction to construct homotopies $F^n: (X, A)^n \times I \to Y$ satisfying

i) $F^{-1} = G$, ii) $F_0^n = f|(X, A)^n$, iii) $F^n|(X, A)^{n-1} \times I = F^{n-1}$.

We then define $F: X \times I \to Y$ by $F(x,t) = F^n(x,t)$ if $x \in (X, A)^n$, $t \in I$. F is well defined because of iii) and continuous because $F|e_\alpha^n \times I = F^n|e_\alpha^n \times I$. $F_0 = f$ by ii) and $F|A \times I = G$ by i). ▯

One application of this result is that collapsing contractible subcomplexes of a CW-complex does not alter its homotopy type.

6.6. Proposition. *If $A \to X$ is a cofibration and A is contractible, then the projection $p: (X, A) \to (X/A, *)$ is a homotopy equivalence.*

Proof: Let $H: A \times I \to A$ be a contracting homotopy; i.e. $H_0 = 1_A$, $H_1 = x_0$, $H(x_0, t) = x_0$, all $t \in I$. Since $i: A \to X$ is a cofibration we can extend $i \circ H$ to a homotopy $K: X \times I \to X$ with $K_0 = 1_X$ and $K|A \times I = i \circ H$. Then $K_1(a) = H_1(a) = x_0$ for all $a \in A$, so K_1 induces a map $k:(X/A,*) \to (X,x_0)$ such that $k \circ p = K_1$. Then K is a homotopy $1_X \simeq k \circ p$. Moreover, since $p \circ K(a,t) = *$ for all $a \in A$, $t \in I$, it follows from 0.8 that $p \circ K$ induces a homotopy $\bar{K}: X/A \times I \to X/A$ such that $\bar{K} \circ (p \times 1) = p \circ K$. Then for every $x \in X$ we have $\bar{K}_0(p(x)) = \bar{K}(p(x),0) = p \circ K(x,0) = p(x)$ and $\bar{K}_1(p(x)) = \bar{K}(p(x),1) = p \circ K(x,1) = p \circ K_1(\check{x}) = p \circ k(p(x))$. Since p is surjective, it follows $\bar{K}_0 = 1_{X/A}$, $\bar{K}_1 = p \circ k$. Thus $p \circ k \simeq 1_{X/A}$. Hence k is a homotopy inverse for p. $\square$

6.7. *Remark.* For any map $f:(X,x_0) \to (Y,y_0)$ the inclusion $\alpha: CY \to (Y \cup_f CX) \cup_j CY$ is a cofibration; for if we are given

$$h:(Y \cup_f CX) \cup_j CY \to Z$$

and $G: CY \times I \to Z$ with $G_0 = h|CY$, then we can extend $G|Y \times I$ over $(Y \cup_f CX) \times I$—to H' say—with $H'_0 = h|Y \cup_f CX$. If we then define $H:[(Y \cup_f CX) \cup_j CY] \times I \to Z$ by $H|(Y \cup_f CX) \times I = H'$, $H|CY \times I = G$, then H is well defined, continuous and satisfies $H_0 = h$ as required. Furthermore, the cone on any space is contractible: define $F: CY \times I \to CY$ by $F([s,y],t) = [s(1-t),y]$. Thus we see that 2.37 is a special case of 6.6.

We now set out to demonstrate that for a CW-complex $\pi_n(X,x_0)$ depends only on the $(n+1)$-skeleton X^{n+1}. The idea of the proof is that S^n is n-dimensional so that the image of S^n in X where it meets an m-cell, $m > n$, ought not to fill the cell and hence can be pushed away radially from any point not in the image, thus deforming it out of the m-cell. Repeating this process will then eventually deform the image of S^n down into X^n. Similarly homotopies ought to be deformable into X^{n+1}. To make this rough idea rigorous requires the simplicial approximation lemma, which says that any map $f: S^n \to X$ can be deformed so that it is linear where it crosses the interior of an m-cell e^m.

Let X be obtained from A by attaching an n-cell e^n. In D^n take two concentric disks $D^n_0 \subset \mathring{D}^n_1 \subset D^n_1 \subset \mathring{D}^n$—e.g. $D^n_0 = \{x \in D^n : |x| \leqslant 1/4\}$, $D^n_1 = \{x \in D^n : |x| \leqslant 1/2\}$. Let $g: D^n \to e^n$ be the characteristic map of e^n and let $e^n_0 = g(D^n_0)$, $e^n_1 = g(D^n_1)$. We can also use g to import a linear structure into $\mathring{e}^n$: for $x, y \in \mathring{e}^n$ and $s, t \in \mathbb{R}$ we define $rx + sy = g(rg^{-1}x + sg^{-1}y)$, provided $rg^{-1}x + sg^{-1}y \in \mathring{D}^n$ (this will be so, for example, if $r + s = 1$, $r, s \geqslant 0$). With these data we can state the simplicial approximation lemma.

6.8. Simplicial approximation lemma. *Suppose $X = A \cup e^n$ as above, (K,L) is a finite simplicial pair and $f:(|K|,|L|) \to (X,A)$ is a map. There exists a subdivision (K',L') of (K,L) and a map $f':(|K|,|L|) \to (X,A)$ such that*

i) *f and f' agree on $f^{-1}(A)$ and $f \simeq f' \operatorname{rel} f^{-1}(A)$;*

ii) *for any simplex σ of K' if $f'(|\sigma|)$ meets e_0^n, then $f'(|\sigma|)$ is contained in $\mathring{e}^n$ and f' is a linear map when restricted to $|\sigma|$.*

Proof: We subdivide K so finely that for any simplex σ of K' if $f(|\sigma|)$ meets $\dot{e}_1^n$, then $f(|\sigma|) \subset \mathring{e}^n$ and the convex hull of $f(|\sigma|)$ does not meet e_0^n. This is possible since $|K|$ is compact, and hence $g^{-1}f$ is uniformly continuous on $f^{-1}e_2^n$, where $D_2^n = \{x \in D^n : |x| \leqslant 3/4\}$, say. There is a $\delta > 0$ such that $d(x,y) < \delta$ in $f^{-1}e_2^n$ implies $d(g^{-1}f(x), g^{-1}f(y)) < 1/4$ in D_2^n. We then subdivide K so finely that no simplex of K' has diameter more than δ.

Now the simplices of K' fall into three disjoint classes:

$$C_1 = \{\sigma \in K' : f(|\sigma|) \subset X - e_1^n\},$$
$$C_2 = \{\sigma \in K' : f(|\sigma|) \subset \mathring{e}_1^n\},$$
$$C_3 = \{\sigma \in K' : f(|\sigma|) \cap \dot{e}_1^n \neq \varnothing\}.$$

We construct f' on the three classes of simplex separately. For $\sigma \in C_1$ we take $f' = f$ on $|\sigma|$. For $\sigma \in C_2$ we define f' to be linear; that is, if $\sigma = (v_0, v_1, \ldots, v_k)$ and if $x = \sum_{i=0}^k r_i v_i$ for $\sum_{i=0}^k r_i = 1$ is a point of $|\sigma|$, then we let $f'(x) = \sum_{i=0}^k r_i f(v_i)$. For $\sigma \in C_3$ we proceed inductively on $\dim \sigma$. If $\dim \sigma = 0$, we let $f'(\sigma) = f(\sigma)$. Suppose f' has been defined on $|\sigma|$ for all $\sigma \in C_3$ with $\dim \sigma < k$ in such a way that $f'(|\sigma|) \subset$ convex hull of $f(|\sigma|)$. Suppose $\sigma = (v_0, \ldots, v_k) \in C_3$ has $\dim \sigma = k$; then f' is defined on $|\dot{\sigma}|$ and $f'(|\dot{\sigma}|) \subset$ convex hull $f(|\sigma|)$. Let $b_\sigma = \sum_{i=0}^k 1/(k+1) v_i$ be the barycenter. Every $x \in |\sigma| - \{b_\sigma\}$ has a unique representation in the form $x = tb_\sigma + (1-t)y_x$ for some $t \in I$ and $y_x \in |\dot{\sigma}|$. Let $f'(b_\sigma) = f(b_\sigma)$ and $f'(x) = tf(b_\sigma) + (1-t)f'(y_x)$. Then f' is continuous on $|\sigma|$ and $f'(|\sigma|) \subset$ convex hull $f(|\sigma|)$. Thus the induction is complete, and we have defined $f' : (|K|, |L|) \to (X, A)$ so that the first half of i) and ii) are satisfied. We define a homotopy $H : |K| \times I \to X$ as follows: on $\sigma \in C_1$ we take H to be stationary. On $\sigma \in C_2 \cup C_3$ we define H by $H(x,t) = (1-t)f(x) + tf'(x)$, $x \in |\sigma|$, $t \in I$. Then H is a homotopy from f to f' rel $f^{-1}A$. □

6.9. Lemma. *If $X = A \cup e^n$ and $x \in \mathring{e}^n$, then A is a strong deformation retract of $X - \{x\}$.*

Proof: First we observe that if $a \in \mathring{D}^n$, then S^{n-1} is a strong deformation retract of $D^n - \{a\}$: every point $y \in D^n - \{a\}$ has a unique representation in the form $y = sa + (1-s)z_y$ for some $z_y \in S^{n-1}$, $s \in [0,1)$. We define $H_a : (D^n - \{a\}) \times I \to D^n - \{a\}$ by

$$H_a(sa + (1-s)z, t) = (1-t)sa + (1 - s(1-t))z$$

for $s \in [0,1)$, $t \in I$, $z \in S^{n-1}$. H_a is the required deformation.

If $f : (D^n, S^{n-1}) \to (X, A)$ is the characteristic map of e^n, we let $a = f^{-1}x$ and define $H : (X - \{x\}) \times I \to X - \{x\}$ by taking H to be stationary on A

and defining H on $(e^n - \{x\}) \times I$ by

$$H(f(y), t) = f \circ H_a(y, t) \quad \text{for} \quad y \in D^n - \{a\}, t \in I. \quad \square$$

The implications of the simplicial approximation lemma (SAL) are wide-reaching. We now begin to deduce some of them.

6.10. Theorem. *If (X, A) is a relative CW-complex then for every $n \geq -1$ the pair $(X, (X, A)^n)$ is n-connected.*

Proof: Suppose $f:(D^r, S^{r-1}) \to (X, (X, A)^n)$ is a map; we must show f is homotopic rel S^{r-1} to a map f' with $f'(D^r) \subset (X, A)^n$ if $r \leq n$. $f(D^r)$ is compact, hence $f(D^r) \subset (X, A)^m$ for some m and $f(D^r)$ meets only finitely many m-cells, say $e_1^m, \ldots, e_q^m$. If $m > n$ we deform f rel $f^{-1}((X, A)^m - \mathring{e}_1^m)$ to a map f' with f' linear where it crosses $(e_1^m)_0$. Since $f'(D^r) \cap \mathring{e}_1^m$ is at most r-dimensional and $r \leq n < m$, it follows there is an $x \in \mathring{e}_1^m - f'(D^r)$. Let $H:[(X, A)^m - \{x\}] \times I \to (X, A)^m - \{x\}$ be a deformation retraction of $(X, A)^m - \{x\}$ onto $(X, A)^m - \mathring{e}_1^m$. Then $K: D^r \times I \to X$ defined by $K(y, t) = H(f'(y), t)$ $y \in D^r$, $t \in I$, is a homotopy rel $f^{-1}((X, A)^m - \mathring{e}_1^m)$ from f' to an f'' with $f''(D^r) \subset (X, A)^m - \mathring{e}_1^m$. One can repeat this process to clear the image of D^r out of the other cells of dimension m. Then one has found an f''' with $f'''(D^r) \subset (X, A)^{m-1}$ and $f''' \simeq f$ rel S^{r-1}. In this way we can push the image of D^r out of one skeleton after another until we have an $\hat{f}$ with $\hat{f} \simeq f$ rel S^{r-1} and $\hat{f}(D^r) \subset (X, A)^n$. This argument works even for $r = 0$. $\square$

6.11. Theorem. *Let X be a CW-complex and let $i: X^n \to X$ be the inclusion of the n-skeleton, $n > 0$. Then $i_*: \pi_r(X^n, x_0) \to \pi_r(X, x_0)$ is an isomorphism for $r < n$ and an epimorphism for $r = n$.*

Proof: The exact homotopy sequence of the pair (X, X^n, x_0) is

$$\cdots \to \pi_{r+1}(X, X^n, x_0) \to \pi_r(X^n, x_0) \xrightarrow{i_*} \pi_r(X, x_0) \to \pi_r(X, X^n, x_0) \to \cdots.$$

Since by 6.10 $\pi_r(X, X^n, x_0) = 0$ for $r \leq n$, the result follows immediately. $\square$

6.12. Corollary. $\pi_r(S^n, s_0) = 0$ *for $r < n$.*

Proof: We give S^n a CW-structure with just one 0-cell and one n-cell. Then $(S^n)^{n-1} = s_0$, so $\pi_r((S^n)^{n-1}, s_0) = 0$ for all r. But by 6.11

$$i_*: \pi_r((S^n)^{n-1}, s_0) \to \pi_r(S^n, s_0)$$

is surjective for $r \leq n - 1$. Hence $\pi_r(S^n, s_0) = 0$ for $r < n$. $\square$

From 6.10 it follows that if (X, A) is a relative CW-complex having no cells of dimensions less than or equal to n—i.e. $(X, A)^n = A$—then the pair (X, A) is n-connected. We also have the following weak converse.

6.13. Proposition. *If (X,A) is an n-connected relative CW-complex, then we can embed (X,A) in a relative CW-complex (X',A') such that*

i) *X is a strong deformation retract of X' and A is a strong deformation retract of A';*
ii) $(X',A')^n = A'$.

Proof: We proceed by induction on n. For $n=-1$ the statement is trivial; we can take $X'=X$, $A'=A$. Suppose the proposition true for $n-1$ and suppose (X,A) is n-connected. Since (X,A) is also $(n-1)$-connected, we may as well assume $(X,A)^{n-1}=A$. Let the n-cells e^n_α have characteristic maps $f^n_\alpha:(D^n,S^{n-1},s_0)\to(X,A,x_0)$. We are going to attach one new $(n+1)$-cell e^{n+1}_α and one new $(n+2)$-cell e^{n+2}_α to X for each $\alpha\in J_n$. Since $\pi_n(X,A,x_0)=0$, we see that $[f^n_\alpha]=0$. Hence f^n_α is homotopic rel S^{n-1} to a map into A. We may regard the homotopy as a map $g_\alpha:D^{n+1}\to X$ such that $g_\alpha|(H^n_+,S^{n-1})=f^n_\alpha$, $g_\alpha(H^n_-)\subset A$. By 6.10 we may assume $g_\alpha(D^{n+1})\subset(X,A)^{n+1}$. We attach e^{n+1}_α by the map $(g_\alpha|S^n):S^n\to X$. Let $h_\alpha:(D^{n+1},S^n)\to(e^{n+1}_\alpha,\dot{e}^{n+1}_\alpha)$ be the characteristic map of e^{n+1}_α. Then we have a map $k_\alpha:S^{n+1}\to X\cup e^{n+1}_\alpha$ such that $k_\alpha|H^{n+1}_+=h_\alpha$, $k_\alpha|H^{n+1}_-=g_\alpha$. We attach e^{n+2}_α by k_α.

Now we take

$$X'=X\cup(\textstyle\bigcup_\alpha e^{n+1}_\alpha)\cup(\bigcup_\alpha e^{n+2}_\alpha)\quad\text{and}\quad A'=A\cup(\bigcup_\alpha e^{n+1}_\alpha).$$

Then clearly $(X,A)\subset(X',A')$. Moreover, since the only n-cells e^n_α are contained in the $(n+1)$-cells e^{n+1}_α, we see that $e^n_\alpha\subset A'$, all α—i.e. $(X',A')^n=A'$. Finally A' is obtained from A by attaching $(n+1)$-disks along the southern hemispheres H^n_- of their boundaries S^n (by the map $g_\alpha|H^n_-$). Since H^n_- is a strong deformation retract of D^{n+1}, we see that A is a strong deformation retract of A'. Similarly X' is obtained from X by attaching $(n+2)$-disks along the southern hemispheres H^{n+1}_- of their boundaries (by the maps g_α). Thus X is a strong deformation retract of X'.

The case $n=0$ is schematically illustrated below.

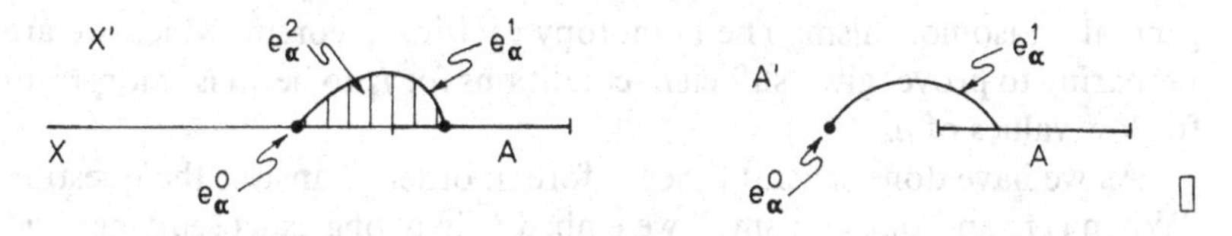

□

For the absolute case we can formulate the result a little differently.

6.14. Corollary. *If (X,x_0) is an n-connected CW-complex then we can find a CW-complex $\bar{X}$ with $(\bar{X})^n=\bar{x}_0$ and a homotopy equivalence $f:(X,x_0)\to(\bar{X},\bar{x}_0)$.*

Proof: We apply 6.13 to the relative CW-complex $(X, \{x_0\})$ and embed the pair in (X', A') with $(X', A')^n = A'$ and X a strong deformation retract of X', $\{x_0\}$ a strong deformation retract of A'. This last means A' is contractible. Thus we may apply 6.6 to conclude $(X', A') \to (X'/A', *)$ is a homotopy equivalence. Clearly $(X'/A')^n = *$, so we take $(\bar{X}, \bar{x}_0) = (X'/A', *)$ and f to be the composite $(X, x_0) \subset (X', A') \to (X'/A', *)$. ▯

6.15. Corollary. *If X is an n-connected CW-complex and Y is an m-connected CW-complex, then $X \wedge Y$ is an $(n+m+1)$-connected CW-complex (we take the weak topology on $X \wedge Y$ if necessary).*

Proof: We may assume $X^n = x_0$, $Y^m = y_0$. Then the cells of $X \times Y$ are of the form, $x_0 \times y_0$, $x_0 \times e^{m+1}$, $e^{n+1} \times y_0$, $e^{n+1} \times e^{m+1}$, etc. The first three types all lie in $X \vee Y$, so we see $(X \times Y, X \vee Y)^{n+m+1} = X \vee Y$. Hence $(X \wedge Y)^{n+m+1} = (X \times Y / X \vee Y)^{n+m+1} = *$. ▯

We have not yet exhausted the full power of the simplicial approximation lemma. In the proof of 6.10 when we were looking at maps $f: D^r \to A \cup e^m$ the only use we made of the fact that $r < m$ was in saying that if f is linear across e_0^m then $f(D^r) \cap e_0^m$ cannot be all of e_0^m. If $r \geqslant m$ we can no longer make this assertion, but we can make a statement about the dimension of the set $f^{-1}x$ for certain points $x \in e_0^m$—in fact we can arrange to have $\dim(f^{-1}x) \leqslant r - m$. This generalizes the situation $f^{-1}x = \varnothing$ if $r < m$. We exploit this fact to obtain further information about homotopy groups. In particular we shall be able to compute $\pi_n(S^n, s_0)$.

6.16. A *triad* $(X; A, B, x_0)$ is a topological space X and two subspaces A, B (with $x_0 \in A \cap B$) such that $X = A \cup B$. If X is a CW-complex and A, B are two subcomplexes, or if $A = X - U$, where $U \subset B$ is an open subset with $\bar{U} \subset \mathring{B}$, then the inclusion

$$j:(A, A \cap B) \to (X, B)$$

induces isomorphisms $j_*: H_n(A, A \cap B) \to H_n(X, B)$ in singular homology for all $n \geqslant 0$, as we shall see later. This is in fact the so-called "excision axiom" for homology. One might ask whether the same holds for homotopy; the answer is that $j_*: \pi_n(A, A \cap B, x_0) \to \pi_n(X, B, x_0)$ is not in general an isomorphism. The homotopy excision theorem, which we are preparing to prove, gives sufficient conditions for j_* to be an isomorphism for low values of n.

As we have done several times before in order to answer the question "When is j_* an isomorphism?" we embed j_* in a long exact sequence and turn the question into one about the vanishing of certain groups.

6.17. Definition. We define the nth *triad homotopy set* $\pi_n(X; A, B, x_0)$, $n \geqslant 2$, of the triad $(X; A, B, x_0)$ by

$$\pi_n(X; A, B, x_0) = \pi_{n-1}(P(X; x_0, B), P(A; x_0, A \cap B), \omega_0).$$

If we recall that $\pi_n(X, B, x_0) = \pi_{n-1}(P(X; x_0, B), \omega_0)$ and

$$\pi_n(A, A \cap B, x_0) = \pi_{n-1}(P(A; x_0, A \cap B), \omega_0),$$

then we see that the exact homotopy sequence of the pair

$$(P(X; x_0, B), P(A; x_0, A \cap B), \omega_0)$$

becomes

6.18.

$$\cdots \longrightarrow \pi_{n+1}(X; A, B, x_0) \xrightarrow{\partial'} \pi_n(A, A \cap B, x_0) \xrightarrow{j_*} \pi_n(X, B, x_0) \xrightarrow{k_*} \pi_n(X; A, B, x_0) \xrightarrow{\partial'} \cdots.$$

6.18 is called the *exact homotopy sequence of the triad* $(X; A, B, x_0)$. The following is immediate.

6.19. Proposition. *For any triad $(X; A, B, x_0)$ the inclusion $j:(A, A \cap B) \to (X, B)$ induces a map $j_*:\pi_r(A, A \cap B, x_0) \to \pi_r(X, B, x_0)$ which is an isomorphism for $2 \leqslant r < n$, an epimorphism for $r = n$ and "monic" ($j_*^{-1}(0) = 0$) for $r = 1$ if and only if $\pi_r(X; A, B, x_0) = 0$, $2 \leqslant r \leqslant n$.*

We will simplify our investigations of $\pi_r(X; A, B, x_0)$ if we use the exponential law to identify elements in $\pi_r(X; A, B, x_0)$ with homotopy classes of maps

$$f:(D^{r-1} \times I, S^{r-2} \times I, D^{r-1} \times \{1\}, D^{r-1} \times \{0\} \cup \{s_0\} \times I) \to (X, A, B, x_0).$$

Then we ask when such maps are null homotopic.

6.20. Lemma. *Let (C, x_0) be any Hausdorff space and suppose A is obtained from C by attaching an n-cell e^n, B is obtained from C by attaching an m-cell e^m and $X = A \cup B$. Then $\pi_r(X; A, B, x_0) = 0$ for $2 \leqslant r \leqslant n + m - 2$.*

Proof: Consider any $f:(D^{r-1} \times I,\ S^{r-2} \times I,\ D^{r-1} \times \{1\},\ D^{r-1} \times \{0\} \cup \{s_0\} \times I) \to (X, A, B, x_0)$. By the SAL we may assume that $D^{r-1} \times I$ is triangulated so finely that for any simplex σ, if $f(|\sigma|)$ meets e_0^n (resp. e_0^m) then $f(|\sigma|) \subset \mathring{e}^n$ (resp. $\subset \mathring{e}^m$) and f is linear on $|\sigma|$. For each σ if $f(|\sigma|)$ meets e_0^m then $f(|\sigma|)$ is a convex set (in fact a cell in the sense of PL topology). Its dimension is the maximum number of affine independent points in $f(|\sigma|)$ minus 1. We divide the set of such simplices into two classes: C_1 those with $\dim f(|\mathring{\sigma}|) < m$, and C_2 those with $\dim f(|\mathring{\sigma}|) = m$. Then the set $\bigcup_{\sigma \in C_1} f(|\mathring{\sigma}|)$ will not cover e_0^m, so we can choose a point $p \in e_0^m$ such that if $p \in f(|\mathring{\sigma}|)$, then $\sigma \in C_2$. Then one readily sees that for this p we have that $f^{-1}p$ is a polyhedron in $D^{r-1} \times I$ of dimension at most $r - m$. Let $\pi: D^{r-1} \times I \to D^{r-1}$ be the projection onto the first factor; then $K = \pi^{-1}(\pi(f^{-1}p))$ is a polyhedron of dimension at most $r - m + 1 \leqslant (n + m - 2) - m + 1 = n - 1$. Hence $f(K)$ does not cover e_0^n, so we can find a $q \in e_0^n$ with $f^{-1}q \cap K = \varnothing$. Thus $\pi(f^{-1}q) \cap \pi K = \varnothing$.

Now D^{r-1} is normal and $\pi(f^{-1}p)$ and $\pi(f^{-1}q) \cup S^{r-2}$ are disjoint closed subsets of D^{r-1}. Hence we can find a continuous function $\chi: D^{r-1} \to I$ such that $\chi(\pi(f^{-1}p)) = 1$, $\chi(\pi(f^{-1}q) \cup S^{r-2}) = 0$. We define a homotopy $H: D^{r-1} \times I \times I \to D^{r-1} \times I$ by

$$H(x,s,t) = (x, s(1 - t\chi(x))), \quad x \in D^{r-1}, \quad s,t \in I.$$

Then i) $H_0 = 1_{D^{r-1} \times I}$, ii) $H_1(D^{r-1} \times I) \subset D^{r-1} - f^{-1}p$, iii) $H(D^{r-1} \times \{1\} \times I) \subset D^{r-1} \times I - f^{-1}q$, iv) H is stationary on $S^{r-2} \times I$. Thus if we regard f as a map

$$(D^{r-1} \times I, S^{r-2} \times I, D^{r-1} \times \{1\}, D^{r-1} \times \{0\} \cup \{s_0\} \times I) \to$$

$$(X, A, X - \{q\}, x_0),$$

then $f \circ H$ is a homotopy of f into a map into $(X - \{p\}, A, X - \{p,q\}, x_0)$.

Consider the diagram

$$\begin{array}{ccc} & & \pi_r(X; A, B, x_0) \\ & & \downarrow j_{1*} \\ \pi_r(X - \{p\}; A, X - \{p,q\}, x_0) & \xrightarrow{j_{2*}} & \pi_r(X; A, X - \{q\}, x_0) \\ \uparrow j_{3*} & & \\ \pi_r(A; A, A - \{q\}, x_0). & & \end{array}$$

What we have proved above is that $j_{1*}[f] = j_{2*}[f']$ for some

$$[f'] \in \pi_r(X - \{p\}; A, X - \{p,q\}, x_0).$$

Now by 6.9 B is a strong deformation retract of $X - \{q\}$, so j_{1*} is easily seen to be an isomorphism. Similarly A is a strong deformation retract of $X - \{p\}$, so j_{3*} is an isomorphism. But $\pi_r(A;\ A, A - \{q\}, x_0) = 0$. Thus $j_{1*}[f] = j_{2*}[f'] = 0$, and hence $[f] = 0$. □

Lemma 6.20 contains all the hard work in proving the following important theorem.

6.21. Homotopy excision theorem. *Let $(X;\ A, B, x_0)$ be a triad such that $(A, A \cap B)$ is an n-connected relative CW-complex, $n \geqslant 1$, and $(B, A \cap B)$ is an m-connected relative CW-complex. Then $j_*: \pi_r(A, A \cap B, x_0) \to \pi_r(X, B, x_0)$ is an isomorphism for $1 \leqslant r < n + m$ and an epimorphism for $r = m + n$.*

Proof: To save words let us say a map of pairs $f: (U, V) \to (Y, Z)$ is an n-equivalence if $f_*: \pi_r(U, V, u_0) \to \pi_r(Y, Z, z_0)$ is a bijection for $1 \leqslant r < n$, surjection for $r = n$. Thus we wish to show that j is an $(m+n)$-equivalence

—or equivalently that $\pi_r(X; A, B, x_0) = 0, 2 \leqslant r \leqslant m + n$. We denote $A \cap B$ by C for brevity. Now we consider a number of cases.

i) $A = C \cup e^{n'}, B = C \cup e^{m'}, n' > n, m' > m$. This is just the situation of 6.20.

ii) $B = C \cup e^{m'}, A = C \cup e^{n_1} \cup e^{n_2} \cup \cdots \cup e^{n_k}$, with $m' > m$, $n_i > n$, $1 \leqslant i \leqslant k$.

We have a sequence of subspaces $A_i = C \cup e^{n_1} \cup e^{n_2} \cup \cdots \cup e^{n_i}$, $1 \leqslant i \leqslant k$. Taking $A_0 = C$ we have $C = A_0 \subset A_1 \subset \cdots \subset A_k = A$. Let $X_i = B \cup A_i$, $0 \leqslant i \leqslant k$. Then $B = X_0 \subset X_1 \subset \cdots \subset X_k = X$. By i) above

$$k(i): (A_i, A_{i-1}, x_0) \to (X_i, X_{i-1}, x_0)$$

is an $(m+n)$-equivalence. We shall prove by induction that

$$j(i): (A_i, C, x_0) \to (X_i, B, x_0)$$

is an $(m+n)$-equivalence for $0 \leqslant i \leqslant k$ (note that $j(k) = j$). The statement is clear for $i = 0$. Suppose it proved for $i - 1$; then we have a commutative diagram of exact homotopy sequences of triples.

$$\begin{array}{ccccc}
\pi_{r+1}(A_i, A_{i-1}, x_0) & \to & \pi_r(A_{i-1}, C, x_0) & \to & \pi_r(A_i, C, x_0) \to \\
\downarrow k(i)_* & & \downarrow j(i-1)_* & & \downarrow j(i)_* \\
\pi_{r+1}(X_i, X_{i-1}, x_0) & \to & \pi_r(X_{i-1}, B, x_0) & \to & \pi_r(X_i, B, x_0) \to
\end{array}$$

$$\begin{array}{ccc}
\pi_r(A_i, A_{i-1}, x_0) & \to & \pi_{r-1}(A_{i-1}, C, x_0) \\
\downarrow k(i)_* & & \downarrow j(i-1)_* \\
\pi_r(X_i, X_{i-1}, x_0) & \to & \pi_{r-1}(X_{i-1}, B, x_0).
\end{array}$$

When $1 < r < m + n$ we have precisely the situation of the 5-lemma, and hence $j(i)_*$ is an isomorphism. For $r = m + n$ the second and fourth vertical morphisms are epimorphisms and the fifth an isomorphism, which is enough to show that $j(i)_*$ is an epimorphism. (Note: for $r = 2$ we need the fact that $\pi_1(A_{i-1}, C, x_0) = \pi_1(X_{i-1}, B, x_0) = 0$ because $n \geqslant 1$.)

iii) $A = C \cup e^{n_1} \cup \cdots \cup e^{n_k}$, $B = C \cup e^{m_1} \cup e^{m_2} \cup \cdots \cup e^{m_l}$, $n_i > n$, $1 \leqslant i \leqslant k$, $m_i > m$, $1 \leqslant i \leqslant l$. We let $B_0 = C$, $B_i = C \cup e^{m_1} \cup \cdots \cup e^{m_i}$, $1 \leqslant i \leqslant l$. Then $B_l = B$. We let $X_i = A \cup B_i$, $0 \leqslant i \leqslant l$; then $X_0 = A, X_l = X$. The map $j: (A, C, x_0) \to (X, B, x_0)$ can be factored as

$$(A, C, x_0) = (X_0, B_0, x_0) \xrightarrow{j_1} (X_1, B_1, x_0) \xrightarrow{j_2} \cdots$$
$$\xrightarrow{j_l} (X_l, B_l, x_0) = (X, B, x_0).$$

By ii) above each j_i is an $(m+n)$-equivalence, $1 \leqslant i \leqslant l$. Hence $j = j_l \circ j_{l-1} \circ \cdots \circ j_1$ is also an $(m+n)$-equivalence.

iv) $(A,C)^n = C$, $(B,C)^m = C$. Any map $f:(D^{r-1} \times I, S^{r-2} \times I, D^{r-1} \times \{1\}, D^{r-1} \times \{0\} \cup \{s_0\} \times I) \to (X,A,B,x_0)$ has compact image, and hence we can find A', B' with $C \subset A' \subset A$, $C \subset B' \subset B$, $A' = C \cup e^{n_1} \cup \cdots \cup e^{n_k}$, $B' = C \cup e^{m_1} \cup \cdots \cup e^{m_l}$ and with $f(D^{r-1} \times I) \subset X' = A' \cup B'$—i.e. $[f]$ lies in the image of $\pi_r(X'; A', B', x_0) \to \pi_r(X; A, B, x_0)$. But (A', C) is still n-connected and (B', C) is still m-connected, so by iii) above

$$\pi_r(X'; A', B', x_0) = 0$$

if $2 \leqslant r \leqslant m+n$.

v) $(B,C)^m = C$. By 6.13 we can embed (A,C) in (A',C') such that (A',C') is a relative CW-complex with $(A',C')^n = C'$ and with C a strong deformation retract of C', A a strong deformation retract of A'. Let $B' = B \cup C'$, $X' = B \cup A'$. Then X is a strong deformation retract of X' and B is a strong deformation retract of B', Thus the inclusion

$$i:(X; A, B, x_0) \to (X'; A', B', x_0)$$

induces an isomorphism $i_*:\pi_r(X; A, B, x_0) \to \pi_r(X'; A', B', x_0)$ of triad homotopy groups for $r \geqslant 2$. But by iv) above $\pi_r(X'; A', B', x_0) = 0$ for $2 \leqslant r \leqslant m+n$.

vi) The general case. We repeat the trick of v), embedding (B,C) in (B',C') and taking $A' = A \cup C'$, $X' = A \cup B'$. ▯

6.22. Corollary. *If (X,A,x_0) is a CW-pair with A m-connected and (X,A) n-connected, $n \geqslant 1$, then the projection $p:(X,A) \to (X/A,*)$ induces a map $p_*:\pi_r(X,A,x_0) \to \pi_r(X/A,\{*\},*) \cong \pi_r(X/A,*)$ which is an isomorphism for $2 \leqslant r \leqslant m+n$ and an epimorphism for $r = m+n+1$.*

Proof: We have a commutative diagram

$$\begin{array}{ccc} (X,A) & \xrightarrow{\ i\ } & (X \cup CA, CA) \\ \big\downarrow p & & \big\downarrow p' \\ (X/A,*) & \xrightarrow{\ \phi\ } & (X \cup CA/CA,*), \end{array}$$

where ϕ is the homeomorphism of 2.38 and p' is a homotopy equivalence by 6.6 ($(CA,*)$ is contractible). Thus it will suffice to show that $i_*:\pi_r(X,A,x_0) \to \pi_r(X \cup CA, CA, *)$ is an isomorphism for $2 \leqslant r \leqslant m+n$ and an epimorphism for $r = m+n+1$. We may apply 6.21 to the triad $(X \cup CA; X, CA, *)$ since we are given that (X,A) is n-connected and $\pi_r(CA,A,*) \cong \pi_{r-1}(A,x_0) = 0$, $1 \leqslant r \leqslant m+1$. ▯

One of the consequences for homology of the excision axiom is that there is an isomorphism $\sigma:H_n(X,\{x_0\}) \cong H_{n+1}(SX,\{*\})$ for every $(X,x_0) \in \mathscr{PT}$. Naturally one would now expect a similar isomorphism in homotopy under very restricted conditions.

6.23. Definition. The homotopy suspension

$$\Sigma:\pi_n(X,x_0) \to \pi_{n+1}(SX,*), \quad n \geqslant 0,$$

is defined by $\Sigma\,[f]=[Sf]=[1_{S^1}\wedge f]$:

$$S^{n+1}\cong S^1\wedge S^n \xrightarrow{1\wedge f} S^1\wedge X = SX.$$

Σ is clearly a natural transformation from the functor π_n to the functor $\pi_{n+1}\circ S$.

6.24. Lemma. *For every pair $(X,x_0)\in\mathscr{PT}$ the diagram*

$$\begin{array}{ccc} \pi_n(X,x_0) & \xrightarrow{\Sigma} & \pi_{n+1}(SX,*)\cong\pi_{n+1}(CX/X,*) \\ {\scriptstyle\partial^{-1}}\downarrow\cong & & \cong\downarrow{\scriptstyle j_*} \\ \pi_{n+1}(CX,X,x_0) & \xrightarrow[p_*]{} & \pi_{n+1}(CX/X,\{*\},*) \end{array}$$

commutes for all $n\geqslant 0$.

Proof: First observe that the following diagram commutes

$$\begin{array}{ccc} CS^n & \xrightarrow{q} & CS^n/S^n\cong S(S^n) \\ {\scriptstyle Cf}\downarrow & & \downarrow{\scriptstyle Sf} \\ CX & \xrightarrow[p]{} & CX/X\cong S(X), \end{array}$$

where $f:(S^n,s_0)\to(X,x_0)$ is any map. Now

$$\begin{aligned} j_*\circ\Sigma[f] &= j_*[Sf]=[Sf\circ q]=[p\circ Cf] \\ &= p_*[Cf]=p_*\,\partial^{-1}[f], \end{aligned}$$

since $\partial[Cf]=[f]$. $\square$

6.25. Corollary. *Σ is a homomorphism for all (X,x_0) and $n\geqslant 1$.*

6.26. Freudenthal suspension theorem. *For every n-connected CW-complex X, $n\geqslant 0$, $\Sigma:\pi_r(X,x_0)\to\pi_{r+1}(SX,*)$ is an isomorphism for $1\leqslant r\leqslant 2n$ and an epimorphism for $r=2n+1$.*

Proof: By 6.24 it suffices to prove the corresponding statements about p_*. But they follow from 6.22 applied to the pair $(CX,X,*)$ since X is n-connected and (CX,X) is $(n+1)$-connected. $\square$

We now wish to apply 6.26 to show that $\pi_n(S^n,s_0)\cong\mathbb{Z}$ for $n\geqslant 1$. We know the result is true for $n=1$ from 4.18. We deal with the case $n=2$ next.

6.27. Proposition. *For each* $n \geqslant 0$ $(CP^n, \eta, S^{2n+1}, S^1)$ *is a fibre bundle.*

Proof: This is just the fibre bundle $V_{1,n+1}(\mathbb{C}) \to G_{1,n+1}(\mathbb{C})$ discussed in 4.14. We can also give a direct proof as follows: For each k, $1 \leqslant k \leqslant n+1$, let $U_k = \{[z_0, \ldots, z_{n+1}] \in CP^n : z_k \neq 0\}$.

Then $\{U_1, \ldots, U_{n+1}\}$ is an open cover for CP^n. Define $\phi_k : U_k \times S^1 \to \eta^{-1} U_k$ by

$$\phi_k([z_1, z_2, \ldots, z_{n+1}], z) = \left(\frac{z_1 \bar{z}_k z}{|z_k| \sqrt{\sum_{i=1}^{n+1} |z_i|^2}}, \ldots, \frac{z_{n+1} \bar{z}_k z}{|z_k| \sqrt{\sum_{i=1}^{n+1} |z_i|^2}} \right).$$

Then clearly $\eta \circ \phi_k = p_{U_k}$. ϕ_k^{-1} is given by

$$\phi_k^{-1}(z_1, z_2, \ldots, z_{n+1}) = \left([z_1, z_2, \ldots, z_{n+1}], \frac{z_k}{|z_k|} \right). \quad \square$$

Clearly also (RP^n, p, S^n, S^0) and (HP^n, v, S^{4n+3}, S^3) are fibre bundles, all $n \geqslant 0$.

Now CP^0 is a single point and CP^1 has one 0-cell, one 2-cell. Thus CP^1 must be homeomorphic to S^2 (the homeomorphism is given by $S^2 \cong D^2/S^1 \xrightarrow{\bar{f}} CP^1/CP^0$, where $\bar{f}$ is induced by the characteristic map $f^2 : (D^2, S^1) \to (CP^1, CP^0)$ of the 2-cell). Thus we have a fibre bundle (S^2, η, S^3, S^1). We look at a piece of the homotopy sequence of the fibration:

$$\cdots \to \pi_2(S^3, s_0) \to \pi_2(S^2, s_0) \to \pi_1(S^1, s_0) \to \pi_1(S^3, s_0) \to \cdots.$$

Since $\pi_1(S^3, s_0) = \pi_2(S^3, s_0) = 0$ by 6.12, we see that

$$\pi_2(S^2, s_0) \cong \pi_1(S^1, s_0) \cong \mathbb{Z}.$$

6.28. Theorem. *For every* $n \geqslant 1$

$$\Sigma : \pi_n(S^n, s_0) \to \pi_{n+1}(S^{n+1}, s_0)$$

is an isomorphism, and hence $\pi_n(S^n, s_0) \cong \mathbb{Z}$, $n \geqslant 1$.

Proof: S^n is $(n-1)$-connected, so by 6.26 $\Sigma : \pi_r(S^n, s_0) \to \pi_{r+1}(S^{n+1}, s_0)$ is an isomorphism for $r < 2n-1$ and an epimorphism for $r = 2n-1$. In particular $\Sigma : \pi_n(S^n, s_0) \to \pi_{n+1}(S^{n+1}, s_0)$ is an isomorphism if $n < 2n-1$ (i.e. $n > 1$) and an epimorphism for $n = 1$. But every epimorphism $\mathbb{Z} \to \mathbb{Z}$ is an isomorphism. $\square$

Let us look at another piece of the exact homotopy sequence of the fibration $\eta : S^3 \to S^2$:

$$\cdots \to \pi_3(S^1, s_0) \to \pi_3(S^3, s_0) \xrightarrow{\eta_*} \pi_3(S^2, s_0) \to \pi_2(S^1, s_0) \to \cdots$$

By 4.18 we know $\pi_3(S^1,s_0) = \pi_2(S^1,s_0) = 0$, so $\eta_*:\pi_3(S^3,s_0) \to \pi_3(S^2,s_0)$ is an isomorphism. Thus $\pi_3(S^2,s_0) \cong \mathbb{Z}$, the first example where $\pi_m(S^n,s_0) \neq 0$ for $m > n$. Since $\pi_3(S^3,s_0)$ is generated by $[1_{S^3}]$, we see that $\pi_3(S^2,s_0)$ is generated by $[\eta]$. η is called the *Hopf* map. It can be shown that $v:S^7 \to HP^1 \cong S^4$ defines a non-zero element $[v] \in \pi_7(S^4,s_0)$ of infinite order.

For each r we have

$$\pi_{2r+2}(S^{r+2},s_0) \xrightarrow[\cong]{\Sigma} \pi_{2r+3}(S^{r+3},s_0) \xrightarrow[\cong]{\Sigma} \cdots \xrightarrow[\cong]{\Sigma} \pi_{k+r}(S^k,s_0) \xrightarrow[\cong]{\Sigma} \cdots.$$

We denote the common group $\pi_{k+r}(S^k,s_0)$, $k \geqslant r+2$, by π_r^S. It is called the rth *stable homotopy group* or the *r-stem*. For example, $\pi_1^S \cong \pi_4(S^3,s_0)$ turns out to be isomorphic to $\mathbb{Z}_2$ and is generated by $\Sigma[\eta]$, since $\Sigma:\pi_3(S^2,s_0) \to \pi_4(S^3,s_0)$ is epimorphic.

We turn now to a proof of the theorem of J. H. C. Whitehead that in the category of CW-complexes a weak homotopy equivalence is a homotopy equivalence. The proof does not depend on the simplicial approximation lemma and is in fact more elementary than the proofs we have just been considering. In Chapter 3 we saw how questions about n-equivalences can be converted into questions about n-connectivity, and that is the method we employ.

In 3.14 we saw that (Y,B) is n-connected if and only if every map $f:(D^r,S^{r-1}) \to (Y,B)$ can be deformed $\operatorname{rel} S^{r-1}$ into a map f' with $f'(D^r) \subset B$ for all r, $0 \leqslant r \leqslant n$ (D^0 = point, $S^{-1} = \varnothing$). Now by using the mapping cylinder as in Chapter 3, we can replace the inclusion $i:B \to Y$ by an n-equivalence $f:Z \to Y$ and find that the corresponding result still remains true.

6.29. Lemma. *Suppose $f:Z \to Y$ is an n-equivalence. Then for any maps $g:S^{r-1} \to Z$ and $h:D^r \to Y$, $r \leqslant n$, such that $h|S^{r-1} = f \circ g$ we can find a map $h':D^r \to Z$ with $h'|S^{r-1} = g$ and $f \circ h' \simeq h \operatorname{rel} S^{r-1}$.*

Proof: The diagram of the maps given is Fig. 1 below.

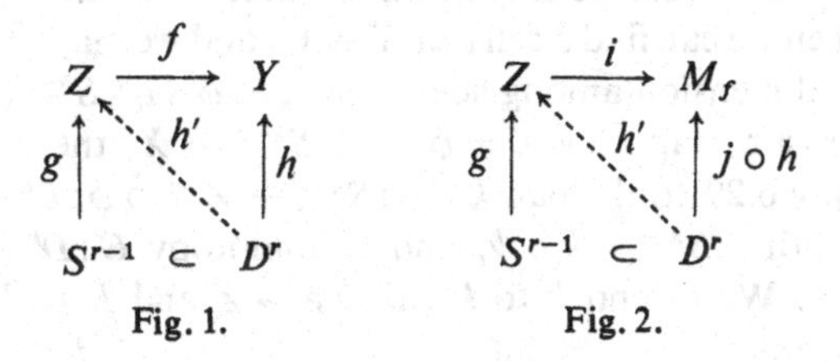

Fig. 1. Fig. 2.

Recall the mapping cylinder M_f of f and its paraphernalia of maps $i:Z \to M_f$, $j:Y \to M_f$, $r:M_f \to Y$ and $H:M_f \times I \to M_f$ (3.15). The idea is to replace $f:Z \to Y$ by $i:Z \to M_f$ and apply 3.14 to the n-connected pair (M_f,Z) as in Fig. 2. The only difficulty is that we do not have $j \circ h|S^{r-1} = i \circ g$; in fact $j \circ h|S^{r-1} = j \circ f \circ g = j \circ r \circ i \circ g \simeq i \circ g$. The homotopy is $H(ig \times \alpha)$, $\alpha(t) = 1 - t$. Since $S^{r-1} \subset D^r$ is a cofibration, we can extend $H \circ (ig \times \alpha)$ to a homotopy $H': D^r \times I \to M_f$ with $H_0' = j \circ h$. Let $\bar{h} = H_1'$; then $\bar{h}|S^{r-1} = H_1'|S^{n-1} = H_0 \circ i \circ g = i \circ g$, so we can apply 3.14 to $\bar{h}:(D^r, S^{r-1}) \to (M_f, Z)$ to find a map $h': D^r \to Z$ which is homotopic to $\bar{h} \operatorname{rel} S^{r-1}$ and such that $h'|S^{r-1} = g$. Then we have $f \circ h' = r \circ i \circ h' \simeq r \circ \bar{h} \simeq r \circ j \circ h = h$, and the homotopy is relative to S^{r-1} since $r \circ H'|S^{r-1} \times I = r \circ H \circ (ig \times \alpha)$ is stationary. ▯

Our experience with CW-complexes up to now might lead us to expect that 6 29 would remain true if we replaced (D^r, S^{r-1}) by a relative CW-complex (X,A) with $\dim(X,A) \leqslant n$. This is in fact true.

6.30. Theorem. *If $f:Z \to Y$ is an n-equivalence and (X,A) is a relative CW-complex with $\dim(X,A) \leqslant n$, then given any maps $g:A \to Z$, $h:X \to Y$ such that $h|A = f \circ g$ we can find a map $h':X \to Z$ with $h'|A = g$ and $f \circ h' \simeq h \operatorname{rel} A$. ($n = \infty$ is allowed.)*

Proof: Let $\mathscr{S}$ be the set of all triples (X',k,K) with $A \subset X' \subset X$, $k:X' \to Z$ a map with $k|A = g$ and $K: X' \times I \to Y$ a homotopy from $f \circ k$ to $h|X'$ which is stationary on A. We give $\mathscr{S}$ a partial ordering by saying $(X',k,K) \leqslant (X'',k',K')$ if and only if $A \subset X' \subset X'' \subset X$, $k'|X' = k$, $K'|X' \times I = K$. Clearly (A,g,K_0) (K_0 the stationary homotopy $A \times I \to Y$) belongs to $\mathscr{S}$, so $\mathscr{S} \neq \varnothing$. We wish to show that $\mathscr{S}$ satisfies the hypothesis of Zorn's lemma.

Suppose $\mathscr{T} \subset \mathscr{S}$ is a linearly ordered subset. We set $W = \cup X'$, where the union is taken over all X' such that there is a $(X',k,K) \in \mathscr{T}$ and define $h': W \to Z$, $H': W \times I \to Y$ by $h'(x) = k(x)$ and $H'(x,t) = K(x,t)$ if $x \in X'$ for some $(X',k,K) \in \mathscr{T}$. h' and H' are then well defined and also continuous, since X has the weak topology. Evidently $(W,h',H') \in \mathscr{S}$ and $(X',k,K) \leqslant (W,h',H')$ for all $(X',k,K) \in \mathscr{T}$.

Thus we may apply Zorn's lemma to conclude that $\mathscr{S}$ has a maximal element (X',k,K). We have only to show that $X' = X$, and we are finished. If $X' \neq X$, then we can find a cell e of $X - X'$, and we may as well assume e has minimal dimension among such cells. Let $\phi:(D^k, S^{k-1}) \to (X, X')$ be the characteristic map of e, $\psi = \phi|S^{k-1}: S^{k-1} \to X'$ the attaching map. We then apply 6.29 to the pair $k \circ \psi: S^{k-1} \to Z$, $h \circ \phi: D^k \to Y$ to find a $\theta: D^k \to Z$ with $\theta|S^{k-1} = k \circ \psi$, and a homotopy $\Theta: D^k \times I \to Y$ from $f \circ \theta$ to $h \circ \phi$. We extend k to $k': X' \cup e \to Z$ and K to $K':(X' \cup e) \times I \to Y$ by setting

$$k'(x) = \begin{cases} h(x) & \text{if } x \in X' \\ \theta(y) & \text{if } x = \phi(y) \quad \text{for} \quad y \in D^k \end{cases}$$

$$K'(x,t) = \begin{cases} K(x,t) & \text{if } x \in X' \\ \Theta(y,t) & \text{if } x = \phi(y) \quad \text{for} \quad y \in D^k. \end{cases}$$

Then $(X' \cup e, k', K') \in \mathscr{S}$ and $(X', k, K) < (X' \cup e, k', K')$, contradicting the maximality of (X', k, K). Thus we must have $X' = X$, so we may take $h' = k$. ▯

Remark. We can of course apply 6.30 to the inclusion $i: B \to Y$, which is an n-equivalence if (Y, B) is n-connected. 6.30 then says that every map $f:(X,A) \to (Y,B)$ of a relative CW-complex (X,A), $\dim(X,A) \leqslant n$, is deformable rel$A$ to a map f' with $f'(X) \subset B$. In particular, if (Y, B) is a relative CW-complex which is n-connected for all n, then we may apply this result to $1:(Y,B) \to (Y,B)$ to show that B is a strong deformation retract of Y. Thus any CW-complex X for which $\pi_r(X, x_0) = 0$, $r \geqslant 0$, is contractible.

6.31. Theorem. *If $f:Z \to Y$ is an n-equivalence ($n = \infty$ is allowed), then for any CW-complex X the function $f_*:[X; Z] \to [X; Y]$ is a surjection if* $\dim X \leqslant n$ *and a bijection if* $\dim X < n$. *Similarly for $f_*:[X, x_0; Z, z_0] \to [X, x_0; Y, y_0]$, where x_0, y_0, z_0 are any base points with $f(z_0) = y_0$.*

Proof: We do the case with base points; the other is similar.

i) f_* is surjective if $\dim X \leqslant n$: given $h:(X,x_0) \to (Y,y_0)$, we apply 6.30 with $A = \{x_0\}$ and $g:A \to Z$ given by $g(x_0) = z_0$. Then there is a map $h':(X,x_0) \to (Z,z_0)$ with $f \circ h' \simeq h \operatorname{rel} x_0$; that is $f_*[h'] = [h]$.

ii) f_* is injective if $\dim X < n$: given maps $g_0, g_1:(X,x_0) \to (Z,z_0)$ and a homotopy $h:X \times I \to Y$ from $f \circ g_0$ to $f \circ g_1 \operatorname{rel} x_0$, we take $X' = X \times I$, $A' = X \times \{0,1\} \cup \{x_0\} \times I$ (i.e. $(X', A') = (X, x_0) \times (I, \dot{I})$) and $g:A' \to Z$ the map defined by $g|X \times \{0\} = g_0$, $g|X \times \{1\} = g_1$, $g(x_0, t) = z_0$, $t \in I$. Then 6.30 gives us a map $h':X \times I \to Z$ with $h'|A' = g$. Thus h' is a homotopy relx_0 from g_0 to g_1—that is, $f_*[g_1] = f_*[g_0]$ implies $[g_1] = [g_0]$. ▯

6.32. Theorem. (J. H. C. Whitehead). *A map $f:X \to Y$ between CW-complexes is a homotopy equivalence if and only if it is a weak homotopy equivalence.*

Proof: The implication homotopy equivalence ⇒ weak homotopy equivalence is clear. Suppose f is a weak homotopy equivalence. Then the two functions

$$[X; X] \xrightarrow{f_*} [X; Y]$$

$$[Y; X] \xrightarrow{f_*} [Y; Y]$$

are bijections. Choose a $g: Y \to X$ such that $f_*[g] = [1_Y]$. Then $[f \circ g] = f_*[g] = [1_Y]$—i.e. $f \circ g \simeq 1_Y$. Also $f_*[g \circ f] = [f \circ g \circ f] = [1_Y \circ f] = [f] =$

$[f \circ 1_X] = f_*[1_X]$, and since f_* is injective it follows $[g \circ f] = [1_X]$—i.e. $g \circ f \simeq 1_X$. □

One's first reaction might be that 6.32 says the homotopy type of a CW-complex is completely determined by its homotopy groups. The existence of the map $f: X \to Y$ in 6.32 however is important; it is not enough to have $\pi_n(X, x_0) \cong \pi_n(Y, y_0)$, $n \geqslant 0$, if the isomorphisms are not induced by a map $f:(X, x_0) \to (Y, y_0)$. The following example illustrates this fact.

6.33. Example. The 3-dimensional lens spaces $L(p,q)$, $p \in \mathbb{N} - \{0\}$, $(p,q) = 1$, are defined as follows. Let $S^3 \subset \mathbb{C}^2$ be the subspace

$$\{(z_0, z_1) \in \mathbb{C}^2 : |z_0|^2 + |z_1|^2 = 1\}.$$

Define $g: S^3 \to S^3$ by $g(z_0, z_1) = (e^{2\pi i/p} z_0,\ e^{2\pi i q/p} z_1)$. if g^k denotes the k-fold iteration $\underbrace{g \circ g \circ \cdots \circ g}_{k}$, then $g^p = 1_{S^3}$. Hence g defines an action of $\mathbb{Z}_p$ on S^3 by $\bar{k} \cdot x = g^k(x)$ for $\bar{k} \in \mathbb{Z}_p$, $x \in S^3$. The action is clearly without fixed points, so by 4.21iii) and 4.20 the projection $q: S^3 \to S^3/\mathbb{Z}_p$ is a covering. We take $L(p,q) = S^3/\mathbb{Z}_p$. Since $\pi_1(S^3, s_0) = 0$, it follows from 4.21i) that $\pi_1(L(p,q), *) \cong \mathbb{Z}_p$, and of course $\pi_n(L(p,q), *) \cong \pi_n(S^3, s_0)$ for $n \geqslant 2$. Moreover the morphism $q_*: \pi_n(S^3, s_0) \to \pi_n(L(p,q), *)$ is a homomorphism of groups with $\mathbb{Z}_p$-action. But $g: S^3 \to S^3$ is homotopic (not rel s_0!) to 1_{S^3} (take $H: S^3 \times I \to S^3$ defined by

$$H(z_0, z_1, t) = (e^{2\pi i t/p} z_0, e^{2\pi i q t/p} z_1))$$

so we find that $\mathbb{Z}_p$ acts trivially on $\pi_n(S^3, s_0)$ and hence on $\pi_n(L(p,q), *)$, $n \geqslant 1$.

Thus we have shown that the homotopy groups of $L(p,q)$ as groups with $\mathbb{Z}_p$-action are independent of q. We might therefore ask whether $L(p,q)$ and $L(p,q')$ have the same homotopy type if $q \neq q'$. The cohomology groups of $L(p,q)$ and $L(p,q')$ are isomorphic as groups but *not as rings* unless $qq' \equiv \pm m^2 \pmod{p}$, some m (cf. [45], for example). Thus although $\pi_*(L(5,1), *) \cong \pi_*(L(5,2), *)$ as $\mathbb{Z}_5$-groups, $L(5,1)$ and $L(5,2)$ cannot have the same homotopy type.

Remark. It is easy to give $L(p,q)$ a CW-structure; we divide S^3 up into cells in such a way that $g: S^3 \to S^3$ carries any cell into another cell. Then $L(p,q) = S^3/\mathbb{Z}_p$ inherits a natural cell structure.

By combining 6.30 with 6.10 we can show that every map between CW-complexes is homotopic to a map which behaves well with respect to the CW-structures.

6.34. Definition. A map $f:(X, A) \to (Y, B)$ between relative CW-complexes is called *cellular* if for every $n \geqslant -1$ we have $f((X, A)^n) \subset (Y, B)^n$. A homotopy is called cellular if it is cellular as a map $(X \times I, A \times I) \to (Y, B)$.

6.35. Proposition. *Every map $f:(X,A)\to(Y,B)$ between relative CW-complexes is homotopic* rel A *to a cellular map and any two cellular maps which are homotopic* rel A *are homotopic* rel A *via a cellular homotopy.*

Proof: By 6.10 $(Y,(Y,B)^n)$ is n-connected for all $n\geqslant -1$. Since $\dim(X,A)^n\leqslant n$, we may apply 6.30 and induction to construct a sequence of homotopies $H^r: X\times I\to Y$, $r\geqslant 0$, satisfying

i) $H_0^0=f$,
ii) $H_0^r=H_1^{r-1}$,
iii) H^r is stationary on $(X,A)^{r-1}$,
iv) $H_1^r((X,Y)^r)\subset(Y,B)^r$.

Now we define $H: X\times I\to Y$ by

$$H(x,t)=\begin{cases} H^{r-1}(x,(r+1)(r(t-1)+1)) & \dfrac{r-1}{r}\leqslant t\leqslant\dfrac{r}{r+1},\ x\in X,\\ H^r(x,1) & t=1,\ x\in(X,A)^r.\end{cases}$$

Then H is a homotopy of f rel A, and by iv) H_1 is cellular.

The second part is proved by applying the first part to a map $(X\times I, A\times I\cup X\times\{0,1\})\to(Y,B)$. □

Thus if (X,x_0), (Y,y_0) are CW-complexes, then whenever we choose a representative $f:(X,x_0)\to(Y,y_0)$ for an element of $[X,x_0;Y,y_0]$ we may assume f is cellular if needed.

Exercise. Show that $Y\cup_f CX$ is a CW-complex if $f:(X,x_0)\to(Y,y_0)$ is cellular.

We next show that we can construct CW-complexes with arbitrarily given homotopy groups.

6.36. Proposition. *If X is an n-connected CW-complex and Y is an m-connected CW-complex, then the maps $i_X:(X,x_0)\to(X\vee Y,*)$, $i_Y:(Y,y_0)\to(X\vee Y,*)$ given by $i_X(x)=(x,y_0)$, $i_Y(y)=(x_0,y)$ induce an isomorphism $(i_{X*},i_{Y*}):\pi_r(X,x_0)\oplus\pi_r(Y,y_0)\to\pi_r(X\vee Y,*)$* for $2\leqslant r\leqslant n+m$, *provided X or Y is locally finite.*

Proof: As we pointed out in the proof of 6.15 $(X\times Y, X\vee Y)^{n+m+1}=X\vee Y$. Hence the exact homotopy sequence for the pair $(X\times Y, X\vee Y,*)$ reduces to the isomorphism

$$j_*:\pi_r(X\vee Y,*)\to\pi_r(X\times Y,*),\ 1\leqslant r\leqslant n+m.$$

We have the isomorphism $(p_{X*},p_{Y*}):\pi_r(X\times Y,*)\to\pi_r(X,x_0)\times\pi_r(Y,y_0)\cong\pi_r(X,x_0)\oplus\pi_r(Y,y_0)$, $r\geqslant 2$. from 4.1. Since $p_X\circ j\circ i_X=1_X$,

$p_Y \circ j \circ i_Y = 1_Y$, it follows $(p_{X*}, p_{Y*}) \circ j_* \circ (i_{X*}, i_{Y*}) = 1$, so that (i_{X*}, i_{Y*}) is indeed an isomorphism. □

6.37. Corollary. *The morphism* $\{i_{\alpha*}\}: \bigoplus_\alpha \pi_n(S^n_\alpha, s_0) \to \pi_n(\bigvee_\alpha S^n_\alpha, *)$ *is an isomorphism for* $n \geqslant 2$.

Proof: If the index set through which the α's run is finite, then the result follows from 6.36 by induction. If the index set is infinite, then we observe that every map $f: (S^n, s_0) \to (\bigvee_\alpha S^n_\alpha, *)$ has compact image, so that there is a finite set $\alpha_1, \alpha_2, \ldots, \alpha_k$ such that $f(S^n) \subset \bigvee_{i=1}^k S^n_{\alpha_i}$. Since we have a factorization

$$\begin{array}{ccccc} & & \bigoplus \pi_n(S^n_\alpha, s_0) & & \\ & \overset{j}{\nearrow} & & \overset{\{i_{\alpha*}\}}{\searrow} & \\ \bigoplus_{i=1}^k \pi_n(S^n_{\alpha_i}, s_0) & & & & \pi_n(\bigvee_\alpha S^n_\alpha, *) \\ & \underset{\{i_{\alpha_i*}\}}{\searrow} \cong & & \nearrow & \\ & & \pi_n(\bigvee_{i=1}^k S^n_{\alpha_i}, *) & & \end{array}$$

we see that $[f] \in \operatorname{im}\{i_{\alpha*}\}$, and hence $\{i_{\alpha*}\}$ is surjective. Similarly any homotopy $H: S^n \times I \to \bigvee_\alpha S^n_\alpha$ has compact image, so $\{i_\alpha *\}$ is injective. □

6.37 can also be stated: $\pi_n(\bigvee_\alpha S^n_\alpha, *)$ is the free abelian group generated by the elements $[i_\alpha]$. For π_1 we have Van Kampen's Theorem, which implies that $\pi_1(X \vee Y, *) \cong \pi_1(X, x_0) * \pi_1(Y, y_0)$, the free product.

6.38. Proposition. $\pi_1(\bigvee_\alpha S^1_\alpha, *)$ *is isomorphic to the free group generated by the* $[i_\alpha]$.

6.39. Theorem: i) *Let G be an abelian group and $n \geqslant 2$ an integer. There is a complex X with*

$$\pi_r(X, x_0) \cong \begin{cases} G & r = n \\ 0 & r \neq n. \end{cases}$$

ii) *Let X be an $(n-1)$-connected CW-complex with $\pi_n(X, x_0) \cong G$ and (Y, y_0) a space with*

$$\pi_r(Y, y_0) \cong \begin{cases} H & r = n \\ 0 & r > n. \end{cases}$$

Let $\phi: G \to H$ be a homomorphism. Then we can find a map $f: (X, x_0) \to (Y, y_0)$ such that the following square commutes

$$\begin{array}{ccc} \pi_n(X, x_0) & \xrightarrow{f_*} & \pi_n(Y, y_0) \\ \| \wr & & \| \wr \\ G & \xrightarrow{\phi} & H. \end{array}$$

f is unique up to homotopy.

Proof: i) We can find free abelian groups F, R and homomorphisms κ, λ such that

$$0 \longrightarrow R \xrightarrow{\kappa} F \xrightarrow{\lambda} G \longrightarrow 0$$

is exact (take F to be the free abelian group generated by the elements of G, λ the obvious map, $R = \ker \lambda$). Let $\{x_\alpha\}_{\alpha \in A}$, $\{r_\gamma\}_{\gamma \in \Gamma}$ be bases for F,R respectively. Take one copy S^n_α of the n-sphere for each $\alpha \in A$ and let $(X^n, x_0) = (\vee_{\alpha \in A} S^n_\alpha, *)$. Then $\pi_n(X^n, x_0) = \pi_n(\vee_\alpha S^n_\alpha, *) \cong \bigoplus_{\alpha \in A} \pi_n(S^n_\alpha, s_0) \cong F$.

For each $\gamma \in \Gamma$ choose a map $h_\gamma : (S^n, s_0) \to (X^n, x_0)$ such that $[h_\gamma] \in \pi_n(X^n, x_0)$ corresponds to $\kappa(r_\gamma) \in F$. Attach an $(n+1)$-cell e^{n+1}_γ by means of h_γ for each $\gamma \in \Gamma$. Let $X^{n+1} = X^n \cup \bigcup_{\gamma \in \Gamma} e^{n+1}_\gamma$. In the exact homotopy sequence

$$\begin{array}{ccccccc}
\cdots \longrightarrow \pi_{n+1}(X^{n+1}, X^n, x_0) & \xrightarrow{\partial} & \pi_n(X^n, x_0) & \xrightarrow{i_*} & \pi_n(X^{n+1}, x_0) & \longrightarrow & \pi_n(X^{n+1}, X^n, x_0) \\
\cong \downarrow p_* & & \| & & \wr & & \| \\
\pi_{n+1}(X^{n+1}/X^n, *) & & \| & & & & 0 \\
\|\wr & & \| & & & & \\
\pi_{n+1}(\vee_{\gamma \in \Gamma} S^{n+1}_\gamma, *) & & \| & & & & \\
\|\wr & & \| & & & & \\
R & \xrightarrow{\kappa} & F & \xrightarrow{\lambda} & G & \longrightarrow & 0
\end{array}$$

we have $\partial[g^{n+1}_\gamma] = [h_\gamma] = \kappa(r_\gamma)$, where $g^{n+1}_\gamma : (D^{n+1}, S^n, s_0) \to (X^{n+1}, X^n, x_0)$ is the characteristic map of e^{n+1}_γ. (That $X^{n+1}/X^n \cong \vee_{\gamma \in \Gamma} S^{n+1}_\gamma$ is 5.16). Thus the sequence above commutes and so $\pi_n(X^{n+1}, x_0) \cong G$.

Suppose we have constructed $X^m (m > n)$ so that

6.40.
$$\pi_r(X^m, x_0) = \begin{cases} G & r = n \\ 0 & 0 \leqslant r < m, r \neq n. \end{cases}$$

Choose a set of generators $\{y_\beta\}_{\beta \in B}$ for $\pi_m(X^m, x_0)$ and maps

$$g_\beta : (S^m, s_0) \to (X^m, x_0)$$

representing y_β, $\beta \in B$. Attach $(m+1)$-cells e^{m+1}_β by means of g_β, $\beta \in B$, and let $X^{m+1} = X^m \cup \bigcup_{\beta \in B} e^{m+1}_\beta$. Let f^{m+1}_β be the characteristic map of e^{m+1}_β. Then in the exact homotopy sequence

$$\cdots \longrightarrow \pi_{m+1}(X^{m+1}, X^m, x_0) \xrightarrow{\partial} \pi_m(X^m, x_0) \longrightarrow \pi_m(X^{m+1}, x_0) \longrightarrow 0$$

we have $\partial[f^{m+1}_\beta] = [g_\beta] = y_\beta$, so ∂ is surjective. Therefore $\pi_m(X^{m+1}, x_0) = 0$. Since $\pi_r(X^{m+1}, x_0) \cong \pi_r(X^m, x_0)$ for $0 \leqslant r < m$, we see that X^{m+1} satisfies

6.40, and hence we may proceed by induction. We take $X = \bigcup_{m \geq n} X^m$ with the weak topology. Then

$$\pi_r(X, x_0) \cong \pi_r(X^{r+1}, x_0) = \begin{cases} G & r = n \\ 0 & r \neq n. \end{cases}$$

ii) If $X^{n-1} \neq x_0$, then replace X by a homotopy equivalent CW-complex for which this is true. We still call the result X. Then X^n is a wedge sum $\bigvee_{\alpha \in A} S^n_\alpha$ of n-spheres; let $i_\alpha : (S^n, s_0) \to (X, x_0)$ be the inclusion map of the αth sphere. Then $\phi[i_\alpha] \in \pi_n(Y, y_0)$, so we can choose a map $f^n_\alpha : (S^n, s_0) \to (Y, y_0)$ with $[f^n_\alpha] = \phi[i_\alpha]$. Define $f|X^n$ by $f|S^n_\alpha = f^n_\alpha$. Then we have constructed $f|X^n$ so that the following diagram commutes

6.41.

$$\begin{array}{ccccccc}
\pi_{n+1}(X^{n+1}, X^n, x_0) & \xrightarrow{\partial} & \pi_n(X^n, x_0) & \xrightarrow{j_*} & \pi_n(X, x_0) & \longrightarrow & 0 \\
 & & \Big\downarrow (f|X^n)_* & & \cong \; G \; \Big\downarrow \phi & & \\
 & & \pi_n(Y, y_0) & \cong & H. & &
\end{array}$$

Let $\{e^{n+1}_\beta\}_{\beta \in B}$ be the $(n+1)$-cells of X and let $g^{n+1}_\beta : (D^{n+1}, S^n, s_0) \to (X^{n+1}, X^n, x_0)$ be the characteristic map of e^{n+1}_β, $\beta \in B$. Then

$$(f|X^n)_* \circ \partial[g^{n+1}_\beta] = \phi \circ j_* \circ \partial[g^{n+1}_\beta] = 0,$$

so we see that $(f|X^n) \circ (g^{n+1}_\beta|S^n)$ is null homotopic. Hence f can be extended to e^{n+1}_β for each $\beta \in B$ (2.34) giving $f|X^{n+1}$.

Suppose $f|X^m$ has been constructed ($m > n$). If $h : (S^m, s_0) \to (X^m, x_0)$ is the attaching map of an $(m+1)$-cell e^{m+1}, then $(f|X^m)_*[h] \in \pi_m(Y, y_0) = 0$, so $(f|X^m) \circ h \simeq 0$. Thus f can be extended over e^{m+1}, and in this way we get $f|X^{m+1}$. Thus by induction f can be extended over all of X, and 6.41 shows that $f_* : \pi_n(X, x_0) \to \pi_n(Y, y_0)$ is ϕ.

Suppose $f, f' : (X, x_0) \to (Y, y_0)$ are two such maps. For each α we have $f'_*[i_\alpha] = \phi[i_\alpha] = f_*[i_\alpha]$—i.e. $f'|S^n_\alpha \simeq f|S^n_\alpha$. Hence we get a homotopy $H^n : X^n \times I \to Y$ from $f'|X^n$ to $f|X^n$. We extend H^n to all of X skeleton by skeleton as above. ▯

6.42. Corollary. *Any two CW-complexes X, X' such that*

$$\pi_r(X, x_0) \cong \pi_r(X', x'_0) \cong \begin{cases} G & r = n \\ 0 & r \neq n \end{cases}$$

are homotopy equivalent.

Proof: Let $\phi : G \to G$ be the identity; by 6.39ii) there are maps $f : (X, x_0) \to (X', x'_0)$ and $g : (X', x'_0) \to (X, x_0)$, unique up to homotopy, such that

$f_*:\pi_n(X,x_0) \to \pi_n(X',x_0')$ and $g_*:\pi_n(X',x_0') \to \pi_n(X,x_0)$ are both ϕ. But $g \circ f:(X,x_0) \to (X,x_0)$ and 1_X are two maps which also induce ϕ on π_n, so by uniqueness $g \circ f \simeq 1_X$. Similarly $f \circ g \simeq 1_Y$. ▯

6.43. Definition. Any CW-complex X with

$$\pi_r(X,x_0) = \begin{cases} G & r = n \\ 0 & r \neq n \end{cases}$$

is called an *Eilenberg–MacLane complex of type* (G,n) and denoted by $H(G,n)$. The symbol $H(G,n)$ really denotes a homotopy type of course.

Note. One generally finds the notation $K(G,n)$ for the Eilenberg–MacLane complex of type (G,n) in the literature. Our reasons for preferring $H(G,n)$ will become clear later.

6.44. We can also construct complexes $H(G,1)$ for any group G (not necessarily abelian). We choose free groups F,R and homomorphisms α,β so that

$$1 \longrightarrow R \xrightarrow{\beta} F \xrightarrow{\alpha} G \longrightarrow 1$$

is exact. We choose free generators $\{x_\alpha\}$ for F and $\{r_\gamma\}$ for R and take $X^1 = \vee_\alpha S^1_\alpha$. We choose maps $h_\gamma:(S^1,s_0) \to (X^1,x_0)$ representing $\beta(r_\gamma)$ and attach 2-cells e^2_γ by the maps h_γ. Then $\pi_1(X^2,x_0) \cong G$, and we proceed to attach higher cells to "kill off" the higher homotopy groups just as before. The analog of 6.39ii) is also true.

6.45. *Remark* Since

$$\pi_r(\Omega H(G,n),\omega_0) \cong \pi_{r+1}(H(G,n),x_0) = \begin{cases} G & r = n-1 \\ 0 & r \neq n-1, \end{cases}$$

we see that there is a weak homotopy equivalence

$$\varepsilon'_{n-1}: H(G,n-1) \to \Omega H(G,n).$$

The adjoints $\varepsilon_{n-1}: SH(G,n-1) \to H(G,n)$ of the ε'_{n-1} will be of importance in Chapter 8.

6.46. Proposition. *Given groups $G_n, n \geqslant 1$, G_n abelian for $n \geqslant 2$, we can find a 0-connected CW-complex X with $\pi_n(X,x_0) \cong G_n$, $n \geqslant 1$.*

Proof: Take $X = \prod_{n\geqslant 1} H(G_n,n)$. Then $\pi_r(X,x_0) \cong \prod_{n\geqslant 1} \pi_r(H(G_n,n),*) = \pi_r(H(G_r,r),*) \cong G_r$, $r \geqslant 1$. ▯

We return briefly to considering how to compute homotopy groups of CW-complexes. We already know that the inclusion $i: X^n \to X$ of the n-skeleton induces an epimorphism $i_*:\pi_n(X^n,x_0) \to \pi_n(X,x_0)$. It thus becomes of interest to know what $\ker i_*$ is. As in the proof of 6.39 we have the exact sequence $\pi_{n+1}(X^{n+1}, X^n, x_0) \xrightarrow{\partial} \pi_n(X^n, x_0) \xrightarrow{i} \pi_n(X, x_0)$ from which it follows $\ker i_* = \operatorname{im} \partial$. In the proof of 6.39 we were able to describe $\pi_{n+1}(X^{n+1}, X^n, x_0)$ immediately, because X^n was 1-connected. We now consider the case where X^n may not be 1-connected.

6.47. Proposition. *Let X be a CW-complex and let the characteristic map of the n-cell e^n_α be denoted by $f^n_\alpha:(D^n, S^{n-1}, s_0) \to (X^n, X^{n-1}, x_0)$. If $n \geqslant 3$ then $\pi_n(X^n, X^{n-1}, x_0)$ is the free abelian group generated by the elements $\gamma \cdot [f^n_\alpha]$ for all $\alpha \in J_n$, $\gamma \in \pi_1(X^{n-1}, x_0) \cong \pi_1(X, x_0)$.*

Proof: We let $\tilde{X} \xrightarrow{p} X^{n-1}$ be the universal covering for X^{n-1}. If $g_\alpha = f^n_\alpha | S^{n-1}$ is the attaching map for e^n_α, then for each $\gamma \in \pi_1(X^{n-1}, x_0)$ there is a lifting $g_{\alpha,\gamma}:(S^{n-1}, s_0) \to (\tilde{X}, \gamma\tilde{x}_0)$ of g_α ($\pi_1(X^{n-1}, x_0)$ acts on $\tilde{X}$, and $p^{-1}(x_0) = \{\gamma\tilde{x}_0; \gamma \in \pi_1(X^{n-1}, x_0)\}$). We attach n-cells $e^n_{\alpha,\gamma}$ to $\tilde{X}$ by the maps $g_{\alpha,\gamma}$ and call the result $\tilde{Y}$. p clearly extends to a map $p: \tilde{Y} \to X^n$ such that $\mathring{e}^n_{\alpha,\gamma}$ is mapped homeomorphically onto $\mathring{e}^n_\alpha$. It is almost immediate that $p: \tilde{Y} \to X^n$ is a covering map. $\tilde{X}$ (and hence also $\tilde{Y}$) can be given a CW-structure. By 4.6 we then have that $p_*:\pi_n(\tilde{Y}, \tilde{X}, \tilde{x}_0) \to \pi_n(X^n, X^{n-1}, x_0)$ is an isomorphism. But $(\tilde{Y}, \tilde{X}, \tilde{x}_0)$ is $(n-1)$-connected and $\tilde{X}$ is 1-connected, so by 6.22 $\pi_n(\tilde{Y}, \tilde{X}, \tilde{x}_0) \cong \pi_n(\tilde{Y}/\tilde{X}, *) \cong \pi_n(\vee_{\alpha,\gamma} S^n, *) \cong$ free abelian group with basis $\pi_1(X, x_0) \times J_n$. ▯

6.48. Proposition. *For $n \geqslant 2$ $\ker[i_*:\pi_n(X^n, x_0) \to \pi_n(X, x_0)]$ is the subgroup of $\pi_n(X^n, x_0)$ generated by the elements $\gamma \cdot [g_\alpha]$, $\gamma \in \pi_1(X, x_0)$, $\alpha \in J_{n+1}$, where $g_\alpha:(S^n, s_0) \to (X^n, x_0)$ is the attaching map of the $(n+1)$-cell e^{n+1}_α. For $n = 1$ $\ker[i_*:\pi_1(X^1, x_0) \to \pi_1(X, x_0)]$ is the normal subgroup N generated by the elements $[g_\alpha]$, $\alpha \in J_2$, where $g_\alpha:(S^1, s_0) \to (X^1, x_0)$ is the attaching map of the 2-cell e^2_α.*

Proof: The statement for $n \geqslant 2$ follows immediately from 6.47. For $n = 1$ we let $\tilde{X} \xrightarrow{p} X^1$ be a covering space of X^1 with $p_*(\pi_1(\tilde{X}, \tilde{x}_0)) = N \subset \pi_1(X^1, x_0)$. Each g_α has a lifting $\tilde{g}_{\alpha,\gamma}:(S^1, s_0) \to (\tilde{X}, \gamma\tilde{x}_0)$, $\gamma \in \pi_1(X^1, x_0)/N$ because $[g_\alpha] \in N$. We let $\tilde{Y}$ be the space obtained from $\tilde{X}$ by attaching 2-cells by the maps $\tilde{g}_{\alpha,\gamma}$. $p:\tilde{X} \to X^1$ has an extension $p: \tilde{Y} \to X^2$ which is a covering projection.

Suppose $f:(I, \dot{I}) \to (X^1, x_0)$ defines an element $[f] \in \ker[i_*]$. f has a lifting $\tilde{f}:(I, 0) \to (\tilde{X}, \tilde{x}_0)$. Because f is null-homotopic in X^2, it follows $\tilde{f}$ is a loop in $\tilde{Y}$ and hence also in $\tilde{X}$. Then $[\tilde{f}] \in \pi_1(\tilde{X}, \tilde{x}_0)$ and $[f] = p_*[\tilde{f}] \in N$. Hence $\ker i_* \subset N$. But clearly $N \subset \ker i_*$. ▯

6.49. Exercise. Show that for any topological space Y one can construct a CW-complex Y' and a weak homotopy equivalence $f: Y' \to Y$. [Hint: Use the methods rather than the results of 6.39.]

6.50. Exercise. Show that if Y', Y'' are CW-complexes and $f': Y' \to Y, f'': Y'' \to Y$ are maps, f' a weak homotopy equivalence, then there is a cellular map $g: Y'' \to Y'$ such that $f' \circ g \simeq f''$ and g is unique up to homotopy.

If Y is a topological space, then a pair (Y', f') as in 6.49 is called a CW-substitute for Y (cf. Dyer [37]). From 6.50 it follows that any two CW-substitutes for Y are homotopy equivalent by an equivalence which is unique up to homotopy. It also follows that given a map $h: X \to Y$ of topological spaces, CW-substitutes (X', f'), (Y', g') for X, Y respectively, we can find a cellular map $h': X' \to Y'$ so that $g' \circ h' \simeq h \circ f'$ and h' is unique up to homotopy.

$$\begin{array}{ccc} X' & \xrightarrow{f'} & X \\ {\scriptstyle h'}\downarrow & & \downarrow{\scriptstyle h} \\ Y' & \xrightarrow[g']{} & Y \end{array}$$

6.51. Exercise. Suppose $f: (X'; A', B', x_0') \to (X; A, B, x_0)$ is a map of triads such that

$$f_*: \pi_*(A' \cap B', x_0') \to \pi_*(A \cap B, x_0)$$

$$f_*: \pi_*(A', x_0') \to \pi_*(A, x_0)$$

and

$$f_*: \pi_*(B', x_0') \to \pi_*(B, x_0)$$

are all isomorphisms. Show that if $X = \mathring{A} \cup \mathring{B}$ then

$$f_*: \pi_*(X'; A', B', x_0') \to \pi_*(X; A, B, x_0)$$

is an isomorphism. [Hint: Replace X by a mapping cylinder, triangulate $D^{n-1} \times I$ sufficiently finely and argue simplex by simplex.]

6.52. Exercise. Show that if $f: (X'; A' B') \to (X; A, B)$ is a map of triads such that $(X'; A', B')$ is a CW-triad, $X = \mathring{A} \cup \mathring{B}$, $f|A' \cap B': A' \cap B' \to A \cap B$, $f|A': A' \to A$, $f|B': B' \to B$ are weak homotopy equivalences and $(A', A' \cap B')$, $(B', A' \cap B')$ are 0-connected, then $f: X' \to X$ is also a weak homotopy equivalence.

6.53. Exercise. Show that the results 6.21–6.26 can be extended to arbitrary spaces with suitable additional hypotheses (such as $X = \mathring{A} \cup \mathring{B}$ in 6.21).

6.54. Exercise. Let $A \subset X$ be a cofibration with (X, A) n-connected, $n \geqslant 1$, A m-connected. Show that

$$[Y, A] \xrightarrow{i_*} [Y, X] \xrightarrow{p_*} [Y, X/A]$$

is exact for all CW-complexes Y with $\dim Y \leqslant m+n$, where $i: A \to X$ is the inclusion and $p: X \to X/A$ the projection.

6.55. Exercise. A *graph* is a 0-connected 1-dimensional CW-complex and a *tree* is a graph which contains no subcomplexes homeomorphic to S^1.

a) Show that trees are contractible.
b) Show that every graph contains a maximal tree.
c) Show that for any graph X with maximal tree T we have $\pi_1(X, x_0) \cong F(T^*)$, the free group generated by the set T^* of all 1-cells not in T.
d) For any 0-connected CW-complex X describe a presentation of $\pi_1(X, x_0)$ in terms of generators and relations.

Comments

The elegant, elementary proof of the homotopy excision theorem given here is due to M. Boardman. There are much more sophisticated proofs in existence (cf. for example [80]) but they use the machinery of spectral sequences, for example, whereas we have used little more than the definitions of the homotopy groups. We shall give a second (sophisticated) proof of the Freudenthal suspension theorem in Chapter 15.

At this point we have enough theorems about homotopy groups that we could begin to do some interesting homotopy theory. The interests of the author are such, however, that we instead turn now to stable homotopy theory and homology theory (which is a branch of stable homotopy theory; see the comments at the end of Chapter 9).

References

1. E. Dyer [37]
2. P. J. Hilton and S. Wylie [45]
3. E. H. Spanier [80]

Chapter 7

Homology and Cohomology Theories

Eilenberg and Steenrod in [40] proved that homology with coefficients G (simplicial or singular) was characterized on the category of finite simplicial complexes by seven axioms now known as the Eilenberg–Steenrod axioms. They showed how to derive many of the properties of homology directly from the axioms. Later others (Atiyah, Hirzebruch and G. W. Whitehead) observed that there were other functors which satisfied six of the axioms (all except the "dimension" axiom) and hence enjoyed many of the properties of homology. Whitehead gave a systematic treatment of these generalized homology theories in [93]. In this chapter we shall give the axioms for a generalized homology theory (and also for cohomology). In Chapter 8 we shall show how to construct a large variety of homology and cohomology theories, and in Chapter 9 we shall show that in Chapter 8 we constructed all cohomology theories on the category of CW-complexes. In Chapters 10, 11 and 12 we shall give three important examples.

7.1. Definition. Let $\mathscr{T}^2$ denote the category of all topological pairs (X, A) ($A \subset X$ a subspace) and maps $f:(X, A) \to (Y, B)$. On $\mathscr{T}^2$ we have the functor $R:\mathscr{T}^2 \to \mathscr{T}^2$ ("restriction") defined by $R(X, A) = (A, \varnothing)$, $R(f) = f|A$. Let $\mathscr{T}^{2\prime}$ be the homotopy category—i.e. objects are as in $\mathscr{T}^2$, morphisms are homotopy classes of maps. We still have $R:\mathscr{T}^{2\prime} \to \mathscr{T}^{2\prime}: R(X, A) = (A, \varnothing)$, $R[f] = [f|A]$.

An *unreduced homology theory* h_* on $\mathscr{T}^{2\prime}$ is a sequence of functors $h_n:\mathscr{T}^{2\prime} \to \mathscr{A}$ for each $n \in \mathbb{Z}$ and natural transformations $\partial_n: h_n \to h_{n-1} \circ R$, $n \in \mathbb{Z}$, satisfying the following two axioms:

i) Exactness: for every pair $(X, A) \in \mathscr{T}^{2\prime}$ the sequence

$$\cdots \longrightarrow h_{n+1}(X, A) \xrightarrow{\partial_{n+1}(X, A)} h_n(A, \varnothing) \xrightarrow{h_n[i]} h_n(X, \varnothing) \xrightarrow{h_n[j]} h_n(X, A) \xrightarrow{\partial_n(X, A)} \cdots$$

is exact, where $i:(A, \varnothing) \to (X, \varnothing)$ and $j:(X, \varnothing) \to (X, A)$ are the inclusions;

ii) Excision: for every pair $(X,A) \in \mathscr{T}^{2'}$ and subset $U \subset A$ with $\bar{U} \subset \mathring{A}$ the inclusion $j:(X-U, A-U) \to (X,A)$ induces an isomorphism $h_n[j]: h_n(X-U, A-U) \to h_n(X,A)$, all $n \in \mathbb{Z}$.

Hereafter we shall abbreviate $h_n[f]$ by f_* and $\partial_n(X,A)$ by ∂.

Remark. Eilenberg–Steenrod Axioms 1 and 2 say h_n is a functor on $\mathscr{T}^2$, Axiom 3 says ∂_n is a natural transformation, Axiom 5 says h_n is actually a functor on $\mathscr{T}^{2'}$, not just on $\mathscr{T}^2$; Axiom 4 is Exactness and Axiom 6 is Excision.

7.2. Proposition. *The excision axiom is equivalent to the following statement:*

7.3. *For every triad $(X; A,B)$ (A,B subspaces of X with $A \cup B = X$) such that $\mathring{A} \cup \mathring{B} = X$ the inclusion $j:(A, A \cap B) \to (X,B)$ induces an isomorphism*

$$j_*: h_n(A, A \cap B) \to h_n(X,B), \quad \textit{all } n \in \mathbb{Z}.$$

Proof: Suppose h_n is a functor satisfying the excision axiom and $(X; A,B)$ is a triad with $\mathring{A} \cup \mathring{B} = X$. Then we can apply excision to the triple $X \supset B \supset X - A$, since $\overline{X-A} = X - \mathring{A} \subset \mathring{B}$. Because $(X-(X-A), B-(X-A)) = (A, A \cap B)$, h_n satisfies 7.3.

Conversely, suppose h_n is a functor satisfying 7.3 and (X,A) is a pair, $U \subset A$ and $\bar{U} \subset \mathring{A}$. Then we may apply 7.3 to the triad $(X; X-U, A)$, since $(X-U)^\circ \cup \mathring{A} = (X-\bar{U}) \cup \mathring{A} \supset (X-\mathring{A}) \cup \mathring{A} = X$. Since

$$(X-U, (X-U) \cap A) = (X-U, A-U),$$

h_n satisfies the excision axiom. □

7.4. Lemma. *If $(X; A,B)$ is a CW-triad then we can find an open set $A' \supset A$ and a homotopy $H: X \times I \to X$ satisfying*

i) $H_0 = 1_X$,
ii) *H is stationary on A,*
iii) $H_1(A') \subset A$,
iv) $H(B \times I) \subset B$.

Proof: Let $A^{-1} = A$ and $H^{-1}: X \times I \to X$ be the stationary homotopy. Suppose we have constructed for each k, $-1 \leqslant k < n$, an open neighborhood A^k of A in $(X,A)^{k+1}$ with $A^k \cap (X,A)^k = A^{k-1}$ and a homotopy $H^k: X \times I \to X$ satisfying

a) $H_0^k = H_1^{k-1}$,
b) H^k stationary on $(X,A)^k$,
c) $H_1^k(A^k) \subset A$,
d) $H^k(B \times I) \subset B$.

Let the $(n+1)$-cells of (X,A) be $\{e_\gamma^{n+1}: \gamma \in \Gamma\}$ and let f_γ^{n+1} be the characteristic map of e_γ^{n+1}. In D^{n+1} let $D_0^{n+1} = \{x \in D^{n+1}: \|x\| \leqslant 1/2\}$. Then $D^{n+1} -$

D_0^{n+1} is an open neighborhood of S^n which can be contracted onto S^n: define $K: D^{n+1} \times I \to D^{n+1}$ by

$$K(x,t) = \begin{cases} (1+t)x & \|x\| \leqslant \dfrac{1}{1+t} \\ \dfrac{x}{\|x\|} & \|x\| \geqslant \dfrac{1}{1+t} \end{cases} \quad x \in D^{n+1},\ t \in I.$$

Now let

$$U_\gamma^{n+1} = \left\{ f_\gamma^{n+1}(y) : f_\gamma^{n+1}\left(\frac{y}{\|y\|}\right) \in A^{n-1}, \quad y \in D^{n+1} - D_0^{n+1} \right\}.$$

U_γ^{n+1} is an open subset of e_γ^{n+1}, so if we take $A^n = A^{n-1} \cup \bigcup_{\gamma \in \Gamma} U_\gamma^{n+1}$, then A^n is an open neighborhood of A in $(X,A)^{n+1}$ with $A^n \cap (X,A)^n = A^{n-1}$. We define $H^n : (X,A)^{n+1} \times I \to X$ by

$$H^n(x,t) = \begin{cases} H_1^{n-1}(x) & x \in (X,A)^n \\ H_1^{n-1}(f_\gamma^{n+1}(K(y,t))) & x = f_\gamma^{n+1}(y), y \in D^{n+1} \end{cases} \quad t \in I.$$

Then H^n is continuous and satisfies $H_0^n = H_1^{n-1}$ on $(X,A)^{n+1}$, so we can extend to X with $H_0^n = H_1^{n-1}$. Clearly a), b) and c) are satisfied. H^n also satisfies d) since H^{n-1} does. Thus we can construct A^k, H^k, all $k \geqslant -1$, by induction.

Now we take $A' = \bigcup_{k \geqslant -1} A^k$; then A' is open because $A' \cap e_\gamma^m = A^{m-1} \cap e_\gamma^m$ for all m, γ. We define $H: X \times I \to X$ by

$$H(x,t) = \begin{cases} H^{r-1}(x,(r+1)(r(t-1)+1)) & \dfrac{r-1}{r} \leqslant t \leqslant \dfrac{r}{r+1}, x \in X, \\ H^r(x,1) & t = 1, x \in (X,A)^r. \end{cases}$$

As in 6.35 H is continuous, $H_0 = H_0^0 = 1_X$; also H is stationary on A, $H_1(A') \subset A$ and $H(B \times I) \subset B$. □

Let $\mathscr{W}^2$ denote the category of CW-pairs (X,A) ($A \subset X$ a subcomplex) and maps of pairs and $\mathscr{W}^{2\prime}$ the corresponding homotopy category.

7.5. Proposition. *For every CW-triad $(X;A,B)$ the inclusion $j:(A, A \cap B) \to (X,B)$ induces an isomorphism*

$$j_* : h_n(A, A \cap B) \to h_n(X,B), \quad \textit{all } n \in \mathbb{Z}.$$

Proof: Given any CW-triad $(X;A,B)$ we choose an open neighborhood A' of A and homotopy H as in 7.4. Let $r = H_1$. Then $r|A' : (A', A' \cap B) \to$

$(A, A \cap B)$ is a homotopy inverse to $i:(A, A \cap B) \to (A', A' \cap B)$ so we have that $(r|A')_*$ is an isomorphism. Since $r \simeq 1_X$, we see that $r_*: h_n(X,B) \to h_n(X,B)$ is 1. We have the commutative diagram

$$\begin{array}{ccc} h_n(A', A' \cap B) & \xrightarrow[\cong]{j'_*} & h_n(X,B) \\ \cong \Big\downarrow (r|A')_* & & \Big\downarrow r_* = 1 \\ h_n(A, A \cap B) & \xrightarrow{j_*} & h_n(X,B), \end{array}$$

and by 7.3 j'_* is an isomorphism, since $X - A$ is an open subset of B and thus $X = A' \cup (X - A) \subset \mathring{A}' \cup \mathring{B}$. Thus j_* is also an isomorphism. ▯

7.6. Thus when we are talking about unreduced homology theories on $\mathscr{W}^{2'}$ we may take 7.5 as the excision axiom.

We proceed now to record some elementary deductions from the axioms for homology theories.

7.7. Proposition. *If $f:(X,A) \to (Y,B)$ is a homotopy equivalence, then $f_*: h_n(X,A) \to h_n(Y,B)$ is an isomorphism, $n \in \mathbb{Z}$.*

7.8. Corollary. *If $A \subset X$ is a deformation retract of X, then $h_n(X,A) = 0$, $n \in \mathbb{Z}$.*

7.9. Corollary. $h_n(X,X) = 0$, $n \in \mathbb{Z}$.

7.10. Proposition. *If $x \in X$ is any point and $i:(\{x\}, \varnothing) \to (X, \varnothing)$ is the inclusion, then i_* injects $h_n(\{x\}, \varnothing)$ as a direct summand in $h_n(X, \varnothing)$ and in fact $h_n(X, \varnothing) \cong h_n(\{x\}, \varnothing) \oplus h_n(X, \{x\})$.*

Proof: Let $c: X \to \{x\}$ be the unique map; then $c \circ i = 1$. We have the exact sequence

$$\cdots \longrightarrow h_n(\{x\}, \varnothing) \underset{c_*}{\overset{i_*}{\rightleftarrows}} h_n(X, \varnothing) \xrightarrow{j_*} h_n(X, \{x\}) \longrightarrow \cdots$$

and because $c_* \circ i_* = (c \circ i)_* = 1_* = 1$, the sequence is split. Thus by 3.22 we have $h_n(X, \varnothing) \cong h_n(\{x\}, \varnothing) \oplus h_n(X, \{x\})$. The isomorphism is given by $a \mapsto (c_*(a), j_*(a))$. ▯

7.11. Proposition. *If (X, A, B) is a triple and $I:(A,B) \to (X,B)$, $J:(X,B) \to (X,A)$ are the inclusions, then we get an exact sequence*

$$\cdots \longrightarrow h_n(A,B) \xrightarrow{I_*} h_n(X,B) \xrightarrow{J_*} h_n(X,A) \xrightarrow{\Delta} h_{n-1}(A,B) \longrightarrow \cdots,$$

where Δ is the composite $h_n(X,A) \xrightarrow{\partial} h_{n-1}(A, \varnothing) \xrightarrow{i_*} h_{n-1}(A,B)$.

Proof: Copy the proof of 3.20. □

7.12. Definition. A topological triad $(X; A, B)$ is called *excisive* with respect to the homology theory h_* if the inclusion $j: (A, A \cap B) \to (X, B)$ induces an isomorphism $j_*: h_n(A, A \cap B) \to h_n(X, B)$ for all $n \in \mathbb{Z}$.

From 7.2 we know that $(X; A, B)$ is excisive if $\mathring{A} \cup \mathring{B} = X$, and from 7.5 we know that every CW-triad is excisive.

7.13. Lemma. *A triad $(X; A, B)$ is excisive if and only if $(X; B, A)$ is.*

Proof: We construct new spaces $C = A \cap B$,

$$\begin{aligned} X' &= (A \times \{0\}) \cup (C \times I) \cup (B \times \{1\}) \subset X \times I, \\ A' &= (A \times \{0\}) \cup (C \times I), \\ B' &= \phantom{(A \times \{0\}) \cup {}} (C \times I) \cup (B \times \{1\}), \\ C' &= \phantom{(A \times \{0\}) \cup {}} C \times I. \end{aligned}$$

The projection $p_X: X \times I \to X$ defines a map $p_X: (X', A', B', C') \to (X, A, B, C)$. In fact $(p_X|A'): (A', C') \to (A, C)$ is a homotopy inverse of the map $(A, C) \to (A', C')$ defined by $a \mapsto (a, 0)$; similarly $(p_X|B'): (B', C') \to (B, C)$ is a homotopy equivalence.

For every $n \in \mathbb{Z}$ we have a commutative square

$$\begin{array}{ccc} h_n(A', C') & \xrightarrow[\cong]{j'_*} & h_n(X', B') \\ \cong \Big\downarrow (p_X|A')_* & & \Big\downarrow p_{X*} \\ h_n(A, C) & \xrightarrow[j_*]{} & h_n(X, B) \end{array}$$

in which j'_* is an isomorphism since $\mathring{A}' \cup \mathring{B}' = X'$. Thus $p_{X*}: h_n(X', B') \to h_n(X, B)$ is an isomorphism for all $n \in \mathbb{Z}$ if and only if $(X; A, B)$ is excisive. Therefore we may apply the 5-lemma to the commutative diagram

$$\begin{array}{ccccccccc} \cdots & \longrightarrow & h_{n+1}(X', B') & \xrightarrow{\Delta} & h_n(B', C') & \xrightarrow{I'_*} & h_n(X', C') & \xrightarrow{J'_*} & \\ & & \Big\downarrow p_{X*} & & \cong \Big\downarrow (p_X|B')_* & & \Big\downarrow p_{X*} & & \\ \cdots & \longrightarrow & h_{n+1}(X, B) & \xrightarrow[\Delta]{} & h_n(B, C) & \xrightarrow[I_*]{} & h_n(X, C) & \xrightarrow[J_*]{} & \end{array}$$

$$\begin{array}{ccccc} h_n(X', B') & \xrightarrow{\Delta} & h_{n-1}(B', C') & \longrightarrow & \cdots \\ \Big\downarrow p_{X*} & & \cong \Big\downarrow (p_X|B')_* & & \\ h_n(X, B) & \xrightarrow[\Delta]{} & h_{n-1}(B, C) & \longrightarrow & \cdots \end{array}$$

to conclude that $p_{X*}: h_n(X', C') \to h_n(X, C)$ is an isomorphism for all $n \in \mathbb{Z}$ if and only if $(X; A, B)$ is excisive. However, $C = B \cap A$ and $C' = B' \cap A'$ also, so we see that $p_{X*}: h_n(X', C') \to h_n(X, C)$ is an isomorphism for all $n \in \mathbb{Z}$ if and only if $(X; B, A)$ is excisive. □

One would expect to be able to take the consequences of the homotopy excision theorem (6.21) and translate them into propositions about homology, but without dimension restrictions. We then get the following propositions.

7.14. Proposition. *If $A \subset X$ is a cofibration, then the projection $p:(X, A) \to (X/A, \{*\})$ induces an isomorphism $p_*: h_n(X, A) \to h_n(X/A, \{*\})$ for all $n \in \mathbb{Z}$.*

Proof: As in the proof of 6.22 the problem reduces to showing that the triad $(X^+ \cup CA^+; X^+, CA^+)$ is excisive. The set

$$X' = X^+ \cup \{[t, a] : a \in A,\ t \in [0, 1/3) \cup (2/3, 1]\}$$

is an open neighborhood of X^+ in $X^+ \cup CA^+$ which deforms onto X^+ as in 7.4.

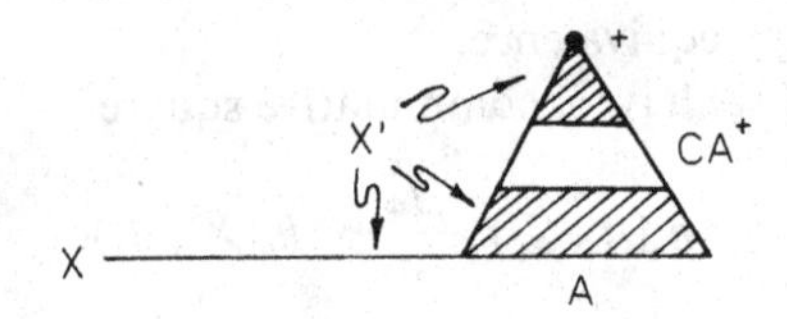

One then completes the proof as in 7.5. □

7.15. For any pointed space (X, x_0) the cone (CX, x_0) is contractible. Hence by 7.8 $h_n(CX, \{x_0\}) = 0$, all $n \in \mathbb{Z}$, and therefore in the exact sequence of the triple $(CX, X, \{x_0\})$, we have an isomorphism $\Delta: h_{n+1}(CX, X) \to h_n(X, \{x_0\})$ for all $n \in \mathbb{Z}$. We define

$$\tilde{\sigma}_n(X, x_0): h_n(X, \{x_0\}) \to h_{n+1}(SX, \{*\})$$

to be the composite

$$h_n(X, \{x_0\}) \underset{\cong}{\overset{\Delta}{\longleftarrow}} h_{n+1}(CX, X) \xrightarrow{p_*} h_{n+1}(CX/X, \{*\}) = h_{n+1}(SX, \{*\}).$$

7.16. Proposition. *If $(X, x_0) \in \mathscr{PT}$ then $\tilde{\sigma}_n(X, x_0)$ is an isomorphism for all $n \in \mathbb{Z}$.*

Proof: By 6.4 $X \subset CX$ is a cofibration, so p_* is an isomorphism by 7.14. □

7.17. Corollary. *For every $n \in \mathbb{Z}$ and non-negative integer k*

$$\tilde{\sigma}_n(S^k, s_0): h_n(S^k, \{s_0\}) \to h_{n+1}(S^{k+1}, \{s_0\})$$

is an isomorphism.

7.18. If $+$ denotes a single point then we have the map $s:+ \to S^0$ given by $s(+) = -1$. Since $(S^0; \{-1\}, \{+1\})$ is an excisive triad, we see that

$$s_* : h_n(+, \varnothing) \to h_n(S^0, \{+1\})$$

is an isomorphism for all $n \in \mathbb{Z}$. Hence $h_n(S^k, \{s_0\}) \cong h_{n-k}(S^0, \{s_0\}) \cong h_{n-k}(+, \varnothing)$, all $n \in \mathbb{Z}$, all $k \geqslant 0$. The groups $h_n(+, \varnothing)$ are called the *coefficient groups* of the homology theory h_*.

7.19. Theorem. *If $(X; A, B)$ is an excisive triad and $C \subset A \cap B$, then there is an exact sequence*

$$\cdots \xrightarrow{\Delta'} h_n(A \cap B, C) \xrightarrow{\alpha} h_n(A, C) \oplus h_n(B, C) \xrightarrow{\beta} h_n(X, C) \xrightarrow{\Delta'} h_{n-1}(A \cap B, C) \longrightarrow \cdots,$$

where $\alpha(x) = (i_{1*}(x), i_{2*}(x))$, $\beta(x,y) = i_{3*}(x) - i_{4*}(y)$ and Δ' is the composite

$$h_n(X, C) \xrightarrow{J_{1*}} h_n(X, B) \underset{\cong}{\xleftarrow{j_{1*}}} h_n(A, A \cap B) \xrightarrow{\Delta_1} h_{n-1}(A \cap B, C).$$

Here i_1, i_2, i_3, i_4, j_1, J_1 are inclusions:

$$\begin{array}{ccccc} & & (A, C) & & \\ & {\scriptstyle i_1}\nearrow & & \searrow{\scriptstyle i_3} & \\ (A \cap B, C) & & & & (X, C), \\ & {\scriptstyle i_2}\searrow & & \nearrow{\scriptstyle i_4} & \\ & & (B, C) & & \end{array} \qquad \begin{array}{l} (A, A \cap B) \xrightarrow{j_1} (X, B), \\ (X, C) \xrightarrow{J_1} (X, B). \end{array}$$

Δ_1 is the boundary for the triple $(A, A \cap B, C)$.

Proof:

$$\beta \circ \alpha(x) = \beta(i_{1*}(x), i_{2*}(x)) = i_{3*} \circ i_{1*}(x) - i_{4*} \circ i_{2*}(x).$$

But clearly $i_3 \circ i_1 = i_4 \circ i_2$, so $\beta \circ \alpha = 0$. The other five facts which must be checked to prove exactness are proved by chasing around the following commutative diagram:

$$\begin{array}{ccccccc} \cdots \longrightarrow & h_{n+1}(A, C) & \longrightarrow & h_{n+1}(A, A \cap B) & \xrightarrow{\Delta_1} & h_n(A \cap B, C) & \xrightarrow{i_{1*}} \\ & \downarrow & & \cong \downarrow j_{1*} & & \downarrow i_{2*} & \\ \cdots \longrightarrow & h_{n+1}(X, C) & \xrightarrow[J_{1*}]{} & h_{n+1}(X, B) & \longrightarrow & h_n(B, C) & \xrightarrow[i_{4*}]{} \end{array}$$

$$\begin{array}{ccccc} h_n(A, C) & \longrightarrow & h_n(A, A \cap B) & \xrightarrow{\Delta_1} & \cdots \\ \downarrow i_{3*} & & \cong \downarrow j_{1*} & & \\ h_n(X, C) & \xrightarrow[J_{1*}]{} & h_n(X, B) & \longrightarrow & \cdots. \end{array}$$

□

7.20. *Remark.* The reader may have noticed a slight arbitrariness in our definition of Δ'; since $(X;\ B, A)$ is also excisive, we could equally well define a Δ'' by

$$h_n(X, C) \xrightarrow{J_{2*}} h_n(X, A) \xleftarrow{j_{2*}} h_n(B, A \cap B) \xrightarrow{\Delta_2} h_{n-1}(A \cap B, C).$$

We would clearly still get an exact sequence if we took Δ'' instead of Δ'. The question then arises: are Δ' and Δ'' equal? The answer is "No"; in fact $\Delta' = -\Delta''$.

7.21. Lemma. *$\Delta'' = -\Delta'$ for all excisive triads $(X;\ A, B)$.*
Proof: We have the following commutative hexagon

7.22.

$$\begin{array}{ccccc}
 & & h_n(X, C) & & \\
 & \swarrow^{J_{2*}} & \downarrow J_* & \searrow^{J_{1*}} & \\
h_n(X, A) & & & & h_n(X, B) \\
\uparrow & \nwarrow^{l_{2*}} & & \nearrow^{l_{1*}} & \uparrow \\
j_{2*} \cong & & h_n(X, A \cap B) & & \cong j_{1*} \\
 & \nearrow^{k_{2*}} & & \nwarrow^{k_{1*}} & \\
h_n(B, A \cap B) & & \downarrow \Delta & & h_n(A, A \cap B) \\
 & \searrow_{\Delta_2} & & \swarrow_{\Delta_1} & \\
 & & h_{n-1}(A \cap B, C) & &
\end{array}$$

in which the vertical sequence and the two diagonal ones are exact. The route from top to bottom around the left side is Δ'', around the right side Δ'.

Given $x \in h_n(X, C)$ choose $y_1 \in h_n(A, A \cap B)$ such that $j_{1*} y_1 = J_{1*} x$. Then $\Delta_1 y_1 = \Delta' x$. Also

$$l_{1*}(J_* x - k_{1*} y_1) = J_{1*} x - j_{1*} y_1 = 0,$$

so by exactness there is a $y_2 \in h_n(B, A \cap B)$ with $k_{2*} y_2 = J_* x - k_{1*} y_1$. Then

$$j_{2*} y_2 = l_{2*} k_{2*} y_2 = l_{2*} J_* x - l_{2*} k_{1*} y_2 = J_{2*} x.$$

Hence $\Delta'' x = \Delta_2 y_2 = \Delta k_{2*} y_2 = \Delta J_* x - \Delta k_{1*} y_1 = -\Delta_1 y_1 = -\Delta' x.$ $\square$

The exact sequence 7.21 is called the *Mayer–Vietoris sequence* and represents one of the most useful properties of homology. Taking $C = \varnothing$ for the moment, we see that if we have a space X which can be written as the union of two subspaces and we have information about the homology of the subspaces and their intersection, then the Mayer–Vietoris sequence may provide us with information about the homology of X. More generally, if $X = X_1 \cup X_2 \cup \cdots \cup X_n$ and we have information about $h_k(X_i, \varnothing)$ and $h_k(X_i \cap (X_1 \cup \cdots \cup X_{i-1}), \varnothing)$, then we can inductively deduce something about $h_k(X, \varnothing)$ in favorable cases.

The following are easy applications of the Mayer–Vietoris sequence.

7.23. Proposition. *If $(X; A, B)$ is an excisive triad, then the inclusions $i_A:(A, A \cap B) \to (X, A \cap B)$, $i_B:(B, A \cap B) \to (X, A \cap B)$ induce an isomorphism*

$$(i_{A*}, i_{B*}): h_n(A, A \cap B) \oplus h_n(B, A \cap B) \to h_n(X, A \cap B), \quad n \in \mathbb{Z}.$$

Proof: If we take $C = A \cap B$ in 7.21, then $h_n(A \cap B, C) = 0$, $n \in \mathbb{Z}$, by 7.9. Hence $\beta = (i_{A*}, -i_{B*})$ is an isomorphism, from which it follows (i_{A*}, i_{B*}) is also an isomorphism. □

7.24. Corollary. *If $X = A \cup B$ where A, B are disjoint open sets, then $(i_{A*}, i_{B*}): h_n(A, \varnothing) \oplus h_n(B, \varnothing) \to h_n(X, \varnothing)$ is an isomorphism for $n \in \mathbb{Z}$.*

Proof: $\mathring{A} \cup \mathring{B} = A \cup B = X$, so $(X; A, B)$ is excisive. □

7.25. Corollary. *For any two pointed CW-complexes (X, x_0), (Y, y_0) the inclusions $i_X:(X, \{x_0\}) \to (X \vee Y, \{*\})$ and $i_Y:(Y, \{y_0\}) \to (X \vee Y, \{*\})$ induce an isomorphism*

$$(i_{X*}, i_{Y*}): h_n(X, \{x_0\}) \oplus h_n(Y, \{y_0\}) \to h_n(X \vee Y, \{*\})$$

for $n \in \mathbb{Z}$.

Proof: $(X \vee Y; X, Y)$ is a CW-triad, hence excisive. □

7.26. Examples. i) The torus $T = S^1 \times S^1$ can be cut into two pieces $T = A \cup B$, each of which is homeomorphic to $S^1 \times I$ and such that

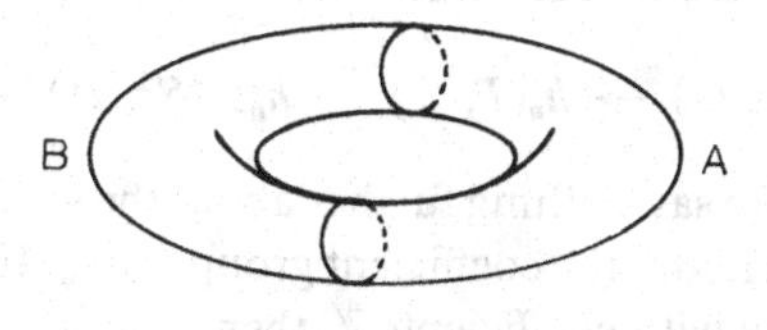

$A \cap B = S^1 \cup S^1$. As A, B have the homotopy type of S^1 (the map $p_A: A \cong S^1 \times I \to S^1$ is a homotopy equivalence) the Mayer–Vietoris sequence looks as follows:

7.27.

$$\cdots \longrightarrow h_{n+1}(T, \varnothing) \xrightarrow{\Delta'}$$

$$\begin{array}{ccccc} h_n(A \cap B, \varnothing) & \xrightarrow{\alpha} & h_n(A, \varnothing) \oplus h_n(B, \varnothing) & \xrightarrow{\beta} & h_n(T, \varnothing) \xrightarrow{\Delta'} \cdots \\ \| \wr & & \cong \downarrow p_{A*} \oplus p_{B*} & & \\ h_n(S^1, \varnothing) \oplus h_n(S^1, \varnothing) & \xrightarrow{\alpha'} & h_n(S^1, \varnothing) \oplus h_n(S^1, \varnothing). & & \end{array}$$

It thus becomes of interest to know how α' looks. If we orient the various circles as shown in Fig. 1, then the two ends of A are

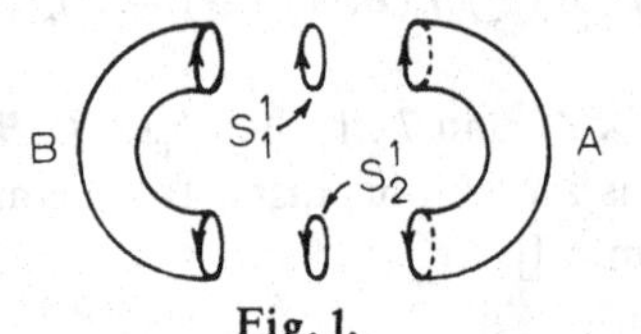

Fig. 1.

consistently oriented (and similarly for B), and the inclusions $j_A: A \cap B \to A$, $j_B: A \cap B \to B$ are such that

$$S_1^1 \xrightarrow{j_A | S_1^1} A \xrightarrow{p_A} S^1$$

is the identity 1_{S^1} and so are $p_A \circ j_A | S_2^1$, $p_B \circ j_B | S_1^1$, $p_B \circ j_B | S_2^1$. Thus α' is the map given by the matrix

$$\begin{pmatrix} 1 & 1 \\ 1 & 1 \end{pmatrix};$$

that is $\alpha'(x, y) = (x + y, x + y)$. The matrix is equivalent to

$$\begin{pmatrix} 1 & 0 \\ 0 & 0 \end{pmatrix},$$

so $\ker \alpha' \cong h_n(S^1, \varnothing)$, $\operatorname{coker} \alpha' \cong h_n(S^1, \varnothing)$. Therefore 7.27 yields a collection of short exact sequences

7.28. $0 \to h_n(S^1, \varnothing) \to h_n(T, \varnothing) \to h_{n-1}(S^1, \varnothing) \to 0, \quad n \in \mathbb{Z}.$

At this point we can say nothing further about the groups $h_n(T, \varnothing)$ without some information about the coefficient groups of h_*. If we look at ordinary singular homology with coefficients $\mathbb{Z}$, then

$$h_n(S^1, \varnothing) = \begin{cases} \mathbb{Z} & n = 0, 1 \\ 0 & n \neq 0, 1. \end{cases}$$

In particular, $h_{n-1}(S^1, \varnothing)$ is free abelian, so the sequence above splits and we get

$$h_n(T, \varnothing) \cong h_{n-1}(S^1, \varnothing) \oplus h_n(S^1, \varnothing) \cong \begin{cases} \mathbb{Z} & n = 0,2, \\ \mathbb{Z} \oplus \mathbb{Z} & n = 1, \\ 0 & n \neq 0,1,2. \end{cases}$$

ii) The Klein bottle K can be cut up in a similar fashion except that this time one finds α' has the matrix

$$\begin{pmatrix} 1 & 1 \\ 1 & v'_* \end{pmatrix},$$

where $v' : S^1 \to S^1$ is the map reversing orientation (as in 2.22). We still have the short exact sequences corresponding to 7.28:

$$0 \to \operatorname{coker} \alpha'_n \to h_n(K, \varnothing) \to \ker \alpha'_{n-1} \to 0.$$

In the case of singular integral homology we find $v'_* : H_1(S^1, \varnothing) \to H_1(S^1, \varnothing)$ is -1, $v'_* : H_0(S^1, \varnothing) \to H_0(S^1, \varnothing)$ is 1. Thus

$$\alpha'_1 = \begin{pmatrix} 1 & 1 \\ 1 & -1 \end{pmatrix} \sim \begin{pmatrix} 2 & 0 \\ 0 & 1 \end{pmatrix} \quad \text{and} \quad \alpha'_0 = \begin{pmatrix} 1 & 1 \\ 1 & 1 \end{pmatrix} \sim \begin{pmatrix} 1 & 0 \\ 0 & 0 \end{pmatrix}.$$

Thus $\ker \alpha'_1 = 0$, $\operatorname{coker} \alpha'_1 = \mathbb{Z}_2$, $\ker \alpha'_0 = \mathbb{Z}$, $\operatorname{coker} \alpha'_0 = \mathbb{Z}$. Hence

$$h_n(K, \varnothing) = \begin{cases} \mathbb{Z} & n = 0, \\ \mathbb{Z} \oplus \mathbb{Z}_2 & n = 1, \\ 0 & n \neq 0, 1. \end{cases}$$

We shall have other applications of the Mayer–Vietoris sequence later (see Chapter 14).

We have defined the notion of an unreduced homology theory h_* on $\mathscr{T}^{2\prime}$; we now turn to the closely related notion of a reduced homology theory on $\mathscr{PT}'$.

7.29. On $\mathscr{PT}'$ we have the suspension functor S defined by $S(X, x_0) = (SX, *)$ and $S[f] = [Sf] = [1_{S^1} \wedge f]$. A *reduced homology theory* k_* on $\mathscr{PT}'$ is a collection of functors $k_n : \mathscr{PT}' \to \mathscr{A}$ and natural equivalences $\sigma_n : h_n \to h_{n+1} \circ S$, $n \in \mathbb{Z}$, satisfying

Exactness: for every pointed pair (X, A, x_0) with inclusions $i : (A, x_0) \to (X, x_0)$ and $j : (X, x_0) \to (X \cup CA, *)$ the sequence

$$k_n(A, x_0) \xrightarrow{k_n[i]} k_n(X, x_0) \xrightarrow{k_n[j]} k_n(X \cup CA, *)$$

is exact.

Hereafter we shall abbreviate $k_n[f]$ by f_* and $\sigma_n(X,x_0)$ by σ. In fact, we shall write $k_n(X)$ rather than $k_n(X,x_0)$, suppressing the base point as we often do.

There are two further axioms which we shall often impose on a reduced homology theory.

7.30. *Wedge axiom:* For every collection $\{(X_\alpha,x_\alpha):\alpha\in A\}$ of pointed spaces the inclusions $i_\alpha:X_\alpha\to\bigvee_{\beta\in A}X_\beta$ induce an isomorphism

$$\{i_{\alpha*}\}:\bigoplus_{\alpha\in A}k_n(X_\alpha)\to k_n(\bigvee_{\alpha\in A}X_\alpha),\quad n\in\mathbb{Z}.$$

7.31. *Weak homotopy equivalence axiom:* If $f:X\to Y$ is a weak homotopy equivalence then $f_*:k_n(X,x_0)\to k_n(Y,f(x_0))$ is an isomorphism for all $n\in\mathbb{Z}$, $x_0\in X$.

We shall see that the wedge axiom is useful for making deductions about CW-complexes which are not finite and the weak homotopy equivalence axiom for making deductions about spaces which are not CW-complexes. There are equivalent axioms for unreduced theories.

7.32. Proposition. *For any single point* $\{x\}$ *we have* $k_n(\{x\})=0$, $n\in\mathbb{Z}$.

Proof: The inclusions $i:(\{x\},x)\to(\{x\},x)$ and $j:(\{x\},x)\to(\{x\}\cup C\{x\},*)$ are both the identity. But the sequence

$$k_n(\{x\})\xrightarrow{1}k_n(\{x\})\xrightarrow{1}k_n(\{x\})$$

can be exact only if $k_n(\{x\})=0$. ▯

7.33. Proposition. *There are natural transformations* $\hat\partial_n$ *and for each pointed pair* (X,A,x_0) *a long exact sequence*

$$\cdots\longrightarrow k_n(A)\xrightarrow{i_*}k_n(X)\xrightarrow{j_*}k_n(X\cup CA)\xrightarrow{\hat\partial_n(X,A,x_0)}k_{n-1}(A)\xrightarrow{i_*}\cdots.$$

Proof: Recall the map $k':X\cup CA\to SA$ given in 2.41 and the homotopy inverse $\nu':SA\to SA$ of the H-cogroup SA as in 2.22. We define $\hat\partial_n(X,A,x_0)$ to be the composite

$$k_n(X\cup CA)\xrightarrow{(\nu'\circ k')_*}k_n(SA)\underset{\cong}{\xleftarrow{\sigma}}k_{n-1}(A).$$

$\hat\partial_n$ is clearly natural and we have a commutative diagram

$$\begin{array}{ccccccc}
k_n(A) & \xrightarrow{i_*} & k_n(X) & \xrightarrow{j_*} & & & \\
 & & k_n(X \cup CA) & \xrightarrow{k_*} & k_n((X \cup CA) \cup CX) & \xrightarrow{l_*} & k_n(((X \cup CA) \cup CX) \cup C(X \cup CA)) \\
 & & & \searrow^{(v' \circ k')_*} & \cong \downarrow (v' \circ \bar{q})_* & & \cong \downarrow \bar{q}'_* \\
 & & & & k_n(SA) & \xrightarrow{Si_*} & k_n(SX) \\
 & & & & \cong \uparrow \sigma & & \sigma \uparrow \cong \\
 & & & & k_{n-1}(A) & \xrightarrow{i_*} & k_{n-1}(X),
\end{array}$$

in which the top line is exact by the exactness axiom. Hence the sequence 7.33 is exact also. □

It is clear that reduced theories are more appropriate to the study of pointed spaces than unreduced ones. Many of the results on homotopy sets in Chapter 2 have their analogs for reduced theories. Even 7.15–7.17 are really statements about reduced theories, as we shall see.

Remark. Eilenberg and Steenrod defined reduced singular homology groups $\tilde{H}_n(X)$ by $\tilde{H}_n(X) = H_n(X, \{x_0\})$. We are about to demonstrate that this trick applied to a general unreduced theory always yields a reduced theory.

7.34. Given an unreduced homology theory h_* denote by $\tilde{h}_*$ the collection of functors $\tilde{h}_n : \mathscr{PT}' \to \mathscr{A}$ defined by $\tilde{h}_n(X, x_0) = h_n(X, \{x_0\})$, $\tilde{h}_n[f] = h_n[f]$, $f : (X, x_0) \to (Y, y_0)$, and natural transformations $\tilde{\sigma}_n : \tilde{h}_n \to \tilde{h}_{n+1} \circ S$ as in 7.15 for all $n \in \mathbb{Z}$. By 7.16 $\tilde{\sigma}_n$ is a natural equivalence for all $n \in \mathbb{Z}$.

To show that $\tilde{h}_*$ is a reduced homology theory it only remains to show that it satisfies the exactness axiom. Let (X, A, x_0) be a pointed pair; we have a commutative diagram

$$\begin{array}{ccccc}
\tilde{h}_n(A, x_0) & \xrightarrow{\tilde{h}_n[i]} & \tilde{h}_n(X, x_0) & \xrightarrow{\tilde{h}_n[j]} & \tilde{h}_n(X \cup CA, *) \\
\| & & \| & & \| \\
 & & & & h_n(X \cup CA, \{*\}) \\
 & & & & j'_* \downarrow \cong \\
 & & & & h_n(X \cup CA, CA) \\
 & & & & i'_* \downarrow \cong \\
h_n(A, \{x_0\}) & \xrightarrow{I_*} & h_n(X, \{x_0\}) & \xrightarrow{J_*} & h_n(X, A),
\end{array}$$

in which the bottom line is part of the exact sequence 7.9 of the triple $(X,A,\{x_0\}), i'_*$ is an isomorphism by excision (see the proof of 7.14) and j'_* is an isomorphism because $(CA,*)$ is contractible. Thus the top line is exact also, proving $\tilde{h}_*$ is a reduced homology theory.

7.35. Given a reduced homology theory k_*, denote by $\hat{k}_*$ the collection of functors $\hat{k}_n:\mathscr{T}^{2'} \to \mathscr{A}$ defined by $\hat{k}_n(X,A) = k_n(X^+ \cup CA^+)$, $\hat{k}_n[f] = k_n[\hat{f}]$, where $\hat{f}:X^+ \cup CA^+ \to Y^+ \cup CB^+$ is induced by $f:(X,A) \to (Y,B)$, and natural transformations $\hat{\partial}_n:\hat{k}_n \to \hat{k}_{n-1} \circ R$ as in 7.33. Then $\hat{k}_*$ satisfies the exactness axiom by 7.33.

Now suppose k_* satisfies the WHE axiom. Let $(X;\ A,B)$ be a triad with $X = \mathring{A} \cup \mathring{B}$. We wish to show that the inclusion $j:(A,A \cap B) \to (X,B)$ induces an isomorphism

$$\hat{k}_n(A,A \cap B) \xrightarrow{\hat{k}_n[j]} \hat{k}_n(X,B), \quad n \in \mathbb{Z}.$$

If $(X;\ A,B)$ is a CW-triad, then this is clear even without the WHE axiom, for we have a commutative diagram

$$\begin{array}{ccc} k_n(A^+ \cup C(A \cap B)^+) & \xrightarrow{\hat{j}_*} & k_n(X^+ \cup CB^+) \\ \cong \downarrow p_{1*} & & \cong \downarrow p_{2*} \\ k_n(A/A \cap B) & \underset{\cong}{\xrightarrow{\bar{j}_*}} & k_n(X/B) \end{array}$$

in which p_1 and p_2 are homotopy equivalences and $\bar{j}:A/A \cap B \to X/B$ is a homeomorphism. Thus $\hat{k}_*$ is an unreduced theory on the category $\mathscr{W}^{2'}$ of CW-pairs without further assumption. Something more seems to be needed on $\mathscr{T}^{2'}$, however.

Given an arbitrary triad $(X;\ A,B)$ with $\mathring{A} \cup \mathring{B} = X$, then, we find CW-substitutes C' for $C = A \cap B$, A' for A, B' for B with $A' \cap B' = C'$ and let all the maps be called f. We take $X' = A' \cup B'$. By 6.52 it follows $f:X' \to X$ is a CW-substitute. From the exactness and WHE axioms it follows that $f_*:k_n(A'^+ \cup C(A' \cap B')^+) \to k_n(A^+ \cup C(A \cap B)^+)$ and $f_*:k_n(X'^+ \cup CB'^+) \to k_n(X^+ \cup CB^+)$ are isomorphisms. We then have a commutative diagram

$$\begin{array}{ccc} k_n(A^+ \cup C(A \cap B)^+) & \xrightarrow{\hat{j}_*} & k_n(X^+ \cup CB^+) \\ \cong \uparrow f_* & & \cong \uparrow f_* \\ k_n(A'^+ \cup C(A' \cap B')^+) & \xrightarrow{\hat{j}'_*} & k_n(X'^+ \cup CB'^+) \end{array}$$

in which $\hat{j}'_*$ is an isomorphism because $(X'; A', B')$ is a CW-triad. Thus $\hat{k}_*$ satisfies the excision axiom and is hence an unreduced homology theory.

We shall presently show that $\tilde{\hat{h}}_*$ is naturally equivalent to h_* and $\tilde{\hat{k}}_*$ to k_*, thus establishing a one–one correspondence between reduced and unreduced theories. It is already clear, however, that to any theorem about an unreduced theory there corresponds one about a reduced theory and conversely.

7.36. Proposition. *For any $(X,x_0), (Y,y_0) \in \mathscr{PW}'$ the inclusions $i_X:(X,x_0) \to (X \vee Y, *)$ and $i_Y:(Y,y_0) \to (X \vee Y, *)$ induce an isomorphism*

$$(i_{X*}, i_{Y*}): k_n(X) \oplus k_n(Y) \to k_n(X \vee Y).$$

Proof: We have a commutative diagram

$$\begin{array}{ccc}
\hat{k}_n(X,\{x_0\}) \oplus \hat{k}_n(Y,\{y_0\}) & \xrightarrow{(i_{X*},\, i_{Y*})} & \hat{k}_n(X \vee Y, \{*\}) \\
\| & & \| \\
k_n(X^+ \cup Cx_0^+) \oplus k_n(Y^+ \cup Cy_0^+) & & k_n((X \vee Y)^+ \cup C*^+) \\
\|\wr & & \|\wr \\
k_n(X) \oplus k_n(Y) & \xrightarrow{(i_{X*},\, i_{Y*})} & k_n(X \vee Y),
\end{array}$$

in which the top morphism is an isomorphism by 7.25. Therefore the bottom morphism is also. ▯

7.37. Lemma. *If (L, l_0) is an H-cogroup with comultiplication $\mu': L \to L \vee L$, homotopy inverse $\nu': L \to L$, L a CW-complex, then*

$$k_n(L) \xrightarrow{\mu'_*} k_n(L \vee L) \xleftarrow[\cong]{(i_{1*},\, i_{2*})} k_n(L) \oplus k_n(L)$$

is given by $x \mapsto (x,x)$ and $\nu'_*: k_n(L) \to k_n(L)$ is given by $x \mapsto -x$, all $x \in k_n(L)$, $n \in \mathbb{Z}$.

Proof: if $l_0: L \to L$ is the constant map ($l_0(l) = l_0$, all $l \in L$), then the following diagram commutes

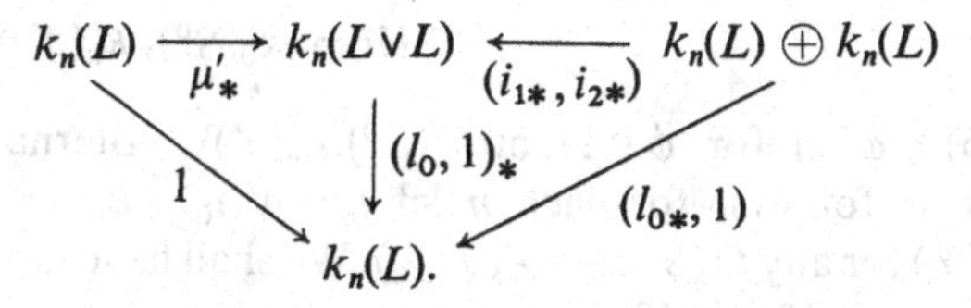

Since $l_0: L \to L$ factors through $c: L \to \{l_0\}$ and since $k_n(\{l_0\}) = 0$ by 7.32, we have $l_{0*} = 0$. If $\alpha = (i_{1*}, i_{2*})^{-1} \circ \mu'_*$, then $x = (0,1) \circ \alpha(x)$, showing that $\alpha(x) = (y,x)$ for some y. Using the fact that $(1, l_0) \circ \mu' \simeq 1$, we get $\alpha(x) = (x, y')$ for some y'. Thus $\alpha(x) = (x,x)$.

Now we also have $(\nu', 1) \circ \mu' \simeq l_0$, so we get $(\nu'_*, 1) \circ \alpha = (\nu', 1)_* \circ \mu'_* = l_{0*} = 0$. That is, $0 = (\nu'_*, 1) \circ \alpha(x) = \nu'_*(x) + x$ for all $x \in k_n(L)$—i.e. $\nu'_*(x) = -x$. □

Remark. We also have the folding map $\Delta': X \vee X \to X$ for any pointed space (X, x_0). Since $\Delta' \circ i_1 = 1_X = \Delta' \circ i_2$, we find that $\Delta'_* \circ (i_{1*}, i_{2*})(x, y) = x + y$, all $x, y \in k_n(X)$.

For any reduced homology theory k_* we have a function

$$\kappa_n : [X, x_0;\ Y, y_0] \to \operatorname{Hom}(k_n(X), k_n(Y))$$

defined by $\kappa_n[f] = f_* = k_n[f]$.

7.38. Proposition. *If (L, l_0) is an H-cogroup and L a CW-complex, then for every $n \in \mathbb{Z}$, $(Y, y_0) \in \mathscr{PT}'$*

$$\kappa_n : [L, l_0;\ Y, y_0] \to \operatorname{Hom}(k_n(L), k_n(Y))$$

is a homomorphism.

Proof: Given $f, g : (L, l_0) \to (Y, y_0)$ we have $[f] + [g] = [\Delta' \circ (f \vee g) \circ \mu']$.

Hence

$$\kappa_n([f] + [g])(x) = \Delta'_* \circ (f \vee g)_* \circ \mu'_*(x) = \Delta'_*(f_*(x), g_*(x)) = f_*(x) + g_*(x) = \kappa_n[f](x) + \kappa_n[g](x),$$

all $x \in k_n(L)$. Thus $\kappa_n([f] + [g]) = \kappa_n[f] + \kappa_n[g]$. □

7.39. Definition. Of particular interest is the case where $(L, l_0) = (S^n, s_0)$ and there is a distinguished element $\iota_0 \in k_0(S^0)$. We define the *Hurewicz homomorphism*

$$h : \pi_n(Y, y_0) \to k_n(Y)$$

to be the composite

$$\pi_n(Y, y_0) \cong [S^n, s_0;\ Y, y_0] \xrightarrow{\kappa_n} \operatorname{Hom}(k_n(S^n), k_n(Y)) \xrightarrow{\operatorname{Hom}(\sigma^n, 1)} \operatorname{Hom}(k_0(S^0), k_n(Y)) \xrightarrow{e} k_n(Y),$$

where $e(\phi) = \phi(\iota_0)$ for $\phi \in \operatorname{Hom}(k_0(S^0), k_n(Y))$. Alternatively we can describe h as follows: for each n let $\iota_n = \sigma^n(\iota_0) \in k_n(S^n)$. Then $h[f] = f_*(\iota_n) \in k_n(Y)$ for any $f : (S^n, s_0) \to (Y, y_0)$. We shall have more to say about h later (Chapters 10, 13, 19).

Let us finish the proof that there is a one–one correspondence between unreduced and reduced theories.

7.40. Definition. A *natural transformation* $T_* : h_* \to h'_*$ between unreduced homology theories h_*, h'_* is a collection of natural transformations

$T_n : h_n \to h'_n$ such that for every pair (X, A) the square

$$\begin{array}{ccc} h_n(X,A) & \xrightarrow{-\partial} & h_{n-1}(A,\varnothing) \\ \Big\downarrow{\scriptstyle T_n(X,A)} & & \Big\downarrow{\scriptstyle T_{n-1}(A,\varnothing)} \\ h'_n(X,A) & \xrightarrow{\partial'} & h'_{n-1}(A,\varnothing) \end{array}$$

commutes. There is a corresponding notion of a natural transformation between reduced theories. T_* is a *natural equivalence of homology theories* if each T_n is a natural equivalence, $n \in \mathbb{Z}$.

7.41. For any reduced homology theory k_* and pointed space (X, x_0) we have

$$\tilde{\hat{k}}_n(X, x_0) = \hat{k}_n(X, \{x_0\}) = k_n(X^+ \cup C\{x_0\}^+).$$

A pointed space (X, x_0) is said to have *non-degenerate base point* if the inclusion $\{x_0\} \subset X$ is a cofibration; in that case the projection $p : (X^+ \cup C\{x_0\}^+, *) \to (X, x_0)$ (collapsing $C\{x_0\}^+$) is a homotopy equivalence by 6.6. If $\mathscr{PT}'_0$ denotes the homotopy category of pointed spaces with non-degenerate base point, then p defines on $\mathscr{PT}'_0$ a natural equivalence $T_n : \tilde{\hat{k}}_n \to k_n$ for each $n \in \mathbb{Z}$.

7.42. Lemma. *The equivalences T_n satisfy $T_{n+1}(SX) \circ \tilde{\sigma} = \sigma \circ T_n(X)$. Thus the collection $T_* = \{T_n\}$ defines a natural equivalence of reduced homology theories on $\mathscr{PT}'_0$ if k_* satisfies the WHE axiom.*

Proof: The following diagram is easily seen to be commutative because of the definitions of $\tilde{\sigma}$, $\hat{\partial}$, $\hat{\Delta}$, $T_n(X)$ and $T_{n+1}(SX)$ and the naturality of σ:

$$\begin{array}{ccccc}
k_n(X) & & \xrightarrow{\quad\sigma\quad} & & k_{n+1}(SX) \\
\cong\Big\uparrow{\scriptstyle p_{1*}} & \nwarrow{\scriptstyle j_*} & & \nearrow{\scriptstyle Sj_*} & \Big\uparrow{\scriptstyle p_{2*}}\cong \\
k_n(X^+ \cup C\{x_0\}^+) & \xleftarrow{\hat{j}_*} k_n(X^+) & \xrightarrow{\sigma} & k_{n+1}(S(X^+)) & \\
\| \quad {\scriptstyle T_n(X)} & \| & & \Big\uparrow{\scriptstyle (v' \circ k')_*} & {\scriptstyle T_{n+1}(SX)} \\
& \hat{k}_n(X,\varnothing) & & k_{n+1}((CX)^+ \cup C(X^+)) \xrightarrow{\hat{p}_{3*}} & k_{n+1}((SX)^+ \cup C\{*\}^+) \\
\| & \swarrow{\scriptstyle j_*} & \nwarrow{\scriptstyle \hat{\partial}} & \| & \| \\
\hat{k}_n(X,\{x_0\}) & \xleftarrow[\cong]{\hat{\Delta}} & \hat{k}_{n+1}(CX,X) & \xrightarrow[p_{3*}]{\cong} & \hat{k}_{n+1}(SX,\{*\}) \\
\| & & & & \| \\
\tilde{\hat{k}}_n(X) & & \xrightarrow{\quad\tilde{\sigma}\quad} & & \tilde{\hat{k}}_{n+1}(SX)
\end{array}$$

,

where $j:(X^+,+) \to (X,x_0)$ is defined by $j|X = 1_X$, $j(+) = x_0$. Every projection has been labelled with a p_i, $i = 1, 2, 3$. Note in particular that the v' is necessary in order that $p_2 \circ \hat{p}_3 \simeq Sj \circ v' \circ \underset{\sim}{k}'$.

Since $\hat{\Delta}$ is an isomorphism for every $x \in \tilde{k}_n(X)$ we can find a $y \in \hat{k}_{n+1}(CX, X)$ with $\hat{\Delta}(y) = x$. Then

$$\sigma \circ T_n(X)(x) = \sigma \circ p_{1*}(x) = \sigma p_{1*} \hat{\Delta}(y) =$$
$$\sigma \circ j_* \circ \hat{\partial}(y) = Sj_* \circ \sigma \circ \hat{\partial}(y) =$$
$$Sj_* \circ (v' \circ k')_*(y) = T_{n+1}(SX) \circ \tilde{\sigma} \circ \hat{\Delta}(y) = T_{n+1}(SX) \circ \tilde{\partial}(x). \quad \square$$

7.43. For any unreduced theory h_* and pair $(X, A) \in \mathcal{T}^{2\prime}$ we have

$$\hat{\tilde{h}}_n(X, A) = \tilde{h}_n(X^+ \cup CA^+) = h_n(X^+ \cup CA^+, \{*\}).$$

The composite

$$h_n(X^+ \cup CA^+, \{*\}) \xrightarrow[\cong]{J_*} h_n(X^+ \cup CA^+, CA^+) \xleftarrow[\cong]{i_*} h_n(X, A)$$

defines a natural equivalence $T_n' : \hat{\tilde{h}}_n \to h_n$ for each $n \in \mathbb{Z}$.

7.44. Lemma. *The equivalences T_n' satisfy $\partial \circ T_n'(X, A) = T_{n-1}'(A, \varnothing) \circ \hat{\tilde{\partial}}$ for all (X, A). Thus the collection $T_*' = \{T_n'\}$ defines a natural equivalence of unreduced homology theories on $\mathcal{T}^{2\prime}$ if h_* satisfies the WHE axiom.*

Proof: If one writes down the definitions of $\hat{\tilde{h}}_n(X, A)$, $\hat{\tilde{h}}_{n-1}(A, \varnothing)$ and $\hat{\tilde{\partial}}$, one finds $\hat{\tilde{\partial}}$ is the following composite

$$h_n(X^+ \cup CA^+, \{*\}) \xrightarrow{(v' \circ k')_*} h_n(SA^+, \{*\}) \xleftarrow[\cong]{p_{3*}} h_n(CA^+, A^+) \xrightarrow[\cong]{\Delta} h_{n-1}(A^+, \{+\}).$$

One then checks that the following diagram commutes

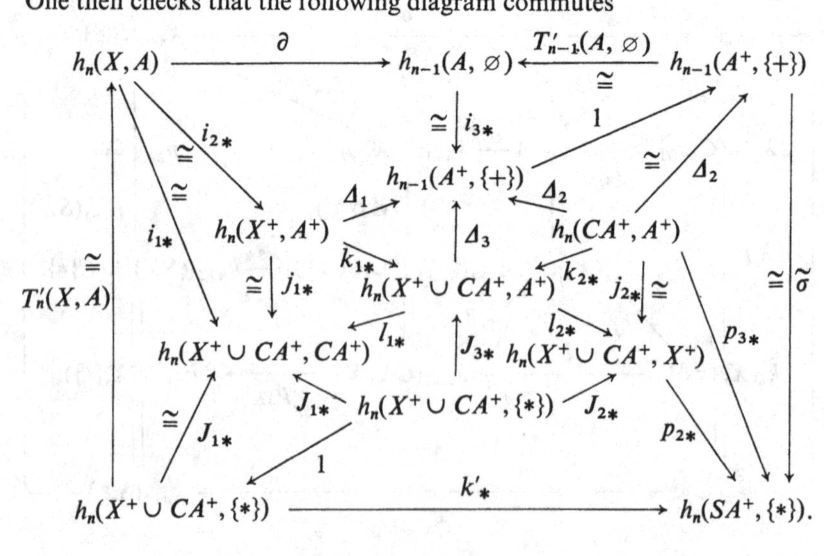

The route around the periphery from lower left corner to lower right corner to upper right corner is $-\hat{\partial}$; the minus sign occurs because v'_* has been left out (7.37). Note that at the center of this diagram is incorporated the hexagon diagram 7.22 of the triad $(X^+ \cup CA^+; X^+, CA^+)$. Therefore $\Delta_1 \circ j_{1*}^{-1} \circ J_{1*} = -\Delta_2 \circ j_{2*}^{-1} \circ J_{2*}$.

Suppose given $x \in \hat{h}_n(X, A) = h_n(X^+ \cup CA^+, \{*\})$;

$\partial \circ T'_n(x) = \partial \circ i_{2*}^{-1} \circ j_{1*}^{-1} \circ J_{1*}(x) =$

$i_{3*}^{-1} \circ \Delta_1 \circ j_{1*}^{-1} \circ J_{1*}(x) = -i_{3*}^{-1} \circ \Delta_2 \circ j_{2*}^{-1} \circ J_{2*}(x) =$

$-T'_{n-1}(A, \varnothing) \circ \tilde{\sigma}^{-1} \circ p_{2*} \circ J_{2*}(x) =$

$-T'_{n-1}(A, \varnothing) \circ \tilde{\sigma}^{-1} \circ k'_*(x) = T'_{n-1}(A, \varnothing) \circ \tilde{\sigma}^{-1} \circ (v' \circ k')_*(x) =$

$T'_{n-1}(A, \varnothing) \circ \hat{\partial}(x)$. □

Remarks. i) Note that if $(X,A) \in \mathscr{T}^2$, then $X^+ \cup CA^+ \in \mathscr{PT}_0$. Hence the one–one correspondence is demonstrated.

ii) If we restrict attention to the categories $\mathscr{W}^{2\prime}$ and $\mathscr{PW}'$ of CW-pairs and pointed CW-complexes, then we get the one–one correspondence unreduced ↔ reduced without assuming the WHE axiom (which holds automatically by 6.32 and 7.7).

It is now clear that it is not essential to consider both reduced and unreduced theories. Reduced theories are evidently somewhat more closely related to homotopy groups, but we shall find unreduced theories more convenient in certain other contexts such as the study of manifolds.

Since for any single point $\{x\}$ the space $\{x\}^+$ is homeomorphic to S^0, we have $k_n(S^0) \cong k_n(\{x\}^+) = \hat{k}_n(\{x\}, \varnothing)$ for all $n \in \mathbb{Z}$ and any reduced theory k_*. Therefore it is reasonable to call the groups $k_n(S^0)$, $n \in \mathbb{Z}$, the *coefficient groups* of the reduced theory k_*. Note that $k_m(S^n) \cong k_{m-n}(S^0)$ for all m,n.

Since for any CW-complex (X,x_0) we have $X^n/X^{n-1} \cong \bigvee_{\alpha \in J_n} S^n$, we see that the exact sequence of the pair (X^n, X^{n-1}) becomes

$$\cdots \longrightarrow k_{q+1}(\vee_\alpha S^n) \xrightarrow{\hat{\partial}} k_q(X^{n-1}) \longrightarrow k_q(X^n) \longrightarrow k_q(\vee_\alpha S^n) \longrightarrow \cdots.$$

If J_n is finite, then applying 7.36 and induction we obtain

$$k_q(\vee_{\alpha \in J_n} S^n) \cong \oplus_{\alpha \in J_n} k_q(S^n_\alpha) \cong \oplus_{\alpha \in J_n} k_{q-n}(S^0).$$

Thus given information about the coefficient groups $k_*(S^0)$, we can often obtain information about $k_q(X^n)$ inductively. If X is not finite, however, an additional assumption is needed. If k_* satisfies the wedge axiom then we can say

$$k_q(X^n/X^{n-1}) \cong k_q(\vee_{\alpha \in J_n} S^n_\alpha) \cong \oplus_{\alpha \in J_n} k_q(S^n) \cong \oplus_{\alpha \in J_n} k_{q-n}(S^0)$$

even if J_n is infinite.

Of course, although this kind of argument now gives us information about $k_q(X^n)$ for each n, it need not a priori tell us anything about $k_q(X)$ if X is infinite dimensional. Recall that for homotopy groups $\pi_q(X,x_0)$ elements were represented by maps $f:S^n \to X$ and $f(S^n)$, being compact, always lay in a finite subcomplex. This kind of argument will clearly be of no help for k_*, so we proceed differently.

The inclusions $j_n^m:X^n \to X^m$, $n \leqslant m$, induce morphisms $j_{n*}^m:k_q(X^n) \to k_q(X^m)$ satisfying $j_{n*}^m \circ j_{p*}^n = j_{p*}^m$. The inclusions $i_n:X^n \to X$ also induce morphisms $i_{n*}:k_q(X^n) \to k_q(X)$ such that $i_{m*} \circ j_{n*}^m = i_{n*}$. The general algebraic situation of which this is a special case is that of a direct system of groups.

7.45. Definition. A *directed set* A is a partially ordered set $(A,\leqslant)$ such that any $\alpha, \beta \in A$ have an upper bound $\gamma \in A: \alpha \leqslant \gamma$, $\beta \leqslant \gamma$. A *direct system* of abelian groups is a collection $\{G_\alpha : \alpha \in A\}$ of abelian groups G_α indexed by a directed set A and homomorphisms $j_\alpha^\beta : G_\alpha \to G_\beta$, $\alpha \leqslant \beta$, such that $j_\beta^\gamma \circ j_\alpha^\beta = j_\alpha^\gamma$, $\alpha \leqslant \beta \leqslant \gamma$, and $j_\alpha^\alpha = 1$, $\alpha \in A$. Given such a direct system $\{G_\alpha, j_\alpha^\beta, A\}$, the *direct limit* dir lim G_α is the group defined as follows: let $R \subset \bigoplus_{\alpha \in A} G_\alpha$ be the subgroup generated by elements of the form $i_\beta \circ j_\alpha^\beta(g) - i_\alpha(g)$ for all $g \in G_\alpha$, all $\alpha \leqslant \beta$, $\alpha, \beta \in A$. Then we take $\operatorname{dir\,lim} G_\alpha = \bigoplus_{\alpha \in A} G_\alpha / R$. The injections $i_\alpha : G_\alpha \to \bigoplus_{\alpha \in A} G_\alpha$ define homomorphisms which we also call $i_\alpha : G_\alpha \to \operatorname{dir\,lim} G_\alpha$. The i_α satisfy $i_\beta \circ j_\alpha^\beta = i_\alpha$, $\alpha \leqslant \beta$.

7.46. Proposition. $(\operatorname{dir\,lim} G_\alpha, \{i_\alpha\})$ *is characterized up to isomorphism by the property that given any group* H *and any homomorphisms* $f_\alpha : G_\alpha \to H$, $\alpha \in A$, *such that* $f_\beta \circ j_\alpha^\beta = f_\alpha$ *for all* $\alpha \leqslant \beta$ *there is a unique homomorphism* $f:\operatorname{dir\,lim} G_\alpha \to H$ *with* $f \circ i_\alpha = f_\alpha$, $\alpha \in A$.

Proof: $\bigoplus_{\alpha \in A} G_\alpha$ is characterized up to isomorphism by the property that for any collection of homomorphisms $f_\alpha : G_\alpha \to H$ there is a unique homomorphism $f' : \bigoplus_\alpha G_\alpha \to H$ with $f' \circ i_\alpha = f_\alpha$, $\alpha \in A$. If the f_α also satisfy $f_\beta \circ j_\alpha^\beta = f_\alpha$, all $\alpha \leqslant \beta$, then $f'(R) = 0$, so f' factors through $\operatorname{dir\,lim} G_\alpha = \bigoplus_\alpha G_\alpha / R$ to give a map $f:\operatorname{dir\,lim} G_\alpha \to H$ with $f \circ i_\alpha = f_\alpha$, $\alpha \in A$.

Suppose $g:\operatorname{dir\,lim} G_\alpha \to H$ is another homomorphism satisfying $g \circ i_\alpha = f_\alpha$, $\alpha \in A$. If $q:\bigoplus_\alpha G_\alpha \to \operatorname{dir\,lim} G_\alpha$ is the projection, then $g \circ q : \bigoplus_\alpha G_\alpha \to H$ satisfies $(g \circ q) \circ i_\alpha = f_\alpha$, $\alpha \in A$, so $g \circ q = f'$. But f is the unique homomorphism such that $f \circ q = f'$. Thus $f = g$.

Finally suppose G is a group and $\bar{\imath}_\alpha : G_\alpha \to G$ are homomorphisms such that for any collection $f_\alpha : G_\alpha \to H$, $\alpha \in A$, with $f_\beta \circ j_\alpha^\beta = f_\alpha$ there is a homomorphism $\phi : G \to H$ with $\phi \circ \bar{\imath}_\alpha = f_\alpha$. Taking the f_α to be $i_\alpha : G_\alpha \to \operatorname{dir\,lim} G_\alpha$, we find a unique $\phi : G \to \operatorname{dir\,lim} G_\alpha$ with $\phi \circ \bar{\imath}_\alpha = i_\alpha$. Similarly we find a unique $\phi' : \operatorname{dir\,lim} G_\alpha \to G$ with $\phi' \circ i_\alpha = \bar{\imath}_\alpha$. Then both $\psi = \phi' \circ \phi$ and $\psi = 1_G$ satisfy $\psi \circ \bar{\imath}_\alpha = \bar{\imath}_\alpha$, so by uniqueness $\phi' \circ \phi = 1_G$. Similarly $\phi \circ \phi' = 1_{\operatorname{dir\,lim} G_\alpha}$. Thus ϕ is an isomorphism and is the unique one such that $\phi \circ \bar{\imath}_\alpha = i_\alpha$. □

The unique map f given by 7.46 will be denoted by $\{f_\alpha\}$: dir lim $G_\alpha \to H$.

7.47. Definition. A *morphism of direct systems* $\{G_\alpha, j_\alpha^\beta, A\} \xrightarrow{\{\phi_\alpha\}} \{G'_\alpha, j'^\beta_\alpha, A\}$ is a collection of homomorphisms $\phi_\alpha: G_\alpha \to G'_\alpha$, $\alpha \in A$, satisfying $j'^\beta_\alpha \circ \phi_\alpha = \phi_\beta \circ j_\alpha^\beta$.

7.48. Corollary. *A morphism $\{\phi_\alpha\}: \{G_\alpha\} \to \{G'_\alpha\}$ of direct systems induces a homomorphism* $\operatorname{dir\,lim} \phi_\alpha: \operatorname{dir\,lim} G_\alpha \to \operatorname{dir\,lim} G'_\alpha$ *such that* $(\operatorname{dir\,lim} \phi_\alpha) \circ i_\beta = i'_\beta \circ \phi_\beta$, *all* $\beta \in A$.

Proof: The maps $f_\beta = i'_\beta \circ \phi_\beta: G_\beta \to \operatorname{dir\,lim} G'_\alpha$ satisfy $f_\beta \circ j_\alpha^\beta = i'_\beta \circ \phi_\beta \circ j_\alpha^\beta = i'_\beta \circ j'^\beta_\alpha \circ \phi_\alpha = i'_\alpha \circ \phi_\alpha = f_\alpha$, so we may take $\operatorname{dir\,lim} \phi_\alpha = \{f_\alpha\}$. $\square$

7.49. Corollary. *Suppose $\{G_\alpha\}$, $\{G'_\alpha\}$ are direct systems and $f_\alpha: G_\alpha \to H$, $f'_\alpha: G'_\alpha \to H'$, $\alpha \in A$, are homomorphisms with $f_\beta \circ j_\alpha^\beta = f_\alpha$, $f'_\beta \circ j'^\beta_\alpha = f'_\alpha$. Suppose $\{\phi_\alpha\}: \{G_\alpha\} \to \{G'_\alpha\}$ is a morphism of direct systems and $\phi: H \to H'$ is a homomorphism such that*

$$\begin{array}{ccc} G_\alpha & \xrightarrow{f_\alpha} & H \\ \phi_\alpha \downarrow & & \downarrow \phi \\ G'_\alpha & \xrightarrow{f'_\alpha} & H' \end{array}$$

commutes for all $\alpha \in A$. Then

$$\begin{array}{ccc} \operatorname{dir\,lim} G_\alpha & \xrightarrow{\{f_\alpha\}} & H \\ \operatorname{dir\,lim} \phi_\alpha \downarrow & & \downarrow \phi \\ \operatorname{dir\,lim} G'_\alpha & \xrightarrow{\{f'_\alpha\}} & H' \end{array}$$

commutes.

Proof: $\phi \circ \{f_\alpha\} \circ i_\beta = \phi \circ f_\beta = f'_\beta \circ \phi_\beta$ and

$$\{f'_\alpha\} \circ \operatorname{dir\,lim} \phi_\alpha \circ i_\beta = \{f'_\alpha\} \circ i'_\beta \circ \phi_\beta = f'_\beta \circ \phi_\beta,$$

all $\beta \in A$. By uniqueness in 7.46 it follows $\phi \circ \{f_\alpha\} = \{f'_\alpha\} \circ \operatorname{dir\,lim} \phi_\alpha$. $\square$

Direct limits also behave well with respect to exact sequences.

7.50. Proposition. *Suppose $\{A_\alpha\}$, $\{B_\alpha\}$, $\{C_\alpha\}$ are direct systems and $\{\phi_\alpha\}: \{A_\alpha\} \to \{B_\alpha\}$, $\{\psi_\alpha\}: \{B_\alpha\} \to \{C_\alpha\}$ are morphisms of direct systems. If the sequence*

$$A_\alpha \xrightarrow{\phi_\alpha} B_\alpha \xrightarrow{\psi_\alpha} C_\alpha$$

is exact for every $\alpha \in A$, then the sequence

$$\operatorname{dir\,lim} A_\alpha \xrightarrow{\operatorname{dir\,lim} \phi_\alpha} \operatorname{dir\,lim} B_\alpha \xrightarrow{\operatorname{dir\,lim} \psi_\alpha} \operatorname{dir\,lim} C_\alpha$$

is also exact.

Proof: Suppose we introduce the relation $\sim$ on $\bigcup_{\alpha \in A} G_\alpha$ as follows: $g \sim g'$, $g \in G_\alpha$, $g' \in G_\beta \Leftrightarrow$ there is a $\gamma \geqslant \alpha$, $\gamma \geqslant \beta$ with $j_\alpha^\gamma g = j_\beta^\gamma g'$. The quotient $\hat{G} = \cup_\alpha G_\alpha/\sim$ has a natural addition ($\{g\} + \{g'\} = \{j_\alpha^\gamma g + j_\beta^\gamma g'\}$, $g \in G_\alpha$, $g' \in G_\beta$) and morphisms $i_\alpha : G_\alpha \to \hat{G}$ defined by $i_\alpha(g) = \{g\}$. $\{\hat{G}, i_\alpha\}$ is easily seen to have the universal property of $\operatorname{dir\,lim} G_\alpha$ and hence $\operatorname{dir\,lim} G_\alpha \cong \hat{G}$. In particular $g \in G_\alpha$ has $i_\alpha(g) = 0 \in \operatorname{dir\,lim} G_\alpha$ if and only if $j_\alpha^\gamma(g) = 0$, some $\gamma \geqslant \alpha$.

Now clearly $(\operatorname{dir\,lim} \psi_\alpha) \circ (\operatorname{dir\,lim} \phi_\alpha) = \operatorname{dir\,lim} (\psi_\alpha \circ \phi_\alpha) = 0$. Suppose $\{b\} \in \hat{B} \cong \operatorname{dir\,lim} B_\alpha$ has $\operatorname{dir\,lim} \psi_\alpha \{b\} = 0$; if $b \in B_\alpha$, this means

$$0 = {}_C j_\alpha^\beta \psi_\alpha(b) = \psi_\beta \circ {}_B j_\alpha^\beta(b)$$

for some $\beta \geqslant \alpha$. Thus ${}_B j_\alpha^\beta(b) = \phi_\beta(a)$, some $a \in A_\beta$. Hence $\operatorname{dir\,lim} \phi_\alpha(\{a\}) = \{b\}$. ▯

7.51. Corollary. *If $\{\phi_\alpha\} : \{G_\alpha\} \to \{G'_\alpha\}$ is a morphism of direct systems such that each ϕ_α is a monomorphism (resp. epimorphism, resp. isomorphism), then the same is true of* $\operatorname{dir\,lim} \phi_\alpha$.

A fact which we implicitly used in Chapter 6 is the following.

7.52. Proposition. *Let (X, x_0) be a pointed space and let $\{X_\alpha : \alpha \in A\}$ be a collection of subspaces ordered by inclusion such that for every compact subset $C \subset X$ there is an X_γ with $C \subset X_\gamma$. Let $j_\alpha^\beta : X_\alpha \to X_\beta$ and $i_\alpha : X_\alpha \to X$ be the inclusions. Then the groups $\pi_n(X_\alpha, x_0)$ and homomorphisms $j_{\alpha *}^\beta$ form a direct system of groups ($n \geqslant 2$) and*

$$\{i_{\alpha *}\} : \operatorname{dir\,lim} \pi_n(X_\alpha, x_0) \to \pi_n(X, x_0)$$

is an isomorphism.

Proof: The first statement is obvious. Given $f : (S^n, s_0) \to (X, x_0)$, there is an $X_\alpha \subset X$ with $f(S^n) \subset X_\alpha$. Thus f defines an element $[f] \in \pi_n(X_\alpha, x_0)$ and clearly $\{i_{\alpha *}\}(\{[f]\}) = [f]$. Hence $\{i_{\alpha *}\}$ is surjective.

Suppose $g_1 : (S^n, s_0) \to (X_\alpha, x_0)$, $g_2 : (S^n, s_0) \to (X_\beta, x_0)$ are two maps such that $\{i_{\alpha *}\}(\{[g_1]\}) = \{i_{\alpha *}\}(\{[g_2]\})$. This means simply that $i_\alpha \circ g_1 \simeq i_\beta \circ g_2$ in X. Let the homotopy be H. Since $H(S^n \times I)$ is compact, we can find an X_γ with $H(S^n \times I) \subset X_\gamma$. In other words, we have shown $j_{\alpha *}^\gamma [g_1] = j_{\beta *}^\gamma [g_2]$, or $\{[g_1]\} = \{[g_2]\}$. Thus $\{i_{\alpha *}\}$ is injective. ▯

We have a corresponding result for homology theories which satisfy the wedge axiom on $\mathscr{PW}$.

7.53. Proposition. *Let (X,x_0) be a CW-complex and let*

$$X^0 \subset \cdots \subset X^n \subset \cdots \subset X$$

be subcomplexes with $\bigcup_{n\geqslant 0} X^n = X$, $j_n^m: X^n \to X^m$, $i_n: X^n \to X$ the inclusions. Then the groups $k_q(X^n)$ and homomorphisms j_{n*}^m form a direct system, and if k_* satisfies the wedge axiom on $\mathscr{PW}$, then

$$\{i_{n*}\}: \operatorname{dir\,lim} k_q(X^n) \to k_q(X)$$

is an isomorphism for all $q \in \mathbb{Z}$.

Proof: We construct the "infinite telescope"

$$X' = \bigcup_{n\geqslant -1} [n-1,n]^+ \wedge X^n \subset [-2,\infty)^+ \wedge X.$$

Then X' is a CW-complex and we have a map $r: X' \to X$ which is the restriction of the projection $p_X: [-2,\infty) \times X \to X$. For each n we let $X_n' = \bigcup_{k=-1}^{n} [k-1,k]^+ \wedge X^k \subset X'$. Then we have inclusions $j_n'^m: X_n' \to X_m'$, $n \leqslant m$, and $i_n': X_n' \to X'$. There are also the maps $r_n: X_n' \to X^n$ (restriction of the projection $[-2,n]^+ \wedge X^n \to X^n$) and $k_n: X^n \to X_n'$ defined by $k_n(x) = [n,x]$, $x \in X^n$. Among these there are the relations

$$r_m \circ j_n'^m = j_n^m \circ r_n \qquad r_n \circ k_n = 1_{X^n}$$
$$i_n \circ r_n = r \circ i_n' \qquad k_n \circ r_n \simeq 1_{X_n'}.$$

The homotopy $k_n \circ r_n \simeq 1_{X_n'}$ is given by

$$K_n([s,x],t) = [(1-t)\,s + tn, x], \quad s \in [k-1,n], x \in X^k, t \in I.$$

Hence r_n is a homotopy equivalence for each $n \geqslant -1$.

$\{\pi_q(X^n,x_0), j_{n*}^m\}$, $\{\pi_q(X_n',*), j_{n*}'^m\}$ are then direct systems and $\{r_{n*}\}$ is a morphism of direct systems. The square

$$\begin{array}{ccc}
\operatorname{dir\,lim} \pi_q(X_n',*) & \xrightarrow[\cong]{\{i_{n*}'\}} & \pi_q(X',x_0) \\
\operatorname{dir\,lim} r_{n*} \downarrow \cong & & \downarrow r_* \\
\operatorname{dir\,lim} \pi_q(X^n,x_0) & \xrightarrow[\cong]{\{i_{n*}\}} & \pi_q(X,x_0)
\end{array}$$

commutes (7.49). $\{i_{n*}'\}$, $\{i_{n*}\}$ are isomorphisms by 7.52. Hence r_* is an isomorphism for each q. Therefore by 6.32 r is a homotopy equivalence. (This is a good example of a case in which it is difficult to imagine how a homotopy inverse of r might look.)

Now let

$$A = \bigcup_{\substack{k\geq -1 \\ k \text{ odd}}} [k-1,k]^+ \wedge X^k \subset X'$$

$$B = \bigcup_{\substack{k\geq 0 \\ k \text{ even}}} [k-1,k]^+ \wedge X^k \subset X'.$$

Then it is easy to see that

i) $A \cup B = X'$,

ii) $A \cap B = \bigvee_{k \geqslant -1} \{k\} \times X^k \simeq \bigvee_{k \geqslant -1} X^k$,

iii) $\bigvee_{\substack{k \geqslant -1 \\ k \text{ odd}}} \{k\} \times X^k \subset A$ is a strong deformation retract,

iv) $\bigvee_{\substack{k \geqslant 0 \\ k \text{ even}}} \{k\} \times X^k \subset B$ is a strong deformation retract.

The Mayer–Vietoris sequence for the triad $(X'; A, B)$ forms the top line of the following commutative diagram

$$\begin{array}{ccccccc}
\cdots \longrightarrow & k_q(A \cap B) & \xrightarrow{\alpha} & k_q(A) \oplus k_q(B) & \xrightarrow{\beta} & k_q(X') & \xrightarrow{\Delta'} \cdots \\
& \uparrow \cong & & \uparrow \cong & \nearrow \beta' & \cong \downarrow r_* & \\
& & & (\oplus_{n \text{ odd}} k_q(X^n)) \oplus (\oplus_{n \text{ even}} k_q(X^n)) & & & \\
& & & \| & & & \\
& \oplus_{n \geqslant -1} k_q(X^n) & \xrightarrow[\alpha']{} & \oplus_{n \geq -1} k_q(X^n) & \xrightarrow{\{(-1)^{n+1} i_{n*}\}} & k_q(X), &
\end{array}$$

where α' is defined by $\alpha'(x) = i_n(x) + i_{n+1} \circ j_{n*}^{n+1}(x)$ for $x \in k_q(X^n)$ (and $i_n : k_q(X^n) \to \oplus_n k_q(X^n)$ is the injection) and β' is defined by $\beta'(x) = (-1)^{n+1} i'_{n*} k_{n*}(x)$, $x \in k_q(X^n)$.

Clearly α' is a monomorphism and hence β' (and therefore $\{(-1)^{n+1} i_{n*}\}$) is an epimorphism. We then have a commutative diagram of exact

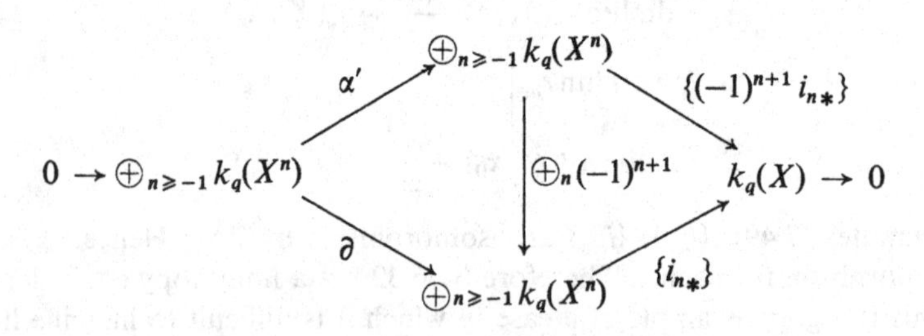

sequences, where $\partial(x) = (-1)^n \{i_{n+1} j_{n*}^{n+1}(x) - i_n(x)\}$ for $x \in k_q(X^n)$. But $\operatorname{dir\,lim} k_q(X^n) = \oplus\, k_q(X^n)/\operatorname{im} \partial$. $\square$

Thus if k_* satisfies the wedge axiom on $\mathscr{PW}$, then information about $k_q(X^n)$, all $n \geqslant -1$, may give information about $\operatorname{dir\,lim} k_q(X^n) \simeq k_q(X)$.

The proof of the following theorem typifies the kind of skeleton-by-skeleton argument outlined above.

7.55. Theorem. *Let $T_*: k_* \to k'_*$ be a natural transformation of homology theories. If*

$$T_q(S^0): k_q(S^0) \to k'_q(S^0)$$

is an isomorphism for $q < N$ and an epimorphism for $q = N$, then for every finite $(n-1)$-connected CW-complex (X, x_0)

$$T_q(X): k_q(X) \to k'_q(X)$$

is an isomorphism for $q < n + N$, an epimorphism for $q = n + N$. If k_*, k'_* also satisfy the wedge axiom on $\mathscr{PW}'$, X may be infinite. If k_*, k'_* also satisfy the WHE axiom, X may be any $(n-1)$-connected space.

Proof: Since T commutes with σ, the square

$$\begin{array}{ccc} k_q(S^0) & \xrightarrow[\cong]{\sigma^n} & k_{q+n}(S^n) \\ T_q(S^0)\downarrow & & \downarrow T_{q+n}(S^n) \\ k'_q(S^0) & \xrightarrow[\cong]{\sigma'^n} & k'_{q+n}(S^n) \end{array}$$

commutes, showing that $T_{q+n}(S^n)$ is an isomorphism for $q < N$ and an epimorphism for $q = N$. By naturality of T the following square commutes:

$$\begin{array}{ccc} \bigoplus_{\alpha \in J_n} k_q(S^n_\alpha) & \xrightarrow{\{i_{\alpha *}\}} & k_q(\bigvee_{\alpha \in J_n} S^n_\alpha) \\ \bigoplus_\alpha T_q(S^n_\alpha)\downarrow & & \downarrow T_q(\bigvee_\alpha S^n_\alpha) \\ \bigoplus_{\alpha \in J_n} k'_q(S^n_\alpha) & \xrightarrow{\{i_{\alpha *}\}} & k'_q(\bigvee_{\alpha \in J_n} S^n_\alpha). \end{array}$$

$\bigoplus_{\alpha \in J_n} T_q(S^n_\alpha)$ is an isomorphism for $q < n + N$, an epimorphism for $q = n + N$. In both theories $\{i_{\alpha *}\}$ is an isomorphism if J_n is finite or k_*, k'_* satisfy the wedge axiom. In either case $T_q(\bigvee_\alpha S^n_\alpha)$ is an isomorphism for $q < n + N$, an epimorphism for $q = n + N$.

We may assume $X^{n-1} = \{x_0\}$, since if not we can replace X by a homotopy equivalent complex for which this is true. Since $k_q(X^{n-1}) = k_q(\{x_0\}) = 0 = k'_q(X^{n-1})$ for all q, it follows $T_q(X^{n-1})$ is always an isomorphism. Suppose we have proved $T_q(X^{m-1})$ is an isomorphism for

$q < n + N$, an epimorphism for $q = n + N$, some $m \geqslant n$. We have a commutative diagram of exact sequences

$$\begin{array}{ccccccc}
\cdots \to & k_{q+1}(\bigvee_{\alpha \in J_m} S^m_\alpha) & \to & k_q(X^{m-1}) & \to & k_q(X^m) & \to \\
& \downarrow T_{q+1}(\bigvee_\alpha S^m_\alpha) & & \downarrow T_q(X^{m-1}) & & \downarrow T_q(X^m) & \\
\cdots \to & k'_{q+1}(\bigvee_{\alpha \in J_m} S^m_\alpha) & \to & k'_q(X^{m-1}) & \to & k'_q(X^m) & \to
\end{array}$$

$$\begin{array}{cccc}
k_q(\bigvee_{\alpha \in J_m} S^m_\alpha) & \to & k_{q-1}(X^{m-1}) & \to \cdots \\
\downarrow T_q(\bigvee_\alpha S^m_\alpha) & & \downarrow T_{q-1}(X^{m-1}) & \\
k'_q(\bigvee_{\alpha \in J_m} S^m_\alpha) & \to & k'_{q-1}(X^{m-1}) & \to \cdots.
\end{array}$$

If $q < n + N$, then $T_q(X^{m-1})$, $T_q(\bigvee_\alpha S^m_\alpha)$ and $T_{q-1}(X^{m-1})$ are isomorphisms and $T_{q+1}(\bigvee_\alpha S^m_\alpha)$ is an epimorphism. Thus by the 5-lemma $T_q(X^m)$ is an isomorphism. If $q = n + N$, then $T_q(X^{m-1})$, $T_q(\bigvee_\alpha S^m_\alpha)$ are epimorphisms and $T_{q-1}(X^{m-1})$ is an isomorphism, so $T_q(X^m)$ is an epimorphism.

Now in the case that X is infinite dimensional and k_*, k'_* satisfy the wedge axiom we have a commutative diagram

$$\begin{array}{ccc}
\operatorname{dir\,lim} k_q(X^m) & \xrightarrow[\cong]{\{i_{m*}\}} & k_q(X) \\
\operatorname{dir\,lim} T_q(X^m) \downarrow & & \downarrow T_q(X) \\
\operatorname{dir\,lim} k'_q(X^m) & \xrightarrow[\cong]{\{i_{m*}\}} & k'_q(X).
\end{array}$$

Thus if $T_q(X^m)$ is an isomorphism for $q < n + N$ and an epimorphism for $q = n + N$, all $m \geqslant -1$, the same will be true of $\operatorname{dir\,lim} T_q(X^m)$ and hence of $T_q(X)$.

Finally, if k_*, k'_* satisfy the WHE axiom, we take CW-substitutes. □

Dual to the notion of homology theory is that of cohomology theory.

7.56. Definition. A *reduced cohomology theory* k^* on $\mathscr{P}\mathscr{T}'$ is a collection of cofunctors $k^n: \mathscr{P}\mathscr{T}' \to \mathscr{A}$ and natural equivalences $\sigma^n: k^{n-1} \circ S \to k^n$, $n \in \mathbb{Z}$, satisfying

Exactness: for every pointed pair (X, A, x_0) with inclusions $i:(A, x_0) \to (X, x_0)$ and $j:(X, x_0) \to (X \cup CA, *)$ the sequence

$$k^n(A, x_0) \xleftarrow{k^n[i]} k^n(X, x_0) \xleftarrow{k^n[j]} k^n(X \cup CA, *)$$

is exact.

There is a corresponding notion of an unreduced cohomology theory on $\mathscr{T}^{2\prime}$.

For every one of the propositions, theorems, etc. from 7.2 through 7.44 there is a dual result for cohomology with a dual proof. Cohomology does not behave so well on limits, however, as homology does, and so the results from 7.45 onward need a separate treatment for cohomology theories.

The inclusions $j_n^m: X^n \to X^m$, $n \leqslant m$, and $i_n: X^n \to X$ induce maps $j_n^{m*}: k^q(X^m) \to k^q(X^n)$ and $i_n^*: k^q(X) \to k^q(X^n)$ satisfying $j_n^{m*} \circ j_m^{p*} = j_n^{p*}$ and $j_n^{m*} \circ i_m^* = i_n^*$. This leads us to the notion dual to that of direct system.

7.57. Definition. An *inverse system* of abelian groups is a collection $\{G_\alpha\}$ of abelian groups indexed by a directed set A and homomorphisms $j_\alpha^\beta: G_\beta \to G_\alpha$ for all $\alpha \leqslant \beta$ satisfying $j_\alpha^\beta \circ j_\beta^\gamma = j_\alpha^\gamma$, $\alpha \leqslant \beta \leqslant \gamma$, $j_\alpha^\alpha = 1$, all α. Given an inverse system $\{G_\alpha, j_\alpha^\beta, A\}$, the *inverse limit* $\lim^0 G_\alpha = \operatorname{inv\,lim} G_\alpha$ is the group defined as follows:

$$\operatorname{inv\,lim} G_\alpha \subset \prod_{\alpha \in A} G_\alpha$$

is the subgroup of all $f \in \prod_{\alpha \in A} G_\alpha$ such that $j_\alpha^\beta(f(\beta)) = f(\alpha)$ for all $\alpha \leqslant \beta$ in A. The restrictions of the projections $p_\alpha: \prod_{\alpha \in A} G_\alpha \to G_\alpha$ to $\operatorname{inv\,lim} G_\alpha$ define homomorphisms

$$p_\alpha: \operatorname{inv\,lim} G_\alpha \to G_\alpha$$

which satisfy $j_\alpha^\beta \circ p_\beta = p_\alpha$ for all $\alpha \leqslant \beta$ in A.

7.58. Proposition. ($\operatorname{inv\,lim} G_\alpha, \{p_\alpha\}$) *is characterized up to unique isomorphism by the property that given any group H and homomorphisms $f_\alpha: H \to G_\alpha$, $\alpha \in A$, such that $j_\alpha^\beta \circ f_\beta = f_\alpha$ for all $\alpha \leqslant \beta$, there is a unique homomorphism $f: H \to \operatorname{inv\,lim} G_\alpha$ such that $p_\alpha \circ f = f_\alpha$ for all $\alpha \in A$.*

Proof: There is the unique map $\{f_\alpha\}: H \to \prod_{\alpha \in A} G_\alpha$ such that $p_\alpha \circ \{f_\alpha\} = f_\alpha$. One has only to check that $\operatorname{im}\{f_\alpha\} \subset \operatorname{inv\,lim} G_\alpha$ under the given assumptions. The rest of the proof is dual to that of 7.46. □

We shall again denote the unique map f by $\{f_\alpha\}: H \to \operatorname{inv\,lim} G_\alpha$.

7.59. There is the obvious definition of a morphism of inverse systems and the dual of 7.48 says a morphism $\{\phi_\alpha\}: \{G_\alpha\} \to \{G'_\alpha\}$ of inverse systems induces a homomorphism

$$\operatorname{inv\,lim} \phi_\alpha: \operatorname{inv\,lim} G_\alpha \to \operatorname{inv\,lim} G'_\alpha$$

such that $p'_\beta \circ \operatorname{inv\,lim} \phi_\alpha = \phi_\beta \circ p_\beta$, all $\beta \in A$.

7.60. The dual of 7.49 says given a morphism of inverse systems

$$\{\phi_\alpha\}: \{G_\alpha\} \to \{G'_\alpha\},$$

homomorphisms $f_\alpha : H \to G_\alpha$, $f'_\alpha : H' \to G'_\alpha$, $\phi : H \to H'$ satisfying $j_\alpha^\beta \circ f_\beta = f_\alpha$, $j_\alpha'^\beta \circ f'_\beta = f'_\alpha$ and $f'_\alpha \circ \phi = \phi_\alpha \circ f_\alpha$, then the diagram

$$\begin{array}{ccc} H & \xrightarrow{\{f_\alpha\}} & \operatorname{inv\,lim} G_\alpha \\ \phi \downarrow & & \downarrow \operatorname{inv\,lim} \phi_\alpha \\ H' & \xrightarrow{\{f'_\alpha\}} & \operatorname{inv\,lim} G'_\alpha \end{array}$$

commutes.

So far everything has been properly dual. Duality breaks down, however, when we come to 7.50: invlim does not preserve exactness.

7.61. Example. The diagram

$$\begin{array}{ccccc} \mathbb{Z} & \xrightarrow{p} & \mathbb{Z}_2 & \longrightarrow & 0 \\ \downarrow \times 3 & & \downarrow 1 & & \downarrow 1 \\ \mathbb{Z} & \xrightarrow{p} & \mathbb{Z}_2 & \longrightarrow & 0 \end{array}$$

commutes if p is the canonical quotient homomorphism, and the rows are exact. Let $\{A_n, j_n^m\}$ be the inverse system with $A_n = \mathbb{Z}$, all integers $n \geqslant 0$, $j_n^m(z) = 3^{m-n} z$, $n \leqslant m$, and $\{B_n, k_n^m\}$ the inverse system with $B_n = \mathbb{Z}_2$, all n, $k_n^m = 1$, all $n \leqslant m$. We take $\{\phi_n\} : \{A_n\} \to \{B_n\}$ and $\{\psi_n\} : \{B_n\} \to \{0\}$ defined by $\phi_n(x) = p(x)$, $x \in A_n$, $\psi_n = 0$. Then $\{\phi_n\}, \{\psi_n\}$ are morphisms of inverse systems and for every n

$$A_n \xrightarrow{\phi_n} B_n \xrightarrow{\psi_n} 0$$

is exact.

Now there is no sequence of integers $\{a_0, a_1, a_2, \ldots\}$ such that $a_i = 3a_{i+1}$ for every $i \geqslant 0$ except the zero sequence. Hence $\operatorname{inv\,lim} A_n = 0$. Clearly $\operatorname{inv\,lim} B_n = \mathbb{Z}_2$. Thus the sequence

$$\operatorname{inv\,lim} A_n \xrightarrow{\operatorname{inv\,lim} \phi_n} \operatorname{inv\,lim} B_n \xrightarrow{\operatorname{inv\,lim} \psi_n} \operatorname{inv\,lim} 0$$

is not exact.

In fact the functor invlim is what is called *left exact* (cf. [56]) and hence generates derived functors $\lim^k$, $k \geqslant 0$. We shall restrict our attention to the case where $A = \mathbb{N} =$ natural numbers, as this case is much simpler than the general case and covers the application we have in mind ($\{k^q(X^n)\}$).

7.62. Consider the map $\prod_n G_n \xrightarrow{\delta} \prod_n G_n$ defined by

$$\delta f(n) = (-1)^n f(n) + (-1)^{n+1} j_n^{n+1}(f(n+1))$$

for $f \in \prod_n G_n$. Note that $\ker \delta \subset \prod_n G_n$ is the set of all f such that $j_n^{n+1}(f(n+1)) = f(n)$, all $n \geqslant 0$. This is just $\lim^0 G_n = \operatorname{inv\,lim} G_n$. We define

$$\lim{}^1 G_n = \operatorname{coker} \delta.$$

7.63. Proposition. *A short exact sequence*

$$0 \longrightarrow \{A_n\} \xrightarrow{\{\phi_n\}} \{B_n\} \xrightarrow{\{\psi_n\}} \{C_n\} \longrightarrow 0$$

of inverse systems (i.e. $0 \to A_n \to B_n \to C_n \to 0$ *is exact for every* $n \geqslant 0$) *gives rise to an exact sequence*

$$0 \to \lim{}^0 A_n \to \lim{}^0 B_n \to \lim{}^0 C_n \to \lim{}^1 A_n \to \lim{}^1 B_n \to \lim{}^1 C_n \to 0.$$

Proof: We can think of

$$\prod_n G_n \xrightarrow{\delta} \prod_n G_n$$

as a cochain complex $C^*\{G_n\}$ with $C^0 = C^1 = \prod_n G_n$, $C^k = 0$, $k \geqslant 2$. In those terms $\lim^0 G_n = H^0(C^*(\{G_n\}))$, $\lim^1 G_n = H^1(C^*(\{G_n\}))$. Here we have a short exact sequence of cochain complexes

$$0 \longrightarrow C^*(\{A_n\}) \xrightarrow{\{\phi_n\}^\#} C^*(\{B_n\}) \xrightarrow{\{\psi_n\}^\#} C^*(\{C_n\}) \longrightarrow 0.$$

The resulting long exact sequence of cohomology groups is just the exact sequence of 7.63. □

7.64. Corollary. *If* $\{\phi_n\}:\{G_n\} \to \{G_n'\}$ *is a morphism of inverse systems such that each* ϕ_n *is an isomorphism then* $\lim^0 \phi_n$ *and* $\lim^1 \phi_n$ *are isomorphisms.*

The following proposition is the nearest we can come to a dual for 7.53.

7.65. *Wedge axiom.* For every collection $\{(X_\alpha, x_\alpha):\alpha \in A\}$ of pointed spaces the inclusions $i_\alpha: X_\alpha \to \bigvee_{\beta \in A} X_\beta$ induce an isomorphism

$$\{i_\alpha^*\}: k^q(\bigvee_\alpha X_\alpha) \to \prod_{\alpha \in A} k^q(X_\alpha).$$

7.66. Proposition. *Let* (X, x_0) *be a CW-complex and let* $X^0 \subset \cdots \subset X^n \subset \cdots \subset X$ *be subcomplexes with* $\bigcup_{n \geqslant 0} X^n = X$, $j_n^m: X^n \to X^m$, $i_n: X^n \to X$ *the inclusions,* $n \geqslant m$. *Then* $\{k^q(X^n), j_n^{m*}, \mathbb{N}\}$ *is an inverse system for every*

$q \in \mathbb{Z}$, and if k^* satisfies the wedge axiom on $\mathscr{PW}'$, then there is an exact sequence

$$0 \longrightarrow \lim^1 k^{q-1}(X^n) \longrightarrow k^q(X) \xrightarrow{\{i_n^*\}} \lim^0 k^q(X^n) \longrightarrow 0\cdot$$

Proof: As in 7.53 we introduce the telescope X'. The proof proceeds as before except that this time the Mayer–Vietoris sequence for $(X'; A, B)$ looks as follows:

$$\begin{array}{ccccc} \cdots \longrightarrow k^{q-1}(A) \oplus k^{q-1}(B) & \xrightarrow{\alpha} & k^{q-1}(A \cap B) & \xrightarrow{\Delta'} \\ \Vert\wr & & \Vert\wr & \\ \prod_{n \geqslant -1} k^{q-1}(X^n) & \xrightarrow[\alpha']{} & \prod_{n \geqslant -1} k^{q-1}(X^n) & \end{array}$$

$$\begin{array}{ccccccc} k^q(X') & \xrightarrow{\beta} & k^q(A) \oplus k^q(B) & \xrightarrow{\alpha} & k^q(A \cap B) & \longrightarrow \cdots \\ r^* \uparrow \cong & \searrow^{\beta'} & \Vert\wr & & \Vert\wr & \\ k^q(X) & \xrightarrow[\{i_n^*\}]{} & \prod_{n \geqslant -1} k^q(X^n) & \xrightarrow[\alpha']{} & \prod_{n \geqslant -1} k^q(X^n), & \end{array}$$

where α' is given by $[\alpha'(f)](n) = (-1)^{n+1}\{f(n) - j_n^{n+1*} f(n+1)\}$ for $f \in \prod_{n \geqslant -1} k^q(X^n)$ and β' is given by $[\beta'(x)](n) = k_n^* i_n'^*(x)$, $x \in k^q(X')$. Thus $\alpha' = -\delta$, and we get an exact sequence

$$\begin{array}{ccccccccc} 0 & \longrightarrow & \operatorname{coker} \alpha' & \longrightarrow & k^q(X) & \xrightarrow{\{i_n^*\}} & \ker \alpha' & \longrightarrow & 0 \\ & & \Vert\wr & & & & \Vert\wr & & \\ & & \lim^1 k^{q-1}(X^n) & & & & \lim^0 k^q(X^n). & & \end{array}$$ □

Therefore in cohomology $\{i_n^*\}: k^q(X) \to \lim^0 k^q(X^n)$ is an isomorphism if and only if $\lim^1 k^{q-1}(X^n) = 0$.

7.57–7.66 are not needed, however, for the appropriate dual of 7.55.

7.67. Theorem. *Let $T^*: k^* \to k'^*$ be a natural transformation of cohomology theories satisfying the wedge axiom on $\mathscr{PW}'$. If $T^q(S^0): k^q(S^0) \to k'^q(S^0)$ is an isomorphism for $q > N$, epimorphism for $q = N$, then for any n-dimensional CW-complex (X, x_0)*

$$T^q(X): k^q(X) \to k'^q(X)$$

is an isomorphism for $q > n + N$, epimorphism for $q = n + N$.

Proof: The argument proceeds exactly as in 7.55, but with everything dualized. ▯

7.68. Now in Chapter 8 we shall show how to construct many reduced homology and cohomology theories on the category $\mathscr{PW}'$. For many purposes, however, it is convenient to have theories which are defined on all topological spaces, so we now show how to extend a reduced homology (or cohomology) theory from $\mathscr{PW}'$ to $\mathscr{PT}'$.

Suppose that k_* is any reduced homology theory on $\mathscr{PW}'$ and suppose (X,x_0) is any pointed space; choose a CW-substitute $f:X' \to X$ for X and a base point x_0' in X' such that $f(x_0')=x_0$. Then we define $k_n(X)=k_n(X')$, $n\in\mathbb{Z}$. Since the CW-substitute is determined up to unique homotopy equivalence, it follows that $k_n(X)$ is a well defined (abstract) group.

Given any map $f:(X,x_0)\to(Y,y_0)$ of pointed spaces, we choose CW-substitutes $g:X'\to X$, $h:Y'\to Y$ and a map $f':(X',x_0')\to(Y',y_0')$ such that $h\circ f'\simeq f\circ g$. Then we take $k_n[f]=k_n[f']$; $k_n[f]$ is well defined since f' is unique up to homotopy.

Suppose (X,x_0) is any pointed space and $f':(X',x_0')\to(X,x_0)$ is a CW-substitute. Let

$$C_1X=\{[t,x]:1/3\leqslant t\leqslant 1,\ x\in X\}\subset SX,$$

$$C_2X=\{[t,x]:0\leqslant t\leqslant 2/3,\ x\in X\}\subset SX.$$

Then $\{SX;\ C_1X,C_2X\}$ is a triad with $SX=C_1X^\circ\cup C_2X^\circ$. Also

$$Sf':(SX';\ C_+X',C_-X')\ \to\ (SX;\ C_1X,C_2X)$$

is a map of triads such that

$$Sf'|C_+X':C_+X'\ \to\ C_1X,\quad Sf'|C_-X':C_-X'\ \to\ C_2X,$$

$$Sf'|C_+X'\cap C_-X':C_+X'\cap C_-X'\ \to\ C_1X\cap C_2X$$

are all weak homotopy equivalences. Hence by 6.52 $Sf':SX'\to SX$ is a CW-substitute for SX. Thus we define $\sigma:k_n(X)\to k_{n+1}(SX)$ by requiring that the square

$$\begin{array}{ccc} k_n(X) & \xrightarrow{\ \sigma\ } & k_{n+1}(SX) \\ \| & & \| \\ k_n(X') & \underset{\cong}{\xrightarrow{\ \sigma\ }} & k_{n+1}(SX') \end{array}$$

commute.

If (X, A, x_0) is any pointed pair then we can choose CW-substitutes $f': A' \to A$ and $g': X' \to X$. Let $h: A' \to X'$ be any cellular map such that

$$\begin{array}{ccc} A' & \xrightarrow{f'} & A \\ h \downarrow & & \cap \\ X' & \xrightarrow{g'} & X \end{array}$$

commutes up to homotopy. Then we may replace X' by the mapping cylinder M_h and g' by $g' \circ r$ ($r: M_h \to X'$ the projection). Thus we may as well assume $A' \subset X'$. Now just as above we show that $\hat{g}': X' \cup CA' \to X \cup CA$ is a weak homotopy equivalence. Hence from the commutative diagram

$$\begin{array}{ccccc} k_n(A) & \xrightarrow{i_*} & k_n(X) & \xrightarrow{j_*} & k_n(X \cup CA) \\ \| & & \| & & \| \\ k_n(A') & \xrightarrow{i'_*} & k_n(X') & \xrightarrow{j'_*} & k_n(X' \cup CA') \end{array}$$

we see that k_* satisfies the exactness axiom. Hence k_* becomes a reduced homology theory on $\mathscr{P}\mathscr{T}'$. k_* obviously satisfies the WHE axiom. Therefore $\hat{k}_*$ is an unreduced homology theory on $\mathscr{T}^{2\prime}$.

7.69. *Remark.* When we consider singular homology theory in Chapter 10 and bordism in Chapter 12 we shall see that each of these homology theories has a very natural definition on all of $\mathscr{P}\mathscr{T}'$. Whenever this occurs we should feel ourselves under an obligation to try to show that the extension from $\mathscr{P}\mathscr{W}'$ to $\mathscr{P}\mathscr{T}'$ which we have just given here agrees with the natural definition on $\mathscr{P}\mathscr{T}'$. This will be so if the natural definition satisfies the WHE axiom. By considering the mapping cylinder M_f of a weak homotopy equivalence $f: X \to Y$ and the resulting exact sequence

$$\cdots \longrightarrow h_{n+1}(M_f, X) \longrightarrow h_n(X, \varnothing) \xrightarrow{f_*} h_n(Y, \varnothing) \longrightarrow h_n(M_f, X) \longrightarrow \cdots$$

for an unreduced theory h_*, we see that the WHE axiom for h_* is equivalent to either of the following.

7.70. If $A \subset X$ is a cofibration and (X, A) is n-connected for all $n \geqslant 0$, then $h_*(X, A) = 0$.

7.71. If X is n-connected for all $n \geqslant 0$, then $h_*(\{x_0\}) \to h_*(X)$ is surjective for all $x_0 \in X$.

7.72. Exercise: Show that $\pi_q^S = \operatorname{dir\,lim} \pi_{q+n}(S^n \wedge -)$ satisfies the WHE axiom.

7.73. Exercise. Suppose $X_1 \subset X_2 \subset \cdots \subset X_n \subset \cdots \subset X$ is a sequence of subspaces of a topological space X such that for each compact subspace $K \subset X$ there is a $k \geqslant 1$ with $K \subset X_k$. Then we can inductively construct CW-substitutes $f^k: X'_k \to X_k$, $k \geqslant 1$, such that X'_k is a subcomplex of X'_{k+1} for $k \geqslant 1$ and $f^{k+1}|X'_k = f^k$. We let $X' = \bigcup_{k \geqslant 1} X'_k$ with the weak topology and $f': X' \to X$ be the map with $f'|X'_k = f^k$.
a) Show that $f': X' \to X$ is a CW-substitute for X.
b) Suppose k_* is a homology theory which satisfies the wedge and weak homotopy equivalence axioms. Show that $\{i_{k*}\}: \operatorname{dir\,lim} k_q(X_k) \to k_q(X)$ is an isomorphism for $q \in \mathbb{Z}$.

Appendix

In this appendix we give a useful criterion for the vanishing of $\lim^1$.

Let

$$G_0 \overset{j_0}{\leftarrow} G_1 \overset{j_1}{\leftarrow} G_2 \overset{j_2}{\leftarrow} \cdots \leftarrow G_n \overset{j_n}{\leftarrow} G_{n+1} \leftarrow \cdots$$

be an inverse system $\{G_n\}$ of abelian groups indexed by the non-negative integers.

7.74. Definition. $\{G_n\}$ satisfies the *Mittag-Leffler condition* (ML) if for each $n \geqslant 0$ there exists an $m(n) \geqslant n$ such that

$$\operatorname{im}[j_n^r: G_r \to G_n] = \operatorname{im}[j_n^{m(n)}: G_{m(n)} \to G_n]$$

for all $r \geqslant m(n)$. In other words the image of G_r in G_n becomes stable for sufficiently large r.

7.75. Theorem. If $\{G_n\}$ satisfies (ML), then $\lim^1 G_n = 0$.

Proof: Without loss of generality we may assume the function $m(n)$ is monotone increasing. Given an arbitrary element $f \in \prod_{n \geqslant 0} G_n$ we must find a $g \in \prod_{n \geqslant 0} G_n$ such that

7.76. $f(n) = (-1)^n g(n) + (-1)^{n+1} j_n g(n+1)$ for all $n \geqslant 0$.

Case I: Suppose $f(n) \in \operatorname{im} j_n^{m(n)}$ for all n. We construct $g(n)$ inductively, starting with $g(0) = 0$. Assume $g(0), g(1), \ldots, g(n)$ have already been chosen satisfying 7.76 and also

7.77. $g(k) \in \operatorname{im} j_k^{m(k)}$ $k = 0, 1, \ldots, n$. Then we also have

$$g(n) - (-1)^n f(n) \in \operatorname{im} j_n^{m(n)} = \operatorname{im} j_n^{m(n+1)},$$

so we choose a c with $g(n) - (-1)^n f(n) = j_n^{m(n+1)}(c)$and define $g(n+1) = j_{n+1}^{m(n+1)}(c)$. Then we have

$$g(n) - (-1)^n f(n) = j_n g(n+1)$$

i.e.

$$f(n) = (-1)^n g(n) + (-1)^{n+1} j_n g(n+1),$$

completing the induction step.

Case II: We now consider a general f and define h by

$$h(n) = (-1)^n f(n) + (-1)^{n+1} j_n^{n+1} f(n+1) + \cdots + (-1)^{m(n)} j_n^{m(n)} f(m(n)).$$

Then

$$(-1)^n h(n) + (-1)^{n+1} j_n h(n+1) \equiv f(n) \bmod \operatorname{im} j_n^{m(n)},$$

so we have reduced Case II to Case I. ▯

Remark 1. Suppose $\{G_n, j_n^m : n, m \in \mathbb{N}\}$ is an inverse system with each G_n finite. Then the subgroups $\operatorname{im}[j_n^r : G_r \to G_n]$, $r \geqslant n$, cannot all be distinct, and in fact there must exist an $m(n) \geqslant n$ with $\operatorname{im}[j_n^r : G_r \to G_n] = \operatorname{im}[j_n^{m(n)} : G_{m(n)} \to G_n]$ for all $r \geqslant m(n)$—that is (ML) is satisfied.

Remark 2. Suppose that X is a space and

$$X_0 \subset X_1 \subset \cdots \subset X_n \subset \cdots \subset X$$

is a filtration by subspaces X_n such that $H^*(X_n)$ is finitely generated. Suppose further that k^* is a cohomology theory such that each $k^{q-s-1}(S^0)$ is finite whenever $\tilde{H}^s(X_n) \neq 0$. Then one quickly sees using the Atiyah–Hirzebruch–Whitehead spectral sequence (Chapter 15) that $k^{q-1}(X_n)$ is finite for all n. By Remark 1 it then follows that $\lim^1 k^{q-1}(X_n) = 0$ and hence $k^q(X) \cong \lim^0 k^q(X_n)$.

References

1. J. F. Adams [10]
2. E. Dyer [37, 38]
3. S. Eilenberg and N. Steenrod [40]
4. J. W. Milnor [65]
5. G. W. Whitehead [93]

Chapter 8

Spectra

We now wish to construct homology and cohomology theories. The idea for what follows comes from Brown's Theorem (Chapter 9) which says that every cofunctor $h:\mathscr{PW}' \to \mathscr{A}$ which satisfies the wedge axiom and something resembling the Mayer–Vietoris theorem is of the form $h(X,x_0) = [X,x_0;\ E,e_0]$ for some fixed $(E,e_0) \in \mathscr{PW}'$. Now in particular, if h^* is a reduced cohomology theory satisfying the wedge axiom, then for every $n \in \mathbb{Z}$ h^n is a cofunctor of the required form, and hence $h^n(-) = [-;\ E_n, *]$ for some $(E_n, *) \in \mathscr{PW}'$. The cofunctors h^n are not unrelated, however; we have natural equivalences

$$\begin{array}{ccc} h^r & \xrightarrow{(\sigma^r)^{-1}} & h^{r+1} \circ S \\ \| & & \| \\ [-;E_r] & & [S(-);E_{r+1}] = [-;\ \Omega E_{r+1}]. \end{array}$$

Then $(\sigma^r)^{-1}[1_{E_r}] \in [E_r;\ \Omega E_{r+1}]$ is represented by a map $\varepsilon_r': E_r \to \Omega E_{r+1}$ such that $(\varepsilon_r')_* = (\sigma^r)^{-1}$. The adjoint of ε_r' is a map $\varepsilon_r: SE_r \to E_{r+1}$. This collection $\{E_r, \varepsilon_r : r \in \mathbb{Z}\}$ suggests the definition of spectrum we are about to give.

We have another aim in constructing the category of spectra. In homology theory the suspension homomorphism $\sigma: h_n(X) \to h_{n+1}(SX)$ is always an isomorphism. For various reasons this suggests trying to embed $\mathscr{PW}'$ in a larger category in which the suspension functor S has an inverse S^{-1}. This the category of spectra will do for us. Otherwise it should have most of the pleasant properties $\mathscr{PW}'$ has: inclusions are always cofibrations, J. H. C. Whitehead theorem, coexact sequences, for example. The category given here was constructed by Boardman; this description of it is due to Adams.

8.1. Definition. A *spectrum* E is a collection $\{(E_n, *) : n \in \mathbb{Z}\}$ of CW-complexes such that SE_n is (or is homeomorphic to) a subcomplex of E_{n+1}, all $n \in \mathbb{Z}$. A *subspectrum* $F \subset E$ consists of subcomplexes $F_n \subset E_n$ such that $SF_n \subset F_{n+1}$.

8.2. Example. If X is any CW-complex, then we can define a spectrum $E(X)$ by taking

$$E(X)_n = \begin{cases} * & n < 0 \\ S^n X & n \geqslant 0. \end{cases}$$

It appears that the collection $\{E_n, \varepsilon_n\}$ constructed from a cohomology theory using Brown's theorem is not necessarily a spectrum; the maps ε_n need not be homeomorphisms of SE_n onto a subcomplex of E_{n+1}. We have seen before, however, that any map can be "turned into" an inclusion.

8.3. Proposition. *Given any collection $\{E_n, \varepsilon_n\}$ of CW-complexes $(E_n, *)$ and cellular maps $\varepsilon_n : SE_n \to E_{n+1}$ we can construct a spectrum $E' = \{E'_n\}$ and homotopy equivalences $r_n : E'_n \to E_n$ such that*

$$r_{n+1} | SE'_n = \varepsilon_n \circ Sr_n.$$

$$\begin{array}{ccc} SE'_n & \subset & E'_{n+1} \\ \downarrow{\scriptstyle Sr_n} & & \downarrow{\scriptstyle r_{n+1}} \\ SE_n & \xrightarrow{\varepsilon_n} & E_{n+1} \end{array}$$

Proof: We employ another version of the "telescope". Let

$$E'_n = E_n \wedge \{n\}^+ \cup \textstyle\bigcup_{k<n} S^{n-k} E_k \wedge [k, k+1]^+, \; n \in \mathbb{Z},$$

with each $[x, k+1] \in S^{n-k} E_k \wedge [k, k+1]^+$ identified with

$$[S^{n-k-1} \varepsilon_k(x), k+1] \in S^{n-k-1} E_{k+1} \wedge [k+1, k+2]^+.$$

This is a CW-complex since each ε_k is a cellular map. We define $r_n : E'_n \to E_n$ by $r_n[x,t] = \varepsilon_{n-1} \circ S\varepsilon_{n-2} \circ \cdots \circ S^{n-k-1} \varepsilon_k(x)$, $(x,t) \in S^{n-k} E_k \times [k, k+1]$. We define $i_n : E_n \to E'_n$ by $i_n(x) = [x, n] \in E_n \times \{n\} \subset E'_n$. Then clearly $r_n \circ i_n = 1_{E_n}$. We define $H_n : E'_n \times I \to E'_n$ by

$$H([x,t], s) = \begin{cases} [S^{n-m} \varepsilon_{m-1} \circ S^{n-m+1} \varepsilon_{m-2} \circ \cdots \circ S^{n-k-1} \varepsilon_k(x), (1-s)t + sn] \\ \quad \text{for } [x,t] \in S^{n-k} E_k \wedge [k, k+1]^+, k < n, m \leqslant (1-s)t + sn \\ \quad \leqslant m+1, s \in I, k \leqslant m < n, \\ \quad [x, n] \text{ for } [x,t] \in E_n \wedge \{n\}^+, s \in I. \end{cases}$$

Then H is continuous and $H_0 = 1_{E'_n}$, $H_1 = i_n \circ r_n$. Hence r_n is a homotopy equivalence. Moreover we clearly have SE'_n contained in E'_{n+1} as subcomplex and $r_{n+1} | SE'_n = \varepsilon_n \circ Sr_n$. $\square$

8.4. We can define the notion of a *cell* of a spectrum E. There is the subspectrum $F \subset E$ such that $F_n = *$ for all n; we denote F by $*$ also and call it a cell of dimension $-\infty$. If e^d_n is any d-cell of E_n (other than $*$, if $d = 0$), then Se_n^d is a $(d+1)$-cell of E_{n+1}, and in general $S^m e_n^d$ is an $(m+d)$-cell of E_{n+m}. The d-cell e_n^d cannot be "desuspended" more than d times, so we can continue the sequence backwards until we reach a cell $e_{n'}^{d'}$ in $E_{n'}$ such that $e_n^d = S^{d-d'} e_{n'}^{d'}$ (in which case $n = n' + d - d'$) but $e_{n'}^{d'}$ is not the suspension of any cell in $E_{n'-1}$. Then the sequence

$$e = \{e_{n'}^{d'}, Se_{n'}^{d'}, S^2 e_{n'}^{d'}, \ldots\}$$

will be called a *cell of dimension* $d' - n'$ in E. Thus each cell in each complex E_n is a member of exactly one cell of E.

8.5. Definition. A spectrum E is called *finite* if it has only finitely many cells. It is called *countable* if it has countably many cells.

Remark. E is finite if and only if there is an integer N such that $E_n = S^{n-N} E_N$ for $n \geqslant N$ and the complex E_N is finite.

8.6. Definition. A *filtration* of a spectrum E is an increasing sequence $\{E^n : n \in \mathbb{Z}\}$ of subspectra of E whose union is E.

8.7. Examples. i) The *skeletal* filtration $\{E^{(n)}\}$ is defined as follows: $E^{(n)}$ = the union of all the cells of E of dimension at most n. Unlike the situation for CW-complexes $E^{(n)}$ may in general be non-trivial for arbitrarily large negative n. This fact prevents the skeletal filtration from being useful for inductive proofs.

ii) The *layer* filtration $\{E^n\}$ is defined as follows: for each cell $e = \{e_n, Se_n, \ldots\}$ of E we can find a finite subspectrum $F \subset E$ of which e is a cell (for example, let $F_n \subset E_n$ be the subcomplex consisting of e_n and all its faces; then take $F_m = *, m < n, F_m = S^{m-n} F_n, m \geqslant n$). Let $l(e)$ = the smallest number of cells in any such F (in fact, $l(e)$ = number of faces of e_n). We then define $E^n = *, n \leqslant 0$, E^n = union of all cells with $l(e) \leqslant n, n > 0$. The terms E^n are called the *layers* of E.

8.8. Lemma. *$\{E^n\}$ is a filtration of E.*

Proof: The subspace $E_m^n \subset E_m$ is the union of all cells e_m such that e_m is a term in a cell $e = \{e_k, Se_k, \ldots\}$ of E with $l(e) \leqslant n$. Choose a finite subspectrum $F \subset E$ with e as a cell and containing no more than n cells. Then e_m is a cell of $F_m \subset E_m$, and clearly $F_m \subset E_m^n$. Thus every face of e_m is in E_m^n, so E_m^n is a subcomplex of E_m. It is obvious that $SE_m^n \subset E_{m+1}^n$, that $E^n \subset E^{n+1}$, all $n \in \mathbb{Z}$, and that $\cup E^n = E$. ▯

Intuitively E is built up layer by layer: one starts with $E^1 = *$, attaches cells of various dimensions at various levels E_m^1, together with all their suspensions, thus obtaining E^2, etc. We shall use the layer filtration for inductive proofs.

In order to construct the category of spectra we need to specify the maps between spectra. The obvious definition turns out not to be adequate for our purposes, so we call these objects "functions".

8.9. Definition. A *function* $f:E \to F$ between spectra is a collection $\{f_n : n \in \mathbb{Z}\}$ of cellular maps $f_n : E_n \to F_n$ such that $f_{n+1}|SE_n = Sf_n$. Composition of functions is defined in the obvious way. The inclusion $i:F \to E$ of a subspectrum $F \subset E$ is a function and if $g:E \to G$ is a function then $g|F = g \circ i$ is also a function.

8.10. Definition. A subspectrum $F \subset E$ is called *cofinal* if each cell in E ultimately lies in F; that is, for any cell $e_n \subset E_n$ there is an m such that $S^m e_n \subset F_{n+m}$. Obviously if F is cofinal and $K_n \subset E_n$ is a finite subcomplex, then there is an m such that $S^m K_n \subset F_{n+m}$.

8.11. Lemma. *The intersection of two cofinal subspectra is cofinal and if $G \subset F \subset E$ are subspectra such that F is cofinal in E and G is cofinal in F, then G is cofinal in E. An arbitrary union of cofinal subspectra is cofinal.*

8.12. Definition. Let E,F be spectra. We consider the set S of all pairs (E',f') such that $E' \subset E$ is a cofinal subspectrum and $f':E' \to F$ is a function. On S we introduce a relation as follows: $(E',f') \sim (E'',f'')$ if and only if there is a pair (E''',f''') with $E''' \subset E' \cap E''$, E''' cofinal and $f'|E''' = f''' = f''|E'''$. From 8.11 it follows $\sim$ is an equivalence relation. We shall call equivalence classes *maps* from E to F; that is, $\hom(E,F) = S/\sim$. Intuitively maps only need to be defined on each cell "eventually".

The following lemma is needed if we are to compose maps.

8.13. Lemma. *Let E and F be spectra and $f:E \to F$ a function. If $F' \subset F$ is a cofinal subspectrum, then there is a cofinal subspectrum $E' \subset E$ with $f(E') \subset F'$.*

Proof: Let S be the set of all subspectra $G \subset E$ such that $f(G) \subset F'$ and let $E' = \bigcup_{G \in S} G$. Then E' is a subspectrum of E and $f(E') \subset F'$. We must show that E' is cofinal. Let $e = \{e_n, Se_n, \ldots\}$ be any cell of E and $L \subset E_n$ a finite subcomplex containing e_n; then $f(L)$ is contained in a finite subcomplex K of F_n, so since $F' \subset F$ is cofinal, there is an N such that $S^N K \subset F'_{n+N}$ and hence $f(S^N L) \subset F'_{n+N}$. Therefore $G = \{S^N L, \ldots\}$ is a subspectrum of E with $f(G) \subset F'$. Thus $\{S^N e_n, \ldots\} \subset E'$, showing that E' is cofinal. $\square$

Composition of maps is now possible; we compose any representatives which can be composed. 8.13 guarantees that there are such representatives. The result is a well defined map. Composition of maps is clearly associative. The inclusion function $i:F \to E$ of a subspectrum defines a map, so for any map $g:E \to G$ the restriction $g|F = g \circ i$ is also a map.

8.14. Note that in the category $\mathcal{SP}$ of spectra and maps any spectrum is equivalent to any cofinal subspectrum of it. For if $E' \subset E$ is cofinal, then the inclusion $i:E' \to E$ defines a map and the identity $1:E' \to E'$ defines a map $f:E \to E'$ such that $f \circ i = 1_{E'}$ and $i \circ f = 1_E$. Therefore any two cofinal subspectra E', $E'' \subset E$ are equivalent to each other, both being equivalent to E.

If $E = \{E_n\}$ is a spectrum and (X,x_0) is a CW-complex, then we can form a new spectrum $E \wedge X$: we take $(E \wedge X)_n = E_n \wedge X$ with the weak topology. This is again a spectrum, as $S(E \wedge X)_n = S(E_n \wedge X) = S^1 \wedge (E_n \wedge X) \cong (S^1 \wedge E_n) \wedge X \subset E_{n+1} \wedge X$. Given a map $f:E \to F$ of spectra represented by (E',f') and a map $g:K \to L$ of CW-complexes, we get a map $f \wedge g: E \wedge K \to F \wedge L$ of spectra represented by $(E' \wedge K, f' \wedge g)$, since $E' \wedge K$ is cofinal in $E \wedge K$.

8.15. We can now define homotopies of spectra. A *homotopy* is a map $h:E \wedge I^+ \to F$. There are two maps $i_0:E \to E \wedge I^+$, $i_1:E \to E \wedge I^+$ induced by the inclusions of 0, 1 in I^+. Then we say two maps of spectra $f_0, f_1:E \to F$ are *homotopic* if there is a homotopy $h:E \wedge I^+ \to F$ with $h \circ i_0 = f_0$, $h \circ i_1 = f_1$. We shall write h_0 for $h \circ i_0$, h_1 for $h \circ i_1$ as with spaces.

In terms of cofinal subspectra we can say that two maps $f_0, f_1:E \to F$ represented by (E_0', f_0'), (E_1', f_1') respectively are homotopic if there is a cofinal subspectrum $E'' \subset E_0' \cap E_1'$ and a function $h'':E'' \wedge I^+ \to F$ such that $h_0'' = f_0' | E''$, $h_1'' = f_1' | E''$.

Homotopy is an equivalence relation, so we may define $[E,F]$ to be the set of equivalence classes of maps $f:E \to F$. Composition passes to homotopy classes.

We have now defined the category $\mathcal{SP}$ of spectra and maps of spectra and also the associated homotopy category $\mathcal{SP}'$. We have the functor $\mathcal{PW} \to \mathcal{SP}$ defined by $(X,x_0) \mapsto E(X)$ for $(X,x_0) \in \mathcal{PW}$ and $f \mapsto E(f)$, where $E(f)_n = S^n f: S^n X \to S^n Y$, $n \geqslant 0$, for any $f:(X,x_0) \to (Y,y_0)$. This embeds $\mathcal{PW}$ in $\mathcal{PS}$. We shall write X for $E(X)$ unless there is a possibility of confusion.

8.16. *Remark.* Although for CW-complexes X, Y the functor described above embeds $\hom_{\mathcal{PW}}(X,Y)$ in $\hom_{\mathcal{SP}}(E(X),E(Y))$, it is not true that the induced functor embeds $\hom_{\mathcal{PW}'}(X,Y) = [X,Y]$ in $\hom_{\mathcal{SP}'}(E(X),E(Y)) = [E(X),E(Y)]$; it may be possible to find a homotopy $H:E(X) \wedge I^+ \to E(Y)$ from $E(f_0)$ to $E(f_1)$ which is not of the form $E(K)$ for any homotopy $K:X \wedge I^+ \to Y$ from f_0 to f_1. For example, it can happen that $Sf_0 \simeq Sf_1$ although $f_0 \not\simeq f_1$.

8.17. Given a map $f:E \to F$ of spectra, we can construct the *mapping cone* $F \cup_f CE$. We give I the base point 0 and define CE to be $E \wedge I$. We take

$F \cup_f CE$ to be the spectrum with $(F \cup_f CE)_n = F_n \cup_{f'_n} (E'_n \wedge I)$, where (E', f') represents f. If (E'', f'') is another representative of f, then $\{F_n \cup_{f'_n} (E'_n \wedge I)\}$ and $\{F_n \cup_{f''_n} (E''_n \wedge I)\}$ have a mutual cofinal subspectrum $\{F_n \cup_{f'''_n} (E'''_n \wedge I)\}$ and hence are equivalent.

For any spectrum $E = \{E_n\}$ we can define ΣE to be the spectrum with $\Sigma E_n = E_{n+1}$, $n \in \mathbb{Z}$. For any function $f: E \to F$ we define the function $\Sigma f: \Sigma E \to \Sigma F$ by $(\Sigma f)_n = f_{n+1}$. Then for any map $f: E \to F$ represented by (E', f') we can define $\Sigma f: \Sigma E \to \Sigma F$ to be the map represented by $(\Sigma E', \Sigma f')$. $\Sigma: \mathscr{SP} \to \mathscr{SP}$ is a functor and induces a functor on $\mathscr{SP}'$, since $f_0 \simeq f_1$ implies $\Sigma f_0 \simeq \Sigma f_1$. We can iterate $\Sigma: \Sigma^n = \Sigma \circ \Sigma^{n-1}, n \geqslant 2$, but Σ also has an inverse Σ^{-1} defined by $(\Sigma^{-1} E)_n = E_{n-1}$, $(\Sigma^{-1} f)_n = f_{n-1}$. Clearly $\Sigma^n \circ \Sigma^m = \Sigma^{m+n}$ for all integers m, n. Later we shall show that ΣE and $E \wedge S^1$ have the same homotopy type, thereby demonstrating that suspension is invertible in $\mathscr{SP}'$. First, however, let us demonstrate that $\mathscr{SP}$ and $\mathscr{SP}'$ have some of the other properties desired.

We have wedge sums in $\mathscr{SP}$; given a collection $\{E^\alpha : \alpha \in A\}$ of spectra, we define $\vee_\alpha E^\alpha$ by $(\vee_\alpha E^\alpha)_n = \vee_\alpha E^\alpha_n$. Since $S(\vee_\alpha E^\alpha_n) = \vee_\alpha S E^\alpha_n \subset \vee_\alpha E^\alpha_{n+1}$, this is a spectrum.

8.18. Proposition. *For any collection $\{E^\alpha : \alpha \in A\}$ of spectra the inclusions $i_\beta : E^\beta \to \vee_\alpha E^\alpha$ induce bijections*

$$\{\hom(i_\alpha, 1)\} : \hom_{\mathscr{SP}}(\vee_\alpha E^\alpha, F) \to \textstyle\prod_\alpha \hom_{\mathscr{SP}}(E^\alpha, F)$$

$$\{i^*_\alpha\} : [\vee_\alpha E^\alpha, F] \to \textstyle\prod_\alpha [E^\alpha, F]$$

for all $F \in \mathscr{SP}$.

Spectra can be built up cell by cell just as CW-complexes can. Let $\{E^n\}$ and $\{E^{(n)}\}$ denote the layer and skeletal filtrations respectively. If $e = \{e_n, Se_n, \ldots\}$ is a d-cell of E, let $f: S^{n+d-1} \to E_n$ be the attaching map of e_n. Let S^0 denote the spectrum $E(S^0)$; then the spectrum F with

$$F_m = \begin{Bmatrix} * & m < n \\ S^{m+d-1} & m \geqslant n \end{Bmatrix}$$

is a cofinal subspectrum of $\Sigma^{d-1} S^0$. Thus the collection $\{f, Sf, \ldots\}$ defines a map $f: \Sigma^{d-1} S^0 \to E$ which we call the attaching map of e.

8.19. Proposition. *Let E be a spectrum.*

a) *For any $n \in \mathbb{Z}$ let $\{e_\alpha : \alpha \in J_n\}$ be the cells of $E^{(n)} - E^{(n-1)}$, and for each $\alpha \in J_n$ let $f_\alpha : \Sigma^{n-1} S^0 \to E$ be the attaching map of e_α. If $g = \{f_\alpha\} : \vee_\alpha \Sigma^{n-1} S^0 \to E$, then g factors through $E^{(n-1)}$ and $E^{(n)} = E^{(n-1)} \cup_g C(\vee_\alpha \Sigma^{n-1} S^0)$.*

b) *For any $n > 0$ let $\{e_\alpha : \alpha \in A_n\}$ be the cells of $E^n - E^{n-1}$, and for each*

$\alpha \in A_n$ let $f_\alpha : \Sigma^{d_\alpha - 1} S^0 \to E$ be the attaching map of e_α. If

$$g = \{f_\alpha\} : \vee_\alpha \Sigma^{d_\alpha - 1} S^0 \to E,$$

then g factors through E^{n-1} and $E^n = E^{n-1} \cup_g C(\vee_\alpha \Sigma^{d_\alpha - 1} S^0)$.
The proof is evident.
Inclusions of spectra possess the homotopy extension property just as with CW-complexes.

8.20. Lemma. *Let E and H be spectra, F a subspectrum of E and G a cofinal subspectrum of $E \wedge \{0\}^+ \cup F \wedge I^+$. Given a function $g : G \to H$, we can find a cofinal subspectrum K of $E \wedge I^+$ containing G and an extension of g to a function $k : K \to H$. Moreover, if $G = E \wedge \{0\}^+ \cup F \wedge I^+$, we can choose $K = E \wedge I^+$.*

Proof: We construct $K^n \subset E^n \wedge I^+$ and $k|K^n$ by induction on n. Since $G^0 = E^0 = *$, we can take $K^0 = E^0 \wedge I^+$ and $k|K^0 = *$. Suppose we have constructed K^n and $k|K^n$ satisfying:

i) K^n is cofinal in $E^n \wedge I^+$,
ii) $G^n \subset K^n$,
iii) $k|G^n = g|G^n$.

For each cell $e = \{e_m, Se_m, \ldots\}$ of $E^{n+1} - (E^n \cup F^{n+1})$ we can find an N large enough so that the cells of $S^N e_m \wedge I^+$ are attached to K^n_{m+N} and g is defined on $S^N e_m \wedge \{0\}^+$. Then we have a map $S^N e_m \wedge \{0\}^+ \cup S^N \dot{e}_m \wedge I^+ \to H_{m+N}$ and by 6.5 we can extend it to a map $S^N e_m \wedge I^+ \to H_{m+N}$. Thus we may add the subcomplex $S^N e_m \wedge I^+$ and all its suspensions to $K^n \cup G^{n+1}$. Carrying out this procedure for all cells of $E^{n+1} - (E^n \cup F^{n+1})$ we obtain K^{n+1} and $k|K^{n+1}$ satisfying i)–iii). This completes the induction. We take $K = \bigcup_{n \geq 0} K^n$ of course. The last statement is obvious; indeed, it is just 6.5 applied to $E_n \wedge I^+$ for each n. □

8.21. We define the homotopy groups of a spectrum by taking $\pi_n(E) = [\Sigma^n S^0, E]$, $n \in \mathbb{Z}$. We can also give an alternate description of $\pi_n(E)$: observe that the cofinal subspectra of $\Sigma^n S^0$ are the spectra $\Sigma^{n-r} E(S^r)$, $r \geq 0$. A function $f : \Sigma^{n-r} E(S^r) \to E$ is just a map $f : S^r \to E_{r-n}$ together with all its suspensions $S^s f : S^{r+s} \to S^s E_{r-n} \subset E_{r+s-n}$. The groups $\pi_{k+n}(E_k, *)$ $(k \geq \min(2, 2-n))$ and morphisms

$$\pi_{k+n}(E_k, *) \xrightarrow{\Sigma} \pi_{k+n+1}(SE_k, *) \longrightarrow \pi_{k+1+n}(E_{k+1}, *)$$

define a direct system of abelian groups, and we can define $\alpha : [\Sigma^n S^0, E] \to \operatorname{dir lim} \pi_{k+n}(E_k, *)$ by $\alpha\{\Sigma^{n-r} E(S^r), f\} = \{[f]\}$. α is clearly surjective, and if $\{[f]\} = 0$ in $\operatorname{dir lim} \pi_{k+n}(E_k, *)$, then this means we can find a null homotopy $H : S^{r+s} \wedge I^+ \to E_{r+s-n}$ of $S^s f$ for some s. Then $\{\Sigma^{n-r-s}(S^{r+s}) \wedge I^+, H\}$ is a null homotopy of $\{\Sigma^{n-r}E(S^r), f\}$; hence α is also injective. Thus we have

$$\pi_n(E) \simeq \operatorname{dir\,lim} \pi_{n+k}(E_k, *), \quad n \in \mathbb{Z}.$$

In 8.27 below we shall see in a different way that $\pi_n(E)$ is a group.

Warning: Here is a case where $E(X)$ must be distinguished from the CW-complex X: $\pi_n(E(X)) \cong \operatorname{dir\,lim} \pi_{n+k}(S^k X, *) = \pi_n^S(X)$ (cf. 6.28 ff). This may be quite different from $\pi_n(X, x_0)$.

Of course any map of spectra $f: E \to F$ induces a homomorphism $f_*: \pi_n(E) \to \pi_n(F)$, $n \in \mathbb{Z}$. We call f a *weak homotopy equivalence* if f_* is an isomorphism for all $n \in \mathbb{Z}$. The proof of the Whitehead Theorem will be formally much the same as that for CW-complexes. The following is the analog of 6.30.

8.22. Lemma. *Suppose given a commutative diagram of spectra and functions as in Fig. 1. If f is a weak homotopy equivalence, then we can find a cofinal*

$$\begin{array}{ccc} C & \xrightarrow{f} & D \\ g \uparrow & & \uparrow h \\ A & \subset & B \end{array}$$

Fig. 1.

subspectrum $B' \subset B$ and functions $h': B' \to C$, $k: B' \wedge I^+ \to D$ such that

i) $A \subset B'$ iii) $(k)_0 = h|B'$ v) *k is stationary on A.*
ii) $h'|A = g$ iv) $(k)_1 = f \circ h'$

Proof: We consider a special case first. Suppose $A \subset \Sigma^n S^0$, $B \subset C\Sigma^n S^0$, $A \subset B$; then $g_k = S^{k+N} g_{-N}$ for all $k \geqslant -N$, some N, and similarly for h. The pair $(S^{k+N} h_{-N}, S^{k+N} g_{-N})$ defines an element of the group $\operatorname{dir\,lim} \pi_{k+n+1}(M_{f_k}, C_k, *)$; since the sequence

$$\ldots \operatorname{dir\,lim} \pi_{k+n}(C_k, *) \xrightarrow{\operatorname{dir\,lim} f_{k*}} \operatorname{dir\,lim} \pi_{k+n}(D_k, *) \to \operatorname{dir\,lim} \pi_{k+n}(M_{f_k}, C_k, *) \to \cdots$$

is exact, it follows $\operatorname{dir\,lim} \pi_{k+n+1}(M_{f_k}, C_k, *) = 0$, and hence $\{(h_k, g_k)\} = 0$. Thus we can find an $h_r': D^{n+r+1} \to C_r$ and a homotopy $k_r: D^{n+r+1} \wedge I^+ \to D_r$ for some r such that $h_r'|S^{n+r} = S^{r+N} g_{-N}$ and k_r is a homotopy rel S^{n+r} from $S^{r+N} h_{-N}$ to $f_r \circ h_r'$. Then we take

$$B_m' = \begin{cases} S^{m+n} & -N \leqslant m < r \\ D^{m+n+1} & m \geqslant r \end{cases}, \qquad h_m' = \begin{cases} g_m & -N \leqslant m < r \\ S^{m-r} h_r' & m \geqslant r \end{cases},$$

and

$$k_m(x,t) = \begin{cases} f_m \circ g_m(x) & -N \leqslant m < r \\ S^{m-r} k_r(x,t) & m \geqslant r. \end{cases}$$

B' is cofinal and $A \subset B'$.

The proof for the general case is now formally the same as that of 6.30: one considers the set $\mathscr{S}$ of triples (B', h', k) satisfying i)–v). One applies Zorn's lemma to find a maximal element (B', h', k) in $\mathscr{S}$ and then applies the result above to show that B' is cofinal in B. ▯

8.23. Corollary. *Suppose given a commutative diagram of spectra and maps*

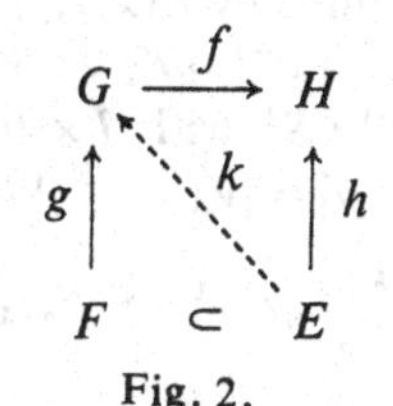

Fig. 2.

as in Fig. 2. If f is a weak homotopy equivalence, then there is a map $k: E \to G$ such that $k|F = g$ and $f \circ k \simeq h \operatorname{rel} F$.

Proof: Choose cofinal subspectra $E' \subset E$, $G' \subset G$, $F' \subset F \cap E'$ and functions $g': F' \to G' \subset G$, $f': G' \to H$, $h': E' \to H$ representing g, f, h respectively. Then apply 8.22 to this configuration to find a cofinal subspectrum $E'' \subset E'$ with $F' \subset E''$ and functions $k': E'' \to G'$, $\hat{k}: E'' \wedge I^+ \to H$ such that $k'|F' = g'$ and $\hat{k}$ is a homotopy from h' to $f' \circ k' \operatorname{rel} F'$. Take $k = \{E'', k'\}$. ▯

8.24. Corollary. *If $f: E \to F$ is a map of spectra which is a weak homotopy equivalence, then $f_*: [G, E] \to [G, F]$ is a bijection for any spectrum G.*

The proof is formally identical to that of 6.31.

8.25. Theorem. *A map of spectra is a weak homotopy equivalence if and only if it is a homotopy equivalence.*

The proof is formally identical to that of 6.32.

Armed with 8.25 we can now prove that ΣE and $E \wedge S^1$ have the same homotopy type.

8.26. Theorem. *There is an equivalence $E \wedge S^1 \to \Sigma E$ which is natural (in $\mathscr{SP}'$) for maps of E.*

Proof: We resurrect our friend the telescope E' from 8.3:

$$E'_n = E_n \wedge \{n\}^+ \cup \bigcup\nolimits_{k<n} S^{n-k} E_k \wedge [k, k+1]^+, \quad n \in \mathbb{Z}.$$

We also have the projections $r_n:E'_n \to E_n$, which define a function and hence a map $r:E' \to E$ of spectra. The injections $i_n:E_n \to E'_n$ do not define a function of spectra, but i_n is a homotopy inverse of r_n for each n, so that $r_*:\pi_n(E') \to \pi_n(E)$ is clearly an isomorphism for each $n \in \mathbb{Z}$. By 8.25 it follows r is a homotopy equivalence.

We shall also construct a homotopy equivalence $f:E' \wedge S^1 \to \Sigma E$. We have our map $\nu':S^1 \to S^1$ such that $(\nu')^2 = 1$ (cf. 2.22). We define a homotopy $H:(S^1 \wedge S^1) \times I \to S^1 \wedge S^1$ as follows: let $\alpha:(I^2,\dot{I}^2) \to (D^2,S^1)$ be the homeomorphism obtained by translating the center of the square to the origin and then projecting radially. Let $K:(D^2 \times I, S^1 \times I) \to (D^2,S^1)$ be the homotopy defined by $K(x,y,t) = (x\cos(\pi t/2) - y\sin(\pi t/2), x\sin(\pi t/2) + y\cos(\pi t/2))$. Then $\alpha^{-1} \circ K_t \circ \alpha$ is a homotopy of $(I^2,\dot{I}^2)$ to itself; and if we identify $S^1 \wedge S^1$ with $I \times I/I \times \dot{I} \cup \dot{I} \times I$, then $\alpha^{-1}K_t\alpha$ induces H. $H_0 = 1$ and $H_1[x,y] = [\nu'(y),x]$. We now define

$$f_n:E'_n \wedge S^1 \to E_{n+1}$$

by

$$f_n[\xi,n,y] = [(\nu')^n(y),\xi] \in S^1 \wedge E_n \subset E_{n+1} \text{ for } \xi \in E_n, y \in S^1,$$
$$f_n[s,x,\xi,m+t,y] = [s, H(x,(\nu')^m(y),t),\xi] \in S^{n-m+1}E_m \subset E_{n+1}$$
$$\text{for } s \in S^{n-m-1}, x,y \in S^1, \xi \in E_m, 0 \leqslant t \leqslant 1.$$

Then f_n is continuous and $Sf_n = f_{n+1}|S(E'_n \wedge S^1)$. Thus we get a function $f:E' \wedge S^1 \to \Sigma E$.

The functions $g_n:SE_n \to E'_n \wedge S^1$ defined by $g_n[x,\xi] = [\xi,n,(\nu')^n(x)]$, $\xi \in E_n$, $x \in S^1$, satisfy $g_n|S^2E_{n-1} \simeq Sg_{n-1}$, $f_n \circ g_n = 1$, $g_n \circ f_n \simeq 1$. Thus $f_*:\pi_n(E' \wedge S^1) \to \pi_n(\Sigma E)$ is an isomorphism for all $n \in \mathbb{Z}$, and therefore by 8.25 f is a homotopy equivalence. Thus we have $E \wedge S^1 \simeq E' \wedge S^1 \simeq \Sigma E$. The maps r, f are clearly natural. □

Remark. It seems difficult to find a homotopy inverse to either r or f above. Thus 8.25 is particularly useful here.

8.27. Corollary. *We can give each set $[E,F]$ the structure of an abelian group so that composition is bilinear.*

Proof: Since the functor Σ is invertible, $[E,F] \to [\Sigma E, \Sigma F]$ is a one–one correspondence. By 8.26 $[\Sigma E, \Sigma F] \to [E \wedge S^1, F \wedge S^1]$ is a one–one correspondence, so the function

$$\sigma:[E,F] \to [E \wedge S^1, F \wedge S^1]$$

defined by $\sigma([f]) = [f \wedge 1_{S^1}]$ is a one–one correspondence. Thus $\sigma^2:[E,F] \to [E \wedge S^2, F \wedge S^2]$ is also bijective. Now we can show that

$E \wedge S^2$ is a homotopy commutative H-cogroup in the spectrum sense: the comultiplication $\mu': S^2 \to S^2 \vee S^2$ gives us $E \wedge S^2 \xrightarrow[1 \wedge \mu']{} E \wedge (S^2 \vee S^2) \simeq (E \wedge S^2) \vee (E \wedge S^2)$. The rest goes as in 2.22. □

8.28. Definition. For any map $f: E \to F$ of spectra we call the sequence

$$E \xrightarrow{f} F \xrightarrow{j} F \cup_f CE$$

a *special cofibre sequence*. A *general cofibre sequence*, or simply a *cofibre sequence*, is any sequence

$$G \xrightarrow{g} H \xrightarrow{h} K$$

for which there is a homotopy commutative diagram

$$\begin{array}{ccccc} G & \xrightarrow{g} & H & \xrightarrow{h} & K \\ \downarrow{\scriptstyle \alpha} & & \downarrow{\scriptstyle \beta} & & \downarrow{\scriptstyle \gamma} \\ E & \xrightarrow{f} & F & \xrightarrow{j} & F \cup_f CE \end{array}$$

in which α, β, γ are homotopy equivalences.

8.29. *Remark.* In the diagram of 8.28 one may even assume that f is an inclusion of spectra: if $\{E', f'\}$ is a representative for f, then the mapping cylinder $M_f = \{F_n \cup E'_n \wedge I^+\}$ is a spectrum having the same homotopy type as F and containing E' as subspectrum.

8.30. Proposition. *In the sequence*

$$E \xrightarrow{f} F \xrightarrow{j} F \cup_f CE \xrightarrow{k'} E \wedge S^1 \xrightarrow{f \wedge 1} F \wedge S^1$$

each pair of consecutive maps forms a cofibre sequence.

The proof is essentially 2.39.

8.31. Lemma. *Given a homotopy commutative diagram of spectra and maps as in Fig.* 3

$$\begin{array}{ccccccc} G & \xrightarrow{g} & H & \xrightarrow{h} & K & \xrightarrow{k} & G \wedge S^1 \\ \downarrow{\scriptstyle \alpha} & & \downarrow{\scriptstyle \beta} & & \downarrow{\scriptstyle \gamma} & & \downarrow{\scriptstyle \alpha \wedge 1} \\ G' & \xrightarrow{g'} & H' & \xrightarrow{h'} & K' & \xrightarrow{k'} & G' \wedge S^1 \end{array}$$

Fig. 3.

in which the rows are cofibre sequences, we can find a map $\gamma: K \to K'$ such that the resulting diagram is homotopy commutative.

Proof: We may assume the rows are special cofibre sequences (i.e. $K = H \cup_g CG$, $K' = H' \cup_{g'} CG'$, h,h' inclusions) and even that g is an inclusion (8.29). We choose representatives (B,β') for β, (A',f') for g' and (A,α') for α so that $g(A) \subset B$ and $\alpha'(A) \subset A'$. Then we have a homotopy commutative diagram

$$\begin{array}{ccccccc} A & \xrightarrow{g} & B & \longrightarrow & B \cup_g CA & \longrightarrow & A \wedge S^1 \\ \downarrow{\scriptstyle \alpha'} & & \downarrow{\scriptstyle \beta'} & & \vdots{\scriptstyle \gamma'} & & \downarrow{\scriptstyle \alpha' \wedge 1} \\ A' & \xrightarrow{f'} & H' & \longrightarrow & H' \cup_{f'} CA' & \longrightarrow & A' \wedge S^1 \end{array}$$

whose rows are cofibre sequences and g an inclusion. By 8.20 we can find a $\beta'': B \to H'$ with $\beta'' \circ g = f' \circ \alpha'$ and $\beta'' \simeq \beta'$. We define γ' by $\gamma'|B = \beta''$, $\gamma'|CA = C\alpha': CA \to CA'$. Then γ' is a function which makes the diagram commute strictly if β' is replaced by β''. Hence the diagram commutes up to homotopy even with β'. □

8.32. Proposition. *If $G \xrightarrow{g} H \xrightarrow{h} K$ is a cofibre sequence, then for any spectrum E the sequences*

$$[E,G] \xrightarrow[g_*]{} [E,H] \xrightarrow[h_*]{} [E,K]$$

$$[G,E] \xleftarrow[g^*]{} [H,E] \xleftarrow[h^*]{} [K,E]$$

are exact.

Proof: i) Since $h \circ g \simeq 0$, it follows $h_* \circ g_* = 0$. Suppose $f: E \to H$ satisfies $h_*[f] = 0$. We apply 8.31 to the diagram

$$\begin{array}{ccccccc} E & \longrightarrow & E \wedge I & \longrightarrow & E \wedge S^1 & \xrightarrow{1} & E \wedge S^1 \\ {\scriptstyle f}\downarrow & & \downarrow{\scriptstyle \bar{h}} & & \downarrow{\scriptstyle k} & & \downarrow{\scriptstyle f \wedge 1} \\ H & \xrightarrow{h} & K & \longrightarrow & G \wedge S^1 & \xrightarrow{g \wedge 1} & H \wedge S^1 \end{array}$$

where $\bar{h}: E \wedge I \to K$ is a null homotopy of $h \circ f$. We obtain a map $k: E \wedge S^1 \to G \wedge S^1$ such that $(g \wedge 1) \circ k \simeq f \wedge 1$. From the natural equivalence $\sigma: [E,G] \to [E \wedge S^1, G \wedge S^1]$ we get a map $k': E \to G$ with $k \simeq k' \wedge 1$. Then $(g \circ k') \wedge 1 = (g \wedge 1) \circ (k' \wedge 1) \simeq (g \wedge 1) \circ k \simeq f \wedge 1$. Since $\sigma: [E,H] \to [E \wedge S^1, H \wedge S^1]$ is injective, it follows $g \circ k' \simeq f$—i.e. $g_*[k'] = [f]$.
ii) Again $g^* \circ h^* = 0$ follows from $h \circ g \simeq 0$. Suppose given $f: H \to E$ such that $g^*[f] = 0$. We apply 8.31 to the diagram

$$\begin{array}{ccccccc} G & \xrightarrow{g} & H & \xrightarrow{h} & K & \longrightarrow & G\wedge S^1 \\ & & \downarrow{\scriptstyle f} & & \downarrow{\scriptstyle f'} & & \\ * & \longrightarrow & E & \xrightarrow{1} & E & \longrightarrow & * \end{array}$$

to obtain a map $f':K \to E$ such that $f' \circ h \simeq f$; that is, $h^*[f'] = [f]$. □

Thus in the category $\mathscr{SP}'$ the two kinds of sequence, the cofibre sequence 2.41 and the fibre sequence 2.52, coincide.

8.33. We can now define the (reduced) homology and cohomology theories associated with any spectrum E. For each $(X,x_0) \in \mathscr{PW}'$ and $n \in \mathbb{Z}$ we take

$$E_n(X) = \pi_n(E\wedge X) = [\Sigma^n S^0, E\wedge X]$$
$$E^n(X) = [E(X), \Sigma^n E] \cong [\Sigma^{-n} S^0 \wedge X, E].$$

For $f:(X,x_0) \to (Y,y_0)$ we take $E_n(f) = (1\wedge f)_*$, $E^n(f) = E(f)^*$. We define $\sigma_n: E_n(X) \to E_{n+1}(SX)$ to be the composite

$$E_n(X) = [\Sigma^n S^0, E\wedge X] \underset{\cong}{\xrightarrow{\Sigma}} [\Sigma^{n+1} S^0, \Sigma E\wedge X[\underset{\cong}{\longrightarrow} [\Sigma^{n+1} S^0, E\wedge S^1 \wedge X]$$
$$= E_{n+1}(SX).$$

σ_n is clearly a natural equivalence. We define $\sigma^n: E^{n+1}(SX) \to E^n(X)$ to be the composite

$$E^{n+1}(SX) = [E(SX), \Sigma^{n+1} E] \underset{\cong}{\xleftarrow{i^*}} [\Sigma E(X), \Sigma^{n+1} E] \underset{\cong}{\xrightarrow{\Sigma^{-1}}} [E(X), \Sigma^n E]$$
$$= E^n(X).$$

Here we have used the fact that $E(SX)$ is a cofinal subspectrum of $\Sigma E(X)$ and hence the inclusion $i: E(SX) \to \Sigma E(X)$ induces an isomorphism i^*. σ^n is clearly a natural equivalence.

Let (X,A,x_0) be any pointed CW-pair. Since $E_n \wedge (X \cup CA) \cong (E_n \wedge X) \cup C(E_n \wedge A)$, $n \in \mathbb{Z}$, we see that

$$E\wedge A \xrightarrow{1\wedge i} E\wedge X \xrightarrow{1\wedge j} E\wedge (X \cup CA)$$

is a cofibre sequence. Therefore by 8.32

$$[\Sigma^n S^0, E\wedge A] \xrightarrow{(1\wedge i)_*} [\Sigma^n S^0, E\wedge X] \xrightarrow{(1\wedge j)_*} [\Sigma^n S^0, E\wedge (X \cup CA)]$$

is exact; but this is just the sequence

$$E_n(A) \xrightarrow{i_*} E_n(X) \xrightarrow{j_*} E_n(X \cup CA).$$

Thus E_* is a homology theory on $\mathscr{PW}'$.

Since $S^n(X \cup CA) \cong S^n X \cup C(S^n A)$, $n \in \mathbb{Z}$, we see that

$$E(A) \xrightarrow{E(i)} E(X) \xrightarrow{E(j)} E(X \cup CA)$$

is a cofibre sequence. Hence by 8.32

$$[E(A), \Sigma^n E] \xleftarrow{E(i)^*} [E(X), \Sigma^n E] \xleftarrow{E(j)^*} [E(X \cup CA), \Sigma^n E]$$

is exact; but this is just the sequence

$$E^n(A) \xleftarrow{i^*} E^n(X) \xleftarrow{j^*} E^n(X \cup CA).$$

Thus E^* is a cohomology theory on $\mathscr{PW}'$.

Since for any collection $\{X_\alpha : \alpha \in A\}$ of CW-complexes we have $S^n(\bigvee_\alpha X_\alpha) \cong \bigvee_\alpha S^n X_\alpha$ and hence $E(\bigvee_\alpha X_\alpha) \cong \bigvee_\alpha E(X_\alpha)$, we conclude from 8.18 that

$$\{i_\alpha^*\} : E^n(\bigvee_\alpha X_\alpha) \to \prod_\alpha E^n(X_\alpha)$$

is an isomorphism for all $n \in \mathbb{Z}$—in other words, E^* satisfies the wedge axiom.

To prove that E_* also satisfies the wedge axiom we need a lemma.

8.34. Lemma. *If E is a finite spectrum and $\{F^\alpha : \alpha \in A\}$ is a directed set of subspectra of a spectrum F ordered by inclusion such that $\bigcup_\alpha F^\alpha = F$, then the inclusions $i_\alpha : F^\alpha \to F$ induce an isomorphism*

$$\{i_{\alpha *}\} : \operatorname{dir\,lim}\, [E, F^\alpha] \to [E, F].$$

Proof: Since E is finite we can find an N such that E_N is a finite complex and $E_m = S^{m-N} E_N$, $m \geqslant N$. If $f : E \to F$ is any map represented by (E', f'), we may assume $E'_N = E_N$ (by increasing N if necessary). There is a finite subcomplex $K \subset F_N$ with $f'(E_N) \subset K$. Since A is directed and $\bigcup_\alpha F^\alpha = F$, we can find an α_0 with $K \subset F_N^{\alpha_0}$. Let E'' be the cofinal subspectrum of E with

$$E''_m = \begin{cases} * & m < N \\ S^{m-N} E_N & m \geqslant N, \end{cases}$$

and $f'' : E'' \to F^{\alpha_0}$ the function defined by $f' | E''$. Then $\{(E'', f'')\}$ is a map $g : E \to F^{\alpha_0}$ such that $i_{\alpha_0}[g] = [f]$. Hence $\{i_{\alpha *}\}$ is epimorphic.

If $g : E \to F^\beta$ is a map such that $\{i_{\alpha *}\}\{[g]\} = 0$, let $h : E \wedge I^+ \to F$ be a null homotopy of $i_\beta \circ g$. $E \wedge I^+$ is also a finite spectrum, so we can repeat the above argument to find a $\gamma \geqslant \beta$ and a null homotopy of $j_\beta^\gamma \circ g$. Hence $\{[g]\} = 0$, so $\{i_{\alpha *}\}$ is monomorphic. $\square$

8.35. Corollary. *For any spectrum E and any directed set $\{Y^\alpha:\alpha\in A\}$ of subcomplexes $Y^\alpha\subset Y$ of a CW-complex Y such that $\bigcup_\alpha Y^\alpha = Y$ the inclusions $i_\alpha: Y^\alpha\to Y$ induce an isomorphism*

$$\{i_{\alpha*}\}:\operatorname{dir\,lim} E_n(Y^\alpha) \to E_n(Y),\quad n\in\mathbb{Z}.$$

Proof: This is just 8.34 with $F^\alpha = E\wedge Y^\alpha$, $\alpha\in A$, since $\Sigma^n S^0$ is a finite spectrum. □

8.36. Corollary. *For any spectrum E the homology theory E_* satisfies the wedge axiom.*

Proof: If the collection $\{X_\alpha:\alpha\in A\}$ is finite, then the fact that

$$\{i_{\alpha*}\}:\bigoplus_{\alpha\in A} E_n(X_\alpha) \to E_n(\bigvee_{\alpha\in A} X_\alpha)$$

is an isomorphism follows from 7.36 plus induction. In the general case we let B be the set of all finite subsets of A directed by inclusion, and for each $\beta\in B$ we let $Y^\beta = \bigvee_{\alpha\in\beta} X_\alpha \subset \bigvee_{\alpha\in A} X_\alpha$. Then we have a commutative diagram

$$\begin{array}{ccc} \operatorname{dir\,lim} E_n(Y^\beta) & \xrightarrow[\cong]{\{i_{\beta*}\}} & E_n(\bigvee_{\alpha\in A} X_\alpha) \\ \cong\Big\uparrow \operatorname{dir\,lim}\{i_{\alpha*}\} & & \Big\uparrow \{i_{\alpha*}\} \\ \operatorname{dir\,lim}(\bigoplus_{\alpha\in\beta} E_n(X_\alpha)) & \xrightarrow[\cong]{\{j_\beta\}} & \bigoplus_{\alpha\in A} E_n(X_\alpha), \end{array}$$

where $j_\beta:\bigoplus_{\alpha\in\beta} E_n(X_\alpha)\to\bigoplus_{\alpha\in A} E_n(X_\alpha)$ is the inclusion for each $\beta\in B$. By the remark above $\{i_{\alpha*}\}:\bigoplus_{\alpha\in\beta} E_n(X_\alpha)\to E_n(Y^\beta)$ is an isomorphism for each $\beta\in B$ and hence by 7.51 $\operatorname{dir\,lim}\{i_{\alpha*}\}$ is also. That $\{j_\beta\}$ is an isomorphism is simple algebra. Hence $\{i_{\alpha*}\}$ is an isomorphism. □

The nearest we have to a dual for 8.35 is the following.

8.37. Proposition. *For any spectrum E and any filtration $\{X^n\}$ of a CW-complex X we have an exact sequence*

$$0 \longrightarrow \lim{}^1 E^{q-1}(X^n) \longrightarrow E^q(X) \xrightarrow{\{i_n^*\}} \lim{}^0 E^q(X^n) \longrightarrow 0.$$

This is just 7.66 applied to E^*.

8.38. The coefficient groups of the homology theory E_* are

$$E_n(S^0) = \pi_n(E\wedge S^0) = \pi_n(E),\quad n\in\mathbb{Z}.$$

The coefficient groups of E^* are

$$E^n(S^0) = [E(S^0),\Sigma^n E] = [S^0,\Sigma^n E] \cong [\Sigma^{-n} S^0, E] = \pi_{-n}(E),\quad n\in\mathbb{Z}.$$

8.39. Any map $f:E \to F$ of spectra induces natural transformations $T_*(f):E_* \to F_*$, $T^*(f):E^* \to F^*$ of homology and cohomology theories. If f is a homotopy equivalence, then $T_*(f)$, $T^*(f)$ are natural equivalences. By 8.25 this is the case if and only if $f_*:\pi_n(E) \to \pi_n(F)$ is an isomorphism for all $n \in \mathbb{Z}$—i.e. $T_*(f)$ is a natural equivalence if and only if $T_*(f)$ is an isomorphism on the coefficient groups. This is a special case of 7.55. Of course a priori there may be natural transformations $T:E_* \to F_*$ of homology theories which are not of the form $T_*(f)$ for any map $f:E \to F$.

8.40. We can extend the cohomology theory E^* to a cohomology theory E^* on the category $\mathcal{SP}'$ by simply taking $E^n(F) = [F, \Sigma^n E]$, $n \in \mathbb{Z}$, $F \in \mathcal{SP}'$. E^* is a cohomology theory in the sense that we have natural equivalences

$$\sigma^n : E^{n+1}(F \wedge S^1) \xrightarrow[\cong]{} E^{n+1}(\Sigma F) \xrightarrow[\cong]{} E^n(F)$$

for all $n \in \mathbb{Z}$, $F \in \mathcal{SP}'$, and E^* satisfies the following exactness axiom: for any cofibre sequence $F \xrightarrow{f} G \xrightarrow{g} H$ the sequence

$$E^n(F) \xleftarrow{f^*} E^n(G) \xleftarrow{g^*} E^n(H)$$

is exact. (This axiom is equivalent to the usual one on $\mathcal{PW}'$.)

Now if $T^*:E^* \to F^*$ is a natural equivalence of cohomology theories on $\mathcal{SP}'$, we can show $T^* = T^*(f)$ for some map $f:E \to F$. In $E^0(E) = [E,E]$ we have $[1_E]$; let $T^0(E)[1_E] \in F^0(E) = [E,F]$ be represented by $f:E \to F$. Then for any spectrum G and $x \in E^n(G)$ represented by $g:G \to \Sigma^n E$ we have

$$\begin{aligned} T^n(G)(x) &= T^n(G)[g] = T^n(G)[1_{\Sigma^n E} \circ g] = T^n(G) \circ g^*[1_{\Sigma^n E}] = g^* T^n(\Sigma^n E)[1_{\Sigma^n E}] \\ &= g^* T^0(E)[1_E] = g^*[f] = [f \circ g] = T^n(f)(G)[g] = (T^n(f)(G))(x). \end{aligned}$$

Thus $T^* = T^*(f)$.

When we have defined the smash product $E \wedge G$ of two spectra we can also extend E_* to $\mathcal{SP}'$: $E_n(G) = \pi_n(E \wedge G) = [\Sigma^n S^0, E \wedge G]$. In this case, however, it is not clear that a natural transformation $T_*:E_* \to F_*$ on $\mathcal{SP}'$ is of the form $T_* = T_*(f)$ for some map $f:E \to F$.

8.41. Definition. If E is a spectrum such that the adjoint $\varepsilon_n':E_n \to \Omega E_{n+1}$ of the inclusion $\varepsilon_n:SE_n \to E_{n+1}$ is always a weak homotopy equivalence, then we call E an *Ω-spectrum.*

8.42. Theorem. *If E is an Ω-spectrum, then for every CW-complex (X,x_0) we have a natural isomorphism $E^n(X) \cong [X,x_0; E_n, *]$.*

Proof: We define $k^n(X)$ to be $[X,x_0;\, E_n,*]$ and $\bar{\sigma}^n: k^{n+1}(SX) \to k^n(X)$ to be the natural equivalence

$$[SX,*;\, E_{n+1},*] \xrightarrow[A]{\cong} [X,x_0;\, \Omega E_{n+1},\omega_0] \xleftarrow[\varepsilon'_{n*}]{\cong} [X,x_0;\, E_n,*].$$

Since for any pair (X,A,x_0) the sequence

$$[A,x_0;\, E_n,*] \xleftarrow{i^*} [X,x_0;\, E_n,*] \xleftarrow{j^*} [X \cup CA,*;\, E_n,*]$$

is exact, it follows k^* is a cohomology theory on $\mathscr{PW}'$. Since for any collection $\{X_\alpha : \alpha \in A\}$

$$\{i_\alpha^*\} : [\vee_\alpha X_\alpha,*;\, E_n,*] \to \prod_\alpha [X_\alpha,x_\alpha;\, E_n,*]$$

is an isomorphism, k^* satisfies the wedge axiom. We define a natural transformation $T^n: k^n \to E^n$ as follows: any map $f:(X,x_0) \to (E_n,*)$ defines a function $f': E'(X) \to E$, where

$$E'(X)_m = \begin{cases} * & m < n \\ S^{m-n} X & m \geqslant n \end{cases}$$

is a cofinal subspectrum of $\Sigma^{-n} E(X)$. Thus we get a map $\bar{f}: \Sigma^{-n} E(X) \to E$ of spectra, and we take

$$T([f]) = [\bar{f}] \in [\Sigma^{-n} E(X), E] \cong [E(X), \Sigma^n E] = E^n(X).$$

Since the diagram

$$\begin{array}{ccccc}
[\Sigma^{-n-1} E(SX), E] & & \xleftarrow{\quad\cong\quad} & & [\Sigma^{-n} E(X), E] \\
\uparrow T^{n+1}(SX) & & [SX,*;\, SE_n,*] & & \uparrow T^n(X) \\
& \swarrow \varepsilon_{n*} & & \nwarrow \Sigma & \\
[SX,*;\, E_{n+1},*] & & & & [X,x_0;\, E_n,*] \\
& \searrow A \ (\cong) & & \swarrow \varepsilon'_{n*}\ (\cong) & \\
& & [X,x_0;\, \Omega E_{n+1},\omega_0] & &
\end{array}$$

commutes, we see that T^* is a natural transformation of cohomology theories.

Now from the commutativity of the square

$$\begin{array}{ccccc}
& & \pi_{k+1}(SE_{n+k},*) & & \\
& \nearrow \Sigma & & \searrow \varepsilon_{n+k*} & \\
\pi_k(E_{n+k},*) & & & & \pi_{k+1}(E_{n+k+1},*) \\
& \searrow \varepsilon'_{n+k*}\ (\cong) & & \nearrow A\ (\cong) & \\
& & \pi_k(\Omega E_{n+k+1},\omega_0) & &
\end{array}$$

we see that $[S^k,s_0;\, E_{n+k},*] \to \operatorname{dirlim}[S^{k+l},s_0;\, E_{n+k+l},*] \cong E^n(S^0)$ is an isomorphism—i.e. $T^n(S^0): k^n(S^0) \to E^n(S^0)$ is an isomorphism for all n. Thus by 7.67 T^* is a natural equivalence. $\square$

8.43. In Chapters 10, 11 and 12 we shall consider some important examples of spectra and their associated homology and cohomology theories. We close this chapter by mentioning an important spectrum which is already known to us: the sphere spectrum $S^0 = E(S^0)$. The associated homology theory S_*^0 is more usually known as *stable homotopy*: $S_*^0(X) = \pi_n(S^0 \wedge X) = \operatorname{dir\,lim} \pi_{n+k}(S^k \wedge X, *)$. The notation $\pi_n^S(X)$ is more usual for $S_n^0(X)$. The associated cohomology theory S^{0*} is called *stable cohomotopy* and denoted by π_S^*. By 7.72 $\pi_n^S(-) = \operatorname{dir\,lim} \pi_{n+q}(S^q \wedge -)$ is equivalent to S_*^0 on $\mathscr{PT}'$.

For any $n \geqslant 2$ we have the natural map

$$i_0 : \pi_n(X, x_0) \to \operatorname{dir\,lim} \pi_{n+k}(S^k X, *) = \pi_n^S(X)$$

sending $x \in \pi_n(X, x_0)$ to $\{x\} \in \operatorname{dir\,lim} \pi_{n+k}(S^k X, *)$. We can also define i_0 as follows: any map $f:(S^n, s_0) \to (X, x_0)$ defines a function $E(f): E(S^n) \to E(X)$; since $E(S^n)$ is a cofinal subspectrum of $\Sigma^n S^0$, we get a map $\{E(f)\}: \Sigma^n S^0 \to E(X)$ and

$$i_0[f] = [\{E(f)\}] \in [\Sigma^n S^0, E(X)] = \pi_n^S(X).$$

This definition of i_0 applies even for $n = 0$ or 1. For $n \geqslant 1$ i_0 is a homomorphism.

The coefficient groups $\pi_n^S(S^0) = \operatorname{dir\,lim} \pi_{n+k}(S^k, s_0)$ are the stable stems π_n^S mentioned following 6.28. These groups are known only through a finite range of $n > 0$ (note $\pi_n^S = 0$, $n < 0$, $\pi_0^S \cong \mathbb{Z}$). The attempt to compute π_*^S represents one of the important areas of research in algebraic topology at present.

Remark. The construction at the end of Chapter 7 allows us to define a reduced homology theory $E_*(-)$ and a reduced cohomology theory $E^*(-)$ on $\mathscr{PT}'$ for every spectrum E. Let us establish here the notational convention which will prevail in all subsequent chapters. For any *space* $X \in \mathscr{PT}'$ we will denote by $\tilde{E}_*(X)$ the reduced homology $E_*(X') = \pi_*(E \wedge X')$, where X' is any CW-substitute for X. For any pair (X, A) we will denote by $E_*(X, A)$ the unreduced homology $\tilde{E}_*(X^+ \cup CA^+)$. For any (unpointed) space X we shall use the abbreviation $E_*(X)$ for $E_*(X, \varnothing)$. This convention at least has the advantage of agreeing with standard conventions long in use.

Comments

After this book was written I received a copy of Adams' Chicago Lecture Notes "Stable Homotopy and Generalised Homology". Influenced by those notes I have thought it worthwhile to rewrite parts of Chapter 13 but have left Chapter 8 as it was. The reader, however, will certainly want to have a look at Adams' treatment of spectra in those notes for comparison.

References

1. J. F. Adams [8]
2. M. Boardman [21]
3. G. W. Whitehead [93]

Chapter 9

Representation Theorems

In Chapter 8 we saw how to associate a homology theory and a cohomology theory (both satisfying the wedge axiom) to a spectrum E. In this chapter we shall prove a converse result: given a cohomology theory k^* satisfying the wedge axiom on $\mathscr{PW}'$ we shall construct a spectrum E and a natural equivalence of cohomology theories $T:\tilde{E}^* \to k^*$ on $\mathscr{PW}'$. In fact, we shall do somewhat more than that; for any cofunctor $F^*:\mathscr{PW}' \to \mathscr{PS}$ satisfying the wedge axiom and a suitable exactness axiom we shall find a CW-complex (Y,y_0) and a natural equivalence $T:[-;\ Y,y_0] \to F^*$. We shall also prove such a theorem for cofunctors F^* defined only on the category $\mathscr{PW}'_F$ of finite CW-complexes provided F^* takes values in $\mathscr{G}$. Y is called a *classifying space* for F^*.

The basic idea is this; if $(Y,y_0) \in \mathscr{PW}'$ and $u \in F^*(Y)$, then we have a natural transformation $T_u:[-;\ Y,y_0] \to F^*$ given by $T_u[f] = f^*(u) \in F^*(X)$ for any $f:(X,x_0) \to (Y,y_0)$. The problem is to construct (Y,y_0) and $u \in F^*(Y)$ so that T_u is an equivalence.

We introduce the following conventions for the remainder of this chapter. Since we are only concerned with cofunctors here, we shall drop the star and simply write F for a cofunctor. If $A \subset X$ with inclusion $i:A \to X$ and $u \in F(X)$, then we shall often denote $i^*(u) \in F(A)$ by $u|A$.

The two axioms we want our cofunctors F to satisfy on $\mathscr{PW}'$ are the following:

W) (*Wedge Axiom.*) For an arbitrary wedge $\vee_\alpha X_\alpha$ in $\mathscr{PW}'$ with inclusions $i_\beta:X_\beta \to \vee_\alpha X_\alpha$ the induced morphism $\{i_\alpha^*\}:F(\vee_\alpha X_\alpha) \to \prod_\alpha F(X_\alpha)$ is a bijection.

MV) (*Mayer–Vietoris Axiom.*) For any CW-triad $(X;\ A_1,A_2)$ (i.e. A_1,A_2 subcomplexes of X with $X = A_1 \cup A_2$) and for any $x_1 \in F(A_1)$, $x_2 \in F(A_2)$ with $x_1|A_1 \cap A_2 = x_2|A_1 \cap A_2$ there is a $y \in F(X)$ with $y|A_1 = x_1$, $y|A_2 = x_2$.

9.1. Proposition. *For any $(Y,y_0) \in \mathscr{PW}$ the cofunctor $F = [-;\ Y,y_0]$ satisfies W) and MV).*

Proof: W) An element of $\prod_\alpha F(X_\alpha)$ is an indexed family $\{[f_\alpha]\}$ of homotopy classes of maps $f_\alpha:(X_\alpha,x_\alpha) \to (Y,y_0)$. The f_α's define a map $f:\vee_\alpha X_\alpha \to Y$ with $f \circ i_\alpha = f_\alpha$; that is $\{i_\alpha^*\}[f] = \{[f_\alpha]\}$. Hence $\{i_\alpha^*\}$ is surjective.

If two elements $[f]$, $[g] \in [\vee_\alpha X_\alpha, *; Y, y_0]$ have the same image under $\{i_\alpha^*\}$, then there is an indexed family of homotopies $H^\alpha: X_\alpha \wedge I^+ \to Y$ with $H_0^\alpha = f \circ i_\alpha$, $H_1^\alpha = g \circ i_\alpha$. Since $\vee_\alpha(X_\alpha \wedge I^+) \cong (\vee_\alpha X_\alpha) \wedge I^+$, we get a homotopy $H:(\vee_\alpha X_\alpha) \wedge I^+ \to Y$ with $H \circ (i_\alpha \wedge 1) = H^\alpha$, and $H_0 = f$, $H_1 = g$. Thus $[f] = [g]$—i.e. $\{i_\alpha^*\}$ is injective.

MV) Suppose $X = A_1 \cup A_2$, A_1, A_2 subcomplexes of X and $[f_1] \in [A_1, x_0; Y, y_0]$, $[f_2] \in [A_2, x_0; Y, y_0]$ are elements with $[f_1]|A_1 \cap A_2 = [f_2]|A_1 \cap A_2$; in other words, $f_1|A_1 \cap A_2 \simeq f_2|A_1 \cap A_2$. The inclusion $A_1 \cap A_2 \to A_1$ is a cofibration, so if $H:(A_1 \cap A_2) \times I \to Y$ is a homotopy from $f_1|A_1 \cap A_2$ to $f_2|A_1 \cap A_2$, then we can extend H to a homotopy $\bar{H}: A_1 \times I \to Y$ with $\bar{H}_0 = f_1$. Then $f_1' = \bar{H}_1$ is another representative for $[f_1] \in [A_1, x_0; Y, y_0]$ and $f_1'|A_1 \cap A_2 = f_2|A_1 \cap A_2$. Hence we have a map $g:(X,x_0) \to (Y,y_0)$ with $g|A_1 = f_1'$, $g|A_2 = f_2$. That is, the element $[g] \in [X, x_0; Y, y_0]$ has the property that $[g]|A_1 = [f_1]$, $[g]|A_2 = [f_2]$. □

Our goal is to prove the converse of 9.1. We assume from now on that $F:\mathscr{PW}' \to \mathscr{PS}$ is a cofunctor satisfying *W)* and *MV)*.

9.2. Lemma. *$F(\{x_0\})$ contains only one element for any one-point space $\{x_0\}$.*

Proof: $F(\{x_0\}) = F(\{x_0\} \vee \{x_0\}) \cong F(\{x_0\}) \times F(\{x_0\})$ and the map is given by $a \mapsto (a,a)$. This is only possible if $F(\{x_0\})$ has but one element. □

9.3. Lemma. If $(X,x_0) \in \mathscr{PW}'$ and $\{x_0\} = X_{-1} \subset X_0 \subset \cdots \subset X_n \subset \cdots \subset X$ is any sequence of subcomplexes with $X = \bigcup_n X_n$, then

$$\{i_n^*\}: F(X) \to \operatorname{inv\,lim} F(X_n)$$

is a surjection.

Remark. One defines the inverse limit for sets in an obvious fashion.

Proof: We employ the telescope $X' = \bigcup_{n \geq -1} [n-1,n]^+ \wedge X_n$, and the proof very much resembles that of 7.66. We set

$$A_1 = \bigcup_{\substack{k \geq -1 \\ k \text{ odd}}} [k-1,k]^+ \wedge X_k \subset X'$$

$$A_2 = \bigcup_{\substack{k \geq 0 \\ k \text{ even}}} [k-1,k]^+ \wedge X_k \subset X',$$

getting

$$X' = A_1 \cup A_2, \quad A_1 \cap A_2 = \vee_k X_k,$$

$$A_1 \simeq \vee_{k \text{ odd}} X_k \quad \text{and} \quad A_2 \simeq \vee_{k \text{ even}} X_k.$$

Given any $\{x_n\} \in \operatorname{invlim} F(X_n)$ we can (using the wedge axiom) find $y_1 \in F(A_1)$ with $y_1|X_k = x_k$, k odd, and $y_2 \in F(A_2)$ with $y_2|X_k = x_k$, k even. Then consider $y_1|A_1 \cap A_2$: for k odd $y_1|X_k = x_k$. For k even $y_1|X_k = j_k^*(y_1|X_{k+1}) = j_k^*(x_{k+1}) = x_k$, $j_k: X_k \to X_{k+1}$ the inclusion. Similarly for $y_2|A_1 \cap A_2$. Thus $y_1|A_1 \cap A_2 = y_2|A_1 \cap A_2$. Hence by MV) there is a $y' \in F(X')$ with $y'|A_1 = y_1$, $y'|A_2 = y_2$. Then $y'|X_k = x_k$, $k \geqslant -1$. But $X' \simeq X$; so there is a $y \in F(X)$ with $y|X_k = x_k$, $k \geqslant -1$. □

9.4. Proposition. *For any map $f:(X,x_0) \to (Y,y_0)$ in $\mathscr{PW}$ the sequence*

$$F(X) \xleftarrow{f^*} F(Y) \xleftarrow{j^*} F(Y \cup_f CX)$$

is exact.

Proof: Clearly $f^* \circ j^* = (j \circ f)^* = 0$. Suppose $y \in F(Y)$ is an element with $f^*(y) = 0$. In $Y \cup_f CX$ we take $A_1 = [0,1/2] \wedge X$, $A_2 = [1/2,1]^+ \wedge X \cup Y$. Then $A_1 \cap A_2 = \{1/2\}^+ \wedge X \cong X$, $A_2 \simeq Y$ and the inclusion $A_1 \cap A_2 \to A_2$ is essentially f. Let $y_1 \in F(A_1)$ be 0, $y_2 \in F(A_2) \cong F(Y)$ be y. Then $y_2|A_1 \cap A_2 = y_1|A_1 \cap A_2$. Hence there is a $z \in F(Y \cup_f CX)$ with $z|Y = y$. □

9.5. *Remark.* From W) it readily follows that $F(K)$ is a group whenever K is an H-cogroup.

9.6. Definition. An element $u \in F(Y)$ is called *n-universal* if

$$T_u: [S^q, s_0;\ Y, y_0] \to F(S^q)$$

is an isomorphism for $q < n$ and an epimorphism for $q = n$. u is called *universal* if it is n-universal for all $n \geqslant 0$.

9.7. Lemma. *If $f:(Y,y_0) \to (Y',y_0')$ is a map in $\mathscr{PW}$ and $u \in F(Y)$, $u' \in F(Y')$ are universal elements such that $f^*u' = u$, then f induces an isomorphism $f_*: \pi_q(Y,y_0) \to \pi_q(Y',y_0')$.*

Proof: We have a commutative diagram

$$\begin{array}{ccc} \pi_q(Y,y_0) & \xrightarrow{f_*} & \pi_q(Y',y_0') \\ & T_u \searrow \cong \quad \cong \swarrow T_{u'} & \\ & F(S^q) & \end{array}$$

for all q. The assertion follows. □

We shall call any element $u \in F(Y)$ (-1)-universal.

9.8. Lemma. *For any $(Y,y_0) \in \mathscr{PW}$ and n-universal element $u_n \in F(Y)$ we* *can find a CW-complex Y' obtained from Y by attaching $(n+1)$-cells and an $(n+1)$-universal element $u_{n+1} \in F(Y')$ with $u_{n+1}| Y = u_n$.*

Proof: For each $\lambda \in F(S^{n+1})$ we take a copy S_λ^{n+1} of S^{n+1} and form $Y \vee (\vee_\lambda S_2^{n+1})$. If $n \geqslant 0$ then for every $\alpha \in \pi_n(Y, y_0)$ with $T_{U_n}(\alpha) = 0$ we choose a representative $f:(S^n, s_0) \to (Y, y_0)$ and attach an $(n+1)$-cell e_α^{n+1} by f, getting Y'. By W) there is a $v \in F(Y \vee \vee_\lambda S_\lambda^{n+1})$ with $v| Y = u_n$, $v|S_\lambda^{n+1} = \lambda$. If $g: \vee_\alpha S_\alpha^n \to Y \vee \vee_\lambda S_\lambda^{n+1}$ is the "big attaching map" for the $(n+1)$-cells e_α^{n+1}, then

$$F(\vee_\alpha S_\alpha^n) \xleftarrow{g^*} F(Y \vee \vee_\lambda S_\lambda^{n+1}) \xleftarrow{j^*} F(Y')$$

is exact. $g^*(v)|S_\alpha^n = T_{u_n}(\alpha) = 0$ for each α; since $\{i_\alpha^*\}: F(\vee_\alpha S_\alpha^n) \to \prod_\alpha F(S_\alpha^n)$ is bijective, it follows $g^*(v) = 0$. Hence there is a $u_{n+1} \in F(Y')$ with $u_{n+1}|(Y \vee \vee_\lambda S_\lambda^{n+1}) = v$. In particular $u_{n+1}| Y = u_n$ and $u_{n+1}|S_\lambda^{n+1} = \lambda$.

Now we must show that u_{n+1} is $(n+1)$-universal. We have a commutative diagram

$$\begin{array}{ccc} \pi_q(Y, y_0) & \xrightarrow{i_*} & \pi_q(Y', y_0) \\ & {\scriptstyle T_{u_n}} \searrow \quad \swarrow {\scriptstyle T_{u_{n+1}}} & \\ & F(S^q) & \end{array}$$

in which i_* is an isomorphism for $q < n$ and an epimorphism for $q = n$ since Y' is obtained from Y by attaching $(n+1)$-cells. It follows $T_{u_{n+1}}$ is an isomorphism for $q < n$ and an epimorphism for $q = n$. Suppose $T_{u_{n+1}}(\beta) = 0$ for some $\beta \in \pi_n(Y', y_0)$; since i_* is surjective for $q = n$, there is an $\alpha \in \pi_n(Y, y_0)$ with $i_*(\alpha) = \beta$. Then $T_{u_n}(\alpha) = T_{u_{n+1}} i_*(\alpha) = T_{u_{n+1}}(\beta) = 0$. Therefore there is a cell e_α^{n+1} in Y', which shows $i_*(\alpha) = 0$—that is, $\beta = 0$. Hence $T_{u_{n+1}}$ is a monomorphism for $q = n$. Finally, for every $\lambda \in F(S^{n+1})$ we have $T_{u_{n+1}}([i_\lambda]) = i_\lambda^*(u_{n+1}) = \lambda$, where $i_\lambda: S_\lambda^{n+1} \to Y'$ is the inclusion. Hence $T_{u_{n+1}}$ is an epimorphism for $q = n+1$. ▯

9.9. Corollary. *For any $(Y, y_0) \in \mathscr{PW}$ and $v \in F(Y)$ we can find a CW-complex Y' containing Y as a subcomplex and a universal element $u \in F(Y')$ with $u| Y = v$.*

Proof: We take $Y_{-1} = Y$, $u_{-1} = v$ and then use the lemma inductively to find a sequence $Y = Y_{-1} \subset Y_0 \subset Y_1 \subset \cdots \subset Y_n \subset \cdots$ of CW-complexes with Y_n obtained from Y_{n-1} by attaching n-cells and elements $u_n \in F(Y_n)$ with $u_n| Y_{n-1} = u_{n-1}$ and u_n n-universal. We take $Y' = \bigcup_{n \geqslant -1} Y_n$ with the weak topology and use 9.3 to find a $u \in F(Y')$ with $u| Y_n = u_n$.

In the commutative diagram

$$\begin{array}{ccc} \pi_q(Y_n, y_0) & \xrightarrow{i_{n*}} & \pi_q(Y', y_0) \\ & T_{u_n} \searrow \quad \swarrow T_u & \\ & F(S^q) & \end{array}$$

i_{n*} is an isomorphism for $q < n$, since Y' is obtained from Y_n by attaching cells of dimensions greater than n. Thus $T_u:\pi_q(Y',y_0) \to F(S^q)$ is an isomorphism. This holds for all q, so u is universal. ▯

9.10. Corollary. *There exists a CW-complex $(Y,y_0) \in \mathcal{PW}$ and universal element $u \in F(Y)$.*

Proof: In 9.9 take $Y = \{y_0\}$, $v =$ base point in $F(Y)$. Then the Y' of 9.9 is the desired CW-complex and u is the universal element. ▯

From this point on the proof of the representation theorem reminds one of the proof of the theorem of J. H. C. Whitehead (6.32).

9.11. Lemma. *Let Y be a space with universal element $u \in F(Y)$, (X,A,x_0) a CW-pair, $g:(A,x_0) \to (Y,y_0)$ a cellular map and $v \in F(X)$ an element with $v|A = g^*(u)$. Then there exists a cellular map $h:(X,x_0) \to (Y,y_0)$ with $h|A = g$ and $v = h^*(u)$.*

Proof: Let T be the space constructed from $(I^+ \wedge A) \vee X \vee Y$ by identifying $[0,a] \in I^+ \wedge A$ with $a \in X$ and $[1,a] \in I^+ \wedge A$ with $g(a) \in Y$. Let A_1, $A_2 \subset T$ be the subcomplexes given by

$$A_1 = ([0,1/2]^+ \wedge A) \cup X, \; A_2 = ([1/2,1]^+ \wedge A) \cup Y.$$

Then $A_1 \cup A_2 = T$, $A_1 \cap A_2 = \{1/2\} \times A \cong A$, there is a strong deformation retraction $f:A_1 \to X$, Y is a strong deformation retract of A_2. Thus there are a $\bar{v} \in F(A_1)$ with $\bar{v}|X = v$ and a $\bar{u} \in F(A_2)$ with $\bar{u}|Y = u$. Clearly $\bar{v}|A_1 \cap A_2 = f^*(v|A) = f^*g^*(u) = \bar{u}|A_1 \cap A_2$, so by MV) there is a $w \in F(T)$ with $w|X = v$, $w|Y = u$.

By 9.8 we can embed T in a CW-complex Y' and find a universal element $u' \in F(Y')$ with $u'|T = w$. Let $j: Y \to Y'$ be the inclusion; then $j^*(u') = u'|Y = w|Y = u$, so by 9.7 $j_*:\pi_*(Y,y_0) \cong \pi_*(Y',y_0')$.

Now let $\bar{g}:X \to Y'$ be the inclusion. We have $\bar{g}|A \simeq j \circ g$, for indeed the composite $I^+ \wedge A \subset T \to Y'$ is a homotopy from $\bar{g}|A$ to $j \circ g$. As $A \subset X$ is a cofibration, we can find a map $\bar{\bar{g}}:X \to Y'$ with $\bar{\bar{g}}|A = j \circ g$ and $\bar{\bar{g}} \simeq \bar{g}$.

$$\begin{array}{ccc} Y & \xrightarrow{j} & Y' \\ g \uparrow & & \uparrow \bar{\bar{g}} \\ A & \subset & X \end{array}$$

From the proofs of 6.29–6.30 it follows there is a map $h:X \to Y$ with $h|A = g, j \circ h \simeq \bar{\bar{g}} \simeq \bar{g}$.

Then we have $h^*(u) = h^*j^*(u') = \bar{g}^*(u') = v$. □

9.12. Theorem (E. H. Brown). *If $F:\mathscr{PW}' \to \mathscr{PS}$ is a cofunctor satisfying W) and MV), then there is a classifying space $(Y,y_0) \in \mathscr{PW}$ and universal element $u \in F(Y)$ such that $T_u:[-;\ Y,y_0] \to F$ is a natural equivalence.*

Proof: By 9.10 we know we can find a space $(Y,y_0) \in \mathscr{PW}$ and a universal element $u \in F(Y)$. We must show that $T_u:[X,x_0;\ Y,y_0] \to F(X)$ is bijective for all $(X,x_0) \in \mathscr{PW}$.

a) Suppose $v \in F(X)$. In 9.11 we take $A = \{x_0\}$, $g:(A,x_0) \to (Y,y_0)$ the unique map. Then we can find a map $h:(X,x_0) \to (Y,y_0)$ with $v = h^*(u) = T_u([h])$. Thus T_u is surjective.

b) Suppose $T_u[g_0] = T_u[g_1]$ for two elements g_0, $g_1:(X,x_0) \to (Y,y_0)$. We may as well assume the representatives g_0, g_1 are cellular. We let $X' = X \wedge I^+$ and $A' = X \wedge \{0,1\}^+$ and define $g:(A',*) \to (Y,y_0)$ by $g[x,0] = g_0(x)$, $g[x,1] = g_1(x)$, $x \in X$. We define $p:X' \to X$ by $p[x,t] = x$, $[x,t] \in X'$ and let $v = p^*g_0^*u \in F(X')$. Then $v|X \wedge \{0\}^+ = g_0^*u = g^*u|X \wedge \{0\}^+$ and $v|X \wedge \{1\}^+ = g_0^*u = T_u[g_0] = T_u[g_1] = g_1^*u = g^*u|X \wedge \{1\}^+$. Then we see that $g^*u = v|A$, so by 9.11 we can find an $h:X' \to Y$ with $h|A = g$. But then h is a homotopy from g_0 to g_1, so in fact T_u is injective. □

9.13. Theorem. *Let $F,F':\mathscr{PW}' \to \mathscr{PS}$ be two cofunctors with classifying spaces (Y,y_0), (Y',y_0') and universal elements u, u' respectively. If $T:F \to F'$ is a natural transformation, then there is a map $f:(Y,y_0) \to (Y',y_0')$, unique up to homotopy, such that the diagram*

$$\begin{array}{ccc}
[X,x_0;\ Y,y_0] & \xrightarrow{f_*} & [X,x_0;\ Y',y_0'] \\
\downarrow T_u(X) & & \downarrow T_{u'}(X) \\
F(X) & \xrightarrow{T(X)} & F'(X)
\end{array}$$

commutes for all $(X,x_0) \in \mathscr{PW}$.

Proof: We choose a class $[f] \in [Y,y_0;\ Y',y_0']$ such that

$$T(Y) \circ T_u(Y)([1_Y]) = T_{u'}(Y)([f]).\ \square$$

In particular it follows the classifying space (Y,y_0) is determined up to homotopy equivalence.

9.14. Exercise. Show that if $[-;\ Y,y_0]$ is a cofunctor to $\mathscr{G}$, then we can give (Y,y_0) the structure of an H-group so that the multiplication in $[-;\ Y,y_0]$ is given by this H-group structure.

We now turn to the case of a cofunctor F defined only on the category $\mathscr{PW}'_F$ of finite CW-complexes and satisfying MV) and the weak wedge axiom W_F):

$$(i_1^*, i_2^*): F(X_1 \vee X_2) \rightarrow F(X_1) \times F(X_2)$$

is bijective for all $(X_1, x_1), (X_2, x_2) \in \mathscr{PW}_F$. In addition we shall assume that F is a group-valued cofunctor—i.e. $F: \mathscr{PW}'_F \rightarrow \mathscr{G}$.

The proof of the representation theorem for cofunctors F on $\mathscr{PW}'_F$ must necessarily be somewhat different from the proof given above, since the classifying spaces Y will generally be infinite CW-complexes, and we can no longer consider $F(Y)$. The crucial idea here is to extend F to $\mathscr{PW}'$ by taking $\hat{F}(X) = \operatorname{inv\,lim} F(X_\alpha)$, where $(X, x_0) \in \mathscr{PW}$ and X_α runs over all finite subcomplexes of X containing x_0. Clearly $\hat{F}(X) = F(X)$ if $(X, x_0) \in \mathscr{PW}_F$. One might at first hope to be able to prove that $\hat{F}$ satisfies the strong wedge axiom and the Mayer–Vietoris axiom and hence apply 9.12 to find a classifying space Y. This is not possible, however; we shall see that $\hat{F}$ satisfies a stronger result than 9.3 and a weaker version of MV). These properties will be enough to permit us to find a classifying space Y for F.

9.15. Lemma. *$\hat{F}$ satisfies W).*

The proof is a tedious but straightforward manipulation with elements $\{x_\alpha\} \in \operatorname{inv\,lim} F(X_\alpha)$.

9.16. Lemma. *For any $(X, x_0) \in \mathscr{PW}$ and any directed system $\{X_\alpha\}$ of subcomplexes of X with $X = \cup X_\alpha$ we have that $\{i_\alpha^*\}: \hat{F}(X) \rightarrow \operatorname{inv\,lim} \hat{F}(X_\alpha)$ is an isomorphism.*

Again the proof is tedious but straightforward.

We are going to prove that F satisfies MV) for any triad $(X; A_1, A_2)$ for which $A_1 \cap A_2$ is a finite complex. The proof of this fact will require some preparation.

Suppose $\mathscr{C}$ is a category with the following properties:

i) the objects of $\mathscr{C}$ are non-empty sets;
ii) the morphisms of $\mathscr{C}$ are surjective functions;
iii) for any $X, Y \in \mathscr{C}$ there is at most one morphism $\alpha: X \rightarrow Y$ in $\mathscr{C}$;
iv) for any $X, Y \in \mathscr{C}$ there are an object $Z \in \mathscr{C}$ and morphisms $\alpha: Z \rightarrow X$, $\beta: Z \rightarrow Y$.

Then we can define $\operatorname{inv\,lim} \mathscr{C}$; we think of the set of objects of $\mathscr{C}$ as having a "partial ordering" given by $X \leqslant Y$ if there is a morphism $\alpha: Y \rightarrow X$ in $\mathscr{C}$ ($X \leqslant Y$ and $Y \leqslant X \nRightarrow X = Y$). Then $\mathscr{C}$ is an "inverse system" of sets and surjections. It is not clear, however, that $\operatorname{inv\,lim} \mathscr{C}$ is non-empty; for that we need another assumption.

We enlarge $\mathscr{C}$ to a category $\overline{\mathscr{C}}$ with the same objects but more morphisms: any function $f:X \to Y$ between objects of $\mathscr{C}$ will be a morphism of $\overline{\mathscr{C}}$ if and only if there are morphisms $\alpha:Z \to X$, $\beta:Z \to Y$ in $\mathscr{C}$ with $f \circ \alpha = \beta$. $\overline{\mathscr{C}}$ still possesses the properties i)–iv) and is clearly the largest category with these properties which has the same objects as $\mathscr{C}$. There is an obvious injection i: $\operatorname{invlim}\overline{\mathscr{C}} \to \operatorname{invlim}\mathscr{C}$, and in fact it is easy to see that i is a bijection. Thus it suffices to find a condition which guarantees that $\operatorname{invlim}\overline{\mathscr{C}} \neq \varnothing$.

9.17. Proposition. *If $\overline{\mathscr{C}}$ has only countably many equivalence classes, then* $\operatorname{invlim}\mathscr{C} = \operatorname{invlim}\overline{\mathscr{C}} \neq \varnothing$.

Proof: Let $Y_0, Y_1, \ldots, Y_n, \ldots$ be a countable collection of objects in $\mathscr{C}$ such that every $Z \in \mathscr{C}$ is equivalent to exactly one Y_i in $\overline{\mathscr{C}}$. By induction we can construct a sequence $X_0, X_1, \ldots$ in $\mathscr{C}$ and elements $x_k \in X_k$ such that
a) $X_k \leqslant X_{k+1}$, $k \geqslant 0$;
b) $Y_i \leqslant X_i$ for each $i \geqslant 0$;
c) if $\alpha:X_{k+1} \to X_k$ is the unique morphism, then $\alpha(x_{k+1}) = x_k$.

Now we define an element $x_Y \in Y$ for every object $Y \in \mathscr{C}$ as follows: there is a unique i such that Y is equivalent to Y_i in $\overline{\mathscr{C}}$; let $f: Y_i \to Y$ be the equivalence. There is a morphism $\alpha:X_i \to Y_i$ in $\mathscr{C}$ and we take $x_Y = f \circ \alpha(x_i)$. We show that the collection $\{x_Y\}$ is an element of $\operatorname{invlim}\mathscr{C}$. Suppose $\alpha:Z \to Y$ is a morphism in $\mathscr{C}$. There are integers i,j and equivalences $f:Y_i \to Y$, $g:Y_j \to Z$ in $\overline{\mathscr{C}}$. There are morphisms $\alpha_i:X_i \to Y_i$, $\alpha_j:X_j \to Y_j$ in $\mathscr{C}$. By definition $x_Y = f \circ \alpha_i(x_i)$, $x_Z = g \circ \alpha_j(x_j)$.

We consider two possible cases:
$i \leqslant j$: then there is a morphism $\alpha_{ij}:X_j \to X_i$ in $\mathscr{C}$ and $f \circ \alpha_i \circ \alpha_{ij}$, $\alpha \circ g \circ \alpha_j$ are two morphisms in $\overline{\mathscr{C}}$ from X_j to Y. Hence $f \circ \alpha_i \circ \alpha_{ij} = \alpha \circ g \circ \alpha_j$, so

$$x_Y = f \circ \alpha_i(x_i) = f \circ \alpha_i \circ \alpha_{ij}(x_j) = \alpha \circ g \circ \alpha_j(x_j) = \alpha(x_Z).$$

$j \leqslant i$: then there is a morphism $\alpha_{ji}:X_i \to X_j$ in $\mathscr{C}$ and $f \circ \alpha_i$, $\alpha \circ g \circ \alpha_j \circ \alpha_{ji}$ are two morphisms in $\overline{\mathscr{C}}$ from X_i to Y. Thus $f \circ \alpha_i = \alpha \circ g \circ \alpha_j \circ \alpha_{ji}$, so

$$x_Y = f \circ \alpha_i(x_i) = \alpha \circ g \circ \alpha_j \circ \alpha_{ji}(x_i) = \alpha \circ g \circ \alpha_j(x_j) = \alpha(x_Z).$$

Thus in either case $\alpha(x_Z) = x_Y$. This shows that $\{x_Y\}$ is an element of $\operatorname{invlim}\mathscr{C}$. □

9.18. Lemma. *$\hat{F}$ satisfies the following axiom: MV_F) For any CW-triad $(X;\, A,B)$ with $A \cap B$ finite and any $a \in \hat{F}(A)$, $b \in \hat{F}(B)$ with $a|A \cap B = b|A \cap B$ there is an $x \in \hat{F}(X)$ with $x|A = a$ and $x|B = b$.*

Proof: The set of all finite subcomplexes of A containing $A \cap B$ is cofinal in the set of all finite complexes of A; that is, for every finite subcomplex C of A there is a finite subcomplex C' with $A \cap B \subset C'$ and $C \subset C'$. It follows that $\hat{F}(A) = \operatorname{invlim} F(A_\alpha)$, where A_α runs over the finite subcomplexes of A containing $A \cap B$. Similarly $\hat{F}(B) = \operatorname{invlim} F(B_\beta)$, where B_β runs over the finite subcomplexes of B containing $A \cap B$, and $\hat{F}(X) = \operatorname{invlim} F(A_\alpha \cup B_\beta)$. Naturally $\hat{F}(A \cap B) = F(A \cap B)$.

Suppose given $a \in \hat{F}(A)$, $b \in \hat{F}(B)$ with $a|A \cap B = b|A \cap B$. If $a = \{a_\alpha\}$, $b = \{b_\beta\}$ then $a_\alpha|A \cap B = b_\beta|A \cap B$ for all α, β. Let $X_{\alpha\beta} \subset F(A_\alpha \cup B_\beta)$ be the set of all elements w with $w|A_\alpha = a_\alpha$, $w|B_\beta = b_\beta$; since F satisfies MV), $X_{\alpha\beta} \neq \varnothing$ for all α, β. Whenever $A_\alpha \subset A_\theta$ and $B_\beta \subset B_\phi$ we have an inclusion $A_\alpha \cup B_\beta \subset A_\theta \cup B_\phi$ which induces a morphism $X_{\theta\phi} \to X_{\alpha\beta}$. We take $\mathscr{C}$ to be the category whose objects are the sets $X_{\alpha\beta}$ and whose morphisms are the induced functions just described. The only one of the properties i)–iv) which is not immediately clear for $\mathscr{C}$ is ii); we must show the morphisms $X_{\theta\phi} \to X_{\alpha\beta}$ are surjections.

Just as in 9.4 the sequence $F(X) \xleftarrow{f^*} F(Y) \xleftarrow{j^*} F(Y \cup_f CX)$ is exact for any map $f:(X,x_0) \to (Y,y_0)$ in $\mathscr{PW}_F$. In the usual way one then shows that $F(X) \xleftarrow{f^*} F(Y) \xleftarrow{j^*} F(Y \cup_f CX) \xleftarrow{k^*} F(SX) \xleftarrow{Sf^*} F(SY) \leftarrow \cdots$ is exact and then deduces the exactness of the long Mayer–Vietoris sequence of a triad $(X; A, B)$ in $\mathscr{PW}_F$:

$$F(A \cap B) \leftarrow F(A) \times F(B) \leftarrow F(X) \leftarrow F(S(A \cap B)) \leftarrow$$
$$F(SA) \times F(SB) \cong F(SA \vee SB).$$

Thus whenever $A_\alpha \subset A_\theta$, $A_\gamma \subset A_\theta$, $B_\beta \subset B_\phi$, $B_\delta \subset B_\phi$ we have a commutative diagram of exact sequences

9.19.

$$\begin{array}{ccccccc}
F(A_\alpha) \times F(B_\beta) & \longleftarrow & F(A_\alpha \cup B_\beta) & \xleftarrow{\Delta_1} & F(S(A \cap B)) & \xleftarrow{g_{\alpha\beta}} & F(S(A_\alpha \vee B_\beta)) \\
\uparrow & & \uparrow j_1^* & & \| & & \uparrow \\
F(A_\theta) \times F(B_\phi) & \longleftarrow & F(A_\theta \cup B_\phi) & \xleftarrow{\Delta_2} & F(S(A \cap B)) & \xleftarrow{g_{\theta\phi}} & F(S(A_\theta \vee B_\phi)) \\
\downarrow & & \downarrow j_2^* & & \| & & \downarrow \\
F(A_\gamma) \times F(B_\delta) & \longleftarrow & F(A_\gamma \cup B_\delta) & \xleftarrow{\Delta_3} & F(S(A \cap B)) & \xleftarrow{g_{\gamma\delta}} & F(S(A_\gamma \vee B_\delta)).
\end{array}$$

Suppose $w \in X_{\alpha\beta}$; let w' be any element of $X_{\theta\phi}$. Then $w - j_1^* w'$ goes to 0 in $F(A_\alpha) \times F(B_\beta)$, so there is a $z \in F(S(A \cap B))$ with $\Delta_1 z = w - j_1^* w'$. Then $j_1^*(\Delta_2 z + w') = w - j_1^* w' + j_1^* w' = w$, showing that $j_1^*(X_{\theta\phi}) = X_{\alpha\beta}$.

Now an element of $\operatorname{invlim} \mathscr{C}$ is an element $x \in \operatorname{invlim} F(A_\alpha \cup B_\beta) = \hat{F}(X)$ such that $x|A = a$, $x|B = b$. Thus it suffices to show $\operatorname{invlim} \mathscr{C} \neq \varnothing$. We wish to apply 9.17, so we must show that $\overline{\mathscr{C}}$ has only finitely many equivalence classes. To this end we now prove the following statement:

there is a morphism $f:X_{\gamma\delta}\to X_{\alpha\beta}$ in $\overline{\mathscr{C}}$ if and only if $\operatorname{img}_{\gamma\delta}\subset\operatorname{img}_{\alpha\beta}$ (see diagram 9.19).

Suppose there is a morphism $f:X_{\gamma\delta}\to X_{\alpha\beta}$ in $\overline{\mathscr{C}}$; this means there is a pair (θ,ϕ) and morphisms $\mu:X_{\theta\phi}\to X_{\gamma\delta}$, $v:X_{\theta\phi}\to X_{\alpha\beta}$ in $\mathscr{C}$ with $f\circ\mu=v$. For any $w\in F(S(A_\gamma\vee B_\delta))$ we have $\Delta_3\circ g_{\gamma\delta}(w)=0$ in $F(A_\gamma\cup B_\delta)$. Let $x\in X_{\theta\phi}$ be arbitrary; then in $F(A_\alpha\cup B_\beta)$ we have $v(x)+\Delta_1 g_{\gamma\delta}(w)=v(x+\Delta_2 g_{\gamma\delta}(w))=f\circ\mu(x+\Delta_2 g_{\gamma\delta}(w))=f(\mu(x)+\Delta_3 g_{\gamma\delta}(w))=f\circ\mu(x)=v(x)$. Thus $\Delta_1 g_{\gamma\delta}(w)=0$, so there is a $w'\in F(S(A_\alpha\vee B_\beta))$ with $g_{\alpha\beta}(w')=g_{\gamma\delta}(w)$—that is, $\operatorname{img}_{\gamma\delta}\subset\operatorname{img}_{\alpha\beta}$.

Conversely suppose $\operatorname{img}_{\gamma\delta}\subset\operatorname{img}_{\alpha\beta}$; we construct $f:X_{\gamma\delta}\to X_{\alpha\beta}$ as follows: take $A_\theta=A_\alpha\cup A_\gamma$, $B_\phi=B_\beta\cup B_\delta$. For any $x\in X_{\gamma\delta}$ choose $x'\in X_{\theta\phi}$ with $j_2^*(x')=x$ and set $f(x)=j_1^*(x')$. We must show f is well defined; if $x''\in X_{\theta\phi}$ also satisfies $j_2^*(x'')=x$, then $x'-x''$ has image 0 in $F(A_\theta)\times F(B_\phi)$ and also $j_2^*(x'-x'')=0$. We can find a $u\in F(S(A\cap B))$ with $\Delta_2 u=x'-x''$. As $\Delta_3 u=j_2^*(x'-x'')=0$, there is a $w\in F(S(A_\gamma\vee B_\delta))$ with $g_{\gamma\delta}(w)=u$. Choose $w'\in F(S(A_\alpha\vee B_\beta))$ with $g_{\alpha\beta}(w')=g_{\gamma\delta}(w)=u$. Then $0=\Delta_1 g_{\alpha\beta}(w')=\Delta_1(u)=j_1^*(x'-x'')=j_1^*(x')-j_1^*(x'')$. This $j_1^*(x'')=j_1^*(x')$—that is, f is well defined. Clearly $f\circ(j_2^*|X_{\theta\phi})=j_1^*|X_{\theta\phi}$.

In particular $X_{\alpha\beta}\simeq X_{\gamma\delta}$ in $\overline{\mathscr{C}}$ if and only if $\operatorname{img}_{\alpha\beta}=\operatorname{img}_{\gamma\delta}$. Now there are only countably many homotopy equivalence classes of finite CW-complexes; for example we can choose a finite simplicial complex in each of these homotopy types. There are only countably many homotopy classes of maps of $S(A\cap B)$ into each of this collection of finite simplicial complexes (e.g. replace $S(A\cap B)$ by a finite simplicial complex; then each homotopy class of maps contains a simplicial map). Hence there can be at most countably many distinct sets $\operatorname{img}_{\alpha\beta}$, which shows that $\overline{\mathscr{C}}$ has only countably many equivalence classes. Thus by 9.17 $\operatorname{invlim}\mathscr{C}\neq\varnothing$. □

We shall call an element $u\in\hat{F}(Y)$ *n-universal* if $T_u:[S^q,s_0;Y,y_0]\to F(S^q)$ is an isomorphism for $q<n$ and an epimorphism for $q=n$. u is called *universal* if it is n-universal for all $n\geqslant 0$.

9.20. Lemma. *For any* $(Y,y_0)\in\mathscr{PW}$ *and n-universal element* $u_n\in\hat{F}(Y)$ *we can find a CW-complex* Y' *obtained from* Y *by attaching* $(n+1)$*-cells and an* $(n+1)$*-universal element* $u_{n+1}\in\hat{F}(Y')$ *with* $u_{n+1}|Y=u_n$.

Proof: We construct Y' as in the proof of 9.8, attaching $(n+1)$-cells e_α^{n+1} for each $\alpha\in\pi_n(Y,y_0)$ with $T_{u_n}(\alpha)=0$ and $(n+1)$-spheres S_λ^{n+1} for each $\lambda\in F(S^{n+1})$. Let S be the set of all pairs (Y'',u) such that Y'' is a subcomplex of Y' containing Y and $u\in\hat{F}(Y'')$ is an element with $u|Y=u_n$. We give S the obvious order relation: $(Y_1'',u_1)\leqslant(Y_2'',u_2)$ if and only if Y_1'' is a subcomplex of Y_2'' and $u_2|Y_1''=u_1$.

From 9.16 it follows S is inductive—that is, satisfies the hypothesis of Zorn's lemma. Hence S contains a maximal element (Y_0'',u_0). We show next that $Y_0''=Y'$.

Suppose Y_0'' did not include one of the $(n+1)$-cells e_α^{n+1} of $Y' - Y$. If $g:(S^n, s_0) \to Y \subset Y_0''$ is the attaching map of e_α^{n+1}, then by 9.18 the sequence $\hat{F}(S^n) \xleftarrow{g^*} \hat{F}(Y_0'') \xleftarrow{j^*} \hat{F}(Y_0'' \cup e_\alpha^{n+1})$ is exact; but $g^*(u_0) = g^*(u_n) = T_{u_n}(\alpha) = 0$, so there is a $u_1 \in \hat{F}(Y_0'' \cup e_\alpha^{n+1})$ with $u_1 | Y_0'' = u_0$, contradicting the maximality of (Y_0'', u_0). Similarly if Y_0'' omits one of the S_λ^{n+1}, then by W) we can find a $u_1 \in \hat{F}(Y_0'' \vee S_\lambda^{n+1})$ with $u_1 | Y_0'' = u_0$, again contradicting the maximality of (Y_0'', u_0). Hence $Y_0'' = Y'$.

We therefore take $u_{n+1} = u_0 \in \hat{F}(Y')$; the proof that u_{n+1} is $(n+1)$-universal goes exactly as in 9.8. $\square$

From this point on the proof of the representation theorem for F runs exactly as before, and we have the following theorem .

9.21. Theorem (J. F. Adams). *If $F:\mathscr{PW}'_F \to \mathscr{G}$ is a cofunctor satisfying W_F) and MV), then there is a classifying space $(Y, y_0) \in \mathscr{PW}$ and a natural equivalence $T:[-;\ Y, y_0] \to F$ on $\mathscr{PW}'_F$.*

In fact there is a universal element $u \in \hat{F}(Y)$ and T is given by

$$T_u:[X, x_0;\ Y, y_0] \to \hat{F}(X) = F(X) \text{ for } (X, x_0) \in \mathscr{PW}_F.$$

There is also a representation theorem for natural transformations on $\mathscr{PW}'_F$, but to formulate the uniqueness part of this theorem we need the following definition.

9.22. Two maps $f, g: X \to Y$ are called *weakly homotopic* if for any finite CW-complex Z and any map $h: Z \to X$ we have $f \circ h \simeq g \circ h$. We write $f \simeq_w g$. $\simeq_w$ is an equivalence relation and we write $[X, Y]_w$ or $[X, x_0;\ Y, y_0]_w$ for the set of all weak homotopy equivalence classes of maps $X \to Y$ or $(X, x_0) \to (Y, y_0)$.

9.23. Proposition. *There are natural equivalences*

$$\begin{aligned} \hat{F}(X) &\cong \mathrm{Nat}_{\mathscr{PW}'_F}([-;\ X, x_0], F) \\ &\cong \mathrm{Nat}_{\mathscr{PW}'}([-;\ X, x_0]_w, \hat{F}), \quad (X, x_0) \in \mathscr{PW}'. \end{aligned}$$

(For any cofunctors $G, H: \mathscr{C} \to \mathscr{D}$ we write $\mathrm{Nat}_{\mathscr{C}}(G, H)$ for the set of all natural transformations $T: G \to H$ on $\mathscr{C}$).

Proof: Given $u \in \hat{F}(X)$ we take $T_u:[-;\ X, x_0] \to F$ in $\mathrm{Nat}_{\mathscr{PW}'_F}([-;\ X, x_0], F)$. Given $T \in \mathrm{Nat}_{\mathscr{PW}'_F}([-;\ X, x_0], F)$ we take the element

$$\{T[i_\alpha]\} \in \mathrm{inv\,lim}\, F(X_\alpha) = \hat{F}(X),$$

where $i_\alpha: X_\alpha \to X$ is the inclusion. Given $u = \{u_\alpha\} \in \hat{F}(X) = \mathrm{inv\,lim}\, F(X_\alpha)$, we find $\{T_u[i_\alpha]\} = \{i_\alpha^*(u)\} = \{u_\alpha\} = u$. Thus $u \mapsto T_u \mapsto \{T_u[i_\alpha]\}$ is the identity on $\hat{F}(X)$. Given $T \in \mathrm{Nat}_{\mathscr{PW}'_F}([-;X, x_0], F)$, let $u = \{T[i_\alpha]\}$. For any $[f] \in [K, k_0; X, x_0]$ (K finite) we can find a β with $f(K) \subset X_\beta$; that is,

$f = i_\beta \circ f'$ for some $f':(K,k_0) \to (X_\beta, x_0)$. Thus $T_u[f] = f^*(u) = f'^* i_\beta^*(u) = f'^*(T[i_\beta]) = T(f'^*[i_\beta]) = T[i_\beta \circ f'] = T[f]$—that is, $T_u = T$. Hence

$$T \mapsto \{T[i_\alpha]\} \mapsto T_u$$

is the identity on $\mathrm{Nat}_{\mathscr{PW}'_F}([-;X,x_0],F)$.

The proof that $\hat{F}(X) \cong \mathrm{Nat}_{\mathscr{PW}'}([-;X,x_0]_w,\hat{F})$ is much simpler. □

This proposition indicates why $\hat{F}$ plays such a crucial role in the representation theorem.

Clearly $[X,x_0;Y,y_0]_w = [X,x_0;Y,y_0]$ if X is finite. If $u \in \hat{F}(Y)$ is the universal element of Thm. 9.21 then the natural transformation $T_u:[-;Y,y_0] \to F$ on $\mathscr{PW}'_F$ has an extension $\hat{T}_u:[-;Y,y_0]_w \to \hat{F}$ to $\mathscr{PW}'$. $\hat{T}_u$ is also a natural equivalence; for if $f,g:(X,x_0) \to (Y,y_0)$ are maps with $\hat{T}_u[f]_w = \hat{T}_u[g]_w$ and $h:(K,k_0) \to (X,x_0)$ is any map of a finite complex K to X, then $h^*(\hat{T}_u[f]_w) = h^*(\hat{T}_u[g]_w)$—that is $T_u[f \circ h] = T_u[g \circ h]$. Since T_u is a bijection, it follows $f \circ h \simeq g \circ h$—that is, $f \simeq_w g$, thus proving that $\hat{T}_u$ is injective. The proof that $\hat{T}_u$ is surjective is not difficult.

9.24. Theorem. *Let $F,F':\mathscr{PW}'_F \to \mathscr{G}$ be two cofunctors with classifying spaces (Y,y_0), (Y',y_0') and universal elements u, u' respectively. If $T:F \to F'$ is a natural transformation, then there is a map $f:(Y,y_0) \to (Y',y_0)$, unique up to weak homotopy, such that the diagram*

$$\begin{array}{ccc} [X,x_0;\ Y,y_0] & \xrightarrow{f_*} & [X,x_0;\ Y',y_0'] \\ \Big\downarrow T_u(X) & & \Big\downarrow T_{u'}(X) \\ F(X) & \xrightarrow{T(X)} & F'(X) \end{array}$$

commutes for all $(X,x_0) \in \mathscr{PW}'_F$.

Proof: There is an obvious extension of T to $\hat{T}:\hat{F} \to \hat{F}'$ on $\mathscr{PW}'$; in fact $\hat{T}(X) = \mathrm{inv\,lim}\, T(X_\alpha)$ for $(X,x_0) \in \mathscr{PW}'$. We choose a class

$$[f]_w \in [Y,y_0;\ Y',y_0']_w$$

so that $\hat{T}(Y) \circ \hat{T}_u(Y)([1_Y]_w) = \hat{T}_{u'}(Y)([f]_w)$. Then the diagram

$$\begin{array}{ccc} [X,x_0;\ Y,y_0]_w & \xrightarrow{f_*} & [X,x_0;\ Y',y_0']_w \\ \Big\downarrow \hat{T}_u(X) & & \Big\downarrow \hat{T}_{u'}(X) \\ \hat{F}(X) & \xrightarrow{\hat{T}(X)} & \hat{F}'(X) \end{array}$$

commutes for all $(X,x_0) \in \mathscr{PW}'$ and reduces to the diagram of 9.24 if X is finite. □

It is clear that we should expect f to be determined only up to weak homotopy.

9.25. Exercise. Suppose $F:\mathscr{PW}' \to \mathscr{G}$ is a cofunctor satisfying W) and MV). Show that the natural morphism $F(X) \to \operatorname{invlim} F(X_\alpha)$ is surjective, where $(X,x_0) \in \mathscr{PW}'$ and X_α runs over all finite subcomplexes of X containing x_0.

9.26. Exercise. Define the notion *weak H-group* and show that the classifying space (Y,y_0) of 9.21 can be given a weak H-group structure so that $T_u:[-;Y,y_0] \to F$ is a natural equivalence of cofunctors from $\mathscr{PW}'_F$ to $\mathscr{G}$.

Now suppose k^* is a reduced cohomology theory defined either on $\mathscr{PW}'$ and satisfying the wedge axiom or on $\mathscr{PW}'_F$.

9.27. Theorem. *There is an Ω-spectrum E and a natural equivalence $T:\tilde{E}^* \to k^*$.*

Proof: Let $F = k^n:\mathscr{PW}' \to \mathscr{A}$ (or $\mathscr{PW}'_F \to \mathscr{A}$). By 7.19* F satisfies MV). By assumption F satisfies W) if its domain is $\mathscr{PW}'$; if the domain is $\mathscr{PW}'_F$, then W_F) follows from 7.36*. Hence we can find a classifying space E_n for $F = k^n$ and a universal element $u_n \in k^n(E_n)$ (or $u_n \in \hat{F}(E_n)$).

Now we have natural equivalences

$$[X,x_0;\Omega E_{n+1},\omega_0] \xrightarrow{A^{-1}} [SX,*;E_{n+1},*] \xrightarrow{T_{u_{n+1}}} k^{n+1}(SX) \xrightarrow{\sigma} k^n(X).$$

Therefore there are maps $\phi_n:E_n \to \Omega E_{n+1}$ and $\psi_n:\Omega E_{n+1} \to E_n$ representing $A \circ T_{u_{n+1}}^{-1} \circ \sigma^{-1}$ and $\sigma \circ T_{u_{n+1}} \circ A^{-1}$ respectively. Then $\psi_n \circ \phi_n$ and $\phi_n \circ \psi_n$ represent the identity transformations, and hence $\psi_n \circ \phi_n \simeq 1$, $\phi_n \circ \psi_n \simeq 1$ (or at least $\phi_n \circ \psi_n \simeq_w 1$, $\psi_n \circ \phi_n \simeq_w 1$).
In the latter case ϕ_n, ψ_n are weak homotopy equivalences. Thus $\{E_n,\phi_n\}$ defines an Ω-spectrum.

The following diagram commutes

$$\begin{array}{ccccc}
k^{n+1}(SX) & & \xrightarrow{\sigma} & & k^n(X) \\
\uparrow T_{u_{n+1}} & & [SX,*;SE_n,*] & & \uparrow T_{u_n} \\
& \swarrow \varepsilon_{n*} & & \nwarrow \Sigma & \\
[SX,*;E_{n+1},*] & & & & [X,x_0;E_n,*] \\
& \searrow A & & \swarrow \phi_{n*} & \\
& & [X,x_0;\Omega E_{n+1},\omega_0] & &
\end{array}$$

if $\varepsilon_n:SE_n \to E_{n+1}$ is the adjoint of ϕ_n. This shows that the natural transformations T_{u_n} do define a natural equivalence of cohomology theories. $\square$

Of course we also have a representation theorem for natural transformations of cohomology theories on $\mathscr{PW}'$.

9.28. Theorem. *If E, E' are Ω-spectra and $T: \tilde{E}^* \to \tilde{E}'^*$ is a natural transformation of cohomology theories on $\mathscr{PW}'$, then there is a map $f: E \to E'$ of spectra such that $T = T_f$.*

Unfortunately, the obvious attempt to prove a corresponding result for natural transformations on $\mathscr{PW}'_F$ fails; we do indeed get maps $f_n: E_n \to E'_n$ inducing $T_n: \tilde{E}^n(-) \to \tilde{E}'^n(-)$, but the diagrams

$$\begin{array}{ccc} E_n & \xrightarrow{f_n} & E'_n \\ \phi_n \downarrow & & \downarrow \phi'_n \\ \Omega E_{n+1} & \xrightarrow{\Omega f_{n+1}} & \Omega E'_{n+1} \end{array}$$

then commute only up to weak homotopy and not up to homotopy as required to give a map of spectra (see 10.4).

As the representation of transformations is important, we attack this problem from a slightly different angle. We can consider cohomology theories k^* defined on the category $\mathscr{SP}_F$ of finite spectra. We define $\hat{k}^*(X) = \operatorname{invlim} k^*(X_\alpha)$ for $X \in \mathscr{SP}$, where X_α runs over the finite subspectra of X. Then we have natural equivalences

$$\hat{k}^*(E) \cong \operatorname{Nat}_{\mathscr{SP}'_F}(E^*, k^*) \cong \operatorname{Nat}_{\mathscr{SP}'}([-;E]^*_w, \hat{k}^*),$$

where weak homotopy of maps between spectra is defined in the obvious way and $[-;E]^*_w$ denotes the collection of cofunctors $[-;E]^n_w = [-, \Sigma^n E]_w$ for $n \in \mathbb{Z}$.

$\hat{k}^*$ is of course not a cohomology theory, but it does satisfy the analogs of 9.16 and MV_F). An element $u \in \hat{k}^0(E)$ is called *n-universal* if $T_u: \pi_q(E) = E^0(S^q) \to k^0(S^q)$ is an isomorphism for $q < n$, epimorphism for $q = n$.

9.29. Lemma. *Suppose $k^q(S^0) = 0$ for $q > 0$. If F is any spectrum and $v \in \hat{k}^0(F)$ is n-universal, $n \geqslant 0$, then we can find a spectrum E containing F as subspectrum and an $(n+1)$-universal element $u \in \hat{k}^0(E)$ with $u|F = v$.*

Proof: For every $\alpha \in F^0(S^n)$ with $T_u(\alpha) = 0$, we choose a representing map $f_\alpha: \Sigma^n S^0 \to F$ and attach a cell e^{n+1}_α to F. If the result is F', then we take $E = F' \vee (\bigvee_\lambda \Sigma^{n+1} S^0)$, where λ runs over $\hat{k}^0(S^{n+1}) = k^0(S^{n+1})$. The construction of u and the proof that it is $(n+1)$-universal goes as in 9.20. $\square$

Now we can prove the analog of 9.11 and hence the existence of a natural equivalence $\hat{T}_u: [-;E]^*_w \to \hat{k}^*$ in the case $k^q(S^0) = 0$, $q > 0$.

9.30. Theorem. If *k^*, k'^* are cohomology theories defined on $\mathscr{S}\mathscr{P}'_F$* with $k^q(S^0) = k'^q(S^0) = 0$, $q > 0$, *and with classifying spectra E, E' and universal elements u, u' respectively, and if $T: k^* \to k'^*$ is a natural transformation, then there is a map of spectra $f: E \to E'$, unique up to weak homotopy, such that the diagram*

$$\begin{array}{ccc} E^* & \xrightarrow{T_f} & E'_* \\ {\scriptstyle T_u}\downarrow & & \downarrow{\scriptstyle T_{u'}} \\ k^* & \xrightarrow{T} & k'^* \end{array}$$

commutes.

Comments

The Brown Representation Theorem represented a turning point in algebraic topology, because it made clear that many of the most important functors in algebraic topology were essentially homotopy functors and hence accessible to the methods of homotopy theory. Theorem 9.27 shows that cohomology theory is to a large extent a branch of stable homotopy theory. We shall encounter specific representing complexes and spectra in Chapters 10, 11 and 12, and in Chapter 14 we shall prove a representation theorem for homology theories.

References

1. J. F. Adams [9]
2. E. H. Brown [27]
3. E. H. Spanier [80]
4. G. W. Whitehead [93]

Chapter 10

Ordinary Homology Theory

By an *ordinary* homology theory k_* we shall mean one with $k_n(S^0) = 0$ unless $n = 0$. If $k_0(S^0) = G$, then k_* will be called an ordinary homology theory with *coefficients* G. Reduced singular homology $\tilde{H}_*(-;G)$ is an ordinary homology theory with coefficients G on the category $\mathscr{PT}'$. We shall show that any two ordinary homology theories with coefficients G satisfying the wedge and WHE axioms are naturally equivalent. We shall also construct the Eilenberg–MacLane spectrum $H(G)$ with

$$\pi_n(H(G)) = \begin{cases} G & n = 0 \\ 0 & n \neq 0. \end{cases}$$

The resulting homology theory $H(G)_*(-)$ is then an ordinary homology theory with coefficients G, so that on $\mathscr{PT}'$ we have a natural equivalence $H(G)_*(-) \cong \tilde{H}_*(-;G)$. Finally we establish a sufficient condition for the Hurewicz homomorphism $h:\pi_n(Y,y_0) \to H_n(Y;\mathbb{Z})$ (cf. 7.39) to be an isomorphism.

10.1. For each $n \geqslant 0$ we take the *standard n-simplex* $\Delta_n \subset \mathbb{R}^{n+1}$ to be the simplex with vertices $e_0 = (1,0,\ldots,0)$, $e_1 = (0,1,\ldots,0)$, ..., $e_n = (0,\ldots,0,1)$. For each $i, 0 \leqslant i \leqslant n$, let $\partial_i:\Delta_{n-1} \to \Delta_n$ be the map defined by

$$\partial_i(e_k) = \begin{cases} e_k & k < i \\ e_{k+1} & k \geqslant i \end{cases}$$

on the vertices and extended to be affine linear, i.e.

$$\partial_i(\textstyle\sum_{j=0}^{n-1} \lambda_j e_j) = \sum_{j=0}^{n-1} \lambda_j \, \partial_i(e_j) \quad \text{for all } \lambda_0, \ldots, \lambda_{n-1} \quad \text{with}$$

$$\lambda_j \geqslant 0,\ 0 \leqslant j \leqslant n-1,\quad \textstyle\sum_{j=0}^{n-1} \lambda_j = 1.$$

A *singular n-simplex* in a topological space X is a map $u:\Delta_n \to X$. The ith *face* $\partial_i u$ of u is the singular $(n-1)$-simplex $u \circ \partial_i:\Delta_{n-1} \to X$.

Let $S_n(X)$ = the free abelian group generated by the singular n-simplices of X. $S_n(X)$ is called the group of *n-chains* of X. Let $d:S_n(X) \to$

$S_{n-1}(X)$ be the homomorphism defined by $du = \sum_{i=0}^{n} (-1)^i \partial_i u$ on the basis. d is called the *differential*. One verifies that $d \circ d = 0$, so that $\{S_n(X), d\}$ becomes a *chain complex*. We define

$$Z_n(X) = \ker\{d: S_n(X) \to S_{n-1}(X)\} \subset S_n(X)$$
$$B_n(X) = \operatorname{im}\{d: S_{n+1}(X) \to S_n(X)\} \subset S_n(X).$$

$Z_n(X)$ is the group of *cycles*, $B_n(X)$ the group of *boundaries*. Since $d \circ d = 0$, $B_n(X) \subset Z_n(X)$. The *singular homology group* $H_n(X)$ is defined to be $H_n(X) = Z_n(X)/B_n(X)$, $n \geqslant 0$, $= 0$ for $n < 0$.

If $f: X \to Y$ is a map, then we can define a homomorphism $f_\#: S_n(X) \to S_n(Y)$ for $n \geqslant 0$ by taking $f_\#(u) = f \circ u: \Delta_n \to Y$ on the basis. Then

$$d \circ f_\#(u) = \textstyle\sum_{i=0}^{n} (-1)^i (f \circ u) \circ \partial_i = \sum_{i=0}^{n} (-1)^i f_\#(\partial_i u)$$
$$= f_\#(\textstyle\sum_{i=0}^{n} (-1)^i \partial_i u) = f_\#(du)$$

for all singular simplices u. Hence $d \circ f_\# = f_\# \circ d$—i.e. $f_\#: \{S_*(X), d\} \to \{S_*(Y), d\}$ is a *chain map*. Since

$$f_\#(Z_n(X)) \subset Z_n(Y)$$
$$f_\#(B_n(X)) \subset B_n(Y)$$

$f_\#$ induces a homomorphism $f_*: H_n(X) \to H_n(Y)$ for all $n \in \mathbb{Z}$.

If $(X, A) \in \mathcal{T}^2$ we define $S_n(X, A)$ to be the quotient $S_n(X)/S_n(A)$. $d: S_n(X) \to S_{n-1}(X)$ induces a differential d on $S_*(X, A)$, and we define

$$H_n(X, A) = \ker d/\operatorname{im} d = \begin{cases} Z_n(X, A)/B_n(X, A) & n \geqslant 0, \\ 0 & n < 0. \end{cases}$$

A map $f: (X, A) \to (Y, B)$ defines $f_\#: (S_n(X), S_n(A)) \to (S_n(Y), S_n(B))$ which in turn induces $f_\#: S_n(X, A) \to S_n(Y, B)$. $f_\#$ induces a homomorphism $f_*: H_n(X, A) \to H_n(Y, B)$, $n \in \mathbb{Z}$.

If

$$0 \longrightarrow A_* \xrightarrow{f_\#} B_* \xrightarrow{g_\#} C_* \longrightarrow 0$$

is an exact sequence of chain complexes and chain maps then we can construct a homomorphism $\partial_n: H_n(C_*) \to H_{n-1}(A_*)$ for every n as follows: given $z \in Z_n(C_*)$ we can find an $x \in B_n$. With $g_n(x) = z$. Then $g_{n-1}(d_B x) = d_C g_n(x) = d_C z = 0$, so by exactness there is a $y \in A_{n-1}$ with $f_{n-1}(y) = d_B x$. Then $f_{n-2}(d_A y) = d_B f_{n-1}(y) = d_B \circ d_B x = 0$, so since $f_\#$ is a monomorphism $d_A y = 0$—i.e. $y \in Z_{n-1}(A)$. We define $\partial_n\{z\} = \{y\}$ and check that ∂_n is well defined and a homomorphism. ∂_n is also natural with respect to morphisms of exact sequences:

$$\begin{array}{ccccccccc} 0 & \longrightarrow & A_* & \xrightarrow{f_\#} & B_* & \xrightarrow{g_\#} & C_* & \longrightarrow & 0 \\ & & \downarrow{\scriptstyle \alpha_\#} & & \downarrow{\scriptstyle \beta_\#} & & \downarrow{\scriptstyle \gamma_\#} & & \\ 0 & \longrightarrow & A'_* & \xrightarrow[f'_\#]{} & B'_* & \xrightarrow[g'_\#]{} & C'_* & \longrightarrow & 0 \end{array}$$

commutative implies $\alpha_* \circ \partial_n = \partial_n \circ \gamma_* : H_n(C_*) \to H_{n-1}(A'_*)$. Moreover, the sequence

$$\cdots \xrightarrow{\partial_{n+1}} H_n(A_*) \xrightarrow{f_*} H_n(B_*) \xrightarrow{g_*} H_n(C_*) \xrightarrow{\partial_n} H_{n-1}(A_*) \longrightarrow \cdots$$

is exact.

If we now apply this general theory to the exact sequence of chain complexes and chain maps

$$0 \longrightarrow S_*(A, \varnothing) \xrightarrow{i_\#} S_*(X, \varnothing) \xrightarrow{j_\#} S_*(X, A) \longrightarrow 0$$

we get a natural transformation

$$\partial_n(X, A) : H_n(X, A) \to H_{n-1}(A, \varnothing)$$

such that the sequence

$$\cdots \xrightarrow{\partial_{n+1}} H_n(A, \varnothing) \xrightarrow{i_*} H_n(X, \varnothing) \xrightarrow{j_*} H_n(X, A) \xrightarrow{\partial_n} H_{n-1}(A, \varnothing) \longrightarrow \cdots$$

is exact.

For any abelian group G and pair $(X, A) \in \mathcal{T}^2$ we can form a new chain complex $S_*(X, A; G) = \{S_n(X, A) \otimes G,\ d \otimes 1\}$. We then define $H_n(X, A; G)$ to be $H_n(S_*(X, A; G))$, $n \geqslant 0$, and 0 for $n < 0$. Since $S_n(X, A)$ is free abelian (being generated by the singular n-simplices $u : \Delta_n \to X$ such that $u(\Delta_n) \not\subset A$), the sequence

$$0 \longrightarrow S_n(A, \varnothing) \otimes G \xrightarrow{i \otimes 1} S_n(X, \varnothing) \otimes G \xrightarrow{j \otimes 1} S_n(X, A) \otimes G \longrightarrow 0$$

is still exact, and hence we get a long exact sequence

$$\cdots \xrightarrow{\partial_{n+1}} H_n(A, \varnothing; G) \xrightarrow{i_*} H_n(X, \varnothing; G) \xrightarrow{j_*} H_n(X, A; G) \xrightarrow{\partial_n} H_{n-1}(A, \varnothing; G) \longrightarrow \cdots,$$

where ∂_n is a natural transformation.

Thus if we show that $f \simeq f' : (X, A) \to (Y, B)$ implies

$$f_* = f'_* \cdot H_*(X, A; G) \to H_*(Y, B; G)$$

and that $H_*(-; G)$ satisfies the excision axiom, then we will have constructed an unreduced homology theory on $\mathscr{T}^{2'}$—*singular homology with coefficients G.*

We regard $\Delta_n \times I$ as a subspace of $\mathbb{R}^{n+1} \times \mathbb{R}$. For points $v_0, v_1, \ldots, v_m$ in $\mathbb{R}^{n+2}$ we let the symbol $[v_0, v_1, \ldots v_m]$ denote the linear singular simplex with vertices $v_0, v_1, \ldots, v_m$; that is, it is given by the unique affine linear map $u: \Delta_m \to \mathbb{R}^{n+2}$ such that $u(e_i) = v_i$, $0 \leqslant i \leqslant m$. Let $\delta_n \in S_n(\Delta_n)$ be the singular simplex $1: \Delta_n \to \Delta_n$. In these terms we define an $(n+1)$-chain $P\delta_n$ in $S_{n+1}(\Delta_n \times I)$:

$$P\delta_n = \sum_{i=0}^n (-1)^i [(e_0, 0), \ldots, (e_i, 0), (e_i, 1), \ldots, (e_n, 1)].$$

We define $P: S_n(X) \to S_{n+1}(X \times I)$ by taking $Pu = (u \times 1)_\#(P\delta_n)$ for any $u: \Delta_n \to X$ and extending linearly. Then P is natural. One calculates that

$$dP\delta_n + Pd\delta_n = i_{1\#}\,\delta_n - i_{0\#}\,\delta_n,$$

where for any space X, $i_0, i_1: X \to X \times I$ are defined by $i_0(x) = (x, 0)$, $i_1(x) = (x, 1)$, all $x \in X$. Applying $(u \times 1)_\#$ we get

$$dPu + Pdu = i_{1\#}(u) - i_{0\#}(u), \quad \text{i.e.}$$

$$dP + Pd = i_{1\#} - i_{0\#}.$$

Now suppose $F: X \times I \to Y$ is a homotopy from f_0 to f_1. Then $F \circ i_0 = f_0$, $F \circ i_1 = f_1$, and hence if we define $H: S_n(X) \to S_{n+1}(Y)$ by $H = F_\# \circ P$ we get

$$dH + Hd = d \circ F_\# \circ P + F_\# \circ P \circ d = F_\# \circ (dP + Pd) = F_\# \circ (i_{1\#} - i_{0\#})$$
$$= f_{1\#} - f_{0\#},$$

which shows that $f_{0\#}$ and $f_{1\#}$ are *chain homotopic* and hence that $f_{0*} = f_{1*}: H_*(X; G) \to H_*(Y; G)$. Thus $f_0 \simeq f_1$ implies $f_{0*} = f_{1*}$.

We shall only sketch the proof of the excision axiom. If $\mathscr{E} = \{U_\alpha : \alpha \in A\}$ is an open covering of a space X, then we denote by $S_*^{\mathscr{E}}(X)$ the chain complex generated by "small" simplices—i.e. those u such that $u(\Delta_n) \subset U_\alpha$ for some α. Clearly the differential d maps $S_*^{\mathscr{E}}(X)$ into itself. The crucial fact is that the inclusion $i: S_*^{\mathscr{E}}(X) \to S_*(X)$ is a chain homotopy equivalence. One proves this by taking a singular simplex $u: \Delta_n \to X$ and considering the open covering $\{u^{-1}(U_\alpha) : \alpha \in A\}$ of Δ_n. As Δ_n is a compact metric space this covering has a Lebesgue number λ. One then applies barycentric subdivision to Δ_n repeatedly until no simplex of the subdivision has diameter more than λ. In terms of this subdivision one then defines a chain homotopy $D: S_*(X) \to S_*(X)$ such that $dD + Dd = 1 - \tau$ where $\tau: S_*(X) \to S_*^{\mathscr{E}}(X)$ and $\tau | S_*^{\mathscr{E}}(X) = 1$.

This result is now applied as follows: suppose $(X; A, B)$ is a triad with $X = \mathring{A} \cup \mathring{B}$. Let $\mathscr{E} = \{\mathring{A}, \mathring{B}\}$. Then $S_*(X)$ is chain homotopy equivalent to $S_*^{\mathscr{E}}(X) \cong S_*(A) + S_*(B)$. Hence $S_*(X,B) = S_*(X)/S_*(B)$ is chain homotopy equivalent to

$$S_*(A) + S_*(B)/S_*(B) \cong S_*(A)/S_*(A) \cap S_*(B) \cong S_*(A)/S_*(A \cap B) = S_*(A, A \cap B).$$

This proves that $j_*: H_*(A, A \cap B;\ G) \to H_*(X, B;\ G)$ is an isomorphism. The reader unfamiliar with this proof can find it in [40] or [80], for example.

For each pair (X, A) and abelian group G we obtain a *cochain complex* $S^*(X, A; G) = \{\mathrm{Hom}(S_n(X, A); G),\ \mathrm{Hom}(d, 1)\}$. The cohomology groups $H^n(S^*(X, A;\ G)) = Z^n(X, A;\ G)/B^n(X, A;\ G) = \ker \mathrm{Hom}(d, 1)/\mathrm{im}\,\mathrm{Hom}(d, 1)$ of this cochain complex are the *singular cohomology groups with coefficients G* of the pair (X, A).

Of course we define the reduced homology and cohomology groups $\tilde{H}_*(-;\ G)$, $\tilde{H}^*(-;\ G)$ on $\mathscr{P}\mathscr{T}'$ by

$$\tilde{H}_n(X;\ G) = H_n(X, \{x_0\};\ G)$$
$$\tilde{H}^n(X;\ G) = H^n(X, \{x_0\};\ G).$$

Since the one point space $\{+\}$ has $S_n(\{+\}) = \mathbb{Z}$, $n \geqslant 0$, and $d: S_n(\{+\}) \to S_{n-1}(\{+\})$ the map $d(m) = m \cdot \sum_{j=0}^{n-1} (-1)^j$, $m \in \mathbb{Z}$, we see that

$$H_n(\{+\}, \varnothing\,;\ G) = \begin{cases} G & n = 0 \\ 0 & n \neq 0. \end{cases}$$

Similarly

$$H^n(\{+\}, \varnothing\,;\ G) = \begin{cases} G & n = 0 \\ 0 & n \neq 0. \end{cases}$$

Therefore $\tilde{H}_n$, $\tilde{H}^n$ are ordinary homology, respectively cohomology, theories.

10.2. Now recall the Eilenberg–MacLane complexes $H(G, n)$ constructed in Chapter 6 such that

$$\pi_r(H(G, n), *) = \begin{cases} G & r = n \\ 0 & r \neq n. \end{cases}$$

From 6.39ii) it follows any homomorphism $\phi: G \to G'$ generates a map $H(\phi, n): H(G, n) \to H(G', n)$ which is unique up to homotopy.

We also observed that since

$$\pi_r(\Omega H(G,n),\omega_0) \cong \pi_{r+1}(H(G,n),*) = \begin{cases} G & r = n-1 \\ 0 & r \neq n-1, \end{cases}$$

we have a map $\varepsilon'_{n-1}: H(G,n-1) \to \Omega H(G,n)$ and its adjoint

$$\varepsilon_{n-1}: SH(G,n-1) \to H(G,n),\ n \geqslant 2.$$

10.3. Lemma. *For any homomorphism $\phi: G \to G'$ and $n \geqslant 2$ the diagram*

$$\begin{array}{ccc} SH(G,n-1) & \xrightarrow{\varepsilon_{n-1}} & H(G,n) \\ \Big\downarrow{\scriptstyle SH(\phi,n-1)} & & \Big\downarrow{\scriptstyle H(\phi,n)} \\ SH(G',n-1) & \xrightarrow{\varepsilon_{n-1}} & H(G',n) \end{array}$$

commutes up to homotopy.

Proof: That the diagram

$$\begin{array}{ccc} H(G,n-1) & \xrightarrow{\varepsilon'_{n-1}} & \Omega H(G,n) \\ \Big\downarrow{\scriptstyle H(\phi,n-1)} & & \Big\downarrow{\scriptstyle \Omega H(\phi,n)} \\ H(G',n-1) & \xrightarrow{\varepsilon'_{n-1}} & \Omega H(G',n) \end{array}$$

commutes up to homotopy follows from the uniqueness statement of 6.39ii), since $\Omega H(\phi,n) \circ \varepsilon'_{n-1}$ and $\varepsilon'_{n-1} \circ H(\phi,n-1)$ are both maps inducing ϕ on $\pi_{n-1}(-)$. That $H(\phi,n) \circ \varepsilon_{n-1} \simeq \varepsilon_{n-1} \circ SH(\phi,n-1)$ is then just the naturality of the exponential correspondence. ▯

We may assume the maps $\varepsilon_n: SH(G,n) \to H(G,n+1)$ are cellular because of 6.35. Therefore by 8.3 there is a spectrum $H(G)$ such that $H(G)_n \simeq H(G,n)$, $n \geqslant 1$, $H(G)_n = *$, $n \leqslant 0$—i.e. $H(G)_n$ is an Eilenberg–MacLane complex of type (G,n) for $n \geqslant 1$. $H(G)$ is an Ω-spectrum. Then

$$\pi_n(H(G)) = \operatorname{dir\,lim} \pi_{n+k}(H(G)_k, *)$$

$$= \begin{cases} G & n = 0 \\ 0 & n \neq 0. \end{cases}$$

Hence the homology and cohomology theories $H(G)_*$, $H(G)^*$ are ordinary homology (cohomology) theories with coefficients G.

Any homomorphism $\phi: G \to G'$ defines a sequence of maps

$$H(\phi)_n: H(G)_n \to H(G')_n$$

such that $H(\phi)_n | SH(G)_{n-1} \simeq SH(\phi)_{n-1}$ by (10.3).

10.4. Lemma. a) *For any E, F with $E_n = *$, $n < N$ for some N, and collection of maps $f_n: E_n \to F_n$ such that $f_n | SE_{n-1} \simeq Sf_{n-1}$ we can find a function $f': E \to F$ of spectra such that $f'_n \simeq f_n$ for all $n \in \mathbb{Z}$.*

b) *If $g_n: E_n \to F_n$ is a second such system of maps with associated function $g': E \to F$ and $g_n \simeq f_n$ for all $n \geqslant N$ and if $\lim^1 \tilde{F}^{n-1}(E_n) = 0$, then $g' \simeq f'$.*

Proof: We take $f'_n = f_n$ for $n \leqslant N$. Suppose we have constructed f'_m for $m < n$ such that $f'_m | SE_{m-1} = Sf'_{m-1}$ and $f'_m \simeq f_m$, all $m < n$. Since $f_n | SE_{n-1} \simeq Sf_{n-1} \simeq Sf'_{n-1}$, we may extend the homotopy to one defined on $E_n \wedge I^+$— say H— with $H_0 = f_n$. Let $f'_n = H_1$. Then $f'_n | SE_{n-1} = H_1 | SE_{n-1} = Sf'_{n-1}$ and clearly $f'_n \simeq f_n$. Thus by induction we can construct f'_n for all n, and f' is a function of spectra.

Next we make the following general observation. Suppose

$$E^0 \subset E^1 \subset \cdots \subset E^n \subset \cdots \subset E$$

is a filtration of a spectrum E; then we can imitate the telescope construction 7.52. using spectra and repeat the proof of 7.66 to get an exact sequence

$$0 \to \lim_n{}^1 F^{q-1}(E^n) \to F^q(E) \to \lim_n{}^0 F^q(E_n) \to 0$$

for every spectrum F.

In the situation of 10.4b) we use this remark as follows: E has a filtration $\{E^n\}$ with

$$E^n_m = \begin{Bmatrix} E_m & m \leqslant n \\ S^{m-n} E_n & m \geqslant n \end{Bmatrix} \quad n \geqslant 0.$$

Thus we have an exact sequence

$$\begin{array}{c} 0 \to \lim{}^1 F^{-1}(E^n) \to F^0(E) \to \lim{}^0 F^0(E^n) \to 0. \\ \| \\ [E, F] \end{array}$$

Thus the uniqueness of $f': E \to F$ up to homotopy is guaranteed by $\lim^1 F^{-1}(E^n) = 0$. Finally we observe that $\Sigma^{-n} E(E_n) \subset E^n$ is a cofinal subspectrum, so $F^{-1}(E^n) \cong F^{-1}(\Sigma^{-n} E(E_n)) \cong \tilde{F}^{n-1}(E_n)$. ▯

Remark. Suppose we define subspectra $\hat{E}^n \subset E$ by

$$\hat{E}^n_m = \begin{cases} (E_m)^{2m} & m \leqslant n \\ S^{m-n}(E_n)^{2n} & m \geqslant n \end{cases} \qquad ((E_m)^{2m} = 2m\text{-skeleton of } E_m).$$

Then $\hat{E} = \bigcup_n \hat{E}^n$ is a cofinal subspectrum of E, so we see it even suffices to show $\lim^1 F^{-1}(\hat{E}^n) \cong \lim^1 \tilde{F}^{n-1}(E^{2n}_n) = 0$. As in the appendix to Chapter 7 it is sufficient to have $H^*(E_n)$ of finite type for every n and $\pi_q(F)$ finite whenever $\tilde{H}^{n+q-1}(E_n^{2n}) \neq 0$.

Hence we may as well assume $H(\phi) = \{H(\phi)_n\}$ is a function and therefore defines a map of spectra $H(\phi): H(G) \to H(G')$. Clearly the diagram

$$\begin{array}{ccc} \pi_0(H(G)) & \xrightarrow{H(\phi)_*} & \pi_0(H(G')) \\ \wr\| & & \wr\| \\ G & \xrightarrow{\phi} & G' \end{array} \quad \text{commutes.}$$

As in 8.39 $H(\phi):H(G) \to H(G')$ induces a natural transformation $T_*(\phi) = T_*(H(\phi)): H(G)_* \to H(G')_*$ of homology theories and also of cohomology theories. In this case $T_*(\phi)$ is a natural equivalence if and only if ϕ is an isomorphism.

10.5. The corresponding construction for singular homology is as follows: given a homomorphism $\phi: G \to G'$ for any pair $(X, A) \in \mathcal{T}^2$ we have $\phi_{\#} = 1 \otimes \phi: S_n(X,A) \otimes G \to S_n(X,A) \otimes G'$ defining a chain map $\phi_{\#}: S_*(X,A;\, G) \to S_*(X,A;\, G')$. This in turn induces a homomorphism $\phi_*: H_*(X,A;\, G) \to H_*(X,A;\, G')$ which is natural with respect to maps $f:(X,A) \to (Y,B)$. A short exact sequence $0 \to G \to H \to K \to 0$ induces a long exact homology sequence $\cdots \to H_n(X,A;\, G) \to H_n(X,A;\, H) \to H_n(X,A;\, K) \xrightarrow{\beta} H_{n-1}(X,A;\, G) \to \cdots$. The connecting homomorphism $\beta: H_n(X,A;\, K) \to H_{n-1}(X,A;\, G)$ is called the *Bockstein* homomorphism associated with the sequence $0 \to G \to H \to K \to 0$.

10.6. We now give a means of computing the ordinary homology groups of any CW-complex which is effective if we have a firm enough grasp on the attaching maps of the complex. Let $\{e_\alpha^n : \alpha \in J_n\}$ be the n-cells of X with characteristic maps $f_\alpha^n:(D^n, S^{n-1}, s_0) \to (X^n, X^{n-1}, x_0)$. Let $C_n(X)$ be the free abelian group generated by J_n, $n \in \mathbb{Z}$. We assume k_* is any ordinary homology theory with coefficients G which satisfies the wedge axiom.

10.7. Lemma *For each $n \in \mathbb{Z}$ there is a natural equivalence*

$$\kappa_n(X): k_n(X^n/X^{n-1}) \to C_n(X) \otimes G. \quad k_m(X^n/X^{n-1}) = 0 \text{ if } m \neq n.$$

(Remember that $X^k = \{x_0\}$ if $k < 0$; also $C_k(X) = 0$ for $k<0$, since $J_k = \phi$.)

Proof: From 5.17 we have the natural isomorphism $\lambda_n: k_m(X^n/X^{n-1}) \to k_m(\vee_{\alpha \in J_n} S_\alpha^n)$.

From the wedge axiom we know $\{i_{\alpha *}\}: \oplus_{\alpha \in J_n} k_m(S_\alpha^n) \to k_m(\vee_{\alpha \in J_n} S_\alpha^n)$ is an isomorphism. Since

$$k_m(S^n) \underset{\cong}{\xrightarrow{\sigma^{-n}}} k_{m-n}(S^0) = \begin{cases} G & m = n \\ 0 & m \neq n, \end{cases}$$

the assertion for $m \neq n$ follows immediately. We define $\mu: C_n(X) \oplus_\alpha G \to \oplus_{\alpha \in J_n} k_0(S^0)$ by

$$\mu(\alpha \otimes g)_\beta = \begin{cases} g & \beta = \alpha \\ 0 & \beta \neq \alpha \end{cases},$$

and extend linearly. Then for $\kappa_n(X)$ we can take the composite

$$k_n(X^n/X^{n-1}) \xrightarrow[\cong]{\lambda_n} k_n(\vee_{\alpha \in J_n} S^n_\alpha) \xleftarrow[\cong]{\{i_{\alpha *}\}} \oplus_{\alpha \in J_n} k_n(S^n) \xrightarrow[\cong]{\oplus \sigma^{-n}} \oplus_{\alpha \in J_n} k_0(S^0)$$

$$\xleftarrow[\cong]{\mu} C_n(X) \otimes G. \quad \square$$

10.8. Corollary. i) $k_m(X^n) = 0$ *if* $m > n$,

ii) $k_n(X^n) \to k_n(X^{n+1})$ *is an epimorphism, and*
iii) $k_m(X^n) \xrightarrow{i_{n*}} k_m(X)$ *is an isomorphism if* $m < n$.

Proof: We have the exact sequence of the pair (X^{n+1}, X^n)

$$\cdots \longrightarrow k_{m+1}(X^{n+1}/X^n) \longrightarrow k_m(X^n) \xrightarrow{i_*} k_m(X^{n+1}) \longrightarrow$$

$$k_m(X^{n+1}/X^n) \longrightarrow \cdots.$$

From 10.7 we see that i_* is
a) an isomorphism for $m \neq n, n+1$;
b) an epimorphism for $m = n$;
c) a monomorphism for $m = n + 1$.

Thus if $m > n$ we have $k_m(X^n) \cong k_m(X^{n-1}) \cong \cdots \cong k_m(X^{-1}) = k_m(\{x_0\}) = 0$ by 7.32. ii) follows from b). If $m < n$, then a) implies

$$k_m(X^n) \cong k_m(X^{n+1}) \cong \cdots \cong k_m(X^r)$$

for all $r \geqslant n$. We have a commutative diagram

$$\begin{array}{ccc} & \xrightarrow{j_n} & \operatorname{dir\,lim} k_m(X^r) \\ k_m(X^n) & & \cong \big\downarrow \{i_{r*}\} \\ & \xrightarrow[i_{n*}]{} & k_m(X) \end{array}$$

in which $\{i_{r*}\}$ is an isomorphism by 7.53 and j_n defined by $j_n x = \{x\}$ is an isomorphism since all morphisms $j^r_{s*}: k_m(X^s) \to k_m(X^r)$ are isomorphisms for $n \leqslant s \leqslant r$. Thus i_{n*} is an isomorphism. $\square$

Let $\Delta: k_n(X^n/X^{n-1}) \to k_{n-1}(X^{n-1}/X^{n-2})$ be the boundary map of the triple (X^n, X^{n-1}, X^{n-2}) and define $d: C_n(X) \otimes G \to C_{n-1}(X) \otimes G$ to be the unique homomorphism such that

$$\begin{array}{ccc} k_n(X^n/X^{n-1}) & \xrightarrow[\cong]{\kappa_n(X)} & C_n(X) \otimes G \\ \downarrow{\scriptstyle \Delta} & & \downarrow{\scriptstyle d} \\ k_{n-1}(X^{n-1}/X^{n-2}) & \xrightarrow[\cong]{\kappa_{n-1}(X)} & C_{n-1}(X) \otimes G \end{array}$$

commutes, $n \geqslant 0$. Since Δ is the composite

$$k_n(X^n/X^{n-1}) \xrightarrow{\ \hat{\partial}\ } k_{n-1}(X^{n-1}) \xrightarrow{\ p_*\ } k_{n-1}(X^{n-1}/X^{n-2}),$$

it follows that $\Delta \circ \Delta = p_* \circ \hat{\partial} \circ p_* \circ \hat{\partial} = 0$, because $\hat{\partial} \circ p_* = 0$ (7.33). Thus $\{k_n(X^n/X^{n-1}), \Delta\}$ and $\{C_n(X) \otimes G, d\}$ are chain complexes and $\kappa_\#$ is a natural chain equivalence.

10.9. Theorem. *There is a natural isomorphism*

$$H_n(\{C_*(X) \otimes G, d\}) \cong k_n(X), n \in \mathbb{Z}.$$

Proof: We shall construct a natural equivalence $H_n(\{k_n(X^n/X^{n-1}), \Delta\}) \cong k_n(X)$; we may then compose this with the natural equivalence κ_*. We have a commutative diagram

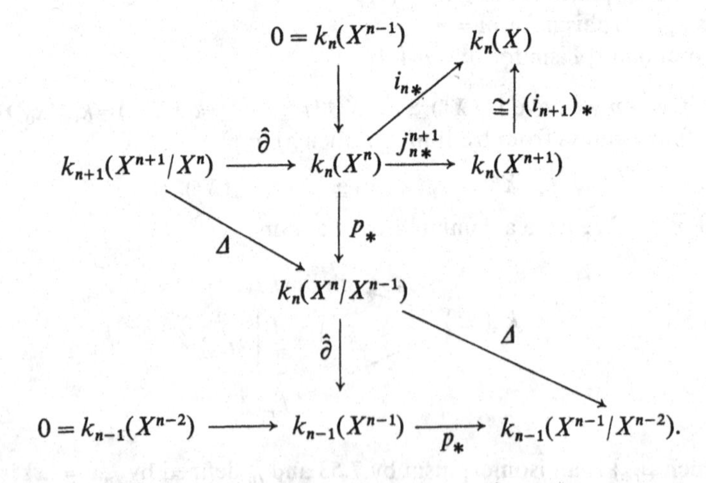

$z \in k_n(X^n/X^{n-1})$ is a cycle $\Leftrightarrow 0 = \Delta z = p_* \hat{\partial} z \Leftrightarrow \hat{\partial} z = 0 \Leftrightarrow z \in \operatorname{im} p_*$. Thus p_* defines an isomorphism $p_*: k_n(X^n) \to Z_n$.

$z \in k_n(X^n/X^{n-1})$ is a boundary $\Leftrightarrow z = \Delta z'$ for some $z' \in k_{n+1}(X^{n+1}/X^n) \Leftrightarrow z = p_* \hat{\partial} z' \Leftrightarrow z \in p_*(\operatorname{im} \hat{\partial}) \Leftrightarrow z \in p_*(\ker j_{n*}^{n+1}) \Leftrightarrow z \in p_*(\ker i_{n*})$.

Thus p_* defines an isomorphism $p_*: \ker i_{n*} \to B_n$. Thus we have a commutative diagram of exact sequence

$$\begin{array}{ccccccccc}
0 & \longrightarrow & \ker i_{n*} & \longrightarrow & k_n(X^n) & \xrightarrow{i_{n*}} & k_n(X) & \longrightarrow & 0 \\
 & & p_* \downarrow \cong & & \cong \downarrow p_* & & & & \\
0 & \longrightarrow & B_n & \longrightarrow & Z_n & \longrightarrow & H_n(\{k_n(X^n/X^{n-1}), \Delta\}) & \longrightarrow & 0,
\end{array}$$

from which it follows there is a natural isomorphism

$$k_n(X) \cong H_n(\{k_n(X^n/X^{n-1}), \Delta\}). \quad \square$$

Fortunately for the sake of calculation the differential $d: C_n(X) \otimes G \to C_{n-1}(X) \otimes G$ can be described in more elementary terms, as we are about to show.

10.10. Definition. For any $n \geqslant 1$ we have $\pi_n(S^n, s_0) \cong \mathbb{Z}$, the generator of $\pi_n(S^n, s_0)$ which corresponds to $1 \in \mathbb{Z}$ being $[1_{S^n}]$. For any $f: (S^n, s_0) \to (S^n, s_0)$ we define the *degree* $\deg(f)$ to be the integer $m \in \mathbb{Z}$ which corresponds to $[f]$ under the isomorphism. In other words, if $\deg(f) = m$, then $[f] = m \cdot [1_{S^n}]$. The case $n = 0$ is special; $\pi_0(S^0, s_0) \cong \mathbb{Z}_2$ as sets. We take $\deg(1) = 1$, $\deg(s_0) = 0$.

For any n-cell e_α^n of X with attaching map $g_\alpha^n: (S^{n-1}, s_0) \to (X^{n-1}, x_0)$ and any $(n-1)$-cell e_β^{n-1} let X_β^{n-1} denote the complex $X^{n-1} - \mathring{e}_\beta^{n-1}$ $(= X^{n-2} \cup$ all $(n-1)$-cells except $e_\beta^{n-1})$ and define $h_\beta^\alpha: (S^{n-1}, s_0) \to (S^{n-1}, s_0)$ to be the composite

$$(S^{n-1}, s_0) \xrightarrow{g_\alpha^n} (X^{n-1}, x_0) \xrightarrow{p_\beta} (X^{n-1}/X_\beta^{n-1}, *) \underset{\cong}{\xleftarrow{f_\beta^{n-1}}} (D^{n-1}/S^{n-2}, *) \cong (S^{n-1}, *).$$

Intuitively h_β^α indicates the way e_α^n is attached across e_β^{n-1}.

10.11. Proposition. *$d: C_n(X) \otimes G \to C_{n-1}(X) \otimes G$ is of the form $d = d' \otimes 1$, where $d': C_n(X) \to C_{n-1}(X)$ is given by $d'(\alpha) = \sum_{\beta \in J_{n-1}} \deg(h_\beta^\alpha) \beta$, $\alpha \in J_n$.*

Proof: For any $\alpha \in J_n$, $g \in G$, we shall have $d(\alpha \otimes g) = \sum_{\gamma \in J_{n-1}} \gamma \otimes r_\gamma^\alpha$ for some $r_\gamma^\alpha \in G$. d is given by $d = \kappa_{n-1}(X) \circ \Delta \circ \kappa_n(X)^{-1}$. The following diagram is commutative

$$
\begin{array}{ccccccc}
 & k_n(S^n) & \xrightarrow{i_{\alpha *}} & k_n(\vee_\delta S^n_\delta) & & & \\
\sigma^n \nearrow & \uparrow \cong & & \lambda_n \uparrow \cong & & & \\
k_0(S^0) & k_n(D^n/S^{n-1}) & \xrightarrow{\bar{f}^n_{\alpha *}} & k_n(X^n/X^{n-1}) & & & \\
\sigma^{n-1} \searrow & \partial \downarrow \cong & & \hat{\partial} \downarrow & \searrow \Delta & & \\
 & k_{n-1}(S^{n-1}) & \xrightarrow{g^n_{\alpha *}} & k_{n-1}(X^{n-1}) \xrightarrow{p_*} & k_{n-1}(X^{n-1}/X^{n-2}) \xrightarrow{\lambda_{n-1}} & k_{n-1}(\vee_\gamma S^{n-1}_\gamma) \xleftarrow{\{i_{\gamma *}\} \circ \sigma^{n-1}} & \oplus_\gamma k_0(S^0) \\
 & & & p_{\beta *} \searrow & \downarrow p'_{\beta *} & p''_{\beta *} \downarrow & \downarrow \pi_\beta \\
 & & & & k_{n-1}(X^{n-1}/X^{n-1}_\beta) \underset{\cong}{\rightarrow} & k_{n-1}(S^{n-1}) \underset{\cong}{\xleftarrow{\sigma^{n-1}}} & k_0(S^0).
\end{array}
$$

Here $p''_\beta : \vee_{\gamma \in J_{n-1}} S^{n-1}_\gamma \to S^{n-1}$ is the projection on the βth summand, as is $\pi_\beta : \oplus_{\gamma \in J_{n-1}} k_0(S^0) \to k_0(S^0)$. Thus

$$
\begin{aligned}
r^\alpha_\beta &= \pi_\beta(\mu(\textstyle\sum \gamma \otimes r^\alpha_\gamma)) = \pi_\beta \circ \mu(d(\alpha \otimes g)) \\
&= \pi_\beta \circ \mu \circ \kappa_{n-1}(X) \circ \Delta \circ \kappa_n(X)^{-1}(\alpha \otimes g) \\
&= \pi_\beta \circ \oplus_\gamma (\sigma^{n-1})^{-1} \circ \{i_{\gamma *}\}^{-1} \circ \lambda_{n-1} \circ \Delta \circ \lambda_n^{-1} \circ i_{\alpha *} \circ \sigma^n(g) \\
&= (\sigma^{n-1})^{-1} \circ h^\alpha_{\beta *} \circ \sigma^{n-1}(g).
\end{aligned}
$$

Now for any element $x \in k_{n-1}(S^{n-1})$ we have $h^\alpha_\beta(x) = (\deg(h^\alpha_\beta) 1_{S^n})_*(x) = \deg(h^\alpha_\beta) x$ (7.38). Therefore

$$r^\alpha_\beta = (\sigma^{n-1})^{-1}(\deg(h^\alpha_\beta)\, \sigma^{n-1}(g)) = \deg(h^\alpha_\beta)(\sigma^{n-1})^{-1}(\sigma^{n-1}(g)) = \deg(h^\alpha_\beta)\, g.$$

Hence we have proved that

$$
\begin{aligned}
d(\alpha \otimes g) &= \textstyle\sum_{\gamma \in J_{n-1}} \deg(h^\alpha_\gamma)\, \gamma \otimes g = (\sum_{\gamma \in J_{n-1}} \deg(h^\alpha_\gamma)\, \gamma) \otimes g = d'(\alpha) \otimes g \\
&= (d' \otimes 1)(\alpha \otimes g).
\end{aligned}
$$

Since the elements $\alpha \otimes g$ generate $C_n(X) \otimes G$, 10.11 follows. □

10.12. We can even determine the homomorphism $f_* : k_n(X) \to k_n(Y)$ induced by a cellular map $f : (X, x_0) \to (Y, y_0)$. f defines a map $\bar{f} : X^n/X^{n-1} \to Y^n/Y^{n-1}$. Let e^n_α be an n-cell of X, e^n_β an n-cell of Y. Let $f^\alpha_\beta : (S^n, s_0) \to (S^n, s_0)$ be the composite

$$
\begin{aligned}
(S^n, s_0) \cong (D^n/S^{n-1}, *) &\xrightarrow{\bar{f}^n_\alpha} (X^n/X^{n-1}, *) \xrightarrow{\bar{f}} (Y^n/Y^n_\beta, *) \\
&\underset{\cong}{\xleftarrow{\bar{f}^n_\beta}} (D^n/S^{n-1}, *) \cong (S^n, s_0).
\end{aligned}
$$

10.13. Proposition. *$f_\# : C_n(X) \otimes G \to C_n(Y) \otimes G$ is of the form $f'_\# \otimes 1$ where $f'_\# : C_n(X) \to C_n(Y)$ is given by $f'_\#(\alpha) = \sum_{\gamma \in J'_n} \deg(f^\alpha_\gamma)\gamma$, $\alpha \in J_n$.*

Proof: Again we have a commutative diagram

$$\begin{array}{ccccccc}
k_0(S^0) \xrightarrow{\sigma^n} k_n(S^n) & \xrightarrow{i_{\alpha *}} & k_n(\vee_\delta S^n_\delta) & & & & \\
\uparrow \cong & & \cong \uparrow \lambda_n & & & & \\
k_n(D^n/S^{n-1}) & \xrightarrow{f^n_{\alpha *}} & k_n(X^n/X^{n-1}) & \xrightarrow{f_*} & k_n(Y^n/Y^{n-1}) & \xrightarrow{\lambda_n} & k_n(\vee_\gamma S^n_\gamma) \\
 & & & & \downarrow p'_{\beta *} & & \downarrow p''_{\beta *} \\
 & & & & k_n(Y^n/Y^n_\beta) & \xleftarrow[\cong]{f^n_{\beta *}} & k_n(S^n) \xleftarrow{\sigma^n} k_0(S^0).
\end{array}$$

We know that $f_\#(\alpha \otimes g) = \sum_{\gamma \in J'_n} \gamma \otimes r^\alpha_\gamma$ for some $r^\alpha_\gamma \in G$. From the diagram we see that $r^\alpha_\beta = \sigma^{-n} \circ f^\alpha_{\beta *} \circ \sigma^n(g) = \deg(f^\alpha_\beta)g$. Thus $f_\#(\alpha \otimes g) = \sum_\gamma \deg(f^\alpha_\gamma)\gamma \otimes g = (f'_\# \otimes 1)(\alpha \otimes g)$. ▯

10.14. Observe that the chain complex $C_*(X)$, the differential d' and the chain map $f'_\#$ all have descriptions which depend only on the geometry of the complexes and maps but not on the homology theory k_*. This suggests that any two ordinary homology theories with G coefficients are naturally equivalent.

10.15. Theorem. *If k_*, k'_* are ordinary homology theories with G, respectively G', coefficients satisfying the wedge axiom on $\mathscr{PW}'$, the WHE axiom on $\mathscr{PT}'$ and $\phi : G \to G'$ is a homomorphism, then there is a natural transformation $T_*(\phi) : k_* \to k'_*$ of homology theories on $\mathscr{PT}'$ such that $T_0(\phi)(S^0) : k_0(S^0) \to k'_0(S^0)$ is ϕ. If k''_* is an ordinary homology theory with G'' coefficients and $\phi' : G' \to G''$ a homomorphism, then $T_*(\phi' \circ \phi) = T_*(\phi') \circ T_*(\phi)$. $T_*(1_G) =$ the identity transformation. In particular, if ϕ is an isomorphism, then $T_*(\phi)$ is a natural equivalence.*

Proof: $1 \otimes \phi$: $C(X) \otimes G \to C(X) \otimes G'$ is a chain map by 10.11 and natural with respect to maps $f : (X, x_0) \to (Y, y_0)$ by 10.13. Hence if X is a CW-complex we can take $T_n(\phi)(X)$ to be the composite

$$k_n(X) \cong H_n(\{C_*(X) \otimes G, d\}) \xrightarrow{(1 \otimes \phi)_*} H_n(\{C_*(X) \otimes G', d\}) \cong k'_n(X).$$

For other spaces we take CW-substitutes. The remaining statements are obvious. ▯

The results 10.6–10.15 have their duals for ordinary cohomology theories.

10.16. Proposition. *$\tilde{H}_*(-;G)$ satisfies the wedge axiom on $\mathcal{PT}'_0$, the category of pointed spaces with non-degenerate base points.*

Proof: First we show that the equivalent axiom for an unreduced theory h_* is the following:

For every collection $\{X_\alpha : \alpha \in A\}$ of topological spaces the inclusions $i_\alpha : X_\alpha \to \bigcup_{\alpha \in A} X_\alpha$ into the disjoint union induce an isomorphism

$$\{i_{\alpha *}\} : \bigoplus_{\alpha \in A} h_n(X_\alpha) \to h_n(\bigcup_{\alpha \in A} X_\alpha), \quad n \in \mathbb{Z}.$$

For if the reduced theory $\tilde{h}_*$ satisfies the wedge axiom then

$$h_n(\bigcup_\alpha X_\alpha) \cong \tilde{h}_n((\bigcup_\alpha X_\alpha)^+) = \tilde{h}_n(\bigvee_\alpha X_\alpha^+) \cong \bigoplus_\alpha \tilde{h}_n(X_\alpha^+) \cong \bigoplus_\alpha h_n(X_\alpha).$$

On the other hand suppose h_* satisfies the axiom above and suppose $\{(X_\alpha, x_\alpha) : \alpha \in A\}$ is a collection of pointed spaces $(X_\alpha, x_\alpha) \in \mathcal{PT}_0$. Let $Y = \{x_\alpha : \alpha \in A\} \subset \bigcup_\alpha X_\alpha$. Then

$$p : ((\bigcup X_\alpha)^+ \cup CY^+, CY^+) \to ((\bigcup X_\alpha)^+ \cup CY^+/CY^+, \{*\}) \cong (\bigvee_\alpha X_\alpha, \{*\})$$

is a homotopy equivalence ($Y \subset \bigcup X_\alpha$ is a cofibration), and the triad $((\bigcup X_\alpha)^+ \cup CY^+; (\bigcup X_\alpha)^+, CY^+)$ is excisive. Hence from the commutative diagram

$$\begin{array}{ccc}
\bigoplus_\alpha \tilde{h}_n(X_\alpha) = \bigoplus_\alpha h_n(X_\alpha, \{x_\alpha\}) & \xrightarrow[\{i_{\alpha *}\}]{\cong} & h_n(\bigcup_\alpha X_\alpha, Y) \\
\downarrow \{i_{\alpha *}\} \qquad \{i_{\alpha *}\} \downarrow & & \downarrow \cong \\
\tilde{h}_n(\bigvee_\alpha X) = h_n(\bigvee_\alpha X_\alpha, \{*\}) & \xleftarrow[p_*]{\cong} & h_n((\bigcup X_\alpha)^+ \cup CY^+, CY^+)
\end{array}$$

it follows that $\{i_{\alpha *}\} : \bigoplus_\alpha \tilde{h}_n(X_\alpha) \to \tilde{h}_n(\bigvee_\alpha X_\alpha)$ is an isomorphism.

Now for a disjoint union $\bigcup X_\alpha$ we clearly have $S_*(\bigcup X_\alpha; G) = \bigoplus_\alpha S_*(X_\alpha; G)$. On the other hand $H_*(\bigoplus_\alpha C_*^\alpha) \cong \bigoplus_\alpha H_*(C_*^\alpha)$ for any collection $\{C_*^\alpha : \alpha \in A\}$ of chain complexes C_*^α. Hence $H_*(-;G)$ satisfies the unreduced "wedge" axiom above. □

Now we may conclude that $\tilde{H}_*(-;G)$ and $H(G)_*(-)$ are naturally equivalent on $\mathcal{PW}'$.

10.17. Proposition. *$\tilde{H}^*(-;G)$ satisfies the wedge axiom for every G.*

Proof: Again it suffices to show that $H^*(\bigcup X_\alpha; G) \cong \prod_\alpha H^*(X_\alpha; G)$ for any disjoint union $\bigcup X_\alpha$. On the one hand we have

$$S^*(\bigcup X_\alpha; G) = \mathrm{Hom}(S_*(\bigcup X_\alpha), G) \cong \mathrm{Hom}(\oplus_\alpha S_*(X_\alpha), G)$$
$$\cong \prod_\alpha \mathrm{Hom}(S_*(X_\alpha), G)$$
$$= \prod_\alpha S^*(X_\alpha; G).$$

On the other hand $H^*(\prod_\alpha C_\alpha^*) \cong \prod_\alpha H^*(C_\alpha^*)$ for any collection $\{C_\alpha^*: \alpha \in A\}$ of cochain complexes. □

Thus $\tilde{H}^*(-; G)$ and $H(G)^*$ are also naturally equivalent on $\mathscr{PW}'$.

We shall next show that $H_*(-; G)$, $H^*(-; G)$ satisfy the weak homotopy equivalence axiom.

10.18. Definition. For each $n \geqslant 0$ and each pair (X, A) let

$$S_*^{(n,A)}(X) \subset S_*(X)$$

be the subcomplex generated by the singular simplices $u: \Delta_m \to X$ which map the $(n-1)$-skeleton of Δ_m into A. The *nth Eilenberg complex* of (X, A) is the subcomplex

$$S_*^{(n)}(X, A) = S_*^{(n,A)}(X)/S_*(A) \subset S_*(X, A).$$

Note that $S_m^{(n)}(X, A) = 0$ for $0 \leqslant m < n$.

10.19. Lemma. *If* (X, A) *is* $(n-1)$*-connected, then the inclusion*

$$i: S^{(n)}(X, A) \to S_*(X, A)$$

is a chain homotopy equivalence and hence defines isomorphisms

$$i_*: H_*(S_*^{(n)}(X, A) \otimes G) \to H_*(S_*(X, A; G))$$

and $i^*: H^*(S^*(X, A; G)) \to H^*(\mathrm{Hom}(S_*^{(n)}(X, A), G))$.

Proof: We wish to construct for each $u: \Delta_m \to X$ a map $ku: \Delta_m \times I \to X$ with the properties

i) $ku \circ i_0 = u$, $ku \circ i_1 \in S^{(n,A)}(X)$;
ii) $ku \circ (\partial_i \times 1) = k\partial_i u$, $0 \leqslant i \leqslant m$;
iii) if $u \in S_m^{(n,A)}(X)$, then $ku(x, t) = u(x)$, all $x \in \Delta_m$, $t \in I$.

Here $i_0: \Delta_m \to \Delta_m \times I$ is given by $i_0(x) = (x, 0)$ and similarly for i_1. We proceed by induction on m; for $m = 0$ we let $ku: \Delta_0 \times I \to X$ be any path from $u(\Delta_0)$ to A, except that if $u(\Delta_0) \in A$, it must be the constant path. Suppose ku has been defined for all $u: \Delta_k \to X$, $k < m$. For a given $u: \Delta_m \to X$ we have ku already defined on $\Delta_m \times \{0\} \cup \dot{\Delta}_m \times I$ such that $ku|\dot{\Delta}_m \times 1 \in S_m^{(n,A)}(X)$. If $m < n$, then we observe that $(\Delta_m \times \{0\} \cup \dot{\Delta}_m \times I, \Delta_m \times \{1\}) \cong (D^m, S^{m-1})$, so since (X, A) is $(n-1)$-connected, we can find a homotopy $ku: \Delta_m \times I \to X$ agreeing with what was already given and such that $ku|\Delta_m \times \{1\} \in S_m^{(n,A)}(X)$. If $u \in S_m^{(n,A)}(X)$, then we take ku to be the stationary homotopy. If $m \geqslant n$ then we take ku to be any extension

$(\Delta_m \times \{0\} \cup \dot{\Delta}_m \times I$ is a retract of $\Delta_m \times I)$, except that again it should satisfy iii) if $u \in S_m^{(n,A)}(X)$.

Now we recall the chain homotopy $P: S_*(\Delta_m) \to S_*(\Delta_m \times I)$ constructed in 10.1. Let $\delta_m \in S_m(\Delta_m)$ be the singular chain $1: \Delta_m \to \Delta_m$ and define a chain homotopy $\bar{P}: S_*(X) \to S_*(X)$ by taking $\bar{P}u = (ku)_\#(P\delta_m)$ and extending linearly. From the properties of P we compute

$$\begin{aligned} d\bar{P}u + \bar{P}du &= ku_\#(dP\delta_m) + \sum (-1)^i (k\partial_i u)_\#(P\delta_{m-1}) \\ &= ku_\#(i_{0\#}(\delta_m) - i_{1\#}(\delta_m) - Pd\delta_m) + \sum (-1)^i ku_\#(\partial_i \times 1)_\#(P\delta_{m-1}) \\ &= u_\#(\delta_m) - (ku \circ i_1)_\#(\delta_m) + ku_\#[\sum (-1)^i (\partial_i \times 1)_\#(P\delta_{m-1}) - Pd\delta_m] \\ &= u - \tau(u) + ku_\#[P(\sum (-1)^i \partial_{i\#}(\delta_{m-1}) - d\delta_m)] \\ &= u - \tau(u), \end{aligned}$$

where $\tau(u) = ku \circ i_1$ is a singular simplex in $S_m^{(n,A)}(X)$. Then τ is a chain map $S_*(X) \to S_*^{(n,A)}(X)$ which is chain homotopic in $S_*(X)$ to the identity. Also by condition iii) $\tau | S_*^{(n,A)}(X) = 1$. Because $\tau(S_*(A)) \subset S_*(A)$ τ defines a chain homotopy inverse $\bar{\tau}: S_*(X,A) \to S_*^{(n)}(X,A)$ to the inclusion i. The other statements are clear. □

10.20. Corollary. *$H_*(-;G)$ and $H^*(-;G)$ satisfy the weak homotopy equivalence axiom.*

Proof: By 7.70 it suffices to show that if (X,A) is n-connected for all n then $H_*(X,A;G) = 0$. But by 10.19 $H_q(X,A;G) \cong H_q(S_*^{(q+1)}(X,A) \otimes G) = 0$, since $S_q^{(q+1)}(X,A) = 0$, all q. Similarly for cohomology. □

10.21. Theorem. $\tilde{H}_*(-;G) = H(G)_*(-)$ *and* $\tilde{H}^*(-;G) = H(G)^*(-)$.

10.22. Examples. i) Let us give S^n the CW-structure with two p-cells e_+^p, e_-^p for every p, $0 \leqslant p \leqslant n$, as in 5.4.2b), except that $e_+^0 = s_0$ is a $(-\infty)$-cell. Then $C_p(S^n) \cong \mathbb{Z} \oplus \mathbb{Z}$, $1 \leqslant p \leqslant n$, $C_0(S^n) \cong \mathbb{Z}$. The attaching maps of e_+^p, e_-^p are the identity $1: S^{p-1} \to S^{p-1}$; the collapsing map $S^{p-1} \to S^{p-1}/e_-^{p-1} \cong S^{p-1}$ is clearly homotopic to 1, while $S^{p-1} \to S^{p-1}/e_+^{p-1} \cong S^{p-1}$ is homotopic to ν'. Thus $d(e_+^p) = e_+^{p-1} - e_-^{p-1}$, $d(e_-^p) = e_+^{p-1} - e_-^{p-1}$, and so d has the matrix $\begin{pmatrix} 1 & -1 \\ 1 & -1 \end{pmatrix}$. This has rank 1, and hence $B_p = Z_p \cong \mathbb{Z}$, $0 < p \leqslant n-1$, $Z_n \cong \mathbb{Z}$, $B_n = 0$. $d: C_0(S^n) \to C_{-1}(S^n)$ is 0, and $e_-^0 = d(e_+^1)$. Therefore we get

$$\tilde{H}_p(S^n; G) = \begin{cases} G & p = n \\ 0 & p \neq n, \end{cases}$$

as we should.

ii) CP^n has a $2p$-cell for each p, $1 \leqslant p \leqslant n$, and one $(-\infty)$-cell. Thus $C_{2p}(CP^n) \cong \mathbb{Z}$, $1 \leqslant p \leqslant n$, $C_m(CP^n) = 0$ for all other m. Clearly

$$d: C_m(CP^n) \to C_{m-1}(CP^n)$$

must be 0 for all m. Hence

$$\tilde{H}_m(CP^n; G) \cong \begin{cases} G & m = 2p, 1 \leqslant p \leqslant n \\ 0 & \text{otherwise.} \end{cases}$$

Passing to the direct limit, we find

$$\tilde{H}_m(CP^\infty; G) \cong \begin{cases} G & m = 2k, k \geqslant 1 \\ 0 & \text{otherwise.} \end{cases}$$

iii) Similarly

$$\tilde{H}_m(HP^n; G) \cong \begin{cases} G & m = 4k, 1 \leqslant k \leqslant n \\ 0 & \text{otherwise,} \end{cases}$$

and

$$\tilde{H}_m(HP^\infty; G) \cong \begin{cases} G & m = 4k, k \geqslant 1 \\ 0 & \text{otherwise.} \end{cases}$$

iv) The calculation for RP^n is slightly more interesting. We have $C_p(RP^n) = \mathbb{Z}$, $1 \leqslant p \leqslant n$. We could write down the composite h^m

$$S^{m-1} \xrightarrow{q^{m-1}} RP^{m-1} \xrightarrow{p} RP^{m-1}/RP^{m-2} \xleftarrow[\cong]{f^{m-1}} D^{m-1}/S^{m-2} \cong S^{m-1}$$

and show that $\deg h^m = 1 + (-1)^m$. Instead we proceed as follows: if we give S^n the CW-structure of i) above, then $q^n: S^n \to RP^n$ is a cellular map. We look at $q^n_{\#}: C_p(S^n) \to C_p(RP^n)$. The composite

$$D^p/S^{p-1} \xrightarrow{\bar{p}_+} S^p/S^{p-1} \xrightarrow{\bar{q}^p} RP^p/RP^{p-1} \xleftarrow[\cong]{\hat{q}_p} H^p_+/S^{p-1} \xleftarrow[\cong]{\bar{p}_+} D^p/S^{p-1}$$

is clearly the identity; and if we replace $\bar{p}_+$ by $\bar{p}_-$ on the left, then the composite becomes the map $\{x_1, x_2, \ldots, x_p\} \to \{-x_1, -x_2, \ldots, -x_p\}$ which has degree $(-1)^p$, as one easily sees. Thus we see that $q^n_{\#}: C_p(S^n) \to C_p(RP^n_{\#})$ is given by $q^n_{\#}(e^p_+) = e^p$, $q^n_{\#}(e^p_-) = (-1)^p e^p$. Therefore $de^p = dq^n_{\#}(e^p_+) =$

$q^n_\#(de^p_+) = q^n_\#(e^{p-1}_+ - e^{p-1}_-) = e^{p-1} + (-1)^p e^{p-1} = [1 + (-1)^p]e^{p-1}$. Hence we find that

$$\tilde{H}_m(RP^n; G) = \begin{cases} G_2 & m \text{ odd}, 1 \leqslant m < n, \\ G^2 & m \text{ even}, 2 \leqslant m \leqslant n, \\ G & m = n, n \text{ odd}, \\ 0 & \text{otherwise}, \end{cases}$$

where $G_2 = G/2G$, G^2 = elements of order 2 in G.

Since $\pi_0(H(\mathbb{Z})) = [S^0, H(\mathbb{Z})] = \mathbb{Z}$, we may choose a map $\imath: S^0 \to H(\mathbb{Z})$ of spectra representing $1 \in \mathbb{Z}$. $\imath$ then induces a natural transformation

$$T_*(\imath): \pi^S_* \to H(\mathbb{Z})_* = \tilde{H}_*(-; \mathbb{Z})$$

as in 8.39. We shall denote $T_*(\imath)$ by h^S and call it the *stable Hurewicz homomorphism*. Since $\pi^S_n(S^0) = 0 = \pi^S_n(H(\mathbb{Z}))$ for $n < 0$ and $\imath_*: \pi^S_0(S^0) \to \pi_0(H(\mathbb{Z}))$ is an isomorphism ($\pi^S_0(S^0)$ is generated by $[1_{S^0}]$, and $\imath_*[1_{S^0}] = [\imath]$), while $\imath_*: \pi^S_1(S^0) \to \pi_1(H(\mathbb{Z}))$ is certainly an epimorphism, we may apply 7.55 with $N = 1$ to obtain the following theorem.

10.23. Theorem. *If X is an $(n-1)$-connected space then $h^S: \pi^S_q(X) \to \tilde{H}_q(X; \mathbb{Z})$ is an isomorphism for $q \leqslant n$ and an epimorphism for $q = n + 1$.*

We also have the *unstable Hurewicz homomorphism*

$$h: \pi_n(X, x_0) \to \tilde{H}_n(X; \mathbb{Z})$$

since $\tilde{H}_0(S^0; \mathbb{Z}) \cong \mathbb{Z}$, (cf. 7.39). In fact $\tilde{H}(\mathbb{Z})_0(S^0) = \pi_0(H(\mathbb{Z}) \wedge S^0) = \pi_0(H(\mathbb{Z}))$, so we can choose $\imath \in \tilde{H}_0(S^0; \mathbb{Z})$ as generator. We let $\imath_n = \sigma^n(\imath) \in \tilde{H}_n(S^n; \mathbb{Z})$ for every $n \geqslant 0$ and then h is defined by $h[f] = f_*(\imath_n) \in \tilde{H}_n(X; \mathbb{Z})$ for any map $f: (S^n, s_0) \to (X, x_0)$.

10.24. Proposition. *For every (X, x_0) and $n \geqslant 0$ the diagram*

$$\begin{array}{ccc} & & \pi^S_n(X) \\ & \overset{i_0}{\nearrow} & \\ \pi_n(X, x_0) & & \downarrow h^S \\ & \underset{h}{\searrow} & \\ & & \tilde{H}_n(X; \mathbb{Z}) \end{array}$$

commutes, where $i_0: \pi_n(X, x_0) \to \pi^S_n(X) \cong \operatorname{dir\,lim} \pi_{n+k}(S^k X, *)$ is the natural homomorphism.

Proof: Recall that 8.26 gives us an equivalence $k(E): \Sigma E \to E \wedge S^1$ for any spectrum E; $k(E)$ is natural in $\mathscr{SP}'$. By iteration we get a natural equivalence $k_n(E): \Sigma^n E \to E \wedge S^n$. Then $\iota_n \in \tilde{H}_n(S^n; \mathbb{Z}) = [\Sigma^n S^0; H(\mathbb{Z}) \wedge S^n]$ is represented by the map

$$\Sigma^n S^0 \xrightarrow{[\Sigma^n \iota]} \Sigma^n H(\mathbb{Z}) \xrightarrow{[k_n(H(\mathbb{Z}))]} H(\mathbb{Z}) \wedge S^n.$$

The following diagram is commutative in $\mathcal{SP}'$

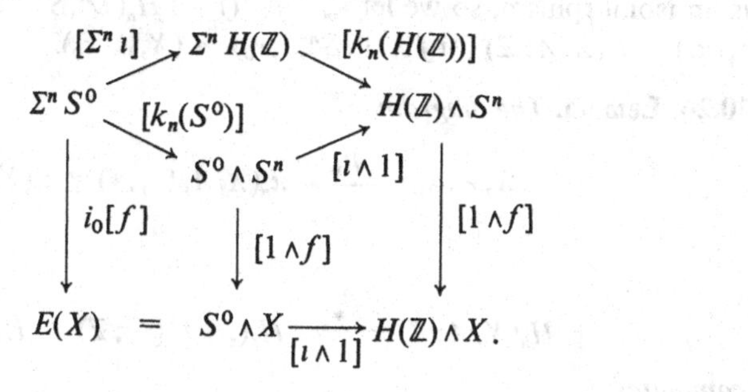

Thus

$$\begin{aligned} h[f] = f_*(\iota_n) &= [1 \wedge f] \circ [k_n(H(\mathbb{Z}))] \circ [\Sigma^n \iota] \\ &= [\iota \wedge 1] \circ i_0[f] = h^S \circ i_0[f]. \quad \square \end{aligned}$$

10.25. Theorem. (Hurewicz Isomorphism Theorem). *If X is an $(n-1)$-connected space, $n \geqslant 2$, then $h: \pi_q(X, x_0) \to \tilde{H}_q(X; \mathbb{Z})$ is an isomorphism for $q \leqslant n$ and an epimorphism for $q = n + 1$. In the case $n = 1$ h is an epimorphism for $q = n$.*

Proof: Combining 10.23 and 10.24, we see that it suffices to show i_0 is an isomorphism for $q \leqslant n$ and an epimorphism for $q = n + 1$ (or that i_0 is an epimorphism for $q = n$ in the case $n = 1$). $\pi_q^S(X) = \operatorname{dir\,lim} \pi_{q+k}(S^k X, *)$. The direct limit is over the system

$$\pi_q(X, x_0) \xrightarrow{\Sigma} \pi_{q+1}(SX, *) \xrightarrow{\Sigma} \pi_{q+2}(S^2 X, *) \xrightarrow{\Sigma} \cdots;$$

since X is $(n-1)$-connected (see 6.53), it follows $S^k X$ is $(n+k-1)$-connected and hence $\Sigma: \pi_{q+k}(S^k X, *) \to \pi_{q+k+1}(S^{k+1} X, *)$ is an isomorphism for $q < 2n + k - 1$ and an epimorphism for $q = 2n + k - 1$. If $q \leqslant n$, then all Σ are epimorphisms—in fact they are all isomorphisms except the first one in the case $n = 1$. If $n \geqslant 2$ and $q = n + 1$, then all Σ are epimorphisms. $\square$

Remark. In the case $n = 1$ we can say $\ker h = \ker\{\Sigma: \pi_1(X, x_0) \to \pi_2(SX, *)\}$. Since $\pi_2(SX, *)$ is abelian, it is clear that $\ker \Sigma$ contains the commutator subgroup $[\pi_1(X, x_0), \pi_1(X, x_0)]$ generated by all elements of the form $xyx^{-1}y^{-1}$ for $x, y \in \pi_1(X, x_0)$. A more detailed study of Σ than we have made would show that $\ker \Sigma$ is precisely $[\pi_1, \pi_1]$. Alternatively one can show directly that $\ker h$ is precisely $[\pi_1, \pi_1]$.

There is also a *relative Hurewicz homomorphism* $h: \pi_n(X, A, x_0) \to H_n(X, A; \mathbb{Z})$ for all $n \geqslant 0$, pairs (X, A, x_0).

$$p_*: H_n(D^n, S^{n-1}; \mathbb{Z}) \xrightarrow{\cong} H_n(S^n, \{*\}; \mathbb{Z}) = \tilde{H}_n(S^n; \mathbb{Z})$$

is an isomorphism, so we let $\kappa_n = p_*^{-1}(\iota_n) \in H_n(D^n, S^{n-1}; \mathbb{Z})$. Then $h[f] = f_*(\kappa_n) \in H_n(X, A; \mathbb{Z})$ for $f: (D^n, S^{n-1}, s_0) \to (X, A, x_0)$.

10.26. Lemma. *The diagram*

$$\begin{array}{ccc} \pi_n(X, A, x_0) & \xrightarrow{p_*} & \pi_n(X/A, \{*\}, *) \cong \pi_n(X/A, *) \\ \downarrow h & & \downarrow h \\ H_n(X, A; \mathbb{Z}) & \xrightarrow{p_*} & H_n(X/A, \{*\}; \mathbb{Z}) = \tilde{H}_n(X/A; \mathbb{Z}) \end{array}$$

commutes.

Proof: The morphism $\pi_n(X, A, x_0) \xrightarrow{p_*} \pi_n(X/A, \{*\}, *) \cong \pi_n(X/A, *)$ sends a class $[f]$ into the class $[\bar{f}]$, where if $f: (D^n, S^{n-1}, s_0) \to (X, A, x_0)$, then $\bar{f}: S^n = D^n/S^{n-1} \to X/A$ is the induced map such that $\bar{f} \circ p = p \circ f$. Then $h \circ p_*[f] = h[\bar{f}] = \bar{f}_*(\iota_n) = \bar{f}_*(p_*(\kappa_n)) = (\bar{f} \circ p)_*(\kappa_n) = p_* \circ f_*(\kappa_n) =$
$= p_* \circ h([f])$. □

10.27. Theorem. (Relative Hurewicz Isomorphism Theorem). *If (X, A) is an $(n-1)$-connected pair, $n \geqslant 2$, and A is 1-connected, then*

$$h: \pi_q(X, A, x_0) \to H_q(X, A; \mathbb{Z}),$$

is an isomorphism for $q \leqslant n$ and an epimorphism for $q = n + 1$. h is an epimorphism for $q = n$ even if A is only 0-connected.

Proof: By 6.22 $p_*: \pi_q(X, A, x_0) \to \pi_q(X/A, *)$ is an isomorphism for $q \leqslant n$, epimorphism for $q = n + 1$. Thus $(X/A, *)$ is $(n-1)$-connected, so by 10.25 $h: \pi_q(X/A, *) \to \tilde{H}_q(X/A)$ is an isomorphism for $q \leqslant n$, epimorphism for $q = n + 1$. If (X, A) is a CW-pair, then $p_*: H_q(X, A) \to \tilde{H}_q(X/A)$ is an isomorphism for all q. Thus the theorem follows for CW-pairs. The general case follows by taking CW-substitutes. □

Remark. Suppose A is not 1-connected. Recall that there is a natural action of $\pi_1(A, x_0)$ on $\pi_n(X, A, x_0)$. If $\alpha \in \pi_1(A, x_0)$ and $x \in \pi_n(X, A, x_0)$ then $p_*(\alpha x) = p_*(\alpha) \cdot p_*(x)$, where $p_*(\alpha) \in \pi_1(\{*\}, *) = 1$. Thus $p_*(\alpha) = 1$, so $p_*(\alpha x) = p_*(x)$. Hence $\ker h = \ker\{p_*: \pi_n(X, A, x_0) \to \pi_n(X/A, \{*\}, *)\}$ certainly contains the smallest normal subgroup $H \subset \pi_n(X, A, x_0)$ containing all elements of the form $\alpha x - x$ for $\alpha \in \pi_1(A, x_0)$, $x \in \pi_n(X, A, x_0)$. One can show that in fact $\ker h = \ker p_* = H$.

10.28. Theorem. (Whitehead). *Let $f:X \to Y$ be a map of spaces which are 0-connected.*

i) *If f is an n-equivalence ($n = \infty$ allowed), then $f_*: \tilde{H}_q(X;\mathbb{Z}) \to \tilde{H}_q(Y;\mathbb{Z})$ is an isomorphism for $q < n$ and an epimorphism for $q = n$.*
ii) *If X, Y are 1-connected and $f_*: \tilde{H}_q(X;\mathbb{Z}) \to \tilde{H}_q(Y;\mathbb{Z})$ is an isomorphism for $q < n$ and an epimorphism for $q = n$, then f is an n-equivalence ($n = \infty$ allowed).*

Proof: If M_f is the mapping cylinder of f, then f is an n-equivalence if an only if (M_f, X) is n-connected (3.18 and 3.32). Since we also have an exact homology sequence

$$\cdots \longrightarrow \tilde{H}_n(X;\mathbb{Z}) \xrightarrow{f_*} \tilde{H}_n(Y;\mathbb{Z}) \longrightarrow H_n(M_f, X;\mathbb{Z}) \xrightarrow{\partial} \tilde{H}_{n-1}(X;\mathbb{Z}) \xrightarrow{f_*} \cdots,$$

it follows $f_*: \tilde{H}_q(X;\mathbb{Z}) \to \tilde{H}_q(Y;\mathbb{Z})$ is an isomorphism for $q < n$ and an epimorphism for $q = n$ if and only if $H_q(M_f, X;\mathbb{Z}) = 0$ for $q \leqslant n$.

Suppose f is an n-equivalence; then $\pi_q(M_f, X, *) = 0$ for $q \leqslant n$, so by 10.27 $H_q(M_f, X;\mathbb{Z}) = 0$ for $q \leqslant n$. Conversely, if $H_q(M_f, X;\mathbb{Z}) = 0$ for $q \leqslant n$ and X, Y are 1-connected, then (M_f, X) is 1-connected, so by 10.27 $h: \pi_2(M_f, X, *) \to H_2(M_f, X;\mathbb{Z})$ is an isomorphism. If $n \geqslant 2$, then it follows (M_f, X) is 2-connected. Repeating this argument we find (M_f, X) is n-connected—i.e. f is an n-equivalence. □

10.29. Corollary. *A map $f:X \to Y$ between 1-connected CW-complexes is a homotopy equivalence if and only if $f_*: \tilde{H}_n(X;\mathbb{Z}) \to \tilde{H}_n(Y;\mathbb{Z})$ is an isomorphism for all $n \geqslant 0$.*

The results 10.23–10.25 have their duals for cohomology. $\iota: S^0 \to H(\mathbb{Z})$ induces a natural transformation $T^*(\iota): \pi_S^* \to H(\mathbb{Z})^*$ from stable cohomotopy to integral cohomology. We shall denote $T^*(\iota)$ by ψ_S and call it the *stable Hopf homomorphism.* Since $\psi_S: \pi_S^n(S^0) \to H(\mathbb{Z})^n(S^0)$ is just $\iota_*: \pi_{-n}(S^0) \to \pi_{-n}(H(\mathbb{Z}))$, we see that ψ_S is an isomorphism for $n > -1$ and an epimorphism for $n = -1$. We may therefore apply 7.67 with $N = -1$ to obtain the following theorem.

10.30. Theorem. *If (X, x_0) is a CW-complex with* $\dim X \leqslant n$, *then*

$$\psi_S: \pi_S^q(X) \to \tilde{H}^q(X;\mathbb{Z})$$

is an isomorphism for $q \geqslant n$ and an epimorphism for $q = n - 1$.

For any space (X, x_0) the nth *cohomotopy set* $\pi^n(X, x_0)$ is defined to be $[X, x_0; S^n, s_0]$. π^n is a cofunctor $\mathscr{PT} \to \mathscr{PS}$.

10.31. Lemma. *If (X,x_0) is a CW-complex with* $\dim X \leq n$ $(n \geq 2)$, *then* $i_0:[X,x_0;\ S^q,s_0] \to [E(X),\Sigma^q S^0] = \pi_S^q(X)$ *is a bijection if* $q \geq n$ *and a surjection if* $q = n-1$.

Proof: i_0 is defined by $i_0[f] = [E(f)]$ for $f:(X,x_0) \to (S^q,s_0)$. i_0 is a natural transformation of cofunctors satisfying W) and MV); since

$$i_0:[S^k,s_0;\ S^q,s_0] \to \pi_S^q(S^k) \cong \pi_S^{q-k}(S^0) = \pi_{k-q}^S(S^0)$$

is an isomorphism for $0 \leq k \leq q$ and an epimorphism for $k = q+1$, we may apply results of Chapter 9. □

In particular we may give the set $\pi^n(X,x_0)$ a group structure for $\dim X \leq n$ so that i_0 is an isomorphism. The group is abelian.

Exercise. Show that $\Sigma:[X,x_0;S^n,s_0] \to [SX,*;S^{n+1},s_0]$ is a bijection for $\dim X \leq 2n-2$. Hence $\pi^n(X,x_0)$ may be given a group structure if $\dim X \leq 2n-2$.

We have the *unstable Hopf function* $\psi:\pi^n(X,x_0) \to \tilde{H}^n(X;\ \mathbb{Z})$ defined to be $\psi_S \circ i_0$. ψ is a homomorphism for $\dim X \leq n$ (or even $\dim X \leq 2n-2$). An alternative description of ψ is clearly the following: $\tilde{H}^n(S^n;\ \mathbb{Z}) \cong \tilde{H}^0(S^0;\mathbb{Z}) = \pi_0(H(\mathbb{Z}))$; let $\iota^n \in \tilde{H}^n(S^n;\mathbb{Z})$ correspond to $\iota \in \pi_0(H(\mathbb{Z}))$. Then $\psi[f] = f^*\iota^n \in \tilde{H}^n(X;\ \mathbb{Z})$.

10.32. Theorem. (Hopf Isomorphism Theorem). *If (X,x_0) is a CW-complex with* $\dim X \leq n$ *then* $\psi:\pi^q(X,x_0) \to \tilde{H}^q(X;\ \mathbb{Z})$ *is an isomorphism for* $q \geq n$ *and an epimorphism for* $q = n-1$.

10.33. Theorem 10.32 has a generalization; if Y is $(n-1)$-connected, then for any CW-substitute Y' for Y there is by 6.39ii) a map

$$f':Y' \to H(\pi_n(Y,y_0),n)$$

which induces an isomorphism on $\pi_n(-)$. $[f'] \in [Y',y_0';H(\pi_n(Y,y_0),n),*]$ defines an element $\iota' \in \tilde{H}^n(Y';G)$, $G = \pi_n(Y,y_0)$. ι' defines an $\iota \in \tilde{H}^n(Y;G)$ which is $(n+1)$-universal in the sense of 9.6: $T_\iota:[S^q,s_0;\ Y,y_0] \to \tilde{H}^n(S^q;G)$ is an isomorphism for $q \leq n$, an epimorphism for $q = n+1$. 6.31 then gives the following.

10.34. Theorem. *For any CW-complex (X,x_0)*

$$T_\iota:[X,x_0;\ Y,y_0] \to \tilde{H}^n(X;\ G)$$

is an isomorphism if $\dim X \leq n$ *and an epimorphism if* $\dim X = n+1$.

10.35. Exercise. Suppose X is a CW-complex with $X^0 = \{x_0\}$. By considering the description of $H_1(X;\ \mathbb{Z})$ given by 10.9–10.11 and of $\pi_1(X,x_0)$ given in 6.55d) show that $h:\pi_1(X,x_0) \to \tilde{H}_1(X;\ \mathbb{Z})$ induces an isomorphism $\pi_1(X,x_0)/[\pi_1] \to H_1(X;\ \mathbb{Z})$, where $[\pi_1]$ denotes the commutator subgroup of $\pi_1(X, x_0)$. Deduce that the same is true for any 0-connected space (X, x_0).

10.36. Suppose $(X^\alpha : \alpha \in A\}$ is a directed set of subspaces of the space X $(\alpha \leqslant \beta \Rightarrow X_\alpha \subset X_\beta)$ such that for all compact subsets $C \subset X$ there is an $\alpha \in A$ with $C \subset X^\alpha$. Show that the inclusions $i_\alpha : X^\alpha \to X$ induce an isomorphism $\{i_{\alpha *}\} : \operatorname{dir\,lim} H_n(X^\alpha; G) \to H_n(X; G)$ for all $n \geqslant 0, G$.

Comments

Of the various homology theories ordinary homology is for most purposes the most useful for the simple reason that it is usually much easier to compute the ordinary homology groups of a given space X than it is to determine $k_*(X)$ for some other homology theory k_*. In fact, the first step in finding $k_*(X)$ usually consists in computing $H*(X)$ (cf. Chapter 15, 16). Thus ordinary homology is in a sense the fundamental homology theory.

Our proof of the Hurewicz isomorphism theorem seems comparatively simple, but it rests of course upon the Freudenthal suspension theorem and thus upon the deep homotopy results of Chapter 6. There are elementary proofs of the Hurewicz isomorphism theorem which employ little more than the definitions of homotopy group and singular homology group; see [80], for example.

References

1. A. Dold [36]
2. S. Eilenberg and N. Steenrod [40]
3. M. Greenberg [41]
4. E. H. Spanier [80]

Chapter 11

Vector Bundles and K-Theory

In Chapter 4 we defined the notion of a fibre bundle (a locally trivial fibration); in this chapter we consider an important class of fibre bundles—those for which every fibre has the structure of a vector space in a way which is compatible on neighboring fibres. We show how equivalence classes of such vector bundles over a CW-complex can be used to define groups $K^*(X)$ in such a way that K^* becomes a cohomology theory.

11.1. Definition Let F denote $\mathbb{R}$, $\mathbb{C}$ or $\mathbb{H}$—the real, complex or quaternionic numbers. An *n-dimensional F-vector bundle* is a fibre bundle $\xi = (B, p, E, F^n)$ in which each fibre $p^{-1}(b)$, $b \in B$, has the structure of a vector space over F such that there is an open covering $\{U_\alpha : \alpha \in A\}$ of B and for each $\alpha \in A$ a homeomorphism $\phi_\alpha : U_\alpha \times F^n \to p^{-1} U_\alpha$ with $p \circ \phi_\alpha = p_{U_\alpha}$ and $(\phi_\alpha | \{b\} \times F^n) : \{b\} \times F^n \to p^{-1}(b)$ a vector space isomorphism for each $b \in U_\alpha$. We speak of *real*, *complex* or *quaternionic* vector bundles according to whether $F = \mathbb{R}$, $\mathbb{C}$ or $\mathbb{H}$.

11.2. Examples. i) For any space B the *trivial* n-dimensional F-vector bundle is $(B, p_B, B \times F^n, F^n)$.

ii) For any $n \geqslant 1$ the *tangent bundle* $\tau(S^n)$ of the n-sphere S^n is the fibre bundle $(S^n, p, E, \mathbb{R}^n)$, where

$$E = \{(x, y) \in \mathbb{R}^{n+1} \times \mathbb{R}^{n+1} : \|x\| = 1,\ x \cdot y = 0\}$$

and $p : E \to S^n$ is defined by $p(x, y) = x$. We take $U_i \subset S^n$ to be the open set

$$U_i = \{x \in \mathbb{R}^{n+1} : \|x\| = 1,\ x_i \neq 0\}, \quad 1 \leqslant i \leqslant n + 1.$$

Then $\{U_i : 1 \leqslant i \leqslant n + 1\}$ is an open covering of S^n. We define $\phi_i : U_i \times \mathbb{R}^n \to p^{-1} U_i$ by

$$\phi_i(x, y) = (x, f_i(y) - (x \cdot f_i(y))\, x),$$

where $f_i : \mathbb{R}^n \to \mathbb{R}^{n+1}$ is defined by $f_i(y_1, \ldots, y_n) = (y_1, \ldots, y_{i-1}, 0, y_i, \ldots, y_n)$. The ϕ_i are homeomorphisms which satisfy $p \circ \phi_i = p_{U_i}$ and are linear on the fibres.

iii) For any $n \geqslant 1$ the *normal bundle* $v(S^n)$ is the fibre bundle $(S^n, p', E', \mathbb{R}^1)$, where

$$E' = \{(x, y) \in \mathbb{R}^{n+1} \times \mathbb{R}^{n+1} : \|x\| = 1, y = \lambda x, \lambda \in \mathbb{R}^1\}$$

and $p' : E' \to S^n$ is defined by $p'(x, y) = x$. Let us define

$$\phi : S^n \times \mathbb{R}^1 \to E' \text{ and } \psi : E' \to S^n \times \mathbb{R}^1$$

by $\phi(x, \lambda) = (x, \lambda x)$ for $(x, \lambda) \in S^n \times \mathbb{R}^1$ and $\psi(x, y) = (x, x \cdot y)$ for $(x, y) \in E'$. Then ϕ, ψ are inverses of one another, so $v(S^n)$ is a trivial bundle.

11.3. Definition. A *morphism* $\phi : \xi \to \xi'$ of F-vector bundles $\xi = (B, p, E, F^n)$ and $\xi' = (B', p', E', F^n)$ is a pair of maps $\phi : E \to E'$, $\bar{\phi} : B \to B'$ with $p' \circ \phi = \bar{\phi} \circ p$ and $(\phi | p^{-1}(b)) : p^{-1}(b) \to p'^{-1}(\bar{\phi}(b))$ a linear isomorphism for each $b \in B$. The identity maps $1 : E \to E$, $1 : B \to B$ define a morphism $1 : \xi \to \xi$. It is clear how to define the composition $\phi' \circ \phi$ of two morphisms so that $\phi' \circ \phi$ is also a morphism. We say that two bundles ξ, ξ' over the same base space B are *equivalent* if there are morphisms $\phi : \xi \to \xi'$, $\psi : \xi' \to \xi$ with $\bar{\phi} = 1 = \bar{\psi}$ and $\phi \circ \psi = 1$, $\psi \circ \phi = 1$. Note that equivalent bundles have the same dimension.

Closely related to the notion of vector bundle is that of principal G-bundle, where G is a topological group.

11.4. Definition. Let G be a topological group. A (locally trivial) *principal G-bundle* is a fibre bundle $\xi = (B, p, E, G)$ with a right G-action (cf. 4.19) $E \times G \to E$ of G on E such that there is an open covering $\{U_\alpha : \alpha \in A\}$ of B and for each $\alpha \in A$ a homeomorphism $\phi_\alpha : U_\alpha \times G \to p^{-1}(U_\alpha)$ satisfying $p \circ \phi_\alpha = p_{U_\alpha}$ and $\phi_\alpha(b, g) = \phi_\alpha(b, 1) \cdot g$, $b \in U_\alpha$, $g \in G$. A *morphism* $\phi : \xi \to \xi'$ of principal G-bundles is a pair of maps $\phi : E \to E'$, $\bar{\phi} : B \to B'$ such that $p' \circ \phi = \bar{\phi} \circ p$ and $\phi(eg) = \phi(e) \cdot g$ for all $g \in G$, $e \in E$. We have a notion of *equivalence* of two G-bundles ξ, ξ' over B as above.

11.5. Definition. A *chart* (U, ϕ) for a principal G-bundle $\xi = (B, p, E, G)$ is an open set $U \subset B$ and a homeomorphism $\phi : U \times G \to p^{-1} U$ such that $p \circ \phi = p_U$ and $\phi(b, g) = \phi(b, 1) \cdot g$, $b \in U$, $g \in G$. An *atlas* is a collection of charts $\{(U_\alpha, \phi_\alpha) : \alpha \in A\}$ such that $\{U_\alpha : \alpha \in A\}$ is an open covering of B.

By definition every principal G-bundle has at least one atlas.

11.6. Definition. A *set of transition functions* ξ for a space B and a topological group G is an open covering $\{U_\alpha : \alpha \in A\}$ for B and a collection of maps $g_{\alpha\beta} : U_\alpha \cap U_\beta \to G$ for $\alpha, \beta \in A$ such that

11.7. $$g_{\alpha\gamma}(b) = g_{\alpha\beta}(b) g_{\beta\gamma}(b) \quad \text{for all } b \in U_\alpha \cap U_\beta \cap U_\gamma.$$

Since in particular $g_{\alpha\alpha}(b) = g_{\alpha\alpha}(b)g_{\alpha\alpha}(b)$ for all $b \in U_\alpha$, it follows that

11.8. $$g_{\alpha\alpha}(b) = 1 \quad \text{for all } b \in U_\alpha.$$

11.7 and 11.8 then imply that

11.9. $$g_{\beta\alpha}(b) = g_{\alpha\beta}(b)^{-1} \quad \text{for all } b \in U_\alpha \cap U_\beta.$$

A *morphism* $r: \bar{\xi} \to \bar{\xi}'$ between sets of transition functions $\bar{\xi} = \{U_\alpha, g_{\alpha\beta} : \alpha, \beta \in A\}$ on B and $\bar{\xi}' = \{U'_\gamma, g'_{\gamma\delta} : \gamma, \delta \in \Gamma\}$ on B' is a map $\bar{r}: B \to B'$ and a collection of maps $r_{\gamma\alpha}: U_\alpha \cap \bar{r}^{-1} U'_\gamma \to G$ satisfying

11.10.
$$r_{\gamma\alpha}(b)\, g_{\alpha\beta}(b) = g'_{\gamma\delta}(\bar{r}(b))\, r_{\delta\beta}(b), \quad \text{all } b \in U_\alpha \cap U_\beta \cap \bar{r}^{-1} U'_\gamma \cap \bar{r}^{-1} U'_\delta.$$

Two sets of transition functions $\{U_\alpha, g_{\alpha\beta} : \alpha, \beta \in A\}$, $\{U'_\gamma, g'_{\gamma\delta} : \gamma, \delta \in \Gamma\}$ on the same space B are called *equivalent* if there are maps $r_{\gamma\alpha}: U_\alpha \cap U'_\gamma \to G$, $\alpha \in A$, $\gamma \in \Gamma$ satisfying

11.11. $$g'_{\gamma\delta}(b) = r_{\gamma\alpha}(b)\, g_{\alpha\beta}(b)\, r_{\delta\beta}(b)^{-1}, \quad \text{all } b \in U_\alpha \cap U_\beta \cap U'_\gamma \cap U'_\delta.$$

What we wish to show is that equivalence classes of G-bundles over a fixed space B are in one–one correspondence (in a natural way) with equivalence classes of sets of transition functions.

11.12. Lemma. i) *Let ξ be a principal G-bundle over B. To every atlas $\{(U_\alpha, \phi_\alpha) : \alpha \in A\}$ for ξ we can associate a unique set of transition functions $\bar{\xi} = \{U_\alpha, g_{\alpha\beta} : \alpha, \beta \in A\}$ such that*

11.13. $$\phi_\beta(b, g) = \phi_\alpha(b, g_{\alpha\beta}(b) g), \quad \textit{all } b \in U_\alpha \cap U_\beta,\ g \in G.$$

ii) *If $\bar{\xi}, \bar{\xi}'$ are sets of transition functions associated to atlases $\{(U_\alpha, \phi_\alpha) : \alpha \in A\}$, $\{(U'_\gamma, \phi'_\gamma) : \gamma \in \Gamma\}$ on bundles $\xi = (B, p, E, G)$, $\xi' = (B', p', E', G)$ as in i) above, and if $\phi: \xi \to \xi'$ is a bundle morphism, then there is a unique morphism of sets of transition functions $r: \bar{\xi} \to \bar{\xi}'$ such that $\bar{r} = \bar{\phi}$ and*

11.14. $$\phi \circ \phi_\alpha(b, g) = \phi'_\gamma(\bar{\phi}(b), r_{\gamma\alpha}(b) g), \quad \text{all } b \in U_\alpha \cap \bar{\phi}^{-1} U'_\gamma,\ g \in G.$$

Proof: i) For any $\alpha, \beta \in A$ the map

$$\psi_{\alpha\beta} = \phi_\alpha^{-1} \circ (\phi_\beta | (U_\alpha \cap U_\beta) \times G) : (U_\alpha \cap U_\beta) \times G \to (U_\alpha \cap U_\beta) \times G$$

satisfies $p_{U_\alpha \cap U_\beta} \circ \psi_{\alpha\beta} = p_{U_\alpha \cap U_\beta}$ and hence must be of the form $\psi_{\alpha\beta}(b, g) = (b, h_{\alpha\beta}(b, g))$ for some map $h_{\alpha\beta} : (U_\alpha \cap U_\beta) \times G \to G$—i.e. we have $\phi_\beta(b, g) = \phi_\alpha(b, h_{\alpha\beta}(b, g))$, all $b \in U_\alpha \cap U_\beta$, $g \in G$. Thus $\phi_\alpha(b, h_{\alpha\beta}(b, g)) = \phi_\beta(b, g) = \phi_\beta(b, 1) \cdot g = \phi_\alpha(b, h_{\alpha\beta}(b, 1)) \cdot g = \phi_\alpha(b, h_{\alpha\beta}(b, 1) \cdot g)$, for all $b \in U_\alpha \cap U_\beta$, $g \in G$. This implies that $h_{\alpha\beta}(b, g) = h_{\alpha\beta}(b, 1) \cdot g$. Therefore 11.13 will be satisfied if we take $g_{\alpha\beta}(b) = h_{\alpha\beta}(b,1)$ for $b \in U_\alpha \cap U_\beta$, $\alpha,\beta \in A$. For $b \in U_\alpha \cap U_\beta \cap U_\gamma$ we then have $\phi_\alpha(b, g_{\alpha\gamma}(b)g) = \phi_\gamma(b,g) = \phi_\beta(b, g_{\beta\gamma}(b)g) = \phi_\alpha(b, g_{\alpha\beta}(b)g_{\beta\gamma}(b)g)$, which implies that $g_{\alpha\gamma}(b) = g_{\alpha\beta}(b)g_{\beta\gamma}(b)$. Thus $\xi = \{U_\alpha, g_{\alpha\beta} : \alpha,\beta \in A\}$ is a set of transition functions.

ii) For any $\alpha \in A$, $\gamma \in \Gamma$ the maps

$$\theta_{\gamma\alpha} = \phi_\gamma'^{-1} \circ \phi \circ (\phi_\alpha | (U_\alpha \cap \bar{\phi}^{-1} U_\gamma') \times G) : (U_\alpha \cap \bar{\phi}^{-1} U_\gamma') \times G \to U_\gamma' \times G$$

satisfy $p_{U_\gamma'} \circ \theta_{\gamma\alpha} = \bar{\phi} \circ p_{U_\alpha \cap \bar{\phi}^{-1} U_\gamma'}$ and hence must be of the form $\theta_{\gamma\alpha}(b,g) = (\bar{\phi}(b), h_{\gamma\alpha}(b,g))$ for some map $h_{\gamma\alpha} : (U_\alpha \cap \bar{\phi}^{-1} U_\gamma') \times G \to G$. Thus $\phi \circ \phi_\alpha(b,g) = \phi_\gamma'(\bar{\phi}(b), h_{\gamma\alpha}(b,g))$, all $b \in U_\alpha \cap \bar{\phi}^{-1} U_\gamma'$, $g \in G$. Then $\phi_\gamma'(\bar{\phi}(b), h_{\gamma\alpha}(b,g)) = \phi \circ \phi_\alpha(b,g) = [\phi \circ \phi_\alpha(b,1)] \cdot g = \phi_\gamma'(\bar{\phi}(b), h_{\gamma\alpha}(b,1)) \cdot g = \phi_\gamma'(\bar{\phi}(b), h_{\gamma\alpha}(b,1)g)$, from which it follows that $h_{\gamma\alpha}(b,g) = h_{\gamma\alpha}(b,1) \cdot g$. Thus 11.14 will be satisfied if we take $r_{\gamma\alpha}(b) = h_{\gamma\alpha}(b,1)$ for $b \in U_\alpha \cap \bar{\phi}^{-1} U_\gamma'$, $\alpha \in A$, $\gamma \in \Gamma$. Suppose $\xi = \{U_\alpha, g_{\alpha\beta} : \alpha,\beta \in A\}$, $\xi' = \{U_\gamma', g_{\gamma\delta}' : \gamma,\delta \in \Gamma\}$. Then for all $b \in U_\alpha \cap U_\beta \cap \bar{\phi}^{-1} U_\gamma' \cap \bar{\phi}^{-1} U_\delta'$ and $g \in G$ we have

$$\phi_\gamma'(\bar{\phi}(b), r_{\gamma\alpha}(b) g_{\alpha\beta}(b) g) = \phi \circ \phi_\alpha(b, g_{\alpha\beta}(b) g) = \phi \circ \phi_\beta(b,g) = \phi_\delta'(\bar{\phi}(b), r_{\delta\beta}(b) g) = \phi_\gamma'(\bar{\phi}(b), g_{\gamma\delta}'(\bar{\phi}(b)) r_{\delta\beta}(b) g),$$

whence it follows that $r_{\gamma\alpha}(b) g_{\alpha\beta}(b) = g_{\gamma\delta}'(\bar{\phi}(b)) r_{\delta\beta}(b)$. Thus $r = \{r_{\gamma\alpha}\}$ is a morphism of sets of transition functions. □

11.15. Proposition. i) *If $\xi = \{U_\alpha, g_{\alpha\beta}\}$ is a set of transition functions for the space B and topological group G, then there is a principal G-bundle $\zeta = (B,p,E,G)$ and an atlas $\{U_\alpha, \phi_\alpha\}$ for ζ such that ξ is the set of transition functions for this atlas.*

ii) *Let $\zeta = (B,p,E,G)$, $\zeta' = (B',p',E',G)$ be two principal G-bundles with atlases $\{U_\alpha, \phi_\alpha\}$, $\{U_\gamma', \phi_\gamma'\}$ and associated sets of transition functions ξ, ξ'. If $r : \xi \to \xi'$ is a morphism of sets of transition functions, then there is a morphism $\phi : \zeta \to \zeta'$ of principal G-bundles inducing r.*

Proof: i) Given $\xi = \{U_\alpha, g_{\alpha\beta}\}$ we let $\bar{E} = \bigcup_{\alpha \in A} U_\alpha \times G \times \{\alpha\}$ and on $\bar{E}$ define the relation $\sim$ by $(b,g,\alpha) \sim (b',g',\beta)$ if and only if $b = b'$, $g = g_{\alpha\beta}(b)g'$. By 11.7–11.9 it follows $\sim$ is an equivalence relation. We take $E = \bar{E}/\!\sim$ and denote by $\{b,g,\alpha\}$ the class of (b,g,α) in E. We define $p : E \to B$ by $p\{b,g,\alpha\} = b$; p is clearly well defined and continuous. We define the action of G on E by $\{b,g,\alpha\}g' = \{b,gg',\alpha\}$, $g' \in G$; since $\{b,(g_{\alpha\beta}(b)g)g',\alpha\} = \{b, g_{\alpha\beta}(b)(gg'),\alpha\} = \{b,gg',\beta\}$, this action is well defined and clearly continuous.

If we define $\phi_\alpha : U_\alpha \times G \to p^{-1} U_\alpha$ by $\phi_\alpha(b,g) = \{b,g,\alpha\}$, then ϕ_α is continuous and satisfies $p \circ \phi_\alpha = p_{U_\alpha}$, $\phi_\alpha(b,g) = \{b, 1 \cdot g, \alpha\} = \{b,1,\alpha\}g = \phi_\alpha(b,1)g$, $b \in U_\alpha$, $g \in G$. ϕ_α is clearly a homeomorphism. Moreover, $\phi_\alpha(b, g_{\alpha\beta}(b)g) = \{b, g_{\alpha\beta}(b)g, \alpha\} = \{b,g,\beta\} = \phi_\beta(b,g)$ for all $b \in U_\alpha \cap U_\beta$,

$g \in G$, so $\{U_\alpha, g_{\alpha\beta}\}$ is the set of transition functions associated with the atlas $\{U_\alpha, \phi_\alpha\}$.

ii) Let us define $\phi: E \to E'$ as follows: let $\bar{\phi} = \bar{r}$ and if $e \in E$ and $p(e) \in U_\alpha \cap \bar{\phi}^{-1}(U'_\gamma)$ let $\phi(e) = \phi'_\gamma(\bar{\phi}(b), r_{\gamma\alpha}(b)g)$, where $e = \phi_\alpha(b,g)$. If $p(e) \in U_\beta \cap \bar{\phi}^{-1}(U'_\delta)$ also, then $e = \phi_\beta(b, g_{\beta\alpha}(b)g)$ and

$$\phi'_\delta(\bar{\phi}(b), r_{\delta\beta}(b)\, g_{\beta\alpha}(b)\, g) = \phi'_\delta(\bar{\phi}(b), g'_{\delta\gamma}(\bar{\phi}(b))\, r_{\gamma\alpha}(b)\, g) = \phi'_\gamma(\bar{\phi}(b), r_{\gamma\alpha}(b)\, g),$$

so $\phi(e)$ is well defined. Since $\phi | p^{-1}(U_\alpha \cap \bar{\phi}^{-1}(U'_\gamma))$ is the composite

$$p^{-1}(U_\alpha \cap \bar{\phi}^{-1}(U'_\gamma)) \xrightarrow{\phi_\alpha^{-1}} (U_\alpha \cap \bar{\phi}^{-1}(U'_\gamma)) \times G \xrightarrow{(\bar{\phi}, r_{\gamma\alpha}) \times 1}$$

$$(\bar{\phi}(U_\alpha) \cap U'_\gamma) \times G \times G \xrightarrow{1 \times \mu} (\bar{\phi}(U_\alpha) \cap U'_\gamma) \times G \xrightarrow{\phi'_\gamma}$$

$$p'^{-1}(\bar{\phi}(U_\alpha) \cap U'_\gamma),$$

we see that ϕ is continuous. We have $p' \circ \phi(e) = p' \circ \phi'_\gamma(\bar{\phi}(b), r_{\gamma\alpha}(b)g) = \bar{\phi}(b) = \bar{\phi} \circ p(e)$. Since $eh = \phi_\alpha(b,g)h = \phi_\alpha(b,gh)$, we have $\phi(eh) = \phi'_\gamma(\bar{\phi}(b), r_{\gamma\alpha}(b)gh) = \phi'_\gamma(\bar{\phi}(b), r_{\gamma\alpha}(b)g)h = \phi(e)h$ for $h \in G$, $e \in E$. Thus ϕ is a morphism of principal G-bundles. Since by definition of ϕ we have $\phi \circ \phi_\alpha(b,g) = \phi'_\gamma(\bar{\phi}(b), r_{\gamma\alpha}(b)g)$, $b \in U_\alpha \cap \bar{\phi}^{-1}(U'_\gamma)$, $g \in G$, we see that ϕ induces the morphism $r = \{r_{\gamma\alpha}\}$ of sets of transition functions. □

11.16. Theorem. *There is a one–one correspondence between equivalence classes of principal G-bundles over B and equivalence classes of sets of transition functions which can be described as follows: if ξ is a G-bundle, then to the equivalence class of ξ we assign the equivalence class $\{\bar{\xi}\}$ of sets of transition functions associated to an atlas of ξ by* 11.12i).

Proof: If $\phi: \xi \to \xi'$ is an equivalence of G-bundles, then 11.12ii) gives us a morphism $r(\phi): \bar{\xi} \to \bar{\xi}'$ which is clearly an equivalence of sets of transition functions since $\bar{\phi} = 1_B$. Thus the correspondence described is well defined.

Suppose $\bar{\xi}$ is any set of transition functions. 11.15 gives us a G-bundle ξ and an atlas $\{(U_\alpha, \phi_\alpha): \alpha \in A\}$ of ξ such that $\bar{\xi}$ is the associated set of transition functions. Thus the correspondence is surjective.

Suppose ξ, ξ' are two G-bundles such that $\bar{\xi}$ and $\bar{\xi}'$ are equivalent. Let $r: \bar{\xi} \to \bar{\xi}'$ be the equivalence; 11.15ii) gives us a morphism $\phi: \xi \to \xi'$ inducing r. In particular $\bar{\phi} = 1_B$. We also have the morphism $r^{-1}: \bar{\xi}' \to \bar{\xi}$ given by $r^{-1} = \{r_{\gamma\alpha}^{-1}\}$. The associated morphism $\phi^{-1}: \xi' \to \xi$ of bundles is the inverse of ϕ, for we have

$$\phi^{-1} \circ \phi \circ \phi_\alpha(b,g) = \phi^{-1} \circ \phi'_\gamma(b, r_{\gamma\alpha}(b)g) = \phi_\alpha(b, r_{\gamma\alpha}^{-1}(b)\, r_{\gamma\alpha}(b)\, g) = \phi_\alpha(b,g)$$

for all $b \in U_\alpha \cap U'_\gamma$, $g \in G$, $\alpha \in A$, $\gamma \in \Gamma$, whence $\phi^{-1} \circ \phi = 1$. Similarly $\phi \circ \phi^{-1} = 1$. Hence $\xi \simeq \xi'$, so the correspondence is injective. □

11.17. Corollary. *If $\phi:\xi \to \xi'$ is a morphism of principal G-bundles over B such that $\bar{\phi} = 1_B$, then ϕ is an equivalence.*

Proof: The associated morphism $r(\phi):\bar{\xi} \to \bar{\xi}'$ of sets of transition functions is obviously an equivalence. □

11.18. Let $GL(n,F)$ denote the group of all nonsingular $n \times n$ matrices over F. This is a topological group, its topology being that obtained from regarding $GL(n,F)$ as a subset of F^{n^2}. We wish to show there is a one–one correspondence between equivalence classes of n-dimensional F-vector bundles over B and equivalence classes of sets of transition functions for B and $GL(n,F)$—and hence also a one–one correspondence with principal $GL(n,F)$-bundles.

11.19. Definition. If $\xi = (B,p,E,F^n)$ is a vector bundle over B, then a *chart* (U,ϕ) for ξ is an open set $U \subset B$ and a homeomorphism $\phi: U \times F^n \to p^{-1}U$ such that $p \circ \phi = p_U$ and ϕ is linear on fibres. An *atlas* is a collection $\{(U_\alpha,\phi_\alpha):\alpha \in A\}$ of charts such that $\{U_\alpha:\alpha \in A\}$ is an open covering for B. ξ has at least one atlas.

Given an atlas $\{(U_\alpha,\phi_\alpha):\alpha \in A\}$ for a vector bundle ξ over B we can construct a set of transition functions $(U_\alpha, g_{\alpha\beta}:\alpha,\beta \in A\}$ for B and $GL(n,F)$ as follows: for any $\alpha,\beta \in A$ the maps

$$\psi_{\alpha\beta} = \phi_\alpha^{-1} \circ (\phi_\beta|(U_\alpha \cap U_\beta) \times F^n):(U_\alpha \cap U_\beta) \times F^n \to (U_\alpha \cap U_\beta) \times F^n$$

are of the form $\psi_{\alpha\beta}(b,v) = (b,h_{\alpha\beta}(b,v))$ for some map $h_{\alpha\beta}:(U_\alpha \cap U_\beta) \times F^n \to F^n$. For fixed $b \in U_\alpha \cap U_\beta$ the map $h_{\alpha\beta}(b,-):F^n \to F^n$ is a linear isomorphism—i.e. $h_{\alpha\beta}(b,-) \in GL(n,F)$. Thus $h_{\alpha\beta}(b,v) = g_{\alpha\beta}(b)\cdot v$, where $g_{\alpha\beta}(b) = h_{\alpha\beta}(b,-)$. Therefore we have

$$\phi_\beta(b,v) = \phi_\alpha(b,g_{\alpha\beta}(b)\cdot v), \quad b \in U_\alpha \cap U_\beta,\ v \in F^n.$$

From here on the argument runs parallel to that given above and proves the following.

11.20. Theorem. *There is a one–one correspondence between equivalence classes of n-dimensional F-vector bundles over B and equivalence classes of sets of transition functions for B and GL(n,F).*

We complete the triangle by describing explicitly the correspondence {vector bundles} $\leftrightarrow$ {$GL(n,F)$-bundles}.

11.21. Suppose $\xi = (B,p,E,G)$ is a principal G-bundle and Y is a topological space on which G acts on the left. We can form the *associated fibre bundle* $\xi[Y]$ with fibre Y as follows: define a right action of G on $E \times Y$ by $(e,y)g = (eg,g^{-1}y)$ for $g \in G$, $e \in E$, $y \in Y$. Let $E_Y = E \times Y/G$. If we denote by $\{e,y\}$ the image of (e,y) in E_Y, then $\{eg,y\}=\{e,gy\}$ for $g\in G$. We define $p_Y:E_Y\to B$ by $p_Y\{e,y\}=p(e)$. Since $p(eg)=p(e)$ for $g\in G$, p_Y is well defined and clearly continuous. If $\{(U_\alpha,\phi_\alpha)\}$ is an atlas for ξ we define an atlas $\{(U_\alpha,\psi_\alpha)\}$ for $\xi[Y]=(B,p_Y,E_Y,Y)$ by

$$\psi_\alpha(b,y)=\ \{\phi_\alpha(b,1),y\},\quad b\in U_\alpha,\ y\in Y.$$

ψ_α is continuous and satisfies $p_Y\circ\psi_\alpha=p_{U\alpha}$. The composite

$$p^{-1}(U_\alpha)\times Y\xrightarrow{\ \phi_\alpha^{-1}\times 1\ }U_\alpha\times G\times Y\xrightarrow{\ 1\times\rho\ }U_\alpha\times Y$$

$(\rho:G\times Y\to Y$ the action map of G on $Y)$ induces a map $p_Y^{-1}(U_\alpha)\to U_\alpha\times Y$ which is the inverse of ψ_α.

If $Y=F^n$, we can give the fibres a vector space structure so that ψ_α is linear on fibres for each α by taking $r\{e,v\}+s\{e',v'\}=\{e,rv+sg^{-1}v'\}$ if $e'g=e$, $g\in GL(n,F)$, $r,s\in F$.

11.22. Proposition. *If η is a vector bundle and ξ is a $GL(n,F)$-bundle associated to η by* 11.16 *and* 11.20 *then $\eta\simeq\xi[F^n]$.*

Proof: ξ is constructed by taking an atlas for η and constructing the principal bundle associated to the set of transition functions of the atlas. But this is the same set of transition functions as that for the resulting atlas on $\xi[F^n]$. □

Transition functions are often useful for proofs, as the following illustrates.

11.23. Proposition. *Suppose $(X;A,B)$ is a triad with $X=\mathring{A}\cup\mathring{B}$ and suppose ξ is an n-dimensional F-vector bundle (resp. principal G-bundle) over A, ξ' an n-dimensional F-vector bundle (resp. principal G-bundle) over B, $\phi:\xi|A\cap B\to\xi'|A\cap B$ an equivalence of the restrictions* (in general if $\eta=(B,p,E,F^n)$ and $A\subset B$, then $\eta|A=(A,p|p^{-1}(A),p^{-1}(A),F^n)$). *Then there is an n-dimensional F-vector bundle (resp. principal G-bundle) η over X with $\eta|A$ equivalent to ξ, $\eta|B$ equivalent to ξ'.*

Proof: In both cases we actually work with sets of transition functions. Let $\{U_\alpha,g_{\beta\alpha}:\alpha,\beta\in M\}$, $\{U'_\gamma,g'_{\delta\gamma}:\gamma,\delta\in N\}$ be sets of transition functions for ξ,ξ'. Let $r_{\gamma\alpha}:U_\alpha\cap U'_\gamma\to G$ (or $GL(n,F)$) be functions defining the equivalence $\xi|A\cap B\simeq\xi'|A\cap B$; i.e.

$$r_{\delta\beta}(b)\,g_{\beta\alpha}(b)=g'_{\delta\gamma}(b)\,r_{\gamma\alpha}(b),\quad \text{all } b\in U_\alpha\cap U_\beta\cap U'_\gamma\cap U'_\delta.$$

Then the sets $\{\mathring{A} \cap U_\alpha, \mathring{B} \cap U'_\gamma : \alpha \in M, \gamma \in N\}$ form an open covering of X and the functions

$$g_{\beta\alpha} : \mathring{A} \cap U_\alpha \cap U_\beta \to G$$

$$g'_{\delta\gamma} : \mathring{B} \cap U'_\gamma \cap U'_\delta \to G$$

$$r_{\gamma\alpha} : \mathring{A} \cap \mathring{B} \cap U_\alpha \cap U'_\gamma \to G$$

form a set of transition functions on X, as we easily check; for example, if $b \in U_\alpha \cap U_\beta \cap U'_\gamma$, then

$$r_{\gamma\beta}(b)\, g_{\beta\alpha}(b) = r_{\gamma\alpha}(b)$$

as we see by taking $\delta = \gamma$ in the equation above. Thus we get a principal G-bundle η (or vector bundle) on X which clearly satisfies $\eta | A \simeq \xi$, $\eta | B \simeq \xi'$. $\square$

11.24. If $\xi = (B, p, E, G)$ is a principal G-bundle over B and $f : B' \to B$ is a map, then we construct a principal G-bundle $f^*\xi$ over B' as follows: let $E' = \{(b', e) \in B' \times E : f(b') = p(e)\}$. We define $p' : E' \to B'$ by $p'(b', e) = b'$. If we define $f' : E' \to E$ by $f'(b', e) = e$, then the diagram

$$\begin{array}{ccc} E' & \xrightarrow{f'} & E \\ {\scriptstyle p'}\downarrow & & \downarrow{\scriptstyle p} \\ B' & \xrightarrow{f} & B \end{array}$$

commutes. The G-action on E' is defined by $(b', e)h = (b', eh)$, $h \in G$. Let $\{(U_\alpha, \phi_\alpha) : \alpha \in A\}$ be an atlas for ξ. Then $(\{f^{-1} U_\alpha, \phi'_\alpha) : \alpha \in A\}$ is an atlas for $f^*\xi = (B', p', E', G)$, where

$$\phi'_\alpha : f^{-1} U_\alpha \times G \to p'^{-1}(f^{-1} U_\alpha)$$

is defined by $\phi'_\alpha(b', g) = (b', \phi_\alpha(f(b'), g))$, $b' \in f^{-1} U_\alpha$, $g \in G$. Clearly if $\xi \simeq \xi'$ then $f^*\xi \simeq f^*\xi'$.

In terms of transition functions we can describe $f^*\xi$ as follows: if $\{U_\alpha, g_{\beta\alpha} : \alpha, \beta \in A\}$ is a set of transition functions for ξ, then $\{f^{-1} U_\alpha, g_{\beta\alpha} \circ f\}$ is a set of transition functions for $f^*\xi$. Thus it is clear that $(f \circ g)^*\xi \simeq g^*(f^*\xi)$ if $g : B'' \to B'$ is another map and also $1_B^*\xi \simeq \xi$.

There is an analogous construction for the *induced bundle* $f^*\xi$ if ξ is a vector bundle.

11.25. Proposition. *If $\phi : \xi \to \eta$ is a morphism of bundles, then there is a morphism $\psi : \xi \to \bar{\phi}^*\eta$ with $\bar{\psi} = 1_B$; hence by 11.17 $\xi \simeq \bar{\phi}^*\eta$.*

Proof: Recall that $\bar{\phi}^*\eta = (B, p', E', G)$, where $E' = \{(b, e) \in B \times E_\eta : \bar{\phi}(b) = p_\eta(e)\}$. We define $\psi : E_\xi \to E'$ by $\psi(e) = (p_\xi(e), \phi(e))$. $\psi(e)$ lies in E', since

$p_\eta \circ \phi(e) = \bar{\phi} \circ p_\xi(e)$ for all $e \in E_\xi$. Clearly $p' \circ \psi = p_\xi$ and $\psi(eh) = (p_\xi(eh), \phi(eh)) = (p_\xi(e), \phi(e)h) = (p_\xi(e), \phi(e))h = \psi(e)h$, $h \in G$. Thus ψ is a morphism of principal G-bundles with $\bar{\psi} = 1_B$. ▯

We now give a criterion for the existence of bundle morphisms.

11.26. Definition. For any map $p: E \to B$ a *section* of p is a map $\lambda: B \to E$ with $p \circ \lambda = 1_B$.

11.27. Proposition. *Let ξ, ξ' be two principal G-bundles. There is a one–one correspondence between morphisms $\phi: \xi \to \xi'$ and sections λ of the fibre bundle $\xi[E']$ given by $\phi \mapsto \lambda_\phi$ with $\lambda_\phi(b) = \{e, \phi(e)\}$ for any $e \in p^{-1}(b)$. (We give a left action of G on E' by $g \cdot e' = e' \cdot g^{-1}$ for $e' \in E'$, $g \in G$.)*

Proof: First we observe that λ_ϕ is well defined, for if $\tilde{e} \in p^{-1}(b)$ is another element in the fibre over b, then there is a $g \in G$ with $\tilde{e} = eg$ and $\{\tilde{e}, \phi(\tilde{e})\} = \{eg, \phi(eg)\} = \{eg, \phi(e)g\} = \{e, \phi(e)\}$. λ_ϕ is continuous, because it is in fact the map $E/G \to E \times E'/G$ induced by $(1, \phi)$. Clearly $p_{E'} \circ \lambda_\phi = 1_B$.

Now suppose $\lambda: B \to E_{E'} = E \times E'/G$ is a section of $p_{E'}$. $\lambda \circ p$ must be of the form $e \mapsto \{e, \phi_\lambda(e)\}$ for some function $\phi_\lambda: E \to E'$. For $g \in G$ we have $\{eg, \phi_\lambda(e)g\} = \{e, \phi_\lambda(e)\} = \lambda \circ p(e) = \lambda \circ p(eg) = \{eg, \phi_\lambda(eg)\}$, from which it follows $\phi_\lambda(eg) = \phi_\lambda(e)g$ for $e \in E$, $g \in G$. We take $\bar{\phi}_\lambda: B \to B'$ to be the induced map $E/G \to E'/G$. $(\phi_\lambda, \bar{\phi}_\lambda)$ will be a morphism of G-bundles if we show that ϕ_λ is continuous.

Suppose $e \in E$ and $U \subset E'$ is an open neighborhood of $\phi_\lambda(e)$. Because the action of G on E' is continuous, we can find an open neighborhood U' of $\phi_\lambda(e)$ in E' and an open neighborhood V of 1 in G with $U' \cdot V \subset U$. Choose a neighborhood V' of 1 in G with $(V')^{-1} V' \subset V$. Now let (W, ψ) be a chart for ξ around $b = p(e)$ and suppose $e = \psi(b, h)$, $h \in G$. $W' = \psi(W \times hV')$ is open in E and $\{W' \times U'\}$ is open in $E \times E'/G$. Thus $O = W' \cap p^{-1} \circ \lambda^{-1}\{W' \times U'\}$ is open in E, $e \in O$, and also

$$\lambda \circ p(O) \subset \{W' \times U'\}.$$

We now show that $\phi_\lambda(O) \subset U$. Suppose $x \in O$; then $\lambda \circ p(x) \in \{W' \times U'\}$—say $\lambda \circ p(x) = \{x', y'\}$ for $x' \in W'$, $y' \in U'$. Since $x, x' \in W'$, it follows $x' = x \cdot g^{-1}$ for some $g \in V$. Thus $\{x, \phi_\lambda(x)\} = \lambda \circ p(x) = \{x', y'\} = \{x \cdot g^{-1}, y'\} = \{x, y' \cdot g\}$. But $y' \in U'$, $g \in V$ means $y' \cdot g \in U$—that is $\phi_\lambda(x) \in U$. ▯

Remark. If $\pi_2: E \times E'/G \to E'/G = B'$ is the map induced by projection on the second factor, then the relations $\pi_2 \circ \lambda_\phi = \bar{\phi}$ and $\pi_2 \circ \lambda = \bar{\phi}_\lambda$ hold.

We wish to show now that if ξ is a principal G-bundle over B and $f_0, f_1: B' \to B$ are homotopic maps, then $f_0^* \xi \simeq f_1^* \xi$. To this end we define $\xi \times I$ for any principal G-bundle $\xi = (B, p, E, G)$ by $\xi \times I = (B \times I, p \times 1,$

$E \times I$, G), where the G-action is defined by $(e,t)h=(eh,t)$ for all $(e,t) \in E \times I$, $h \in G$. If $\{(U_\alpha, \phi_\alpha)\}$ is an atlas for ξ, then

$$\{(U_\alpha \times I, (\phi_\alpha \times 1) \circ (1 \times \tau))\}$$

is an atlas for $\xi \times I$.

11.28. Lemma. *Suppose $\xi = (D^n, p, E, G)$, $\xi' = (B', p', E', G)$ are principal G-bundles and $\phi:(p \times 1)^{-1}(D^n \times \{0\} \cup S^{n-1} \times I) \to E'$ is a bundle morphism, $F: D^n \times I \to B'$ an extension of $\bar{\phi}$. Then there is a bundle morphism $\Phi: \xi \times I \to \xi'$ extending ϕ with $\bar{\Phi} = F$.*

Proof: The proof is similar to that of 4.10. We choose atlases $\{(U_\alpha, \phi_\alpha)\}$, $\{(U'_\gamma, \phi'_\gamma)\}$ for ξ, ξ'. The sets $(U_\alpha \times I) \cap F^{-1}(U'_\gamma)$ form an open cover of $D^n \times I$, so since $D^n \times I$ is a compact metric space, we can find a number $\lambda > 0$ such that every subset $S \subset D^n \times I$ of diameter $< \lambda$ is contained in some $(U_\alpha \times I) \cap F^{-1}(U'_\gamma)$. We triangulate (D^n, S^{n-1}) so finely that every simplex has diameter $< \lambda/2$ and then subdivide I by $0 = t_0 < t_1 < \cdots < t_k = 1$ so that each set $\sigma \times [t_i, t_{i+1}]$ has diameter $< \lambda$, σ any simplex of D^n, $0 \leqslant i < k$.

Suppose Φ has already been constructed on $E \times [t_0, t_i] \cup p^{-1}(S^{n-1}) \times I$ ($\Phi | E \times \{0\} \cup p^{-1}(S^{n-1}) \times I = \phi$). We shall construct Φ on $E \times [t_i, t_{i+1}]$ simplex by simplex by induction over $\dim \sigma$. If $\dim \sigma = 0$, $\sigma \not\subset S^{n-1}$, then we choose α and γ with $\sigma \times [t_i, t_{i+1}] \subset (U_\alpha \times I) \cap F^{-1}(U'_\gamma)$. Since $p' \circ \Phi(\phi_\alpha(\sigma, g), t_i) = F(\sigma, t_i) \in U'_\gamma$ for every $g \in G$, there is a map $f_\sigma: G \to G$ such that

$$\Phi(\phi_\alpha(\sigma, g), t_i) = \phi'_\gamma(F(\sigma, t_i), f_\sigma(g)).$$

Since $\Phi(\phi_\alpha(\sigma, g), t_i) = \Phi(\phi_\alpha(\sigma, 1), t_i)g$, we find that $f_\sigma(g) = f_\sigma(1)g$ for all $g \in G$. Let $g_\sigma = f_\sigma(1)$; then $\Phi(\phi_\alpha(\sigma, g), t_i) = \phi'_\gamma(F(\sigma, t_i), g_\sigma g)$. We define Φ on $(p \times 1)^{-1}(\sigma \times [t_i, t_{i+1}])$ by

$$\Phi(\phi_\alpha(\sigma, g), t) = \phi'_\gamma(F(\sigma, t), g_\sigma g), \quad t \in [t_i, t_{i+1}].$$

Then Φ is continuous, extends $\Phi | (p \times 1)^{-1}(\sigma \times \{t_i\})$ and satisfies $p' \circ \Phi = F \circ (p \times 1)$, $\Phi((e,t)h) = \Phi(e,t)h$ for $(e,t) \in (p \times 1)^{-1}(\sigma \times [t_i, t_{i+1}])$, $h \in G$.

Suppose Φ has been constructed on $(p \times 1)^{-1}(\sigma' \times [0, t_{i+1}])$ for all σ' with $\dim \sigma' < m$ and suppose $\dim \sigma = m$, $\sigma \not\subset S^{n-1}$. Again choose α, γ with $\sigma \times [t_i, t_{i+1}] \subset (U_\alpha \times I) \cap F^{-1}(U'_\gamma)$. Φ is defined on $(p \times 1)^{-1}(\sigma \times \{t_i\} \cup \dot{\sigma} \times [t_i, t_{i+1}])$ and must be extended over $(p \times 1)^{-1}(\sigma \times [t_i, t_{i+1}])$. As before we find that on $(p \times 1)^{-1}(\sigma \times \{t_i\} \cup \dot{\sigma} \times [t_i, t_{i+1}])$ Φ has the form

$$\Phi(\phi_\alpha(x, g), t) = \phi'_\gamma(F(x, t), f_\sigma(x, t) g)$$

for some map $f_\sigma: \sigma \times \{t_i\} \cup \dot{\sigma} \times [t_i, t_{i+1}] \to G$. Since $\sigma \times \{t_i\} \cup \dot{\sigma} \times [t_i, t_{i+1}]$ is a retract of $\sigma \times [t_i, t_{i+1}]$ we can find an extension $\bar{f}_\sigma: \sigma \times$

$[t_i, t_{i+1}] \to G$. We then define Φ on $(p \times 1)^{-1}(\sigma \times [t_i, t_{i+1}])$ by

$$\Phi(\phi_\alpha(x,g),t) = \phi'_\gamma(F(x,t), f_\sigma(x,t)g),$$

$(x,t) \in \sigma \times [t_i, t_{i+1}]$, $g \in G$. Then Φ is continuous, extends

$$\Phi|(p \times 1)^{-1}(\sigma \times \{t_i\} \cup \dot{\sigma} \times [t_i, t_{i+1}])$$

and satisfies $p' \circ \Phi = F \circ (p \times 1)$, $\Phi((e,t)h) = \Phi(e,t)h$ for

$$(e,t) \in (p \times 1)^{-1}(\sigma \times [t_i, t_{i+1}]),\ h \in G.$$

This completes the induction. □

11.29. Lemma. *Let $\phi: \xi \to \xi'$ be a bundle morphism and $F: B \times I \to B'$ a homotopy with $F_0 = \bar{\phi}$. If B is a CW-complex, then there is a bundle morphism $\Phi: \xi \times I \to \xi'$ with $\bar{\Phi} = F$ and $\Phi|E \times \{0\} = \phi$.*

Proof: The proof will be the familiar skeleton by skeleton induction argument: if Φ is defined on $(p \times 1)^{-1}(B^{n-1} \times I)$, then by 11.27 it defines a section $\lambda: B^{n-1} \times I \to E \times I \times E'/G$. We wish to extend this section over $B^n \times I$. ($\lambda|(B^{-1} \times I)$ is easy whether $B^{-1} = \varnothing$ or $B^{-1} = \{b_0\}$ and F is rel b_0.) Let $f^n_\alpha: (D^n, S^{n-1}) \to (B^n, B^{n-1})$ be the characteristic map of the αth n-cell e^n_α. For each α $(y,t) \mapsto (y, \lambda(f^n_\alpha(y),t))$, $(y,t) \in D^n \times \{0\} \cup S^{n-1} \times I$, defines a section of $(f^{n*}_\alpha \xi \times I)[E']$. By 11.27 and 11.28 it follows there is an extension $\lambda_\alpha: D^n \times I \to E_{f^{n*}_\alpha \xi} \times I \times E'/G$ with $\pi_2 \circ \lambda_\alpha = F \circ (f^n_\alpha \times 1)$. If we extend λ over $e^n_\alpha \times I$ by $\lambda(f^n_\alpha(y),t) = (f^{n\prime}_\alpha \times_G 1) \circ \lambda_\alpha(y,t)$ for $y \in D^n$, $t \in I$, then λ is continuous and a section. λ in turn defines a morphism $\Phi: (p \times 1)^{-1}(B^n \times I) \to E'$. □

11.30. Theorem. *Let $\xi = (B,p,E,G)$ be a principal G-bundle and $f_0, f_1: B' \to B$ two maps of a CW-complex B' into B which are homotopic. Then $f_0^*\xi \simeq f_1^*\xi$.*

Proof: Let $F: B' \times I \to B$ be the homotopy. By 11.25 and 11.29 $f_0^*\xi \times I \simeq F^*\xi$, because there is a bundle morphism $\Phi: f_0^*\xi \times I \to \xi$ covering F. If $i_1: B' \to B' \times I$ is defined by $i_1(b') = (b',1)$, $b' \in B'$, then $f_1 = F \circ i_1$, so

$$f_1^*\xi = (F \circ i_1)^*\xi \simeq i_1^*(F^*\xi) \simeq i_1^*(f_0^*\xi \times I) = (f_0^*\xi \times I)|(B' \times \{1\}) \simeq f_0^*\xi. \quad \square$$

Remark. In everything which has been said so far no mention has been made of base points. We can equally well discuss bundles (B,p,E,G) in which B, E have base points b_0, e_0 and all maps are required to preserve base points. This alters slightly the notion of equivalence: two bundles $\xi = (B,p,E,G)$, $\xi' = (B,p',E',G)$ over B are equivalent if and only if there is a map $\phi: (E,e_0) \to (E',e_0')$ which satisfies $\phi(eh) = \phi(e)h$, $h \in G$, $e \in E$, and $p' \circ \phi = p$. The reader should check that all results proved so far apply equally well with base points.

11.31. For any topological group G we can define a cofunctor $k_G : \mathscr{PW}' \to \mathscr{PS}$ as follows: let $k_G(X)$ be the set of all (pointed) equivalence classes of principal G-bundles over X for each $(X, x_0) \in \mathscr{PW}'$. For any homotopy class $[f]$ of maps $f:(X,x_0) \to (Y,y_0)$ we define $k_G[f]:k_G(Y) \to k_G(X)$ by $k_G[f](\{\xi\}) = \{f^*\xi\}$. This function is well defined because of 11.30. The distinguished element of $k_G(X)$ is the equivalence class of the trivial bundle $(X, p \times 1, X \times G, G)$.

11.32. Proposition. *The cofunctor k_G satisfies the Wedge and Mayer–Vietoris axioms of Chapter 9.*

Proof: Suppose $(X; A, B)$ is a CW-triad. By 7.4 we can find an open neighborhood A' of A and a deformation $H: X \times I \to X$ of A' onto A carrying B into itself. If $j:(A, A \cap B) \to (A', A' \cap B)$ is the inclusion, then $j^*: k_G(A') \to k_G(A)$ and $(j|A \cap B)^*: k_G(A' \cap B) \to k_G(A \cap B)$ are bijections. Suppose given $x_1 \in k_G(A)$, $x_2 \in k_G(B)$ such that $i_1^*(x_1) = i_2^*(x_2) \in k_G(A \cap B)$. We can choose $x_1' \in k_G(A')$ so that $j^*(x_1') = x_1$. Hence $i_1'^*(x_1') = i_2'^*(x_2) \in k_G(A' \cap B)$ ($i_1': A' \cap B \to A'$, $i_2': A' \cap B \to B$ the inclusions). Thus x_1', x_2 are represented by bundles ξ_1 on A', ξ_2 on B such that $\xi_1|A' \cap B \simeq \xi_2|A' \cap B$. By 11.23 ($\mathring{A}' \cup \mathring{B} = X$) there is a bundle η over X with $\eta|A' \simeq \xi_1$, $\eta|B \simeq \xi_2$. If $y = \{\eta\} \in k_G(X)$, then $j_1'^* y = x_1'$, $j_2^* y = x_2$, $j_1': A' \to X$, $j_2: B \to X$ the inclusions. Thus $j_1^* y = j^* j_1'^* y = j^* x_1' = x_1$ ($j_1: A \to X$ the inclusion). Hence k_G satisfies MV).

Suppose $\{(X_\alpha, x_\alpha): \alpha \in A\}$ is a collection of pointed CW-complexes. Given $y_\alpha \in k_G(X_\alpha)$, $\alpha \in A$, choose a representing bundle $\xi_\alpha = (X_\alpha, p_\alpha, E_\alpha, G)$. We define a bundle $\xi = (\vee_\alpha X_\alpha, p, E, G)$ over $\vee_\alpha X_\alpha$ as follows: in $\vee_\alpha E_\alpha$ we identify e and e' if $e \in E_\alpha$, $e' \in E_\beta$ and there is a $g \in G$ such that $e = e_\alpha g$, $e' = e_\beta g$ ($e_\alpha \in E_\alpha$, $e_\beta \in E_\beta$ are the base points). We let E be the result of this identification and $p: E \to \vee_\alpha X_\alpha$ be the obvious projection. If we define a G-action on E by $\{e\}h = \{eh\}$, $e \in E_\alpha$, $h \in G$, then the action is well defined and $p(\{e\}) = p(\{e'\})$ if and only if $\{e'\} = \{e\}h$ for some $h \in G$. If $\{(U_\beta^\alpha, \phi_\beta^\alpha): \beta \in B_\alpha\}$ is an atlas for ξ_α, $\alpha \in A$, such that $\phi_\beta^\alpha(x_\alpha, 1) = e_\alpha$ whenever $x_\alpha \in U_\beta^\alpha$ (this last condition is always imposed for pointed spaces) then out of $\{(U_\beta^\alpha, \phi_\beta^\alpha): \beta \in B_\alpha, \alpha \in A\}$ we can make an atlas for ξ. It is clear that $i_\alpha^*: k_G(\vee_\alpha X_\alpha) \to k_G(X_\alpha)$ sends $\{\xi\}$ to y_α. Thus $\{i_\alpha^*\}: k_G(\vee_\alpha X_\alpha) \to \prod_\alpha k_G(X_\alpha)$ is surjective. Similarly if ξ, η are two bundles over $\vee_\alpha X_\alpha$ such that $\xi|X_\alpha \simeq \eta|X_\alpha$ for all $\alpha \in A$, then we can glue equivalences together on $\vee_\alpha X_\alpha$ (they preserve base points) to obtain an equivalence $\xi \simeq \eta$. Thus $\{i_\alpha^*\}$ is injective, and hence k_G satisfies W). $\square$

11.33. Therefore by 9.12 for each topological group G there is a CW-complex $(BG, *)$ (determined up to homotopy type) and a principal

G-bundle $\xi_G = (BG, p_G, EG, G)$ such that the natural transformation on $\mathscr{P}\mathscr{W}'$

$$T_G: [-; BG, *] \to k_G(-)$$

defined by $T[f] = \{f^* \xi_G\}$ is a natural equivalence. Thus principal G-bundles over a CW-complex X are *classified* by homotopy classes of maps $f:(X, x_0) \to (BG, *)$. It follows that n-dimensional F-vector bundles over X are classified by homotopy classes of maps $(X, x_0) \to (BGL(n, F), *)$.

We are going to give an explicit construction for $BO(n)$, $EO(n)$, $BU(n)$, $EU(n)$ shortly. We need a criterion for recognizing which principal G-bundles ξ can serve as universal bundles ξ_G. ξ_G must have the property that for every principal G-bundle ξ there is a morphism $\phi: \xi \to \xi_G$—or by 11.27 a section $\lambda: B \to E \times E_G/G$. There are several useful criteria for the existence of sections, of which the following is an important example.

11.34. Theorem. *Suppose (X, A) is a relative CW-complex and $\xi = (X, p, E, F)$ a fibre bundle over X. Let $\lambda': A \to E$ be a section of $\xi|A$. If F is $(n-1)$-connected and $\dim(X, A) \leqslant n$, then λ' has an extension to a section $\lambda: X \to E$ of p.*

Proof: From the exact sequence of the weak fibration $p: E \to X$ we see that p is an n-equivalence. From Lemma 6.30 it follows we can find a map $\lambda'': X \to E$ extending λ' and with $p \circ \lambda'' \simeq 1_X \operatorname{rel} A$.

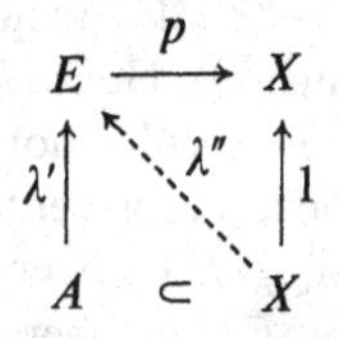

Since p is a weak fibration and (X, A) is a relative CW-complex, it follows (5.15) that the homotopy $p \circ \lambda'' \simeq 1_X \operatorname{rel} A$ can be lifted to a homotopy of $\lambda'' \operatorname{rel} A$. Thus we find a $\lambda: X \to E$ with $\lambda|A = \lambda'$ and $p \circ \lambda = 1_X$. ▯

11.35. Theorem. *If $\hat{\xi} = (\hat{B}, \hat{p}, \hat{E}, G)$ is a principal G-bundle with $\hat{E}$ n-connected, then the natural transformation*

$$T: [X, x_0; \hat{B}, \hat{b}_0] \to k_G(X)$$

given by $T[f] = \{f^* \hat{\xi}\}$ is a bijection for all CW-complexes X of dimension at most n.

Proof: Suppose ξ is a principal G-bundle over X; the fibre of $\xi[\hat{E}]$ is n-connected, so there is a section λ of $\xi[\hat{E}]$: we apply 11.34 with $A = \{x_0\}$ and $\lambda(x_0) = \{e_0, \hat{e}_0\}$. By 11.27 it follows there is a morphism $\phi: \xi \to \hat{\xi}$; therefore $\xi \simeq \bar{\phi}^* \hat{\xi} = T[\bar{\phi}]$, which shows T is surjective.

Suppose $f_0, f_1:(X,x_0) \to (\hat{B},\hat{b}_0)$ are maps with $f_0^*\hat{\xi} \simeq f_1^*\hat{\xi}$. We apply 11.34 to the pair $(X \times I, X \times \dot{I} \cup x_0 \times I)$; we take $\xi = f_0^*\xi$ and then have a section λ' of $((\xi \times I)|X \times \dot{I} \cup \{x_0\} \times I)[\hat{E}]$ corresponding to the morphisms $((\xi \times I)|X \times 0) = f_0^*\hat{\xi} \to \hat{\xi}$ and $((\xi \times I)|X \times 1) \simeq f_1^*\hat{\xi} \to \hat{\xi}$. If X is at most n-dimensional, then $(X \times I, X \times \dot{I} \cup x_0 \times I)$ is at most $(n+1)$-dimensional, so there is an extension of λ' to a section of $(\xi \times I)[\hat{E}]$ and hence a morphism $\phi:\xi \times I \to \hat{\xi}$ extending the given one on the ends. But then $\bar{\phi}:X \times I \to B$ is a homotopy from f_0 to f_1. This proves that T is injective. ▯

11.36. With the aid of 11.35 we can easily construct classifying spaces for $O(n)$-bundles. In 4.14 we constructed the Stiefel manifolds $V_{k,n} \cong O(n)/O(n-k)$ and the Grassman manifolds $G_{k,n} \cong O(n)/O(n-k) \times O(k)$ and various associated fibrations. From the exact homotopy sequence of the fibration $O(n-1) \to O(n) \to O(n)/O(n-1) \cong S^{n-1}$ we see that

$$\pi_q(O(n-1),*) \to \pi_q(O(n),*)$$

is an epimorphism for $q = n-2$ and $\pi_q(O(n-1),*) \cong \pi_q(O(n),*)$ for $q < n-2$. Thus $\pi_q(O(n),*) \to \pi_q(O(n+k),*)$ is an epimorphism for $q \leqslant n-1$ and an isomorphism for $q < n-1$, which shows that $\pi_q(V_{k,k+n},*) = 0$ for $q \leqslant n-1$. It follows that the principal $O(k)$-bundle $\xi_{k,k+n} = (G_{k,k+n}, \pi, V_{k,k+n}, O(k))$ is universal for $O(k)$-bundles over CW-complexes of dimension at most $n-1$. The inclusions

$$\mathbb{R}^k \subset \mathbb{R}^{k+1} \subset \cdots \subset \mathbb{R}^{k+n} \subset \cdots$$

given by $(x_1,\ldots,x_{k+1}) \to (0,x_1,\ldots,x_{k+1})$ induce inclusions

$$G_{k,k} \xrightarrow{j} G_{k,k+1} \xrightarrow{j} \cdots \longrightarrow G_{k,k+n} \longrightarrow \cdots;$$

we take $BO(k) = \bigcup_{n \geqslant 0} G_{k,k+n}$ with the weak topology—$U \subset BO(k)$ is open if and only if $U \cap G_{k,k+n}$ is open in $G_{k,k+n}$ for all $n \geqslant 0$. We take $EO(k) = \bigcup_{n \geqslant 0} V_{k,k+n}$ and $\pi:EO(k) \to BO(k)$ the obvious map. Then $\xi_{O(k)} = (BO(k), \pi, EO(k), O(k))$ is a universal $O(k)$-bundle for all CW-complexes.

In a similar fashion one sees that the bundle $V_{k,k+n}(\mathbb{C}) \to G_{k,k+n}(\mathbb{C})$ is universal for principal $U(k)$-bundles over CW-complexes of dimension at most $2n-1$; we take $BU(k) = \bigcup_{n \geqslant 0} G_{k,k+n}(\mathbb{C})$, $EU(k) = \bigcup_{n \geqslant 0} V_{k,k+n}(\mathbb{C})$. Likewise $BSp(k)$ can be taken to be $BSp(k) = \bigcup_{n \geqslant 0} G_{k,k+n}(\mathbb{H})$. If we let $G_{k,n}^+$ denote the space of *oriented* k-planes in $\mathbb{R}^n$, then

$$G_{k,n}^+ \cong O(n)/O(n-k) \times SO(k),$$

and we can take $BSO(k) = \bigcup_{n \geqslant 0} G_{k,k+n}^+$. Similarly

$$BSU(k) = \bigcup_{n \geqslant 0} G_{k,k+n}^+(\mathbb{C}).$$

11.37. The natural inclusion $O(k) \subset O(k+1)$ described in 4.14 induces a map $O(n+k)/O(n) \times O(k) \xrightarrow{i} O(n+k+1)/O(n) \times O(k+1)$ which commutes with the inclusions j of the last paragraph and hence induces $Bi: BO(k) \to BO(k+1)$. Similarly the natural inclusions $U(k) \subset O(2k)$ (think of $\mathbb{C}^k$ as $\mathbb{R}^{2k}$) induce maps $BU(k) \to BO(2k)$. More generally if $h: G \to G'$ is any continuous homomorphism of topological groups and if ξ is a principal G-bundle over B with transition functions $\{U_\alpha, g_{\beta\alpha}: \alpha, \beta \in A\}$, then $\{U_\alpha, h \circ g_{\beta\alpha}: \alpha, \beta \in A\}$ is a set of transition functions for a principal G'-bundle, which we denote by $h_* \xi$. Note that if $\xi \simeq \xi'$, then $h_* \xi \simeq h_* \xi'$, and if $f: B' \to B$ is a map, then $f^*(h_* \xi) \simeq h_*(f^* \xi)$, because both have transition functions $\{U_\alpha, h \circ g_{\beta\alpha} \circ f: \alpha, \beta \in A\}$. Thus h defines a natural transformation $T(h): k_G \to k_{G'}$ with $T(h)(\{\xi\}) = \{h_* \xi\}$. By 9.13 there is a map $Bh: BG \to BG'$ such that the diagram

$$\begin{array}{ccc} [-; BG, *] & \xrightarrow{T_G} & k_G \\ \big\downarrow Bh_* & & \big\downarrow T(h) \\ [-; BG', *] & \xrightarrow{T_{G'}} & k_{G'} \end{array}$$

commutes. If $h': G' \to G''$ is another continuous homomorphism, then $B(h' \circ h) \simeq Bh' \circ Bh$.

11.38. Let $H \subset G$ be a closed subgroup of the topological group G. We say that a principal G-bundle ξ has a *reduction to* H if there is a principal H-bundle η such that $i_* \eta$ and ξ are equivalent as G-bundles ($i: H \to G$ the inclusion).

11.39. Proposition. *Let ξ be a principal G-bundle over X with classifying map $:(X, x_0) \to (BG, *)$. Then ξ has a reduction to H if and only if there is a map $g:(X, x_0) \to (BH, *)$ so that the diagram*

$$\begin{array}{ccc} & & BH \\ & \overset{g}{\nearrow} & \big\downarrow Bi \\ X & \underset{f}{\searrow} & BG \end{array}$$

commutes up to homotopy.

The proof is obvious.

Remark. If $H \subset GL(n, F)$ is a subgroup, then the inclusion $i: H \to GL(n, F)$ induces an action of H on F^n. For any H-bundle η the fibre bundle $\eta[F^n]$ has a natural vector bundle structure, and if η is the reduction to H of some $GL(n, F)$-bundle ξ, then $\eta[F^n]$ and $\xi[F^n]$ are equivalent as F-vector bundles. For example, if $\{U_\alpha, g_{\beta\alpha}: \alpha, \beta \in A\}$ is a set of transition functions for η, then $\{U_\alpha, i \circ g_{\beta\alpha}: \alpha, \beta \in A\}$ is a set of transition functions for ξ and is also a set of transition functions for both the vector bundles $\eta[F^n]$ and $\xi[F^n]$.

11.40. Equivalence classes of G-bundles over a suspension SX are in one–one correspondence with the homotopy set $[SX,*;\ BG,*] \cong [X,x_0;\ \Omega BG,\omega_0]$. We can, however, describe $k_G(SX)$ as a set of homotopy classes in a different way. A *clutching function* for SX is a map $f:(X,x_0) \to (G,1)$. We can construct a bundle $\xi(f)$ on SX as follows: Let $C_-X = [0,3/4) \wedge X$, $C_+X = (1/4,1] \wedge X$, where we give $[0,3/4)$ the base point 0 and $(1/4,1]$ base point 1. Then $\{C_-X, C_+X\}$ is an open covering of SX with $C_-X \cap C_+X = (1/4,3/4)^+ \wedge X$, which has $\{1/2\} \times X$ as a strong deformation retract. We define $\bar{f}: C_-X \cap C_+X \to G$ by $\bar{f}[t,x] = f(x)$, $x \in X$, $t \in (1/4,3/4)$. Then $\{C_-X, C_+X, \bar{f}\}$ together with the functions $C_-X \to G$, $C_+X \to G$ which are identically 1 is a set of transition functions on SX and hence defines a bundle $\xi(f)$ by 11.15. Clearly if $g:(X',x_0') \to (X,x_0)$ is a map, then $Sg^*(\xi(f)) \simeq \xi(f \circ g)$.

11.41. Lemma. *If X is a CW-complex and $f_0, f_1:(X,x_0) \to (G,1)$ are two clutching functions, then $f_0 \simeq f_1 \operatorname{rel} x_0$ if and only if $\xi(f_0) \simeq \xi(f_1)$.*

Proof: Suppose $F: X \wedge I^+ \to G$ is a homotopy from f_0 to f_1. Then $\xi(F)$ is a bundle over $S(X \wedge I^+) \cong SX \wedge I^+$. If $i_0, i_1: SX \to SX \wedge I^+$ are the injections $i_\varepsilon(y) = [y,\varepsilon]$, $\varepsilon = 0,1$, $y \in SX$, then $i_0^*(\xi(F)) \simeq \xi(F \circ i_0) = \xi(f_0)$, $i_1^*(\xi(F)) \simeq \xi(F \circ i_1) = \xi(f_1)$. Now the projection $SX \times I \to SX \wedge I^+$ is a homotopy from i_0 to i_1, so by 11.30 $i_0^*(\xi(F)) \simeq i_1^*(\xi(F))$.

Conversely, if $\xi(f_0) \simeq \xi(f_1)$ then there are functions $r_-:(C_-X,*) \to (G,1)$, $r_+:(C_+X,*) \to (G,1)$ with $\bar{f}_1[t,x] = r_-[t,x] \cdot \bar{f}_0[t,x] \cdot r_+[t,x]$ for $x \in X$, $t \in (1/4,3/4)$. We define $H: X \times I \to G$ by

$$H(x,t) = r_-[t/2,x] \cdot f_0(x) \cdot r_+[1-t/2,x]$$

for $(x,t) \in X \times I$. Then $H_0(x) = r_-[0,x] \cdot f_0(x) \cdot r_+[1,x] = 1 \cdot f_0(x) \cdot 1 = f_0(x)$ and

$$\begin{aligned} H_1(x) = r_-[1/2,x] \cdot f_0(x) \cdot r_+[1/2,x] = r_-[1/2,x] \cdot \bar{f}_0[1/2\ x] \cdot r_+[1/2,x] = \\ \bar{f}_1[1/2,x] = f_1(x). \quad \square \end{aligned}$$

Thus we get a natural transformation $T': [-;\ G,1] \to k_G \circ S$ defined by $T'[f] = \{\xi(f)\}$. $T'(X)$ is an injection for every X.

11.42. Lemma. *T' is a natural equivalence on $\mathscr{PW}'$.*

Proof: We must show that every G-bundle ξ over SX is of the form $\xi(f)$ for some clutching function $f: X \to G$—at least up to equivalence. Since C_-X, C_+X are contractible $k_G(C_-X) = 0 = k_G(C_+X)$, and hence $\xi|C_-X$, $\xi|C_+X$ are trivial. We therefore have an atlas $\{(C_-X, \phi_-), (C_+X, \phi_+)\}$ for ξ, where ϕ_-, ϕ_+ are any trivializations of $\xi|C_-X$, $\xi|C_+X$. This atlas yields a set of transition functions $\{C_-X, C_+X, g\}$, $g: C_-X \cap C_+X \to G$. We define $f:(X, x_0) \to (G, 1)$ by $f(x) = g[1/2, x]$. We shall show $\xi \simeq \xi(f)$.

Clearly $\bar{f} \simeq g \operatorname{rel} *$; let $H:(C_-X \cap C_+X) \wedge I^+ \to G$ be the homotopy. Then $\{C_-X \wedge I^+, C_+X \wedge I^+, H\}$ is a set of transition functions on $SX \wedge I^+$ and hence defines a bundle η over $SX \wedge I^+$ such that $i_0^*\eta \simeq \xi(f)$, $i_1^*\eta \simeq \xi$. But $\eta \simeq \pi^* i_0^* \eta \simeq i_0^*\eta \times I$ ($\pi: SX \times I \to SX \times 0$ the projection), so $i_1^*\eta \simeq i_1^*(i_0^*\eta \times I) \simeq i_0^*\eta$. □

11.43. Suppose that G has the homotopy type of a CW-complex. Since the cofunctors $[-; G, 1]$ and $[-; \Omega BG, \omega_0]$ are naturally equivalent on $\mathscr{PW}'$, it follows from 9.13 that there is a homotopy equivalence

$$\mu_G:(G, 1) \simeq (\Omega BG, \omega_0).$$

It therefore follows, in particular, that $\pi_n(BG, *) \cong \pi_{n-1}(G, 1)$, $n \geqslant 1$. μ_G is natural for homomorphisms $h: G \to G'$.

Let $O(n, F)$ denote $O(n)$, $U(n)$ or $Sp(n)$ according to whether $F = \mathbb{R}$, $\mathbb{C}$ or $\mathbb{H}$.

11.44. Proposition. *$O(n, F)$ is a strong deformation retract of $GL(n, F)$ for $F = \mathbb{R}$, $\mathbb{C}$ or $\mathbb{H}$ and all $n \geqslant 1$.*

Proof: Let $H(n, F) \subset GL(n, F)$ denote the group of upper triangular matrices. An immediate consequence of the Gram–Schmidt theorem is that every matrix $A \in GL(n, F)$ has a unique representation in the form $A = OH$, where $O \in O(n, F)$, $H \in H(n, F)$. Thus the map $\theta: O(n, F) \times H(n, F) \to GL(n, F)$ defined by $\theta(O, H) = OH$ is a bijection. θ is in fact a homeomorphism. Now $H(n, F)$ is contractible: we define $G: H(nF) \times I \to H(n, F)$ by

$$G\left(\begin{pmatrix} 1 & a_{12} & \cdots & a_{1n} \\ & 1 & \cdots & \cdots \\ & & \cdots & \cdots \\ & & 1 & a_{n-1,n} \\ & & & 1 \end{pmatrix}, t\right) = \begin{pmatrix} 1 & (1-t)a_{12} & \cdots & (1-t)a_{1n} \\ & 1 & \cdots & \cdots \\ & & \cdots & \cdots \\ & & 1 & (1-t)a_{n-1,n} \\ & & & 1 \end{pmatrix}.$$

Then $1 \times G: O(n, F) \times H(n, F) \times I \to O(n, F) \times H(n, F)$ is a deformation of $O(n, F) \times H(n, F)$ onto $O(n, F) \times \{I\}$. □

11.45. Proposition. *Every $GL(n, F)$-bundle over a CW-complex has a reduction to an $O(n, F)$-bundle. Thus we may regard any F-vector bundle as being of the form $\eta[F^n]$ for some $O(n, F)$-bundle η.*

Proof: The diagram

$$\begin{array}{ccc}
\pi_n(BO(n,F),*) & \xrightarrow{Bi_*} & \pi_n(BGL(n,F),*) \\
\| \wr & & \| \wr \\
\pi_{n-1}(\Omega BO(n,F),\omega_0) & \xrightarrow{\Omega Bi_*} & \pi_{n-1}(\Omega BGL(n,F),\omega_0) \\
\cong \uparrow \mu_{O*} & & \cong \uparrow \mu_{G*} \\
\pi_{n-1}(O(n,F),1) & \underset{\cong}{\xrightarrow{i_*}} & \pi_{n-1}(GL(n,F),1)
\end{array}$$

commutes, showing that Bi is a weak homotopy equivalence. Hence by 6.31

$$[X,x_0;BO(n,F),*] \xrightarrow{Bi_*} [X,x_0;BGL(n,F),*]$$

is a bijection for all CW-complexes X. Therefore any map $f:(X,x_0)\to(BGL(n,F),*)$ has a lifting to $BO(n,F)$. ▯

We may thus regard n-dimensional F-vector bundles as being classified by homotopy classes of maps into $BO(n,F)$.

11.46. We can construct the *product* $\xi\times\xi'$ of two fibre bundles $\xi=(B,p,E,F)$ and $\xi'=(B',p',E',F')$; we take $\xi\times\xi'=(B\times B,'\ p\times p',$ $E\times E',\ F\times F')$. The product of a principal G-bundle with a principal G'-bundle is in a natural fashion a $G\times G'$-bundle; for example, if $\bar{\xi}=\{U_\alpha,g_{\beta\alpha}:\alpha,\beta\in A\}$ and $\bar{\xi}'=\{V_\gamma,g'_{\delta\gamma}:\gamma,\delta\in\Gamma\}$ are sets of transition functions for ξ and ξ' respectively, then

$$\bar{\xi}\times\bar{\xi}'=\{U_\alpha\times V_\gamma,g_{\beta\alpha}\times g'_{\delta\gamma}:\alpha,\beta\in A,\gamma,\delta\in\Gamma\}$$

is a set of transition functions for $\xi\times\xi'$. Similarly, if ξ,ξ' are F-vector bundles with fibres F^n,F^m respectively, then $\xi\times\xi'$ can be given a vector bundle structure in a natural way so that the fibre is F^{n+m}; for example, if $\{U_\alpha,\phi_\alpha:\alpha\in A\}$, $\{V_\gamma,\psi_\gamma:\gamma\in\Gamma\}$ are atlases for ξ and ξ' respectively, then $\{U_\alpha\times V_\gamma,(\phi_\alpha\times\psi_\gamma)\circ(1\times\tau\times1):\alpha\in A,\gamma\in\Gamma\}$ is an atlas for $\xi\times\xi'$. Moreover, for a $GL(n,F)$-bundle ξ and a $GL(m,F)$-bundle ξ' it is clear that $\xi[F^n]\times\xi'[F^m]\simeq(\xi\times\xi')[F^{n+m}]$.

We can also define the *Whitney sum* $\xi\oplus\xi'$ of two vector bundles over the same space B. It is the vector bundle $\Delta^*(\xi\times\xi')$, where $\Delta:B\to B\times B$ is the diagonal map. Thus $\xi\oplus\xi'$ is again a vector bundle over B. The fibre of $\xi\oplus\xi'$ over $b\in B$ is the vector space direct sum of the fibre of ξ over b and the fibre of ξ' over b.

Let ξ_n denote the universal $O(n,F)$-bundle over $BO(n,F)$. Let $h:O(n,F)\times O(m,F)\to O(n+m,F)$ be the inclusion of 4.14. Then

$h_*(\xi_n \times \xi_m)$ over $BO(n,F) \times BO(m,F)$ is classified by a map

$$\omega_{n,m}: BO(n,F) \times BO(m,F) \to BO(n+m,F)$$

which is unique up to homotopy.

11.47. Lemma. *For any three fibre bundles ξ_1, ξ_2, ξ_3 there is an equivalence $(\xi_1 \times \xi_2) \times \xi_3 \simeq \xi_1 \times (\xi_2 \times \xi_3)$. If ξ_1, ξ_2, ξ_3 are principal bundles or vector bundles, then the equivalence is one of principal bundles, respectively vector bundles.*

11.48. Corollary. *The diagram*

$$\begin{array}{ccc} BO(n,F) \times BO(m,F) \times BO(q,F) & \xrightarrow[1 \times \omega_{m,q}]{} & BO(n,F) \times BO(m+q,F) \\ \downarrow \omega_{n,m} \times 1 & & \downarrow \omega_{n,m+q} \\ BO(n+m,F) \times BO(q,F) & \xrightarrow[\omega_{n+m,q}]{} & BO(n+m+q,F) \end{array}$$

commutes up to homotopy for all n, m, q.

The proofs are evident.

11.49. Lemma. *If ξ_1, ξ_2 are fibre bundles over B_1, B_2 respectively and $\tau: B_1 \times B_2 \to B_2 \times B_1$ is the "switch" map, then $\tau^*(\xi_2 \times \xi_1) \simeq \xi_1 \times \xi_2$. If ξ_1, ξ_2 are principal bundles or vector bundles, then the equivalence is one of principal bundles, respectively vector bundles.*

11.50. Corollary. *If ξ_1, ξ_2 are vector bundles over the same space B then $\xi_1 \oplus \xi_2 \simeq \xi_2 \oplus \xi_1$.*

Proof: $\tau \circ \Delta = \Delta$; therefore

$$\xi_1 \oplus \xi_2 = \Delta^*(\xi_1 \times \xi_2) \simeq \Delta^* \tau^*(\xi_2 \times \xi_1) = \Delta^*(\xi_2 \times \xi_1) = \xi_2 \oplus \xi_1. \quad \square$$

11.51. Corollary. *The diagram*

$$\begin{array}{ccc} BO(n,F) \times BO(m,F) & \xrightarrow{\tau} & BO(m,F) \times BO(n,F) \\ & \searrow \omega_{n,m} \quad \omega_{m,n} \swarrow & \\ & BO(n+m,F) & \end{array}$$

commutes up to homotopy for all n, m.

For each n let ε_n denote the trivial $O(n,F)$-bundle:

$$\varepsilon_n = (*, p_n, O(n,F), O(n,F)).$$

We shall also write ε_n for the trivial $O(n,F)$-bundle over B: $\varepsilon_n = (B, p_B, B \times O(n,F), O(n,F))$. One readily verifies the following.

11.52. Lemma. *If $Bi: BO(n,F) \to BO(n+m,F)$ is the map of classifying spaces induced by the inclusion $O(n,F) \subset O(n+m,F)$ and $h: O(n,F) \times O(m,F) \to O(n+m,F)$ is the inclusion of 4.14, then $h_*(\xi_n \times \varepsilon_m) \simeq Bi^*(\xi_{n+m})$ as bundles over $BO(n,F) \times \{*\} \cong BO(n,F)$.*

11.53. Corollary. *The diagram*

$$\begin{array}{ccc} & BO(n,F) \times BO(m,F) & \\ {\scriptstyle (1,*)} \nearrow & & \searrow {\scriptstyle \omega_{n,m}} \\ BO(n,F) & \xrightarrow{Bi} & BO(n+m,F) \end{array}$$

commutes up to homotopy for all n, m.

11.54. Definition. In what follows we allow a vector bundle to have different dimensions over different components of a space. We say that two vector bundles ξ, η over the same base space X (but not necessarily of the same dimension) are *stably equivalent* (notation: $\xi \simeq_s \eta$) if and only if there are trivial bundles $\varepsilon', \varepsilon''$ over X with $\xi \oplus \varepsilon' \simeq \eta \oplus \varepsilon''$. We denote by $\widetilde{KO}(X)$ the set of all stable equivalence classes of real vector bundles over X. The stable class of a vector bundle ξ will be denoted by $\{\xi\}_s$.

We introduce an addition $+$ on $\widetilde{KO}(X)$ by taking $\{\xi\}_s + \{\eta\}_s = \{\xi \oplus \eta\}_s$. $+$ is clearly well defined, and from 11.47 and 11.50 it follows that $+$ is associative and commutative. The class $\{\varepsilon\}_s$ of any trivial bundle is a zero element for $+$. We wish to show that $\widetilde{KO}(X)$ is a group—that is, that negatives $-\{\xi\}_s$ exist.

11.55. Proposition. *If ξ is a vector bundle over a finite dimensional CW-complex X, then there is a vector bundle η over X such that $\xi \oplus \eta \simeq \varepsilon^N$ for some N.*

Proof: Suppose ξ is a k-dimensional bundle and X is n-dimensional. From 11.36 it follows that k-dimensional bundles over X are classified by homotopy classes of maps of X into $G_{k,k+n+1}$. The universal k-plane bundle $\xi_k = (G_{k,k+n+1}, \pi, E_k, \mathbb{R}^k)$ can be described as follows:

$$E_k = \{(W,x) \in G_{k,k+n+1} \times \mathbb{R}^{n+k+1} : x \in W\}$$

and $\pi: E_k \to G_{k,k+n+1}$ is the map given by $\pi(W,x) = W$.

We construct the complementary bundle $\bar{\xi}_k = (G_{k,k+n+1}, \bar{\pi}, \bar{E}_k, \mathbb{R}^{n+1})$ with $\bar{E}_k = \{(W,x) \in G_{k,k+n+1} \times \mathbb{R}^{n+k+1} : x \perp W\}$ and $\bar{\pi}(W,x) = W$. Now we can define a bundle map $\phi: \xi_k \oplus \bar{\xi}_k \to \varepsilon^{n+k+1} = G_{k,k+n+1} \times \mathbb{R}^{n+k+1}$ by $\phi((W,x),(W,x')) = (W, x+x')$. ϕ is in fact a bundle equivalence.

Now suppose our given ξ is represented by $f: X \to G_{k,k+n+1}$, so that $\xi \simeq f^*\xi_k$. Let $\eta = f^*\bar{\xi}_k$; then $\xi \oplus \eta \simeq f^*\xi_k \oplus f^*\bar{\xi}_k \simeq f^*(\xi_k \oplus \bar{\xi}_k) \simeq \varepsilon^{n+k+1}$. $\square$

$\widetilde{KO}$ is a representable cofunctor on the category $\mathscr{PCW}'_F$ of finite, connected pointed CW-complexes. There are natural inclusions $G_{k,k+n} \subset G_{k+1,k+n+1}$ given by $W \mapsto \langle W, e_{k+n+1}\rangle =$ the $(k+1)$-plane spanned by W and e_{k+n+1}. These inclusions induce inclusions $BO(k) \subset BO(k+1)$ for all $k \geqslant 1$. We let $BO = \bigcup_{k\geqslant 1} BO(k)$ with the weak topology. Let ξ_k be the universal k-dimensional vector bundle over $BO(k)$. If X is a finite CW-complex, then X is compact, so any map $f:X \to BO$ factors through $BO(k)$ for some k. We set $T[f] = \{f^*\xi_k\}_s \in \widetilde{KO}(X)$. We must show T is well defined. Suppose $g:X \to BO$ is another map with $g \simeq f$. If g factors through $BO(l)$, then $T[g] = \{g^*\xi_l\}_s$. Let $H:X \times I \to BO$ be a homotopy from f to g; then H factors through $BO(n)$, say. Since $f = H \circ i_0$, $g = H \circ i_1$, we may as well assume $n \geqslant k$, $n \geqslant l$. By 11.30 $f^*\xi_n \simeq g^*\xi_n$. By 11.52 we have $g^*\xi_l \oplus \varepsilon^{n-l} \simeq g^*\xi_n \simeq f^*\xi_n \simeq f^*\xi_k \oplus \varepsilon^{n-k}$. In other words $g^*\xi_l \simeq_s f^*\xi_k$, or $T[g] = T[f]$.

11.56. Theorem. $T:[-; BO, *] \to \widetilde{KO}$ *is a natural equivalence on* $\mathscr{PCW}'_F$.

Proof: Suppose ξ is any vector bundle over X; then there is a map $f:X \to BO(k)$ with $f^*\xi_k \simeq \xi$. If $i_k: BO(k) \to BO$ is the inclusion, then $T[i_k \circ f] = \{\xi\}_s$, so T is surjective.

Suppose $T[f] = T[g]$ for two maps $f, g: X \to BO$. Then there is an n such that f and g both factor through $BO(n)$ and $\{f^*\xi_n\}_s = \{g^*\xi_n\}_s$ or in other words $f^*\xi_n \oplus \varepsilon^k \simeq g^*\xi_n \oplus \varepsilon^k$ for some k. Thus $f^*\xi_{n+k} \simeq g^*\xi_{n+k}$, so by the universality of ξ_{n+k} we have $f \simeq g$ in $BO(n+k)$ and hence in BO. Thus $[f] = [g]$. □

In similar fashion we can define $\widetilde{K}(X) =$ stable equivalence classes of complex vector bundles over X and $\widetilde{KSp}(X) =$ stable equivalence classes of quaternionic vector bundles over X. We have natural equivalences

$$[-; BU, *] \simeq \widetilde{K}$$

$$[-; BSp, *] \simeq \widetilde{KSp}$$

on $\mathscr{PCW}'_F$. By 9.14 we know we can give BO, BU and BSp the structure of weak H-groups, but they are in fact more—namely H-groups (even infinite loop spaces; cf. [22]). To show this we make the following observations first made by Boardman and Vogt [22].

Let $\mathscr{C}$ be the category whose objects are inner product spaces of finite or countable dimension and whose morphisms are linear isometries. We define $B:\mathscr{C} \to \mathscr{T}$ as follows: for any k-dimensional inner product space V we let $B(V) = G_k(V \otimes \mathbb{R}^\infty)$, the Grassmannian of k-planes in $V \otimes \mathbb{R}^\infty$ (the topology is defined essentially as in 4.14). If $f: V \to W$ is a linear isometry of a k-dimensional inner product space V into an m-dimensional inner product space W, then the orthogonal complement

$U = f(V)^{\perp}$ of $f(V)$ in W is defined; it is an $(m-k)$-dimensional plane. Given a k-plane P in $V \otimes \mathbb{R}^{\infty}$, we let

$$B(f)(P) = (f \otimes 1)(P) + U \otimes e_1 \quad (e_1 = (1, 0, 0, \ldots) \in \mathbb{R}^{\infty}).$$

Then

$$B(f)(P) \in G_m(W \otimes \mathbb{R}^{\infty}) = B(W),$$

and $B(f)$ is easily seen to be continuous. Now for an infinite dimensional inner product space $V \in \mathscr{C}$ we note that V is the direct limit of its finite dimensional subspaces, so we define $B(V) = \bigcup_{\dim W < \infty} B(W)$ with the weak topology. If $f: V \to V'$ is a linear isometry, $W \subset V$ finite dimensional, $x \in B(W)$, then $B(f|W)(x) \in B(f(W))$ defines an element of $B(V')$ and thus f defines $B(f): B(V) \to B(V')$. $B: \mathscr{C} \to \mathscr{T}$ is then a functor. One readily checks that if $f_t: V \to W$ is a homotopy through linear isometries, then $B(f_t): B(V) \to B(W)$ is also a homotopy. Note that $B(\mathbb{R}^0) = *$, $B(\mathbb{R}^n) = BO(n)$, $B(\mathbb{R}^{\infty}) = BO$. We can make similar considerations for BU, BSp and indeed even for BSO, BSU.

We define a "Whitney sum"

$$w: B(V) \times B(W) \to B(V \oplus W)$$

for all finite dimensional $V, W \in \mathscr{C}$ by $w(P, P') = P \oplus P'$ for P a k-plane in $V \otimes \mathbb{R}^{\infty}$, P' an m-plane in $W \otimes \mathbb{R}^{\infty}$. For then $P \oplus P'$ is a $(k+m)$-plane in $V \otimes \mathbb{R}^{\infty} \oplus W \otimes \mathbb{R}^{\infty} = (V \oplus W) \otimes \mathbb{R}^{\infty}$. We then extend w to arbitrary $V, W \in \mathscr{C}$ by taking limits. It is immediate that w makes the following diagrams commute (strictly)

11.57.

$$\begin{array}{ccc} B(U) \times B(V) \times B(W) & \xrightarrow{w \times 1} & B(U \oplus V) \times B(W) \\ \Big\downarrow{\scriptstyle 1 \times w} & & \Big\downarrow{\scriptstyle w} \\ B(U) \times B(V \oplus W) & \xrightarrow{w} & B(U \oplus V \oplus W) \end{array}$$

$$\begin{array}{ccc} B(V) \times B(W) & \xrightarrow{\tau} & B(W) \times B(V) \\ & {\scriptstyle w}\searrow \quad \swarrow{\scriptstyle w} & \\ & B(V \oplus W) & \end{array} \qquad \begin{array}{ccc} B(\mathbb{R}^0) \times B(V) & \xrightarrow{w} & B(\mathbb{R}^0 \oplus V) \\ & \searrow{\scriptstyle \cong} & \Big\downarrow{\scriptstyle \cong} \\ & & B(V) \end{array}$$

$$\begin{array}{ccc} B(V) \times B(W) & \xrightarrow{w} & B(V \oplus W) \\ \Big\downarrow{\scriptstyle B(f) \times B(g)} & & \Big\downarrow{\scriptstyle B(f \oplus g)} \\ B(V') \times B(W') & \xrightarrow{w} & B(V' \oplus W') \end{array}$$

$f: V \to V'$, $g: W \to W'$ linear isometries.

Now we wish to define our product $\omega: BO \times BO \to BO$ as follows:

$$B(\mathbb{R}^\infty) \times B(\mathbb{R}^\infty) \xrightarrow{w} B(\mathbb{R}^\infty \oplus \mathbb{R}^\infty) \xrightarrow{B(f)} B(\mathbb{R}^\infty)$$

where f is some appropriate isometric isomorphism $f: \mathbb{R}^\infty \oplus \mathbb{R}^\infty \to \mathbb{R}^\infty$. But which one? The clever observation of Boardman and Vogt is that it makes no difference which, for any two such—say f,g—are homotopic through linear isometries.

11.58. Lemma. *Any two linear isometries $f,g: V \to \mathbb{R}^\infty$ are homotopic through linear isometries.*

Proof: Let $i_1, i_2: \mathbb{R}^\infty \to \mathbb{R}^\infty \oplus \mathbb{R}^\infty$ be the inclusions of the two summands and let $a: \mathbb{R}^\infty \to \mathbb{R}^\infty \oplus \mathbb{R}^\infty$ be defined by $a(e_{2n}) = (e_n, 0)$, $a(e_{2n-1}) = (0, e_n)$ for $n \geqslant 1$.

i) $i_1 \simeq i_2$: for $\sqrt{1-t} \cdot i_1 + \sqrt{t} \cdot i_2 = H_t$ is such a homotopy through isometries.
ii) $i_1 \simeq a$: for if $b: \mathbb{R}^\infty \to \mathbb{R}^\infty$ is defined by $b(e_n) = e_{2n}$, then $b \simeq 1_{\mathbb{R}^\infty}$, because we can apply Gram–Schmidt to the homotopy $(1-t) \cdot 1_{\mathbb{R}^\infty} + t \cdot b$ to get a homotopy through isometries. Thus $a = a \circ 1_{\mathbb{R}^\infty} \simeq a \circ b = i_1$. Let K denote the homotopy $i_1 \simeq a$.

Now let h be any linear isometry $h: V \to \mathbb{R}^\infty$ and define $F: V \times I \to \mathbb{R}^\infty$ by

$$F_t = \begin{cases} a^{-1} \circ K_{1-4t} \circ f & 0 \leqslant t \leqslant 1/4 \\ a^{-1} \circ (f \oplus h) \circ H_{4t-1} & 1/4 \leqslant t \leqslant 1/2 \\ a^{-1} \circ (g \oplus h) \circ H_{3-4t} & 1/2 \leqslant t \leqslant 3/4 \\ a^{-1} \circ K_{4t-3} \circ g & 3/4 \leqslant t \leqslant 1. \end{cases}$$

(In fact one sees from the proof that the *space* of linear isometries $V \to \mathbb{R}^\infty$ is contractible.) □

The proof that (BO, ω) is a homotopy-commutative H-group is now relatively simple. To show that $* = B(\mathbb{R}^0) \in B(\mathbb{R}^\infty)$ is a homotopy unit it suffices (because of the properties 11.57 of w) to show that

$$\begin{array}{ccc} B(\mathbb{R}^0 \oplus \mathbb{R}^\infty) & \xrightarrow{B(i \oplus 1)} & B(\mathbb{R}^\infty \oplus \mathbb{R}^\infty) \\ & {\scriptstyle B(g)} \searrow & \downarrow {\scriptstyle B(f)} \\ & & B(\mathbb{R}^\infty) \end{array}$$

commutes up to homotopy, where $i: \mathbb{R}^0 \to \mathbb{R}^\infty$ is the inclusion, $f: \mathbb{R}^\infty \oplus \mathbb{R}^\infty \to \mathbb{R}^\infty$ is the isometric isomorphism used to define ω and $g: \mathbb{R}^0 \oplus \mathbb{R}^\infty \to$

$\mathbb{R}^\infty$ is the obvious isomorphism. For this, however, it suffices to have $f \circ (i \oplus 1) \simeq g: \mathbb{R}^0 \oplus \mathbb{R}^\infty \to \mathbb{R}^\infty$ through linear isometries, which in turn follows from 11.58. To prove homotopy associativity of ω it suffices (again because of 11.57) to show that

$$\begin{array}{ccc} B(\mathbb{R}^\infty \oplus \mathbb{R}^\infty \oplus \mathbb{R}^\infty) & \xrightarrow{B(f \oplus 1)} & B(\mathbb{R}^\infty \oplus \mathbb{R}^\infty) \\ \downarrow{\scriptstyle B(1 \oplus f)} & & \downarrow{\scriptstyle B(f)} \\ B(\mathbb{R}^\infty \oplus \mathbb{R}^\infty) & \xrightarrow{B(f)} & B(\mathbb{R}^\infty) \end{array}$$

commutes up to homotopy. Again it suffices to prove $f \circ (f \oplus 1) \simeq f \circ (1 \oplus f): \mathbb{R}^\infty \oplus \mathbb{R}^\infty \oplus \mathbb{R}^\infty \to \mathbb{R}^\infty$ through linear isometries, and again this follows from 11.58. Homotopy commutativity is proved similarly.

Analogous considerations show that BU, BSp, BSO and BSU are homotopy commutative H-groups.

We can give a slightly different description of $\widetilde{KO}(X)$. Let $V_{\mathbb{R}}(X)$ denote the set of all equivalence classes of real vector bundles over X, $X \in \mathscr{W}'_F$. We can define an addition $+$ on $V_{\mathbb{R}}(X)$ by taking $\{\xi\} + \{\eta\} = \{\xi \oplus \eta\}$. $V_{\mathbb{R}}(X)$ then becomes an abelian semigroup. Now there is a standard procedure for extending an abelian semigroup to an abelian group.

11.59. Proposition. *Let S be an abelian semigroup. There exist an abelian group $K(S)$ and a homomorphism $\phi: S \to K(S)$ of semigroups with the following universal property: for any abelian group A and homomorphism $\psi: S \to A$ of semigroups there is a unique homomorphism $\theta: K(S) \to A$ with $\theta \circ \phi = \psi$. $K(S)$ is determined up to isomorphism by this property.*

Proof: The uniqueness of $K(S)$ is clear. We proceed to construct such a $K(S)$. Let $\sim$ be the relation on $S \times S$ defined by $(x, x') \sim (y, y')$ if and only if there is a $z \in S$ with $x + y' + z = x' + y + z$ (one thinks of (x, x') as being $x - x'$). $\sim$ is an equivalence relation; we let $K(S) = S \times S/\sim$ and denote by $\{x, x'\}$ the equivalence class of (x, x'). We introduce a sum $+$ in $K(S)$ by taking $\{x, x'\} + \{y, y'\} = \{x + y, x' + y'\}$ and checking that this is well defined. With $+$ $K(S)$ becomes an abelian group; the zero element is $\{0, 0\}$ and $-\{x, x'\} = \{x', x\}$. If we define $\phi: S \to K(S)$ by $\phi(x) = \{x, 0\}$, then ϕ is a homomorphism.

Now suppose given a homomorphism $\psi: S \to A$, A an abelian group. If $\theta: K(S) \to A$ exists such that $\theta \circ \phi = \psi$ then we must have $\theta(\{x, x'\}) = \theta(\{x, 0\} - \{x', 0\}) = \theta(\phi(x) - \phi(x')) = \theta(\phi(x)) - \theta(\phi(x')) = \psi(x) - \psi(x')$. On the other hand if we define θ by $\theta(\{x, x'\}) = \psi(x) - \psi(x')$, then θ is well defined and a homomorphism such that $\theta \circ \phi = \psi$. $\square$

We now let $KO(X) = K(V_{\mathbb{R}}(X))$ for $X \in \mathscr{W}'_F$. The elements of $KO(X)$ can be written in the form $\{\xi\} - \{\eta\}$ for ξ, η real vector bundles over X. We have the homomorphism $\psi: V_{\mathbb{R}}(X) \to \widetilde{KO}(X)$ for $(X, x_0) \in \mathscr{PW}'_F$ given by $\psi\{\xi\} = \{\xi\}_s$, the stable equivalence class of ξ. Thus there is a homomorphism $\theta: KO(X) \to \widetilde{KO}(X)$ such that $\theta\{\xi\} = \{\xi\}_s$ for $(X, x_0) \in \mathscr{PW}'_F$. We also have a homomorphism $\kappa: \widetilde{KO}(X) \to KO(X)$ given by $\kappa(\{\xi\}_s) = \{\xi\} - \{\varepsilon^n\}$ if $\dim \xi = n$; for if $\{\xi\}_s = \{\eta\}_s$, then $\xi \oplus \varepsilon^k \simeq \eta \oplus \varepsilon^l$ for some k, l. If $\dim \eta = m$, then $n + k = m + l$, or in other words $n - l = m - k$. Thus also $\xi \oplus \varepsilon^m \simeq \eta \oplus \varepsilon^n$, which implies that $\{\xi\} - \{\varepsilon^n\} = \{\eta\} - \{\varepsilon^m\}$ in $KO(X)$. Clearly $\theta \circ \kappa = 1_{\widetilde{KO}(X)}$.

KO is a cofunctor on $\mathscr{W}'_F$. For $(X, x_0) \in \mathscr{PCW}'_F$ we can consider the maps $c: X \to \{x_0\}$, $i: \{x_0\} \to X$ and the induced maps $i^*: KO(X) \to KO(\{x_0\})$, $c^*: KO(\{x_0\}) \to KO(X)$. The sequence

$$0 \longrightarrow \widetilde{KO}(X) \xrightarrow{\ \kappa\ } KO(X) \underset{c^*}{\overset{i^*}{\rightleftarrows}} KO(\{x_0\}) \longrightarrow 0$$

is exact, so we may identify $\widetilde{KO}(X)$ with $\ker[i^*: KO(X) \to KO(\{x_0\})]$ for all $(X, x_0) \in \mathscr{PCW}'_F$. We then have a direct sum decomposition

$$KO(X) \cong \widetilde{KO}(X) \oplus KO(\{x_0\}).$$

The elements $\{\xi\} \in V_{\mathbb{R}}(\{x_0\})$ are clearly determined by $\dim \xi$, so $V_{\mathbb{R}}(\{x_0\}) \cong \mathbb{N}$, the natural numbers. Thus $KO(\{x_0\}) = K(V_{\mathbb{R}}\{x_0\}) = K(\mathbb{N}) = \mathbb{Z}$. We have a natural equivalence

$$[-; \mathbb{Z} \times BO] \to KO$$

on $\mathscr{PCW}'_F$, where $\mathbb{Z}$ is given the discrete topology.

In an analogous fashion we can define $V_{\mathbb{C}}(X)$ and $K(X) = K(V_{\mathbb{C}}(X))$, $V_{\mathbb{H}}(X)$ and $KSp(X) = K(V_{\mathbb{H}}(X))$. We get

$$K(X) \cong \widetilde{K}(X) \oplus K(\{x_0\})$$

$$KSp(X) \cong \widetilde{KSp}(X) \oplus KSp(\{x_0\}).$$

$K(\{x_0\}) = KSp(\{x_0\}) = \mathbb{Z}$, and

$$[-; \mathbb{Z} \times BU] \cong K, \quad [-; \mathbb{Z} \times BSp] \cong KSp.$$

Having constructed cofunctors $\widetilde{KO}, \widetilde{K}, \widetilde{KSp}$ on $\mathscr{PW}'_F$, we may naturally ask about the "coefficient groups" $\widetilde{KO}(S^q)$, $\widetilde{K}(S^q)$, $\widetilde{KSp}(S^q)$, $q \geqslant 0$. By 11.56 we have

$$\widetilde{KO}(S^q) = [S^q, s_0; BO, *] = \pi_q(BO, *), q \geqslant 1.$$

For $q = 0$ we have

$$\mathbb{Z} \oplus \mathbb{Z} \cong KO(S^0) \cong \widetilde{KO}(S^0) \oplus \mathbb{Z},$$

so clearly $\widetilde{KO}(S^0) \cong \mathbb{Z}$. Similarly

$$\tilde{K}(S^0) \cong \mathbb{Z}, \quad \tilde{K}(S^q) \cong \pi_q(BU, *), \quad q \geqslant 1,$$
$$\widetilde{KSp}(S^0) \cong \mathbb{Z}, \quad \widetilde{KSp}(S^q) \cong \pi_q(BSp, *), \quad q \geqslant 1.$$

In computing these coefficient groups we will be greatly helped by the following very important theorem.

11.60. Theorem (Bott). *There are homotopy equivalences*

$$\mathbb{Z} \times BU \simeq \Omega^2 BU$$
$$\mathbb{Z} \times BO \simeq \Omega^4 BSp$$
$$\mathbb{Z} \times BSp \simeq \Omega^4 BO.$$

We shall prove $\mathbb{Z} \times BU \simeq \Omega^2 BU$ in Chapter 16. The proofs of the other two statements are more complicated but the same in principle.

In particular we have

$$\pi_{q+2}(BU, *) \cong \pi_q(\Omega^2 BU, \omega_0) \cong \pi_q(\mathbb{Z} \times BU, *) \cong \begin{cases} \mathbb{Z} & q = 0 \\ \pi_q(BU, *) & q \geqslant 1, \end{cases}$$

$$\pi_{q+4}(BO, *) \cong \pi_q(\Omega^4 BO, \omega_0) \cong \pi_q(\mathbb{Z} \times BSp, *) \cong \begin{cases} \mathbb{Z} & q = 0 \\ \pi_q(BSp, *) & q \geqslant 1, \end{cases}$$

$$\pi_{q+4}(BSp, *) \cong \pi_q(\Omega^4 BSp, \omega_0) \cong \pi_q(\mathbb{Z} \times BO, *) \cong \begin{cases} \mathbb{Z} & q = 0 \\ \pi_q(BO, *) & q \geqslant 1. \end{cases}$$

Thus the groups $\pi_q(BU, *)$ repeat with period 2, the groups $\pi_q(BO, *)$ repeat with period 8 and the groups $\pi_q(BSp, *)$ are the same as the groups $\pi_q(BO, *)$ with a shift of 4 in dimension. For this reason 11.60 is called the "Bott Periodicity Theorem".

Thus in order to obtain all the coefficient groups $\widetilde{KO}(S^q)$, $\tilde{K}(S^q)$, $\widetilde{KSp}(S^q)$ it suffices to compute $\pi_1(BU, *)$ and $\pi_q(BO, *)$, $\pi_q(BSp, *)$ for $q = 1, 2, 3$. From 11.43 we have $\pi_q(BG, *) \simeq \pi_{q-1}(G, 1)$ for $q \geqslant 1$ and any topological group G. Hence it suffices to compute $\pi_0(U, 1)$, $\pi_q(O, 1)$, $\pi_q(Sp, 1)$, $q = 0, 1, 2$. From the remarks in 11.36 it suffices to compute $\pi_0(U(1), 1)$, $\pi_q(Sp(1), 1)$, $\pi_q(O(q+2), 1)$, $q = 0, 1, 2$.

Since $U(1) \cong S^1$ and $Sp(1) \cong S^3$, we immediately have $\pi_0(U(1), 1) = 0$, $\pi_q(Sp(1), 1) = 0$, $q = 0, 1, 2$. From the fibration $SO(2) \to O(2) \to Z_2$ we see that $\pi_0(O(2),1) \cong Z_2$. We turn to the computation of $\pi_1(O(3),1) \cong \pi_1(SO(3),1)$ and $\pi_2(O(4),1) \cong \pi_2(SO(4),1)$.

11.61. Proposition. $SO(3) \cong RP^3$.

Proof: We regard RP^3 as being D^3 with antipodal points of S^2 identified, $SO(3)$ as the group of rotations in $\mathbb{R}^3$; hence an element $A \in SO(3)$ is a rotation about a line l through an angle θ, $-\pi \leqslant \theta \leqslant \pi$. Let x be the point of the upper hemisphere H_+^2 through which l passes and let $f(A) = [(\theta/\pi) \cdot x] \in RP^3$. Then f is continuous and bijective, hence a homeomorphism. ▯

11.62. Corollary. $\pi_1(SO(3),1) \cong \mathbb{Z}_2$ *and* $\pi_2(SO(3),1) \cong \pi_2(S^3,s_0) = 0$.

Now consider the fibration $SO(3) \to SO(4) \to S^3$ and its associated homotopy sequence. Since $\pi_2(SO(3),1) = 0 = \pi_2(S^3,s_0)$, it follows that $\pi_2(SO(4),1) = 0$.

We summarize these calculations in the following table.

$q \bmod 8$	0	1	2	3	4	5	6	7
$\widetilde{KO}(S^q)$	$\mathbb{Z}$	$\mathbb{Z}_2$	$\mathbb{Z}_2$	0	$\mathbb{Z}$	0	0	0
$\tilde{K}(S^q)$	$\mathbb{Z}$	0	$\mathbb{Z}$	0	$\mathbb{Z}$	0	$\mathbb{Z}$	0
$\widetilde{KSp}(S^q)$	$\mathbb{Z}$	0	0	0	$\mathbb{Z}$	$\mathbb{Z}_2$	$\mathbb{Z}_2$	0

By virtue of Theorem 11.60 we can construct two Ω-spectra KO and K as follows: we take KO to be the spectrum of period 8 (i.e. $KO_{q+8} = KO_q$ for all $q \in \mathbb{Z}$) as follows

$$\begin{array}{ccccccccc} \cdots KO_0 & KO_1 & KO_2 & KO_3 & KO_4 & KO_5 & KO_6 & KO_7 & KO_8 \cdots \\ \| & \| & \| & \| & \| & \| & \| & \| & \| \\ \mathbb{Z} \times BO & \Omega^3 BSp & \Omega^2 BSp & \Omega BSp & \mathbb{Z} \times BSp & \Omega^3 BO & \Omega^2 BO & \Omega BO & \mathbb{Z} \times BO. \end{array}$$

The map $\varepsilon'_{8q}: KO_{8q} \to \Omega KO_{8q+1}$ is to be the homotopy equivalence $\mathbb{Z} \times BO \to \Omega^4 BSp$ of 11.60. Similarly the map $\varepsilon'_{8q+4}: KO_{8q+4} \to \Omega KO_{8q+5}$ is to be the homotopy equivalence $\mathbb{Z} \times BSp \to \Omega_4 BO$ of 11.60. The other maps ε'_q are all to be the identity. For K we take the spectrum of period 2 with $K_{2q} = \mathbb{Z} \times BU$, $K_{2q+1} = \Omega BU$, $\varepsilon'_{2q}: K_{2q} \to \Omega K_{2q+1}$ the homotopy equivalence $\mathbb{Z} \times BU \to \Omega^2 BU$ of 11.60 and $\varepsilon'_{2q+1} = 1$.

We then have homology and cohomology theories K_*, KO_* and K^*, KO^* defined by these spectra. Since they are Ω-spectra, we have

$$\widetilde{KO}^0(X) \cong [X, x_0; \mathbb{Z} \times BO, (0,*)] \cong \widetilde{KO}(X)$$
$$\tilde{K}^0(X) \cong [X, x_0; \mathbb{Z} \times BU, (0,*)] \cong \tilde{K}(X)$$

on the category $\mathscr{PCW}'_F$. Since $\Sigma^2 K = K$ and $\Sigma^8 KO = KO$, the coefficient groups $\tilde{K}^q(S^0)$ have period 2 and the coefficient groups $\widetilde{KO}^q(S^0)$ have period 8. They are given in the table above.

The relative groups $KO^0(X, A)$ have a geometric description in terms of bundles over X with a trivialization over A.

Comments.

The groups $K(X)$, $KO(X)$ were first introduced by Atiyah and Hirzebruch in [18] to study manifolds. Since then their usefulness for the investigation of manifolds has rapidly grown (cf. [19], for example). In Chapter 20 we shall see an important example of their application in this connection (the Stong–Hattori Theorem 20.34). The cohomology theories $K^*(X)$, $KO^*(X)$ have also proved useful for solving problems in algebraic topology; good examples are Adams' solution of the vector fields on spheres problem [2] and his e-invariant for investigating homotopy groups of spheres (Chapter 19).

References

1. J. F. Adams [2]
2. M. F. Atiyah [16]
3. M. F. Atiyah and F. Hirzebruch [18]
4. M. F. Atiyah and I. M. Singer [19]
5. J. M. Boardman and R. M. Vogt [22]
6. R. Bott [25]
7. D. Husemoller [49]

Chapter 12

Manifolds and Bordism

In this chapter we consider another collection of homology theories, the bordism theories, which arise from manipulations with manifolds, although in a different sense from that in which the K-theories have their origins in manipulations with bundles. We begin with the definition of a manifold, but we shall not prove all the theorems about manifolds which we shall employ, because to do so would lead to a chapter out of all proportion with the rest of the book (it is recommended that the reader unfamiliar with the theory of manifolds see [69] or [52]). Instead we go on to define the Thom complex of a vector bundle and the various Thom spectra MG. Then we sketch the proof of Thom that the homology theories associated with these spectra can be described in terms of singular manifolds.

12.1. Definition. A Hausdorff space M is called a (topological) *manifold of dimension n* if there is a collection

$$\{(U_\alpha, \phi_\alpha) : \alpha \in A,\ U_\alpha \subset M,\ \phi_\alpha : U_\alpha \to \mathbb{R}^n\}$$

such that $\{U_\alpha\}_{\alpha \in A}$ is an open covering of M and each ϕ_α is a homeomorphism of U_α onto an open subset of $\mathbb{R}^n_+ = \{x \in \mathbb{R}^n : x_n \geqslant 0\}$. The pairs (U_α, ϕ_α) are called *charts* for M and a collection $\{(U_\alpha, \phi_\alpha) : \alpha \in A\}$ of charts which cover M is called an *atlas*.

Suppose $\{(U_\alpha, \phi_\alpha) : \alpha \in A\}$ is an atlas for M. With respect to this atlas we define a subset ∂M in M as follows: $x \in \partial M$ if there is a chart (U_α, ϕ_α) with $x \in U_\alpha$ and $\phi_\alpha(x) \in \mathbb{R}^{n-1} = \{x \in \mathbb{R}^n : x_n = 0\}$.

12.2. Lemma. *For any* $x \in \partial M$ *we have* $H_n(M, M - \{x\}; \mathbb{Z}) = 0$, *whereas for* $x \in M - \partial M$ *we have* $H_n(M, M - \{x\}; \mathbb{Z}) \cong \mathbb{Z}$.

Proof: Suppose $x \in \partial M$ and (U_α, ϕ_α) is a chart around x with $\phi_\alpha(x) \in \mathbb{R}^{n-1}$. Let D^n denote a small disk around $\phi_\alpha(x)$ in $\mathbb{R}^n$; then $V = \phi_\alpha^{-1}(D^n \cap R^n_+)$ is a neighborhood of x in M. Now the inclusion

$$j : (V, V - \{x\}) \to (M, M - \{x\})$$

induces an isomorphism

$$j_*: H_n(V, V - \{x\}) \to H_n(M, M - \{x\})$$

by the excision axiom. But $V, V - \{x\}$ can both be deformation retracted onto $\dot{V} = \phi_\alpha^{-1}(S^{n-1} \cap \mathbb{R}^n_+)$, so $H_n(V, V - \{x\}; \mathbb{Z}) = 0$.

If $x \in M - \partial M$ we choose a chart (U_α, ϕ_α) around x and let D^n denote a small disk around $\phi_\alpha(x)$ in $\mathbb{R}^n$; then $V = \phi_\alpha^{-1}(D^n)$ is a neighborhood of x in M and again we have $H_n(M, M - \{x\}) \cong H_n(V, V - \{x\}) \cong H_n(V, \dot{V})$ with $\dot{V} = \phi_\alpha^{-1}(S^{n-1})$. But $(V, \dot{V}) \cong (D^n, S^{n-1})$, so we have $H_n(V, \dot{V}) \cong H_n(D^n, S^{n-1}) \cong \mathbb{Z}$. □

Now the characterization of ∂M as the set of those points $x \in M$ with $H_n(M, M - \{x\}; \mathbb{Z}) = 0$ is independent of the choice of chart. Thus ∂M is an invariantly determined subset of M.

12.3. Definition. ∂M is called the *boundary* of M. If $\partial M = \varnothing$, then M is called a manifold *without boundary*. A compact manifold without boundary is called *closed*.

12.4. Proposition. *∂M is a manifold without boundary of dimension $n - 1$.*
Proof: Let $\{(U_\alpha, \phi_\alpha): \alpha \in A\}$ be an atlas for M and let $A' \subset A$ be the set of indices such that $U_\alpha \cap \partial M \neq \varnothing$ if $\alpha \in A'$. Then clearly

$$\{(U_\alpha \cap \partial M, \phi_\alpha | U_\alpha \cap \partial M) : \alpha \in A'\}$$

can be made into an atlas for ∂M. □

12.5. Definition. *A differentiable atlas of class C^k* on a manifold M is an atlas $\{(U_\alpha, \phi_\alpha) : \alpha \in A\}$ such that for all $\alpha, \beta \in A$ the function

$$\phi_\beta \circ (\phi_\alpha^{-1} | \phi_\alpha(U_\alpha \cap U_\beta)) : \phi_\alpha(U_\alpha \cap U_\beta) \to \mathbb{R}^n$$

has continuous partial derivatives of order k. Two differentiable atlases $\{(U_\alpha, \phi_\alpha) : \alpha \in A\}$ and $\{(V_\beta, \psi_\beta) : \beta \in B\}$ of class C^k are called *equivalent* if

$$\{(U_\alpha, \phi_\alpha) : \alpha \in A\} \cup \{(V_\beta, \psi_\beta) : \beta \in B\}$$

is again a differentiable atlas of class C^k (this is an equivalence relation). A *differentiable structure of class C^k* on M is an equivalence class of differentiable atlases of class C^k on M. A *differentiable manifold of class C^k* is a manifold M together with a differentiable structure of class C^k. A *smooth* manifold is a differentiable manifold of class C^∞.

Remark 1. The union of atlases in an equivalence class is the largest atlas in the equivalence class.

Remark 2. In general a topological manifold M may have no differentiable structures at all, and if it does have any, then it may have several distinct ones. For example, Milnor and Kervaire [50] have shown that the sphere S^7 has 28 distinct oriented smooth structures.

In what follows we shall consider only smooth structures and smooth manifolds, although many of the results have analogs for C^k manifolds.

12.6. Let M be a smooth manifold of dimension n. We can define a vector bundle $\tau(M)$ of dimension n as follows: choose a smooth atlas $\{(U_\alpha, \phi_\alpha) : \alpha \in A\}$ in the smooth structure of M. For $\alpha, \beta \in A$ we write $\phi_\beta \circ \phi_\alpha^{-1}$ for $\phi_\beta \circ \phi_\alpha^{-1} | \phi_\alpha(U_\alpha \cap U_\beta)$. Then the Jacobian matrix

$$M_{\beta\alpha}(x) = \left(\frac{\partial(\phi_\beta \circ \phi_\alpha^{-1})_i}{\partial x_j} \bigg|_{\phi_\alpha(x)} \right)$$

is invertible for all $x \in U_\alpha \cap U_\beta$. Thus $M_{\beta\alpha}(x) \in GL(n, R)$ for all $x \in U_\alpha \cap U_\beta$, $\alpha, \beta \in A$.

$$\begin{aligned} M_{\gamma\beta}(x) \cdot M_{\beta\alpha}(x) &= \left(\frac{\partial(\phi_\gamma \circ \phi_\beta^{-1})_i}{\partial x_j} \bigg|_{\phi_\beta(x)} \right) \left(\frac{\partial(\phi_\beta \circ \phi_\alpha^{-1})_j}{\partial x_k} \bigg|_{\phi_\alpha(x)} \right) \\ &= \left(\frac{\partial(\phi_\gamma \circ \phi_\alpha^{-1})_i}{\partial x_k} \bigg|_{\phi_\alpha(x)} \right) \end{aligned}$$

by the chain rule. Thus $\{U_\alpha, M_{\alpha\beta} : \alpha, \beta \in A\}$ is a set of transition functions and hence defines a vector bundle $\tau(M)$ (or rather an equivalence class).

Suppose $\{(V_\gamma, \psi_\gamma) : \gamma \in \Gamma\}$ is a smooth atlas equivalent to $\{(U_\alpha, \phi_\alpha) : \alpha \in A\}$. Let

$$N_{\delta\gamma}(x) = \left(\frac{\partial(\psi_\delta \circ \psi_\gamma^{-1})_i}{\partial x_j} \bigg|_{\psi_\gamma(x)} \right)$$

be the Jacobian matrix for $x \in V_\gamma \cap V_\delta$. For $x \in U_\alpha \cap V_\gamma$ let

$$R_{\gamma\alpha}(x) = \left(\frac{\partial(\psi_\gamma \circ \phi_\alpha^{-1})_i}{\partial x_j} \bigg|_{\phi_\alpha(x)} \right).$$

Again $R_{\gamma\alpha}$ is invertible, and we have

$$N_{\delta\gamma}(x) = R_{\delta\beta}(x) \cdot M_{\beta\alpha}(x) \cdot R_{\gamma\alpha}(x)^{-1}$$

for all $x \in U_\alpha \cap U_\beta \cap V_\gamma \cap V_\delta$. Thus the sets of transition functions

$$\{U_\alpha, M_{\beta\alpha} : \alpha, \beta \in A\}, \quad \{V_\gamma, N_{\delta\gamma} : \gamma, \delta \in \Gamma\}$$

are equivalent, and hence the equivalence class of $\tau(M)$ is well defined.

12.7. Definition. The uniquely defined equivalence class $\tau(M)$ is called the *tangent bundle* of M, although it is not a bundle. If one wants an honest bundle, he must take the largest atlas and carry out the glueing process described in 11.15. M is *orientable* if $\det(M_{\beta\alpha}(x)) > 0$ for all x, α, β.

12.8. Definition. Let M, M' be smooth manifolds of dimensions n, n' and $f: M \to M'$ a map between them. f is called *smooth* if for some smooth atlases $\{(U_\alpha, \phi_\alpha): \alpha \in A\}$ for M and $\{(V_\beta, \psi_\beta): \beta \in B\}$ for M' the functions

$$\psi_\beta \circ f \circ \phi_\alpha^{-1} | \phi_\alpha(f^{-1}(V_\beta) \cap U_\alpha): \phi_\alpha(f^{-1}(V_\beta) \cap U_\alpha) \to \mathbb{R}^{n'}$$

are of class C^∞.

12.9. Lemma. *If $f: M \to M'$ is smooth with respect to atlases*

$$\{(U_\alpha, \phi_\alpha): \alpha \in A\}, \quad \{(V_\beta, \psi_\beta): \beta \in B\}$$

for M, M' then it is also smooth with respect to equivalent atlases

$$\{(U'_\delta, \theta_\delta): \delta \in \Delta\}, \quad \{(V'_\gamma, \eta_\gamma): \gamma \in \Gamma\}.$$

Proof: Since $\{(U'_\delta, \theta_\delta): \delta \in \Delta\}$ is equivalent to $\{(U_\alpha, \phi_\alpha): \alpha \in A\}$ we know that $\phi_\alpha \circ (\theta_\delta^{-1} | \theta_\delta(U_\alpha \cap U'_\delta))$ is of class C^∞ for all $\alpha \in A$, $\gamma \in \Gamma$. Similarly $\eta_\gamma \circ (\psi_\beta^{-1} | \psi_\beta(V_\beta \cap V'_\gamma))$ is of class C^∞ for $\beta \in B$, $\gamma \in \Gamma$. Then for $\delta \in \Delta$, $\gamma \in \Gamma$ and $y \in \theta_\delta(f^{-1}(V'_\gamma) \cap U'_\delta)$ we choose $\alpha \in A$, $\beta \in B$ such that $y \in \theta_\delta(f^{-1}(V_\beta) \cap U_\alpha)$ and we have

$$\eta_\gamma \circ f \circ \theta_\delta^{-1} | \theta_\delta(f^{-1}(V'_\gamma) \cap U'_\delta) = (\eta_\gamma \circ \psi_\beta^{-1}) \circ (\psi_\beta \circ f \circ \phi_\alpha^{-1}) \circ (\phi_\alpha \circ \theta_\delta^{-1})$$

on $\theta_\delta(f^{-1}(V_\beta \cap V'_\gamma) \cap U_\alpha \cap U'_\delta)$. But $\theta_\delta(f^{-1}(V_\beta \cap V'_\gamma) \cap U_\alpha \cap U'_\delta)$ is an open neighborhood of y, so it follows that $\eta_\gamma \circ f \circ \theta_\delta^{-1} | \theta_\delta(f^{-1}(V'_\gamma) \cap U'_\delta)$ is of class C^∞. ▯

Thus the definition of smooth map between two smooth manifolds is independent of the choice of atlases.

12.10. Definition. If $f: M \to M'$ is a homeomorphism such that both f and f^{-1} are smooth, then f is called a *diffeomorphism*. M and M' are said to be *diffeomorphic* if there is a diffeomorphism between them. In that case they have the same dimension, of course.

12.11. Suppose $f: M \to M'$ is a smooth map and $\{(U_\alpha, \phi_\alpha): \alpha \in A\}$, $\{(V_\gamma, \psi_\gamma): \gamma \in \Gamma\}$ are atlases for M, resp. M'; then the Jacobian

$$R_{\gamma\alpha}(x) = \left(\frac{\partial(\psi_\gamma \circ f \circ \phi_\alpha^{-1})_i}{\partial x_j} \bigg|_{\phi_\alpha(x)} \right), \quad x \in f^{-1}(V_\gamma) \cap U_\alpha,$$

is an $n' \times n$ matrix. If

$$N_{\alpha\beta}(x) = \left(\frac{\partial(\phi_\alpha \circ \phi_\beta^{-1})_i}{\partial x_j} \bigg|_{\phi_\beta(x)} \right), \quad x \in U_\alpha \cap U_\beta,$$

$$N'_{\gamma\delta}(y) = \left(\frac{\partial(\psi_\gamma \circ \psi_\delta^{-1})_i}{\partial x_j} \bigg|_{\psi_\delta(y)} \right), \quad y \in V_\gamma \cap V_\delta,$$

are the transition matrices for $\tau(M)$, $\tau(M')$ then we have

$$R_{\gamma\alpha}(x) \cdot N_{\alpha\beta}(x) = N'_{\gamma\delta}(f(x)) \cdot R_{\delta\beta}(x) \quad \text{for all } x \in f^{-1}(V_\gamma \cap V_\delta) \cap U_\alpha \cap U_\beta.$$

This equation would be exactly the definition of a morphism of sets of transition functions (11.10) if $R_{\gamma\alpha}(x)$ were an invertible square matrix. In 11.3 we defined a morphism of vector bundles to be a commutative diagram

$$\begin{array}{ccc} E & \xrightarrow{\phi} & E' \\ \downarrow{\scriptstyle p} & & \downarrow{\scriptstyle p'} \\ B & \xrightarrow{\bar{\phi}} & B' \end{array}$$

in which ϕ is a linear isomorphism on each fibre. If we relax this last condition a little and allow morphisms in which ϕ is merely linear on each fibre (and the bundles may have different dimensions), then we see that the $R_{\gamma\alpha}$ above can be regarded as defining a morphism $\tau(f):\tau(M) \to \tau(M')$. If $g:M' \to M''$ is another smooth map, then $\tau(g \circ f) = \tau(g) \circ \tau(f)$ and clearly $\tau(1) = 1_{\tau(M)}$. $\tau(f)$ is the *differential* of f.

For a smooth map $f:M \to M'$ and a point $x \in M$ we define $(\operatorname{rank} f)_x$ to be the rank of the linear map $\tau(f)|p^{-1}(x):p^{-1}(x) \to p'^{-1}(f(x))$. Clearly $(\operatorname{rank} f)_x = \operatorname{rank} R_{\gamma\alpha}(x)$ for some choice of atlases $\{(U_\alpha, \phi_\alpha):\alpha \in A\}$ for M and $\{(V_\gamma, \psi_\gamma):\gamma \in \Gamma\}$ for M' and α, γ with $x \in U_\alpha, f(x) \in V_\gamma$.

12.12. Definition. A smooth map $f:M \to M'$ from a manifold M of dimension n to a manifold M' of dimension $n' \geqslant n$ is called an *immersion* if $(\operatorname{rank} f)_x = n$ for all $x \in M$. An *embedding* is an immersion which is a homeomorphism of M onto $f(M)$. The following picture is supposed to indicate an immersion of $\mathbb{R}^1$ in $\mathbb{R}^2$ which is injective but nevertheless not an embedding.

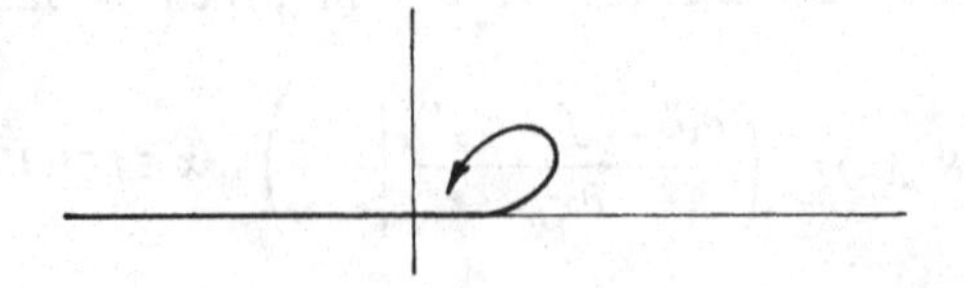

The arrow is supposed to indicate that the curve "approaches but does not reach" the x-axis.

Exercise. Show that an immersion is always locally injective.

12.13. In 11.45 we remarked that every vector bundle over a CW-complex can be regarded as of the form $\xi[\mathbb{R}^n]$ for ξ an $O(n)$-bundle. Thus if we assume that M has a CW-structure (which is certainly true if M is compact) then we may assume there is an inner product in the fibres of $\tau(M)$ which varies continuously from fibre to fibre in a suitable fashion (this is called a Riemann metric on M). If $f:M \to N$ is an immersion, then $\tau(f):\tau(M) \to \tau(N)$ induces a bundle map $\phi:\tau(M) \to f^*\tau(N)$ (cf. 11.25) with $\bar{\phi} = 1_M$. We let $v_f =$ the orthogonal complement of $\phi(\tau(M))$ in $f^*\tau(N)$; that is, the fibre of v_f over $x \in M$ is the orthogonal complement of $\tau(M)_x$ in $(f^*\tau(N))_x$. Then clearly $\tau(M) \oplus v_f \simeq f^*\tau(N)$. In particular if $N = \mathbb{R}^{n+k}$, then since $\tau(N) = \varepsilon^{n+k}$ we have $\tau(M) \oplus v_f \simeq \varepsilon^{n+k}$; that is, $\{v_f\}_s$ is the negative of $\{\tau(M)\}_s$ in $\widetilde{KO}(M)$. In particular the stable equivalence class of v_f depends only on M and not on the particular immersion $f:M \to \mathbb{R}^{n+k}$. This stable equivalence class is called the *stable normal bundle* of M; any representative v in this class is also frequently called a stable normal bundle of M.

12.14. Theorem. *If M is a compact manifold then there is an embedding $f:M \to \mathbb{R}^{n+k}$ for sufficiently large k and a homeomorphism $h:D(v_f) \to U$ from the disk bundle $D(v_f)$ of all vectors of length $\leqslant 1$ in v_f to a neighborhood U of $f(M)$ in $\mathbb{R}^{n+k}$ such that if $i:M \to E(v_f)$ is the zero-section of v_f ($i(x) = 0$ in $(v_f)_x$) then $h \circ i = f$. Moreover, if $f':M \to \mathbb{R}^{n+k}$ and $h':D(v_{f'}) \to U'$ is another such pair of embedding plus homeomorphism, then there is an* isotopy *$H:\mathbb{R}^{n+k} \times I \to \mathbb{R}^{n+k}$ (each H_t is a diffeomorphism, $t \in I$) with $H_0 = 1_{\mathbb{R}^{n+k}}$, $H_1 \circ f = f'$ and $H_1(U) = U'$.*

The neighborhood U which occurs in this theorem is called a *tubular neighborhood* of $f(M)$ in $\mathbb{R}^{n+k}$; it has the property that there is a deformation retraction $r:U \to f(M)$ such that $r \circ h = f \circ p$ ($p:E(v_f) \to M$ the bundle projection). Thus locally U looks like $V \times D^k$, V open in $f(M)$.

The proof of the results stated in 12.14 can be found in [52] as can the proof of the following theorem (12.17).

12.15. Definition. Let M be a manifold and $N \subset M$ a subspace such that for each point $x \in N$ we can find a smooth chart (U, ϕ) in the maximal atlas of M around x in M with $N \cap U = \phi^{-1}(\mathbb{R}^k \subset \mathbb{R}^n)$. Then N is called a *submanifold* of *dimension* k (or *codimension* $n - k$) in M. Clearly the set $\{(U \cap N, \phi|U \cap N)\}$ forms an atlas for N if we regard $\phi|U \cap N$ as a map into $\mathbb{R}^k$, so N is a smooth manifold of dimension k. Clearly the inclusion $i:N \to M$ is an embedding, and more generally if $f:M' \to M$ is an embedding, then $f(M')$ is a submanifold of M.

12.16. Definition. Suppose $N \subset M'$ is a submanifold and $f:M \to M'$ is a smooth map, $\partial M = \varnothing$. We say f is *transverse regular* on N if for every $x \in M$ such that $f(x) \in N$ we have $[\tau(f)(\tau(M)) + \tau(N)]_{f(x)} = \tau(M')_{f(x)}$.

For example, if $\dim M + \dim N < \dim M'$, then f is transverse regular on N if and only if $f(M) \cap N = \varnothing$. If M has a boundary, then we require transverse regularity on $M - \partial M$ and ∂M separately.

12.17. Theorem. *If $f: M \to M'$ is transverse regular on N then $f^{-1}(N)$ is a submanifold of M and* $\operatorname{codim} f^{-1}(N) = \operatorname{codim} N$*—that is,*

$$\dim M - \dim f^{-1}(N) = \dim M' - \dim N.$$

12.18. Theorem. *If $f: M \to M'$ is a continuous map and $N \subset M'$ is a submanifold, then there is a smooth map $f': M \to M'$ which is transverse regular on N and homotopic to f.*

12.14, 12.17 and 12.18 are the essential ingredients in the proof of the Thom theorem (12.30).

In 1946 Steenrod proposed the following problem: given a topological space X and a homology class $x \in H_n(X; \mathbb{Z})$ do there exist a smooth n-dimensional oriented closed manifold M with fundamental class $z_M \in H_n(M; \mathbb{Z})$ and a continuous map $f: M \to X$ with $f_*(z_M) = x$? In other words one would like to represent cycles by continuous images of closed smooth manifolds. One can ask the same question with $\mathbb{Z}_2$-homology and unoriented manifolds. Steenrod also asked which closed smooth (oriented) manifolds M are boundaries of smooth (oriented) manifolds with boundary. Thom laid the foundations for the solution of these problems by introducing the relation of *cobordism* on the set of closed smooth (oriented) manifolds [84]. He completely classified unoriented manifolds up to cobordism.

One says two closed smooth n-dimensional manifolds M, M' are *cobordant* if there is a compact smooth $(n+1)$-dimensional manifold W with boundary ∂W such that $\partial W = M \cup M'$, the disjoint union of M

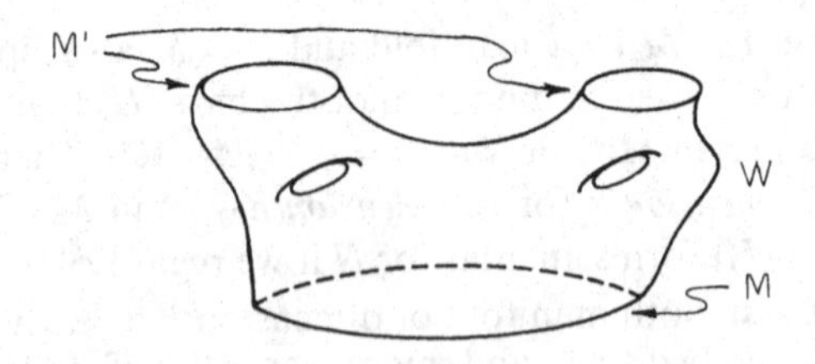

and M'. Subsequently others (notably Milnor) applied Thom's idea of cobordism to manifolds with various additional structures, obtaining other classification theorems. We shall describe some of these.

We consider a collection of spaces X_n and strictly commutative diagrams

$$\begin{array}{ccc} X_n & \xrightarrow{f_n} & BO(n) \\ {\scriptstyle g_n}\downarrow & & \downarrow{\scriptstyle Bi_n} \\ X_{n+1} & \xrightarrow{f_{n+1}} & BO(n+1), \end{array}$$

where the maps f_n are to be fibrations. Such a system will be denoted by the letter X.

12.19. Definition. An *X-structure* on a smooth manifold M is a pair $(h, \tilde{\nu})$ such that $h: M \to \mathbb{R}^{n+k}$ is an embedding with normal bundle classified by ν and $\tilde{\nu}$ is a lifting $\tilde{\nu}: M \to X_k$ of ν:

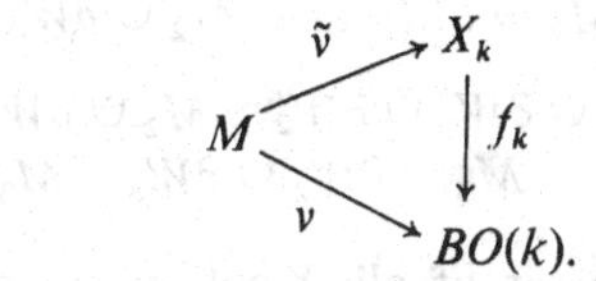

(An embedding $h: M \to \mathbb{R}^{n+k}$ defines an explicit $\nu_h: M \to G_{k,n+k} \subset BO(k)$ by parallel transport of the normal hyperplanes. If $T: \mathbb{R}^{n+k} \to \mathbb{R}^{n+k}$ is a translation, then $\nu_{T\circ h} = \nu_h$.) An X-structure $(h, \tilde{\nu})$ defines a family of pairs $(h_m, \tilde{\nu}_m)$, $m \geqslant k$, with $\nu_m = Bi_{m-1} \circ Bi_{m-2} \circ \cdots \circ Bi_k \circ \nu: M \to BO(m)$, $\tilde{\nu}_m = g_{m-1} \circ \cdots \circ g_k \circ \tilde{\nu}: M \to X_m$ and $h_m = i \circ h: M \to \mathbb{R}^{n+m}$. Two X-structures $(h, \tilde{\nu})$ $(\nu: M \to BO(k))$ and $(h', \tilde{\nu}')$ $(\nu': M \to BO(k'))$ are called *equivalent* if there is some $k'' \geqslant \max(k, k')$ such that $h'_{k''} = T \circ h_{k''}$ for some translation $T: \mathbb{R}^{n+k''} \to \mathbb{R}^{n+k''}$ and $\tilde{\nu}'_{k''} \simeq \tilde{\nu}_{k''}$, where the homotopy $H: M \times I \to X_{k''}$ is required to satisfy $f_{k''} \circ H_t = \nu_{k''}$ for all $t \in I$—i.e. H is a homotopy through liftings. An *X-manifold* is a smooth manifold M with an equivalence class of X-structures.

Remark. The specific maps f_k definitely play a rôle here; if we replace them by homotopic maps f'_k we get a different notion of X-structure.

We shall require that a manifold W with boundary shall always be embedded in $\mathbb{R}^{n+k+1}$ in such a way that W lies in the upper half-space $\mathbb{R}_+^{n+k+1} = \{x \in R^{n+k+1}: x_1 \geqslant 0\}$ and $\partial W \subset \mathbb{R}^{n+k} = \{x \in \mathbb{R}^{n+k+1}: x_1 = 0\}$. If $(h, \tilde{\nu})$ is an X-structure on W and $i: \partial W \to W$ is the inclusion, then $\nu \circ i$ classifies the normal bundle of ∂W in $\mathbb{R}^{n+k}$. Thus $(h \circ i, \tilde{\nu} \circ i)$ is an X-structure on ∂W called the *induced* X-structure.

12.20. Definition. A *map of X-manifolds* is a smooth map $f: M \to M'$ between manifolds M, M' with X-structures $(h, \tilde{\nu})$, $(h', \tilde{\nu}')$ respectively such that $h' \circ f = T \circ h$ for some translation T and $\tilde{\nu}' \circ f \simeq \tilde{\nu}$ (homotopy over $\nu' \circ f = \nu$).

Convention: The empty set $\varnothing$ will be regarded as an n-dimensional smooth manifold for all $n \geqslant 0$. $\varnothing$ has a unique X-structure.

12.21. Definition. Let M_1, M_2 be two closed n-dimensional X-manifolds. We say M_1, M_2 are *X-cobordant* (written $M_1 \sim_X M_2$) if there are $(n+1)$-dimensional compact X-manifolds W_1, W_2 with $M_1 \dot{\cup} \partial W_1 \cong M_2 \dot{\cup} \partial W_2$, where ∂W_1, ∂W_2 have the induced X-structures and "$U \cong V$" means there is an X-diffeomorphism $f: U \to V$.

12.22. Lemma. *$\sim_X$ is an equivalence relation.*

Proof: i) Clearly $M_1 \dot{\cup} \partial\varnothing \cong M_1 \dot{\cup} \partial\varnothing$, so $M_1 \sim_X M_1$.

ii) $M_1 \dot{\cup} \partial W_1 \cong M_2 \dot{\cup} \partial W_2$ implies $M_2 \dot{\cup} \partial W_2 \cong M_1 \dot{\cup} \partial W_1$.

iii) If $M_1 \dot{\cup} \partial W_1 \cong M_2 \dot{\cup} \partial W_2$ and $M_2 \dot{\cup} \partial W_2' \cong M_3 \dot{\cup} \partial W_3$, then

$$M_1 \dot{\cup} \partial(W_1 \dot{\cup} W_2') = M_1 \dot{\cup} \partial W_1 \dot{\cup} \partial W_2' \cong M_2 \dot{\cup} \partial W_2 \dot{\cup} \partial W_2' \cong M_3 \dot{\cup} \partial W_2 \dot{\cup} \partial W_3 = M_3 \dot{\cup} \partial(W_2 \dot{\cup} W_3). \quad \square$$

We write Ω_n^X for the set of all X-cobordism classes of closed n-dimensional X-manifolds, $n \geqslant 0$.

12.23. Lemma. *The operation of taking disjoint union $\dot{\cup}$ defines a sum $+$ on Ω_n^X such that Ω_n^X becomes an abelian group, $n \geqslant 0$. The class $\{\varnothing\}$ defines the zero element.*

Proof: Suppose $M_1 \dot{\cup} \partial W_1 \cong M_2 \dot{\cup} \partial W_2$ and $N_1 \dot{\cup} \partial U_1 \cong N_2 \dot{\cup} \partial U_2$. Then

$$M_1 \dot{\cup} N_1 \dot{\cup} \partial(W_1 \dot{\cup} U_1) = (M_1 \dot{\cup} \partial W_1) \dot{\cup} (N_1 \dot{\cup} \partial U_1) \cong (M_2 \dot{\cup} \partial W_2) \dot{\cup} (N_2 \dot{\cup} \partial U_2) = M_2 \dot{\cup} N_2 \dot{\cup} \partial(W_2 \dot{\cup} U_2).$$

Hence $+$ is well defined if we set $\{M\} + \{N\} = \{M \dot{\cup} N\}$. $+$ is clearly associative and commutative. Also $\{M\} + \{\varnothing\} = \{M\}$, all M.

Given an X-manifold M with embedding $h: M \to \mathbb{R}^{n+k}$, we embed $M \times 1$ in $\mathbb{R}^{n+k}$ by taking a suitable translation of the given embedding so that the images of $M (= M \times 0)$ and $M \times 1$ are disjoint. We can extend these embeddings to embeddings of $M \times [0, 1/4]$, $M \times [3/4, 1]$ into $\mathbb{R}_+^{n+k+1}$ in the obvious way. The two embeddings $M \times 1/4 \to \mathbb{R}_+^{n+k+1}$, $M \times 3/4 \to \mathbb{R}_+^{n+k+1}$ are isotopic, so we can find a map $M \times [1/4, 3/4] \to \mathbb{R}_+^{n+k+1}$ which is an isotopy between them. This isotopy can be deformed rel $M \times \{1/4, 3/4\}$ to be an embedding, so we get an embedding of $M \times I$ into $\mathbb{R}_+^{n+k+1}$ which agrees with the given embedding on $M \times 0$. The normal map $\nu': M \times I \to BO(k)$ has a lifting $\tilde{\nu}': M \times I \to X_k$ because $f_k: X_k \to$

$BO(k)$ is a fibration. Then $M \times 0 \cup M \times 1 \cong \partial(M \times I)$, or in other words $\{M\} + \{M \times 1) = 0$. $\square$

12.24. Examples. 1) Let $X = BO = \{BO(n)\}$ with maps $f_n: X_n \to BO(n)$ the identity and $g_n = Bi_n$. The result is the *unoriented cobordism group* Ω_*^0, also denoted by $\mathcal{N}_*$ by Thom. Every element of Ω_n^0 except 0 is of order 2, for in this case $\partial(M \times I) \cong M \cup M$.

2) Let $X = BSO = \{BSO(n)\}$ with maps $f_n: X_n \to BO(n)$ induced by the inclusions $j_n: SO(n) \to O(n)$ and chosen to be fibrations. The result is *oriented cobordism groups* Ω_*^{SO} (denoted by Ω_* by Thom). The additive inverse of any class $\{M\}$ is $\{\bar{M}\}$, where $\bar{M}$ denotes M with the opposite orientation; for the two ends of $M \times I$ get opposite orientations.

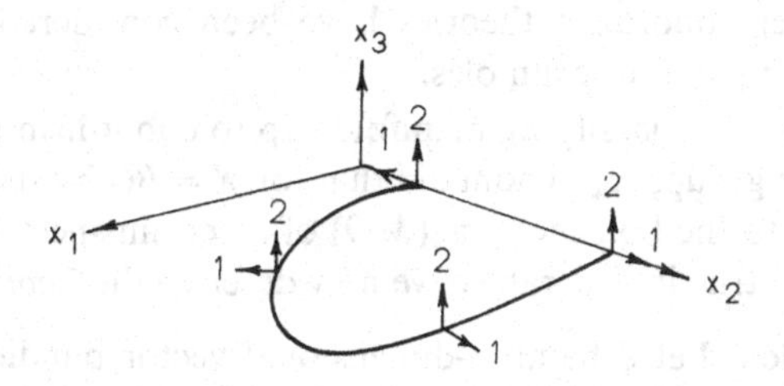

3) $SO(n)$ is not simply connected: $\pi_1(SO(n), 1) \cong \mathbb{Z}_2$. Therefore $SO(n)$ has a double covering (which is also a Lie group) called *Spin*(n). The projection $p_n: Spin(n) \to SO(n)$ induces $Bp_n: BSpin(n) \to BSO(n)$, which can be taken to be a fibration. We let $X_n = BSpin(n)$ and $f_n: X_n \to BO(n)$ be $Bj_n \circ Bp_n$. The result is the *Spin cobordism groups* Ω_*^{Spin}.

4) Let $X = BU$ where $X_{2n} = BU(n) = X_{2n+1}$, $n \geqslant 0$, and $f_{2n}: X_{2n} \to BO(2n)$ is the classifying map for the universal bundle $\gamma_n \to BU(n)$ regarded as real bundle and $f_{2n+1}: X_{2n+1} \to BO(2n+1)$ classifies $\gamma_n \oplus \varepsilon^1$ as an $O(2n+1)$-bundle. The result is the *unitary cobordism groups* Ω_*^U.

5) In analogous fashion we take $X_{2n} = BSU(n) = X_{2n+1}$ and get Ω_*^{SU}.

6) Similarly with $X_{4n} = X_{4n+1} = X_{4n+2} = X_{4n+3} = BSp(n)$ we get Ω_*^{Sp}, the *symplectic cobordism groups*.

7) The smallest subgroup of $O(n)$ is $G = \{1\}$. A G-structure on a bundle is a trivialization or *framing* of the bundle. For $BG(n)$ we could take $EO(n)$, the total space of the universal principal $O(n)$-bundle, since $EO(n)$ is contractible. With $X_n = EO(n)$, $f_n: X_n \to BO(n)$ the bundle projection $p_n: EO(n) \to BO(n)$, we get Ω_*^{fr}, the *framed cobordism groups*. This example illustrates very clearly the necessity of taking homotopies through liftings in the definition 12.19 of X-structure. Since $EO(n)$ is contractible, we would have that all framings were equivalent if we allowed arbitrary homotopies—and hence $\Omega_*^{fr} = 0$. We shall see, however, that Ω_*^{fr} has a very complicated structure.

8) Let $x \in H^q(BO; \mathbb{Z}_2)$ denote any cohomology class,

$$x_n \in H^q(BO(n); \mathbb{Z}_2)$$

its restriction to $BO(n)$. Let $h_n : BO(n) \to H(\mathbb{Z}_2, q)$ be a map representing the class x_n and $f_n : B\langle x\rangle_n \to BO(n)$ the fibration induced from the path fibration $PH(\mathbb{Z}_2,q) \to H(\mathbb{Z}_2,q)$ by h_n.

$$\begin{array}{ccc} B\langle x\rangle_n & \longrightarrow & PH(\mathbb{Z}_2,q) \\ {\scriptstyle f_n}\downarrow & & \downarrow{\scriptstyle p} \\ BO(n) & \xrightarrow{h_n} & H(\mathbb{Z}_2,q) \end{array}$$

Taking $X_n = B\langle x\rangle_n$ gives rise to cobordism groups $\Omega_*^{B\langle x\rangle}$ considered by Browder in [26].

Many other cobordism theories have been considered as well. See Stong [83] for numerous examples.

The problem of classifying manifolds up to cobordism is now that of computing the groups Ω_*^X. Thom did that for $X = BO$ by showing that Ω_*^O is isomorphic to the homotopy $\pi_*(MO)$ of a certain spectrum MO. This turns out to be true in general, so we now describe the *Thom spectra* MX.

12.25. Definition. Let ξ be an n-dimensional vector bundle over a CW-complex X. We can regard ξ as being an $O(n)$-bundle and hence as having an inner product on the fibres. We let $D(\xi)$ denote the associated disk bundle consisting of those vectors of length at most 1 in each fibre. We let $S(\xi)$ denote the associated sphere bundle. If we write $\xi = \eta[\mathbb{R}^n]$ for some principal $O(n)$-bundle η, then $D(\xi) \cong \eta[D^n]$, $S(\xi) \cong \eta[S^{n-1}]$. If $ED(\xi)$ and $ES(\xi)$ denote the total spaces of these fibre bundles, then $ES(\xi) \subset ED(\xi)$ and we can form

$$M(\xi) = ED(\xi)/ES(\xi).$$

We let $\pi : ED(\xi) \to M(\xi)$ be the projection.

12.26. Lemma. *If X is a CW-complex with cells $\{e_\alpha^m : \alpha \in J_m,\ m \geqslant -1\}$, then $M(\xi)$ can be given the structure of a CW-complex with cells $\{\hat{e}_\alpha^{m+n} : \alpha \in J_m,\ m \geqslant -1\}$.*

Proof: Suppose we have already constructed the $(k+n)$-skeleton $M(\xi)^{k+n}$ with $M(\xi)^{k+n} = \pi p^{-1}(X^k)$ $(M(\xi)^{n-1} = *)$. Let $f_\alpha^{k+1} : (D^{k+1}, S^k) \to (X^{k+1}, X^k)$ be the characteristic map of e_α^{k+1}, $\alpha \in J_{k+1}$. We can consider the disk bundle $(f_\alpha^{k+1})^*(D(\xi))$ over D^{k+1}. Since D^{k+1} is contractible, all $O(n)$-bundles over D^{k+1} are trivial, so we can find a trivialization of $(f_\alpha^{k+1})^*(D(\xi))$ and hence a map $\tilde{f}_\alpha^{k+1} : D^{k+1} \times D^n \to ED(\xi)$ with $p \circ \tilde{f}_\alpha^{k+1} = f_\alpha^{k+1} \circ p_{D^{k+1}}$ such that $\tilde{f}_\alpha^{k+1}$ is a homeomorphism on each fibre. If we denote by $\hat{f}_\alpha^{k+1}$ the composite

$$D^{k+n+1} \cong D^{k+1} \times D^n \xrightarrow{\tilde{f}_\alpha^{k+1}} ED(\xi) \xrightarrow{\pi} M(\xi),$$

then $\hat{f}_\alpha^{k+1}(S^{k+n}) \subset M(\xi)^{k+n}$, $\hat{f}_\alpha^{k+1}(D^{k+n+1}) \subset \pi p^{-1}(X^{k+1})$ and $\hat{f}_\alpha^{k+1}$ is a homeomorphism on $\mathring{D}^{k+n+1}$. Thus we can take $\hat{e}_\alpha^{k+n+1} = \hat{f}_\alpha^{k+n+1}(D^{k+n+1})$. One readily verifies that the interiors of the $\hat{e}_\alpha^{k+n+1}$ are disjoint from one another and from $M(\xi)^{k+n}$ and that $M(\xi)^{k+n+1} = M(\xi)^{k+n} \cup \bigcup_{\alpha \in J_{k+1}} \hat{e}_\alpha^{k+n+1}$ has the weak topology. Thus the CW-structure on $M(\xi)$ may be constructed inductively. □

12.27. Definition. $M(\xi)$ is called the *Thom space* of ξ or the *Thom complex* if X is a CW-complex. If $f:\xi \to \eta$ is a map of $O(n)$-bundles, then $f(ED(\xi)) \subset ED(\eta)$ and $f(ES(\xi)) \subset ES(\eta)$, so f defines a map $M(f): M(\xi) \to M(\eta)$. It is clear that $M(f \circ g) = M(f) \circ M(g)$, $M(1) = 1$. If $\bar{f}: X \to Y$ is the associated map of base spaces, then we sometimes write $M(\bar{f})$ rather than $M(f)$, although this is an abuse of notation.

12.28. Proposition. *If ξ, η are vector bundles over X, Y respectively, then there is a natural homeomorphism*

$$M(\xi) \wedge M(\eta) \to M(\xi \times \eta).$$

Proof: We have a homeomorphism $D^n \times D^m \simeq D^{n+m}$. The fibre of $\xi \times \eta$ over $(x,y) \in X \times Y$ is $\xi_x \times \eta_y$ and we have a homeomorphism $D(\xi)_x \times D(\eta)_y \simeq D(\xi \times \eta)_{(x,y)}$ which in fact defines a homeomorphism $D(\xi) \times D(\eta) \simeq D(\xi \times \eta)$ so that $D(\xi) \times S(\eta) \cup S(\xi) \times D(\eta)$ is mapped onto $S(\xi \times \eta)$. Thus we get a homeomorphism

$$M(\xi) \wedge M(\eta) = \frac{D(\xi) \times D(\eta)}{D(\xi) \times S(\eta) \cup S(\xi) \times D(\eta)} \cong \frac{D(\xi \times \eta)}{S(\xi \times \eta)} = M(\xi \times \eta). \quad \square$$

Now if ε^n denotes the trivial n-dimensional bundle over a point x_0, then clearly $M(\varepsilon^n) \simeq S^n$. Since for any bundle ξ over X the bundle $\xi \oplus \varepsilon^n$ can be regarded as $\xi \times \varepsilon^n$ over $X \times x_0$, we see that

12.29.

$$M(\xi \oplus \varepsilon^n) = M(\xi \times \varepsilon^n) \cong M(\xi) \wedge M(\varepsilon^n) \simeq M(\xi) \wedge S^n \simeq S^n M(\xi).$$

Now let $X = \{X_n, g_n, f_n\}$ be such a system as we discussed in 12.19 and let ω_n denote the universal $O(n)$-bundle over $BO(n)$. Over X_n we have the bundle $\gamma_n = f_n^* \omega_n$ and

$$g_n^* \gamma_{n+1} = g_n^* f_{n+1}^* \omega_{n+1} \simeq f_n^*(Bi_n)^* \omega_{n+1} \simeq f_n^*(\omega_n \oplus \varepsilon^1) \simeq f_n^* \omega_n \oplus \varepsilon^1 = \gamma_n \oplus \varepsilon^1.$$

Thus g_n induces a bundle map

$$\gamma_n \oplus \varepsilon^1 \to \gamma_{n+1}$$

and hence a map

$$M(g_n): SM(\gamma_n) \to M(\gamma_{n+1})$$

of Thom complexes. The complexes $MX_n = M(\gamma_n)$ and maps $SMX_n \to MX_{n+1}$ define a spectrum (cf. 8.3), which we denote by MX. In particular we have the Thom spectra MO, MSO, $MSpin$, MU, MSU, MSp and $MB\langle x\rangle$, $x \in H^q(BO; \mathbb{Z}_2)$. For $G = \{1\} \subset O$ we find that MG is homotopy equivalent to the sphere spectrum S^0.

12.30. Theorem (Thom). $\Omega_*^X \cong \pi_*(MX)$.

Proof sketch: We describe the map $\Phi: \Omega_n^X \to \pi_n(MX)$ as follows: given a closed smooth n-dimensional manifold M with X-structure $(h, \tilde{v})$, $h: M \to \mathbb{R}^{n+k}$ an embedding, we must construct a homotopy class of maps $f_M: \Sigma^n S^0 \to MX$. We regard S^{n+k} as the one-point compactification of $\mathbb{R}^{n+k}$ and the normal disk bundle $D(v)$ of M in $\mathbb{R}^{n+k}$ as a tubular neighborhood of M in $S^{n+k} - \{s_0\}$. Then we can define a map $g: S^{n+k} \to M(v)$ by taking $g|D(v)$ to be the projection $\pi: D(v) \to M(v)$ and $g(S^{n+k} - \mathring{D}(v)) = *$. We also have $M(\tilde{v}): M(v) \to MX_k$ and the composite $M(\tilde{v}) \circ g$ is a map $f_M: (S^{n+k}, s_0) \to (MX_k, *)$; we take $\Phi(M) = \{f_M\} \in \pi_n(MX)$. Φ is at least a well defined function on the set of X-diffeomorphism classes of X-manifolds.

Suppose $(M, h, \tilde{v})$, $(M', h', \tilde{v}')$ are two X-manifolds. If $T: \mathbb{R}^{n+k} \to \mathbb{R}^{n+k}$ is a translation, then $T \circ h$ has the same classifying map v for its normal bundle, and since $(M, T \circ h, \tilde{v})$ is equivalent to $(M, h, \tilde{v})$, we replace h by $T \circ h$ whenever this seems desirable. In particular we may assume $h'(M') \cap h(M) = \varnothing$—indeed even that $h'(M')$ is contained in the lower hemisphere $\mathring{H}_-^{n+k}$, $h(M)$ in the upper hemisphere $\mathring{H}_+^{n+k}$ of S^{n+k}. If we now carry out the above construction to get $f_{M \cup M'}$, then the equator of S^{n+k} gets pinched to a point, and we see that $\{f_{M \cup M'}\} = \{f_M\} + \{f_{M'}\}$.

Now suppose W is an X-manifold with boundary; we can regard the embedding $W \to \mathbb{R}_+^{n+k+1}$ as an embedding in $S^{n+k} \times [0, 1)$ with ∂W embedded in $S^{n+k} \times 0$. We carry out the analogous construction to that above, collapsing everything outside a tubular neighborhood of W in $S^{n+k} \times I$ to a point and obtain a map $H: S^{n+k} \times I \to MX_k$ such that $H_0 = f_{\partial W}$, $H_1 = *$; thus $\{f_{\partial W}\} = 0 \in \pi_n(MX)$.

In particular it follows that $\{f_{M \cup \partial W}\} = \{f_M\} + \{f_{\partial W}\} = \{f_M\}$. Hence if $M \sim_X M'$, or in other words $M \cup \partial W \cong M' \cup \partial W'$ for some W, W', then $\{f_M\} = \{f_{M \cup \partial W}\} = \{f_{M' \cup \partial W'}\} = \{f_{M'}\}$. Thus we may regard Φ as a homomorphism from Ω_*^X to $\pi_*(MX)$.

To show that Φ is surjective one argues as follows: the disk bundle $ED(\omega_k)$ can be given the structure of a smooth manifold and we regard $BO(k)$ as a closed submanifold in $E\mathring{D}(\omega_k)$ (the 0-section). Suppose $f: (S^{n+k}, s_0) \to (MX_k, *)$ represents $\alpha \in \pi_n(MX)$. $MO_k - \{*\}$ is the image in MO_k of $ED(\omega_k)^\circ$; on $U = f^{-1} \circ (Mf_k)^{-1}(MO_k - \{*\})$ we can deform $Mf_k \circ f$ a little bit so that it becomes transverse regular on $BO(k) \subset ED(\omega_k)^\circ$. We lift this homotopy to $E\mathring{D}(\gamma_k) \subset MX_k$ and extend the result of this deformation to S^{n+k} by sending $S^{n+k}-U$ to $*$; we call the result g. $g^{-1}\circ(Mf_k)^{-1}(BO(k))$ is a submanifold M of S^{n+k} of codimension n. We may even arrange that M has a tubular neighborhood T in S^{n+k} such that $g|T$ maps T in a fibre-preserving fashion into $M(\gamma_n)-\{*\}$. Thus g induces an X-structure on M and one can show that $\Phi(\{M\})=\{f\}$.

The proof that Φ is injective goes similarly. For details see [83] or [84]. □

In Chapter 20 we shall compute the homotopy groups $\pi_*(MO)$, $\pi_*(MU)$ and $\pi_*(MSO)$. We state the structure of the cobordism groups here for completeness. In each case we shall describe Ω_*^X as a graded ring; the products are defined in Chapter 13.

12.31. Theorem. *$\Omega_*^O \cong \mathbb{Z}_2[x_2,x_4,x_5,\ldots]$, a polynomial ring with one generator x_n in dimension n for each n not of the form 2^t-1 for some $t\geqslant 0$. x_{2k} may be taken to be the class $\{RP^{2k}\}$, $k\geqslant 1$.*

The description of manifolds representing odd dimensional generators is more complicated; Dold has done this in [33].

12.32. Theorem. *$\Omega_*^U \cong \mathbb{Z}[x_2,x_4,x_6,\ldots]$, a polynomial ring with one generator x_{2k} in dimension $2k$ for every $k\geqslant 1$. x_{2k} may be taken to be the class $\{CP^k\}$ if k is of the form $p-1$ for some prime p.*

12.33. Theorem. *Ω_*^{SO} has only 2-torsion and $\Omega_*^{SO}/torsion \cong \mathbb{Z}[x_4,x_8,\ldots]$, a polynomial ring with one generator x_{4k} in dimension $4k$ for every $k\geqslant 1$.*

Ω_*^{SU}, Ω_*^{Spin} are also known, but only partial results about Ω^{Sp} are known at this time. $\Omega_*^{fr}\cong\pi_*(S^0)=\pi_*^S$ is the stable homotopy groups of the spheres. This was observed by Pontrjagin even before Thom defined Ω_*^O.

Of course, as soon as one has a spectrum one gets homology and cohomology theories. Thus we get the so-called *bordism* theories $MX_*(-)$ and *cobordism* theories $MX^*(-)$ for the various spectra MX. We have just given the coefficient groups for these theories in the cases $X=BO$, BU and BSO.

For any space Y, the homology groups $MX_n(Y)$ have a geometrical description.

12.34. Definition. A *singular X-manifold* in Y is a pair (M,f), where M is a closed X-manifold and $f:M\to Y$ is a continuous map (cf. singular simplex). Two singular manifolds (M,f), (M',f') in Y are called *X-cobordant* if there is a pair (W,g) such that W is a compact X-manifold with boundary $\partial W\cong M\,\dot\cup\,(-M')$ ($-M'$ in the sense of negative in Ω_*^X) and g is a continuous map $g:W\to Y$ such that $g|M=f$, $g|M'=f'$.

12.35. Proposition. *The group $\Omega_n^X(Y)$ of X-cobordism classes of singular n-dimensional X-manifolds in Y is isomorphic to* $MX_n(Y) \cong \pi_n(MX \wedge Y^+)$.

The proof is an adaptation of that of 12.30.

12.36. *Remark.* In 7.69 we put ourselves under the obligation to show that Ω_*^X satisfies the WHE axiom, which by 7.71 is equivalent to showing that if Y is n-connected for all $n \geqslant 0$ then $\Omega_*^X(y_0) \to \Omega_*^X(Y)$ is surjective for all $y_0 \in Y$. But every compact smooth manifold M has the homotopy type of a CW-complex, so by 6.31 any singular X-manifold $f: M \to Y$ is null-homotopic and hence X-cobordant to a singular manifold $M \to y_0 \to Y$.

Exercise. Show that the definition of $\sim_X$ in 12.21 is equivalent to the definition of "cobordant" on the previous page if $X = BO$.

References

1. W. Browder [26]
2. A. Dold [33]
3. M. A. Kervaire and J. W. Milnor [50]
4. S. Lang [52]
5. A. Liulevicius [55]
6. J. W. Milnor [60, 64]
7. J. R. Munkres [69]
8. R. E. Stong [83]
9. R. Thom [84, 85]
10. C. T. C. Wall [89]

Chapter 13

Products

We have now described homology theories in general and three important particular examples. Let us not loose sight of one of the reasons for studying homology functors: one wants to investigate the existence or nonexistence of maps $f:X \to Y$ by looking at the corresponding algebraic morphisms $f_*:k_*(X) \to k_*(Y)$. As we have said before, the richer the algebraic structure on $k_*(X)$, the more useful k_* will be for these investigations. In this chapter we introduce products, so that under appropriate assumptions $k^*(X)$ will be a ring for all X. In Chapters 17, 18 and 19 we introduce another very useful algebraic structure: we make $E_*(X)$ into a comodule over a certain Hopf algebra (and $E^*(X)$ into a module over the dual Hopf algebra).

We begin this chapter by showing how to describe $H_*(X;G)$ and $H^*(X;G)$ in terms of $H_*(X;\mathbb{Z})$ (the universal coefficient theorems) and how to describe $H_*(X \times Y;\mathbb{Z})$ in terms of $H_*(X;\mathbb{Z})$ and $H_*(Y;\mathbb{Z})$ (cf. 4.1). The machinery we set up in the process will then provide us with products on $H_*(-,R)$ and $H^*(-;R)$ for any ring R. Then comes a digression to sketch the construction of smash products of spectra, after which we will be able to show how to introduce products on $E_*(-)$ and $E^*(-)$ for any ring spectrum E.

We defined $H_*(X;G)$ with the help of the chain complex $S_*(X) \otimes G$. The first naive conjecture, then, might be that $H_*(X;G) \cong H_*(X;\mathbb{Z}) \otimes G$. This is certainly true for several favorable cases such as $X = S^n$, $X = CP^n$ or $X = HP^n$, $n \geqslant 0$. However, for $X = RP^\infty$, for example, we had

$$H_q(RP^\infty;\mathbb{Z}) \cong \begin{cases} \mathbb{Z} & q=0 \\ \mathbb{Z}_2 & q \text{ odd} \\ 0 & q \text{ even}, q \neq 0, \end{cases}$$

but $H_q(RP^\infty;\mathbb{Z}_2) \cong \mathbb{Z}_2$, all $q \geqslant 0$. There is a natural homomorphism

$$\mu: H_*(X) \otimes G \to H_*(X;G)$$

given by $\{a\} \otimes g \mapsto \{a \otimes g\}$ for $a \in Z_*(X)$, $g \in G$, and we shall see that this homomorphism is always a monomorphism. Its cokernel, however, is not trivial in general. To describe the cokernel we need a little homological algebra.

Let G be an abelian group; if

$$0 \longrightarrow A \xrightarrow{\phi} B \xrightarrow{\psi} C \longrightarrow 0$$

is an exact sequence of abelian groups, then

$$0 \longrightarrow A \otimes G \xrightarrow{\phi \otimes 1} B \otimes G \xrightarrow{\psi \otimes 1} C \otimes G \longrightarrow 0$$

need not be exact. Specifically, $\phi \otimes 1$ may fail to be injective. For example

$$0 \longrightarrow \mathbb{Z} \xrightarrow{2} \mathbb{Z} \longrightarrow \mathbb{Z}_2 \longrightarrow 0$$

is exact, but

$$\begin{array}{ccccccccc} 0 & \longrightarrow & \mathbb{Z} \otimes \mathbb{Z}_2 & \xrightarrow{2} & \mathbb{Z} \otimes \mathbb{Z}_2 & \longrightarrow & \mathbb{Z}_2 \otimes \mathbb{Z}_2 & \longrightarrow & 0 \\ & & \wr\| & & \wr\| & & & & \\ & & \mathbb{Z}_2 & & \mathbb{Z}_2 & & & & \end{array}$$

is not exact, because $Z_2 \xrightarrow{2} Z_2$ is not monomorphic. Investigation of the kernel of $\phi \otimes 1$ leads to a new functor Tor.

Given any abelian group A we can find an exact sequence

$$0 \longrightarrow R \xrightarrow{\alpha} F \xrightarrow{\beta} A \longrightarrow 0$$

with R, F free abelian; we could for example take F to be the free abelian group with the elements $a \in A$ as generators and β the unique homomorphism with $\beta(a) = a$ for all $a \in A$. Then we would take $R = \ker \beta$ and α the inclusion. We now define $\operatorname{Tor}(A, G)$ to be $\ker(\alpha \otimes 1)$; that is, the sequence

$$0 \longrightarrow \operatorname{Tor}(A, G) \longrightarrow R \otimes G \xrightarrow{\alpha \otimes 1} F \otimes G \xrightarrow{\beta \otimes 1} A \otimes G \longrightarrow 0$$

is exact. Unfortunately it is not immediately clear that $\operatorname{Tor}(A, G)$ is independent of the choices R, F, α, β. This we must show.

13.1. Lemma. i) Suppose $\phi: A \to A'$ is a homomorphism and $0 \to R \xrightarrow{\alpha} F \xrightarrow{\beta} A \to 0$, $0 \to R' \xrightarrow{\alpha'} F' \xrightarrow{\beta'} A' \to 0$ are exact sequences with R, F free abelian. Then there exist homomorphisms $\psi: F \to F'$, $\theta: R \to R'$ such that the diagram

$$\begin{array}{ccccccccc} 0 & \longrightarrow & R & \xrightarrow{\alpha} & F & \xrightarrow{\beta} & A & \longrightarrow & 0 \\ & & \downarrow{\scriptstyle\theta} & & \downarrow{\scriptstyle\psi} & & \downarrow{\scriptstyle\phi} & & \\ 0 & \longrightarrow & R' & \xrightarrow{\alpha'} & F' & \xrightarrow{\beta'} & A' & \longrightarrow & 0 \end{array}$$

commutes.

ii) If (θ',ψ') is another such pair, then there exists a homomorphism $\gamma:F \to R'$ such that $\alpha' \circ \gamma = \psi' - \psi$ and $\gamma \circ \alpha = \theta' - \theta$.

Proof: i) Let $\{f_\mu : \mu \in M\}$, $\{r_\nu : \nu \in N\}$ be bases for F, R respectively. Then for each $\mu \in M$ we can find an $f'_\mu \in F'$ with $\beta'(f'_\mu) = \phi\beta(f_\mu)$ since β' is surjective. If we let ψ be the unique homomorphism with $\psi(f_\mu) = f'_\mu$, $\mu \in M$, then the right-hand square commutes. For each $\nu \in N$ $\beta'\psi\alpha(r_\nu) = \phi\beta\alpha(r_\nu) = 0$, so by exactness there is an $r'_\nu \in R'$ with $\alpha'(r'_\nu) = \psi\alpha(r_\nu)$. If we let θ be the unique homomorphism with $\theta(r_\nu) = r'_\nu$, $\nu \in N$, then the left-hand square commutes.

ii) Suppose (θ,ψ), (θ',ψ') are two such pairs, $\{f_\mu\}$ a basis for F as above. For each $\mu \in M$ we have $\beta'(\psi' - \psi)(f_\mu) = \beta'\psi'(f_\mu) - \beta'\psi(f_\mu) = \phi\alpha(f_\mu) - \phi\alpha(f_\mu) = 0$, so there exists an $r'_\mu \in R'$ with $\alpha'(r'_\mu) = (\psi' - \psi)(f_\mu)$. If we let $\gamma:F \to R'$ be the unique homomorphism with $\gamma(f_\mu) = r'_\mu$, $\mu \in M$, then $\alpha' \circ \gamma = \psi' - \psi$. We also have $\alpha' \circ \gamma \circ \alpha = (\psi' - \psi) \circ \alpha = \alpha' \circ (\theta' - \theta)$, so since α' is a monomorphism it follows $\gamma \circ \alpha = \theta' - \theta$. ▯

13.2. We can think of $\cdots \to 0 \to 0 \to R \to F$ as a chain complex $C_*(A)$ with

$$C_q(A) = \begin{cases} F & q = 0 \\ R & q = 1 \\ 0 & q \neq 0, 1. \end{cases}$$

In these terms $H_0(C_*(A) \otimes G) \cong A \otimes G$ and $H_1(C_*(A) \otimes G) \cong \operatorname{Tor}(A, G)$. Lemma 13.1 says that every homomorphism $\phi: A \to A'$ induces a chain map $C_*(\phi): C_*(A) \to C_*(A')$ which is not unique but is unique up to chain homotopy. Thus ϕ induces a unique homomorphism

$$\operatorname{Tor}(\phi, 1): \operatorname{Tor}(A, G) \to \operatorname{Tor}(A', G)$$

—namely $(C_*(\phi) \otimes 1)_*: H_1(C_*(A) \otimes G) \to H_1(C_*(A') \otimes G)$. From the definition it is clear that $\operatorname{Tor}(1_A, 1) = 1_{\operatorname{Tor}(A, G)}$ and for a homomorphism $\phi': A' \to A''$ we have

$$\operatorname{Tor}(\phi', 1) \circ \operatorname{Tor}(\phi, 1) = \operatorname{Tor}(\phi' \circ \phi, 1).$$

Taking $\phi = 1_A = \phi'$ we see that $\operatorname{Tor}(A,G)$ is independent (up to isomorphism) of the *free resolution* $C_*(A)$. $\operatorname{Tor}(-, G)$ is a functor for fixed G.

Now any homomorphism $\psi: G \to G'$ induces a "coefficient homomorphism" (cf. 10.5) $(1 \otimes \psi)_*: H_1(C_*(A) \otimes G) \to H_1(C_*(A) \otimes G')$, which we shall denote by

$$\mathrm{Tor}(1,\psi):\mathrm{Tor}(A,G) \to \mathrm{Tor}(A,G').$$

$\mathrm{Tor}(A,-)$ is a functor for fixed A.

13.3. Proposition. Tor *has the following properties:*

i) *if A is free abelian, then* $\mathrm{Tor}(A,G)=0$;
ii) *if G is free abelian, then* $\mathrm{Tor}(A,G)=0$;
iii) $\mathrm{Tor}(A,G)\cong\mathrm{Tor}(G,A)$ *for all A,G;*
iv) *if $0\to A\to B\to C\to 0$ is a short exact sequence, then for every G there is a long exact sequence*

$$0 \to \mathrm{Tor}(A,G) \to \mathrm{Tor}(B,G) \to \mathrm{Tor}(C,G) \to A\otimes G \to B\otimes G \to C\otimes G \to 0.$$

Proof: i) If A is free abelian, then $0\to 0 \xrightarrow{\alpha} A \xrightarrow{1} A \to 0$ is a free resolution of A and $\mathrm{Tor}(A,G)=\ker(\alpha\otimes 1)=0$.

ii) G free abelian implies $A\otimes G\cong\bigoplus A$, etc. But a direct sum of exact sequences is exact.

iii) Let $0\to R\to F\to A\to 0$ and $0\to K\to H\to G\to 0$ be free resolutions of A,G respectively. Since $A\otimes G\cong G\otimes A$, etc., it follows that the following commutative diagram has exact rows and columns

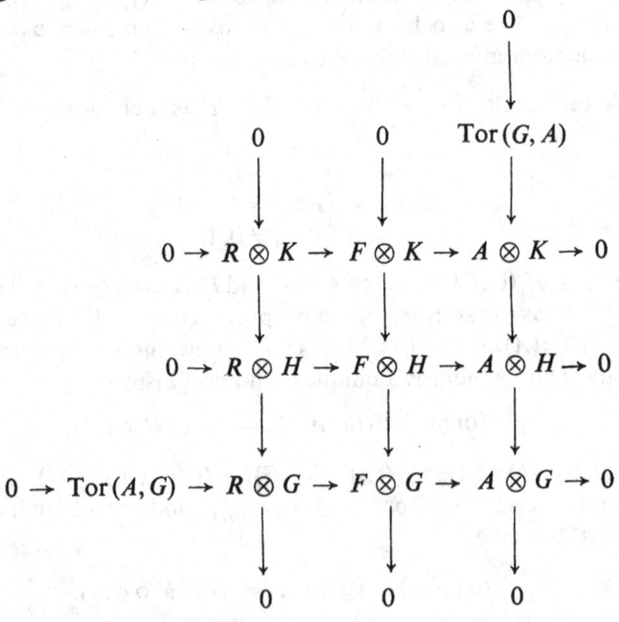

Diagram chasing in this diagram yields homomorphisms

$$\phi:\mathrm{Tor}(A,G) \to \mathrm{Tor}(G,A), \quad \psi:\mathrm{Tor}(G,A) \to \mathrm{Tor}(A,G)$$

which are inverses of one another.

iv) Let $C_*(G)$ be the chain complex corresponding to a free resolution of G. Because $C_*(G)$ is a free chain complex, it follows from ii) that

$$0 \to C_*(G) \otimes A \to C_*(G) \otimes B \to C_*(G) \otimes C \to 0$$

is a short exact sequence of chain complexes. The exact homology sequence together with the isomorphism of iii) yields the long exact sequence of iv). □

13.4. Exercise. Show that Tor commutes with dirlim and $\oplus$ and deduce that $\operatorname{Tor}(A,B) \cong \operatorname{Tor}(T(A),T(B))$, where $T(A)$ denotes the torsion subgroup of A.

13.5. Theorem (Universal Coefficient Theorem). *For any free chain complex C_* and any abelian group G there is a natural exact sequence*

$$0 \to H_n(C_*) \otimes G \to H_n(C_* \otimes G) \to \operatorname{Tor}(H_{n-1}(C_*), G) \to 0,$$

for all $n \in \mathbb{Z}$. The sequence splits but not naturally.
Proof: We have two exact sequences

$$0 \longrightarrow Z_n(C_*) \longrightarrow C_n \xrightarrow{d} B_{n-1}(C_*) \longrightarrow 0,$$

13.7.

$$0 \longrightarrow B_n(C_*) \xrightarrow{i} Z_n(C_*) \longrightarrow H_n(C_*) \longrightarrow 0, \quad n \in \mathbb{Z}.$$

$B_{n-1} \subset C_{n-1}$ is free abelian, so by 13.3i) and iv) the sequence

$$0 \longrightarrow Z_n \otimes G \longrightarrow C_n \otimes G \xrightarrow{d \otimes 1} B_{n-1} \otimes G \longrightarrow 0$$

is also exact; that is, we have a short exact sequence of chain complexes

$$0 \longrightarrow Z_* \otimes G \longrightarrow C_* \otimes G \xrightarrow{d \otimes 1} B_* \otimes G \longrightarrow 0,$$

where $d \otimes 1$ lowers degree by 1 and the differentials for $Z_* \otimes G$, $B_* \otimes G$ are 0. The resulting long exact sequence of homology groups is then

$$\cdots \longrightarrow H_{n+1}(C_* \otimes G) \longrightarrow$$

$$\begin{array}{ccccc} H_n(B_* \otimes G) & \xrightarrow{\partial} & H_n(Z_* \otimes G) & \longrightarrow & H_n(C_* \otimes G) \longrightarrow \cdots, \\ \wr\Vert & & \wr\Vert & & \\ B_n \otimes G & & Z_n \otimes G & & \end{array}$$

and examination of the definition of ∂ for this case shows that it is nothing but $i \otimes 1$. Thus we obtain short exact sequences

$$0 \to \operatorname{coker}(i \otimes 1)_n \to H_n(C_* \otimes G) \to \ker(i \otimes 1)_{n-1} \to 0, \quad n \in \mathbb{Z}.$$

Now 13.7 can be regarded as a free resolution of $H_n(C_*)$, so we have the exact sequence

$$0 \longrightarrow \operatorname{Tor}(H_n(C_*), G) \longrightarrow B_n \otimes G \xrightarrow{i \otimes 1} Z_n \otimes G \longrightarrow H_n(C_*) \otimes G \longrightarrow 0,$$

from which we immediately have

$$\operatorname{coker}(i \otimes 1)_n \cong H_n(C_*) \otimes G$$

$$\ker(i \otimes 1)_{n-1} \cong \operatorname{Tor}(H_{n-1}(C_*), G).$$

One easily sees that the map $H_n(C_*) \otimes G \to H_n(C_* \otimes G)$ is the μ defined earlier. Naturality is also clear.

To construct a splitting we take any splitting $\kappa: C_* \to Z_*$ of 13.6 (such exist because B_* is free). Then the composite

$$Z(C_* \otimes G) \xrightarrow{(\kappa \otimes 1)|Z} Z_* \otimes G \to H_*(C_*) \otimes G$$

maps $B(C_* \otimes G)$ to 0 and hence defines a homomorphism

$$\kappa': H_*(C_* \otimes G) \to H_*(C_*) \otimes G$$

which is a left inverse of μ. ▯

13.8. Corollary. *For every topological pair (X, A) and abelian group G there is a natural exact sequence*

$$0 \to H_n(X, A) \otimes G \to H_n(X, A; G) \to \operatorname{Tor}(H_{n-1}(X, A), G) \to 0$$

and an unnatural splitting

$$H_n(X, A; G) \cong H_n(X, A) \otimes G \oplus \operatorname{Tor}(H_{n-1}(X, A), G).$$

Now we also want to express $H^*(X; G)$ in terms of $H_*(X)$. Cohomology is constructed with the help of the cofunctor $\operatorname{Hom}(-, G)$, and $\operatorname{Hom}(-, G)$ does not preserve exact sequences, as the following example shows:

$$0 \longrightarrow \mathbb{Z} \xrightarrow{2} \mathbb{Z} \longrightarrow \mathbb{Z}_2 \longrightarrow 0$$

is exact but

$$\begin{array}{ccccccc} 0 \longleftarrow & \operatorname{Hom}(\mathbb{Z}, \mathbb{Z}_2) & \xleftarrow{2} & \operatorname{Hom}(\mathbb{Z}, \mathbb{Z}_2) & \longleftarrow & \operatorname{Hom}(\mathbb{Z}_2, \mathbb{Z}_2) & \longleftarrow 0 \\ & \| \wr & & \| \wr & & & \\ & \mathbb{Z}_2 & & \mathbb{Z}_2 & & & \end{array}$$

is not. After our experience with $\otimes$ and Tor we might now expect Hom to generate a derived functor, and indeed the method of construction already lies at hand. We take a free resolution

$$0 \longrightarrow R \xrightarrow{\alpha} F \xrightarrow{\beta} A \longrightarrow 0$$

of any abelian group A and regard

$$0 \leftarrow \mathrm{Hom}(R, G) \leftarrow \mathrm{Hom}(F, G)$$

as a cochain complex

$$C^q(A; G) = \begin{cases} \mathrm{Hom}(F, G) & q=0 \\ \mathrm{Hom}(R, G) & q=1 \\ 0 & q \neq 0, 1. \end{cases}$$

Then we have $H^0(C^*(A;G)) \cong \mathrm{Hom}(A, G)$ and we define $\mathrm{Ext}(A, G) = H^1(C^*(A;G))$. That is $\mathrm{Ext}(A, G) = \mathrm{coker}\,\mathrm{Hom}(\alpha, 1)$. The proof that Ext does not depend on the free resolution of A should be clear.

Exercise 13.4 is supposed to suggest why the derived functor of $\otimes$ is called Tor. The derived functor of Hom is called Ext because it has to do with group extensions: if A, B are abelian groups, then an *extension* of A by B is an exact sequence

$$0 \longrightarrow B \xrightarrow{\rho} C \xrightarrow{\sigma} A \longrightarrow 0$$

with C abelian (one sometimes calls C an extension of A by B, but this is incorrect; ρ, σ are important too). Two extensions

$$0 \longrightarrow B \xrightarrow{\rho} C \xrightarrow{\sigma} A \longrightarrow 0$$

$$0 \longrightarrow B \xrightarrow{\rho'} C' \xrightarrow{\sigma'} A \longrightarrow 0$$

are *equivalent* if there is a commutative diagram

$$\begin{array}{ccccccccc} 0 & \longrightarrow & B & \xrightarrow{\rho} & C & \xrightarrow{\sigma} & A & \longrightarrow & 0 \\ & & \downarrow{\scriptstyle 1} & & \downarrow{\scriptstyle \phi} & & \downarrow{\scriptstyle 1} & & \\ 0 & \longrightarrow & B & \xrightarrow{\rho'} & C' & \xrightarrow{\sigma'} & A & \longrightarrow & 0 \end{array}$$

(ϕ is, of course, an isomorphism). We let $E(A, B)$ denote the set of equivalence classes of extensions of A by B. One can introduce a sum on $E(A, B)$ so that $E(A, B)$ becomes an abelian group.

Suppose $x \in E(A, B)$ is represented by $0 \to B \xrightarrow{\rho} C \xrightarrow{\sigma} A \to 0$ and suppose $0 \to R \xrightarrow{\alpha} F \xrightarrow{\beta} A \to 0$ is a free resolution of A. Then by 13.1 we can find ϕ_1, ϕ_2 so that

$$\begin{array}{ccccccccc} 0 & \longrightarrow & R & \xrightarrow{\alpha} & F & \xrightarrow{\beta} & A & \longrightarrow & 0 \\ & & \downarrow{\scriptstyle \phi_2} & & \downarrow{\scriptstyle \phi_1} & & \downarrow{\scriptstyle 1} & & \\ 0 & \longrightarrow & B & \xrightarrow{\rho} & C & \xrightarrow{\sigma} & A & \longrightarrow & 0 \end{array}$$

commutes. $\phi_2 \in \mathrm{Hom}(R,B)$ is determined modulo $\mathrm{im}[\mathrm{Hom}(F,B) \to \mathrm{Hom}(R,B)]$, hence defines an element $\kappa(x) \in \mathrm{Ext}(A,B)$.

13.9. Exercise. Show that $\kappa(x)$ is well defined and that $\kappa: E(A,B) \to \mathrm{Ext}(A,B)$ is bijective.

13.10. Theorem (Universal Coefficient Theorem). *For any free chain complex C and any abelian group G there is a natural exact sequence*

$$0 \longrightarrow \mathrm{Ext}(H_{n-1}(C_*), G) \longrightarrow H^n(\mathrm{Hom}(C_*, G)) \xrightarrow{\mu^*} \mathrm{Hom}(H_n(C_*), G) \longrightarrow 0$$

for all $n \in \mathbb{Z}$, where μ^* is given by

$$\mu^* \{\phi\}(\{z\}) = \phi(z) \quad \{\phi\} \in H^n(\mathrm{Hom}(C_*, G)), \{z\} \in H_n(C_*).$$

The sequence splits but not naturally.

Proof: Because B_* is free applying $\mathrm{Hom}(-,G)$ to 13.6 gives an exact sequence

$$0 \leftarrow \mathrm{Hom}(Z_*, G) \leftarrow \mathrm{Hom}(C_*, G) \leftarrow \mathrm{Hom}(B_*, G) \leftarrow 0$$

(the sequence is even split). We can regard this as a short exact sequence of cochain complexes if we give $\mathrm{Hom}(Z_*, G)$, $\mathrm{Hom}(B_*, G)$ zero differentials. The resulting long exact sequence of cohomology groups looks as follows:

$$\cdots \longrightarrow H^n(\mathrm{Hom}(C_*, G)) \longrightarrow \mathrm{Hom}(Z_n, G) \xrightarrow{\mathrm{Hom}(i,1)} \mathrm{Hom}(B_n, G) \longrightarrow H^{n+1}(\mathrm{Hom}(C_*, G)) \longrightarrow \cdots$$

from which we get the short exact sequences

$$0 \to \mathrm{coker}\,\mathrm{Hom}(i,1)_{n-1} \to H^n(\mathrm{Hom}(C_*, G)) \to \ker \mathrm{Hom}(i,1)_n \to 0.$$

On the other hand from 13.7 we get the exact sequence

$$0 \longleftarrow \mathrm{Ext}(H_n(C_*), G) \longleftarrow \mathrm{Hom}(B_n, G) \xleftarrow{\mathrm{Hom}(i,1)} \mathrm{Hom}(Z_n, G) \longleftarrow \mathrm{Hom}(H_n(C_*), G) \longleftarrow 0,$$

from which it follows

$$\mathrm{coker}\,\mathrm{Hom}(i,1)_{n-1} \cong \mathrm{Ext}(H_{n-1}(C_*), G)$$
$$\ker \mathrm{Hom}(i,1)_n \cong \mathrm{Hom}(H_n(C_*), G).$$

The other statements of the theorem are easily checked. □

13.11. Corollary. *For every topological pair (X,A) and abelian group G there is a natural exact sequence*

$$0 \longrightarrow \operatorname{Ext}(H_{n-1}(X,A),G) \longrightarrow H^n(X,A;G) \xrightarrow{\mu^*} \operatorname{Hom}(H_n(X,A),G) \longrightarrow 0$$

and an unnatural splitting

$$H^n(X,A;G) \simeq \operatorname{Hom}(H_n(X,A),G) \oplus \operatorname{Ext}(H_{n-1}(X,A),G).$$

The student would do well to determine $\operatorname{Tor}(\mathbb{Z}_n,\mathbb{Z}_m)$, $\operatorname{Ext}(\mathbb{Z}_n,\mathbb{Z}_m)$, $\operatorname{Ext}(\mathbb{Z}_m,\mathbb{Z})$ for all n,m and then apply 13.8 and 13.11 to find $H_*(RP^n;\mathbb{Z}_m)$, $H^*(RP^n;\mathbb{Z}_m)$ and check whether the result agrees with 10.22iv).

We turn now to the problem of describing $H_*(X \times Y)$ in terms of $H_*(X)$ and $H_*(Y)$. The discussion here naturally falls into a "geometric" part ($S_*(X \times Y) \simeq S_*(X) \otimes S_*(Y)$) and an algebraic part

$$(H_*(C_* \otimes C'_*) \simeq \cdots).$$

We do the algebraic part first.

13.12. Definition. Suppose A_*, B_* are two graded abelian groups. We can define the tensor product $A_* \otimes B_*$: it is the graded abelian group with $(A_* \otimes B_*)_n = \bigoplus_{i+j=n} A_i \otimes B_j$, $n \in \mathbb{Z}$. In a similar fashion we define the Tor product: $\operatorname{Tor}(A_*,B_*)_n = \bigoplus_{i+j=n} \operatorname{Tor}(A_i,B_j)$, $n \in \mathbb{Z}$. If (C_*,d), (C'_*,d') are two chain complexes, then we can define a new chain complex $(C_* \otimes C'_*, d_\otimes)$: we define $d_\otimes$ on $C_* \otimes C'_*$ by $d_\otimes(a \otimes b) = da \otimes b + (-1)^p a \otimes d'b$ if $a \in C_p$, $b \in C'_*$. Note that in the second term of the expression for $d_\otimes(a \otimes b)$ we have "interchanged a and d"; a has degree p and d has degree -1, so we put a sign $(-1)^{p(-1)} = (-1)^p$ in front of this term. The reader will easily check that $d_\otimes$ so defined is a differential ($d_\otimes \circ d_\otimes = 0$) but would not be a differential if we left this sign out. In this chapter and in all following ones we shall often employ the convention that whenever something of degree p is "shifted past" something of degree q a sign $(-1)^{pq}$ occurs before the term in question.

The following theorem (algebraic Künneth theorem) expresses $H_*(C_* \otimes C'_*)$ in terms of $H_*(C_*)$ and $H_*(C'_*)$.

13.13. Theorem. *Suppose C_*, C'_* are two chain complexes, one of which is free. Then there is a natural exact sequence*

$$0 \longrightarrow [H_*(C_*) \otimes H_*(C'_*)]_n \xrightarrow{\times} H_n(C_* \otimes C'_*) \longrightarrow [\operatorname{Tor}(H_*(C_*), H_*(C'_*))]_{n-1} \longrightarrow 0.$$

If C_*, C'_* are both free, then the sequence even splits but not naturally.

Proof: Let us suppose C_* is free; the proof for the case C'_* free is similar. We have exact sequences

13.14. $$0 \longrightarrow Z_n(C_*) \longrightarrow C_n \xrightarrow{d} B_{n-1}(C_*) \longrightarrow 0$$

13.15. $$0 \longrightarrow B_n(C_*) \xrightarrow{i} Z_n(C_*) \longrightarrow H_n(C_*) \longrightarrow 0$$

13.16. $$0 \longrightarrow Z_n(C'_*) \longrightarrow C'_n \xrightarrow{d'} B_{n-1}(C'_*) \longrightarrow 0$$

13.17. $$0 \longrightarrow B_n(C'_*) \longrightarrow Z_n(C'_*) \longrightarrow H_n(C'_*) \longrightarrow 0,$$

and 13.14 is split because $B_{n-1}(C_*)$ is free. Therefore 13.14 remains exact if we tensor with C'_*:

13.18.

$$0 \longrightarrow Z(C_*) \otimes C'_* \longrightarrow C_* \otimes C'_* \xrightarrow{d \otimes 1} B(C_*) \otimes C'_* \longrightarrow 0.$$

We may regard 13.18 as an exact sequence of chain complexes if we give $Z(C_*) \otimes C'_*$ and $B(C_*) \otimes C'_*$ the differential $\omega \otimes d'$, $-\omega \otimes d'$ respectively, where $\omega(a) = (-1)^p a$ for $a \in Z_p$ or B_p. We then get a long exact sequence of homology groups

13.19.

$$\cdots \longrightarrow H_{n+1}(C_* \otimes C'_*) \longrightarrow H_n(B(C_*) \otimes C'_*) \xrightarrow{\partial}$$
$$H_n(Z(C_*) \otimes C'_*) \longrightarrow H_n(C_* \otimes C'_*) \longrightarrow \cdots.$$

We wish to identify $H_*(B(C_*) \otimes C'_*)$, $H_*(Z(C_*) \otimes C'_*)$ and ∂ with more familiar objects.

Since $Z(C_*) \subset C_*$ is free, the sequence 13.16 remains exact if we tensor with $Z(C_*)$:

13.20

$$0 \longrightarrow Z(C_*) \otimes Z(C'_*) \longrightarrow Z(C_*) \otimes C'_* \xrightarrow{1 \otimes d'}$$
$$Z(C_*) \otimes B(C'_*) \longrightarrow 0.$$

We also have the exact sequence

13.21.

$$0 \to Z(C_*) \otimes B(C'_*) \to Z(C_*) \otimes Z(C'_*) \to Z(C_*) \otimes H_*(C'_*) \to 0.$$

From 13.20 it then follows that

$$\operatorname{im}(\omega \otimes d') = \operatorname{im}(1 \otimes d') \cong Z(C_*) \otimes B(C'_*)$$
$$\ker(\omega \otimes d') = \ker(1 \otimes d') \cong Z(C_*) \otimes Z(C'_*)$$

and from 13.21

$$H_*(Z(C_*) \otimes C'_*) = \frac{\ker(\omega \otimes d')}{\operatorname{im}(\omega \otimes d')} \cong \frac{Z(C_*) \otimes Z(C'_*)}{Z(C_*) \otimes B(C'_*)} \cong Z(C_*) \otimes H_*(C'_*).$$

Analogously it follows $H_*(B(C_*) \otimes C'_*) \cong B(C_*) \otimes H_*(C'_*)$. Moreover, under these identifications $\partial: H_*(B(C_*) \otimes C'_*) \to H_*(Z(C_*) \otimes C'_*)$ becomes

$$i \otimes 1: B(C_*) \otimes H_*(C'_*) \to Z(C_*) \otimes H_*(C'_*),$$

as one quickly checks. Thus from 13.19 we get short exact sequences

13.22.

$$0 \to \operatorname{coker}(i \otimes 1)_n \to H_n(C_* \otimes C'_*) \to \ker(i \otimes 1)_{n-1} \to 0.$$

Now we may regard 13.15 as a free resolution of $H_n(C_*)$, so we have an exact sequence

13.23.

$$0 \longrightarrow \operatorname{Tor}(H_n(C_*), H_m(C'_*)) \longrightarrow B_n(C_*) \otimes H_m(C'_*) \xrightarrow{i \otimes 1}$$
$$Z_n(C_*) \otimes H_m(C'_*) \longrightarrow H_n(C_*) \otimes H_m(C'_*) \longrightarrow 0.$$

From 13.23 we have immediately

$$\operatorname{coker}(i \otimes 1)_n \cong [H_*(C_*) \otimes H_*(C'_*)]_n$$
$$\ker(i \otimes 1)_{n-1} \cong [\operatorname{Tor}(H_*(C_*), H_*(C'_*))]_{n-1}.$$

Now suppose C'_* is also free and let $\kappa: C_* \to Z(C_*)$, $\kappa': C'_* \to Z(C'_*)$ be splittings of 13.14 and 13.16. Then the restriction of $\kappa \otimes \kappa'$ to $Z(C_* \otimes C'_*)$ defines a homomorphism

$$\lambda: Z(C_* \otimes C'_*) \to Z(C_*) \otimes Z(C'_*)$$

such that $\lambda(B(C_* \otimes C_*)) \subset \operatorname{im}[B(C_*) \otimes Z(C'_*) + Z(C_*) \otimes B(C'_*)]$. Thus λ defines a homomorphism

$$\bar{\lambda}: H_*(C_* \otimes C'_*) \to \frac{Z(C_*) \otimes Z(C'_*)}{\operatorname{im}[B(C_*) \otimes Z(C'_*) + Z(C_*) \otimes B(C'_*)]}.$$

On the other hand from the commutative diagram

$$\begin{array}{ccccccc}
B(C_*) \otimes B(C'_*) & \to & B(C_*) \otimes Z(C'_*) & \to & B(C_*) \otimes H(C'_*) & \to & 0 \\
\downarrow & & \downarrow & & \downarrow & & \\
Z(C_*) \otimes B(C'_*) & \to & Z(C_*) \otimes Z(C'_*) & \to & Z(C_*) \otimes H(C'_*) & \to & 0 \\
\downarrow & & \downarrow & & \downarrow & & \\
H(C_*) \otimes B(C'_*) & \to & H(C_*) \otimes Z(C'_*) & \to & H(C_*) \otimes H(C'_*) & \to & 0 \\
\downarrow & & \downarrow & & \downarrow & & \\
0 & & 0 & & 0 & &
\end{array}$$

with exact rows and columns we get the exact sequence

$$B(C_*) \otimes Z(C'_*) + Z(C_*) \otimes B(C'_*) \to Z(C_*) \otimes Z(C'_*) \to H(C_*) \otimes H(C'_*) \to 0$$

from which it follows that

$$\frac{Z(C_*) \otimes Z(C'_*)}{\operatorname{im}\,[B(C_*) \otimes Z(C'_*) + Z(C_*) \otimes B(C'_*)]} \cong H(C_*) \otimes H(C'_*).$$

(This result does not depend on C_*, C'_* being free.) Thus we may regard $\bar{\lambda}$ as a map

$$H_*(C_* \otimes C'_*) \to H_*(C_*) \otimes H_*(C'_*),$$

and as such it is a left inverse for $\times$. □

Exercise. Prove a corresponding result for cohomology; you will need some finiteness condition.

Now we turn to the geometric part of our investigation—namely the proof that $S_*(X \times Y)$ is chain homotopy equivalent to $S_*(X) \otimes S_*(Y)$. To this end we must construct chain maps

$$\rho: S_*(X \times Y) \to S_*(X) \otimes S_*(Y)$$

$$\sigma: S_*(X) \otimes S_*(Y) \to S_*(X \times Y)$$

which are chain homotopy inverses of one another. We shall prove a much more general theorem about the existence of such chain maps, a theorem which will be of use to us later when we discuss products.

13.24. Definition. Let $\mathscr{K}$ denote the category of chain complexes and chain maps and $\mathscr{C}$ any category. Let M be a set of objects from $\mathscr{C}$ (the elements of M will be called *models*). A functor $F:\mathscr{C} \to \mathscr{K}$ is *acyclic on the models M* if $H_q(F(X)) = 0$ for all $q > 0$ and all $X \in M$. In other words

$$\cdots \xrightarrow{d} F_q(X) \xrightarrow{d} F_{q-1}(X) \longrightarrow \cdots \xrightarrow{d} F_0(X) \xrightarrow{\varepsilon} H_0(F(X)) \longrightarrow 0$$

is an exact sequence. F is called *free on the models M* if there is an indexed collection $B = \{b_j \in F(M_j)\}_{j \in J}$, where $M_j \in M$, such that for every $Y \in \mathscr{C}$ the indexed set

$$\{F(f)(b_j)\}_{j \in J,\, f \in \hom(M_j, Y)} \subset F(Y)$$

forms a basis for $F(Y)$.

13.25. Examples. i) Let $\mathscr{C} = \mathscr{T}$, the category of all topological spaces and continuous functions and $F = S_*:\mathscr{T} \to \mathscr{K}$, the singular chain complex functor. Let $M = \{\Delta_n : n \geqslant 0\}$. Because Δ_n is contractible we have $H_q(F(\Delta_n)) = H_q(\Delta_n) = 0, q > 0, n \geqslant 0$, so $F = S_*$ is acyclic on the models M. Let $\delta_n : \Delta_n \to \Delta_n$ be the identity and $B = \{\delta_n\}_{n \geqslant 0}$. Then for every $Y \in \mathscr{T}$ the set

$$\begin{aligned}
&\{F(f)(\delta_n) : f \in \hom_{\mathscr{T}}(\Delta_n, Y),\, n \geqslant 0\} \\
&= \{f_\#(\delta_n) : f : \Delta_n \to Y \text{ a map},\, n \geqslant 0\} \\
&= \{f : \Delta_n \to Y,\, n \geqslant 0\} \subset S_*(Y)
\end{aligned}$$

is a basis for $S_*(Y)$, so $F = S_*$ is also free on the models M.

ii) Let $\mathscr{C} = \mathscr{T} \times \mathscr{T}$, the category of ordered pairs of topological spaces and define $F:\mathscr{C} \to \mathscr{K}$ by $F(X, Y) = S_*(X \times Y)$. Let

$$M = \{(\Delta_n, \Delta_m) : n, m \geqslant 0\}.$$

Since $\Delta_n \times \Delta_m$ is contractible $H_q(F(\Delta_n, \Delta_m)) = H_q(\Delta_n \times \Delta_m) = 0$, $q > 0$, $n, m \geqslant 0$, so F is acyclic on models M. Let $\hat{\delta}_n : \Delta_n \to \Delta_n \times \Delta_n$ be the diagonal map; then $\hat{\delta}_n \in S_n(\Delta_n \times \Delta_n)$ and for every pair $(X, Y) \in \mathscr{C}$ the set

$$\begin{aligned}
&\{F(f, g)(\hat{\delta}_n) : (f, g) \in \hom_{\mathscr{C}}((\Delta_n, \Delta_n), (X, Y)),\, n \geqslant 0\} \\
&= \{(f \times g)_\#(\hat{\delta}_n) : (f, g) \in \hom_{\mathscr{C}}((\Delta_n, \Delta_n), (X, Y)),\, n \geqslant 0\} \\
&= \{(f, g) : \Delta_n \to X \times Y,\, n \geqslant 0\} \subset S_*(X \times Y)
\end{aligned}$$

is a basis for $S_*(X \times Y)$, so F is also free on the models M.

iii) Let $\mathscr{C}$ be as in ii) and let $G:\mathscr{C} \to \mathscr{K}$ be given by $G(X,Y) = S_*(X) \otimes S_*(Y)$. With M as in ii) we have

$$H_q(G(\Delta_n, \Delta_m)) = H_q(S_*(\Delta_n) \otimes S_*(\Delta_m)) \cong \bigoplus_{r+s=q} H_r(\Delta_n) \otimes H_s(\Delta_m) \oplus \bigoplus_{r+s=q-1} \operatorname{Tor}(H_r(\Delta_n), H_s(\Delta_m)) = 0 \text{ for } q > 0,\ n,m \geqslant 0$$

(here we have used Theorem 13.13). Thus G is acyclic on the models M. Also for each pair (X,Y) the set

$$\begin{aligned}&\{G(f,g)(\delta_n \otimes \delta_m) : (f,g) \in \hom_{\mathscr{C}}((\Delta_n,\Delta_m),(X,Y)),\ n,m \geqslant 0\}\\ &\quad = \{(f_\# \otimes g_\#)(\delta_n \otimes \delta_m) : (f,g) \in \hom_{\mathscr{C}}((\Delta_n,\Delta_m),(X,Y)),\ n,m \geqslant 0\}\\ &\quad = \{f \otimes g : f:\Delta_n \to X, g:\Delta_m \to Y,\ n,m \geqslant 0\} \subset S_*(X) \otimes S_*(Y)\end{aligned}$$

forms a basis for $S_*(X) \otimes S_*(Y)$, so G is free on the models M.

13.26. Theorem. *Let $\mathscr{C}$ be a category and $F,G:\mathscr{C} \to \mathscr{K}$ be two functors such that F is free on models $M \subset \mathscr{C}$ and G is acyclic on models M.*

i) *For every natural transformation $\phi: H_0(F(-)) \to H_0(G(-))$ there exists a natural transformation $\phi_\#: F \to G$ such that*

$$\begin{array}{ccccccccc} \cdots \to & F_2(X) & \to & F_1(X) & \to & F_0(X) & \to & H_0(F(X)) \\ & \downarrow \phi_2(X) & & \downarrow \phi_1(X) & & \downarrow \phi_0(X) & & \downarrow \phi(X) \\ \cdots \to & G_2(X) & \to & G_1(X) & \to & G_0(X) & \to & H_0(G(X)) \end{array}$$

commutes.

ii) *If $\phi'_\#$ is another such natural transformation, then there is a natural chain homotopy $\Phi_q: F_q(-) \to G_{q+1}(-)$, $q \geqslant 0$, such that $d \circ \Phi_q + \Phi_{q-1} \circ d = \phi'_q - \phi_q$, $q \geqslant 0$ ($\Phi_{-1} = 0$).*

Proof: i) The actual construction takes place on the models and naturality then takes care of the rest (cf. the proof of the homotopy axiom for singular homology). Let $b_j \in B$ be such that $b_j \in F_0(M_j)$. Since $G_0(M_j) \xrightarrow{\varepsilon} H_0(G(M_j)) \to 0$ is exact, we can find a $b'_j \in G_0(M_j)$ with $\varepsilon(b'_j) = \phi(M_j)(\varepsilon(b_j))$. We set $\phi_0(M_j)(b_j) = b'_j$ and we make such a choice for every $b_j \in B$ with $b_j \in F_0(M_j)$.

Now if ϕ_0 is to be natural, then for every $f: M_j \to X$ we must have

13.27.

$$\phi_0(X)(F(f)(b_j)) = G(f)\phi_0(M_j)(b_j) = G(f)(b'_j) \quad \text{for} \quad b_j \in B \cap F_0(M_j).$$

On the other hand $\{F(f)(b_j) : j \in J, b_j \in F_0(M_j), f \in \hom(M_j, X)\}$ is a basis for $F_0(X)$, so 13.27 defines a unique homomorphism $\phi_0(X): F_0(X) \to G_0(X)$. From the choice of b'_j we have

$$\begin{aligned}\varepsilon \circ \phi_0(X)(F(f)(b_j)) &= \varepsilon \circ G(f)(b_j') = H_0(G(f)) \circ \varepsilon(b_j') \\ &= H_0(G(f))\, \phi(M_j)(\varepsilon(b_j)) = \phi(X)\, H_0(F(f))(\varepsilon(b_j)) \\ &= \phi(X) \circ \varepsilon(F(f)(b_j)),\end{aligned}$$

so it follows $\varepsilon \circ \phi_0(X) = \phi(X) \circ \varepsilon$ for all $X \in \mathscr{C}$ as desired. The naturality of ϕ_0 follows immediately from the definition of ϕ_0.

Suppose we have already constructed $\phi_0, \phi_1, \ldots, \phi_{k-1}$ satisfying i). For any b_j with $b_j \in F_k(M_j)$ we have $d\phi_{k-1}(M_j)(db_j) = \phi_{k-2}(M_j)(ddb_j) = 0$. Since $G_k(M_j) \to G_{k-1}(M_j) \to G_{k-2}(M_j)$ is exact, we can find a $b_j' \in G_k(M_j)$ with $db_j' = \phi_{k-1}(M_j)(db_j)$. We set $\phi_k(M_j)(b_j) = b_j'$, and we make such a choice for each b_j with $b_j \in F_k(M_j)$.

The naturality requirement on ϕ_k then forces us to set

13.28. $$\phi_k(X)(F(f)(b_j)) = G(f)(b_j')$$

for every $X \in \mathscr{C}$, b_j with $b_j \in F_k(M_j)$ and $f: M_j \to X$. On the other hand 13.28 defines a unique homomorphism $\phi_k(X): F_k(X) \to G_k(X)$ for every $X \in \mathscr{C}$ and ϕ_k is automatically natural. Moreover from the choice of b_j' it follows

$$\begin{aligned}d\phi_k(X)(F(f)(b_j)) &= dG(f)(b_j') = G(f)(db_j') = G(f)\, \phi_{k-1}(M_j)(db_j) \\ &= \phi_{k-1}(X)\, F(f)(db_j) = \phi_{k-1}(X) \circ d(F(f)(b_j)),\end{aligned}$$

so it follows $d\phi_k(X) = \phi_{k-1}(X) \circ d$ for all $X \in \mathscr{C}$ as desired.

We hope the idea of the proof is by now clear, so that the reader can prove ii) for himself. ▯

13.29. Examples. i) For every $X \in \mathscr{T}$ we can regard $H_0(X)$ as the free abelian group with the 0-components of X as basis. Thus $H_0(X \times Y)$ has a basis $\{C \times C' : C$ 0-component of X, C' 0-component of $Y\}$, while $H_0(S_*(X) \otimes S_*(Y)) \cong H_0(X) \otimes H_0(Y)$ has a basis $\{C \otimes C' : C$ 0-component of X, C' 0-component of $Y\}$. Therefore we can define

$$\rho_{-1}: H_0(X \times Y) \to H_0(S_*(X) \otimes S_*(Y))$$

$$\sigma_{-1}: H_0(S_*(X) \otimes S_*(Y)) \to H_0(X \times Y)$$

by $\rho_{-1}(C \times C') = C \otimes C'$, $\sigma_{-1}(C \otimes C') = C \times C'$. Clearly $\sigma_{-1} \circ \rho_{-1} = 1$, $\rho_{-1} \circ \sigma_{-1} = 1$.

From Theorem 13.26i) and Examples 13.25ii), iii) it then follows there are natural chain maps

$$\rho: S_*(X \times Y) \to S_*(X) \otimes S_*(Y)$$

$$\sigma: S_*(X) \otimes S_*(Y) \to S_*(X \times Y)$$

inducing ρ_{-1}, σ_{-1} on H_0. Moreover $\sigma \circ \rho$ and 1 both induce 1 on $H_0(X \times Y)$, so by 13.26ii) $\sigma \circ \rho$ is chain homotopic to 1. Similarly $\rho \circ \sigma$ is chain homotopic to 1. Thus we have proved the following theorem.

13.30. Theorem (Eilenberg–Zilber). *There is a natural chain homotopy equivalence*

$$\rho : S_*(X \times Y) \to S_*(X) \otimes S_*(Y).$$

13.31. Theorem (Künneth). *For any topological spaces X, Y there is a natural exact sequence*

$$0 \longrightarrow [H_*(X) \otimes H_*(Y)]_n \xrightarrow{\times} H_n(X \times Y) \longrightarrow [\operatorname{Tor}(H_*(X), H_*(Y))]_{n-1} \longrightarrow 0.$$

The sequence splits but not naturally.

The homomorphism $\times$ in 13.31 is the composition of the algebraic $\times$

$$H_*(X) \otimes H_*(Y) \xrightarrow{\times} H_*(S_*(X) \otimes S_*(Y))$$

with the induced map

$$H_*(S_*(X) \otimes S_*(Y)) \xrightarrow{\sigma_*} H_*(X \times Y).$$

Exercise. Use 13.31 to compute the homology groups of the torus and compare the result with the values obtained by other metohds (Mayer–Vietoris, CW-decompositions).

There is also a relative version of 13.30. For any pairs $(X, A), (Y, B)$ we have a commutative diagram with exact rows and columns

$$\begin{array}{ccccccccc}
 & & 0 & & 0 & & 0 & & \\
 & & \downarrow & & \downarrow & & \downarrow & & \\
0 & \to & S_*(A) \otimes S_*(B) & \to & S_*(A) \otimes S_*(Y) & \to & S_*(A) \otimes S_*(Y, B) & \to & 0 \\
 & & \downarrow & & \downarrow & & \downarrow & & \\
0 & \to & S_*(X) \otimes S_*(B) & \to & S_*(X) \otimes S_*(Y) & \to & S_*(X) \otimes S_*(Y, B) & \to & 0 \\
 & & \downarrow & & \downarrow & & \downarrow & & \\
0 & \to & S_*(X, A) \otimes S_*(B) & \to & S_*(X, A) \otimes S_*(Y) & \to & S_*(X, A) \otimes S_*(Y, B) & \to & 0 \\
 & & \downarrow & & \downarrow & & \downarrow & & \\
 & & 0 & & 0 & & 0 & &
\end{array}$$

and as in the proof of 13.13 this leads to an exact sequence

$$0 \to S_*(A) \otimes S_*(Y) + S_*(X) \otimes S_*(B) \to S_*(X) \otimes S_*(Y) \to S_*(X,A) \otimes S_*(Y,B) \to 0,$$

or in other words

13.32. $$S_*(X,A) \otimes S_*(Y,B) \cong \frac{S_*(X) \otimes S_*(Y)}{S_*(A) \otimes S_*(Y) + S_*(X) \otimes S_*(B)}.$$

We also have a commutative diagram with exact rows

$$\begin{array}{ccccccc} 0 & \longrightarrow & S_*(A \times Y) + S_*(X \times B) & \longrightarrow & S_*(X \times Y) & \longrightarrow & \dfrac{S_*(X \times Y)}{S_*(A \times Y) + S_*(X \times B)} \to 0 \\ & & \rho \downarrow \quad \uparrow \sigma & & \rho \downarrow \quad \uparrow \sigma & & \downarrow \ \uparrow \\ 0 & \to & S_*(A) \otimes S_*(Y) + S_*(X) \otimes S_*(B) & \to & S_*(X) \otimes S_*(Y) & \to & \dfrac{S_*(X) \otimes S_*(Y)}{S_*(A) \otimes S_*(Y) + S_*(X) \otimes S_*(B)} \to 0 \end{array}$$

from which it follows ρ, σ induce chain homotopy equivalences

13.33.

$$\frac{S_*(X) \otimes S_*(Y)}{S_*(A) \otimes S_*(Y) + S_*(X) \otimes S_*(B)} \simeq \frac{S_*(X \times Y)}{S_*(A \times Y) + S_*(X \times B)}.$$

Finally we have the exact sequence of chain complexes

$$0 \to \frac{S_*(A \times Y) + S_*(X \times B)}{S_*(A \times Y)} \to \frac{S_*(A \times Y \cup X \times B)}{S_*(A \times Y)} \to \frac{S_*(A \times Y \cup X \times B)}{S_*(A \times Y) + S_*(X \times B)} \to 0,$$

$$\frac{S_*(A \times Y) + S_*(X \times B)}{S_*(A \times Y)} \cong \frac{S_*(X \times B)}{S_*(A \times Y \cap X \times B)}$$

which gives a homology sequence

$$\cdots \to H_{n+1}(A \times Y \cup X \times B, A \times Y) \to H_{n+1}\left(\frac{S_*(A \times Y \cup X \times B)}{S_*(A \times Y) + S_*(X \times B)}\right)$$

$$\to H_n(X \times B, A \times Y \cap X \times B) \xrightarrow{j_*} H_n(A \times Y \cup X \times B, A \times Y) \to \cdots.$$

Thus if the triad $(A \times Y \cup X \times B;\ X \times B, A \times Y)$ is excisive (i.e. j_* is an isomorphism) then

$$H_*\left(\frac{S_*(A \times Y \cup X \times B)}{S_*(A \times Y) + S_*(X \times B)}\right) = 0,$$

or in other words

13.34. $$S_*(A \times Y) + S_*(X \times B) \simeq S_*(A \times Y \cup X \times B).$$

Combining 13.32, 13.33 and 13.34 we get the following proposition.

13.35. Proposition. *For any two topological pairs (X, A), (Y, B) such that $(A \times Y \cup X \times B;\ A \times Y, X \times B)$ is excisive there is a chain homotopy equivalence*

$$\rho : S_*(X \times Y, A \times Y \cup X \times B) \simeq S_*(X, A) \otimes S_*(Y, B).$$

We often write $(X, A) \times (Y, B)$ for the pair $(X \times Y, A \times Y \cup X \times B)$. Note that $X \times Y / A \times Y \cup X \times B \cong (X/A) \wedge (Y/B)$.

13.36. Theorem (Relative Künneth Theorem). *For any topological pairs (X, A), (Y, B) such that $(A \times Y \cup X \times B;\ A \times Y, X \times B)$ is excisive there is a natural exact sequence*

$$0 \to [H_*(X, A) \otimes H_*(Y, B)]_n \to H_n((X, A) \times (Y, B)) \to$$

$$[\operatorname{Tor}(H_*(X, A), H_*(Y, B))]_{n-1} \to 0.$$

We could prove a cohomology analog of 13.36 (cf. [80]), but instead we merely describe $\times : H^*(X; \mathbb{Z}) \times H^*(Y; \mathbb{Z}) \to H^*(X \times Y; \mathbb{Z})$. More generally we can consider three abelian groups G, H, K and a homomorphism $\mu : G \otimes H \to K$; then we can define

$$\times : H^*(X; G) \otimes H^*(Y; H) \to H^*(X \times Y; K).$$

For any two chain complexes C_*, C'_* we have the algebraic $\times$-homomorphism

$$H^p(\operatorname{Hom}(C_*, G)) \otimes H^q(\operatorname{Hom}(C'_*, H)) \xrightarrow{\times} H^{p+q}(\operatorname{Hom}(C_* \otimes C'_*, G \otimes H))$$

defined as follows: if $\phi \in \operatorname{Hom}(C_p, G)$, $\psi \in \operatorname{Hom}(C'_q, H)$ are cocycles, then $\phi \otimes \psi \in \operatorname{Hom}(C_p \otimes C'_q, G \otimes H)$ and for all $a \in C_p$, $b \in C'_q$ we have

$$\begin{aligned}\delta(\phi \otimes \psi)(a \otimes b) &= (\phi \otimes \psi)(d_\otimes(a \otimes b)) \\ &= \phi \otimes \psi(da \otimes b + (-1)^{|a|} a \otimes db) \\ &= \phi(da) \otimes \psi(b) + (-1)^{|a|} \phi(a) \otimes \psi(db) = 0.\end{aligned}$$

If we extend $\phi \otimes \psi$ over $(C_* \otimes C'_*)_{p+q}$ be letting $\phi \otimes \psi | C_i \otimes C'_j = 0$ for $(i,j) \neq (p,q)$, then $\phi \otimes \psi$ is a cocycle, and we set

$$\{\phi\} \times \{\psi\} = \{\phi \otimes \psi\} \in H^{p+q}(\operatorname{Hom}(C_* \otimes C'_*, G \otimes H)).$$

We also have $\rho: S_*(X \times Y) \to S_*(X) \otimes S_*(Y)$ and hence

$$\operatorname{Hom}(\rho, 1)^*: H^*(\operatorname{Hom}(S_*(X) \otimes S_*(Y), G \otimes H)) \to H^*(X \times Y; G \otimes H).$$

Finally μ induces a coefficient homomorphism μ_* and the $\times$-homomorphism we want is the composite

$$H^p(X; G) \otimes H^q(Y; H) \xrightarrow{\times} H^{p+q}(\operatorname{Hom}(S_*(X) \otimes S_*(Y), G \otimes H)) \longrightarrow$$
$$\xrightarrow{\operatorname{Hom}(\rho, 1)^*} H^{p+q}(X \times Y; G \otimes H) \xrightarrow{\mu_*} H^{p+q}(X \times Y; K).$$

The homology $\times$-homomorphism can be similarly generalized:

$$H_p(X; G) \otimes H_q(Y; H) \xrightarrow{\times} H_{p+q}(S_*(X) \otimes S_*(Y) \otimes G \otimes H) \longrightarrow$$
$$\xrightarrow{(\sigma \otimes 1)_*} H_{p+q}(X \times Y; G \otimes H) \xrightarrow{\mu_*} H_{p+q}(X \times Y; K).$$

Moreover, if (X, A), (Y, B) are pairs such that

$$(A \times Y \cup X \times B; A \times Y, X \times B)$$

is excisive, then we have

$$\times: H_p(X, A; G) \otimes H_q(Y, B; H) \to H_{p+q}((X, A) \times (Y, B); K)$$

$$\times: H^p(X, A; G) \otimes H^q(Y, B; H) \to H^{p+q}((X, A) \times (Y, B); K).$$

The homomorphism $\times$ is called the *homology* (resp. *cohomology*) $\times$*-product* associated to $\mu: G \otimes H \to K$. Of particular interest is the case where R is a ring and $\mu: R \otimes R \to R$ is the product or M is an R-module and

$\mu: R \otimes M \to M$ is the R-action. The next proposition lists some of the properties of $\times$; the proposition is stated for the absolute case, but analogous results hold for the relative case whenever the $\times$-products are defined.

13.37. Proposition. *Let X, Y, Z be topological spaces, G_1, G_2, G_3, H, K, M abelian groups and $\mu_{12}: G_1 \otimes G_2 \to H$, $\mu_{21}: G_2 \otimes G_1 \to H$, $\mu_{23}: G_2 \otimes G_3 \to K$, $\nu_1: G_1 \otimes K \to M$, $\nu_2: H \otimes G_3 \to M$ homomorphisms.*

i) *If*

$$\begin{array}{ccc} G_1 \otimes G_2 \otimes G_3 & \xrightarrow{\mu_{12} \otimes 1} & H \otimes G_3 \\ \downarrow{\scriptstyle 1 \otimes \mu_{23}} & & \downarrow{\scriptstyle \nu_2} \\ G_1 \otimes K & \xrightarrow{\nu_1} & M \end{array}$$

commutes, then so does

$$\begin{array}{ccc} H_p(X; G_1) \otimes H_q(Y; G_2) \otimes H_r(Z; G_3) & \xrightarrow{\times \otimes 1} & H_{p+q}(X \times Y; H) \otimes H_r(Z; G_3) \\ {\scriptstyle 1 \otimes \times}\downarrow & & \downarrow{\scriptstyle \times} \\ H_p(X; G_1) \otimes H_{q+r}(Y \times Z; K) & \xrightarrow{\times} & H_{p+q+r}(X \times Y \times Z; M) \end{array}$$

ii) *If*

$$\begin{array}{ccc} G_1 \otimes G_2 & \xrightarrow{T} & G_2 \otimes G_1 \\ & {\scriptstyle \mu_{12}} \searrow \quad \swarrow {\scriptstyle \mu_{21}} & \\ & H & \end{array}$$

commutes, then so does

$$\begin{array}{ccc} H_p(X; G_1) \otimes H_q(Y; G_2) & \xrightarrow{T} & H_q(Y; G_2) \otimes H_p(X; G_1) \\ \downarrow{\scriptstyle \times} & & \downarrow{\scriptstyle \times} \\ H_{p+q}(X \times Y; H) & \xrightarrow{\tau_*} & H_{p+q}(Y \times X; H), \end{array}$$

where $\tau: X \times Y \to Y \times X$ is the switch map and T is given by $T(a \otimes b) = (-1)^{pq} b \otimes a$, $a \in H_p(X; G_1)$, $b \in H_q(Y, G_2)$.

iii) *If $f: X \to X'$ and $g: Y \to Y'$ are maps, then*

$$\begin{array}{ccc} H_p(X; G_1) \otimes H_q(Y; G_2) & \xrightarrow{\times} & H_{p+q}(X \times Y; H) \\ \Big\downarrow f_* \otimes g_* & & \Big\downarrow (f \times g)_* \\ H_p(X'; G_1) \otimes H_q(Y'; G_2) & \xrightarrow{\times} & H_{p+q}(X' \times Y'; H) \end{array}$$

commutes.

There are analogous results for the cohomology $\times$-product.

Proof: i) It is completely straightforward to check that

$$\begin{array}{ccc} H_p(C_1) \otimes H_q(C_2) \otimes H_r(C_3) & \xrightarrow{\times \otimes 1} & H_{p+q}(C_1 \otimes C_2) \otimes H_r(C_3) \\ \Big\downarrow 1 \otimes \times & & \Big\downarrow \times \\ H_p(C_1) \otimes H_{q+r}(C_2 \otimes C_3) & \xrightarrow{\times} & H_{p+q+r}(C_1 \otimes C_2 \otimes C_3) \end{array}$$

commutes for any chain complexes C_1, C_2, C_3. The only non-trivial step, then, is to verify that

$$\begin{array}{ccc} S_*(X) \otimes S_*(Y) \otimes S_*(Z) & \xrightarrow{\sigma \otimes 1} & S_*(X \times Y) \otimes S_*(Z) \\ \Big\downarrow 1 \otimes \sigma & & \Big\downarrow \sigma \\ S_*(X) \otimes S_*(Y \times Z) & \xrightarrow{\sigma} & S_*(X \times Y \times Z) \end{array}$$

is chain homotopy commutative. But $\sigma \circ (\sigma \otimes 1)$ and $\sigma \circ (1 \otimes \sigma)$ induce the same homomorphism on $H_0: C_1 \otimes C_2 \otimes C_3 \mapsto C_1 \times C_2 \times C_3$ for 0-components C_1, C_2, C_3 of X, Y, Z respectively. Hence by 13.26ii) $\sigma \circ (\sigma \otimes 1) \simeq \sigma \circ (1 \otimes \sigma)$.

ii) The proof is similar to that of i); the crux of the matter is that

$$\begin{array}{ccc} S_*(X) \otimes S_*(Y) & \xrightarrow{T} & S_*(Y) \otimes S_*(X) \\ \Big\downarrow \sigma & & \Big\downarrow \sigma \\ S_*(X \times Y) & \xrightarrow{\tau_\#} & S_*(Y \times X) \end{array}$$

is chain homotopy commutative, which again follows by 13.26ii).

iii) The three ingredients of $\times$ are natural. □

There are two other important "external" products—the slant products

$$/: H^p(X \times Y; G_1) \otimes H_q(Y; G_2) \to H^{p-q}(X; H)$$
$$\backslash: H^p(X; G_1) \otimes H_q(X \times Y; G_2) \to H_{q-p}(Y; H).$$

$/$ can be defined as follows: given a cocycle $\phi \in \operatorname{Hom}(C \otimes C', G_1)$ and a cycle $z = \sum_i c_i' \otimes g_i \in C' \otimes G_2$, we form the cochain $\phi/z \in \operatorname{Hom}(C, H)$ by taking

$$(\phi/z)(c) = \textstyle\sum_i \mu_{12}(\phi(c \otimes c_i') \otimes g_i) \in H$$

for all $c \in C$. One checks immediately that ϕ/z is a cocycle, and hence we can set

$$\{\phi\}/\{z\} = \{\phi/z\} \in H^*(C; H).$$

One then composes this algebraic $/$-product with

$$H^p(X \times Y; G_1) \otimes H_q(Y; G_2) \xrightarrow{\operatorname{Hom}(\sigma, 1)^* \otimes 1}$$
$$H^p(\operatorname{Hom}(S_*(X) \otimes S_*(Y), G_1)) \otimes H_q(Y; G_2)$$

to get the "geometric" $/$-product. $\backslash$ is defined similarly. $/$ and $\backslash$ are natural, and there are appropriate associativity relations connecting $\times$, $\times$, $/$ and $\backslash$.

We can also define "internal" products $\cup$ and $\cap$. Let $\Delta: X \to X \times X$ be the diagonal and let R be a ring with product $\mu: R \otimes R \to R$. Then the $\cup$-product on $H^*(X; R)$ is the composite

$$H^p(X; R) \otimes H^q(X; R) \xrightarrow{\times} H^{p+q}(X \times X; R) \xrightarrow{\Delta^*} H^{p+q}(X; R).$$

With $\cup$ as product $H^*(X; R)$ becomes a ring which is graded commutative if R is commutative and has a 1 if R has a 1. We also have the $\cap$-product

$$\cap: H^p(X; R) \otimes H_q(X; R) \to H_{q-p}(X; R)$$

defined by $a \cap b = a \backslash \Delta_*(b)$, $a \in H^p(X; R)$, $b \in H_q(X; R)$. The $\cup$- and $\cap$-product are natural and enjoy appropriate associativity relations. One can also discuss the behavior of the boundary homomorphisms ∂ and δ with respect to these products. We are not going into detail now, because we want to discuss all these products in the more general setting of generalized homology and cohomology theories after we have constructed smash products of spectra.

The construction of the smash product $E \wedge F$ of two spectra E, F is unfortunately somewhat complicated; the applications of the smash product which we make in Chapters 14–20, however, provide ample evidence that the construction is worthwhile. Naturally we expect $E \wedge F$ to be in some sense a limit of $E_i \wedge F_j$ as i, j tend to ∞($E_i \wedge F_j$ gets the weak topology). Therefore the following construction—called the *naive smash product* by Boardman—seems very natural: let A be some ordered set order isomorphic to $\mathbb{N} = \{0,1,2,\ldots\}$. For any subset $B \subset A$ we have a monotonic function $\beta: A \to \mathbb{N}$ defined by $\beta(a) =$ the number of elements $b \in B$ with $b < a$. In particular we have $\alpha: A \to \mathbb{N}$ corresponding to the subset $A \subset A$; α is an order isomorphism. Now we suppose A partitioned as the union of two subsets: $A = B \cup C$, $B \cap C = \varnothing$. We define the naive smash product $E \wedge_{BC} F$ by

$$(E \wedge_{BC} F)_{\alpha(a)} = E_{\beta(a)} \wedge F_{\gamma(a)}, \quad a \in A,$$

where $\beta, \gamma: A \to \mathbb{N}$ are the monotonic functions associated to B, C respectively. Note that $\beta(a) + \gamma(a) = \alpha(a)$ for all $a \in A$. If $a \in B$, then

$$(E \wedge_{BC} F)_{\alpha(a)+1} = E_{\beta(a)+1} \wedge F_{\gamma(a)},$$

and we identify

$$[t, e, f] \in S^1 \wedge E_{\beta(a)} \wedge F_{\gamma(a)} = S^1 \wedge [E \wedge_{BC} F]_{\alpha(a)}$$

with

$$[[t, e], f] \in E_{\beta(a)+1} \wedge F_{\gamma(a)} = [E \wedge_{BC} F]_{\alpha(a)+1}.$$

If $a \in C$, then $(E \wedge_{BC} F)_{\alpha(a)+1} = E_{\beta(a)} \wedge F_{\gamma(a)+1}$, and we identify

$$[t, e, f] \in S^1 \wedge E_{\beta(a)} \wedge F_{\gamma(a)} = S^1 \wedge [E \wedge_{BC} F]_{\alpha(a)}$$

with

$$[e, [v'^{\beta(a)}(t), f]] \in E_{\beta(a)} \wedge F_{\gamma(a)+1} = [E \wedge_{BC} F]_{\alpha(a)+1},$$

where $v': S^1 \to S^1$ is the homotopy inverse of the H-cogroup S^1.

Example. If X is a CW-complex and $E(X)$ the suspension spectrum of X, then $F \wedge_{A, \varnothing} E(X) = F \wedge X$.

It is clear from the definition that $E \wedge_{BC} F$ is natural with respect to functions of spectra. With maps we must be a little careful; if E' is cofinal in E, then $E' \wedge_{BC} F$ is cofinal in $E \wedge_{BC} F$ if B is infinite but not in general. For example, if B has n elements and $E' \subset E$ is a cofinal subspectrum such that $E'_q = *$ for $q \leqslant n$, then clearly $E' \wedge_{BC} F = *$. However, if B is infinite then $E \wedge_{BC} F$ is natural with respect to maps $f: E \to G$. Similarly if C is infinite $E \wedge_{BC} F$ is natural with respect to maps $g: F \to H$. Moreover $(E \wedge_{BC} F) \wedge I^+ \cong E \wedge_{BC} (F \wedge I^+) \cong (E \wedge I^+) \wedge_{BC} F$, so the homotopy class of $f \wedge g$ depends only on the homotopy class of f if B is infinite and only on the homotopy class of g if C is infinite.

Let us agree to work in the homotopy category $\mathscr{SP}'$, so that "equivalence" means "homotopy equivalence" and diagrams commute if and only if the corresponding diagrams of maps commute up to homotopy.

Now there is clearly a naive smash product $E \wedge_{BC} F$ for every partition $A = B \cup C$; which should we take for $E \wedge F$? Of course, we want $E \wedge F$ to have properties analogous to those of the smash product of spaces, such as $E \wedge F \simeq F \wedge E$, $(E \wedge F) \wedge G \simeq E \wedge (F \wedge G)$, $S^0 \wedge E \simeq E \simeq E \wedge S^0$. Some of the naive smash products are commutative ($E \wedge_{BC} F \simeq F \wedge_{CB} E$ provided $\beta(a)\gamma(a)$ is always even), some are associative ($(E \wedge_{BC} F) \wedge_{B \cup C, D} G = E \wedge_{B, C \cup D} (F \wedge_{CD} G)$ for any partition $A = B \cup C \cup D$ with $B \cup C$ and $C \cup D$ infinite). For some naive smash products S^0 is a unit—notably $S^0 \wedge_{\varnothing, A} E = E = E \wedge_{A, \varnothing} S^0$. But we want all these properties on $E \wedge F$, so we see that we cannot simply choose one partition $A = B \cup C$ and take $E \wedge F = E \wedge_{BC} F$. Instead we glue all the various $E \wedge_{BC} F$ together appropriately to create $E \wedge F$.

We recall our infinite telescope of Chapters 7, 8. We can take the telescope $T(E)$ of a spectrum E to be the following:

$$T(E)_n = (\vee_{i=0}^{n} S^{n-i} \wedge E_i \wedge [i]^+) \vee (\vee_{i=0}^{n-1} S^{n-i} \wedge E_i \wedge [i, i+1]^+),$$

where we identify $[t, e, i] \in S^{n-i} \wedge E_i \wedge [i, i+1]^+$ with $[t, e, i] \in S^{n-i} \wedge E_i \wedge [i]^+$ and $[s, t, e, i+1] \in S^{n-i-1} \wedge S^1 \wedge E_i \wedge [i, i+1]^+$ with

$$[s, [t, e], i+1] \in S^{n-i-1} \wedge E_{i+1} \wedge [i+1]^+.$$

$T(E)$ is then a spectrum (we identify $[s, t, e, u] \in S^1 \wedge S^{n-i} \wedge E_i \wedge [i, i+1]^+ \subset S^1 \wedge T(E)_n$ with $[[s, t], e, u] \in S^{n+1-i} \wedge E_i \wedge [i, i+1]^+ \subset T(E)_{n+1}$, for example) and we have our homotopy equivalence $\rho: T(E) \to E$ which collapses $T(E)_n$ onto $E_n \wedge [n]^+ = E_n$, $n \geqslant 0$.

The telescope construction amounts to giving $[0, +\infty)$ a CW-structure with 0-cells $[n]$ and 1-cells $[n, n+1]$. Then one constructs certain spaces over each cell and makes appropriate identifications on the boundaries. Given two spectra E, F we construct a sort of "2-dimensional telescope" $E \wedge F$ as follows. We give the quarter-plane $[0, +\infty) \times [0, +\infty)$ a CW-structure with 0-cells $[n, m]$, 1-cells $[n, n+1] \times [m]$ and $[n] \times [m, m+1]$, and 2-cells $[n, n+1] \times [m, m+1]$, $n, m \geqslant 0$. To construct $(E \wedge F)_n$ we use only the cells e with lower left corner (i, j) such that $i + j + \dim(e) \leqslant n$.

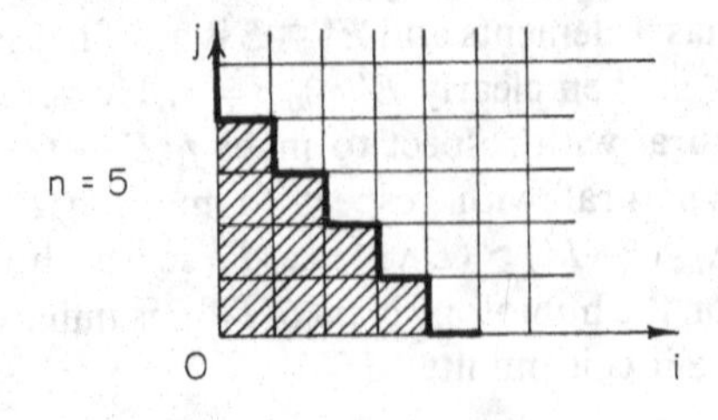

Over (i,j) we take $S^{n-i-j} \wedge E_i \wedge F_j \wedge (i,j)^+$; over $[i,i+1] \times [j]$ we take $S^{n-i-j} \wedge E_i \wedge F_j \wedge ([i,i+1] \times [j])^+$; and over $[i] \times [j,j+1]$ we take $S^{n-i-j} \wedge E_i \wedge F_j \wedge ([i] \times [j,j+1])^+$. The required identifications are clear:

$$[s,e,f,(i,j)] \in S^{n-i-j} \wedge E_i \wedge F_j \wedge ([i,i+1] \times [j])^+$$

or

$$\in S^{n-i-j} \wedge E_i \wedge F_j \wedge ([i] \times [j,j+1])^+$$

with

$$[s,e,f,(i,j)] \in S^{n-i-j} \wedge E_i \wedge F_j \wedge (i,j)^+;$$

$$[s,t,e,f,(i+1,j)] \in S^{n-i-j-1} \wedge S^1 \wedge E_i \wedge F_j \wedge ([i,i+1] \times [j])^+$$

with

$$[s,[t,e],f,(i+1,j)] \in S^{n-i-j-1} \wedge E_{i+1} \wedge F_j \wedge (i+1,j)^+;$$

and

$$[s,t,e,f,(i,j+1)] \in S^{n-i-j-1} \wedge S^1 \wedge E_i \wedge F_j \wedge ([i] \times [j,j+1])^+$$

with

$$[s,e,[v'^i(t),f],(i,j+1)] \in S^{n-i-j-1} \wedge E_i \wedge F_{j+1} \wedge (i,j+1)^+.$$

We must still describe the space over a 2-cell $e = [i,i+1] \times [j,j+1]$; the definition must agree of course with what we already have over ∂e.

We have two paths in ∂e from (i,j) to $(i+1,j+1)$; if we go by way of $(i+1,j)$ then the point $[s,t,u,e,f,(i,j)]$ goes to $[s,t,[u,e],f,(i+1,j)]$ and then to $[s,[u,e],[v'^{i+1}(t),f],\ (i+1,j+1)]$. On the other hand if we go by way of $(i,j+1)$, then $[s,t,u,e,f,(i,j)]$ goes first to $[s,t,e,[v'^i(u),f],(i,j+1)]$ and then to $[s,[t,e],[v'^i(u),f],(i+1,j+1)]$. Thus in the factor $S^1 \wedge S^1$ we have in effect the transition function $(t,u) \mapsto (u,v'(t))$, or if we regard $S^1 \wedge S^1$ as $[-1,+1] \times [-1,+1]$ with appropriate identifications, then the transition function has matrix

$$\begin{pmatrix} 0 & 1 \\ -1 & 0 \end{pmatrix}.$$

Hence we see that the space we have constructed over ∂e is of the form

$$S^{n-i-j-2} \wedge M(\xi) \wedge E_i \wedge F_j,$$

where $M(\xi)$ is the Thom complex of a certain $SO(2)$-bundle ξ over ∂e. We could therefore take the part of $(E \wedge F)_n$ over e to be

$$S^{n-i-j-2} \wedge M(\xi) \wedge E_i \wedge F_j$$

for ξ some $SO(2)$-bundle over e extending $\xi|\partial e$. But $\pi_1(BSO(2),*) \cong \pi_0(SO(2),1) = 0$, so ξ does have an extension over e. Since the bundle $\xi|\partial e$ does not essentially depend on i,j,n,E,F, we choose a ξ on e which is independent of i,j,n,E,F. Thus we have constructed $(E \wedge F)_n$ for $n \geqslant 0$; we set $(E \wedge F)_n = *$ for $n < 0$. It is clear that $E \wedge F$ is a spectrum.

Now we must show that $E \wedge F$ has various desired properties such as naturality, associativity and commutativity. We have remarked that each of these properties is possessed by appropriate naive smash products $E \wedge_{BC} F$, so we set about showing $E \wedge F$ is homotopy equivalent to many of the naive smash products. The main idea is that we can embed the telescope $T(E \wedge_{BC} F)$ in $E \wedge F$ for all partitions $A = B \cup C$. We define

$$\omega: [0, +\infty) \to [0, +\infty) \times [0, +\infty)$$

by

$$\omega(t) = \left.\begin{cases} (\beta(a) - \alpha(a) + t, \gamma(a)) & a \in B \\ (\beta(a), \gamma(a) - \alpha(a) + t) & a \in C \end{cases}\right\} \quad t \in [\alpha(a), \alpha(a) + 1].$$

Example.

$B = (0, 1, 3, 6, 7, \ldots)$
$C = (2, 4, 5, 8, \ldots)$

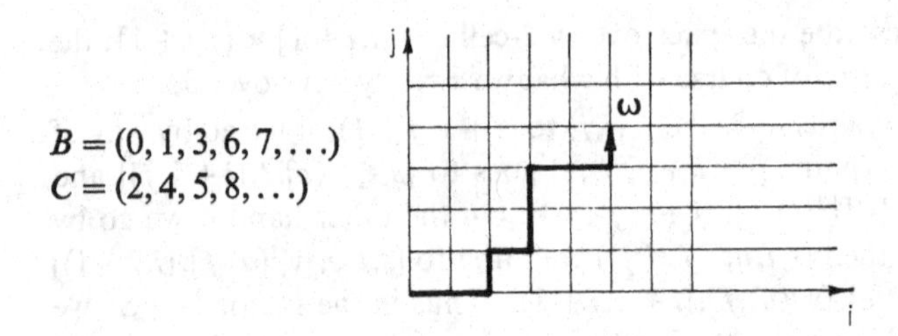

Then we can embed $T(E \wedge_{BC} F)$ in $E \wedge F$ by identifying

$$[s, e, f, t] \in S^{n-\alpha(a)} \wedge E_{\beta(a)} \wedge F_{\gamma(a)} \wedge [\alpha(a), \alpha(a) + 1], \quad a \in A,$$

with

$$[s, e, f, \omega(t)] \in S^{n-\alpha(a)} \wedge E_{\beta(a)} \wedge F_{\gamma(a)} \wedge ([\beta(a), \beta(a) + 1] \times [\gamma(a)])^+, \quad a \in B,$$

or

$$\in S^{n-\alpha(a)} \wedge E_{\beta(a)} \wedge F_{\gamma(a)} \wedge ([\beta(a)] \times [\gamma(a), \gamma(a) + 1])^+, \quad a \in C.$$

We then define $eq_{BC}: E \wedge_{BC} F \to E \wedge F$ to be the composite

$$E \wedge_{BC} F \xrightarrow{\rho^{-1}} T(E \wedge_{BC} F) \subset E \wedge F,$$

where ρ^{-1} is the homotopy inverse of $\rho: T(E \wedge_{BC} F) \to E \wedge_{BC} F$. (Remember that ρ^{-1} has no simple description (cf. 8.26).)

13.38. Lemma. *$eq_{BC}: E \wedge_{BC} F \to E \wedge F$ is a homotopy equivalence if any one of the following is satisfied.*

a) *B and C are both infinite;*
b) *B has d elements and $SE_r = E_{r+1}$ for $r \geqslant d$;*
c) *C has d elements and $SF_r = F_{r+1}$ for $r \geqslant d$.*

Proof: Suppose a) is satisfied. For each $a \in A$ let $G_{\alpha(a)} \subset (E \wedge F)_{\alpha(a)}$ be the subcomplex over the cells e_{ij} with $i \leqslant \beta(a), j \leqslant \gamma(a)$.

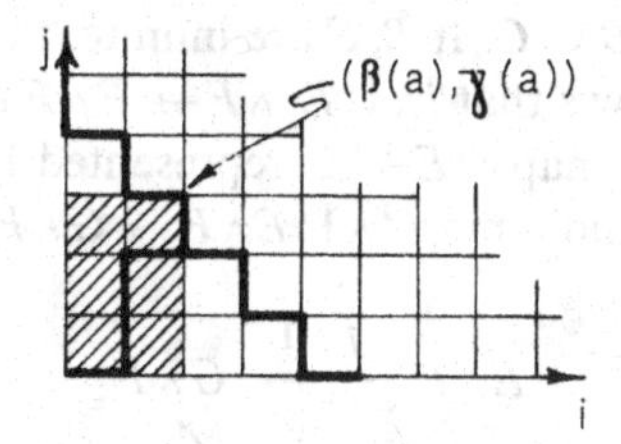

From the assumption a) it follows that $G = \{G_n\}$ is a cofinal subspectrum of $E \wedge F$. On the other hand there is a deformation retraction of $G_{\alpha(a)}$ onto $E_{\beta(a)} \wedge F_{\gamma(a)}$ for all $a \in A$. Thus in the diagram

$$\begin{array}{ccc} & T(E \wedge_{BC} F)_{\alpha(a)} & \\ {\scriptstyle\simeq}\nearrow & & \searrow \\ E_{\beta(a)} \wedge F_{\gamma(a)} & \xrightarrow[\simeq]{} & G_{\alpha(a)} \end{array}$$

two of the three maps induce isomorphisms of homotopy groups. Hence the third does also, and therefore by 8.25

$$T(E \wedge_{BC} F) \to G \to E \wedge F$$

is a homotopy equivalence. Thus eq_{BC} is also.

If assumption b) holds instead, then we construct $G_{\alpha(a)}$ a little differently; we take the subcomplex of $(E \wedge F)_{\alpha(a)}$ over the cells e_{ij} with $i + j \leqslant \alpha(a)$, $j \leqslant \gamma(a)$. Then $G = \{G_n\}$ is still a cofinal subspectrum of $E \wedge F$ and $E_{\beta(a)} \wedge F_{\gamma(a)}$ is still a deformation retract of $G_{\alpha(a)}$ for $\beta(a) = d$, since $S^{n-d} \wedge E_d \wedge Y_j \to S^{n-d-s} \wedge E_{d+s} \wedge Y_j$ is a homeomorphism for all $s \geqslant 0$.

The proof when c) holds is analogous. □

13.39. Corollary. *If X is a CW-complex, $E(X)$ the suspension spectrum and F any spectrum, then $F \wedge E(X) \simeq F \wedge X$.*

We can also demonstrate the naturality of $E \wedge F$. Functions $f: E \to G$, $g: F \to H$ clearly induce a function $f \wedge g: E \wedge F \to G \wedge H$ in an obvious manner. If $E' \subset E$ is a cofinal subspectrum, however, $E' \wedge F$ need not be cofinal in $E \wedge F$. But if $i: E' \to E$ is the inclusion, then i is an equivalence in $\mathscr{SP}'$ and we have a commutative diagram

$$\begin{array}{ccc} E' \wedge F & \xrightarrow{i \wedge 1} & E \wedge F \\ \uparrow{\scriptstyle eq_{BC}} & & \uparrow{\scriptstyle eq_{BC}} \\ E' \wedge_{BC} F & \underset{\simeq}{\xrightarrow{i \wedge 1}} & E \wedge_{BC} F \end{array}$$

for all partitions $A = B \cup C$. If B, C are infinite, then eq_{BC} is a homotopy equivalence, so it follows that $i \wedge 1: E' \wedge F \to E \wedge F$ is a homotopy equivalence. Thus given any map $f: E \to G$ represented by a function $f': E' \to G$, E' cofinal, we can find a map $f \wedge 1: E \wedge F \to G \wedge F$ so that

$$\begin{array}{ccc} E \wedge F & \xrightarrow{f \wedge 1} & G \wedge F \\ & {\scriptstyle i \wedge 1} \nwarrow \quad \nearrow {\scriptstyle f' \wedge 1} & \\ & E' \wedge F & \end{array}$$

commutes. Similarly a map $g: F \to H$ induces $1 \wedge g: E \wedge F \to E \wedge H$. Since $(E \wedge I^+) \wedge F \cong E \wedge (F \wedge I^+) \cong (E \wedge F) \wedge I^+$, the homotopy class of $f \wedge 1$ depends only on the homotopy class of f. Similarly for $1 \wedge g$.

From the construction of $f \wedge 1$ and $1 \wedge g$ it follows that

i) if B is infinite and $f: E \to G$ is a map, then the diagram

$$\begin{array}{ccc} E \wedge_{BC} F & \xrightarrow{eq_{BC}} & E \wedge F \\ \downarrow{\scriptstyle f \wedge 1} & & \downarrow{\scriptstyle f \wedge 1} \\ G \wedge_{BC} F & \xrightarrow{eq_{BC}} & G \wedge F \end{array}$$

commutes;

ii) if C is infinite and $g: F \to H$ is a map, then the diagram

$$\begin{array}{ccc} E \wedge_{BC} F & \xrightarrow{eq_{BC}} & E \wedge F \\ \downarrow{\scriptstyle 1 \wedge g} & & \downarrow{\scriptstyle 1 \wedge g} \\ E \wedge_{BC} H & \xrightarrow{eq_{BC}} & E \wedge H \end{array}$$

commutes.

We now state our main theorem about $E \wedge F$.

13.40. Theorem. a) *$E\wedge F$ is a functor of two variables.* b) *There are natural homotopy equivalences (natural in the homotopy category $\mathscr{S}\mathscr{P}'$)*

$$\begin{aligned} a = a(E,F,G) &: (E\wedge F)\wedge G \to E\wedge(F\wedge G) \\ \tau = \tau(E,F) &: E\wedge F \to F\wedge E \\ l = l(E) &: S^0\wedge E \to E \\ r = r(E) &: E\wedge S^0 \to E \end{aligned}$$

such that the following diagrams commute (in $\mathscr{S}\mathscr{P}'$):

i)

$$\begin{array}{ccccc} & & (E\wedge F)\wedge(G\wedge H) & \xrightarrow{a} & E\wedge(F\wedge(G\wedge H)) \\ & \overset{a}{\nearrow} & & & \uparrow{\scriptstyle 1\wedge a} \\ ((E\wedge F)\wedge G)\wedge H & & & & \\ & \underset{a\wedge 1}{\searrow} & & & \\ & & (E\wedge(F\wedge G))\wedge H & \xrightarrow{a} & E\wedge((F\wedge G)\wedge H) \end{array}$$

ii)

$$\begin{array}{ccc} & F\wedge E & \\ {\scriptstyle\tau}\nearrow & & \searrow{\scriptstyle\tau} \\ E\wedge F & \xrightarrow[1]{} & E\wedge F \end{array}$$

iii)

$$\begin{array}{ccc} & (F\wedge E)\wedge G & \\ {\scriptstyle\tau\wedge 1}\nearrow & & \searrow{\scriptstyle a} \\ (E\wedge F)\wedge G & & F\wedge(E\wedge G) \\ {\scriptstyle a}\downarrow & & \downarrow{\scriptstyle 1\wedge\tau} \\ E\wedge(F\wedge G) & & F\wedge(G\wedge E) \\ {\scriptstyle\tau}\searrow & & \nearrow{\scriptstyle a} \\ & (F\wedge G)\wedge E & \end{array}$$

iv)

$$\begin{array}{ccc} (S^0\wedge E)\wedge F & \xrightarrow{a} & S^0\wedge(E\wedge F) \\ {\scriptstyle l\wedge 1}\searrow & & \swarrow{\scriptstyle l} \\ & E\wedge F & \end{array}$$

v)

$$\begin{array}{ccc} (E\wedge S^0)\wedge F & \xrightarrow{a} & E\wedge(S^0\wedge F) \\ {\scriptstyle r\wedge 1}\searrow & & \swarrow{\scriptstyle 1\wedge l} \\ & E\wedge F & \end{array}$$

vi)

$$\begin{array}{ccc} (E\wedge F)\wedge S^0 & \xrightarrow{a} & E\wedge(F\wedge S^0) \\ & {\scriptstyle r}\searrow \quad \swarrow{\scriptstyle 1\wedge r} & \\ & E\wedge F & \end{array}$$

vii)

$$\begin{array}{ccc} S^0\wedge E & \xrightarrow{\tau} & E\wedge S^0 \\ & {\scriptstyle l}\searrow \quad \swarrow{\scriptstyle r} & \\ & E & \end{array}$$

viii)

$$S^0\wedge S^0 \underset{1}{\overset{\tau}{\rightrightarrows}} S^0\wedge S^0.$$

Remark. One can prove the commutativity of any other diagrams involving a, τ, l and r which one may happen to need (provided they do commute!) by using these eight (cf. MacLane [57]).

Proof: We have already proved a). The construction of l and r is easy; for l we take

$$S^0\wedge E \xrightarrow{eq^{-1}_{\varnothing,A}} S^0\wedge_{\varnothing,A}E\simeq E,$$

and for r we take

$$E\wedge S^0 \xrightarrow{eq^{-1}_{A,\varnothing}} E\wedge_{A,\varnothing}S^0\simeq E.$$

$eq_{\varnothing,A}$ and $eq_{A,\varnothing}$ are homotopy equivalences by 13.38.

We next consider τ. It is tempting to construct τ by requiring the diagram

13.41.

$$\begin{array}{ccc} E\wedge F & \xrightarrow{\tau} & F\wedge E \\ {\scriptstyle eq_{BC}}\uparrow & & \uparrow{\scriptstyle eq_{CB}} \\ E\wedge_{BC}F & \xrightarrow[\tau_{BC}]{} & F\wedge_{CB}E \end{array}$$

to commute up to homotopy, where B and C are both infinite and τ_{BC} is given by $\tau_{BC}[e,f]=[f,e]$ for $e\in E_{\beta(a)}$, $f\in F_{\gamma(a)}$. There are two difficulties to this approach:

i) τ_{BC} so defined is not a function of spectra. We really need $\tau_{BC}[e,f]=(-1)^{\beta(a)\gamma(a)}[f,e]$, but there is no suspension with which to make sense of

$(-1)^{\beta(a)\gamma(a)}$. A simple solution is to consider only such partitions $A = B \cup C$ such that for all $n \geqslant 0$ $\alpha^{-1}(2n+1)$ and $\alpha^{-1}(2n)$ lie either both in B or both in C; then $\beta(a)\gamma(a)$ is always even. In fact we shall make the following restriction on all our partitions:

13.42. $\alpha^{-1}(4n)$, $\alpha^{-1}(4n+1)$, $\alpha^{-1}(4n+2)$, $\alpha^{-1}(4n+3)$ lie either all in B or all in C for $n \geqslant 0$.

The reason for taking four at a time becomes clear shortly. Note that $A = \varnothing \cup A$ satisfies 13.42.

ii) The other difficulty is that we cannot get by with partitions $A = B \cup C$, B and C both infinite, if we want to prove vii), for example. Thus we must work a little harder.

13.43. Lemma. *There exist a spectrum Q and homotopy equivalences*

$$i_0: E \wedge F \to Q, i_1: F \wedge E \to Q$$

such that for every partition $A = B \cup C$ satisfying 13.42 *the diagram*

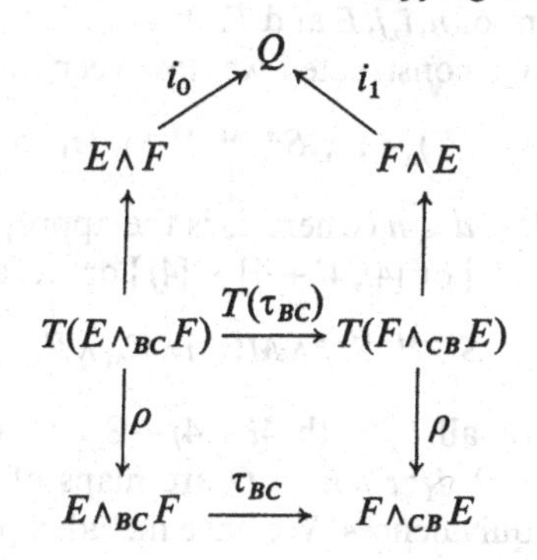

commutes.

Of course if we define $\tau: E \wedge F \to F \wedge E$ to be $i_1^{-1} \circ i_0$, then 13.41 commutes for all partitions $A = B \cup C$ satisfying 13.42.

Proof of 13.43: Q will be a construction over $[0,+\infty) \times [0,+\infty) \times I$ such that the part over $[0,+\infty) \times [0,+\infty) \times 0$ is $E \wedge F$ and that over $[0,+\infty) \times [0,+\infty) \times 1$ is $F \wedge E$. Over cells $e_{ij}^+ \wedge I^+$ for e_{ij} of the form $[i] \times [j]$ for i or j even, $[i] \times [j, j+1]$ for i even and $[i, i+1] \times [j]$ for j even we take the appropriate part of $(E \wedge F)_n \wedge I^+$ and identify the 0 end of the cylinder with the corresponding part of $(E \wedge F)_n$, while at the 1 end we make the identification:

$$[u, e, f, s, t] \in S^{n-i-j} \wedge E_i \wedge F_j \wedge e_{ij}^+$$

is identified with

$$[u, f, e, t, s] \in S^{n-i-j} \wedge F_j \wedge E_i \wedge e_{ji}^+$$

These identifications are consistent (recall 13.42).

We have already described the part of Q_n over ∂e for $e = [2i, 2i+2] \times [2j, 2j+2] \times I$, and it contains a subcomplex

$$S^{n-2i-2j-4} \wedge M(\tau') \wedge E_{2i} \wedge F_{2j},$$

where τ' is a 4-plane bundle over ∂e. Again this bundle τ' does not depend on n, i, j, E or F but only on the choice made earlier in the construction of $E \wedge F$; it is classified by an element $\alpha \in \pi_1(SO, 1) \cong Z_2$.

Now in $e' = [4i, 4i+4] \times [4j, 4j+4] \times I$ there are four cells of the type just considered—say e_1, e_2, e_3, e_4. Q has already been constructed over $\partial e_1 \cup \partial e_2 \cup \partial e_3 \cup \partial e_4$ and there it contains a subcomplex of the form

$$S^{n-4i-4j-8} \wedge M(\tau'') \wedge E_{4i} \wedge F_{4j},$$

where τ'' is an $SO(8)$-bundle over $\partial e_1 \cup \partial e_2 \cup \partial e_3 \cup \partial e_4$. Over ∂e_i we see τ'' is of the form $\tau''|\partial e_i \simeq \tau' \oplus \varepsilon^4$, τ' as above. It thus follows that $\tau''|\partial e'$ is classified by $4\alpha = 0$, and hence τ'' can be extended over e'. We choose an extension independent of n, i, j, E and F.

From what we have constructed we now keep only the following:

$$(E \wedge F)_n \wedge 0^+, (F \wedge E)_n \wedge 1^+, S^{n-4i-4j-d} \wedge M(\tau_d) \wedge E_{4i} \wedge F_{4j} \wedge I^+$$

for all i, j with $4i + 4j + d \leqslant n$ (where τ_d is the appropriate d-plane bundle over the d-cell $[4i] \times [4j]$ or $[4i, 4i+4] \times [4j]$ or ...) and finally

$$S^{n-4i-4j-8} \wedge M(\tau'') \wedge E_{4i} \wedge F_{4j}$$

(with τ'' as above) for all i, j with $4i + 4j + 8 \leqslant n$. This we call Q_n; the inclusions $i_0: E \wedge F \to Q$, $i_1: F \wedge E \to Q$ are maps of spectra and by 8.25 they are homotopy equivalences. We have included precisely enough in Q that the diagram of 13.43 commutes. □

Now we can prove 13.40b)ii). By 13.41 we only need to check the commutativity of

$$\begin{array}{ccc} & F \wedge_{CB} E & \\ {}^{\tau_{BC}}\nearrow & & \searrow^{\tau_{CB}} \\ E \wedge_{BC} F & \xrightarrow[1]{} & E \wedge_{BC} F; \end{array}$$

but this diagram is even commutative as a diagram of functions. Similarly to prove 13.40b)vii) we need only to check the commutativity of

$$\begin{array}{ccc} S^0 \wedge_{\varnothing, A} E & \xrightarrow{\tau_{\varnothing, A}} & E \wedge_{A, \varnothing} S^0 \\ {}_{\cong}\searrow & & \swarrow_{\cong} \\ & E & \end{array}$$

and again this diagram is commutative as a diagram of functions. 13.40b)viii) follows from the fact that

$$S^0 \simeq S^0 \wedge S^0 \xrightarrow{\tau} S^0 \wedge S^0 \simeq S^0$$

was constructed to have degree 1.

We turn to the construction of a. Again it is tempting to define a by requiring the diagram

13.44.

$$\begin{array}{ccccccc}
 & (E\wedge F)\wedge G & & \xrightarrow{\quad a \quad} & & E\wedge(F\wedge G) & \\
{\scriptstyle eq_{B\cup C,D}}\nearrow & & \nwarrow{\scriptstyle eq_{BC}\wedge 1} & & {\scriptstyle 1\wedge eq_{CD}}\nearrow & & \nwarrow{\scriptstyle eq_{B,C\cup D}} \\
(E\wedge F)\wedge_{B\cup C,D}G & & (E\wedge_{BC}F)\wedge G & & E\wedge(F\wedge_{CD}G) & & E\wedge_{B,C\cup D}(F\wedge G) \\
{\scriptstyle eq_{BC}\wedge 1}\nwarrow & & \nearrow{\scriptstyle eq_{B\cup C,D}} & & {\scriptstyle eq_{B,C\cup D}}\nwarrow & & \nearrow{\scriptstyle 1\wedge eq_{CD}} \\
 & (E\wedge_{BC}F)\wedge_{B\cup C,D}G & & \xrightarrow[\quad 1 \quad]{} & & E\wedge_{B,C\cup D}(F\wedge_{CD}G) &
\end{array}$$

to commute for all partitions $A = B\cup C\cup D$ with B, C, D infinite (observe that $(E\wedge_{BC}F)\wedge_{B\cup C,D}G$ and $E\wedge_{B,C\cup D}(F\wedge_{CD}G)$ are in fact the same spectrum). But again unfortunately we cannot content ourselves with only such partitions $A = B\cup C\cup D$ such that B, C, D are infinite if we want to prove 13.40b)iv)–vi).

13.45. Lemma. *For any spectra E, F, G there are spectra P', P'' and for all partitions $A = B\cup C\cup D$ with $B\cup C$, $C\cup D$ infinite commutative diagrams*

$$\begin{array}{ccccc}
(E\wedge_{BC}F)\wedge G & \xleftarrow{\rho\wedge 1} & (T(E\wedge_{BC}F))\wedge G & \xrightarrow{i\wedge 1} & (E\wedge F)\wedge G \\
\uparrow{\scriptstyle i} & & \nearrow{\scriptstyle j'} & & \uparrow{\scriptstyle i'} \\
T((E\wedge_{BC}F)\wedge_{B\cup C,D}G) & & \xrightarrow{\qquad k' \qquad} & & P'
\end{array}$$

$$\begin{array}{ccccc}
E\wedge(F\wedge_{CD}G) & \xleftarrow{1\wedge\rho} & E\wedge T(F\wedge_{CD}G) & \xrightarrow{1\wedge i} & E\wedge(F\wedge G) \\
\uparrow{\scriptstyle i} & & \nearrow{\scriptstyle j''} & & \uparrow{\scriptstyle i''} \\
T(E\wedge_{B,C\cup D}(F\wedge_{CD}G)) & & \xrightarrow{\qquad k'' \qquad} & & P''
\end{array}$$

such that P', P'', i', i'' are independent of B, C, D and i', i'' are equivalences. Moreover, there is an equivalence $e: P' \to P''$ such that

$$\begin{array}{ccc} P' & \xrightarrow{e} & P'' \\ {\scriptstyle k'}\uparrow & & \uparrow{\scriptstyle k''} \\ T((E \wedge_{BC} F) \wedge_{B \cup C, D} G) & = & T(E \wedge_{B, C \cup D} (F \wedge_{CD} G)) \end{array}$$

commutes.

We of course take a to be $i'' \circ e \circ (i')^{-1}$; then 13.44 commutes. We give only the idea for the proof of 13.45. $(E \wedge F) \wedge G$ is a "4-dimensional telescope"—that is, $(E \wedge F) \wedge G$ is constructed over cells of the positive cone in 4-space. One constructs a cellular map θ of the 3-cone into the 4-cone and pulls the constructions (bundles, etc.) of $(E \wedge F) \wedge G$ back to the 3-cone by means of this map. Thus if e_{ijk} is a cell of the 3-cone, then the part of $((E \wedge F) \wedge G)_n$ over $\theta(e_{ijk})$ contains a subcomplex of the form

$$S^{n-i-j-k-d} \wedge M(\tau_d) \wedge E_i \wedge F_j \wedge G_k,$$

and we define the corresponding part of P' to be

$$S^{n-i-j-k-d} \wedge M(\theta^* \tau_d) \wedge E_i \wedge F_j \wedge G_k,$$

where $\theta^* \tau_d$ is the induced bundle over e_{ijk}. P'' is constructed analogously, and one finds that P' and P'' have precisely the same description; this provides $e: P' \to P''$. More details are given in [8].

Now we can prove 13.40b)i); we take any partition

$$A = B_1 \cup B_2 \cup B_3 \cup B_4$$

with B_i infinite, $1 \leqslant i \leqslant 4$. The commutativity of 13.40b)i) then follows from the fact that

$$\begin{array}{ccc} & (E \wedge_{B_1 B_2} F) \wedge_{B_1 \cup B_2, B_3 \cup B_4} (G \wedge_{B_3 B_4} H) & \\ {\scriptstyle 1}\nearrow & & \searrow{\scriptstyle 1} \\ ((E \wedge_{B_1 B_2} F) \wedge_{B_1 \cup B_2, B_3} G) \wedge_{B_1 \cup B_2 \cup B_3, B_4} H & & E \wedge_{B_1, B_2 \cup B_3 \cup B_4} (F \wedge_{B_2, B_3 \cup B_4} (G \wedge_{B_3 B_4} H)) \\ {\scriptstyle 1}\downarrow & & \uparrow{\scriptstyle 1} \\ (E \wedge_{B_1, B_2 \cup B_3} (F \wedge_{B_2 B_3} G)) \wedge_{B_1 \cup B_2 \cup B_3, B_4} H & \xrightarrow{1} & E \wedge_{B_1, B_2 \cup B_3 \cup B_4} ((F \wedge_{B_2 B_3} G) \wedge_{B_2 \cup B_3, B_4} H) \end{array}$$

commutes as a diagram of functions. The commutativity of 13.40b)iii) follows from the commutativity of

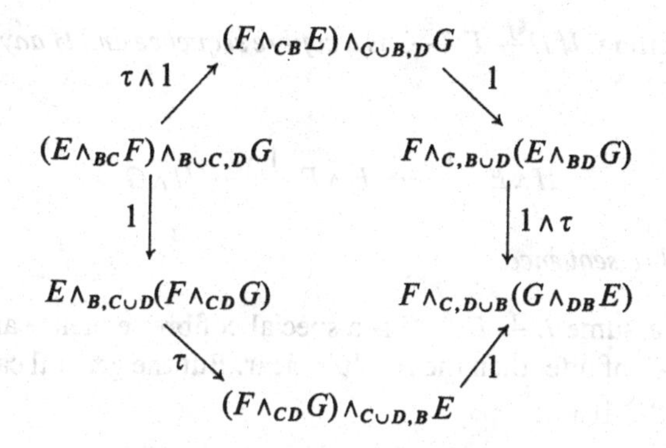

The proofs of iv), v) and vi) are similar. Hence the proof of 13.40 is complete. □

13.46. Proposition. *For all spectra E, F there are natural homotopy equivalences*

$$\Sigma E \wedge F \simeq \Sigma(E \wedge F) \simeq E \wedge \Sigma F.$$

Proof: For $\Sigma E \wedge F \simeq \Sigma(E \wedge F)$ we take

$$\Sigma E \wedge F \simeq (E \wedge S^1) \wedge F \underset{a}{\simeq} E \wedge (S^1 \wedge F) \underset{1 \wedge \tau}{\simeq} E \wedge (F \wedge S^1) \underset{a}{\simeq} (E \wedge F) \wedge S^1 \simeq \Sigma(E \wedge F)$$

and for $E \wedge \Sigma F \simeq \Sigma(E \wedge F)$

$$E \wedge \Sigma F \simeq E \wedge (F \wedge S^1) \underset{a}{\simeq} (E \wedge F) \wedge S^1 \simeq \Sigma(E \wedge F). \quad \square$$

Note:

$$\begin{array}{ccc} & & \Sigma S^n \\ & \overset{\simeq}{\nearrow} & \downarrow \simeq \\ S^{n+1} & \underset{\simeq}{\searrow} & S^n \wedge S^1 \end{array}$$

is homotopy commutative up to a sign $(-1)^n$. Also note that the diagram

$$\begin{array}{ccc} \Sigma^2(X \wedge Y) & \rightarrow & \Sigma(\Sigma X \wedge Y) \\ \downarrow & & \downarrow \\ \Sigma(X \wedge \Sigma Y) & \rightarrow & \Sigma X \wedge \Sigma Y \end{array}$$

commutes up to sign -1.

13.47. Proposition. *If $E \xrightarrow{f} F \xrightarrow{g} G$ is a cofibre sequence and H any spectrum, then*

$$H \wedge E \xrightarrow{1 \wedge f} H \wedge F \xrightarrow{1 \wedge g} H \wedge G$$

is also a cofibre sequence.

Proof: If we assume $E \xrightarrow{f} F \xrightarrow{g} G$ is a special cofibre sequence and replace $\wedge$ by $\wedge_{BC}$, B, C infinite, then the result is clear. But the general case follows from this one. ▯

13.48. Proposition. *For any spectra E, X_α, $\alpha \in A$, we have a natural homotopy equivalence*

$$E \wedge (\vee_\alpha X_\alpha) \simeq \vee_\alpha (E \wedge X_\alpha).$$

Proof: Again the result is clear if we replace $\wedge$ by $\wedge_{BC}$, B, C infinite. But eq_{BC} is a natural homotopy equivalence. ▯

We have had to work hard to construct $E \wedge F$ and verify its most important properties; now we can begin to enjoy the fruits of our labors. For example we can immediately extend the homology theory E_* associated to a spectrum E over the category $\mathscr{SP}$ of spectra by setting

$$E_n(X) = \pi_n(E \wedge X) = [S^n, E \wedge X]$$

for $n \in \mathbb{Z}$, $X \in \mathscr{SP}$. Since for any CW-complex X we have $E \wedge X \simeq E \wedge E(X)$, the new definition agrees with the old one on the category of CW-complexes. The axioms for a homology theory on $\mathscr{SP}'$ follow directly from 13.46–13.48.

We could, however, have defined $E_n(X)$ in terms of direct limits, as the following proposition shows.

13.49. Proposition. *For any two spectra E and X we have a natural isomorphism*

$$E_n(X) \cong \operatorname{dirlim}_q \tilde{E}_{n+q}(X_q),$$

the morphisms of the direct system being given by

$$\tilde{E}_{n+q}(X_q) \underset{\sigma}{\cong} \tilde{E}_{n+q+1}(SX_q) \xrightarrow{\varepsilon_{q*}} \tilde{E}_{n+q+1}(X_{q+1}),$$

$\varepsilon_q : SX_q \to X_{q+1}$ the inclusion.

Proof: The proposition will be a consequence of the following algebraic fact, whose proof we leave to the reader. Let $\{G_{ij}, \phi_{ij}^{kl}:(i,j),(k,l) \in \mathbb{N} \times \mathbb{N}\}$ be a direct system of abelian groups G_{ij} and homomorphisms

$$\phi_{ij}^{kl}: G_{ij} \to G_{kl}$$

for $i \leqslant k, j \leqslant l$ where $\mathbb{N} \times \mathbb{N}$ has the order relation $(i,j) \leqslant (k,l)$ if $i \leqslant k, j \leqslant l$. We can form

$$\operatorname{dir\,lim}_i(\operatorname{dir\,lim}_j G_{ij})$$

and we have natural homomorphisms

$$f_{kl}: G_{kl} \to \operatorname{dir\,lim}_i(\operatorname{dir\,lim}_j G_{ij})$$

for each $k, l \in \mathbb{N}$ such that $f_{mn} \circ \phi_{kl}^{mn} = f_{kl}$. Let $\mathbb{N} = B \cup C$ be a partition and define $\beta, \gamma: \mathbb{N} \to \mathbb{N}$ as in the construction of naive smash products. Then the homomorphisms

$$f_{\beta(n),\,\gamma(n)}: G_{\beta(n),\,\gamma(n)} \to \operatorname{dir\,lim}_i(\operatorname{dir\,lim}_j G_{ij}), \quad n \in \mathbb{N},$$

define a homomorphism

$$f_{BC} = \{f_{\beta(n),\,\gamma(n)}\}: \operatorname{dir\,lim}_n G_{\beta(n),\,\gamma(n)} \to \operatorname{dir\,lim}_i(\operatorname{dir\,lim}_j G_{ij}),$$

and we claim f_{BC} is an isomorphism if B and C are both infinite.

Now let $\mathbb{N} = B \cup C$ be any such partition with B, C infinite. Then we have the sequence of natural isomorphisms

$$\begin{aligned} E_n(X) = \pi_n(E \wedge X) &\xleftarrow[\cong]{eq_{BC}*} \pi_n'(E \wedge_{BC} X) \cong \operatorname{dir\,lim}_r \pi_{n+r}((E \wedge_{BC} X)_r) \\ &= \operatorname{dir\,lim}_r \pi_{n+\beta(r)+\gamma(r)}(E_{\beta(r)} \wedge X_{\gamma(r)}) \to \\ &\xrightarrow{f_{BC}} \operatorname{dir\,lim}_i(\operatorname{dir\,lim}_j \pi_{n+i+j}(E_j \wedge X_i)) \cong \operatorname{dir\,lim}_i \pi_{n+i}(E \wedge X_i), \end{aligned}$$

where we have used 8.21 twice. □

With the smash product we can also show how to construct products in general homology and cohomology theories.

13.50. Definition. A *ring spectrum* is a spectrum E with *product* $\mu: E \wedge E \to E$ (a map of spectra) and *identity* $\iota: S^0 \to E$ such that the diagrams

$$\begin{array}{ccc} E \wedge E \wedge E & \xrightarrow{\mu \wedge 1} & E \wedge E \\ \downarrow{\scriptstyle 1 \wedge \mu} & & \downarrow{\scriptstyle \mu} \\ E \wedge E & \xrightarrow{\mu} & E \end{array} \qquad \begin{array}{ccccc} S^0 \wedge E & \xrightarrow{\iota \wedge 1} & E \wedge E & \xleftarrow{1 \wedge \iota} & E \wedge S^0 \\ & {\scriptstyle l} \searrow & \downarrow{\scriptstyle \mu} & \swarrow {\scriptstyle r} & \\ & & E & & \end{array}$$

commute up to homotopy. μ is *commutative* if the diagram

$$\begin{array}{ccc} E\wedge E & \xrightarrow{\tau} & E\wedge E \\ & \mu \searrow \quad \swarrow \mu & \\ & E & \end{array}$$

also commutes up to homotopy.

For any ring spectrum E we can now define external products:

i) $\wedge: E_n(X) \otimes E_m(Y) \to E_{n+m}(X\wedge Y)$
ii) $\wedge: E^n(X) \otimes E^m(Y) \to E^{n+m}(X\wedge Y)$
iii) $/: E^p(X\wedge Y) \otimes E_q(Y) \to E^{p-q}(X)$
iv) $\backslash: E^p(X) \otimes E_q(X\wedge Y) \to E_{q-p}(Y)$.

i) For $x\in E_n(X)$, $y\in E_m(Y)$ represented by $f: S^n \to E\wedge X$, $g: S^m \to E\wedge Y$ respectively we take $x\wedge y$ to be represented by

$$S^{n+m} \simeq S^n \wedge S^m \xrightarrow{f\wedge g} (E\wedge X)\wedge(E\wedge Y) \xrightarrow{1\wedge\tau\wedge 1} E\wedge E\wedge X\wedge Y \xrightarrow{\mu\wedge 1} E\wedge X\wedge Y.$$

One readily checks that $\wedge$ is bilinear, so that we genuinely do have a homomorphism from the tensor product.

ii) For $x\in E^n(X)$, $y\in E^m(Y)$ represented by $f: X\to \Sigma^n E, g: Y\to \Sigma^m E$ respectively we take $x\wedge y$ to be the element represented by

$$X\wedge Y \xrightarrow{f\wedge g} \Sigma^n E\wedge \Sigma^m E \simeq \Sigma^{n+m}(E\wedge E) \xrightarrow{\Sigma^{n+m}\mu} \Sigma^{n+m} E.$$

Again $\wedge$ is bilinear.

iii) For $w\in E^p(X\wedge Y)$, $y\in E_q(Y)$ represented by $f: X\wedge Y\to \Sigma^p E$, $g: S^q \to E\wedge Y$ respectively we take w/y to be the element represented by

$$X\simeq X\wedge S^0 \xrightarrow{1\wedge\Sigma^{-q}g} X\wedge \Sigma^{-q}E\wedge Y \xrightarrow{1\wedge\tau} X\wedge Y\wedge\Sigma^{-q}E \xrightarrow{f\wedge 1} \Sigma^p E\wedge \Sigma^{-q}E \simeq \Sigma^{p-q}E\wedge E \xrightarrow{\Sigma^{p-q}\mu} \Sigma^{p-q}E.$$

iv) For $x\in E^p(X)$, $w\in E_q(X\wedge Y)$ represented by $f: X\to \Sigma^p E$, $g: S^q \to E\wedge X\wedge Y$ respectively we take $x\backslash w$ to be the element represented by

$$S^{q-p} \xrightarrow{\Sigma^{-p}g} \Sigma^{-p}E\wedge X\wedge Y \xrightarrow{\tau\wedge 1} X\wedge\Sigma^{-p}E\wedge Y \xrightarrow{f\wedge 1} \Sigma^p E\wedge \Sigma^{-p}E\wedge Y \simeq E\wedge E\wedge Y \xrightarrow{\mu\wedge 1} E\wedge Y.$$

Note: Because of the remark following 13.46 we must have a convention about such equivalences as $\Sigma^{n+m}(X\wedge Y) \simeq \Sigma^n X\wedge\Sigma^m Y$. We shall mean the composition $\Sigma^{n+m}(X\wedge Y) \simeq \Sigma^n(X\wedge \Sigma^m Y) \simeq \Sigma^n X\wedge \Sigma^m Y$.

13.51. *Remarks.* i) We have the following mnemonic device for the slant products:
a) the cohomology variable always comes on the left and the homology variable on the right of the slant;
b) the slant products work like "cancelling in fractions" — e.g. "$XY/Y \to X$".

ii) More generally we could define products

$$E_n(X) \otimes F_m(Y) \to (E \wedge F)_{m+n}(X \wedge Y), \text{ etc.}$$

for any two spectra E, F (cf. the $\times$-product

$$H_n(X; G) \otimes H_m(Y; G') \to H_{n+m}(X \times Y; G \otimes G')$$

for any two abelian groups G, G').

iii) We have the obvious notion of a *module spectrum* F over a ring spectrum E: i.e. there is an action map $\alpha: E \wedge F \to F$ such that appropriate diagrams commute up to homotopy. Combining α with ii) above we could then have products $E_n(X) \otimes F_m(Y) \to F_{n+m}(X \wedge Y)$, etc. for any E-module spectrum F.

For CW-pairs (X, A) we also have the relative groups $E_*(X, A)$ and corresponding external products which result from the fact that

$$(X/A) \wedge (Y/B) \simeq X \times Y/X \times B \cup A \times Y.$$

Thus we get products

i) $\times: E_n(X, A) \otimes E_m(Y, B) \to E_{n+m}(X \times Y, X \times B \cup A \times Y)$
ii) $\times: E^n(X, A) \otimes E^m(Y, B) \to E^{n+m}(X \times Y, X \times B \cup A \times Y)$
iii) $/: E^p(X \times Y, X \times B \cup A \times Y) \otimes E_q(Y, B) \to E^{p-q}(X, A)$
iv) $\backslash: E^p(X, A) \otimes E_q(X \times Y, X \times B \cup A \times Y) \to E_{q-p}(Y, B)$.

Since $E_*(X, A)$ for general pairs is defined by taking CW-substitutes, we can extend these products to arbitrary pairs (X, A) and (Y, B) such that $(X \times B \cup A \times Y; X \times B, A \times Y)$ is excisive with respect to E.

The appropriate naturality properties of these products are readily deduced from the definitions. We state them in the following proposition.

13.52. Proposition. *Let $f: X \to X'$, $g: Y \to Y'$ be maps.*

i) *For $x \in E_n(X)$, $y \in E_m(Y)$ we have $(f \wedge g)_*(x \wedge y) = f_*(x) \wedge g_*(y)$.*
ii) *For $x \in E^n(X')$, $y \in E^m(Y')$ we have $(f \wedge g)^*(x \wedge y) = f^*(x) \wedge g^*(y)$.*
iii) *For $u \in E^p(X' \wedge Y')$, $y \in E_q(Y)$ we have $((f \wedge g)^* u)/y = f^*(u/g_* y)$.*
iv) *For $x \in E^p(X')$, $v \in E_q(X \wedge Y)$ we have $x \backslash (f \wedge g)_*(v) = g_*(f^* x \backslash v)$.*

If the product μ is commutative, then we have a commutativity relation for $\wedge$.

13.53. Proposition. *If $\mu: E \wedge E \to E$ is commutative and $x \in E_n(X)$, $y \in E_m(Y)$, then $\tau_*(x \wedge y) = (-1)^{nm} y \wedge x$. An analogous result holds for cohomology.*

Question: How should the commutativity relation be formulated for the more general product $\wedge : E_n(X) \otimes F_m(Y) \to (E \wedge F)_{n+m}(X \wedge Y)$?

Proof: Let x be represented by $f : S^n \to E \wedge X$, y by $g : S^m \to E \wedge Y$. The result follows from the homotopy-commutativity of the diagram

$$\begin{array}{ccccc}
S^{n+m} \simeq S^n \wedge S^m & \xrightarrow{f \wedge g} & (E \wedge X) \wedge (E \wedge Y) & \xrightarrow{1 \wedge \tau \wedge 1} & \\
\downarrow (-1)^{nm} \quad \downarrow \tau & & \downarrow \tau & & \\
S^{n+m} \simeq S^m \wedge S^n & \xrightarrow{g \wedge f} & (E \wedge Y) \wedge (E \wedge X) & \xrightarrow{1 \wedge \tau \wedge 1} &
\end{array}$$

$$\begin{array}{ccc}
E \wedge E \wedge X \wedge Y & \xrightarrow{\mu \wedge 1} & E \wedge X \wedge Y \\
\downarrow \tau \wedge \tau & & \downarrow 1 \wedge \tau \\
E \wedge E \wedge Y \wedge X & \xrightarrow{\mu \wedge 1} & E \wedge Y \wedge X.
\end{array}$$

The only point where homotopy-commutativity is not completely trivial is the left-most square. The latter follows by double induction from the fact that $\tau : S^1 \wedge S^1 \to S^1 \wedge S^1$ and $\nu' \wedge 1 : S^1 \wedge S^1 \to S^1 \wedge S^1$ are homotopic, which we proved in the process of showing $E \wedge S^1 \simeq \Sigma E$ (8.26). □

Naturally we also expect associativity relations; in fact there are eight associativity relations among the four products.

13.54. Proposition.

i) *For* $x \in E_n(X)$, $y \in E_m(Y)$, $z \in E_p(Z)$ *we have*
$$(x \wedge y) \wedge z = x \wedge (y \wedge z) \in E_{n+m+p}(X \wedge Y \wedge Z).$$

ii) *For* $x \in E^n(X)$, $y \in E^m(Y)$, $z \in E^p(Z)$ *we have*
$$(x \wedge y) \wedge z = x \wedge (y \wedge z) \in E^{n+m+p}(X \wedge Y \wedge Z).$$

iii) *For* $x \in E^n(X)$, $u \in E_m(X \wedge Y)$, $z \in E_p(Z)$ *we have*
$$(x \backslash u) \wedge z = x \backslash (u \wedge z) \in E_{m+p-n}(Y \wedge Z).$$

iv) *For* $x \in E^n(X)$, $u \in E^m(Y \wedge Z)$, $z \in E_p(Z)$ *we have*
$$x \wedge (u/z) = (x \wedge u)/z \in E^{n+m-p}(X \wedge Y).$$

v) *For* $w \in E^m(X \wedge Y \wedge Z)$, $y \in E_n(Y)$, $z \in E_p(Z)$ *we have*
$$(w/z)/y = w/(\tau_*(z \wedge y)) \in E^{m-n-p}(X).$$

vi) *For* $x \in E^m(X)$, $y \in E^n(Y)$, $w \in E_p(X \wedge Y \wedge Z)$ *we have*
$$y \backslash (x \backslash w) = \tau^*(y \wedge x) \backslash w \in E_{p-m-n}(Z).$$

vii) *For* $u \in E^m(X \wedge Z)$, $y \in E^n(Y)$, $v \in E_p(Y \wedge Z)$ *we have*

$$u/(y\backslash v) = [(1 \wedge \tau)^*(u \wedge y)]/v \in E^{n+m-p}(X).$$

viii) *For* $u \in E^m(X \wedge Y)$, $y \in E_n(Y)$, $v \in E_p(X \wedge Z)$ *we have*

$$(u/y)\backslash v = u\backslash[(\tau \wedge 1)_*(y \wedge v)] \in E_{n+p-m}(Z).$$

The proof involves writing down the appropriate diagrams and checking that they commute up to homotopy. Naturally the associativity of μ plays a rôle.

Question: How should one formulate these relations for the more general product of 13.51ii)?

Remark. If μ is commutative, then we can use 13.53 to rewrite v) and vi) in the form

$$(w/z)/y = (-1)^{np} w/(y \wedge z)$$

$$y\backslash(x\backslash w) = (-1)^{mn} x \wedge y\backslash w$$

if we wish.

For any ring spectrum E we have in particular the product

$$\wedge : E_n(S^0) \otimes E_m(S^0) \rightarrow E_{n+m}(S^0 \wedge S^0) = E_{n+m}(S^0)$$

—that is, we have a product

$$\pi_n(E) \otimes \pi_m(E) \rightarrow \pi_{n+m}(E),\ n, m \in \mathbb{Z},$$

which makes $\pi_*(E)$ into a *graded ring*. The associativity of this graded ring follows from 13.54i) for $X = Y = Z = S^0$. The element $1 = [S^0 \xrightarrow{\iota} E] \in \pi_0(E)$ is a 1 for the ring, and the ring is (graded) commutative if E is commutative—that is, for $x \in \pi_n(E)$, $y \in \pi_m(E)$ we have $xy = (-1)^{nm} yx$. Moreover for an arbitrary X we have products

$$\wedge : E_n(S^0) \otimes E_m(X) \rightarrow E_{n+m}(S^0 \wedge X) = E_{n+m}(X)$$

$$\wedge : E^n(S^0) \otimes E^m(X) \rightarrow E^{n+m}(S^0 \wedge X) = E^{n+m}(X)$$

making $E_*(X)$ and $E^*(X)$ into *graded (left) modules* over the graded ring $\pi_*(E)$ (note $E^n(S^0) = \pi_{-n}(E)$). The axioms for a module follow from 13.54i), ii) and the assumption on $\iota : S^0 \rightarrow E$. Naturally there are also right module structures.

We must also ask how these products behave with respect to suspension. We have eight relations of the type $\sigma(ab) = (\sigma a)b$ or $\sigma(ab) = a(\sigma b)$ given by the following proposition.

13.55. Proposition. *For any spectra X, Y the following diagrams commute up to the sign indicated.*

i)

$$\begin{array}{ccccc} E_n(X)\otimes E_m(Y) & \xrightarrow{\sigma\otimes 1} & & & E_{n+1}(\Sigma X)\otimes E_m(Y)\\ \downarrow{\scriptstyle\wedge} & & & & \downarrow{\scriptstyle\wedge} \\ E_{n+m}(X\wedge Y) & \xrightarrow{\sigma} & E_{n+m+1}(\Sigma(X\wedge Y)) & \cong & E_{n+m+1}(\Sigma X\wedge Y)\end{array}$$

ii)

$$\begin{array}{ccccc} E_n(X)\otimes E_m(Y) & \xrightarrow{1\otimes \sigma} & & & E_n(X)\otimes E_{m+1}(\Sigma Y)\\ \downarrow{\scriptstyle\wedge} & & (-1)^n & & \downarrow{\scriptstyle\wedge} \\ E_{n+m}(X\wedge Y) & \xrightarrow{\sigma} & E_{n+m+1}(\Sigma(X\wedge Y)) & \cong & E_{n+m+1}(X\wedge \Sigma Y)\end{array}$$

iii)

$$\begin{array}{ccccc} E^n(X)\otimes E^m(Y) & \xleftarrow{\sigma\otimes 1} & & & E^{n+1}(\Sigma X)\otimes E^m(Y)\\ \downarrow{\scriptstyle\wedge} & & & & \downarrow{\scriptstyle\wedge} \\ E^{n+m}(X\wedge Y) & \xleftarrow{\sigma} & E^{n+m+1}(\Sigma(X\wedge Y)) & \cong & E^{n+m+1}(\Sigma X\wedge Y)\end{array}$$

iv)

$$\begin{array}{ccccc} E^n(X)\otimes E^m(Y) & \xleftarrow{1\otimes \sigma} & & & E^n(X)\otimes E^{m+1}(\Sigma Y)\\ \downarrow{\scriptstyle\wedge} & & (-1)^n & & \downarrow{\scriptstyle\wedge} \\ E^{n+m}(X\wedge Y) & \xleftarrow{\sigma} & E^{n+m+1}(\Sigma(X\wedge Y)) & \cong & E^{n+m+1}(X\wedge \Sigma Y)\end{array}$$

v)

$$\begin{array}{ccccc} E^n(X\wedge Y)\otimes E_m(Y) & \xleftarrow{\sigma\otimes 1} & E^{n+1}(\Sigma(X\wedge Y))\otimes E_m(Y) & \cong & E^{n+1}(\Sigma X\wedge Y)\otimes E_m(Y)\\ \downarrow{\scriptstyle /} & & & & \downarrow{\scriptstyle /} \\ E^{n-m}(X) & & \xleftarrow{\sigma} & & E^{n-m+1}(\Sigma X)\end{array}$$

vi)

$$\begin{array}{ccc}
\begin{array}{c}E^{n+1}(\Sigma(X\wedge Y))\\ \otimes E_{m-1}(Y)\end{array} & \xrightarrow{1\otimes\sigma} E^{n+1}(\Sigma(X\wedge Y))\otimes E_m(\Sigma Y)\cong & \begin{array}{c}E^{n+1}(X\wedge\Sigma Y)\\ \otimes E_m(\Sigma Y)\end{array}\\
\sigma\otimes 1\big\downarrow & (-1)^{n+m} & \big\downarrow /\\
E^n(X\wedge Y)\otimes E_{m-1}(Y) & \xrightarrow{\qquad\qquad /\qquad\qquad} & E^{n-m+1}(X)
\end{array}$$

vii)

$$\begin{array}{ccc}
\begin{array}{c}E^n(\Sigma X)\otimes\\ E_m(X\wedge Y)\end{array} & \xrightarrow{1\otimes\sigma} E^n(\Sigma X)\otimes E_{m+1}(\Sigma(X\wedge Y))\cong & \begin{array}{c}E^n(\Sigma X)\otimes\\ E_{m+1}(\Sigma X\wedge Y)\end{array}\\
\sigma\otimes 1\big\downarrow & (-1) & \big\downarrow \backslash\\
E^{n-1}(X)\otimes E_m(X\wedge Y) & \xrightarrow{\qquad\qquad \backslash\qquad\qquad} & E_{m-n+1}(Y)
\end{array}$$

viii)

$$\begin{array}{ccc}
\begin{array}{c}E^n(X)\otimes\\ E_m(X\wedge Y)\end{array} & \xrightarrow{1\otimes\sigma} E^n(X)\otimes E_{m+1}(\Sigma(X\wedge Y))\cong & E^n(X)\otimes E_{m+1}(X\wedge\Sigma Y)\\
\big\downarrow\backslash & (-1)^n & \big\downarrow \backslash\\
E_{m-n}(Y) & \xrightarrow{\qquad\qquad \sigma\qquad\qquad} & E_{m-n+1}(\Sigma Y).
\end{array}$$

Proof: Again the proof consists of writing down representatives and checking that an appropriate diagram commutes up to homotopy. For example the proof of ii) involves checking the homotopy commutativity of the following diagram for any $f\colon S^n\to E\wedge X$, $g\colon S^m\to E\wedge Y$:

$$\begin{array}{ccccc}
 & S^n\wedge\Sigma S^m & \xrightarrow{f\wedge\Sigma g} & (E\wedge X)\wedge\Sigma(E\wedge Y)\cong E\wedge X\wedge E\wedge\Sigma Y & \xrightarrow{1\wedge\tau\wedge 1}\\
S^{n+m+1}\nearrow\!\!\!\searrow & (-1)^n\ \| & & \| & \\
 & \Sigma(S^n\wedge S^m) & \xrightarrow{\Sigma(f\wedge g)} & \Sigma(E\wedge X\wedge E\wedge Y) & \xrightarrow{\Sigma(1\wedge\tau\wedge 1)}
\end{array}$$

$$\begin{array}{ccc}
E\wedge E\wedge X\wedge\Sigma Y & \xrightarrow{\mu\wedge 1} & E\wedge X\wedge\Sigma Y\\
\| & & \|\\
\Sigma(E\wedge E\wedge X\wedge Y) & \xrightarrow{\Sigma(\mu\wedge 1)} & \Sigma(E\wedge X\wedge Y)
\end{array}$$

(see note following 13.46). ∎

Once we know how the products behave with respect to σ we can also deduce their behavior with respect to the boundary homomorphisms $\partial: E_n(X,A) \to E_{n-1}(A)$, $\Delta: E_n(X,A) \to E_{n-1}(A,B)$, etc. We list twelve relations which will follow from 13.55 and the naturality of our products once we have remarked that the diagram on page 277 commutes for any CW-pairs (X,A), (Y,B). That is, $\Delta \circ \cong$ is essentially $(1 \wedge k')^* \circ \sigma^{-1}$. We get three other such diagrams by replacing $(X,A) \times B$ by $A \times (Y,B)$ and E^* by E_*.

13.56. Proposition. *Let (X,A), (Y,B) be topological pairs such that the triad $(X \times B \cup A \times Y; X \times B, A \times Y)$ is excisive with respect to E. Then the following diagrams commute up to the sign indicated.*

i)

$$\begin{array}{ccc}
E_n(X,A) \otimes E_m(Y,B) & \xrightarrow{\partial \otimes 1} & E_{n-1}(A) \otimes E_m(Y,B) \\
\Big\downarrow \times & & \Big\downarrow \times \\
 & & E_{n+m-1}(A \times Y, A \times B) \\
 & & \Big\downarrow \cong \\
E_{n+m}((X,A) \times (Y,B)) & \xrightarrow{\Delta} & E_{n+m-1}(X \times B \cup A \times Y, X \times B),
\end{array}$$

where Δ is the boundary homomorphism of the triple

$(X \times Y, X \times B \cup A \times Y, X \times B)$.

ii)

$$\begin{array}{ccc}
E_n(X,A) \otimes E_m(Y,B) & \xrightarrow{1 \otimes \partial} & E_n(X,A) \otimes E_{m-1}(B) \\
\Big\downarrow \times & & \Big\downarrow \times \\
 & (-1)^n & E_{n+m-1}(X \times B, A \times B) \\
 & & \Big\downarrow \cong \\
E_{n+m}((X,A) \times (Y,B)) & \xrightarrow{\Delta} & E_{n+m-1}(X \times B \cup A \times Y, A \times Y)
\end{array}$$

where Δ is the boundary homomorphism of the triple

$(X \times Y, X \times B \cup A \times Y, A \times Y)$.

$$
\begin{array}{ccccccc}
E^n((X,A)\times B) & \xleftarrow{\cong} & E^n(X\times B\cup A\times Y, A\times Y) & \xrightarrow{j^*} & E^n(X\times B\cup A\times Y) & \xrightarrow{\delta} & E^{n+1}(X\times Y, X\times B\cup A\times Y) \\
\| \wr & & \| \wr & & \| \wr & & \| \wr \\
\tilde{E}^n\left(\dfrac{X\times B}{A\times B}\right) & \xleftarrow{\cong} & \tilde{E}^n\left(\dfrac{X\times B\cup A\times Y}{A\times Y}\right) & \xrightarrow{p^*} & \tilde{E}^n(X\times B\cup A\times Y^+) \underset{\cong}{\xleftarrow{\sigma}} \tilde{E}^{n+1}(S(X\times B\cup A\times Y^+)) & \xrightarrow{k'^*} & \tilde{E}^{n+1}\left(\dfrac{X\times Y}{X\times B\cup A\times Y}\right) \\
\cong\uparrow\sigma & & \uparrow\sigma & \nearrow Sp^* & & & \| \\
\tilde{E}^{n+1}\left(S\left(\dfrac{X\times B}{A\times B}\right)\right) & \cong & \tilde{E}^{n+1}\left(S\left(\dfrac{X\times B\cup A\times Y}{A\times Y}\right)\right) & & & & \| \\
\| \wr & & & & & & \| \\
\tilde{E}^{n+1}(X/A\wedge SB^+) & & & \xrightarrow{(1\wedge k')^*} & & & \tilde{E}^{n+1}(X/A\wedge Y/B).
\end{array}
$$

iii)

$$\begin{array}{ccc}
E^n(A) \otimes E^m(Y,B) & \xrightarrow{\delta \otimes 1} & E^{n+1}(X,A) \otimes E^m(Y,B) \\
\times \downarrow & & \\
E^{n+m}(A \times Y, A \times B) & & \downarrow \times \\
\cong \uparrow & & \\
E^{n+m}(X \times B \cup A \times Y, X \times B) & \xrightarrow{\Delta} & E^{n+m+1}((X,A) \times (Y,B)),
\end{array}$$

where Δ is the boundary homomorphism of the triple $(X \times Y, X \times B \cup A \times Y, X \times B)$.

iv)

$$\begin{array}{ccc}
E^n(X,A) \otimes E^m(B) & \xrightarrow{1 \otimes \delta} & E^n(X,A) \otimes E^{m+1}(Y,B) \\
\times \downarrow & & \\
E^{n+m}(X \times B, A \times B) & (-1)^n & \downarrow \times \\
\cong \uparrow & & \\
E^{n+m}(X \times B \cup A \times Y, A \times Y) & \xrightarrow{\Delta} & E^{n+m+1}((X,A) \times (Y,B)),
\end{array}$$

where Δ is the boundary homomorphism of the triple $(X \times Y, X \times B \cup A \times Y, A \times Y)$.

v)

$$\begin{array}{ccc}
E^n(X \times B \cup A \times Y, X \times B) \otimes E_m(Y,B) & \xrightarrow{\Delta \otimes 1} & E^{n+1}((X,A) \times (Y,B)) \otimes E_m(Y,B) \\
\cong \otimes 1 \downarrow & & \\
E^n(A \times Y, A \times B) \otimes E_m(Y,B) & & \downarrow / \\
/ \downarrow & & \\
E^{n-m}(A) & \xrightarrow{\delta} & E^{n-m+1}(X,A).
\end{array}$$

vi)

$$\begin{array}{ccc}
E^n(X \times B \cup A \times Y, A \times Y) \otimes E_m(Y,B) & \xrightarrow{\Delta \otimes 1} & E^{n+1}((X,A) \times (Y,B)) \otimes E_m(Y,B) \\
\Big\downarrow{\scriptstyle \cong \otimes 1} & & \\
E^n(X \times B, A \times B) \otimes E_m(Y,B) & (-1)^{n+m} & \Big\downarrow{\scriptstyle /} \\
\Big\downarrow{\scriptstyle 1 \otimes \partial} & & \\
E^n((X,A) \times B) \otimes E_{m-1}(B) & \xrightarrow{\quad / \quad} & E^{n-m+1}(X,A).
\end{array}$$

vii)

$$\begin{array}{ccc}
E^n(A) \otimes E_m((X,A) \times (Y,B)) & \xrightarrow{1 \otimes \Delta} & E^n(A) \otimes E_{m-1}(X \times B \cup A \times Y, X \times B) \\
 & & \Big\uparrow{\scriptstyle 1 \otimes \cong} \\
\Big\downarrow{\scriptstyle \delta \otimes 1} & (-1) & E^n(A) \otimes E_{m-1}(A \times Y, A \times B) \\
 & & \Big\downarrow{\scriptstyle \backslash} \\
E^{n+1}(X,A) \otimes E_m((X,A) \times (Y,B)) & \xrightarrow{\quad \backslash \quad} & E_{m-n-1}(Y,B).
\end{array}$$

viii)

$$\begin{array}{ccc}
E^n(X,A) \otimes E_m((X,A) \times (Y,B)) & \xrightarrow{1 \otimes \Delta} & E^n(X,A) \otimes E_{m-1}(X \times B \cup A \times Y, A \times Y) \\
 & & \Big\uparrow{\scriptstyle 1 \otimes \cong} \\
\Big\downarrow{\scriptstyle \backslash} & (-1)^n & E^n(X,A) \times E_{m-1}(X \times B, A \times B) \\
 & & \Big\downarrow{\scriptstyle \backslash} \\
E_{m-n}(Y,B) & \xrightarrow{\quad \partial \quad} & E_{m-n-1}(B).
\end{array}$$

ix)

$$\begin{array}{ccccc}
 & \overset{i_1^* \otimes 1}{\nearrow} & E^n(A \times (Y,B)) \otimes E_m(Y,B) & \xrightarrow{\quad / \quad} & E^{n-m}(A) \\
E^n(X \times Y, A \times B) \otimes E_m(Y,B) & & (-1)^{n+m+1} & & \Big\downarrow{\scriptstyle \delta} \\
 & \underset{i_2^* \otimes \partial}{\searrow} & E^n((X,A) \times B) \otimes E_{m-1}(B) & \xrightarrow{\quad / \quad} & E^{n-m+1}(X,A).
\end{array}$$

x)

$$\begin{array}{ccc}
 & E^n(A) \otimes E_m(X \times B \cup A \times Y, X \times B) & \\
 \nearrow^{1 \otimes j_{1*}} & \| \wr & \\
 & E^n(A) \otimes E_m(A \times (Y, B)) \xrightarrow{\;\backslash\;} & E_{m-n}(Y, B) \\
E^n(A) \otimes E_m(X \times B \cup A \times Y) & (-1)^n & \downarrow \partial \\
 \searrow_{\delta \otimes j_{2*}} & & \\
 & E^{n+1}(X, A) \otimes E_m(X \times B \cup A \times Y, A \times Y) & \\
 & \| \wr & \\
 & E^{n+1}(X, A) \otimes E_m((X, A) \times B) \xrightarrow{\;\backslash\;} & E_{m-n-1}(B).
\end{array}$$

xi)

$$\begin{array}{ccc}
 & E_n(X, A) \otimes E_m(Y, B) & \\
 \swarrow^{\partial \otimes 1} & & \searrow^{1 \otimes \partial} \\
E_{n-1}(A) \otimes E_m(Y, B) & & E_n(X, A) \otimes E_{m-1}(B) \\
\downarrow \times & (-1)^{n+1} & \downarrow \times \\
E_{n+m-1}(A \times (Y, B)) & & E_{n+m-1}((X, A) \times B) \\
 \searrow & & \swarrow \\
 & E_{n+m-1}(X \times Y, A \times B). &
\end{array}$$

xii)

$$\begin{array}{ccc}
 & E^n(A) \otimes E^m(B) & \\
 \swarrow^{\delta \otimes 1} & & \searrow^{1 \otimes \delta} \\
E^{n+1}(X, A) \otimes E^m(B) & & E^n(A) \otimes E^{m+1}(Y, B) \\
\downarrow \times & & \downarrow \times \\
E^{n+m+1}((X, A) \times B) & (-1)^{n+1} & E^{n+m+1}(A \times (Y, B)) \\
\uparrow \cong & & \uparrow \cong \\
E^{n+m+1}(X \times B \cup A \times Y, A \times Y) & & E^{n+m+1}(X \times B \cup A \times Y, X \times B) \\
 \searrow & & \swarrow \\
 & E^{n+m+1}(X \times B \cup A \times Y). &
\end{array}$$

Proof: The proofs of i)–viii) are all similar; we prove vi) as an illustration. We suppose (X, A), (Y, B) are CW-pairs; the general case is then handled by choosing CW-substitutes. $\Delta \circ \cong$ is $(1 \wedge k')^* \circ \sigma^{-1}$, so for $u \in E^n((X, A) \times B)$ and $y \in E_m(Y, B)$ we have

$$\begin{aligned}\Delta u/y &= (1 \wedge k')^* \sigma^{-1} u/y = \sigma^{-1} u/k'_* y \quad \text{by 13.52iii)}\\ &= (-1)^{n+m} u/\sigma^{-1} k'_* y \quad \text{by 13.55vi)}\\ &= (-1)^{n+m} u/\partial y.\end{aligned}$$

ix)–xii) follow simply by filling in the hexagon diagrams 7.22 for the triad $(X \times B \cup A \times Y; X \times B, A \times Y)$ and using i)–viii). □

Now that we have external products we can use the diagonal map to define internal products $\cup$ and $\cap$ as with singular homology. For any space X we define

$$\cup : E^n(X) \otimes E^m(X) \to E^{n+m}(X)$$

by $\cup = \Delta^* \circ \times$. (Note that we do not have a diagonal map for spectra in general, so we cannot define $\cup$ on $E^*(X)$ for a general spectrum X.) For topological pairs (X, A), (X, B) we also have a $\cup$-product

$$E^n(X, A) \otimes E^m(X, B) \xrightarrow{\times} E^{n+m}((X, A) \times (X, B)) \xrightarrow{\Delta^*} E^{n+m}(X, A \cup B),$$

where we regard Δ as a map

$$\Delta : (X, A \cup B) \to (X \times X, X \times B \cup A \times X) = (X, A) \times (X, B).$$

We can also define the $\cap$-product:

$$\cap : E^n(X, A) \otimes E_m(X, A \cup B) \to E_{m-n}(X, B)$$

is the composite

$$E^n(X, A) \otimes E_m(X, A \cup B) \xrightarrow{1 \otimes \Delta_*} E^n(X, A) \otimes E_m((X, A) \times (X, B)) \xrightarrow{\backslash} E_{m-n}(X, B).$$

From the definitions and 13.52 we immediately have the following naturality formulae.

13.57. Proposition. i) *For any map* $f : (X, A, B) \to (X', A', B')$, $x \in E^n(X', A')$, $y \in E^m(X', B')$ *we have*

$$f^*(x \cup y) = f^* x \cup f^* y \in E^{n+m}(X, A \cup B).$$

ii) *For any map*

$$f : (X, A, B) \to (X', A', B'),\ x \in E^n(X', A'),\ y \in E_m(X, A \cup B)$$

we have

$$x \cap f_*(y) = f_*(f^* x \cap y).$$

From the commutativity relation for $\wedge$ we get a commutativity relation for $\cup$.

13.58. Proposition. *If E is a commutative ring spectrum, then for $x \in E^n(X,A)$, $y \in E^m(X,B)$ we have*

$$y \cup x = (-1)^{nm} x \cup y \in E^{n+m}(X, A \cup B).$$

The behavior of $\cup$ with respect to

$$1 = [X^+ \simeq X^+ \wedge S^0 \xrightarrow{c \wedge \iota} (pt)^+ \wedge E \simeq E] \in E^0(X^+)$$

follows from the next lemma.

13.59. Lemma. *Let $\pi_1 : X \times Y \to X$, $\pi_2 : X \times Y \to Y$ be the projections. Then for $x \in E^n(X)$, $y \in E^m(Y)$ we have*

$$\pi_1^*(x) = x \times 1, \quad \pi_2^*(y) = 1 \times y.$$

The proof is trivial.

13.60. Corollary. *For all $x \in E^n(X,A)$, $y \in E_m(X,A)$ we have $1 \cup x = x = x \cup 1$ and $1 \cap y = y$.*

Proof: $1 \cup x = \Delta^*(1 \times x) = \Delta^* \pi_2^* x = (\pi_2 \circ \Delta)^* x = x, x \cup 1 = \Delta^*(x \times 1) = \Delta^* \pi_1^* x = (\pi_1 \circ \Delta)^* x = x$. The statement $1 \cap y = y$ can be verified by writing down representatives. ▯

From the associativity relations for the external products we can generate a large number of associativity relations among $\times$, $\times$, $/$, $\backslash$, $\cup$ and $\cap$. The following proposition lists those we shall need later.

13.61. Proposition.

i) *For $x \in E^n(X,A)$, $y \in E^m(X,B)$, $z \in E^p(X,C)$ we have*

$$(x \cup y) \cup z = x \cup (y \cup z) \in E^{n+m+p}(X, A \cup B \cup C).$$

ii) *For $x \in E^n(X,A)$, $x' \in E^m(X,A')$, $y \in E^p(Y,B)$, $y' \in E^q(Y,B')$ we have*

$$(x \cup x') \times (y \cup y') = (-1)^{mp}(x \times y) \cup (x' \times y') \in E^{n+m+p+q}((X, A \cup A') \times (Y, B \cup B')),$$

provided E is commutative.

iii) *For $x \in E^n(X,A)$, $y \in E^m(X,B)$, $z \in E_p(X, A \cup B \cup C)$ we have*

$$(x \cup y) \cap z = x \cap (y \cap z) \in E_{p-m-n}(X,C).$$

iv) *For $a \in E^n((X,A) \times (Y,B))$, $y \in E^m(Y,C)$, $z \in E_p(Y, B \cup C)$ we have*

$$a/y \cap z = [a \cup (1 \times y)]/z \in E^{n+m-p}(X,A).$$

v) *For $a \in E^n((X,A) \times (Y,B))$, $y \in E^m(X,C)$, $z \in E_p(Y,B)$ we have*

$$y \cup (a/z) = [(y \times 1) \cup a]/z \in E^{n+m-p}(X, A \cup C).$$

vi) *For* $a \in E^n((X,A)\times(Y,B))$, $y \in E_m(Y,B)$, $z \in E_p(X,A\cup C)$ *we have*

$$(a/y)\cap z = \pi_{1*}[a\cap(z\times y)] \in E_{p+m-n}(X,C),$$

where $\pi_1: X\times Y \to X$ *is the projection on the first factor.*

Proof: i) and ii) are evident. For iii) we have

$$\begin{aligned} x\cap(y\cap z) = x\backslash\Delta_*(y\cap z) = x\backslash\Delta_*(y\backslash\Delta_* z) &= x\backslash(y\backslash(1\times\Delta)_*\Delta_* z) && (13.52)\\ &= \tau^*(x\times y)\backslash(\Delta\times 1)_*\Delta_* z && (13.54)\\ &= \Delta^*\tau^*(x\times y)\backslash\Delta_* z && (13.52)\\ &= \Delta^*(x\times y)\backslash\Delta_* z && (\tau\circ\Delta=\Delta)\\ &= (x\cup y)\cap z. \end{aligned}$$

For iv) we have

$$\begin{aligned} [a\cup(1\times y)]/z &= [\Delta^*_{X\times Y}\circ(1\times\pi_2^*)(a\times y)]/z\\ &= [(1\times\Delta_Y)^*(a\times y)]/z = a\times y/\Delta_{Y*}(z)\\ &= (a\times y)/\tau_*\Delta_{Y*}z = [(1\times\tau)^*(a\times y)]/\Delta_* z\\ &= a/(y\backslash\Delta_* z) && (13.54)\\ &= a/y\cap z. \end{aligned}$$

The proofs of v) and vi) are left as an exercise for the reader. □

The behavior of ∪ and ∩ with respect to boundary homomorphisms can be deduced from 13.56.

There is yet another product—the *Kronecker* product—which we shall need in the sequel. This is a product

$$\langle\, ,\rangle: E^n(X)\otimes E_m(X) \to \tilde{E}_{m-n}(S^0) = \pi_{m-n}(E)$$

defined as follows: given $x\in E^n(X)$, $y\in E_m(X)$ we can regard x as an element of $E^n(S^0\wedge X)$ or y as an element of $E_m(X\wedge S^0)$ and then we set

$$\langle x,y\rangle = x/y = x\backslash y \in \tilde{E}_{m-n}(S^0).$$

Given representatives $f: X\to\Sigma^n E$, $g: S^m\to E\wedge X$ for x,y, we get a representative for $\langle x,y\rangle$ by taking

$$S^{m-n}\xrightarrow{\Sigma^{-n}g}\Sigma^{-n}E\wedge X\xrightarrow{\tau}X\wedge\Sigma^{-n}E\xrightarrow{f\wedge 1}\Sigma^n E\wedge\Sigma^{-n}E\xrightarrow{\mu}E.$$

(In singular homology theory the Kronecker product can be defined by evaluating a representing cocycle on a representing cycle.) We have the following obvious naturality and associativity properties of $\langle\, ,\rangle$.

13.62. Proposition.

i) *For any map* $f: X \to X'$, *any* $x \in E^n(X')$, $y \in E_m(X)$ *we have*

$$\langle x, f_* y\rangle = \langle f^* x, y\rangle.$$

ii) *For* $a \in E^n(X \wedge Y)$, $y \in E_m(Y)$, $x \in E_p(X)$ *we have*

$$\langle a/y, x\rangle = \langle a, \tau\ (y \wedge x)\rangle \in \pi_{m+p-n}(E).$$

iii) *For* $y \in E^n(Y)$, $x \in E^m(X)$, $a \in E_p(X \wedge Y)$ *we have*

$$\langle y, x\backslash a\rangle = \langle \tau^*(y \wedge x), a\rangle \in \pi_{p-n-m}(E).$$

iv) *For* $y \in E^n(X, B)$, $x \in E^m(X, A)$, $z \in E_p(X, A \cup B)$ *we have*

$$\langle y, x \cap z\rangle = \langle y \cup x, z\rangle \in \pi_{p-n-m}(E).$$

v) *For* $x \in E^n(X)$, $y \in E^m(Y)$, $a \in E_p(X)$, $b \in E_q(Y)$ *we have*

$$\langle x \wedge y, a \wedge b\rangle = (-1)^{mp}\langle x, a\rangle\langle y, b\rangle.$$

vi) *For* $1 \in E^0(X)$, $y \in E^n(Y)$, $a \in E_m(Y)$ *we have*

$$(1 \times y)/a = \langle y, a\rangle \cdot 1 \in E^{m-n}(X).$$

13.63. From the associativity relations for $\cup$ and $\cap$ and from 13.60 it follows that $\cup$ defines a graded ring structure on $E^*(X, A)$ for all topological pairs (X, A), defines the structure of a graded $E^*(X)$-module on $E^*(X, A)$ and $\cap$ defines the structure of a graded $E^*(X)$-module on $E_*(X, A)$. More is true: $E^*(X, A)$ is not only a graded ring but even a *graded algebra* over the graded ring $\pi_*(E) = \tilde{E}^*(S^0)$. That is, $\cup$ defines a product

$$\cup: E^*(X, A) \otimes E^*(X, A) \to E^*(X, A)$$

which is a homomorphism of graded $\pi_*(E)$-modules. Moreover, these ring and module structures are natural with respect to maps $f: (X, A) \to (Y, B)$. Thus we have succeeded in enriching the natural structure on $E^*(X, A)$, for example—one of our major goals. We give an example to show how this enrichment helps us.

13.64. Example. Let $X = S^2 \times S^3$, $Y = S^2 \vee S^3 \vee S^6$. X and Y both have cell structures with one 0-cell, one 2-cell, one 3-cell and one 6-cell. For any abelian group G we have

$$H^q(X; G) = H^q(S^2 \times S^3; G) \cong \begin{cases} G & q = 0, 2, 3, 6 \\ 0 & q \neq 0, 2, 3, 6 \end{cases}$$

by the Künneth and universal coefficient theorems. We also have

$$H^q(Y;G) \cong \begin{cases} G & q = 0,2,3,6 \\ 0 & q \neq 0,2,3,6. \end{cases}$$

Thus the cohomology *groups* cannot distinguish between the spaces X, Y. Now let us look at the cohomology *rings* $H^*(X;\mathbb{Z})$, $H^*(Y;\mathbb{Z})$. Let $g_i \in H^i(S^i;\mathbb{Z})$ denote a generator, $\pi_1: X \to S^2$, $\pi_2: X \to S^3$ the projections and $x = \pi_1^* g_2 \in H^2(X;\mathbb{Z})$, $y = \pi_2^* g_3 \in H^3(X;\mathbb{Z})$. Then

$$\begin{aligned} x \cup y = \pi_1^* g_2 \cup \pi_2^* g_3 &= (g_2 \times 1) \cup (1 \times g_3) = (g_2 \cup 1) \times (1 \cup g_3) \\ &= g_2 \times g_3. \end{aligned}$$

But by the Künneth theorem for cohomology

$$\times: H^*(S^2;\mathbb{Z}) \otimes H^*(S^3;\mathbb{Z}) \to H^*(X;\mathbb{Z})$$

is an isomorphism, so $x \cup y = g_2 \times g_3 \neq 0$.

Now let $i: S^6 \to Y$ be the inclusion and $\pi: Y \to S^6$ the projection $(Y/Y^5 \cong S^6)$. Then since $\pi \circ i = 1_{S^6}$, we see that

$$i^*: H^6(Y;\mathbb{Z}) \to H^6(S^6;\mathbb{Z})$$

is an epimorphism and hence an isomorphism. But for any $x \in H^2(Y;\mathbb{Z})$, $y \in H^3(Y;\mathbb{Z})$ we have $i^*(x \cup y) = i^*(x) \cup i^*(y) = 0$, since

$$i^*(x) \in H^2(S^6;\mathbb{Z}) = 0.$$

Thus $x \cup y = 0$—or in other words $\tilde{H}^*(Y;\mathbb{Z})$ has no non-trivial products. Thus $H^*(X;\mathbb{Z})$ and $H^*(Y;\mathbb{Z})$ are not isomorphic as rings, and it follows $S^2 \times S^3 \not\simeq S^2 \vee S^3 \vee S^6$.

$\tilde{H}^*(S^n;R) = \tilde{H}^n(S^n;R)$ and for $x, y \in H^n(S^n;R)$ we have

$$x \cup y \in \tilde{H}^{2n}(S^n;R) = 0,$$

so $\tilde{H}^*(S^n;R)$ has a trivial product structure for all $n > 0$ and all rings R. It is not immediately clear, however, that the same is true of $\tilde{E}^*(S^n)$ for all ring spectra E. In fact we now show $\tilde{E}^*(SX)$ has trivial ring structure for all spaces X.

13.65. Proposition. *Suppose $X = A \cup B$ with both A and B contractible to $x_0 \in A \cap B$. Then for all $x, y \in \tilde{E}^*(X) = E^*(X,\{x_0\})$ we have $x \cup y = 0$.*

Proof: Let $i_A: (X,\{x_0\}) \to (X,A)$, $i_B: (X,\{x_0\}) \to (X,B)$, $i: (X,\{x_0\}) \to (X,X)$ denote the inclusions. Then $i_A^*: E^*(X,A) \to \tilde{E}^*(X)$ is an isomorphism and so is i_B^*. Thus we can choose $x' \in E^*(X,A)$, $y' \in E^*(X,B)$ with $i_A^*(x') = x$, $i_B^*(y') = y$. Then we have $x \cup y = i_A^*(x') \cup i_B^*(y') = i^*(x' \cup y')$, but $x' \cup y' \in E^*(X, A \cup B) = E^*(X,X) = 0$. ▯

13.66. Corollary. *$\tilde{E}^*(SX)$ has a trivial product structure.*

Proof: $SX = C^+X \cup C^-X$ and C^+X, C^-X are contractible. □

A similar sort of argument shows that if X is a finite-dimensional CW-complex then all elements of $\tilde{E}^*(X)$ are nilpotent.

13.67. Proposition. *Let X be a CW-complex with r-skeleton X^r and let $F^r E^*(X) = \ker[E^*(X) \to E^*(X^{r-1})]$. Then for $a \in F^p E^*(X)$, $b \in F^q E^*(X)$ we have $a \cup b \in F^{p+q} E^*(X)$.*

Proof: Let $j_1: X \to (X, X^{p-1})$, $j_2: X \to (X, X^{q-1})$, $j_3: X \to (X, X^{p+q-1})$ be the inclusions. Then a can be written $a = j_1^*(a')$ for some $a' \in E^*(X, X^{p-1})$ and similarly $b = j_2^*(b')$ for some $b' \in E^*(X, X^{q-1})$. We have a homotopy commutative diagram

$$\begin{array}{ccc} X & \xrightarrow{\quad\quad\quad\Delta\quad\quad\quad} & X \times X \\ \downarrow{\scriptstyle j_3} & & \downarrow{\scriptstyle j_1 \times j_2} \\ (X, X^{p+q-1}) \xrightarrow{\Delta'} (X \times X, (X \times X)^{p+q-1}) & \subset & (X \times X, X \times X^{q-1} \cup X^{p-1} \times X), \end{array}$$

where Δ' is a cellular map homotopic to Δ (6.35). Thus we find

$$a \cup b = \Delta^*(a \times b) = \Delta^*(j_1 \times j_2)^*(a' \times b') = j_3^* \Delta'^*(a' \times b')$$

—i.e. $a \cup b \in F^{p+q} E^*(X)$. □

13.68. Corollary. *If X is a 0-connected CW-complex of dimension n, then for all $a \in \tilde{E}^*(X)$ we have $a^{n+1} = 0$.*

Proof: We may assume $X^0 = \{x_0\}$. Then $\tilde{E}^*(X) = E^*(X, \{x_0\}) = F^1 E^*(X)$. Thus for $a \in \tilde{E}^*(X)$ we have $a^{n+1} \in F^{n+1} E^*(X) = \ker[E^*(X) \to E^*(X^n)] = \ker[E^*(X) \to E^*(X)] = 0$. □

13.69. So far we have only talked about a ring structure on $E^*(X)$, X a space. The reader may have asked himself why we did not simultaneously define a ring structure on $E_*(X)$, but then he will have noticed that the diagonal $\Delta: X \to X \times X$ goes in the wrong direction for homology. What we really need is a map $X \times X \to X$. Thus we see that although we cannot define a ring structure on $E_*(X)$ for general X, we can if X is an H-group. Specifically, if $\mu: X \times X \to X$ is an H-group multiplication on X and E is a ring spectrum, then we have a *Pontrjagin product*

$$E_n(X) \otimes E_m(X) \xrightarrow{\times} E_{n+m}(X \times X) \xrightarrow{\mu_*} E_{n+m}(X)$$

which makes $E_*(X)$ into a graded ring—in fact, into a graded algebra over $\pi_*(E)$. For example, for the classical Lie groups $G = O$, SO, $Spin$, U, SU, Sp the classifying space BG has an H-group structure inducing the Whitney sum in $k_G = [-;BG]$, so $E_*(BG)$ is a Pontrjagin ring. We shall discuss these Pontrjagin rings further in Chapter 16.

Similarly for any ring spectrum F with product $\mu: F \wedge F \to F$ we have a Pontrjagin product

$$E_n(F) \otimes E_m(F) \xrightarrow{\wedge} E_{n+m}(F \wedge F) \xrightarrow{\mu_*} E_{n+m}(F).$$

13.70. *Remark.* One can define the abstract notion of homology or cohomology theory *with products*. For example if h_* is a homology theory on $\mathcal{SP}'$, then a product structure for h_* is a natural transformation

$$\wedge: h_n(X) \otimes h_m(Y) \to h_{n+m}(X \wedge Y)$$

for $n, m \in \mathbb{Z}$, $X, Y \in \mathcal{SP}$ which satisfies 13.54i), 13.55i) and ii) and an element $1 \in h_0(S^0)$ such that

$$h_0(S^0) \otimes h_n(X) \xrightarrow{\wedge} h_n(S^0 \wedge X) \cong h_n(X)$$

maps $1 \otimes x$ to x for all $x \in h_n(X)$. The definition for a cohomology theory h^* is analogous and there one can define a cup product on $h^*(X)$ for all spaces X.

13.71. Exercise. Suppose h^* is a cohomology theory on $\mathcal{SP}'$ with products and $T: E^* \to h^*$ is a natural equivalence on $\mathcal{SP}'$. Show that E has a product $\mu: E \wedge E \to E$ and identity $\iota: S^0 \to E$ which induce the given product structure on h^*.

The reader has probably asked himself already whether the universal coefficient and Künneth theorems can also be carried over to E_* and E^*. The answer is that they can, but in general they involve a spectral sequence. The interested reader should first read Chapter 15 and then look at the appropriate parts of [6] and [8]. We shall consider only a special case in which the Künneth theorem reduces to an isomorphism; this is the most important case for applications anyway.

13.72. Suppose R is a ring with 1, M a right R-module, N a left R-module. Then we define the tensor product $M \otimes_R N$ as follows: in the tensor product $M \otimes N$ (M, N regarded as abelian groups) we take the subgroup K generated by all elements of the form $mr \otimes n - m \otimes rn$ for $m \in M$, $r \in R$, $n \in N$. Then we set

$$M \otimes_R N = M \otimes N / K.$$

Thus we have an exact sequence

$$M \otimes R \otimes N \xrightarrow{\alpha_M \otimes 1 - 1 \otimes \alpha_N} M \otimes N \to M \otimes_R N \to 0,$$

where $\alpha_M: M \otimes R \to M$, $\alpha_N: R \otimes N \to N$ are the action maps of M, N. Intuitively, we make the relation $mr \otimes n = m \otimes rn$ true in $M \otimes_R N$.

Exercise. Show that $M \otimes_R R \cong M$, $R \otimes_R N \cong N$.

Observe that the associativity relations for $\wedge$ and $\langle\, ,\rangle$ imply that $\wedge$ and $\langle\, ,\rangle$ induce homomorphisms

$$\wedge: E_n(X) \otimes_{\tilde{E}_*(S^0)} E_m(Y) \to E_{n+m}(X \wedge Y)$$
$$\wedge: E^n(X) \otimes_{\tilde{E}^*(S^0)} E^m(Y) \to E^{n+m}(X \wedge Y)$$
$$\langle\, ,\rangle: E^n(X) \otimes_{\pi_*(E)} E_m(X) \to \tilde{E}_{m-n}(S^0) = \pi_{m-n}(E).$$

We now give ourselves a sufficient condition that $\wedge$ and $\wedge$ be isomorphisms.

13.73. Definition. A left R-module N is *flat* if for all exact sequences

$$0 \to A \to B \to C \to 0$$

of right R-modules

$$0 \to A \otimes_R N \to B \otimes_R N \to C \otimes_R N \to 0$$

is also exact. There is an analogous definition for right R-modules.

13.74. Exercise. Show that every free R-module is flat.

13.75. Theorem. *Suppose E is a ring spectrum and X, Y are spectra.* i) *If either $E_*(X)$ is a flat right $\tilde{E}_*(S^0)$-module or $E_*(Y)$ is a flat left $\tilde{E}_*(S^0)$-module, then*

$$\wedge: E_*(X) \otimes_{\tilde{E}_*(S^0)} E_*(Y) \to E_*(X \wedge Y)$$

is an isomorphism.

ii) *If either $E^*(X)$ is a finitely-generated free right $\tilde{E}^*(S^0)$-module or $E^*(Y)$ is a finitely-generated free left $\tilde{E}^*(S^0)$-module, then*

$$\wedge: E^*(X) \otimes_{\tilde{E}^*(S^0)} E^*(Y) \to E^*(X \wedge Y)$$

is an isomorphism.

[In Chapter 17 we shall need to know that

$$\wedge: \pi_*(X \wedge E) \otimes_{\pi_*(E)} \pi_*(E \wedge Y) \to \pi_*(X \wedge E \wedge Y)$$

is an isomorphism if $\pi_*(X \wedge E)$ is a flat right $\pi_*(E)$-module. The proof is the same as for i).]

Proof: i) We assume $E_*(Y)$ is a flat left $\tilde{E}_*(S^0)$-module; the argument is similar if $E_*(X)$ is a flat right $\tilde{E}_*(S^0)$-module. We then regard Y as fixed and X as variable. We have two functors

$$F_1 = E_*(-) \otimes_{\tilde{E}_*(S^0)} E_*(Y) \quad \text{and} \quad F_2 = E_*(- \wedge Y)$$

from $\mathscr{SP}'$ to the category of graded abelian groups and both satisfy the axioms of a homology theory on $\mathscr{SP}'$: F_1 is exact on cofibrations because $E_*(Y)$ is flat and F_2 because if $A \to B \to C$ is a cofibration, then so is $A \wedge Y \to B \wedge Y \to C \wedge Y$ (13.47). We have the natural equivalences of degree 1

$$E_*(X) \otimes_{\tilde{E}_*(S^0)} E_*(Y) \xrightarrow{\sigma \otimes 1} E_*(\Sigma X) \otimes_{\tilde{E}_*(S^0)} E_*(Y)$$

and

$$E_*(X \wedge Y) \xrightarrow{\sigma} E_*(\Sigma(X \wedge Y)) \cong E_*(\Sigma X \wedge Y).$$

F_1 satisfies the wedge axiom because $\otimes_R$ commutes with direct sum (and E_* satisfies the wedge axiom); and F_2 satisfies the wedge axiom because $(\bigvee_\alpha X_\alpha) \wedge Y \simeq \bigvee_\alpha (X_\alpha \wedge Y)$ (13.48). Now $\wedge$ is a natural transformation $F_1 \to F_2$ of homology theories, and clearly $\wedge : F_1(S^0) \to F_2(S^0)$ is an isomorphism. Thus by 7.55 $\wedge$ is a natural equivalence.

ii) The argument goes as above, except that we cannot say that $\otimes_R$ commutes with direct product in general. Therefore we assume that $E^*(Y)$ (or $E^*(X)$) is finitely-generated free (i.e. a finite direct sum of copies of $R = \pi_*(E)$). Then the wedge axiom is satisfied. □

13.76. Example. Suppose F is a field; then $H_*(X;F)$ is always flat as a module over $\tilde{H}_*(S^0;F) \cong F$, so we always have an isomorphism

$$\tilde{H}_*(X;F) \otimes_F \tilde{H}_*(Y;F) \xrightarrow{\wedge} \tilde{H}_*(X \wedge Y;F)$$

or for relative groups

$$H_*(X,A;F) \otimes_F H_*(Y,B;F) \xrightarrow{\times} H_*((X,A) \times (Y,B);F).$$

For cohomology we must also assume that $H^*(X,A;F)$ or $H^*(Y,B;F)$ is a finite-dimensional graded F-vector space (it actually suffices to have $H^n(X,A;F)$ finite-dimensional for all $n \in \mathbb{Z}$).

Now $\langle\, , \rangle : E^n(X) \otimes_{\pi_*(E)} E_m(X) \to \pi_{m-n}(E)$ induces a homomorphism $\kappa : E^n(X) \to \operatorname{Hom}^{-n}_{\pi_*(E)}(E_*(X), \pi_*(E))$, where for graded R-modules A_*, B_* $\operatorname{Hom}^r_R(A_*, B_*)$ means the group of R-homomorphisms $\phi : A_* \to B_*$ of degree r ($\phi(A_q) \subset B_{q+r}$ for all $q \in \mathbb{Z}$). Explicitly, $[\kappa(x)](y) = \langle x, y \rangle$ for all $x \in E^n(X)$, $y \in E_m(X)$. We can also ask when κ is an isomorphism. From

Adams' universal coefficient theorem [8] it follows that κ is an isomorphism if $E_*(X)$ is a free $\pi_*(E)$-module. For $X = S^0$, for example, the result is evident, and if $E_*(X) \cong E_*(S^0)$ by an isomorphism which in some sense is natural (but possibly of degree other than 0) then the result will hold for X also. In Chapter 16 we shall see further examples where

$$E^*(X) \to \operatorname{Hom}^*_{\pi_*(E)}(E_*(X), \pi_*(E))$$

is an isomorphism.

13.77. Exercise. Let R be a ring, M a right R-module and A an abelian group. Show that there is a natural isomorphism

$$\theta : \operatorname{Hom}_R(A \otimes R, M) \to \operatorname{Hom}(A, M)$$

given by $\theta(\phi) = \phi'$ with $\phi'(a) = \phi(a \otimes 1)$ for $\phi \in \operatorname{Hom}_R(A \otimes R, M)$, $a \in A$. Thus for a field F and a topological pair (X, A) we have

$$\begin{aligned} S^n(X, A; F) &= \operatorname{Hom}(S_n(X, A), F) \cong \operatorname{Hom}_F(S_n(X, A) \otimes F, F) \\ &= (S_n(X, A) \otimes F)^* \\ &= \text{the vector space dual of } S_n(X, A) \otimes F. \end{aligned}$$

The boundary homomorphism $d : S^n(X, A; F) \to S^{n+1}(X, A; F)$ is the dual of $d \otimes 1 : S_{n+1}(X, A) \otimes F \to S_n(X, A) \otimes F$. It follows that $H^n(X, A; F) \cong H_n(X, A; F)^*$—and indeed this isomorphism is precisely the

$$\kappa : H^*(X, A; F) \to \operatorname{Hom}_F(H_*(X, A; F), F)$$

above.

In 7.39 we defined a Hurewicz homomorphism $h : \pi_n(Y, y_0) \to k_n(Y)$ for all reduced homology theories k_* with a distinguished element $\iota_0 \in k_0(S^0)$. In particular if E is a ring spectrum with identity $\iota : S^0 \to E$, then we have the element

$$\iota_0 = [\iota] \in \pi_0(E) = \tilde{E}_0(S^0);$$

in this case the Hurewicz homomorphism is just

$$\pi_n(X, x_0) \longrightarrow \pi_n^S(X) \xrightarrow{(\iota \wedge 1)_*} \pi_n(E \wedge X) = E_n(X).$$

More generally, if F is any other spectrum, then we can define a Hurewicz homomorphism

$$F_n(X) = \pi_n(F \wedge X) \cong \pi_n(S^0 \wedge F \wedge X) \xrightarrow{(\iota \wedge 1 \wedge 1)_*} \pi_n(E \wedge F \wedge X) = (E \wedge F)_n(X).$$

We shall denote this homomorphism by H. If in addition $E_*(F)$ is flat as a $\pi_*(E)$-module, then we have the isomorphism

$$\pi_*(E\wedge F)\otimes_{\pi_*(E)}\pi_*(E\wedge X)\to\pi_*(E\wedge F\wedge X)$$

of 13.75, so we can regard H as a homomorphism

$$F_*(X)\xrightarrow{H}E_*(F)\otimes_{\tilde{E}_*(S^0)}E_*(X).$$

Dual to H (cf. 13.76) is the *Boardman homomorphism B*:

$$F^n(X)=[X,\Sigma^nF]=[X,\Sigma^n(S^0\wedge F)]\xrightarrow{\Sigma^n(\iota\wedge 1)_*}[X,\Sigma^n(E\wedge F)]$$
$$=(E\wedge F)^n(X).$$

We have a commutative diagram

$$\begin{array}{ccc} & \xrightarrow{B} & (E\wedge F)^n(X)\\ F^n(X) & & \downarrow\tau\\ & \xrightarrow{\alpha} & \operatorname{Hom}^{-n}_{\tilde{E}_*(S^0)}(E_*(X),E_*(F))\end{array}$$

if we define α by $\alpha[f]=\sigma^nf_*:E_*(X)\to E_*(F)$ for $f:X\to\Sigma^nF$ and τ by

$$(\tau[g])[h]=[S^{q-n}\xrightarrow{\Sigma^{-n}h}\Sigma^{-n}E\wedge X\xrightarrow{1\wedge g}\Sigma^{-n}E\wedge\Sigma^nE\wedge F\xrightarrow{\mu\wedge 1}E\wedge F]$$

for $h:S^q\to E\wedge X, g:X\to\Sigma^n(E\wedge F)$. If $E_*(X)$ is a free $\tilde{E}_*(S^0)$-module with only finitely many generators in each dimension, then we have an isomorphism

$$\operatorname{Hom}^*_{\tilde{E}_*(S^0)}(E_*(X),E_*(F))\cong(E_*(X))^*\hat{\otimes}_{\tilde{E}_*(S^0)}E_*(F)$$

of $\operatorname{Hom}^*_{\tilde{E}_*(S^0)}(E_*(X),E_*(F))$ with the completed tensor product of the dual $(E_*(X))^*=\operatorname{Hom}^*_{\tilde{E}_*(S^0)}(E_*(X),\tilde{E}_*(S^0))$ and $E_*(F)$ ($A\hat{\otimes}B$ means in effect we allow infinite sums $\sum_i a_i\otimes b_i$). The sense in which H and B are dual is the following.

13.78. Theorem. *For any ring spectra E, F and spectrum X the following diagram commutes:*

$$\begin{array}{ccc} F^n(X)\otimes F_m(X) & \xrightarrow{\langle\,,\,\rangle} & \pi_{m-n}(F)\\ \downarrow B\otimes H & & \downarrow H\\ (E\wedge F)^n(X)\otimes(E\wedge F)_m(X) & \xrightarrow{\langle\,,\,\rangle} & \pi_{m-n}(E\wedge F).\end{array}$$

If we replace B by α and if $E_*(F)$ is flat, then the diagram looks as follows:

$$\begin{array}{ccc} F^n(X) \otimes F_m(X) & \xrightarrow{\langle\,,\,\rangle} & \pi_{m-n}(F) \\ \Big\downarrow \alpha \otimes H & & \Big\downarrow H \\ \mathrm{Hom}^{-n}_{\tilde{E}_*(S^0)}(E_*(X), E_*(F)) \otimes E_*(F) \otimes_{\tilde{E}_*(S^0)} E_*(X) & \xrightarrow{\kappa} & E_{m-n}(F) \end{array}$$

where κ is defined by $\kappa(\phi \otimes f \otimes x) = \phi(x)\cdot f$ for $f \in E_q(F)$, $x \in E_{m-q}(X)$ and $\phi \in \mathrm{Hom}^n_{\tilde{E}_*(S^0)}(E_*(X), E_*(F))$ (here $\cdot$ means Pontrjagin product in $E_*(F)$).

Proof: B, H are both natural transformations of theories induced by maps of representing spectra (namely $F \xrightarrow{\iota\wedge 1} E \wedge F$), and $\langle\,,\,\rangle$ is natural, so the commutativity of the first diagram follows. One then need only check that $\kappa \circ (\tau \otimes (\wedge)^{-1}) = \langle\,,\,\rangle$. □

We shall have applications of the Hurewicz and Boardman maps in Chapters 17 and 19.

We now conclude this chapter by constructing explicit products $\mu: E \wedge E \to E$ for $E = H(R)$, R a ring, for $E = K, KO$ and for $E = MG$, G one of the classical Lie groups.

13.79. We have already made a beginning with $H(R)$ for R a ring. Let $g_n = [1] \in [H(R,n),*; H(R,n),*] \cong \tilde{H}^n(H(R,n); R)$. Then we have the element

$$g_n \wedge g_m \in \tilde{H}^{n+m}(H(R,n) \wedge H(R,m); R)$$

and it is represented by a map

$$\bar{\mu}_{nm}: H(R,n) \wedge H(R,m) \to H(R,n+m).$$

From these maps, which obviously satisfy appropriate homotopy associativity and commutativity relations, we want to construct a map $\mu: H(R) \wedge H(R) \to H(R)$. The next proposition provides us with the general method.

13.80. Proposition. *Let E be a spectrum and $\bar{\iota}_n: S^n \to E_n$, $\bar{\mu}_{nm}: E_n \wedge E_m \to E_{n+m}$ be a system of maps such that the following diagrams commute up to homotopy for all m, n, p:*

13.81.

$$\begin{array}{ccc} S^1 \wedge S^n & \xrightarrow{1\wedge \bar{\iota}_n} & S^1 \wedge E_n \\ \wr\wr & & \Big\downarrow \varepsilon_{1,n} \\ S^{n+1} & \xrightarrow{\bar{\iota}_{n+1}} & E_{n+1} \end{array}$$

13.82.

$$\begin{array}{ccccc} S^n \wedge E_m & \xrightarrow{\bar{\imath}_n \wedge 1} & E_n \wedge E_m & \xleftarrow{1 \wedge \bar{\imath}_m} & E_n \wedge S^m \\ & \searrow^{\varepsilon_{nm}} & \downarrow \bar{\mu}_{nm} & & \uparrow \tau' \\ & & E_{n+m} & \xleftarrow{\varepsilon_{mn}} & S^m \wedge E_n \end{array}$$

$$\tau'[t,x] = [x, \nu'^{nm}(t)]$$

13.83.

$$\begin{array}{ccc} E_n \wedge E_m \wedge E_p & \xrightarrow{1 \wedge \bar{\mu}_{mp}} & E_n \wedge E_{m+p} \\ \downarrow \bar{\mu}_{nm} \wedge 1 & & \downarrow \bar{\mu}_{n,m+p} \\ E_{n+m} \wedge E_p & \xrightarrow{\bar{\mu}_{n+m,p}} & E_{n+m+p} \end{array}$$

where $\varepsilon_{nm}: S^n \wedge E_m \to E_{n+m}$ is the inclusion. Suppose further that

$$\lim{}^1 \tilde{E}^{n-1}(E_n) = \lim{}^1 \tilde{E}^{\alpha(a)-1}(E_{\beta(a)} \wedge E_{\gamma(a)}) =$$

$$\lim{}^1 \tilde{E}^{\alpha(a)-1}(E_{\beta(a)} \wedge E_{\gamma(a)} \wedge E_{\delta(a)}) = 0$$

for any partitions $A = B \cup C$ or $A = B \cup C \cup D$. Then there is a map $\iota: S^0 \to E$ and for every partition $A = B \cup C$ with B, C infinite a map $\mu_{BC}: E \wedge E \to E$ such that the homotopy class $[\mu] = [\mu_{BC}]$ is independent of B, C and such that the diagrams

13.84.

$$\begin{array}{ccc} S^n \subset \Sigma^n S^0 \\ \bar{\imath}_n \downarrow \quad \downarrow \Sigma^n \iota \\ E_n \subset \Sigma^n E \end{array} \qquad \begin{array}{ccc} E_{\beta(a)} \wedge E_{\gamma(a)} \subset \Sigma^{\alpha(a)}(E \wedge_{BC} E) & \xrightarrow{\Sigma^{\alpha(a)} eq_{BC}} & \Sigma^{\alpha(a)}(E \wedge E) \\ \downarrow \bar{\mu}_{\beta(a),\gamma(a)} & & \downarrow \Sigma^{\alpha(a)} \mu \\ E_{\alpha(a)} & \longrightarrow & \Sigma^{\alpha(a)} E \end{array}$$

commute (up to homotopy). (E, μ, ι) is a ring spectrum. If, moreover, the diagrams

13.85.

$$\begin{array}{ccc} E_{4n} \wedge E_{4m} & \xrightarrow{\tau} & E_{4m} \wedge E_{4n} \\ & \searrow^{\bar{\mu}_{4n,4m}} \quad \swarrow^{\bar{\mu}_{4m,4n}} & \\ & E_{4n+4m} & \end{array}$$

commute up to homotopy, then (E, μ, ι) is a commutative ring spectrum.

Proof: From 13.82 and 13.83 follows the homotopy commutativity of

$$\begin{array}{ccccc} & & SE_n \wedge E_m & \xrightarrow{\varepsilon_n \wedge 1} & E_{n+1} \wedge E_m \\ & \overset{\kappa}{\nearrow} & & & \downarrow \bar{\mu}_{n+1,\,m} \\ S(E_n \wedge E_m) & \xrightarrow{S\bar{\mu}_{nm}} & SE_{n+m} & \xrightarrow{\varepsilon_{n+m}} & E_{n+m+1} \\ & \underset{\lambda}{\searrow} & & & \uparrow \bar{\mu}_{n,\,m+1} \\ & & E_n \wedge SE_m & \xrightarrow{1 \wedge \varepsilon_m} & E_n \wedge E_{m+1} \end{array}$$

$$\kappa[t,[x,y]] = [[t,x],y], \quad \lambda[t,[x,y]] = [x,[v'^n(t),y]].$$

From 10.4 it follows there are functions $\iota : S^0 \to E$ and $\bar{\mu}_{BC} : E \wedge_{BC} E \to E$ with $\iota_n \simeq \bar{\iota}_n$ and $(\bar{\mu}_{BC})_{\alpha(a)} \simeq \bar{\mu}_{\beta(a),\gamma(a)}$ for $n \in \mathbb{N}$, $a \in A$. We define μ_{BC} to be the composite

$$E \wedge E \xrightarrow[eq_{BC}^{-1}]{} E \wedge_{BC} E \xrightarrow[\bar{\mu}_{BC}]{} E.$$

The diagrams 13.84 evidently commute if we take μ_{BC} for μ.

To prove the associativity of μ we would like to know that the diagram

13.86.

$$\begin{array}{ccc} (E \wedge_{BC} E) \wedge_{B\cup C,D} E & \xrightarrow{\bar{\mu}_{BC} \wedge 1} & E \wedge_{B \cup C,D} E \\ \| & & \\ E \wedge_{B,C\cup D} (E \wedge_{CD} E) & & \downarrow \bar{\mu}_{B\cup C,D} \\ 1 \wedge \bar{\mu}_{CD} \downarrow & & \\ E \wedge_{B,C\cup D} E & \xrightarrow{\bar{\mu}_{B,C\cup D}} & E \end{array}$$

commutes (up to homotopy) for all partitions $A = B \cup C \cup D$ with B, C, D infinite. But this follows directly from 10.4b) because of 13.83 and the assumption $\lim^1 \tilde{E}^{\alpha(a)-1}(E_{\beta(a)} \wedge E_{\gamma(a)} \wedge E_{\delta(a)}) = 0$.

We combine 13.86 with the associativity of eq_{BC} (13.44) and find that

$$\begin{array}{ccc} (E \wedge E) \wedge E & \xrightarrow{\mu_{BC} \wedge 1} & E \wedge E \\ \downarrow a & & \\ E \wedge (E \wedge E) & & \downarrow \mu_{B\cup C,D} \\ \downarrow 1 \wedge \mu_{CD} & & \\ E \wedge E & \xrightarrow{\mu_{B,C\cup D}} & E \end{array}$$

commutes for all partitions $A = B \cup C \cup D$ with B, C, D infinite.

The inclusions $\varepsilon_{nm}: S^n \wedge E_m \to E_{n+m}$ induce a map $\varepsilon_{BC}: S^0 \wedge_{BC} E \to E$ such that $(\varepsilon_{BC})_{\alpha(a)} \simeq \varepsilon_{\beta(a),\gamma(a)}: S^{\beta(a)} \wedge E_{\gamma(a)} \to E_{\alpha(a)}$ for all $a \in A$ and all partitions $A = B \cup C$. One checks that

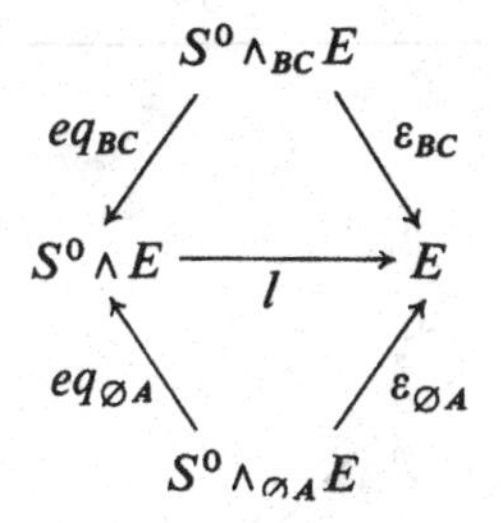

commutes. It follows that

13.87.

$$\begin{array}{ccc} S^0 \wedge_{BC} E & \xrightarrow{\iota \wedge 1} & E \wedge_{BC} E \\ {\scriptstyle eq_{BC}} \downarrow & \searrow^{\varepsilon_{BC}} & \downarrow {\scriptstyle \bar{\mu}_{BC}} \\ S^0 \wedge E & \xrightarrow{l} & E \end{array}$$

commutes (the upper triangle commutes by 13.82). We thus have that

$$\begin{array}{ccc} S^0 \wedge E & \xrightarrow{\iota \wedge 1} & E \wedge E \\ & {}_{l} \searrow & \downarrow {\scriptstyle \mu_{BC}} \\ & & E \end{array}$$

commutes for all partitions $A = B \cup C$ with B, C infinite. The commutativity of

$$\begin{array}{ccc} E \wedge E & \xleftarrow{1 \wedge \iota} & E \wedge S^0 \\ {\scriptstyle \mu_{BC}} \downarrow & \swarrow_{r} & \\ E & & \end{array}$$

follows analogously.

Our proof is complete once we show that $[\mu_{BC}]$ is independent of B, C. To this end we regard the awesome diagram on the following page (in which B and D are taken to be infinite) in which ① commutes by 13.40b)v), ② and ③ by 13.87 and its analog, ④, ⑤, ⑥ and ⑦ by the naturality of $eq_{B,C \cup D}$ and $eq_{B \cup C, D}$, ⑧ by 13.44, ⑨ trivially and ⑩ by 13.86. Thus if B and D are infinite, we see that $\mu_{B \cup C, D} \simeq \mu_{B, C \cup D}$.

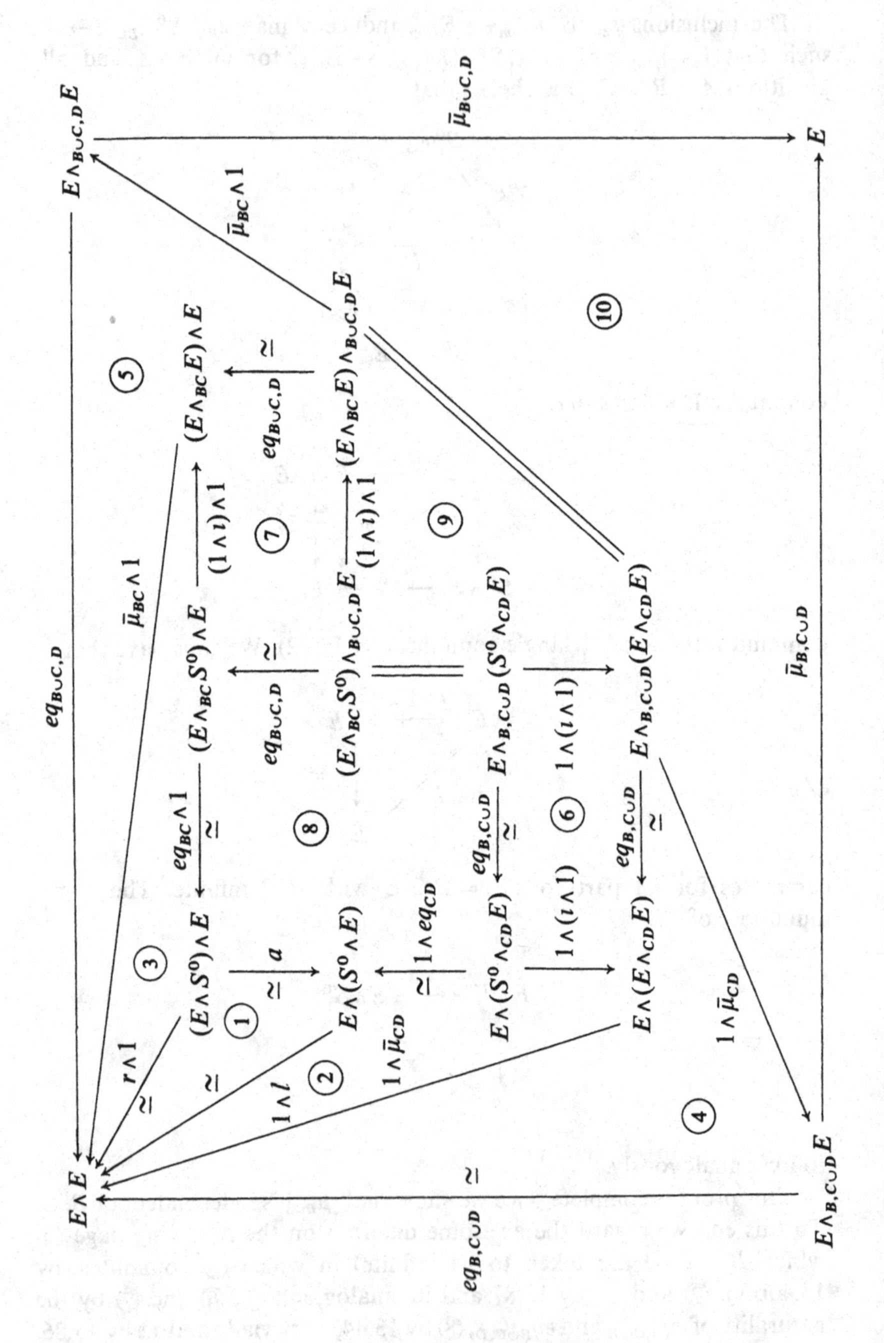
E∧E
E∧B∪C,D E
E
E∧B,C∪D E
(E∧S0)∧E
(E∧BC S0)∧E
(E∧BC E)∧E
E∧(S0∧E)
(E∧BC S0)∧B∪C,D E
(E∧BC E)∧B∪C,D E
E∧(S0∧CD E)
E∧B,C∪D (S0∧CD E)
E∧(E∧CD E)
E∧B,C∪D (E∧CD E)
eqB∪C,D
μ̄B∪C,D
μ̄BC∧1
r∧1
1∧l
1∧μ̄CD
a
eqBC∧1
(1∧ι)∧1
1∧eqCD
eqB,C∪D
1∧(ι∧1)
μ̄B,C∪D
≃
1
2
3
4
5
6
7
8
9
10

Now suppose $A = B_1 \cup C_1 = B_2 \cup C_2$ are two partitions with B_1, B_2, C_1, C_2 infinite. Then $B_1 = (B_1 \cap B_2) \cup (B_1 \cap C_2)$ and $B_1 \cap B_2$ or $B_1 \cap C_2$ must be infinite—say $B_1 \cap B_2$. Similarly $C_1 = (C_1 \cap B_2) \cup (C_1 \cap C_2)$ and one of the latter—say $C_1 \cap C_2$—is infinite. Then we have

$$\begin{aligned}\mu_{B_1,C_1} &= \mu_{(B_1\cap B_2)\cup(B_1\cap C_2),C_1}\\ &\simeq \mu_{B_1\cap B_2,(B_1\cap C_2)\cup(C_1\cap B_2)\cup(C_1\cap C_2)}\\ &\simeq \mu_{(B_1\cap B_2)\cup(C_1\cap B_2),(B_1\cap C_2)\cup(C_1\cap C_2)}\\ &= \mu_{B_2,C_2}.\end{aligned}$$

The other three cases are similar. Thus we have shown $[\mu_{BC}]$ does not depend on B, C. □

Thus to give a spectrum E a ring spectrum structure we only need to construct maps $\bar{\iota}_n: S^n \to E_n$, $\bar{\mu}_{nm}: E_n \wedge E_m \to E_{n+m}$ satisfying 13.81–13.83 and verify the conditions on $\lim^1$. This we now do for $E = H(R)$, R a ring, and $E = K, KO, MG$, G one of the classical Lie groups.

13.88. In 13.79 we constructed

$$\bar{\mu}_{nm}: H(R,n) \wedge H(R,m) \to H(R, n+m)$$

such that $\bar{\mu}^*_{nm}(g_{n+m}) = g_n \wedge g_m$. From the relation $(g_n \wedge g_m) \wedge g_p = g_n \wedge (g_m \wedge g_p)$ (which follows from 13.37) it follows that the $\bar{\mu}_{nm}$ satisfy 13.83. For each n we have our canonical generator $\iota_n \in \tilde{H}^n(S^n; R)$ $(\iota_n = \sigma^{-n}(\iota_0))$, which is represented by a map $\bar{\iota}_n: S^n \to H(R,n)$. The maps $\bar{\iota}_n$ satisfy 13.81 because $\iota_{n+1} = \sigma^{-1}(\iota_n)$. For the $\times$- and $\wedge$-products in singular cohomology which we constructed following 13.36 we can directly verify 13.56iii) and iv) and hence also 13.55iii) and iv). Thus we have

$$\begin{aligned}\varepsilon^*_{nm}(g_{n+m}) = \sigma^{-n}(g_m) = \sigma^{-n}(\iota_0 \wedge g_m) = \sigma^{-n}(\iota_0) \wedge g_m = \iota_n \wedge g_m =\\ (\bar{\iota}_n \wedge 1)^* \bar{\mu}^*_{nm}(g_{n+m}) = [\bar{\mu}_{nm} \circ (\bar{\iota}_n \wedge 1)]^*(g_{n+m}),\end{aligned}$$

from which it follows $\varepsilon_{nm} \simeq \bar{\mu}_{nm} \circ (\bar{\iota}_n \wedge 1)$. Similarly $\varepsilon_{mn} \simeq \bar{\mu}_{nm} \circ (1 \wedge \bar{\iota}_m) \circ \tau'$. Thus $\bar{\mu}_{nm}$, $\bar{\iota}_n$ also satisfy 13.82. From the relation $g_n \wedge g_m = (-1)^{nm} g_m \wedge g_n$ it follows the diagram 13.85 also commutes up to homotopy if R is commutative. The conditions on $\lim^1$ are trivially satisfied in this case;

$$H(R,n) \wedge H(R,m) \wedge H(R,p)$$

is for example $(n+m+p-1)$-connected. Thus we have a ring spectrum $(H(R), \mu, \iota)$ which is commutative if R is commutative and whose $\times$-product agrees with the $\times$-product constructed algebraically following 13.36.

13.89. We consider MG next as the product is particularly easy to construct in this case. In Chapter 11 we had maps

$$\omega_{nm}: BG(n) \times BG(m) \to BG(n+m)$$

such that if ξ_n is the universal $G(n)$-bundle over $BG(n)$, then $\omega_{nm}^*\xi_{n+m} \simeq \xi_n \times \xi_m$; the construction was made for $G = O$, U or Sp but goes equally well for $G = SO$, SU or $Spin$. The induced maps of Thom complexes then have the form

$$\bar{\mu}_{nm} = M(\omega_{nm}) \colon MG(n) \wedge MG(m) \cong M(\xi_n \times \xi_m) \to M(\xi_{n+m}) = MG(n+m).$$

The inclusion $* \to BG(n)$ induces $\bar{\imath}_n : S^n = M(\varepsilon^n) \to MG(n)$. 13.81 follows from the relation $\varepsilon^1 \times \varepsilon^n \simeq \varepsilon^{n+1}$; 13.82, 13.83 and 13.85 follow from 11.53, 11.48 and 11.51 respectively. By the remark following 10.4 one can see that the conditions on $\lim^1$ are satisfied if one can show that $H^{n+q-1}(E_n^{2n}; \mathbb{Q}) \otimes H_q(E; \mathbb{Q}) = 0$ for all q and infinitely many n. In our case this involves showing that $H^{\varepsilon n+q-1}(MG(n); \mathbb{Q})$, $H^{\varepsilon(n+m)+q-1}(MG(n) \wedge MG(m); \mathbb{Q})$, $H^{\varepsilon(n+m+p)+q-1}(MG(n) \wedge MG(m) \wedge MG(p); \mathbb{Q})$ are 0 whenever $H_q(MG; \mathbb{Q}) \neq 0$ for all q and enough n, m, p, where $\varepsilon = 1$, 2, or 4 according to the case. This will follow from the results of Chapter 16 (without using the fact that MG is a ring spectrum!) for G one of the classical groups.

The $\times$-product $MG_*(X) \otimes MG_*(Y) \to MG_*(X \times Y)$ for spaces X, Y has a geometric description in terms of singular manifolds. If $f : M^m \to X$ and $g : N^n \to Y$ are singular G-manifolds, then $M^m \times N^n$ has a natural $G(n+m)$-structure and $f \times g : M^m \times N^n \to X \times Y$ is a singular G-manifold. Then $[M^m, f] \times [N^n, g] = [M^m \times N^n, f \times g] \in MG_{n+m}(X \times Y)$. In particular in Ω_*^G we have $[M^m] \cdot [N^n] = [M^m \times N^n]$, as we see from the Pontrjagin–Thom construction. We embed M^m in $\mathbb{R}^{m+k}$ with tubular neighborhood $N(M)$, N^n in $\mathbb{R}^{n+l}$ with tubular neighborhood $N(N)$. If we regard S^q as $\mathbb{R}^q$ with a point at ∞, then $S^{m+n+k+l} = (\mathbb{R}^{m+k} \times \mathbb{R}^{n+l}) \cup \infty$ and the projection $\pi_{M \times N} : S^{n+m+k+l} \to M(\nu_M \times \nu_N) \cong M(\nu_M) \wedge M(\nu_N)$ is the same as

$$\pi_M \wedge \pi_N : S^{m+k} \wedge S^{n+l} \to M(\nu_M) \wedge M(\nu_N).$$

Hence $[M^m \times N^n]$ is represented by

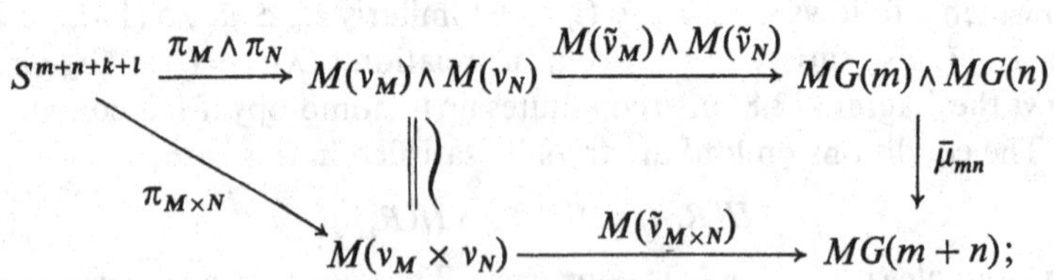

$$\begin{array}{ccccc}
S^{m+n+k+l} & \xrightarrow{\pi_M \wedge \pi_N} & M(\nu_M) \wedge M(\nu_N) & \xrightarrow{M(\tilde{\nu}_M) \wedge M(\tilde{\nu}_N)} & MG(m) \wedge MG(n) \\
& \searrow^{\pi_{M \times N}} & \| \wr & & \downarrow \bar{\mu}_{mn} \\
& & M(\nu_M \times \nu_N) & \xrightarrow{M(\tilde{\nu}_{M \times N})} & MG(m+n);
\end{array}$$

that is $[M^m \times N^n] = [M^m] \cdot [N^n]$.

The graded rings Ω_*^G are described for $G = O$, SO and U in Chapter 12 and computed in Chapter 20.

13.90. Now we construct products for K-theory. K-theory is defined in terms of vector bundles, and we shall construct our product on K-theory using the tensor product of vector bundles.

$GL(n,F)$ can be identified with $\mathrm{Aut}(F^n)$, the group of linear automorphisms of F^n, once we have fixed a basis in F^n—say the standard basis $e_1, e_2, \ldots, e_n$. We define the tensor product homomorphism

$$t_{nm}: GL(n,F) \times GL(m,F) \rightarrow GL(nm,F)$$

by identifying $F^n \otimes F^m$ with F^{nm} by the unique isomorphism which sends $e_i \otimes e_j$ to $e_{(i-1)m+j}$ for $1 \leqslant i \leqslant n$, $1 \leqslant j \leqslant m$, and then taking $t_{nm}(A,B)$ to be the automorphism $A \otimes B: F^n \otimes F^m \rightarrow F^n \otimes F^m$ for $A \in GL(n,F)$, $B \in GL(m,F)$. For $n = m = 2$, for example, we have

$$t_{2,2}\left(\begin{pmatrix} a & c \\ b & d \end{pmatrix}, \begin{pmatrix} a' & c' \\ b' & d' \end{pmatrix}\right) = \begin{pmatrix} aa' & ac' & ca' & cc' \\ ab' & ad' & cb' & cd' \\ ba' & bc' & da' & dc' \\ bb' & bd' & db' & dd' \end{pmatrix}.$$

One readily verifies that $t_{nm}(O(n,F) \times O(m,F)) \subset O(nm,F)$.

Now let F be either $\mathbb{R}$ or $\mathbb{C}$. Given two F-vector bundles $\xi \rightarrow X$, $\eta \rightarrow Y$ of dimensions n,m respectively, we could construct an explicit vector bundle $\xi \otimes_F \eta \rightarrow X \times Y$ of dimension nm with fibre $\xi_x \otimes_F \eta_y$ over (x,y) as follows: we take the largest atlases $\{(U_\alpha, h_\alpha): \alpha \in A\}$, $\{(V_\gamma, h'_\gamma): \gamma \in \Gamma\}$ for ξ, η with corresponding systems

$$\{U_\alpha, g_{\alpha\beta}: \alpha, \beta \in A\}, \quad \{V_\gamma, g'_{\gamma\delta}; \gamma, \delta \in \Gamma\}$$

of transition functions. Then

$$\{U_\alpha \times V_\gamma, t_{nm} \circ (g_{\alpha\beta} \times g'_{\gamma\delta}): \alpha, \beta \in A, \gamma, \delta \in \Gamma\}$$

is a system of transition functions for $GL(nm,F)$ on $X \times Y$. We then carry out the explicit glueing-together procedure of 11.15 and get an nm-dimensional bundle over $X \times Y$, which we call $\xi \otimes_F \eta$.

We can give a more elegant description of the isomorphism class of $\xi \otimes_F \eta$ as follows: we write ξ, η in the form $\xi = \xi'[F^n]$, $\eta = \eta'[F^m]$, where ξ' is a principal $GL(n,F)$-bundle and η' is a principal $GL(m,F)$-bundle (only the isomorphism classes of ξ', η' are determined). Then $(t_{nm})_*(\xi' \times \eta')$ is a $GL(nm,F)$-bundle and we have

$$\xi \otimes_F \eta \simeq ((t_{nm})_*(\xi' \times \eta'))[F^n \otimes_F F^m].$$

From this formulation the following properties of $\otimes_F$ are immediate.

13.91. Proposition.

i) $\otimes_F$ *is a functor of two variables.*

ii) $(\xi \otimes_F \eta) \otimes_F \zeta \simeq \xi \otimes_F (\eta \otimes_F \zeta)$.

iii) *If* $\tau: X \times Y \rightarrow Y \times X$ *is the switch map, then* $\tau^*(\eta \otimes_F \xi) \simeq \xi \otimes_F \eta$.

iv) $\xi \otimes_F (\eta \times \zeta) \simeq (\xi \otimes_F \eta) \times (\xi \otimes_F \zeta)$.

v) $(f \times g)^*(\xi \otimes_F \eta) \simeq f^*\xi \otimes_F g^*\eta$ *for maps* $f: X' \rightarrow X$, $g: Y' \rightarrow Y$.

Naturally given two F-vector bundles ξ, η over the same space X we can apply the diagonal map Δ to get an internal tensor product $\xi \otimes_F \eta$ with $(\xi \otimes_F \eta)_x = \xi_x \otimes_F \eta_x$ for all $x \in X$. The properties of this internal product corresponding to 13.91 are easily deduced.

Remark. The tensor product $V \otimes_H W$ of two quaternionic vector spaces is not a quaternionic vector space but only a real one. We can then define the tensor product $\xi \otimes_H \eta$ of two H-vector bundles; it will be a real vector bundle. Or if V is a real vector space and W is a quaternionic vector space, then $V \otimes_R W$ is a quaternionic vector space. The analogous statement holds for vector bundles.

Now recall that for finite CW-complexes X the group $K(X)$ was constructed by taking the abelian semigroup $V_{\mathbb{C}}(X)$ of isomorphism classes of complex vector bundles over X and completing it to an abelian group $K(X)$ with mapping $\phi: V_{\mathbb{C}}(X) \to K(X)$. Thus $K(X)$ consists of equivalence classes of pairs $(\{\xi\}, \{\eta\})$, ξ, η complex vector bundles over X. One can readily show that for any three semigroups S, S', S'' a bilinear map $\theta: S \times S' \to S''$ induces a unique bilinear map $\theta': K(S) \times K(S') \to K(S'')$ such that

$$\begin{array}{ccc} S \times S' & \xrightarrow{\theta} & S'' \\ \downarrow{\scriptstyle \phi \times \phi} & & \downarrow{\scriptstyle \phi} \\ KS \times KS' & \xrightarrow{\theta'} & KS'' \end{array}$$

commutes. Therefore the map

$$V_{\mathbb{C}}(X) \times V_{\mathbb{C}}(Y) \xrightarrow{\theta} V_{\mathbb{C}}(X \times Y)$$

defined by $(\{\xi\}, \{\eta\}) \mapsto \{\xi \otimes_{\mathbb{C}} \eta\}$ induces a unique map

$$\times : K(X) \otimes K(Y) \to K(X \times Y).$$

Naturality, associativity and commutativity of $\times$ also follow from 13.91. From $\times$ we also get a product

$$\wedge : \tilde{K}(X) \otimes \tilde{K}(Y) \to \tilde{K}(X \wedge Y)$$

for pointed finite CW-complexes X, Y. $\wedge$ is natural, associative and commutative.

We wish to extend $\times$ and $\wedge$ to infinite complexes and spectra. Thus we need a product $K \wedge K \to K$ (or by 13.80 a map $\bar{\mu}: BU \wedge BU \to BU$ with appropriate properties) inducing $\wedge$ on finite complexes.

In Chapter 11 we defined BU to be $BU = \bigcup_{k \geq 1} BU(k)$, where $BU(k) = \bigcup_{n \geq k} G_{k,n}(\mathbb{C})$. In fact, one easily sees that $\bigcup_{n \geq 1} G_{n,2n}(\mathbb{C})$ is homeomorphic to $\bigcup_{k \geq 1} BU(k)$, so for our current purposes we take

$$BU = \bigcup_{n \geq 1} G_{n,2n}(\mathbb{C}).$$

Here we regard $\mathbb{C}^{2n+2}$ as $\mathbb{C} \oplus \mathbb{C}^{2n} \oplus \mathbb{C}$ and for any n-plane $W \in \mathbb{C}^{2n}$ we take the $(n+1)$-plane $W \oplus [e_{2n+2}]$, thus defining our "inclusion"

$$G_{n,2n} \subset G_{n+1,2n+2}.$$

Let ξ_n denote the universal and ε^n the trivial n-bundle over $G_{n,2n}$, which we now call G_n. We let $u_n = \xi_n - \varepsilon^n \in \tilde{K}(G_n)$; if $i_n : G_n \to G_{n+1}$ is the inclusion, then $i_n^* u_{n+1} = u_n$. Thus $\{u_n\}$ is an element of $\lim^0 \tilde{K}(G_n)$ and is in fact the class $u = [1_{BU}] \in [BU, BU] = \tilde{K}(BU)$. We wish to construct $u \wedge u \in \tilde{K}(BU \wedge BU)$; since $\tilde{K}(BU \wedge BU) \to \lim^0 \tilde{K}(G_n \wedge G_n)$ is surjective, there is certainly an element $u \wedge u \in \tilde{K}(BU \wedge BU)$ such that $u \wedge u | G_n \wedge G_n = u_n \wedge u_n$ for all $n \geq 1$. Of course we would like to know that $u \wedge u$ is unique; for this it will suffice to show that $\lim^1 \tilde{K}^{-1}(G_n \wedge G_n) = 0$, since we have the exact sequence 7.66.

$$0 \to \lim{}^1 \tilde{K}^{-1}(G_n \wedge G_n) \to \tilde{K}(BU \wedge BU) \to \lim{}^0 \tilde{K}(G_n \wedge G_n) \to 0.$$

But from the Atiyah–Hirzebruch spectral sequence which we construct in Chapter 15 it will follow easily that $\tilde{K}^{-1}(G_n \wedge G_n) = 0$ for all n, so that $\lim^1 \tilde{K}^{-1}(G_n \wedge G_n) = 0$ trivially.

Now for $x \in \tilde{K}(X)$, $y \in \tilde{K}(Y)$ represented by $f: X \to BU$, $g: Y \to BU$ respectively we define $x \wedge y$ by $x \wedge y = (f \wedge g)^*(u \wedge u)$. To prove associativity of $\wedge$, then, it suffices to show $(u \wedge u) \wedge u = u \wedge (u \wedge u)$. But the restriction of both to $G_n \wedge G_n \wedge G_n$ is $(u_n \wedge u_n) \wedge u_n = u_n \wedge (u_n \wedge u_n)$, so it follows $(u \wedge u) \wedge u = u \wedge (u \wedge u)$ provided $\lim^1 \tilde{K}^{-1}(G_n \wedge G_n \wedge G_n) = 0$. Again $\tilde{K}^{-1}(G_n \wedge G_n \wedge G_n) = 0$.

Let $\iota_2 : S^2 \to BU$ classify $\eta - 1 \in \tilde{K}(S^2)$, where $\eta \to S^2 \cong CP^1$ is the Hopf line bundle. The Bott periodicity isomorphism

$$\tilde{K}(X) \xrightarrow{\cong} \tilde{K}(S^2 \wedge X)$$

is given by $x \to (\eta - 1) \wedge x$ for $x \in \tilde{K}(X)$. That means that the structure map

$$S^2 \wedge BU \to BU$$

of our spectrum K is the composition

$$S^2 \wedge BU \xrightarrow{\iota_2 \wedge 1} BU \wedge BU \xrightarrow{\mu} BU$$

where μ represents $u \wedge u \in \tilde{K}(BU \wedge BU)$. Thus if we define

$$\mu_{2n,2m} : K_{2n} \wedge K_{2m} \to K_{2n+2m}$$

to be $\mu: BU \wedge BU \to BU$ and

$$\iota_{2n}: S^{2n} \to K_{2n}$$

to be

$$S^{2n} \cong S^2 \wedge S^{2n-2} \xrightarrow{\iota_2 \wedge \iota_{2n-2}} BU \wedge BU \xrightarrow{\mu} BU$$

for $n \geqslant 2$, then the diagrams 13.81–13.83, 13.85 commute up to homotopy and the conditions on $\lim^1$ are satisfied because of the remark following 10.4. Thus we get a commutative ring spectrum (K, μ, ι) such that the product $K(X) \otimes K(Y) \to K(X \times Y)$ is induced by the tensor product of bundles.

In analogous fashion tensor product of vector bundles provides us with products

$$\begin{aligned}
\wedge_1: \widetilde{KO}(X) \otimes \widetilde{KO}(Y) &\to \widetilde{KO}(X \wedge Y) \\
\wedge_2: \widetilde{KO}(X) \otimes \widetilde{KSp}(Y) &\to \widetilde{KSp}(X \wedge Y) \\
\wedge_3: \widetilde{KSp}(X) \otimes \widetilde{KO}(Y) &\to \widetilde{KSp}(X \wedge Y) \\
\wedge_4: \widetilde{KSp}(X) \otimes \widetilde{KSp}(Y) &\to \widetilde{KO}(X \wedge Y)
\end{aligned}$$

for finite X, Y and hence with maps

$$\begin{aligned}
\mu_1: BO \wedge BO &\to BO \\
\mu_2: BO \wedge BSp &\to BSp \\
\mu_3: BSp \wedge BO &\to BSp \\
\mu_4: BSp \wedge BSp &\to BO
\end{aligned}$$

from which we construct a product $\mu: KO \wedge KO \to KO$. Again the fact that $\lim^1 \widetilde{KO}^{-1}(G_{n,2n}(\mathbb{H}) \wedge G_{n,2n}(\mathbb{H})) = 0$ etc. follows from the remark following 10.4 and the results of Chapter 16 and suffices to show that μ is unique and associative. We choose $\iota_4: S^4 \to BSp$ to classify $(\xi - 1) \in \widetilde{KSp}(S^4)$, where $\xi \to S^4 \cong HP^1$ is the symplectic Hopf bundle. The Bott periodicity isomorphism

$$\widetilde{KO}(X) \xrightarrow{\cong} \widetilde{KSp}(S^4 \wedge X)$$

is given by $x \to (\xi - 1) \wedge_3 x$ for $x \in \widetilde{KO}(X)$. That means that the structure map

$$S^4 \wedge BO \to BSp$$

of our spectrum KO is the composition

$$S^4 \wedge BO \xrightarrow{\iota_4 \wedge 1} BSp \wedge BO \xrightarrow{\mu_3} BSp.$$

Similarly $S^4 \wedge BSp \to BO$ is the composition

$$S^4 \wedge BSp \xrightarrow{\iota_4 \wedge 1} BSp \wedge BSp \xrightarrow{\mu_4} BO.$$

We then have a ring spectrum (KO, μ, ι), whose product is induced by tensor product of bundles. μ is commutative.

We can now describe the coefficient rings $\tilde{K}^*(S^0) = \pi_*(K)$ and $\widetilde{KO}^*(S^0) = \pi_*(KO)$. We know that

$$\pi_q(K) = \pi_q(BU, *) = \begin{cases} \mathbb{Z} & q \text{ even} \\ 0 & q \text{ odd.} \end{cases}$$

If $t = [\iota_2] \in \pi_2(BU) \cong \pi_2(K)$, then multiplication by t is an isomorphism $\pi_q(K) \xrightarrow{\cong} \pi_{q+2}(K)$, so it follows $t^q \in \pi_{2q}(K)$ is a generator for all $q \in \mathbb{Z}$. Thus it follows that

13.92. $$\pi_*(K) \cong \mathbb{Z}[t, t^{-1}],$$

the graded ring of finite Laurent series in t with integer coefficients.

Now we have certain evident operations on vector bundles.

i) Given a complex vector bundle η we can take the conjugate bundle $\Psi^{-1}(\eta)$.

ii) Given a real vector bundle ξ we can define a complex vector bundle $c(\xi) = \xi \otimes_{\mathbb{R}} \mathbb{C}$.

iii) Given a complex vector bundle η we can forget the complex structure, getting a real vector bundle $r(\eta)$.

iv) Given a complex vector bundle η we can define a quaternionic vector bundle $s(\eta) = \eta \otimes_{\mathbb{C}} \mathbb{H}$.

v) Given a quaternionic vector bundle ξ we can forget the quaternionic structure, getting a complex vector bundle $c'(\xi)$.

These operations define natural transformations $\Psi^{-1}: \tilde{K}^* \to \tilde{K}^*$, $c: \widetilde{KO}^* \to \tilde{K}^*$, $r: \tilde{K}^* \to \widetilde{KO}^*$, $s: \tilde{K}^* \to \widetilde{KSp}^*$, $c': \widetilde{KSp}^* \to \tilde{K}^*$, and indeed even maps of spectra $\Psi^{-1}: K \to K$, $c: KO \to K$, $r: K \to KO$, $s: K \to \Sigma^4 KO$, $c': \Sigma^4 KO \to K$. One easily verifies that they satisfy the following relations.

13.93. i) $rc = 2$, $sc' = 2$;

ii) $cr = 1 + \Psi^{-1} = c's$;

iii) $\Psi^{-1} c = c$, $\Psi^{-1} c' = c'$, $r \Psi^{-1} = r$, $s \Psi^{-1} = s$, $\Psi^{-1} \circ \Psi^{-1} = 1$.

We already know that

$$\pi_q(KO) = \pi_q(BO, *) = \begin{cases} \mathbb{Z} & q \equiv 0 \bmod 4 \\ \mathbb{Z}_2 & q \equiv 1, 2 \bmod 8 \\ 0 & q \equiv 3, 5, 6, 7 \bmod 8. \end{cases}$$

We wish to describe the multiplicative structure of $\pi_*(KO)$. For every pair $G \subset H$ of topological groups we have an exact (fibre) sequence

$$\cdots \longrightarrow G \longrightarrow H \longrightarrow H/G \longrightarrow BG \xrightarrow{Bi} BH$$

(cf. 11.43). In our case then we have the fibration

$$U/O \longrightarrow BO \xrightarrow{c} BU,$$

where $O \to U$ is the map which regards an orthogonal real matrix as a unitary matrix. Now one of the stages of the real periodicity theorem $\mathbb{Z} \times BO \simeq \Omega^4 BSp$ is a homotopy equivalence $\mathbb{Z} \times BO \simeq \Omega(U/O)$ (cf. Chapter 16). Thus we get an exact homotopy sequence

$$\cdots \longrightarrow \pi_{i-1}(BO) \longrightarrow \pi_i(BO) \xrightarrow{c} \pi_i(BU) \longrightarrow \pi_{i-2}(BO) \longrightarrow \cdots.$$

Taking $i = 8$ and 4 we see that

$$c : \pi_8(BO) \to \pi_8(BU) \quad \text{is an isomorphism}$$

and

$$c : \pi_4(BO) \to \pi_4(BU) \quad \text{has image } 2\mathbb{Z}.$$

We choose $x \in \pi_4(BO)$ with $c(x) = 2t^2$ and $y \in \pi_8(BO)$ with $c(y) = t^4$. Since c is a ring homomorphism, it follows $x^2 = 4y$.

Let $\omega \to S^1 \cong RP^1$ denote the real Hopf line bundle and

$$\alpha = \omega - 1 \in \widetilde{KO}(S^1) = \pi_1(BO) = \pi_1(KO).$$

Then α is the non-zero element and one can show that $\alpha^2 \neq 0 \in \pi_2(KO)$. Of course $\alpha^3 = \alpha \cdot x = 0$ since $\alpha^3 \in \pi_3(KO)$ and $\alpha \cdot x \in \pi_5(KO)$. Multiplication by y is the Bott periodicity isomorphism $\pi_q(KO) \cong \pi_{q+8}(KO)$, so we have proved the following:

13.94. $\pi_*(KO)$ is generated as a ring by the elements $\alpha \in \pi_1(KO)$, $x \in \pi_4(KO)$, $y \in \pi_8(KO)$, $y^{-1} \in \pi_{-8}(KO)$ with the relations $2\alpha = \alpha^3 = \alpha \cdot x = 0$, $x^2 = 4y$, $yy^{-1} = 1$.

13.95. Exercise. By using 13.93 describe $r : \pi_*(BU) \to \pi_*(BO)$.

13.96. Exercise. Show that one can construct the maps

$$\bar{\mu}_{nm} : H(R, n) \wedge H(R, m) \to H(R, n+m)$$

using only homotopy theory and without first constructing the $\times$-product with cochains.

Comments

The presentation of products given here is theoretically satisfying, but if the student now tries to compute the cohomology ring $H^*(X;R)$ of some concrete space X (e.g. $X = RP^n$), he will find his task is not simple. In Chapter 15 we shall determine some of these cohomology rings, but only by using spherical fibrations and associated exact sequences of cohomology groups. After our success in Chapter 10 in computing the homology and cohomology groups of CW-complexes using only a knowledge of the cell structure, one might hope to find similar "cellular" methods to determine the ring structure of $H^*(X;R)$ also. The difficulty here is that the diagonal map $\Delta: X \to X \times X$ is not cellular. It is of course homotopic to a cellular map Δ', but one cannot in general say how Δ' looks. In the few cases where one can describe a Δ' explicitly one can indeed use cellular methods to find the ring structure of $H^*(X;R)$.

The historical development of products in homology was quite different from that given in this chapter. In the early days of the subject one worked almost exclusively with manifolds and tried, for example, to represent cycles with embedded submanifolds. One way to define $a \cup b$, $a, b \in H^*(M;R)$, is to find dual cycles $N_a \subset M$, $N_b \subset M$ in M, form their intersection (N_a, N_b should be in "general position") and then dualize $N_a \cap N_b$. The theory of "intersection numbers" was the precursor of the modern products and still provides an effective method for determining the product structure of $H^*(M;R)$ for many manifolds M, especially those which turn up in algebraic geometry. A very instructive exercise for the good student would be to read Chapter 14, then have a look at one of the older books on algebraic topology (e.g. Seifert and Threlfall) and try to translate the theory of intersection numbers into more modern language.

References

1. J. F. Adams [6, 8]
2. E. H. Spanier [80]
3. G. W. Whitehead [93]

Chapter 14

Orientation and Duality

We have seen how to construct homology and cohomology theories E_*, E^* out of a spectrum E, and in the last chapter we showed how to construct various products connecting E_* and E^* if E is a ring spectrum. In many cases, however, we can establish a much stronger connection between $E_*(X)$ and $E^*(X)$ for suitable spaces X. One of the early discoveries of algebraic topology was that if M is a closed n-dimensional orientable manifold, then $H_r(M;\mathbb{Z}) \cong H^{n-r}(M;\mathbb{Z})$ for all r, $0 \leqslant r \leqslant n$. We shall prove this Poincaré duality theorem for general homology theories.

In Chapter 12 we saw that every differentiable manifold had a tangent vector bundle. That is not true in general for topological manifolds, but we shall need some appropriate generalization of the notion of tangent bundle. We therefore turn to Milnor's notion of microbundle.

14.1. Definition. An *n-dimensional microbundle* is a quadruple $\underline{x} = (B, p, X, i)$, where B, X are topological spaces and $i: B \to X$, $p: X \to B$ are maps satisfying

i) $p \circ i = 1_B$,

ii) for every $b \in B$ there are open neighborhoods U of b and V of $i(b)$ and a homeomorphism

$$h_b: U \times \mathbb{R}^n \to V \cap p^{-1}(U)$$

such that $p \circ h_b(u,v) = u$ for all $(u,v) \in U \times \mathbb{R}^n$ and $h_b(u,0) = i(u)$ for all $u \in U$. The pairs (U,h) in ii) are called *charts* for $\underline{x}$.

14.2. Examples. i) The trivial microbundle over B is $(B, p_B, B \times \mathbb{R}^n, i)$, where $i(b) = (b,0)$, $b \in B$.

ii) If $\xi = (B, p, E, \mathbb{R}^n)$ is an n-dimensional vector bundle, then (B, p, E, i) is an n-dimensional microbundle, where i is the 0-section: $i(b) = 0$ in the fibre $p^{-1}(b)$, all $b \in B$.

iii) If M is a topological n-manifold without boundary ($\partial M = \dot{M} = \varnothing$), then the *tangent microbundle of* M is $(M, \pi_1, M \times M, \Delta)$, where π_1 is the projection on the first factor and Δ is the diagonal map. For any $x \in M$ choose a chart (U,f) around x with $f(U) = \mathbb{R}^n$ and define $h_x: U \times \mathbb{R}^n \to U \times U$ by $h_x(y,v) = (y, f^{-1}(f(y) - v))$. Then h_x is a homeomorphism of the required sort. If M has a boundary $\dot{M}$, then the tangent microbundle of M is that of $M \cup C$ restricted to M, where $C = \dot{M} \times [0,1)$ is a collar, and we identify $x \in \dot{M}$ with $(x,0) \in C$.

14.3. Definition. Two microbundles (B,p,X,i) and (B,p',X',i') are *equivalent* or *isomorphic* if there are neighborhoods V of $i(B)$ in X and V' of $i'(B)$ in X' and a homeomorphism $h: V \to V'$ such that

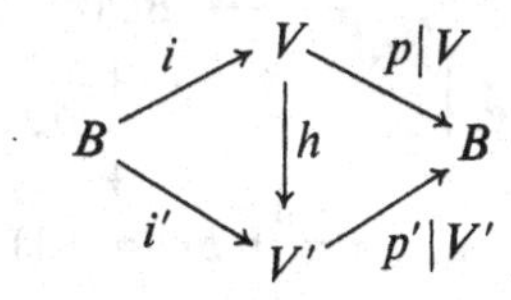

commutes.

14.4. Exercise. Show that if M is a differentiable manifold, then its tangent bundle and its tangent microbundle are microbundle equivalent.

14.5. Definition. Let $\underset{\sim}{x} = (B,p,X,i)$ be an n-dimensional microbundle and E a ring spectrum. For any $b \in B$ we have a chart $h: U \times \mathbb{R}^n \to p^{-1}(U) \cap V$, U a neighborhood of b in B, V a neighborhood of $i(b)$ in X, and an isomorphism

$$E^*(p^{-1}(b), p^{-1}(b) - i(b)) \xrightarrow{\cong}$$

$$E^*(p^{-1}(b) \cap V, p^{-1}(b) \cap V - i(b)) \underset{\cong}{\xrightarrow{h^*}} E^*(b \times (\mathbb{R}^n, \mathbb{R}^n - \{0\})).$$

An element $t \in E^n(X, X - i(b))$ is called an *orientation* or *Thom class* for $\underset{\sim}{x}$ if

$$j_b^*(t) \in E^n(p^{-1}(b), p^{-1}(b) - i(b))$$

is a generator of $E^*(p^{-1}(b), p^{-1}(b) - i(b))$ as a module over $E^*(pt)$ for all $b \in B$. Here $j_b: (p^{-1}(b), p^{-1}(b) - i(b)) \to (X, X - i(b))$ is the inclusion. $\underset{\sim}{x}$ is called *orientable* with respect to E if it has a Thom class. A topological manifold (with or without boundary) is called *orientable* with respect to E if its tangent microbundle has a Thom class.

Examples. i) Suppose $\xi = (B,p,X,\mathbb{R}^n)$ is an $O(n)$-bundle. Then we have an isomorphism

$$E^n(X, X - i(B)) \xrightarrow{\cong} E^n(D(\xi), D(\xi) - i(B)) \xrightarrow{\cong}$$

$$E^n(D(\xi), S(\xi)) \underset{p^*}{\xleftarrow{\cong}} \tilde{E}^n(M(\xi)).$$

An element $t \in \tilde{E}^n(M(\xi))$ is called a Thom class for ξ if and only if its preimage in $E^n(X, X - i(B))$ is a Thom class of the associated microbundle. For any $b \in B$ the inclusion $i_b:\{b\} \to B$ induces $M(i_b): S^n = M(\varepsilon^n) \to M(\xi)$, and clearly $t \in \tilde{E}^n(M(\xi))$ is a Thom class if and only if

$$M(i_b)^*(t) \in \tilde{E}^n(S^n)$$

is a generator of $\tilde{E}^*(S^n)$ as $\tilde{E}^*(S^0)$-module for all $b \in B$.

Suppose

$$\begin{array}{ccc} X_n & \xrightarrow{f_n} & BO(n) \\ {\scriptstyle g_n}\downarrow & & \downarrow{\scriptstyle Bi_n} \\ X_{n+1} & \xrightarrow{f_{n+1}} & BO(n+1) \end{array}$$

is a system of maps and fibrations as in 12.19 with the X_n 0-connected and ξ is a bundle with *X-structure*: that is, ξ has a classifying map $f: B \to BO(n)$ with a lifting $\tilde{f}: B \to X_n$. $\tilde{f}$ induces $M(\tilde{f}): M(\xi) \to MX_n$ and hence an element $t_\xi = \{M(\tilde{f})\} \in MX^n(M(\xi))$. For any $b \in B$ the diagram

$$\begin{array}{ccc} S^n & \xrightarrow{M(i_b)} & M(\xi) \\ & {\scriptstyle \iota_n}\searrow & \downarrow{\scriptstyle M(\tilde{f})} \\ & & MX_n \end{array}$$

commutes up to homotopy, which shows that $M(i_b)^*(t_\xi) = \iota_n \in \widetilde{MX}^n(S^n)$. Thus t_ξ is a Thom class for ξ.

ii) Suppose ξ is an $O(n)$-bundle and $\xi \oplus \varepsilon^q$ has a Thom class

$$t \in \tilde{E}^{n+q}(M(\xi \oplus \varepsilon^q)) \cong \tilde{E}^{n+q}(S^q M(\xi)).$$

Then $t' = \sigma^q(t) \in \tilde{E}^n(M(\xi))$ is easily seen to be a Thom class for ξ.

iii) Suppose M is an X-manifold, X as above. Then the stable normal bundle of M has an X-structure and hence is orientable with respect to MX. For the classical Lie groups $G(n) = O(n), SO(n), U(n), SU(n)$ or $Sp(n)$ we also have maps $\rho: G_{k,n+k} \to G_{n,n+k}$ taking a k-plane into its orthogonal-complementary n-plane. If $\tilde{\nu}: M \to G_{k,n+k}$ defines a G-structure on M, then $\rho \circ \tilde{\nu}$ defines a G-structure on $\tau(M)$. Thus $\tau(M)$ is MG-orientable. Hence every G-manifold is orientable with respect to the cohomology theory $MG^* = \Omega_G^*$.

iv) In Chapter 16 we shall construct Thom classes for other spectra. In particular we shall see that $U(n)$-bundles are always orientable with respect to K^*, while $Sp(n)$-bundles are always orientable with respect to KO^*. Surprisingly it turns out that not all $O(n)$-bundles are orientable with respect to KO^*: Atiyah, Bott and Shapiro [17] showed that an $O(n)$-bundle ξ is KO^*-orientable if and only if ξ has a reduction to *Spin* (stably).

14.6. Theorem (Thom Isomorphism Theorem). *Let $\underset{\sim}{x} = (B,p,X,i)$ be an n-dimensional microbundle with Thom class $t \in E^n(X,X-i(B))$ and B 0-connected and paracompact. Suppose that either B is a finite-dimensional CW-complex or $\pi_q(E) = 0$ for $q < 0$. Then the homomorphisms*

$$\phi_t^*: E^r(B) \to E^{n+r}(X,X-i(B)) \quad \phi_t^*(x) = p^*(x) \cup t, \quad x \in E^r(B)$$

$$\phi_t: E_{n+r}(X,X-i(B)) \to E_r(B) \quad \phi_t(x) = p_*(t \cap x), \quad x \in E_{n+r}(X,X-i(B))$$

are isomorphisms.

Proof: We begin by assuming $\underset{\sim}{x}$ has an atlas with m charts and proceed by induction on m. Suppose $m = 1$—i.e. there is a homeomorphism $h: B \times \mathbb{R}^n \to V$ such that $h(b,0) = i(b)$, $p(h(b,v)) = b$, all $b \in B$, $v \in \mathbb{R}^n$, where V is a neighborhood of $i(B)$. In other words $\underset{\sim}{x}$ is equivalent to the trivial microbundle, and

$$E^*(X,X-i(B)) \cong E^*(V,V-i(B)) \xrightarrow[h^*]{\cong} E^*(B \times \mathbb{R}^n, B \times (\mathbb{R}^n - \{0\})).$$

Therefore we may as well assume $\underset{\sim}{x} = (B,p_B, B \times \mathbb{R}^n, i)$ and $t \in E^n(B \times (\mathbb{R}^n, \mathbb{R}^n - \{0\}))$. Since $E^*(\mathbb{R}^n, \mathbb{R}^n - \{0\})$ is a free module over $E^*(pt)$, it follows by 13.75 that

$$E^*(B) \otimes_{E^*(pt)} E^*(\mathbb{R}^n, \mathbb{R}^n - \{0\}) \xrightarrow{\times} E^*(B \times (\mathbb{R}^n, \mathbb{R}^n - \{0\}))$$

is an isomorphism. Since t is a Thom class, it has the form $(1 + \alpha) \times g$, where $g \in E^n(\mathbb{R}^n, \mathbb{R}^n - \{0\})$ is a generator of the $E^*(pt)$-module, and $\alpha \in \tilde{E}^0(B)$. If B is an n-dimensional CW-complex, then $\alpha^{n+1} = 0$, and it follows $1 + \alpha$ is a unit. If $\pi_q(E) = 0$ for $q < 0$ then $\tilde{E}^0(B) = 0$ and so $\alpha = 0$. In either case

$$\phi_t^*(x) = p_B^*(x) \cup t = (x \times 1) \cup ((1 + \alpha) \times g) = (1 + \alpha)x \times g = \times((1 + \alpha)x \otimes g)$$

for $x \in E^*(B)$, which shows that ϕ_t^* is an isomorphism.

Suppose the theorem proved whenever B has a covering of fewer than m charts and suppose $\underset{\sim}{x} = (B,p,X,i)$ is a microbundle with a covering $U_1, U_2, \ldots, U_m$ of charts. Let $U = U_1 \cup U_2 \cup \cdots \cup U_{m-1}$, $V = U_m$ and $X_1 = p^{-1}U$, $X_2 = p^{-1}V$. $(U,p|X_1,X_1,i|U)$, $(V,p|X_2,X_2,i|V)$ and

$$(U \cap V, p|X_1 \cap X_2, X_1 \cap X_2, i|U \cap V)$$

are microbundles and

$$t_1 = t|(X_1, X_1 - i(U)), \quad t_2 = t|(X_2, X_2 - i(V)), \quad t_3 = t|(X_1 \cap X_2, X_1 \cap X_2 - i(U \cap V))$$

are Thom classes. Since $U, V, U \cap V$ have coverings by fewer than m charts, $\phi_{t_1}^*, \phi_{t_2}^*, \phi_{t_3}^*$ are isomorphisms.

We now wish to use a Mayer–Vietoris sequence argument, but we need a slight variant of the usual Mayer–Vietoris sequence 7.19; if $(X; A, B)$ is a triad with $X = \mathring{A} \cup \mathring{B}$ and $C \subset X$ is a subspace with $A \cup C = \mathring{A} \cup \mathring{C}$, $B \cup C = \mathring{B} \cup \mathring{C}$, then we have an exact sequence in cohomology

$$\cdots \to h^q(X, C) \to h^q(A, A \cap C) \oplus h^q(B, B \cap C) \to$$

$$h^q(A \cap B,\ A \cap B \cap C) \to h^{q+1}(X, C) \to \cdots.$$

One simply takes the sequence 7.19 for the triad $(X; A', B')$ with $A' = A \cup C$, $B' = B \cup C$ and observes that $A' \cap B' = (A \cap B) \cup C$, $h^*(A', C) \cong h^*(A, A \cap C), h^*(B', C) \cong h^*(B, B \cap C)$ and $h^*(A' \cap B', C) \cong h^*(A \cap B, A \cap B \cap C)$. In our case we take $h^* = E^*$, $A = X_1, B = X_2$ and $C = X - i(B)$. We get a commutative diagram of exact sequences

$$\begin{array}{ccc} \cdots \longrightarrow & E^{r-1}(U) \oplus E^{r-1}(V) & \longrightarrow \\ & \big\downarrow \phi_{t_1}^* \oplus \phi_{t_2}^* & \\ \cdots \to & E^{n+r-1}(X_1, X_1 - iU) \oplus E^{n+r-1}(X_2, X_2 - iV) & \to \end{array}$$

$$\begin{array}{cc} E^{r-1}(U \cap V) & \xrightarrow{\Delta'} \\ \big\downarrow \phi_{t_3}^* & \\ E^{n+r-1}(X_1 \cap X_2, X_1 \cap X_2 - i(U \cap V)) & \xrightarrow{\Delta'} \end{array}$$

$$\begin{array}{ccccc} E^r(B) & \longrightarrow & E^r(U) \oplus E^r(V) & \longrightarrow \\ \big\downarrow \phi_t^* & & \big\downarrow \phi_{t_1}^* \oplus \phi_{t_2}^* & \\ E^{n+r}(X, X - iB) & \to & E^{n+r}(X_1, X_1 - iU) \oplus E^{n+r}(X_2, X_2 - iV) & \to \end{array}$$

$$\begin{array}{cc} E^r(U \cap V) & \longrightarrow \cdots \\ \big\downarrow \phi_{t_3}^* & \\ E^{n+r}(X_1 \cap X_2, X_1 \cap X_2 - i(U \cap V)) & \to \cdots. \end{array}$$

By the 5-lemma it follows that ϕ_t^* is an isomorphism.

Now if ξ does not have a finite atlas, then one can easily show that there is at least a countable atlas $\{U_i\}_{i \geq 0}$ (cf. [49], p. 30, for example). Let

$V_j = \cup_{i=0}^{j} U_i$, $X_j = p^{-1}V_j$, $t_j = t|(X_j, X_j - i(V_j))$ for $j \geq 0$. Then we have a commutative diagram

$$\begin{array}{ccccc}
0 \longrightarrow & \lim\limits_{j}{}^{1} E^{r-1}(V_j) & \longrightarrow & E^r(B) & \longrightarrow \\
& \cong \Big\downarrow \lim^1 \phi^*_{t_j} & & \Big\downarrow \phi^*_t & \\
0 \to & \lim\limits_{j}{}^{1} E^{n+r-1}(X_j, X_j - i(V_j)) & \to & E^{n+r}(X, X - i(B)) & \to
\end{array}$$

$$\begin{array}{cc}
\lim\limits_{j}{}^{0} E^r(V_j) & \longrightarrow 0 \\
\cong \Big\downarrow \lim^0 \phi^*_{t_j} & \\
\lim\limits_{j}{}^{0} E^{n+r}(X_j, X_j - i(V_j)) & \to 0
\end{array}$$

in which the rows are exact by the dual of 7.73 and $\lim^0 \phi^*_t$, $\lim^1 \phi^*_t$ are isomorphisms because each V_j has a finite covering by charts. It follows that ϕ^*_t is an isomorphism. The proof for ϕ_t is similar. □

The reader should verify for himself that the diagram of Mayer–Vietoris sequences is commutative, particularly that $\phi^*_t \circ \Delta' = \Delta' \circ \phi^*_{t_3}$.

14.7. Corollary. *If $\underline{x} = (B,p,X,i)$ is a microbundle with B 0-connected and paracompact and either B a finite-dimensional CW-complex or $\pi_q(E) = 0$ for $q < 0$, then for any two Thom classes $t, t' \in E^n(X, X - i(B))$ we have $t' = p^*(u) \cup t$, where $u \in E^0(B)$ is a unit. If moreover*

$$j_b^*(t) = j_b^*(t') \in E^n(p^{-1}(b), p^{-1}(b) - i(b))$$

for some point $b \in B$, then $u = 1 + \alpha$ for some $\alpha \in \tilde{E}^0(B) = E^0(B, \{b\})$.

Proof: Since $\phi^*_t : E^0(B) \to E^n(X, X - i(B))$ is an isomorphism by 14.6, it follows $t' = p^*(u) \cup t$ for some $u \in E^0(B)$. Similarly $t = p^*(u') \cup t' = p^*(u') \cup p^*(u) \cup t = p^*(u'u) \cup t$. In other words $\phi^*_t(u'u) = \phi^*_t(1)$; therefore $u'u = 1$—i.e. u is a unit.

$j_b^* : E^n(X, X - i(B)) \to E^n(p^{-1}(b), p^{-1}(b) - i(b))$ sends $p^*(u) \cup t$ to $i^*(u) \cdot j_b^*(t)$, where $i^*(u) \in E^0(b)$. Thus if $j_b^*(t) = j_b^*(t')$, we have $1 \cdot j_b^*(t) = i^*(u) j_b^*(t)$; since $E^n(p^{-1}(b), p^{-1}(b) - i(b))$ is a free $E^0(pt)$-module, it follows $i^*(u) = 1$ or $i^*(1 - u) = 0$. Hence $-\alpha = 1 - u$ is in $\ker i^* = \tilde{E}^0(B)$. □

14.8. Let us agree to call a spectrum E connected if $\pi_q(E) = 0$ for $q < 0$. For connected E and B 14.7 says the Thom class t is uniquely determined by $j_b^*(t)$ for any $b \in B$.

14.9. Proposition. *Let E be a connected spectrum and $\underline{x} = (B,p,X,i)$ a microbundle. Let $\{U_i : 1 \leq i \leq m\}$ be a finite covering of B by open sets and suppose*

$$t_i \in E^n(p^{-1}(U_i), p^{-1}(U_i) - i(U_i))$$

are elements such that

i) $j_b^* t_i \in E^n(p^{-1}(b), p^{-1}(b) - i(b))$ *is a generator for every* $b \in U_i$;

ii) $j_b^* t_i = j_b^* t_k$, *for every* $b \in U_i \cap U_k$ *and all* i, k.

Then there is a unique Thom class $t \in E^n(X, X - i(B))$ *such that* $t|p^{-1}U_i = t_i$, $1 \leqslant i \leqslant m$.

Proof: We proceed by induction on m. If $m = 1$, then t_1 is the Thom class we are seeking. Suppose the proposition proved for $m' < m$ and suppose $\underset{\sim}{x}$ has a covering by m sets U_i as above. Let $U = U_1 \cup U_2 \cup \cdots \cup U_{m-1}$, $V = U_m$ and let $X_1 = p^{-1}(U)$, $X_2 = p^{-1}(V)$. By the inductive hypothesis there is a Thom class $t_U \in E^n(X_1, X_1 - i(U))$ such that $t_U|p^{-1}(U_i) = t_i$, $1 \leqslant i \leqslant m-1$; let $t_V = t_m$. By 14.8 and ii) above the restrictions of t_U, t_V to $X_1 \cap X_2$ agree. Thus from the exactness of the Mayer–Vietoris sequence

$$\cdots \to E^n(X, X - i(B)) \to E^n(X_1, X_1 - iU) \oplus E^n(X_2, X_2 - iV) \to E^n(X_1 \cap X_2, X_1 \cap X_2 - i(U \cap V)) \to \cdots$$

we see that there is a $t \in E^n(X, X - i(B))$ with $t|X_1 = t_U$, $t|X_2 = t_V$. t is our desired Thom class. If t' were another then t, t' would agree at any $b \in B$ and hence be equal by 14.8. Therefore t is unique. □

14.10. Suppose the U_i of 14.9 were charts for $\underset{\sim}{x}$—that is, there were neighborhoods V_i of $i(U_i)$ in X and homeomorphisms $h_i: U_i \times \mathbb{R}^n \to p^{-1}(U_i) \cap V_i$ with $p \circ h_i(u, v) = u$, $h_i(u, 0) = i(u)$, $u \in U_i$. For each i, $1 \leqslant i \leqslant m$, choose a generator $g_i \in E^n(\mathbb{R}^n, \mathbb{R}^n - \{0\})$ and let t_i be the pre-image of $1 \times g_i$ under

$$E^n(p^{-1}U_i, p^{-1}U_i - i(U_i)) \xrightarrow{\cong} E^n(p^{-1}U_i \cap V_i, p^{-1}U_i \cap V_i - i(U_i)) \underset{\cong}{\xrightarrow{h_i^*}} E^n(U_i \times (\mathbb{R}^n, \mathbb{R}^n - \{0\})).$$

Then the t_i's satisfy i) of 14.9, but they need not satisfy ii). In general for every component N of $U_i \cap U_k$ there will be a unit $u_N \in E^0(pt)$ with $t_i|p^{-1}(N) = u_N \cdot (t_k|p^{-1}(N))$. Of course, in the case $E = H(\mathbb{Z}_2)$, $E^0(pt)$ has only one unit, 1, so in that case we do get a Thom class—i.e. with respect to $H(\mathbb{Z}_2)$ every microbundle is orientable.

Exercise. Suppose M is a manifold with an atlas $\{(U_\alpha, \phi_\alpha)\}$ such that $\phi_\alpha \circ \phi_\beta^{-1}: \phi_\beta(U_\alpha \cup U_\beta) \to \mathbb{R}^n$ is orientation-preserving for all α, β. Show that the tangent microbundle t_M has an orientation for any connected ring spectrum E.

We turn now to the duality theorems for manifolds; they will be consequences of the Alexander duality theorem, which we prove first.

Suppose $B \subset A \subset M - \dot{M}$ are subspaces of an n-manifold M. The inclusion $A \times (M-B) \to M \times M$ induces a map

$$j:(A,B) \times (M-B, M-A) \to (M \times M, M \times M - \Delta).$$

If $t \in E^n(M \times M, M \times M - \Delta)$, then we can define a homomorphism

$$\gamma_t: E_r(M-B, M-A) \to E^{n-r}(A, B)$$

by $\gamma_t(x) = j^* t/x$ for all $x \in E_r(M-B, M-A)$. γ_t is natural with respect to inclusions $(A', B') \subset (A, B)$.

14.11. Theorem. (Alexander Duality). *If M is an n-manifold with boundary $\dot{M}$, $t \in E^n(M \times M, M \times M - \Delta)$ a Thom class for M and $B \subset A \subset M - \dot{M}$ compact polyhedra in $M - \dot{M}$, then*

$$\gamma_t: E_r(M-B, M-A) \to E^{n-r}(A,B)$$

is an isomorphism for all $r \in \mathbb{Z}$. Similarly

$$\gamma_t: E_r(A,B) \to E^{n-r}(M-B, M-A)$$

is an isomorphism.

Proof: We consider various cases.

i) $(A,B) = (\{x\}, \varnothing)$, $x \in M - \dot{M}$. Then $(A,B) \times (M-B, M-A) = \{x\} \times (M, M-\{x\})$; because t is a Thom class, $j^*t = 1 \times g$, where $g \in E^n(M, M-\{x\})$ is a generator over $E^*(pt)$

$$(E^n(M, M-\{x\}) \cong E^n(U, U-\{x\}) \xrightarrow[h^*]{\cong} E^n(\mathbb{R}^n, \mathbb{R}^n - \{0\})$$

if (U,h) is a chart around x). Thus for $y \in E_r(M, M-\{x\})$ we have $\gamma_t(y) = (1 \times g)/y = \langle g, y \rangle \cdot 1$, which shows γ_t is an isomorphism.

ii) $(A,B) = (\sigma^q, \varnothing)$, where $\sigma^q \subset M - \dot{M}$ is an embedded q-simplex which is so small that $\sigma^q \subset U$ for some chart (U,h). Choose a vertex v of σ^q; then we have a commutative diagram

$$\begin{array}{ccccccc}
E^{n-r}(\sigma^q) & \xleftarrow[\gamma_t]{} & E_r(M, M-\sigma^q) & \xleftarrow[\cong]{} & E_r(U, U-\sigma^q) & \xrightarrow[\cong]{h_*} & \\
\cong\downarrow & & \downarrow & & \downarrow & & \\
E^{n-r}(v) & \xleftarrow[\gamma_t]{\cong} & E_r(M, M-\{v\}) & \xleftarrow{\cong} & E_r(U, U-\{v\}) & \xrightarrow[h_*]{\cong} &
\end{array}$$

$$\begin{array}{ccc}
E_r(\mathbb{R}^n, \mathbb{R}^n - h(\sigma^q)) & \xrightarrow[\cong]{k_*} & E_r(S^n, S^n - kh(\sigma^q)) \\
\downarrow & & \downarrow \\
E_r(\mathbb{R}^n, \mathbb{R}^n - h(v)) & \xrightarrow[k_*]{\cong} & E_r(S^n, S^n - kh(v))
\end{array}$$

in which $k:\mathbb{R}^n \to S^n$ embeds $\mathbb{R}^n$ as $S^n - H_+^n$, say. Thus $\gamma_t: E_r(M, M - \sigma^q) \to E^{n-r}(\sigma^q)$ is an isomorphism if and only if

$$E_r(S^n, S^n - kh(\sigma^q)) \to E_r(S^n, S^n - kh(v))$$

is an isomorphism. By applying the 5-lemma to the exact sequences of the pairs $(S^n, S^n - kh(\sigma^q))$ and $(S^n, S^n - kh(v))$ we see that this latter is true if and only if $\tilde{E}_r(S^n - kh(\sigma^q)) \cong \tilde{E}_r(S^n - kh(v)) = 0$. Thus it suffices to show $\tilde{E}_r(S^n - \phi(\sigma^q)) = 0$, $r \in \mathbb{Z}$, for all embeddings $\phi: \sigma^q \to S^n$, $q \geqslant 0$, or equivalently $\tilde{E}_r(S^n - \phi(I^q)) = 0$ for all embeddings $\phi: I^q \to S^n$, $q \geqslant 0$.

We proceed by induction on q. The case $q = 0$ is trivial. We suppose $\tilde{E}_*(S^n - \phi(I^{q-1})) = 0$ for all $\phi: I^{q-1} \to S^n$ and consider an embedding $\phi: I^q \to S^n$. We let $A_0 = \phi(I^q)$, $B_1 = \phi(I^{q-1} \times [0, 1/2])$, $B_2 = \phi(I^{q-1} \times [1/2, 1])$. Then we have a Mayer–Vietoris sequence

$$\cdots \longrightarrow \tilde{E}_{r+1}(S^n - B_1 \cap B_2) \xrightarrow{\Delta'} \tilde{E}_r(S^n - A_0) \xrightarrow{\alpha} \tilde{E}_r(S^n - B_1) \oplus \tilde{E}_r(S^n - B_2) \longrightarrow \cdots.$$

Since $B_1 \cap B_2 = \phi(I^{q-1} \times \{1/2\})$, it follows from the inductive hypothesis that $\tilde{E}_*(S^n - B_1 \cap B_2) = 0$ and hence $\alpha: \tilde{E}_*(S^n - A_0) \to \tilde{E}_*(S^n - B_1) \oplus \tilde{E}_*(S^n - B_2)$ is an isomorphism. If there is a non-zero element

$$x \in \tilde{E}_r(S^n - A_0)$$

for some $r \in \mathbb{Z}$, then either $i_{1*}(x) \in \tilde{E}_r(S^n - B_1)$ or $i_{2*}(x) \in \tilde{E}_r(S^n - B_2)$ is non-zero—say $i_{1*}(x)$. We take $A_1 = B_1$, $x_1 = i_{1*}(x)$. Repeating this process, we find subspaces

$$A_0 \supset A_1 \supset A_2 \supset \cdots \supset A_k \supset \cdots$$

and non-zero elements $x_k \in \tilde{E}_r(S^n - A_k)$ such that $\tilde{E}_r(S^n - A_k) \to \tilde{E}_r(S^n - A_{k+1})$ takes x_k to x_{k+1} and $\bigcap_{k \geqslant 0} A_k = \phi(I^{q-1} \times \{s\})$ for some $s \in I$. By 7.73 and the inductive hypothesis

$$\operatorname{dir\,lim} \tilde{E}_r(S^n - A_k) = \tilde{E}_r(S^n - \textstyle\bigcap_{k \geqslant 0} A_k) = 0;$$

but on the other hand $\{x_k\}$ defines a non-zero element in $\operatorname{dir\,lim} \tilde{E}_r(S^n - A_k)$. This is a contradiction, so it must be the case that $\tilde{E}_*(S^n - \phi(I^q)) = 0$.

iii) Now every compact polyhedron is a compact metric space, so any open covering of it has a Lebesgue number. Hence we always suppose our compact polyhedron A is so finely triangulated that every simplex lies in some chart of a given atlas on M. Such a triangulation of A will be called a *fine* triangulation.

Suppose the theorem proved for all pairs $(A', \varnothing)$, where A' has a fine triangulation with fewer than m simplices. If $A \subset M$ has a fine triangulation with m simplices, we may write $A = A' \cup \sigma^q$, where $A' \subset A$ is a proper subcomplex and σ^q a q-simplex of A. Then A' and $\dot{\sigma}^q$ both have fewer than

m simplices. We have triads $(A;\sigma,A')$ and $(M-\dot\sigma;M-\sigma,M-A')$ with $\sigma\cap A'=\dot\sigma$ and $(M-\sigma)\cap(M-A')=M-A$.

We now employ yet another variant of the Mayer–Vietoris sequence 7.19; if $(X;A,B)$ is an excisive triad for the homology theory h_* and $Y\supset X$, then we have an exact sequence

$$\cdots\to h_r(Y,A\cap B)\xrightarrow{\alpha}h_r(Y,A)\oplus h_r(Y,B)\xrightarrow{\beta}h_r(Y,X)\xrightarrow{\Delta'}h_{r-1}(Y,A\cap B)\to\cdots;$$

the proof is very similar to that of 7.19. In our case, then, we get a commutative diagram of exact sequences

$$\begin{array}{ccccc}
E_{q+1}(M,M-\sigma)\oplus E_{q+1}(M,M-A') & \to & E_{q+1}(M,M-\dot\sigma) & \xrightarrow{\Delta'} \\
\cong\downarrow\gamma_t\oplus\gamma_t & & \cong\downarrow\gamma_t & \\
E^{n-q-1}(\sigma)\oplus E^{n-q-1}(A') & \longrightarrow & E^{n-q-1}(\dot\sigma) & \longrightarrow
\end{array}$$

$$\begin{array}{ccccccc}
E_q(M,M-A) & \to & E_q(M,M-\sigma)\oplus E_q(M,M-A') & \to & E_q(M,M-\dot\sigma) \\
\downarrow\gamma_t & & \cong\downarrow\gamma_t\oplus\gamma_t & & \cong\downarrow\gamma_t \\
E^{n-q}(A) & \longrightarrow & E^{n-q}(\sigma)\oplus E^{n-q}(A') & \longrightarrow & E^{n-q}(\dot\sigma).
\end{array}$$

The 5-lemma then implies that $\gamma_t:E_q(M,M-A)\to E^{n-q}(A)$ is an isomorphism for all q. Therefore by induction γ_t is an isomorphism for all pairs $(A,\varnothing)$. (Again the reader should verify for himself that the diagram above commutes—especially that $\Delta'\circ\gamma_t=\gamma_t\circ\Delta'$.)

iv) For any pair (A,B) we have a commutative diagram of exact sequences

$$\begin{array}{ccccccc}
E_{q+1}(M,M-A) & \to & E_{q+1}(M,M-B) & \to & E_q(M-B,M-A) & \to \\
\cong\downarrow\gamma_t & & \cong\downarrow\gamma_t & & \downarrow\gamma_t & \\
E^{n-q-1}(A) & \longrightarrow & E^{n-q-1}(B) & \longrightarrow & E^{n-q}(A,B) & \longrightarrow
\end{array}$$

$$\begin{array}{ccc}
E_q(M,M-A) & \to & E_q(M,M-B) \\
\cong\downarrow\gamma_t & & \cong\downarrow\gamma_t \\
E^{n-q}(A) & \longrightarrow & E^{n-q}(B).
\end{array}$$

The 5-lemma then implies that $\gamma_t:E_q(M-B,M-A)\to E^{n-q}(A,B)$ is an isomorphism for all $q\in\mathbb{Z}$. The proof that $E_r(A,B)\to E^{n-r}(M-B,M-A)$ is an isomorphism for all $q\in\mathbb{Z}$ is similar. □

As immediate consequences we get the following duality theorems.

14.12. Theorem (Lefschetz Duality). *If M is a compact triangulable n-manifold with boundary $\dot{M}$ and t is a Thom class for M, then there are isomorphisms*

$$E_r(M, \dot{M}) \xrightarrow[\cong]{\gamma_t} E^{n-r}(M - \dot{M}) \xleftarrow[\cong]{i^*} E^{n-r}(M)$$

and

$$E_r(M) \xleftarrow[\cong]{i_*} E_r(M - \dot{M}) \xrightarrow[\cong]{\gamma_t} E^{n-r}(M, \dot{M}), \quad r \in \mathbb{Z}.$$

Proof: Choose a collar neighborhood $N \cong \dot{M} \times [0,1)$ of $\dot{M}$ in M (cf. [88, App. II]). Then we have a commutative diagram

$$\begin{array}{ccccc} E_r(M, \dot{M}) & \xrightarrow{\gamma_t} & E^{n-r}(M - \dot{M}) & & \\ \downarrow \cong & & \downarrow \cong & \nwarrow i^* & \\ & & & & E^{n-r}(M) \\ & & & \swarrow i'^* \cong & \\ E_r(M, N) & \xrightarrow[\cong]{\gamma_t} & E^{n-r}(M - N) & & \end{array}$$

We can apply 14.11 to the pair $(M - N, \varnothing)$ to show that $\gamma_t : E_r(M, N) \to E^{n-r}(M - N)$ is an isomorphism. Clearly $\dot{M}$ is a strong deformation retract of N, $M - N$ is a strong deformation retract of $M - \dot{M}$, and also of M. Thus i'^* and the two vertical arrows are also isomorphisms. The other isomorphism is demonstrated similarly. □

14.13. Theorem (Poincaré Duality). *If M is a compact, triangulable n-manifold without boundary and t is a Thom class for M, then there is an isomorphism*

$$\gamma_t : E_r(M) \to E^{n-r}(M), \quad r \in \mathbb{Z}.$$

14.14. At the time this is written it is an open question whether every topological manifold is triangulable. If the answer were in the affirmative, of course, we could state 14.12 and 14.13 for all compact, oriented topological manifolds. If $E = H(R)$ for some ring R, then 14.12 and 14.13 can be shown to hold for all such M in any case.

14.12 and 14.13 are not in the forms in which the Lefschetz and Poincaré duality theorems usually appear. The usual statements involve the notion of "fundamental class".

14.15. Definition. For any n-manifold M an element $z \in E_n(M, \dot{M})$ is called a *fundamental class* for M if and only if for every $x \in M - \dot{M}$ we have that $j_*(z) \in E_n(M, M - \{x\})$ $(\cong E_n(U, U - \{x\}) \cong E_n(\mathbb{R}^n, \mathbb{R}^n - \{0\}))$ is a generator of $E_*(M, M - \{x\})$ as a module over $E_*(pt)$.

We wish to show that the inverse of the Lefschetz duality isomorphism is given (up to sign) by $y \mapsto y \cap z$ for an appropriate fundamental class z. We first need a lemma from Spanier.

14.16. Lemma. *Let M be a compact, orientable, triangulable n-manifold and let $\pi_1, \pi_2 : M \times M \to M$ be the projections. For any*

$$a \in E^p(M \times M, M \times M - \Delta), \quad b \in E_q(M \times M, M \times M - \Delta)$$

and $c \in E^r(M)$ we have

$$\pi_{1*}(a \cap b) = \pi_{2*}(a \cap b) \quad \text{in } E_{q-p}(M)$$

$$a \cup \pi_1^* c = a \cup \pi_2^* c \quad \text{in} \quad E^{p+r}(M \times M, M \times M - \Delta).$$

Proof: The switch map $1 \times \tau \times 1 : M \times M \times M \times M \to M \times M \times M \times M$ defines a map $T : (M^4, M^4 - \Delta(M^2)) \to (M^2, M^2 - \Delta) \times (M^2, M^2 - \Delta)$, and it is easy to see that if $t \in E^n(M^2, M^2 - \Delta)$ is a Thom class for M, then

$$t' = T^*(t \times t) \in E^{2n}(M^4, M^4 - \Delta(M^2))$$

is a Thom class for $M \times M$. Since $T \circ (\tau \times \tau) = \tau' \circ T$, where

$$\tau'(x_1, x_2, x_3, x_4) = (x_3, x_4, x_1, x_2),$$

it follows $(\tau \times \tau)^*(t') = (-1)^n t'$. Thus the diagram

$$\begin{array}{ccc} E_q(M^2, M^2 - \Delta) & \xrightarrow{\tau_*} & E_q(M^2, M^2 - \Delta) \\ & {\scriptstyle \gamma_{t'}} \searrow \cong \quad \cong \swarrow {\scriptstyle (-1)^n \gamma_{t'}} & \\ & E^{2n-q}(\Delta) & \end{array}$$

commutes, and $\gamma_{t'}$ is an isomorphism by 14.11. Thus $\tau_*(b) = (-1)^n b$ for all $b \in E_q(M^2, M^2 - \Delta)$, and similarly $\tau^*(a) = (-1)^n a$ for all

$$a \in E^p(M^2, M^2 - \Delta).$$

Therefore $\pi_{1*}(a \cap b) = \pi_{2*}\tau_*(a \cap b) = \pi_{2*}(\tau^* a \cap \tau_* b) = \pi_{2*}(a \cap b)$ and $a \cup \pi_1^* c = (-1)^n \tau^*(a \cup \pi_1^* c) = a \cup \tau^* \pi_1^* c = a \cup \pi_2^* c$. □

14.17. Proposition. *Let M be a compact, triangulable n-manifold with Thom class t. Then there is a unique fundamental class $z \in E_n(M, \dot{M})$ such that $\gamma_t : E_n(M, \dot{M}) \to E^0(M - \dot{M})$ sends z to 1. Moreover the inverses of the two Lefschetz duality isomorphisms are given (up to sign) by taking cap product with z.*

Proof: From 14.12 we know that γ_t is an isomorphism, so there is a z with $\gamma_t(z) = 1$. For any $x \in M - \dot{M}$ we have a commutative diagram

$$\begin{array}{ccc} E_n(M,\dot{M}) & \xrightarrow{j_*} & E_n(M, M-\{x\}) \\ \cong \downarrow \gamma_t & & \cong \downarrow \gamma_t \\ E^0(M-\dot{M}) & \xrightarrow{i^*} & E^0(\{x\}). \end{array}$$

Since $i^*(1)=1$, we see that $j_*(z)=\gamma_t^{-1}(1)$. But γ_t is an isomorphism of $E_*(pt)$-modules, so $\gamma_t^{-1}(1)$ is a generator of $E_*(M, M-\{x\})$. Thus z is a fundamental class.

We wish to show that the two triangles

$$E^{n-q}(M) \xrightarrow{\cap z} E_q(M,\dot{M}), \quad i^*: E^{n-q}(M) \to E^{n-q}(M-\dot{M}), \quad \gamma_t: E_q(M,\dot{M}) \to E^{n-q}(M-\dot{M})$$

$$E_q(M-\dot{M}) \xrightarrow{\gamma_t} E^{n-q}(M,\dot{M}), \quad i_*: E_q(M-\dot{M}) \to E_q(M), \quad \cap z: E^{n-q}(M,\dot{M}) \to E_q(M)$$

commute up to sign. Let

$$j_1:(M-\dot{M}) \times (M,\dot{M}) \to (M \times M, M \times M - \Delta), \qquad k_1:(M-\dot{M}) \times M \to M \times M,$$

$$j_2:(M,\dot{M}) \times (M-\dot{M}) \to (M \times M, M \times M - \Delta), \qquad k_2: M \times (M-\dot{M}) \to M \times M$$

be the inclusions. Then for $y \in E^{n-q}(M)$ we have

$$\begin{aligned}
\gamma_t(y \cap z) &= j_1^* t/(y \cap z) = [j_1^* t \cup (1 \times y)]/z \quad \text{(by 13.61)} \\
&= (j_1^*[t \cup \pi_2^* y])/z = (j_1^*[t \cup \pi_1^* y])/z \quad \text{(by 14.16)} \\
&= (-1)^{n(n-q)}(j_1^*[\pi_1^* y \cup t])/z = (-1)^{n+nq}[(i^* y \times 1) \cup j_1^* t]/z \\
&= (-1)^{n+nq}\, i^* y \cup (j_1^* t/z) \quad \text{(by 13.61)} \\
&= (-1)^{n+nq}\, i^* y \cup 1 \quad \text{(by choice of } z\text{)} \\
&= (-1)^{n+nq}\, i^* y.
\end{aligned}$$

Similarly for $x \in E_q(M-\dot{M})$ we have

$$\begin{aligned}
\gamma_t(x) \cap z &= (j_2^* t/x) \cap z = \pi_{1*}[j_2^* t \cap (z \times x)] \quad \text{(by 13.61)} \\
&= \pi_{1*} k_{2*}[j_2^* t \cap (z \times x)] = \pi_{1*}[t \cap j_{2*}(z \times x)] \\
&= \pi_{2*}[t \cap j_{2*}(z \times x)] \quad \text{(by 14.16)} \\
&= \pi_{1*} \tau_*[t \cap j_{2*}(z \times x)] = (-1)^n \pi_{1*}[t \cap j_{1*} \tau_*(z \times x)] \\
&= (-1)^{n+nq} \pi_{1*}[t \cap j_{1*}(x \times z)] \\
&= (-1)^{n+nq} \pi_{1*} k_{1*}[j_1^* t \cap (x \times z)] \\
&= (-1)^{n+nq} i_* \pi_{1*}[j_1^* t \cap (x \times z)] \\
&= (-1)^{n+nq} i_*[(j_1^* t/z) \cap x] \quad \text{(by 13.61)} \\
&= (-1)^{n+nq} i_*(x). \quad \square
\end{aligned}$$

Of course it follows that the dual of the Poincaré isomorphism is also given (up to sign) by taking cap product with the fundamental class.

14.18. Proposition. *Let E be a connected ring spectrum. Then a compact, triangulable n-manifold is orientable if and only if it has a fundamental class.*

Proof: We have just seen that if M is orientable then M has a fundamental class. Suppose z is a fundamental class for M; we must show how to construct a Thom class t. We have for each $b \in M$ isomorphisms

$$E_n(M, M-\{b\}) \xleftarrow{\cong} E_n(U, U-\{b\}) \xrightarrow[h_*]{\cong} E_n(\mathbb{R}^n, \mathbb{R}^n - \{0\})$$

$$E^n(M, M-\{b\}) \xrightarrow{\cong} E^n(U, U-\{b\}) \xleftarrow[h^*]{\cong} E^n(\mathbb{R}^n, \mathbb{R}^n - \{0\}),$$

where $h: U \to \mathbb{R}^n$ is a chart at b with $h(b) = 0$. Let $i_b: (M, \dot{M}) \to (M, M-\{b\})$ be the inclusion and let $z_b = i_{b*}(z) \in E_n(M, M-\{b\})$. Denote by $t_b \in E^n(M, M-\{b\})$ the unique element such that $\langle t_b, z_b \rangle = 1$.

Now choose a finite atlas $\{V_i, h_i\}$ for M. An atlas for the tangent microbundle of M is given by $\{V_i, H_i\}$, where $H_i: V_i \times \mathbb{R}^n \to V_i \times V_i$ is the map with $H_i(v,x) = (v, h_i^{-1}(h_i(v) - x))$ for $v \in V_i$, $x \in \mathbb{R}^n$. Suppose (V,h) is a typical chart, and let us establish the following notation: let $b_0 = h^{-1}(0) \in V$ and for any $b \in V$ let

$$j_b: (V, V-\{b\}) \to (V \times V, V \times V - \Delta(V))$$

be the inclusion of the fibre over b; let

$$h_b: (\mathbb{R}^n, \mathbb{R}^n - \{0\}) \to (V, V - \{b\})$$

be given by $h_b(x) = \pi_2 \circ H(b,x)$, $x \in \mathbb{R}^n$, and

$$g_{b_1, b_2} = h_{b_1} \circ h_{b_2}^{-1}: (V, V-\{b_2\}) \to (V, V-\{b_1\}), \quad b_1, b_2 \in V.$$

Let $t_V = (H^{-1})^*(1 \times h_{b_0}^*(t_{b_0}|(V, V-\{b_0\}))) \in E^n(V \times V, V \times V - \Delta(V))$ (cf. 14.10). We wish to show that $j_b^*(t_V) = t_b|(V, V-\{b\})$ for all $b \in V$. Since the diagram

$$\begin{array}{ccc} & (V, V-\{b\}) & \\ {\scriptstyle b \times h_b^{-1}} \swarrow & & \searrow {\scriptstyle j_b} \\ V \times (\mathbb{R}^n, \mathbb{R}^n - \{0\}) & \xrightarrow[H]{} & (V \times V, V \times V - \Delta(V)) \end{array}$$

commutes, this amounts to showing that

$$\begin{aligned} t_b|(V, V-\{b\}) &= (b \times h_b^{-1})^*(1 \times h_{b_0}^*(t_{b_0}|(V, V-\{b_0\}))) \\ &= (h_b^{-1})^* \circ h_{b_0}^*(t_{b_0}|(V, V-\{b_0\})) \\ &= g_{b_0, b}^*(t_{b_0}|(V, V-\{b_0\})). \end{aligned}$$

Now $g_{b_0,b}$ fits into a commutative diagram

$$\begin{array}{ccc}
(V, V-\{b\}) & \xrightarrow{g_{b_0,b}} & (V, V-\{b_0\}) \\
h \downarrow & & \downarrow h \\
(\mathbb{R}^n, \mathbb{R}^n - \{h(b)\}) & \xrightarrow[T_b]{} & (\mathbb{R}^n, \mathbb{R}^n - \{0\}),
\end{array}$$

where T_b is translation by $-h(b)$. T_b is homotopic to a homeomorphism $s: \mathbb{R}^n \to \mathbb{R}^n$ which is the identity outside a ball B containing $h(b)$ and 0 and sends $h(b)$ to 0. Therefore $g_{b_0,b}$ is homotopic to a map $G_{b_0,b}$ which is the identity outside $h^{-1}(B)$. We may therefore extend $G_{b_0,b}$ to a map $G_{b_0,b}:(M,\dot{M}) \to (M,\dot{M})$ which is in fact homotopic to 1_M. We therefore have a homotopy commutative diagram

$$\begin{array}{ccc}
(V, V-\{b\}) & \xrightarrow{g_{b_0,b}} & (V, V-\{b_0\}) \\
\cap J_b & & \cap J_{b_0} \\
(M, M-\{b\}) & \xrightarrow{G_{b_0,b}} & (M, M-\{b_0\}) \\
 & {}_{i_b}\nwarrow \quad (M, \dot{M}) \quad \nearrow_{i_{b_0}} & .
\end{array}$$

We then calculate $g^*_{b_0,b}(t_{b_0}|(V,V-\{b_0\})) = g^*_{b_0,b} \circ J^*_{b_0}(t_{b_0}) = J^*_b(G^*_{b_0,b}(t_{b_0}))$. Hence

$$\begin{aligned}
\langle g^*_{b_0,b}(t_{b_0}|(V,V-\{b_0\})), J^{-1}_{b*}(z_b)\rangle &= \langle J^*_b \circ G^*_{b_0,b}(t_{b_0}), J^{-1}_{b*}(z_b)\rangle \\
&= \langle t_{b_0}, (G_{b_0,b})_* i_{b*}(z)\rangle = \langle t_{b_0}, z_{b_0}\rangle = 1.
\end{aligned}$$

This establishes that $j^*_b(t_V) = t_b|(V, V-\{b\})$ for all $b \in V$, each chart V. Hence the elements $t_i \in E^n(p^{-1}(V_i), p^{-1}(V_i) - i(V_i))$ which correspond to the t_{V_i} satisfy the hypotheses of 14.9, and therefore there is a Thom class t with $t|(V_i \times V_i, V_i \times V_i - \Delta(V_i)) = t_{V_i}$. □

Remark. It seems regrettable that we need to assume E connected, since KO and K are not.

The Alexander duality theorem says that if X, Y are complementary finite subcomplexes of Euclidean n-space (i.e. $Y \simeq \mathbb{R}^n - X$) then there is a duality isomorphism between $E^*(X)$ and $E_*(Y)$ for any ring spectrum E:

$$E_r(Y) \xrightarrow{\cong} E_r(\mathbb{R}^n - X) \xrightarrow{\gamma_t} E^{n-r}(\mathbb{R}^n, X) \underset{\cong}{\xleftarrow{\Delta}} \tilde{E}^{n-r-1}(X).$$

Alternatively we can define a map $\mu : X \times Y \to S^{n-1}$ by

$$\mu(u, v) = \frac{u - v}{\|u - v\|}.$$

Then the duality isomorphism is given by

$$\gamma(y) = \mu^*(g)/y, \quad y \in E_*(Y),$$

where $g \in \tilde{E}^{n-1}(S^{n-1})$ is the generator of $\tilde{E}^*(S^{n-1})$ as $\tilde{E}^*(S^0)$-module. This observation led Spanier and Whitehead to define a more general kind of duality.

14.19. Suppose X, X^* are spectra and $\mu : X^* \wedge X \to S^0$ is a map. We construct functions

$$D_\mu : [U, V \wedge X^*] \to [U \wedge X, V]$$

and

$${}_\mu D : [U, X \wedge V] \to [X^* \wedge U, V]$$

for any two spectra U, V:

if $f : U \to V \wedge X^*$ is a map, then as representative for $D_\mu[f]$ we take

$$U \wedge X \xrightarrow{f \wedge 1} V \wedge X^* \wedge X \xrightarrow{1 \wedge \mu} V \wedge S^0 \simeq V.$$

If $g : U \to X \wedge V$ is a map, then as representative for ${}_\mu D[g]$ we take

$$X^* \wedge U \xrightarrow{1 \wedge g} X^* \wedge X \wedge V \xrightarrow{\mu \wedge 1} S^0 \wedge V \simeq V.$$

D_μ and ${}_\mu D$ are natural with respect to maps $U' \to U$ and $V \to V'$ and hence are homomorphisms. In like manner, given a map $\rho : S^0 \to X^* \wedge X$ we can define homomorphisms

$$D^\rho : [U \wedge X^*, V] \to [U, V \wedge X]$$
$${}^\rho D : [X \wedge U, V] \to [U, X^* \wedge V].$$

14.20. Definition. We say that μ is an *S-duality map* if D_μ and ${}_\mu D$ are isomorphisms for all spectra U, V. X^* is called an *S-dual* of X.

14.21. Example. If we take $X = \Sigma^p S^0$, $X^* = \Sigma^{-p} S^0$ and $\mu : X^* \wedge X \to S^0$ the standard homotopy equivalence $\Sigma^{-p} S^0 \wedge \Sigma^p S^0 \simeq S^0$, then μ is an S-duality map.

14.22. Proposition. *If X, X^* are finite spectra, then $\mu : X^* \wedge X \to S^0$ is an S-duality map if and only if D_μ, ${}_\mu D$ are isomorphisms for $U = \Sigma^p S^0$, $V = \Sigma^q S^0$ for all $p, q \in \mathbb{Z}$.*

Proof: First assume $U = \Sigma^p S^0$ and $\{V^n\}$ is the layer filtration of V. Then $V^1 = \vee_{\alpha \in A_1} \Sigma^{d_\alpha} S^0$, and we have a commutative diagram

$$\begin{array}{ccc} [U, V^1 \wedge X^*] & \xrightarrow{D_\mu} & [U \wedge X, V^1] \\ \uparrow \{(i_\alpha \wedge 1)_*\} & & \uparrow \{i_{\alpha *}\} \\ \oplus_{\alpha \in A_1} [U, \Sigma^{d_\alpha} S^0 \wedge X^*] & \underset{\oplus_\alpha D_\mu}{\xrightarrow{\cong}} & \oplus_{\alpha \in A_1} [U \wedge X, \Sigma^{d_\alpha} S^0]. \end{array}$$

Now in general $\{i_{\alpha *}\} : \oplus_{\alpha \in A} [W, X_\alpha] \to [W, \vee_{\alpha \in A} X_\alpha]$ is an isomorphism for all finite W and any collection $\{X_\alpha\}$. The result is true for $W = \Sigma^d S^0$ by 8.36. If $W = W' \cup C(\Sigma^d S^0)$, then we have a commutative diagram of exact sequences

$$\begin{array}{ccccccccc} \oplus_\alpha [\Sigma^d S^0, X_\alpha] & \leftarrow & \oplus_\alpha [W', X_\alpha] & \leftarrow & \oplus_\alpha [W, X_\alpha] & \leftarrow & \oplus_\alpha [\Sigma^{d+1} S^0, X_\alpha] & \leftarrow & \oplus_\alpha [\Sigma W', X_\alpha] \\ \downarrow \{i_{\alpha *}\} & & \downarrow \{i_{\alpha *}\} & & \downarrow \{i_{\alpha *}\} & & \downarrow \{i_{\alpha *}\} & & \downarrow \{i_{\alpha *}\} \\ [\Sigma^d S^0, \vee_\alpha X_\alpha] & \leftarrow & [W', \vee_\alpha X_\alpha] & \leftarrow & [W, \vee_\alpha X_\alpha] & \leftarrow & [\Sigma^{d+1} S^0, \vee_\alpha X_\alpha] & \leftarrow & [\Sigma W', \vee_\alpha X_\alpha], \end{array}$$

so that we may argue by induction on the number of cells in W. Hence in the diagram above $\{(i_\alpha \wedge 1)_*\}$ and $\{i_{\alpha *}\}$ are isomorphisms, and hence D_μ is also.

Suppose D_μ has been proved an isomorphism for $U = \Sigma^p S^0$ and V^{n-1}. Then we have a commutative diagram of exact sequences

$$\begin{array}{ccccccccc} \cdots \to & [U, \vee_{\beta \in A_n} \Sigma^{d_\beta} S^0 \wedge X^*] & \to & [U, V^{n-1} \wedge X^*] & \to & [U, V^n \wedge X^*] & \to & [U, \vee_\beta \Sigma^{d_\beta + 1} S^0 \wedge X^*] & \to \cdots \\ & \cong \downarrow D_\mu & & \cong \downarrow D_\mu & & \downarrow D_\mu & & \cong \downarrow D_\mu & \\ \cdots \to & [U \wedge X, \vee_{\beta \in A_n} \Sigma^{d_\beta} S^0] & \to & [U \wedge X, V^{n-1}] & \to & [U \wedge X, V^n] & \to & [U \wedge X, \vee_\beta \Sigma^{d_\beta + 1} S^0] & \to \cdots. \end{array}$$

Thus by the 5-lemma D_μ is an isomorphism for $U = \Sigma^p S^0, V^n$; hence by induction D_μ is an isomorphism for $U = \Sigma^p S^0$, V^n, all $n \geqslant 0$.

Then we have another commutative diagram

$$\begin{array}{ccc} [U,V\wedge X^*] & \xrightarrow[D_\mu]{} & [U\wedge X,V] \\ \cong \Big\uparrow \{(i_n\wedge 1)_*\} & & \cong \Big\uparrow \{i_{n*}\} \\ \operatorname{dir\,lim}\,[U,V^n\wedge X^*] & \xrightarrow[\operatorname{dir\,lim} D_\mu]{\cong} & \operatorname{dir\,lim}\,[U\wedge X,V^n], \end{array}$$

in which $\{(i_n\wedge 1)_*\}$, $\{i_{n*}\}$ are isomorphisms by 8.34. Thus D_μ is an isomorphism for $U=\Sigma^p S^0$, any V.

Now suppose V to be an arbitrary but fixed spectrum and let U have layer filtration $\{U^n\}$. An inductive argument similar to the one above shows $D_\mu:[U^n,V\wedge X^*]\to[U^n\wedge X,V]$ is an isomorphism for all $n\geqslant 0$. Finally we have a commutative diagram of exact sequences

$$\begin{array}{ccccccccc} 0 & \to & \lim^1[U^n,\Sigma^{-1}V\wedge X^*] & \to & [U,V\wedge X^*] & \to & \lim^0[U^n,V\wedge X^*] & \to & 0 \\ & & \cong\Big\downarrow \lim^1 D_\mu & & \Big\downarrow D_\mu & & \cong\Big\downarrow \lim^0 D_\mu & & \\ 0 & \to & \lim^1[U^n\wedge X,\Sigma^{-1}V] & \to & [U\wedge X,V] & \to & \lim^0[U^n\wedge X,V] & \to & 0, \end{array}$$

from which it follows that D_μ is an isomorphism for all U, V.

The proof that ${}_\mu D$ is an isomorphism is similar. ▯

Remark. A similar proof shows D^ρ, ${}^\rho D$ are isomorphisms for all U,V if and only if they are so for $U=\Sigma^p S^0$, $V=\Sigma^q S^0$, $p,q\in\mathbb{Z}$.

14.23. Proposition. *If $\mu:X^*\wedge X\to S^0$ is an S-duality map, then so is $\mu\circ\tau:X\wedge X^*\to S^0$.*

Proof: The diagram

$$\begin{array}{ccc} [U,V\wedge X^*] & \xrightarrow{D_\mu} & [U\wedge X,V] \\ \cong\Big\downarrow \tau_* & & \cong\Big\downarrow \tau^* \\ [U,X^*\wedge V] & \xrightarrow{{}_{\mu\tau}D} & [X\wedge U,V] \end{array}$$

commutes, showing that ${}_{\mu\tau}D$ is an isomorphism if and only if D_μ is. Similarly $D_{\mu\tau}$ is an isomorphism if and only if ${}_\mu D$ is. ▯

Thus if X^* is an S-dual of X, then X is an S-dual of X^*.

If $\mu:X^*\wedge X\to S^0$, $\nu:Y^*\wedge Y\to S^0$ are S-duality maps, then the composite $\theta(\mu,\nu)$

$$[X,Y]\ \underset{\cong}{\xrightarrow{{}_\nu D}}\ [Y^*\wedge X,S^0]\ \underset{\cong}{\xleftarrow{D_\mu}}\ [Y^*,X^*]$$

is an isomorphism. If $f:X \to Y$ is any map, then $[f^*] = \theta(\mu, \nu)[f]$ is characterized by the fact that the diagram

$$\begin{array}{ccc} Y^* \wedge X & \xrightarrow{f^* \wedge 1} & X^* \wedge X \\ {\scriptstyle 1 \wedge f}\downarrow & & \downarrow{\scriptstyle \mu} \\ Y^* \wedge Y & \xrightarrow{\nu} & S^0 \end{array}$$

commutes up to homotopy.

14.24. Proposition. *If $\mu: X^* \wedge X \to S^0$, $\nu: Y^* \wedge Y \to S^0$ and $\pi: Z^* \wedge Z \to S^0$ are S-duality maps and $f: X \to Y$, $g: Y \to Z$ are any maps, then* i) $f^{**} \simeq f$, ii) $(g \circ f)^* \simeq f^* \circ g^*$ *and* iii) $1^* \simeq 1$.

Proof: By 14.23 we can take X for X^{**}, Y for Y^{**}. The diagram

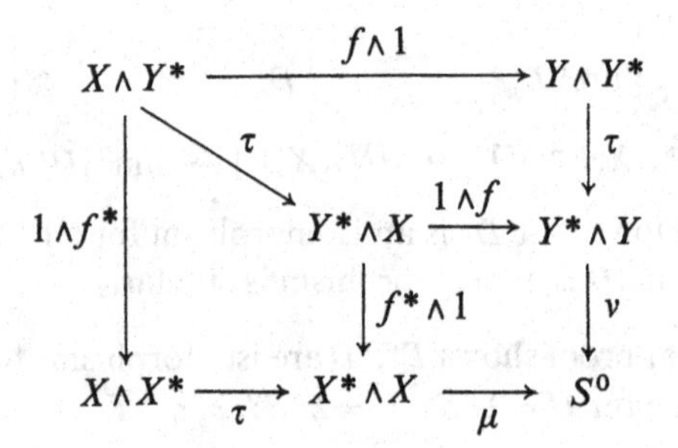

commutes up to homotopy by definition of f^*, which shows that f will do as f^{**}.

The diagram

$$\begin{array}{ccccc} Z^* \wedge X & \xrightarrow{1 \wedge f} & Z^* \wedge Y & \xrightarrow{1 \wedge g} & Z^* \wedge Z \\ {\scriptstyle g^* \wedge 1}\downarrow & & \downarrow{\scriptstyle g^* \wedge 1} & & \\ Y^* \wedge X & \xrightarrow{1 \wedge f} & Y^* \wedge Y & & \downarrow{\scriptstyle \pi} \\ {\scriptstyle f^* \wedge 1}\downarrow & & & \searrow{\scriptstyle \nu} & \\ X^* \wedge X & & \xrightarrow{\mu} & & S^0 \end{array}$$

also commutes up to homotopy by definition of f^* and g^*, which shows $f^* \circ g^*$ will do as $(g \circ f)^*$. iii) is clear. ▯

14.25. Corollary. *If $\mu: X^* \wedge X \to S^0$ and $\hat{\mu}: \hat{X}^* \wedge X \to S^0$ are both S-duality maps, then there is a homotopy equivalence $h: X^* \to \hat{X}^*$ such that $\hat{\mu} \circ (h \wedge 1) \simeq \mu$, and h is unique up to homotopy.*

Proof: $1:X \to X$ induces maps $1^* = h:X^* \to \hat{X}^*$ and $1^* = k:\hat{X}^* \to X^*$, unique up to homotopy, such that $\hat{\mu} \circ (h \wedge 1) \simeq \mu$, $\mu \circ (k \wedge 1) \simeq \hat{\mu}$. Then $h \circ k$ and $1_{\hat{X}^*}$ are both maps which can stand in for $1^*:\hat{X}^* \to \hat{X}^*$, so $h \circ k \simeq 1_{\hat{X}^*}$. Similarly $k \circ h \simeq 1_{X^*}$. □

Thus any two S-duals of X are homotopy equivalent by a canonical homotopy equivalence. We shall show shortly that every finite spectrum has an S-dual.

14.26. Lemma. *If $\mu:X^* \wedge X \to S^0$ and $v:Y^* \wedge Y \to S^0$ are S-duality maps, then so is the composite (μ, v) given by*

$$X^* \wedge Y^* \wedge Y \wedge X \xrightarrow{1 \wedge v \wedge 1} X^* \wedge S^0 \wedge X \simeq X^* \wedge X \xrightarrow{\mu} S^0.$$

Proof: One simply checks that the following diagram commutes for all U,V:

$$\begin{array}{ccc} [U, V \wedge X^* \wedge Y^*] & \xrightarrow{D_{(\mu, v)}} & [U \wedge Y \wedge X, V] \\ & {}_{D_v}\searrow^{\cong} \quad\quad {}^{\cong}\nearrow_{D_\mu} & \\ & [U \wedge Y, V \wedge X^*]. & \end{array}$$

□

Suppose $\mu:X^* \wedge X \to S^0$, $v:Y^* \wedge Y \to S^0$ are S-duality maps as in 14.26 and $\rho: Y \wedge X \to S^0$ is any map. Since $(Y \wedge X)^* = X^* \wedge Y^*$, there is a map $\rho^*: S^0 \to X^* \wedge Y^*$ such that

$$\begin{array}{ccc} S^0 \wedge Y \wedge X & \xrightarrow{\rho^* \wedge 1} & X^* \wedge Y^* \wedge Y \wedge X \\ \downarrow{\scriptstyle 1 \wedge \rho} & & \downarrow{\scriptstyle (\mu, v)} \\ S^0 \wedge S^0 & \cong & S^0 \end{array}$$

commutes up to homotopy.

14.27. Lemma. *ρ is an S-duality map if and only if D^{ρ^*} and ${}^{\rho^*}D$ are isomorphisms for all U,V.*

Proof: One has only to check that the diagram

$$\begin{array}{ccccc} [U, V \wedge Y] & \underset{\cong}{\xrightarrow{\tau_*}} & [U, Y \wedge V] & \underset{\cong}{\xrightarrow{{}_vD}} & [Y^* \wedge U, V] \\ \downarrow{\scriptstyle D_\rho} & & & & \downarrow{\scriptstyle {}^{\rho^*}D} \\ [U \wedge X, V] & \underset{\cong}{\xleftarrow{D_\mu}} & [U, V \wedge X^*] & \underset{\cong}{\xleftarrow{\tau_*}} & [U, X^* \wedge V] \end{array}$$

commutes, and hence D_ρ is an isomorphism if and only if ${}^{\rho^*}D$ is. Similarly ${}_\rho D$ is an isomorphism if and only if D^{ρ^*} is. □

Hence we shall also speak of maps $\rho: S^0 \to X^* \wedge X$ as being S-duality maps if D^ρ and ${}^\rho D$ are isomorphisms for all U,V. 14.27 assures us that there is a one–one correspondence between S-duality maps of this type and those of the type defined in 14.20.

14.28. Lemma. *If $\mu: X^* \wedge X \to S^0$, $\nu: Y^* \wedge Y \to S^0$ are S-duality maps and $f: X \to Y$ is a map, then $(\Sigma^p f)^* \simeq \Sigma^{-p}(f^*)$.*

Proof: The diagram

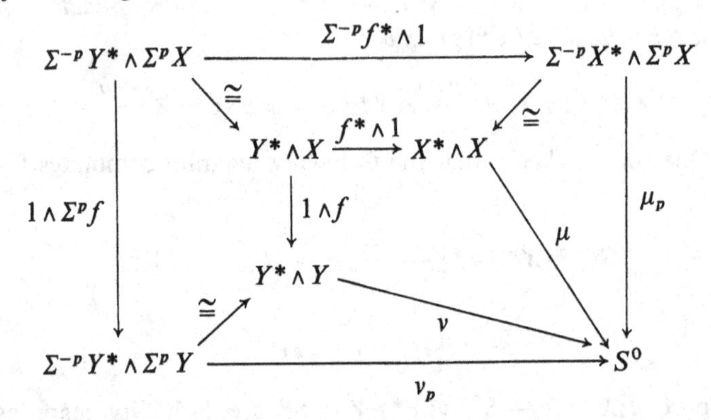

commutes up to homotopy by definition of f^*. Here μ_p, ν_p are the obvious S-duality maps. Thus $\Sigma^{-p} f^*$ will do as $(\Sigma^p f)^*$. ▯

14.29. Lemma. *If $\mu: X^* \wedge X \to S^0$, $\nu: Y^* \wedge Y \to S^0$ are S-duality maps and $f: X \to Y$ is any map, then*

$$\begin{array}{ccc} [U, V \wedge X^*] & \xrightarrow{D_\mu} & [U \wedge X, V] \\ \uparrow (1 \wedge f^*)_* & & \uparrow (1 \wedge f)^* \\ [U, V \wedge Y^*] & \xrightarrow{D_\nu} & [U \wedge Y, V] \end{array}$$

commutes.

14.30. Lemma. *If $f: X \to Y$, $g: U \to V$ are maps of spaces and $Z = Y \cup_f CX$, $W = V \cup_g CU$, then*

$$Z \wedge W / Y \wedge V \cong [(SX \wedge V) \vee (Y \wedge SU)] \cup_h CS[X \wedge U],$$

where h is the composite

$$S(X \wedge U) \xrightarrow{\mu'} S(X \wedge U) \vee S(X \wedge U) \cong (SX \wedge U) \vee (X \wedge SU) \xrightarrow{(1 \wedge g, f \wedge \nu')} (SX \wedge V) \vee (Y \wedge SU).$$

Proof: We have a homeomorphism

$$\frac{I \times I}{I \times \{0,1\} \cup \{0\} \times I} \to \frac{I \times I}{I \times \{0\} \cup \{0\} \times I}$$

given by

$$[s, t] \mapsto \begin{cases} [2st, s] & 0 \leqslant t \leqslant \frac{1}{2} \\ [s, 2s - 2st] & \frac{1}{2} \leqslant t \leqslant 1. \end{cases}$$

We define $\phi: [(SX \wedge V) \vee (Y \wedge SU)] \cup_h CS[X \wedge U] \to Z \wedge W / Y \wedge V$ by

$$\phi[t, x, v] = [t, x, v], \quad [t, x, v] \in SX \wedge V,$$

$$\phi[y, t, u] = [y, t, u], \quad [y, t, u] \in Y \wedge SU,$$

$$\phi[s, t, x, u] = \begin{cases} [2st, x, s, u] & 0 \leqslant t \leqslant \frac{1}{2} \\ [s, x, 2s - 2st, u] & \frac{1}{2} \leqslant t \leqslant 1. \end{cases}$$

Then ϕ is continuous, and one easily writes down ϕ^{-1} also. □

14.31. Lemma. *Let* $X \xrightarrow{f} Y \xrightarrow{g} Z \xrightarrow{h} SX \xrightarrow{Sf} SY$ *and* $Y^* \xrightarrow{f^*} X^* \xrightarrow{h^*} Z^* \xrightarrow{g^*} SY^* \xrightarrow{Sf^*} SX^*$ *be cofibre sequences of spaces. Let* $\mu: X^* \wedge X \to S^n$, $\nu: Y^* \wedge Y \to S^n$ *be maps such that*

$$\begin{array}{ccc} Y^* \wedge X & \xrightarrow{f^* \wedge 1} & X^* \wedge X \\ \downarrow{\scriptstyle 1 \wedge f} & & \downarrow{\scriptstyle \mu} \\ Y^* \wedge Y & \xrightarrow{\nu} & S^n \end{array}$$

commutes up to homotopy. Then we can find a map $\pi: Z^* \wedge Z \to S^{n+1}$ *such that*

14.32.

$$\begin{array}{ccc} Z^* \wedge Y & \xrightarrow{g^* \wedge 1} & SY^* \wedge Y \\ \downarrow{\scriptstyle 1 \wedge g} & & \downarrow{\scriptstyle S\nu} \\ Z^* \wedge Z & \xrightarrow{\pi} & S^{n+1} \end{array} \quad \text{and} \quad \begin{array}{ccc} X^* \wedge Z & \xrightarrow{h^* \wedge 1} & Z^* \wedge Z \\ \downarrow{\scriptstyle 1 \wedge h} & & \downarrow{\scriptstyle \pi} \\ X^* \wedge SX \cong S(X^* \wedge X) & \xrightarrow{S\mu} & S^{n+1} \end{array}$$

commute up to homotopy.

Proof: We may as well assume $Z = Y \cup_f CX$, $Z^* = X^* \cup_{f^*} CY^*$. The composite

$$S(Y^* \wedge X) \xrightarrow{k} (SY^* \wedge Y) \vee (X^* \wedge SX) \xrightarrow{(S\nu, S\mu)} S^{n+1}$$

is null homotopic if k is defined by

$$S(Y^* \wedge X) \xrightarrow{\mu'} (SY^* \wedge X) \vee (Y^* \wedge SX) \xrightarrow{(1 \wedge f, f^* \wedge \nu')} (SY^* \wedge Y) \vee (X^* \wedge SX)$$

as in 14.30. Therefore by 14.30 there is a map $\rho: Z^* \wedge Z / X^* \wedge Y \to S^{n+1}$ with $\rho | [(SY^* \wedge Y) \vee (X^* \wedge SX)] \simeq (S\nu, S\mu)$. We take π to be the composite $Z^* \wedge Z \to Z^* \wedge Z / X^* \wedge Y \xrightarrow{\rho} S^{n+1}$. Since the diagram

$$\begin{array}{ccc}
(Z^* \wedge Y) \vee (X^* \wedge Z) & \xrightarrow{(1 \wedge g, h^* \wedge 1)} & Z^* \wedge Z \\
\downarrow{\scriptstyle (g^* \wedge 1) \vee (1 \wedge h)} & & \downarrow \\
(SY^* \wedge Y) \vee (X^* \wedge SX) & \longrightarrow & Z^* \wedge Z / X^* \wedge Y \\
& {\scriptstyle (S\nu, S\mu)} \searrow \quad S^{n+1} \quad \swarrow {\scriptstyle \rho} &
\end{array}$$

commutes up to homotopy, it follows that the two squares of 14.32 also commute up to homotopy. ▯

14.33. Corollary. *Let $X \xrightarrow{f} Y \xrightarrow{g} Z \xrightarrow{h} \Sigma X \xrightarrow{\Sigma f} \Sigma Y$ be a cofibre sequence and let $\mu: X^* \wedge X \to S^0$, $\nu: Y^* \wedge Y \to S^0$ be S-duality maps. If X, X^*, Y, Y^* are finite spectra, then we can find an S-duality map $\pi: Z^* \wedge Z \to S^0$ and a cofibre sequence*

$$X^* \xleftarrow{f^*} Y^* \xleftarrow{g^*} Z^* \xleftarrow{h^*} \Sigma^{-1} X^* \xleftarrow{\Sigma^{-1} f^*} \Sigma^{-1} Y^*.$$

Proof: First we remark that every finite spectrum E has a cofinal subspectrum $\Sigma^d E(X)$ for some integer d and some finite CW-complex X. For there is an N such that $E_m = S^{m-N} E_N$ for $m \geqslant N$ and E_N is finite; we take $X = E_N$ and $d = -N$.

We now assume $\mu: X^* \wedge X \to S^0$ is of the form $\Sigma^{-d} E(\bar{\mu}): \Sigma^{-d} E(\bar{X}^* \wedge \bar{X}) \to \Sigma^{-d} E(S^d)$ for some map $\bar{\mu}: \bar{X}^* \wedge \bar{X} \to S^d$ of finite complexes $\bar{X}, \bar{X}^*$. Similarly ν is induced by $\bar{\nu}: \bar{Y}^* \wedge \bar{Y} \to S^{d'}$, and by suspending if necessary we can arrange that $d' = d$. Similarly f is induced by $\bar{f}: \bar{X} \to \bar{Y}$, f^* by $\bar{f}^*: \bar{Y}^* \to \bar{X}^*$ (where again we replace any of the spaces $\bar{X}, \bar{Y}$, etc. by an appropriate suspension if necessary). Then we apply 14.31 to the cofibre sequences

$$\bar{X} \xrightarrow{\bar{f}} \bar{Y} \xrightarrow{\bar{g}} (\bar{Y} \cup_{\bar{f}} C\bar{X}) \xrightarrow{\bar{h}} S\bar{X} \xrightarrow{S\bar{f}} S\bar{Y}$$

and

$$\bar{Y}^* \xrightarrow{\bar{f}^*} \bar{X}^* \xrightarrow{\bar{h}^*} (\bar{X}^* \cup_{\bar{f}^*} C\bar{Y}^*) \xrightarrow{\bar{g}^*} S\bar{Y}^* \xrightarrow{S\bar{f}^*} S\bar{X}^*,$$

getting a map $\bar{\pi}: \bar{Z}^* \wedge \bar{Z} \to S^{d+1}$, where $\bar{Z} = \bar{Y} \cup_{\bar{f}} C\bar{X}$, $\bar{Z}^* = \bar{X}^* \cup_{\bar{f}^*} C\bar{Y}^*$. We may assume $Z = \Sigma^{-a} E(\bar{Z})$ and take $Z^* = \Sigma^{a-d-1} E(\bar{Z}^*)$. Then $\bar{\pi}$ gives a function $\Sigma^{-d-1} E(\bar{Z}^* \wedge \bar{Z}) \to \Sigma^{-d-1} E(S^{d+1})$, which defines a map $\pi: Z^* \wedge Z \to S^0$. We can define g^*, h^* in terms of $\bar{g}$, $\bar{h}$. Finally, we have a commutative diagram of exact sequences

$$\begin{array}{ccccc}
[U, V \wedge X^*] & \xleftarrow{(1 \wedge f^*)_*} & [U, V \wedge Y^*] & \xleftarrow{(1 \wedge g^*)_*} & [U, V \wedge Z^*] \\
\cong \downarrow D_\mu & & \cong \downarrow D_\nu & & \downarrow D_\pi \\
[U \wedge X, V] & \xleftarrow{(1 \wedge f)^*} & [U \wedge Y, V] & \xleftarrow{(1 \wedge g)^*} & [U \wedge Z, V]
\end{array}$$

$$\begin{array}{ccccc}
\xleftarrow{(1 \wedge h^*)_*} & [U, V \wedge \Sigma^{-1} X^*] & \longleftarrow & [U, V \wedge \Sigma^{-1} Y^*] \\
& \cong \downarrow D_\mu & & \cong \downarrow D_\nu \\
\xleftarrow{(1 \wedge h)^*} & [U \wedge \Sigma X, V] & \longleftarrow & [U \wedge \Sigma Y, V]
\end{array}$$

for all spectra U, V. By the 5-lemma it follows D_π is an isomorphism. Similarly ${}_\pi D$ is an isomorphism. ▯

14.34. Theorem. *Every finite spectrum has an S-dual.*

Proof: We know $\Sigma^p S^0$ has an S-dual $(\Sigma^{-p} S^0)$, for all $p \in \mathbb{Z}$. We may then argue by induction on the number of cells in the spectrum using 14.33. ▯

As an application of S-duality we can prove a representation theorem for homology theories. Note that for any spectrum E and S-duality map $\mu: X^* \wedge X \to S^0$ we have isomorphisms

$$D_\mu: E_p(X^*) \to E^{-p}(X)$$

$$D_{\mu\tau}: E_p(X) \to E^{-p}(X^*), \quad \text{all } p \in \mathbb{Z}.$$

14.35. Theorem. *If k_* is a homology theory on the category $\mathscr{PW}'_F$ of finite pointed CW-complexes or $\mathscr{SP}'_F$ (finite spectra), then there is a spectrum E and a natural equivalence $T_*: E_* \to k_*$ on $\mathscr{PW}'_F$ or $\mathscr{SP}'_F$.*

Proof: We may as well assume k_* is defined on $\mathscr{SP}'_F$, since we can always extend it to finite spectra F by taking $k_n(F) = \operatorname{dir\,lim} k_{n+p}(F_p)$. Then if $F = \Sigma^d E(X)$ for some finite CW-complex X, clearly $k_n(F) = k_{n-d}(X)$. If $f: F \to F'$ is a function, then $\{f_{p*}: k_{n+p}(F_p) \to k_{n+p}(F'_p)\}$ is a morphism of direct systems and hence defines $f_*: k_n(F) \to k_n(F')$. We take $\sigma: k_n(F) \to$

$k_{n+1}(\Sigma F)$ to be the natural equivalence

$$k_n(F) = \operatorname{dir\,lim} k_{n+p}(F_p) \cong \operatorname{dir\,lim} k_{n+p+1}(F_{p+1}) = k_{n+1}(\Sigma F).$$

Finally if $F \xrightarrow{f} G \xrightarrow{g} H$ is a special cofibre sequence, then for each p we have $k_{n+p}(F_p') \xrightarrow{f_{p*}'} k_{n+p}(G_p') \xrightarrow{g_{p*}'} k_{n+p}(H_p)$ is exact, and hence the sequence remains exact if we take direct limits (7.50). Therefore for any cofibre sequence $F \xrightarrow{f} G \xrightarrow{g} H$ the sequence

$$k_n(F) \xrightarrow{f_*} k_n(G) \xrightarrow{g_*} k_n(H)$$

is exact for all $n \in \mathbb{Z}$. Therefore k_* defines a homology theory on $\mathscr{SP}_F'$.

Now we define cofunctors $k^n: \mathscr{SP}_F' \to \mathscr{A}$ by $k^n(X) = k_{-n}(X^*)$ for $X \in \mathscr{SP}_F'$, X^* any finite S-dual of X. Since X^* is defined up to homotopy type, $k^n(X)$ is well defined. Given $f: X \to Y$, we take

$$k^n[f] = k_{-n}[f^*]: k^n(Y) \to k^n(X).$$

By 14.24 k^n is a cofunctor. We define $\sigma^n: k^{n+1}(\Sigma X) \to k^n(X)$ to be $\sigma_{-n}: k_{-n-1}(\Sigma^{-1} X^*) \to k_{-n}(X^*)$; then σ^n is a natural equivalence. If $X \xrightarrow{f} Y \xrightarrow{g} Z$ is a cofibre sequence, then so also is $X^* \xleftarrow{f^*} Y^* \xleftarrow{g^*} Z^*$; hence the sequence

$$\begin{array}{ccccc} k^n(X) & \xleftarrow{f^*} & k^n(Y) & \xleftarrow{g^*} & k^n(Z) \\ \| & & \| & & \| \\ k_{-n}(X^*) & \xleftarrow{(f^*)_*} & k_{-n}(Y^*) & \xleftarrow{(g^*)_*} & k_{-n}(Z^*) \end{array}$$

is exact. Thus k^* is a cohomology theory on $\mathscr{SP}_F'$.

Now we may apply 9.27 to k^*, obtaining a spectrum E and a natural equivalence $T^*: E^* \to k^*$ on $\mathscr{SP}_F'$. We define $T_*: E_* \to k_*$ to be the unique natural equivalence such that the square

$$\begin{array}{ccc} E_n(X) & \xrightarrow{T_n(X)} & k_n(X) \\ \downarrow{\scriptstyle D_{\mu\tau}} & & \| \\ E^{-n}(X^*) & \xrightarrow{T^{-n}(X^*)} & k^{-n}(X^*) \end{array}$$

commutes. □

14.36. Corollary. *If k_* is a homology theory on $\mathscr{PW}'$ (or $\mathscr{SP}'$) such that $\{i_{\alpha *}\}: \operatorname{dir\,lim} k_n(X_\alpha) \to k_n(X)$ is an isomorphism for all $n \in \mathbb{Z}$, where $\{X_\alpha\}$ runs over the finite subcomplexes (or subspectra) of X, then there is a spectrum E and a natural equivalence $T_*: E_* \to k_*$ on $\mathscr{PW}'$ (or $\mathscr{SP}'$).*

Proof: We take the spectrum E and natural equivalence $T_*: E_* \to k_*$ on $\mathscr{PW}_F'$ (or $\mathscr{SP}_F'$) given by 14.35. We extend T_* to $\mathscr{PW}'$ (or $\mathscr{SP}'$) by taking $T_n(X)$ to be the unique isomorphism such that

$$\begin{array}{ccc} E_n(X) & \xrightarrow[T_n(X)]{} & k_n(X) \\ \{i_{\alpha *}\} \downarrow \cong & & \cong \downarrow \{i_{\alpha *}\} \\ \operatorname{dir\,lim} E_n(X_\alpha) & \xrightarrow[\operatorname{dir\,lim} T_n(X_\alpha)]{\cong} & \operatorname{dir\,lim} k_n(X_\alpha) \end{array}$$

commutes, where $\{X_\alpha\}$ runs over all finite subcomplexes (subspectra) of X. The vertical arrows are isomorphisms because of 8.35 or 8.34 and the hypothesis on k_*. ▯

Remark 1. One can show that the wedge axiom for k_* implies the hypothesis of 14.36; the proof is a generalization of 7.53 using a "generalized telescope".

Remark 2. Suppose we have two homology theories k_*, k'_* on $\mathscr{SP}'_F$ with $k_q(S^0) = k'_q(S^0) = 0$ for $q < 0$ and a natural transformation $T: k_* \to k'_*$ between them. In an obvious fashion we get a natural transformation $T^*: k^* \to k'^*$ between the corresponding cohomology theories and hence by 9.30 a map $f: E \to E'$ between the classifying spectra, unique up to weak homotopy. If T is a natural equivalence, then f is a homotopy equivalence; thus E is determined up to homotopy type. It is not clear what happens for general k_*, k'_*.

As another application of 14.34 we can give a simpler criterion than 14.22 for deciding whether a map $\mu: X^* \wedge X \to S^0$ (X, X^* finite) is an S-duality map.

14.37. Proposition. *Let X, X^* be finite spectra. Then a map $\mu: X^* \wedge X \to S^0$ is an S-duality map if and only if*

$$D_\mu: H_p(X^*; \mathbb{Z}) \to H^{-p}(X; \mathbb{Z})$$

and

$$D_{\mu\tau}: H_p(X; \mathbb{Z}) \to H^{-p}(X^*; \mathbb{Z})$$

are isomorphisms for all $p \in \mathbb{Z}$.

Proof: Since X is finite, 14.34 assures us that there is an S-duality map $\nu: X' \wedge X \to S^0$ for some spectrum X'. Then we have the homomorphism $\theta(\nu, \mu)$

$$[X, X] \xrightarrow{\ {}_\mu D\ } [X^* \wedge X, S^0] \underset{\cong}{\xleftarrow{\ D_\nu\ }} [X^*, X']$$

and we choose a map $f:X^* \to X'$ such that $\theta(\nu,\mu)[1_X] = [f]$—in other words the diagram

$$\begin{array}{ccc} X^* \wedge X & \xrightarrow{f \wedge 1} & X' \wedge X \\ & {}_{\mu}\searrow \quad \swarrow_{\nu} & \\ & S^0 & \end{array}$$

commutes up to homotopy. Hence the diagrams

$$\begin{array}{ccc} \pi_p^S(X') & \xrightarrow[\cong]{D_\nu} & \pi_S^{-p}(X) \\ {}^{f_*}\nwarrow & & \nearrow{}^{D_\mu} \\ & \pi_p^S(X^*) & \end{array} \qquad \begin{array}{ccc} H_p(X';\mathbb{Z}) & \xrightarrow[\cong]{D_\nu} & H^{-p}(X;\mathbb{Z}) \\ {}^{f_*}\nwarrow & & \nearrow{}^{D_\mu} \\ & H_p(X^*;\mathbb{Z}) & \end{array}$$

also commute. If $D_\mu: H_p(X^*;\mathbb{Z}) \to H^{-p}(X;\mathbb{Z})$ is an isomorphism for all $p \in \mathbb{Z}$, then by the stable version of Thm. 10.28 (Whitehead Thm.) f is a homotopy equivalence, and hence $D_\mu: \pi_p^S(X^*) \to \pi_S^{-p}(X)$ is an isomorphism for all $p \in \mathbb{Z}$. A similar proof shows that $D_{\mu\tau}: \pi_p^S(X) \to \pi_S^{-p}(X^*)$ is an isomorphism for all $p \in \mathbb{Z}$. Hence by 14.22 μ is an S-duality map. □

14.38. Exercise. Show that $D_{\mu\tau}: H_p(X;\mathbb{Z}) \to H^{-p}(X^*;\mathbb{Z})$ is an isomorphism for all $p \in \mathbb{Z}$ if and only if ${}_\mu D \circ \tau_*: H_p(X;\mathbb{Z}) \to H^{-p}(X^*;\mathbb{Z})$ is an isomorphism for all $p \in \mathbb{Z}$. Similarly D_μ is an isomorphism if and only if ${}_{\mu\tau}D \circ \tau_*$ is.

The map ${}_\mu D \circ \tau_*$ is easily seen to be given by $x \mapsto \mu^*(\iota^0)/x$ for $x \in H_*(X;\mathbb{Z})$, where $\iota^0 \in \tilde{H}^0(S^0;\mathbb{Z})$ is the generator. Thus for a given $\mu: X^* \wedge X \to S^0$ we need only check that $x \mapsto \mu^*(\iota^0)/x$ is an isomorphism. Alternatively, if $\rho: S^0 \to X^* \wedge X$ is the map dual to μ, then one need only check that $y \mapsto y\backslash\rho_*(\iota_0)$, $y \in H^p(X^*;\mathbb{Z})$, ($\iota_0 \in \tilde{H}_0(S^0;\mathbb{Z})$ the generator) is an isomorphism.

We can now establish a connection between the duality theorems (Alexander, Lefschetz, Poincaré) at the beginning of the chapter and the S-duality just discussed.

14.39. Let M^n be a smooth closed manifold. We can embed M^n in a high-dimensional sphere S^{n+k} with normal bundle ν and tubular neighborhood N, which we identify with the total space of the normal disk bundle $D(\nu)$. As in Chapter 12 we have the map $\pi: S^{n+k} \to M(\nu)$ which is the projection $D(\nu) \to M(\nu)$ on N and sends all of $S^{n+k} - \mathring{N}$ to the base point of $M(\nu)$. We also have the composite ρ given by

$$S^{n+k} \xrightarrow{\pi} M(\nu) \xrightarrow{\Delta'} M(\nu) \wedge D(\nu)^+ \xrightarrow{1 \wedge p} M(\nu) \wedge M^+,$$

where Δ' is induced by the diagonal

$$\Delta.(D(\nu), S(\nu)) \to (D(\nu) \times D(\nu), S(\nu) \times D(\nu))$$

and $p: D(\nu) \to M$ is the bundle projection. We are going to show that ρ is an S-duality map, at least in the case where M is orientable with respect to $H(\mathbb{Z})$—i.e. orientable in the traditional sense.

If we regard the normal bundle ν as a microbundle, then a Thom class is an element $t \in E^k(E(\nu), E(\nu) - i(M))$, where $i(M)$ denotes the 0-section of the bundle. But we have isomorphisms

$$E^k(E(\nu), E(\nu) - i(M)) \cong E^k(D(\nu), D(\nu) - i(M)) \cong$$

$$E^k(D(\nu), S(\nu)) \cong \tilde{E}^k(M(\nu)),$$

so we may regard t as an element of $E^k(D(\nu), S(\nu)) \cong \tilde{E}^k(M(\nu))$. The Thom isomorphism ϕ_t can then be regarded as a morphism

$$\phi_t: \tilde{E}_{k+r}(M(\nu)) \to E_r(M).$$

14.40. Lemma. *If $t \in \tilde{E}^k(M(\nu))$ is a Thom class for ν, then the element $z = \phi_t(\pi_*(\iota_{n+k})) \in E_n(M)$ is a fundamental class for M, where*

$$\iota_{n+k} \in \tilde{E}_{n+k}(S^{n+k})$$

is the generator.

Proof: We must show that for every point $x \in M$ the image of z in $E_n(M, M - \{x\})$ is a generator of $E_*(M, M - \{x\})$ as $\tilde{E}_*(S^0)$-module. We can choose a coordinate chart (U, h) around x such that ν is trivial over $V = h^{-1}(D^n)$. Then we have a commutative diagram

$$\begin{array}{ccccc}
\tilde{E}_{n+k}(M(\nu)) & \xleftarrow{\pi_*} & \tilde{E}_{n+k}(S^{n+k}) & \xrightarrow{\pi'_*} & \tilde{E}_{n+k}(S^{n+k}) \\
\| & & & & \wr\| \\
E_{n+k}(D(\nu), S(\nu)) & \searrow & & & \\
& & E_{n+k}(D(\nu), S(\nu) \cup D(\nu)|(M - \mathring{V})) & \leftarrow & E_{n+k}((V, \dot{V}) \times (D^k, S^{k-1})) \\
\cong \downarrow \phi_t & & \downarrow \phi_t & & \cong \downarrow \phi_{t|V \times D^k} \\
E_n(M) & \longrightarrow & E_n(M, M - \mathring{V}) & \xleftarrow{\cong} & E_n(V, \dot{V}) \\
& {\scriptstyle j_*} \searrow & & \swarrow \cong & \\
& & E_n(M, M - \{x\}), & &
\end{array}$$

where $\pi': S^{n+k} \to V \times D^k / \dot{V} \times D^k \cup V \times S^{k-1} \cong S^{n+k}$ is the projection on $D(\nu)|V$ and maps all of $S^{n+k} - (D(\nu)|V)^\circ$ to the base point. π' is clearly a map of degree 1—that is homotopic to $1_{S^{n+k}}$, and hence sends ι_{n+k} to ι_{n+k}. Therefore $j_*(z)$ is clearly a generator of $E_*(M, M - \{x\})$ as $\tilde{E}_*(S^0)$-module. $\square$

14.41. Lemma. *If $\rho: S^{n+k} \to M(\nu) \wedge M^+$ is the map defined in* 14.39 *and $z = \phi_t(\pi_*(\iota_{n+k}))$ then the diagrams*

$$\begin{array}{ccc} \tilde{E}^{p+k}(M(\nu)) & \xrightarrow{D^\rho} & E_{n-p}(M) \\ & \phi_t^* \nwarrow \quad \nearrow \cap z & \\ & E^p(M) & \end{array} \qquad \begin{array}{ccc} E^p(M) & \xrightarrow{D^{\tau\rho}} & \tilde{E}_{n+k-p}(M(\nu)) \\ & \cap z \searrow \quad \swarrow \phi_t & \\ & E_{n-p}(M) & \end{array}$$

commute up to sign for all $p \in \mathbb{Z}$.

Proof: We may think of ρ as a map $\rho: S^0 \to \Sigma^{-n-k} E(M(\nu)) \wedge E(M^+)$ if we wish.

$$\begin{aligned} D^\rho \circ \phi_t^*(x) &= \phi_t^*(x) \backslash \rho_*(\iota_{n+k}) = \phi_t^*(x) \backslash (1 \wedge p)_* \circ \Delta'_* \circ \pi_*(\iota_{n+k}) \\ &= p_*(\phi_t^*(x) \backslash \Delta'_* \pi_*(\iota_{n+k})) = p_*(\phi_t^*(x) \cap \pi_*(\iota_{n+k})) \\ &= p_*[(p^*(x) \cup t) \cap \pi_*(\iota_{n+k})] \\ &= p_*[p^*(x) \cap (t \cap \pi_*(\iota_{n+k}))] \\ &= x \cap p_*(t \cap \pi_*(\iota_{n+k})) = x \cap \phi_t(\pi_*(\iota_{n+k})) = x \cap z. \end{aligned}$$

The other half is similar. ▯

14.42. Proposition. *If M is a closed smooth manifold which is orientable with respect to $H(\mathbb{Z})$, then the map $\rho: S^0 \to \Sigma^{-n-k} E(M(\nu)) \wedge E(M^+)$ defined above is an S-duality map, and hence $M(\nu)$ is an S-dual for M^+.*

Proof: Since the stable normal bundle ν of M and tangent bundle τ satisfy $\tau \oplus \nu \simeq \varepsilon^{n+k}$ (ε^{n+k} a trivial bundle), it is easy to see M is orientable if and only if ν is. Hence ν has a Thom class $t \in \tilde{H}^k(M(\nu))$ and the diagrams

$$\begin{array}{ccc} \tilde{H}^{p+k}(M(\nu)) & \xrightarrow{D^\rho} & H_{n-p}(M) \\ & \phi_t^* \nwarrow \cong \quad \cong \nearrow \cap z & \\ & H^p(M) & \end{array} \qquad \begin{array}{ccc} H^p(M) & \xrightarrow{D^{\tau\rho}} & \tilde{H}_{n+k-p}(M(\nu)) \\ & \cap z \searrow \cong \quad \cong \swarrow \phi_t & \\ & H_{n-p}(M) & \end{array}$$

commute up to sign for all p. Thus D^ρ and $D^{\tau\rho}$ are isomorphisms, which by 14.37 and 14.38 implies ρ is an S-duality map. ▯

The statement that $M(\nu)$ is an S-dual of M^+ is true even if M is not orientable.

14.43. Theorem (Spanier, Milnor). *If M is a closed smooth manifold, then the Thom complex $M(\nu)$ of the stable normal bundle ν of M is an S-dual of M^+.*

Proof: We shall show that $M(\nu)$ has the same homotopy type as $S(\mathbb{R}^m - M)$. We embed M in $\mathbb{R}^m$ for some large m with tubular neighborhood N, which we identify with the total space of the normal disk bundle $D(\nu)$. $\mathbb{R}^m - \mathring{N}$ is a strong deformation retract of $\mathbb{R}^m - M$, hence an S-dual of M^+ (cf. remarks preceding 14.19).

Now for any vector bundle $\xi \to X$ we can regard $D(\xi)$ as the mapping cylinder of the projection $S(\xi) \to X$ (regard X as the 0-section). But for any map $f: X \to Y$ the inclusion $X \subset M_f$ is a cofibration, so for any vector bundle ξ the inclusion $S(\xi) \subset D(\xi)$ is a cofibration. In particular $\dot{N} \subset N$ is a cofibration and hence so is $\mathbb{R}^m - \mathring{N} \subset \mathbb{R}^m$. Since $\mathbb{R}^m$ is contractible, it follows that $S(\mathbb{R}^m - \mathring{N})$ has the same homotopy type as $\mathbb{R}^m/\mathbb{R}^m - \mathring{N}$. But we have homeomorphisms $\mathbb{R}^m/\mathbb{R}^m - \mathring{N} \cong N/\dot{N} \cong M(\nu)$. Thus $M(\nu)$ is an S-dual of M. □

Atiyah in [15] proved the more general result that if M is a compact smooth manifold with boundary $\dot{M}$, then $M(\nu)$ is an S-dual of $M/\dot{M}$. The proof is scarcely more difficult than that of 14.43.

14.44. Exercise. What are the analogs of 14.39–14.42 for manifolds with boundary?

14.45. *Remark.* Suppose M is an X-manifold (X as in 12.19), $\nu: M \to BO(k)$ the normal bundle map, $\nu: M \to X_k$ a lifting. The identity $1: M \to M$ can be regarded as a singular X-manifold in M and as such defines an element $z_M \in MX_n(M)$ (cf. 12.34). If we apply the Pontrjagin–Thom construction (12.35) in order to represent z_M as a homotopy class in $\pi_n(MX \wedge M^+)$, we find z_M is represented by

$$S^{n+k} \xrightarrow{\rho} M(\nu) \wedge M^+ \xrightarrow{M(\tilde{\nu}) \wedge 1} MX_k \wedge M^+,$$

ρ as in 14.39. Of course $\{M(\tilde{\nu}): M(\nu) \to MX_k\} \in \widetilde{MX}^k(M(\nu))$ is a Thom class t for ν (14.5i)). If we now chase $1 \in MX^0(M)$ around the diagram 14.41 we find $z = 1 \cap z = D^\rho \phi_t^*(1) = D^\rho(t) = \{(M(\tilde{\nu}) \wedge 1) \circ \rho\} = z_M$. Thus the singular X-manifold $1: M \to M$ defines a fundamental class z_M for M with respect to MX and $z_M = \phi_t(\pi_*(\iota_{n+k}))$.

References

1. J. F. Adams [8]
2. M. F. Atiyah [15]
3. E. H. Spanier [79, 80]

Chapter 15

Spectral Sequences

In Chapter 10 we saw that if k_* is an ordinary homology theory with coefficients G and X is a CW-complex, then we can often compute $k_*(X)$ by the following prescription: the groups $k_q(X^q/X^{q-1})$ and the boundary operators $\Delta: k_q(X^q/X^{q-1}) \to k_{q-1}(X^{q-1}/X^{q-2})$ of the triple (X^q, X^{q-1}, X^{q-2}) together form a chain complex $\{k_q(X^q/X^{q-1}), \Delta\}$, and the homology of this chain complex turns out to be $k_*(X)$. This is true essentially because $k_n(X^q/X^{q-1}) = 0$ if $n \neq q$, from which it follows that $k_n(X^q) = 0$ if $n > q$ and $k_n(X^q) \cong k_n(X)$ if $n < q$, among other things.

Now suppose we let k_* be an arbitrary homology theory satisfying the wedge and WHE axioms; what can we say then? $k_n(X^q/X^{q-1})$ will no longer be 0 for all $n \neq q$, so many other related groups will be non-zero too. It will turn out that we can still give a prescription from which $k_*(X)$ is computable in favorable cases, but as one might expect the method will be more complicated than the one described above. The only real difficulty is organizing in a useful way a mass of related groups and homomorphisms, most of which are zero or isomorphic to each other in the case where k_* is ordinary homology. Half the organizational battle is won by naming these groups in a useful way.

Let us assume that h_* is an arbitrary unreduced homology theory (satisfying the wedge axiom). Rather than assuming that X is a CW complex with skeleta X^q we simply assume for the moment that X is a topological space with a *filtration* $\{X^q\}$: that is, the X^q are closed subspaces of X such that $X = \cup_q X^q$, $X^q \subset X^{q+1}$, all q, $X^q = \varnothing$ for $q < 0$ and every compact subset of X is contained in some X^q. We now begin the naming of groups as follows:

$$Z^r_{pq} = \operatorname{im}[j_*: h_{p+q}(X^p, X^{p-r}) \to h_{p+q}(X^p, X^{p-1})]$$

$$B^r_{pq} = \operatorname{im}[\Delta: h_{p+q+1}(X^{p+r-1}, X^p) \to h_{p+q}(X^p, X^{p-1})]$$

$$Z^\infty_{pq} = \operatorname{im}[j_*: h_{p+q}(X^p) \to h_{p+q}(X^p, X^{p-1})]$$

$$B^\infty_{pq} = \operatorname{im}[\Delta: h_{p+q+1}(X, X^p) \to h_{p+q}(X^p, X^{p-1})]$$

and

$$F_{pq} = \operatorname{im}[i_*: h_{p+q}(X^p) \to h_{p+q}(X)].$$

15.1. Proposition. *The groups B^r_{pq}, Z^r_{pq}, B^∞_{pq}, Z^∞_{pq} for fixed p and q are all subgroups of $h_{p+q}(X^p, X^{p-1})$ and satisfy the following inclusion relations:*

$$0 = B^1_{pq} \subset B^2_{pq} \subset \cdots \subset B^r_{pq} \subset B^{r+1}_{pq} \subset \cdots \subset B^\infty_{pq} \subset Z^\infty_{pq} \subset \cdots$$
$$\subset Z^{r+1}_{pq} \subset Z^r_{pq} \cdots Z^1_{pq}.$$

Proof: Clearly $B^1_{pq} = 0$ and $Z^1_{pq} = h_{p+q}(X^p, X^{p-1})$. To prove the inclusion $B^r_{pq} \subset B^{r+1}_{pq}$ we have the commutative diagram

$$\begin{array}{ccc} h_{p+q+1}(X^{p+r}, X^p) & \xrightarrow{\ \Delta\ } & \\ \uparrow{\scriptstyle j_*} & & h_{p+q}(X^p, X^{p-1}). \\ h_{p+q+1}(X^{p+r-1}, X^p) & \xrightarrow{\ \Delta\ } & \end{array}$$

The other inclusions are proved similarly. □

We set $E^r_{pq} = Z^r_{pq}/B^r_{pq}$ and $E^\infty_{pq} = Z^\infty_{pq}/B^\infty_{pq}$.

15.2. Proposition. *There is a natural isomorphism*

$$Z^r_{pq}/Z^{r+1}_{pq} \cong B^{r+1}_{p-r,q+r-1}/B^r_{p-r,q+r-1}.$$

Proof: One simply stares at the commutative diagram

$$\begin{array}{ccccc} & & h_{p+q}(X^p, X^{p-r-1}) & & \\ & & \downarrow & \searrow{\scriptstyle j_{1*}} & \\ h_{p+q}(X^{p-1}, X^{p-r}) & \longrightarrow & h_{p+q}(X^p, X^{p-r}) & \xrightarrow[j_{2*}]{} & h_{p+q}(X^p, X^{p-1}) \\ & {\scriptstyle \Delta_1}\searrow & \downarrow{\scriptstyle \Delta_2} & & \\ & & h_{p+q-1}(X^{p-r}, X^{p-r-1}) & & \end{array}$$

got by putting together the exact sequences of the triples (X^p, X^{p-1}, X^{p-r}) and $(X^p, X^{p-r}, X^{p-r-1})$.

The isomorphism of 15.2 is induced by $\Delta_2 \circ j^{-1}_{2*}$. □

We use this isomorphism to define a map $d^r: E^r_{pq} \to E^r_{p-r,q+r-1}$ for all p, q:

$$\begin{array}{ccccc} Z^r_{pq}/B^r_{pq} & \xrightarrow{\text{epic}} & Z^r_{pq}/Z^{r+1}_{pq} \cong B^{r+1}_{p-r,q+r-1}/B^r_{p-r,q+r-1} & \xrightarrow{\text{monic}} & Z^r_{p-r,q+r-1}/B^r_{p-r,q+r-1} \\ \| & & & & \| \\ E^r_{pq} & & \xrightarrow{\qquad d^r \qquad} & & E^r_{p-r,q+r-1}. \end{array}$$

From the definition of d^r it is clear that $\ker d^r = Z^{r+1}_{pq}/B^r_{pq}$,

$$\operatorname{im} d^r = B^{r+1}_{p-r,q+r-1}/B^r_{p-r,q+r-1}.$$

Thus

$$\operatorname{im}[d^r : E^r_{p+r,q-r+1} \to E^r_{pq}] = B^{r+1}_{pq}/B^r_{pq}.$$

Hence $\operatorname{im} d^r \subset \ker d^r$, so $d^r \circ d^r = 0$. In other words, for fixed r

$$\cdots \longrightarrow E^r_{p+r,q-r+1} \xrightarrow{d^r} E^r_{pq} \longrightarrow E^r_{p-r,q+r-1} \longrightarrow \cdots$$

is a chain complex. Moreover

$$\ker d^r/\operatorname{im} d^r = (Z^{r+1}_{pq}/B^r_{pq})/(B^{r+1}_{pq}/B^r_{pq}) \cong Z^{r+1}_{pq}/B^{r+1}_{pq} = E^{r+1}_{pq}.$$

Therefore we have proved the following.

15.3. Proposition. $E^{r+1}_{**} \cong H(E^r_{**}, d^r)$.

Now we turn to the groups $F_{pq} \subset h_{p+q}(X)$.

15.4. Proposition. *We have inclusion relations*

$$h_{p+q}(X) \supset \cdots \supset F_{pq} \supset F_{p-1,q+1} \supset \cdots \supset F_{-1,p+q+1} = 0$$

and $\bigcup_{r+s=p+q} F_{rs} = h_{p+q}(X)$. *Moreover, there is a natural isomorphism* $F_{pq}/F_{p-1,q+1} \cong E^\infty_{pq}$, *all* p,q.

Thus the subgroups $F_{p,n-p} \subset h_n(X)$ form an infinite *filtration* of $h_n(X)$ and the quotient groups of the filtration are the groups $E^\infty_{p,n-p}$.

Proof: The inclusions are obvious, and since $\bigcup_p X^p = X$ we have $\{i_{p*}\}: \operatorname{dirlim}_p\, h_n(X^p) \cong h_n(X)$ by 7.73. But $\operatorname{im}\{i_{p*}\} = \bigcup_{r+s=n} F_{rs}$. Finally the isomorphism comes from staring at the commutative diagram

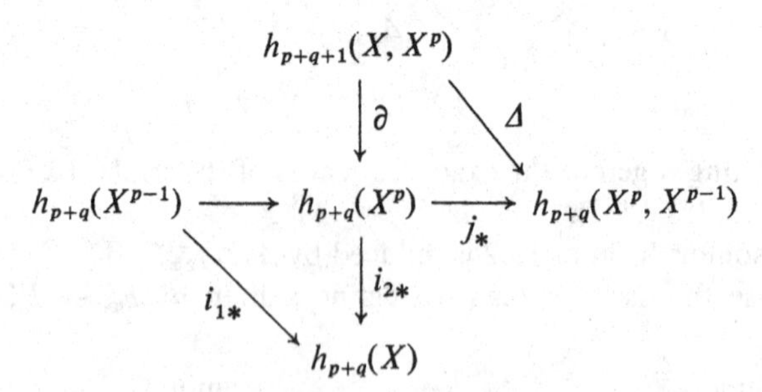

got by putting together the exact sequences of the pairs (X,X^p) and (X^p,X^{p-1}). The isomorphism of 15.4 is induced by $i_{2*} \circ j_*^{-1}$. □

Now it only remains to bridge the gap between E^∞_{**} and E^r_{**} for finite r. The notation already suggests the following result.

15.5. Proposition. *$Z^r_{pq} = Z^\infty_{pq}$ for $r > p$ and $B^\infty_{pq} = \bigcup_{r\geq 1} B^r_{pq}$. Thus we have an epimorphism $E^r_{pq} \to E^{r+1}_{pq}$ for $r > p$ and $E^\infty_{pq} = \operatorname{dir\,lim}_r E^r_{pq}$.*

Proof: For $r > p$ we have

$$\begin{aligned} Z^r_{pq} &= \operatorname{im}[j_*: h_{p+q}(X^p, X^{p-r}) \to h_{p+q}(X^p, X^{p-1})] \\ &= \operatorname{im}[i_*: h_{p+q}(X^p, \varnothing) \to h_{p+q}(X^p, X^{p-1})] \\ &= Z^\infty_{p+q}. \end{aligned}$$

Let $j_r: (X^{p+r-1}, X^p) \to (X, X^p)$ be the inclusion. Since $\bigcup X^n = X$,

$$\{j_{r*}\}: \operatorname{dir\,lim}_r h_{p+q+1}(X^{p+r-1}, X^p) \to h_{p+q+1}(X, X^p)$$

is an isomorphism (7.73). For each r let $\Delta_r: h_{p+q+1}(X^{p+r-1}, X^p) \to h_{p+q}(X^p, X^{p-1})$ be the boundary for the triple $(X^{p+r-1}, X^p, X^{p-1})$. These induce a homomorphism

$$\bar{\Delta}: \operatorname{dir\,lim}_r h_{p+q+1}(X^{p+r-1}, X^p) \to h_{p+q}(X^p, X^{p-1}).$$

Moreover the diagram

$$\begin{array}{ccc} \operatorname{dir\,lim}_r h_{p+q+1}(X^{p+r-1}, X^p) & \xrightarrow{\bar{\Delta}} & h_{p+q}(X^p, X^{p-1}) \\ {\scriptstyle \{j_{r*}\}} \searrow \cong & & \nearrow {\scriptstyle \Delta} \\ & h_{p+q+1}(X, X^p) & \end{array}$$

commutes; hence $B^\infty_{pq} = \operatorname{im} \Delta = \operatorname{im} \bar{\Delta} = \bigcup_{r\geq 1} B^r_{pq}$. The last two statements are then obvious. $\square$

15.6. Let us now summarize what we have achieved so far; we have arranged the job of calculating $h_*(X)$ as follows:

1) Identify $E^1_{pq} = h_{p+q}(X^p, X^{p-1})$ as something familiar (this we shall do for two important cases; compare with 10.7).
2) Identify $d^1: E^1_{pq} \to E^1_{p-1,q}$ as something familiar (this we do also; compare with 10.8ff).
3) Compute $E^2_{pq} = H(E^1_{pq}, d^1)$.
4) Try to work out d^2.
5) Compute $E^3_{pq} = H(E^2_{pq}, d^2)$.
6) Repeat this process, finding E^r_{pq}, $r \geq 1$.
7) Take $E^\infty_{pq} = \operatorname{dir\,lim}_r E^r_{pq}$.
8) $F_{0,n} = E^\infty_{0,n}$. Try to solve the extension problem $0 \to F_{p-1,n-p+1} \to F_{p,n-p} \to E^\infty_{p,n-p} \to 0$ for each $p > 0$, thus inductively determining $F_{p,n-p}$ for $p \geq 0$.
9) $h_n(X) = \bigcup_{p\geq 0} F_{p,n-p}$.

Note: Steps 4) and 8) are in general difficult but can be carried out in many important special cases.

This apparatus $\{E^r_{pq}, d^r\}$ is what one calls a *spectral sequence:* an abstract spectral sequence is a sequence E^1_{**}, E^2_{**}, ..., E^∞_{**} of chain complexes (let's not worry for the moment about the fact that they have double indexing) such that $H(E^n_{**}) \cong E^{n+1}_{**}$. They occur in other contexts as well. In the typical situation one has a good grasp on E^1_{**} (orE^2_{**}) and E^∞_{**} is related to something one would like to know. Thus the in-between terms E^n_{**} form a bridge from the already-known to the not-yet-known, a bridge one can cross if one can work out what the differentials are. This is a rather rough description, however, for it does often occur that one starts with partial information about E^2_{**} and complete knowledge of E^∞_{**} and tries to work backwards to determine E^2_{**} completely (see Thoerem 15.57, for example).

In working with spectral sequences we will often find it helpful to draw diagrams in which the groups E^r_{pq} for fixed r are associated to the integer lattice points (p,q) in the plane and the differentials are represented as arrows as in the figure below.

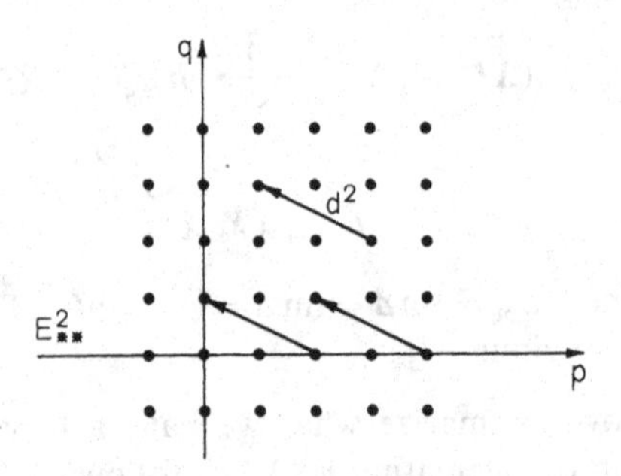

One must draw one such diagram for each r. Often one divides the plane into boxes and writes into the appropriate boxes those groups E^r_{pq} which are not zero—or simply appropriate generators of the non-zero groups. The diagram accompanying 15.38 and the ones on p. 447 are examples of such schemata.

Our spectral sequence has the additional property that the groups E^∞_{**} are the quotients of a filtration

$$\{0\} = F_{0,*} \subset F_{1,*} \subset F_{2,*} \subset \cdots \subset F_{n,*} \subset \cdots \subset \cdots \subset G_*$$

of a graded group $G_*(=h_*(X))$. In this situation if $E^\infty_{pq} = \operatorname{dir\,lim}_r E^r_{pq}$, then one says that the spectral sequence $\{E^r_{pq}, d^r\}$ *converges* to the graded group G_*. Thus our particular spectral sequence converges to $h_*(X)$.

We turn now to steps 1) and 2). Let us consider first the case where X is a CW-complex and $X^n \subset X$ is the n-skeleton. Then $X^p/X^{p-1} \cong$

$\vee_\alpha S_\alpha^p$ where α indexes the p-cells of X. Thus

$$E_{pq}^1 = h_{p+q}(X^p, X^{p-1}) \cong h_{p+q}(X^p/X^{p-1}, \{*\}) \cong h_{p+q}(\vee_\alpha S_\alpha^p, \{*\}) \cong$$

$$\oplus_\alpha h_{p+q}(S^p, \{*\}) \cong \oplus_\alpha h_q(S^0, \{*\}) \cong C_p(X) \otimes h_q(pt)$$

where $C_p(X) = \tilde{H}_p(X^p/X^{p-1})$ is the cellular chain group defined in Chapter 10. Moreover d^1 turns out to be just the boundary

$$h_{p+q}(X^p, X^{p-1}) \xrightarrow{\Delta} h_{p+q-1}(X^{p-1}, X^{p-2}),$$

and one verifies that this is of the form $d \otimes 1: C_p(X) \otimes h_q(pt) \to C_{p-1}(X) \otimes h_q(pt)$ exactly as in 10.11. Therefore we find that $E_{pq}^2 \cong H_p(X; h_q(pt))$, which gives a very satisfactory answer to 1), 2) and 3). Note that E_{pq}^1 depends on the CW-decomposition of X but E_{pq}^2 does not.

15.7. Theorem (Atiyah–Hirzebruch–Whitehead). *For any homology theory h_* and CW-complex X there is a spectral sequence $\{E_{pq}^r, d^r\}$ with $E_{pq}^2 \cong H_p(X; h_q(pt))$ and converging to $h_*(X)$.*

In other words we begin with a knowledge of the ordinary homology of X plus the coefficient groups of h_*, and with a little luck (remember steps 4) and 8)!) we end up with $h_*(X)$.

We turn now to another case of major importance. Suppose $p: E \to B$ is a fibration and B is a CW-complex with skeleta B^n. Then the subspaces $E^n = p^{-1}(B^n)$ form a filtration of E, so we get a spectral sequence $\{E_{pq}^r, d^r\}$ converging to $h_*(E)$ with $E_{pq}^1 = h_{p+q}(p^{-1}B^p, p^{-1}B^{p-1})$. What we wish to prove is that $E_{pq}^1 \cong C_p(B) \otimes h_q(F)$ and $d^1 = d \otimes 1$, so that $E_{pq}^2 \cong H_p(B; h_q(F))$. To do this, however, we shall need to delve a little more deeply into the theory of fibrations.

15.8. Suppose $p: E \to B$ is a fibration and $f: X \to B$ is a map. Let $E_f = \{(x, e) \in X \times E: f(x) = p(e)\} \subset X \times E$. The projections $\pi_1: X \times E \to X$ and $\pi_2: X \times E \to E$ when restricted to E_f give maps $p_f: E_f \to X$ and $\bar{f}: E_f \to E$ such that the square

$$\begin{array}{ccc} E_f & \xrightarrow{\bar{f}} & E \\ \downarrow{\scriptstyle p_f} & & \downarrow{\scriptstyle p} \\ X & \xrightarrow[f]{} & B \end{array}$$

commutes.

15.9. Proposition. *$p_f: E_f \to X$ is also a fibration with fibre homeomorphic to $F = p^{-1}(b_0) = p^{-1}(f(x_0))$.*

Proof: Suppose $g:Y \to E_f$ is a map and $H:Y \times I \to X$ a homotopy of $p_f \circ g$. Then $f \circ H:Y \times I \to B$ is a homotopy of $p \circ \bar{f} \circ g$, so there is a homotopy $H':Y \times I \to E$ of $\bar{f} \circ g$ lifting $f \circ H$—that is, $p \circ H' = f \circ H$. This last equation simply says that for all $(y,t) \in Y \times I$ we have

$$(H(y,t), H'(y,t)) \in E_f.$$

Thus if we define $K:Y \times I \to E_f$ by $K = (H, H')$, then we shall have $K_0 = (H_0, H'_0) = (p_f \circ g,\ \bar{f} \circ g) = g$ and $p_f \circ K = H$. Thus p_f is a fibration. The fibre is $p_f^{-1}(x_0) = \{(x_0, e): f(x_0) = p(e)\} \cong p^{-1}(b_0) = F$. ▯

The fibration $p_f:E_f \to X$ is called the fibration *induced* from $p:E \to B$ by f. We have already encountered an example in 2.52, the fibration $\pi:P_f \to X$ induced from the path fibration $PY \to Y$ by the map $f:X \to Y$. We also had induced vector bundles $f^*\xi$ in Chapter 11.

15.10. Definition. If $p:E \to B$ is a fibration, then a homotopy $H:X \times I \to E$ is called a *fibre homotopy* if and only if $p \circ H$ is a stationary homotopy (in other words, H only moves points around in their fibre). If $f, f':X \to E$ are maps such that $p \circ f = p \circ f'$ then f, f' are called *fibre homotopic* if there is a fibre homotopy between them; we write $f \simeq_p f'$. Two fibrations $p:E \to B$ and $p':E' \to B'$ are called *fibre homotopy equivalent* if there are fibre maps

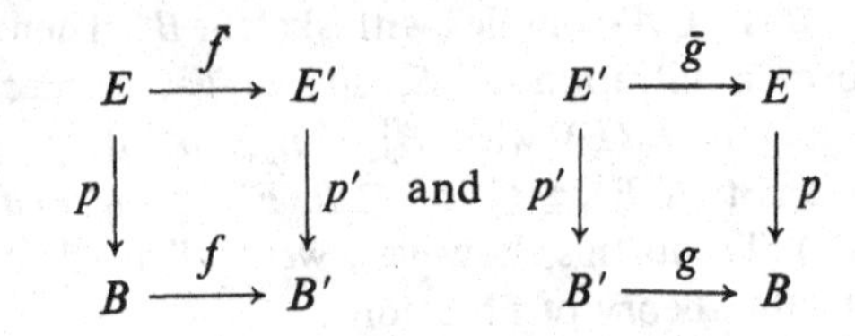

such that $\bar{f} \circ \bar{g} \simeq_{p'} 1_{E'}$, $\bar{g} \circ \bar{f} \simeq_p 1_E$. Two fibrations $p:E \to B$ and $p':E' \to B$ over the same base space B are called *strongly* fibre homotopy equivalent if we can take $f = g = 1$.

15.11. Proposition. *Let $p:E \to B$ be a fibration.*
a) *Let $f_0, f_1:X \to B$ be two maps and let $H:X \times I \to B$ be a homotopy between them. Then there is a fibre map $g_H:E_{f_0} \to E_{f_1}$ such that $\bar{f}_1 \circ g_H \simeq \bar{f}_0$ through a homotopy lifting $H \circ (p_{f_0} \times 1)$.*
b) *Let K be a second homotopy between f_0 and f_1 and let $M:X \times I \times I \to B$ be a homotopy from H to K rel $X \times \dot{I}$. Then there is a fibre homotopy $g_K \simeq_{p_{f_1}} g_H$.*

Proof: a) Let $H':E_{f_0} \times I \to E$ be a lifting of $H \circ (p_{f_0} \times 1)$ which satisfies $H'_0 = \bar{f}_0$. Let g_H be defined by $g_H = (p_{f_0}, H'_1)$; since $p \circ H'_1 = H_1 \circ p_{f_0} = f_1 \circ p_{f_0}$, it follows that $g_H(E_{f_0}) \subset E_{f_1}$. Clearly g_H is a fibre map and $\bar{f}_1 \circ g_H = H'_1 \simeq H'_0 = \bar{f}_0$.

b) Let $K':E_{f_0} \times I \to E$ be a lifting of $K \circ (p_{f_0} \times 1)$ with $K'_0 = \tilde{f}_0$ and let g_K be defined by $g_K = (p_{f_0}, K'_1)$. Let $A \subset I \times I$ be the subset $0 \times I \cup I \times \dot{I}$ and define $M':E_{f_0} \times A \to E$ by $M'(x,t,0) = H'(x,t)$, $M'(x,t,1) = K'(x,t)$, $M'(x,0,s) = \tilde{f}_0(x)$, all $x \in E_{f_0}$, s, $t \in I$. Then $p \circ M' = M \circ (p_{f_0} \times 1) | E_{f_0} \times A$. By using the standard homeomorphism $I \times I \to I \times I$ which carries A to $0 \times I$, we may interpret the homotopy lifting property for p as saying there is a lifting $M':E_{f_0} \times I \times I \to E$ of $M \circ (p_{f_0} \times 1 \times 1)$ which extends the definition already given on $E_{f_0} \times A$. Let $L:E_{f_0} \times I \to E_{f_1}$ be defined by

$$L(e', t) = (p_{f_0}(e'), M'(e', 1, t))$$

for $e' \in E_{f_0}$, $t \in I$. Then $L(e',0) = g_H(e')$, $L(e',1) = g_K(e')$ and L is a fibre homotopy from g_H to g_K. □

Remark. Only the fibre homotopy class $[g_H]$ in 15.11a) is uniquely defined.

We have several important applications of 15.11.

15.12. For any $b \in B$ let F_b denote the fibre $p^{-1}(b)$. If $*$ denotes the space consisting of a single point and $f:* \to B$ the unique map with $f(*) = b$, then $E_f \cong F_b$. Any path w from b_1 to b_2 in B can be interpreted as a homotopy from the map $* \to b_2$ to the map $* \to b_1$ and thus gives rise to a map $h_w:F_{b_2} \to F_{b_1}$. If w' is another path from b_1 to b_2 homotopic to w rel $\dot{I}$, then a homotopy from w to w' is a homotopy of homotopies as in b) above, so there is a homotopy $h_w \simeq h_{w'}$. Recall our notation in Chapter 3 (following 3.30) that $PB(b_1,b_2)$ denotes the set of homotopy classes of paths from b_1 to b_2 in B; then we have a well defined function $\alpha:PB(b_1,b_2) \to [F_{b_2}, F_{b_1}]$ defined by $\alpha[w] = [h_w]$. It follows from the constructions that α satisfies

i) $\alpha[w * w'] = \alpha[w] \circ \alpha[w']$, $[w] \in PB(b_1, b_2)$, $[w'] \in PB(b_2, b_3)$;

ii) $\alpha[w_0] = [1_{F_b}]$ if w_0 is the constant path at b.

In particular it follows $\alpha[w]$ is always a homotopy equivalence and hence any two fibres have the same homotopy type if B is 0-connected.

15.13. Definition. A map $\phi:F_{b_2} \to F_{b_1}$ is called *admissible* if $[\phi] = \alpha[w]$ for some $[w] \in PB(b_1,b_2)$. It follows from i) that the composition of two admissible maps is admissible and from i) and ii) that any admissible map is a homotopy equivalence whose homotopy inverse can be taken to be admissible also. There are admissible maps between any two fibres if B is 0-connected.

15.14. Definition. Let h_* be a homology theory. For any fibration $p:E \to B$ with fibre $F = p^{-1}(b_0)$ we get a pairing

$$\pi_1(B, b_0) \times h_*(F) \xrightarrow{\theta} h_*(F)$$

defined by $\theta([w],x) = \alpha[w]_*(x)$. 15.12i) and ii) imply that this is a group action of $\pi_1(B,b_0)$ on $h_*(F)$. We say that $p:E \to B$ is *orientable* with respect to h if this group action is trivial for all $b_0 \in B$—i.e. $\gamma x = x$ for all $\gamma \in \pi_1(B,b_0)$, $x \in h_*(F)$. If B is 0-connected it suffices that the action be trivial for any one $b_0 \in B$.

15.16. Proposition. *If $f_0, f_1 : X \to B$ are (freely) homotopic maps, then there is a strong fibre homotopy equivalence $g : E_{f_0} \to E_{f_1}$ such that $f_1 \circ g \simeq f_0$ and $(g|F_{f_0(x_0)}) : F_{f_0(x_0)} \to F_{f_1(x_0)}$ is admissible.* ("Freely" means not rel base point.)

Proof: Let H be a homotopy from f_0 to f_1 and define $\bar{H}$ by $\bar{H}(x,t) = H(x,1-t)$, $x \in X$, $t \in I$. Let H', $\bar{H}'$ be liftings as in 15.11 and take $g_H = (p_{f_0}, H'_1)$, $g_{\bar{H}} = (p_{f_1}, \bar{H}'_1)$. Let $K : X \times I \to B$ be defined by

$$K(x,t) = \begin{cases} H(x,2t) & 0 \leqslant t \leqslant 1/2 \\ H(x,2-2t) & 1/2 \leqslant t \leqslant 1, \end{cases}$$

and $M : X \times I \times I \to B$ by

$$M(x,t,s) = \begin{cases} f_0(x) & 0 \leqslant t \leqslant s/2 \\ H(x,2t-s) & s/2 \leqslant t \leqslant 1/2 \\ H(x,2-2t-s) & 1/2 \leqslant t \leqslant 1-s/2 \\ f_0(x) & 1-s/2 \leqslant t \leqslant 1. \end{cases}$$

K' defined by

$$K'(e',t) = \left\{ \begin{matrix} H'(e',2t) & 0 \leqslant t \leqslant 1/2 \\ \bar{H}'(g_H(e'),2t-1) & 1/2 \leqslant t \leqslant 1 \end{matrix} \right\} \; e' \in E_{f_0},$$

is a lifting of $K \circ (p_{f_0} \times 1)$ with $K'_0 = \bar{f}_0$, $(p_{f_0}, K'_1) = g_{\bar{H}} \circ g_H$, so we can take $g_{\bar{H}} \circ g_H$ as g_K. Now M is a homotopy as in 15.11b) from K to the stationary homotopy of f_0, so we have $g_{\bar{H}} \circ g_H = g_K \simeq_{p_{f_0}} 1_{E_{f_0}}$. By symmetry $g_H \circ g_{\bar{H}} \simeq_{p_{f_1}} 1_{E_{f_1}}$. Finally, if $w(t) = H(x_0, 1-t)$, then clearly $[g_H | F_{f_0(x_0)}] = \alpha([w])$. $\square$

15.17. Definition. If $p : E \to B$ is a fibration with fibre $F = p^{-1}(b_0)$ then a *homotopy trivialization* of p is a strong fibre homotopy equivalence $g : B \times F \to E$ such that $g|b_0 \times F$ is admissible.

15.18. Corollary. *If X is contractible to x_0 then every fibration over X has a homotopy trivialization.*

Proof: Let $H : X \times I \to X$ be a homotopy with $H_1 = 1_X$, $H_0(x) = x_0$, all $x \in X$. Note that $E_{1_X} \cong E$ and $E_{H_0} = X \times F$, so by 15.16 there is a strong fibre homotopy equivalence $X \times F = E_{H_0} \xrightarrow{g} E_{H_1} \cong E$ which is admissible on the fibre. □

15.19. Corollary. *Suppose $f_0, f_1 : X \to B$ are (freely) homotopic by a homotopy H, that X is contractible and that*

$$g_0 : X \times F_{f_0(x_0)} \to E_{f_0},\ g_1 : X \times F_{f_1(x_0)} \to E_{f_1}$$

are homotopy trivializations. Then there is an admissible map $\phi : F_{f_0(x_0)} \to F_{f_1(x_0)}$ such that

$$\begin{array}{ccc} X \times F_{f_0(x_0)} & \xrightarrow{g_0} & E_{f_0} \\ {\scriptstyle 1 \times \phi}\downarrow & & \downarrow{\scriptstyle g_H} \\ X \times F_{f_1(x_0)} & \xrightarrow{g_1} & E_{f_1} \end{array}$$

commutes up to fibre homotopy.

Proof: We have strong fibre homotopy equivalences

$$X \times F_{f_0(x_0)} \xrightarrow[g_0]{} E_{f_0} \xrightarrow[g_H]{} E_{f_1} \underset{g_1^{-1}}{\overset{g_1}{\rightleftarrows}} X \times F_{f_1(x_0)},$$

such that each, when restricted to the fibre, is admissible. The composite $g_1^{-1} \circ g_H \circ g_0$ is a map of the form $(x, f) \mapsto (x, \theta(x, f))$ for some map $\theta : X \times F_{f_0(x_0)} \to F_{f_1(x_0)}$ such that $\phi = \theta(x_0, -) : F_{f_0(x_0)} \to F_{f_1(x_0)}$ is admissible. Let $K : X \times I \to X$ be a contraction of X and let $\lambda : X \times F_{f_0(x_0)} \times I \to E_{f_1}$ be defined by $\lambda(x, f, t) = g_1(x, \theta(K(x, t), f))$. Then λ is a fibre homotopy from $g_1 \circ (1 \times \phi)$ to $g_1 \circ g_1^{-1} \circ g_H \circ g_0$ which is fibre homotopic to $g_H \circ g_0$. □

15.20. *Remark.* In the situation of 15.19 since X is contractible it is 0-connected. Thus there is always an admissible map $\phi : F \to F_{f(x_0)}$. We shall also call the composite $\bar{g}_f = g_f \circ (1 \times \phi) : X \times F \to E_f$ a homotopy trivialization. Suppose $f' : X \to B$ is another map homotopic to f, $g_{f'} : X \times F_{f'(x_0)} \to E_{f'}$ a homotopy trivialization and $\phi' : F \to F_{f'(x_0)}$ an admissible map. By 15.19 there is an admissible map $\phi'' : F_{f(x_0)} \to F_{f'(x_0)}$ such that $g_H \circ g_f \simeq_{p_{f'}} g_{f'} \circ (1 \times \phi'')$. Then

$$\phi''' = \phi'^{-1} \circ \phi'' \circ \phi : F \to F$$

is an admissible map such that $g_H \circ \bar{g}_f \simeq_{p_{f'}} \bar{g}_{f'} \circ (1 \times \phi''')$.

15.21. Corollary. *If $p : E \to B$ is a fibration and $f : X \to B$ is a homotopy equivalence, then $\bar{f} : E_f \to E$ is also a homotopy equivalence.*

Proof: Let $g : B \to X$ be a homotopy inverse for f. Then we have the induced bundle $(E_f)_g \cong E_{f \circ g}$ over B. $f \circ g \simeq 1_B$, so by 15.16 there is a strong fibre homotopy equivalence $h:E(=E_{1_B}) \to E_{f\circ g}$ such that $\bar f \circ \bar g \circ h \simeq \bar 1_B = 1_E$. We also have $E_{f\circ g\circ f} \xrightarrow{\bar f} E_{f\circ g} \xrightarrow{\bar g} E_f$, and since $g\circ f \simeq 1_X$, there is a strong fibre homotopy equivalence $k:E_f \to E_{f\circ g\circ f}$ such that $\bar g \circ \bar f \circ k \simeq \bar 1_X = 1_{E_f}$. Thus we have the following string of homotopies:

$$\bar g h \bar f = \bar g h \bar f 1_{E_f} \simeq (\bar g h \bar f)(\bar g \bar f k) \simeq \bar g h \bar f \bar g (hh^{-1}) \bar f k =$$
$$\bar g h(\bar f \bar g h) h^{-1} \bar f k \simeq \bar g h h^{-1} \bar f k \simeq \bar g \bar f k \simeq 1_{E_f}.$$

Thus $\bar f: E_f \to E$ and $\bar g \circ h: E \to E_f$ are homotopy inverses of one another. □

15.22. Now at last we can turn to the proof that $E^2_{pq} \cong H_p(B; h_q(F))$. If $p:E \to B$ is a fibration and $X \subset B$ is any subspace then we let $E|X$ denote $p^{-1}(X) \subset E$. Suppose $Y \subset X \subset B$ are subspaces and

$$f:(D^p, S^{p-1}) \to (X, Y)$$

is a map. D^p is contractible, so by 15.18 there is a homotopy trivialization $g_f: D^p \times F \to E_f$. Then we have the composite κ_f given by

$$h_q(F) \xrightarrow{\alpha} h_{p+q}((D^p, S^{p-1}) \times F) \xrightarrow{g_{f*}}$$
$$h_{p+q}(E_f, E_f|S^{p-1}) \xrightarrow{\bar f_*} h_{p+q}(E|X, E|Y)$$

where α is given by $\alpha(x) = \kappa_p \times x$, $\kappa_p \in h_p(D^p, S^{p-1})$ the canonical generator of $h_*(D^p, S^{p-1})$ as a module over $h_*(pt)$; we assume at this point that h_* is a homology theory with products. Suppose

$$f':(D^p, S^{p-1}) \to (X, Y)$$

is another map freely homotopic to f. By 15.16 there is a strong fibre homotopy equivalence $g: E_f \simeq E_{f'}$ with $\bar f' \circ g \simeq \bar f$. If $g_{f'}: D^p \times F \to E_{f'}$ is a homotopy trivialization, then by 15.19 there is an admissible map $\phi: F \to F$ such that $g \circ g_f \simeq_{p_{f'}} g_{f'} \circ (1 \times \phi)$. Then we have

$$\begin{aligned} \bar f_* \circ g_{f*} \circ \alpha &= \bar f'_* \circ g_* \circ g_{f*} \circ \alpha \\ &= \bar f'_* \circ g_{f'*} \circ (1 \times \phi)_* \circ \alpha \\ &= \bar f'_* \circ g_{f'*} \circ \alpha \circ \phi_*. \end{aligned}$$

If we assume that $p:E \to B$ is orientable with respect to h_*, then $\phi_* = 1$, so we have that $\kappa_f = \kappa_{f'}$. We may therefore write $\kappa_{[f]}$.

Let us introduce the following notation; for any pointed space X let $\pi_1^*(X, x_0) = \pi_1(X, x_0)/[\pi_1, \pi_1]$, where $[\pi_1, \pi_1]$ denotes the commutator subgroup of $\pi_1(X, x_0)$. For any subspace $Y \subset X$ with $x_0 \in Y$ let $\pi_n^*(X, Y) = \pi_n(X, Y, x_0)/G$ where G is the subgroup generated by all elements of the form $\gamma x - x$ for $\gamma \in \pi_1(Y, x_0)$, $x \in \pi_n(X, Y, x_0)$, $n \geqslant 2$. (We suppress the base point in the notation; $\pi_n^*(X,Y)$ can be shown to be independent of the base point anyway).

15.23. Lemma. *Let h_* be a homology theory with products and $p:E \to B$ an oriented fibration. Then κ defines a bilinear pairing $\pi_p^*(X,Y) \times h_q(F) \to h_{p+q}(E|X, E|Y)$, $p \geqslant 1$, which is natural with respect to inclusions of pairs $Y \subset X \subset B$. We assume $Y = \{x_0\}$ if $p = 1$.*

Proof: We clearly have a mapping

$$\pi_p(X, Y, x_0) \times h_q(F) \to h_{p+q}(E|X, E|Y)$$

which is linear in the second variable. Recall that the sum in $\pi_p(X, Y, x_0)$ is defined using the coproduct $\mu': CS^{p-1} \to CS^{p-1} \vee CS^{p-1}$: if $[f], [h] \in \pi_p(X, Y, x_0)$, then $[f] + [h]$ is the class $[\Delta' \circ (f \vee h) \circ \mu']$. E_f and E_h have the same fibre F and $E_{\Delta' \circ (f \vee h)}$ is the space over $CS^{p-1} \vee CS^{p-1}$ obtained from E_f and E_h by identifying corresponding points of F. We can choose homotopy trivializations $g_f: CS^{p-1} \times F \to E_f, g_h: CS^{p-1} \times F \to E_h$ which agree on $* \times F$, so by "glueing" we get a homotopy trivialization

$$g_f \sqcup g_h: (CS^{p-1} \vee CS^{p-1}) \times F \to E_{\Delta' \circ (f \vee h)}.$$

We can then choose a homotopy trivialization

$$g_{f+h}: CS^{p-1} \times F \to E_{f+h}$$

such that

$$\begin{array}{ccc} CS^{p-1} \times F & \xrightarrow{\mu' \times 1} & (CS^{p-1} \vee CS^{p-1}) \times F \\ \Big\downarrow g_{f+h} & & \Big\downarrow g_f \sqcup g_h \\ E_{f+h} & \xrightarrow[\bar{\mu}']{} & E_{\Delta' \circ (f \vee h)} \end{array}$$

commutes up to fibre homotopy.

Also

$$h_*(E_{\Delta' \circ (f \vee h)}, E_{\Delta' \circ (f \vee h)}|(S^{p-1} \vee S^{p-1})) \cong h_*(E_f, E_f|S^{p-1}) \oplus h_*(E_h, E_h|S^{p-1})$$

so we calculate

$$\bar{\Delta}'_* \circ \overline{(f \vee h)}_* \circ \bar{\mu}'_* \circ (g_{f+h})_* \circ \alpha(x) =$$

$$\bar{\Delta}'_* \circ (\bar{f} \sqcup \bar{h})_* \circ (g_f \sqcup g_h)_* \circ (\mu' \times 1)_* \circ \alpha(x) =$$

$$\bar{f}_* \circ g_{f*}(\kappa_p \times x) + \bar{h}_* \circ g_{h*}(\kappa_p \times x) = \kappa_{[f]}(x) + \kappa_{[h]}(x) \text{ for all } x \in h_q(F).$$

This proves "linearity" in the first variable ($\pi_2(X, Y, x_0)$, $\pi_1(X, x_0)$ are not necessarily abelian). Now since representatives of γx and x are freely homotopic, we see that κ factors through $\pi_p^*(X,Y) \times h_q(F)$. □

Thus κ induces a homomorphism

$$\kappa : \pi_p^*(X, Y) \otimes h_q(F) \to h_{p+q}(E|X, E|Y)$$

which is natural with respect to inclusions of pairs (X, Y).

15.24. Lemma. *κ commutes with the boundary operator of the triple (X,Y,Z) in the sense that the following diagram commutes for $p \geqslant 2$ ($Z = \{x_0\}$ if $p = 2$)*

$$\begin{array}{ccccc} \pi_p(X, Y, x_0) \otimes h_q(F) & \xrightarrow{k \otimes 1} & \pi_p^*(X, Y) \otimes h_q(F) & \xrightarrow{\kappa} & h_{p+q}(E|X, E|Y) \\ \downarrow{\scriptstyle \Delta \otimes 1} & & & & \downarrow{\scriptstyle \Delta} \\ \pi_{p-1}(Y, Z, x_0) \otimes h_q(F) & \xrightarrow{k \otimes 1} & \pi_{p-1}^*(Y, Z) \otimes h_q(F) & \xrightarrow{\kappa} & h_{p+q-1}(E|Y, E|Z). \end{array}$$

Here $k : \pi_p(X, Y, x_0) \to \pi_p^*(X, Y)$ is the natural projection.

Proof: Let $x \in h_q(F)$, $[f] \in \pi_p(X, Y, x_0)$. Then

$$\begin{aligned} \Delta \circ \kappa \circ k \otimes 1([f] \otimes x) &= j_* \circ \partial \circ \bar{f}_* \circ g_{f*}(\kappa_p \times x) \\ &= j_* \circ (\bar{f}|(E_f|S^{p-1}))_* \circ (g_f|S^{p-1} \times F)_*(\partial \kappa_p \times x). \end{aligned}$$

Let $\rho : (CS^{p-2}, S^{p-2}) \to (S^{p-1}, *)$ be the standard projection; then $\Delta[f]$ is the class of the composite

$$(CS^{p-2}, S^{p-2}) \xrightarrow{\rho} (S^{p-1}, *) \xrightarrow{f|S^{p-1}} (Y, x_0) \xrightarrow{j} (Y, X).$$

Call this h. Then we can choose a homotopy trivialization $g_h : CS^{p-2} \times F \to E_h$ so that the diagram

$$\begin{array}{ccc} CS^{p-2} \times F & \xrightarrow{g_h} & E_h \\ \downarrow{\scriptstyle \rho \times 1} & & \downarrow{\scriptstyle \bar{\rho}} \\ S^{p-1} \times F & \xrightarrow[(g_f|S^{p-1} \times F)]{} & E_f|S^{p-1} \end{array}$$

commutes up to fibre homotopy. Then we have

$$\begin{aligned} &\kappa \circ k \otimes 1 \circ \Delta \otimes 1([f] \otimes x) \\ &= \bar{h}_* \circ g_{h*}(\kappa_{p-1} \times x) = j_* \circ (\bar{f}|(E_f|S^{p-1}))_* \circ \bar{\rho}_* \circ g_{h*}(\kappa_{p-1} \times x). \\ &= j_* \circ (\bar{f}|(E_f|S^{p-1}))_* \circ (g_f|S^{p-1} \times F)_* \circ (\rho \times 1)_*(\kappa_{p-1} \times x) \\ &= j_* \circ (\bar{f}|(E_f|S^{p-1}))_* \circ (g_f|S^{p-1} \times F)_*(\partial \kappa_p \times x) \\ &= \Delta \circ \kappa \circ (k \otimes 1)([f] \otimes x). \quad \square \end{aligned}$$

15.25. Lemma. *If $p: E \to D^p$ is a fibration then $\kappa: \pi_p^*(D^p, S^{p-1}) \otimes h_q(F) \to h_{p+q}(E, E|S^{p-1})$ is an isomorphism for $p \geq 2$. Similarly*

$$\kappa: \pi_1(S^1, *) \otimes h_q(F) \to h_{q+1}(E, F)$$

is an isomorphism for a fibration $p: E \to S^1$.

Proof: D^p is contractible, so we can find a homotopy trivialization $g: D^p \times F \to E$; in fact $g = g_f$ where $f: (D^p, S^{p-1}) \to (D^p, S^{p-1})$ is the identity. Thus the following square commutes

$$\begin{array}{ccc} \pi_p^*(D^p, S^{p-1}) \otimes h_q(F) & \xrightarrow{\kappa} & h_{p+q}(E, E|S^{p-1}) \\ \cong \uparrow \beta & & \cong \uparrow g_* \\ h_q(F) & \xrightarrow[\cong]{\alpha} & h_{p+q}((D^p, S^{p-1}) \times F), \end{array}$$

where $\beta(x) = [1] \otimes x$. β is an isomorphism because $\pi_p^*(D^p, S^{p-1}) \cong \mathbb{Z}$ and α is an isomorphism by 13.75 because $h_*(D^p, S^{p-1})$ is a free $h_*(pt)$-module with κ_p as generator. Thus κ is an isomorphism. The case $p: E \to S^1$ is similar. □

15.26. Lemma. *If $p: E \to B$ is an orientable fibration and B is a CW-complex in $\mathscr{PW}$ with $B^0 = \{b_0\}$, then*

$$\kappa: \pi_p^*(B^p, B^{p-1}) \otimes h_q(F) \to h_{p+q}(E|B^p, E|B^{p-1})$$

is an isomorphism for all $p \geq 1$.

Proof: Let $\chi: (\vee_\alpha D_\alpha^p, \vee_\alpha S_\alpha^{p-1}) \to (B^p, B^{p-1})$ be the "big characteristic map" for the p-cells of B. First we show that $\bar{\chi}_*: h_*(E_\chi, E_\chi|\vee_\alpha S_\alpha^{p-1}) \to h_*(E|B^p, E|B^{p-1})$ is an isomorphism. Let $N \subset B^p$ be a closed neighborhood of B^{p-1} of which B^{p-1} is a strong deformation retract; then the inclusion $i: B^{p-1} \to N$ is a homotopy equivalence, so by 15.21 $\bar{i}: E|B^{p-1} \to E|N$ is a homotopy equivalence. From the 5-lemma it follows that $\bar{j}_*: h_*(E|B^p, E|B^{p-1}) \to h_*(E|B^p, E|N)$ is an isomorphism. Let $N' = \chi^{-1}(N) \subset \vee_\alpha D_\alpha^p$; again $\vee_\alpha S_\alpha^{p-1}$ is a strong deformation retract of N', so we have an isomorphism $h_*(E_\chi, E_\chi|\vee_\alpha S_\alpha^{p-1}) \to h_*(E_\chi, E_\chi|N')$. If $B^{p-1} \subset N_1 \subset \bar{N}_1 \subset \mathring{N}$, then we have a commutative diagram

$$\begin{array}{ccccc} h_*(E_\chi, E_\chi|\vee_\alpha S_\alpha^{p-1}) & \xrightarrow[\cong]{} & h_*(E_\chi, E_\chi|N') & \xleftarrow[\cong]{} & h_*(E_\chi|(\vee_\alpha D_\alpha^p - N_1'), E_\chi|(N' - N_1')) \\ \downarrow \bar{\chi}_* & & \downarrow \bar{\chi}_* & & \cong \downarrow \bar{\chi}_* \\ h_*(E|B^p, E|B^{p-1}) & \xrightarrow{\cong} & h_*(E|B^p, E|N) & \xleftarrow{\cong} & h_*(E|(B^p - N_1), E|(N - N_1)). \end{array}$$

The two horizontal arrows on the right are isomorphisms by excision and the last $\bar{\chi}$ on the right is a homeomorphism. Thus

$$\bar{\chi}_*: h_*(E_\chi, E_\chi|\vee_\alpha S^{p-1}_\alpha) \to h_*(E|B^p, E|B^{p-1})$$

is an isomorphism as claimed. Note that the proof holds for any homology theory and in particular for ordinary homology.

The result now follows from the commutative diagram

$$\begin{array}{ccccc}
H_p(B^p, B^{p-1}) \otimes h_q(F) & \xleftarrow[\cong]{\chi_* \otimes 1} & H_p(\vee_\alpha D^p_\alpha, \vee_\alpha S^{p-1}_\alpha) \otimes h_q(F) & \xleftarrow[\cong]{\{i_{\alpha*} \otimes 1\}} & \oplus_\alpha H_p(D^p, S^{p-1}) \otimes h_q(F) \\
\cong \uparrow h \otimes 1 & & \uparrow h \otimes 1 & & \cong \uparrow h \otimes 1 \\
\pi^*_p(B^p, B^{p-1}) \otimes h_q(F) & \xleftarrow{\chi_* \otimes 1} & \pi^*_p(\vee_\alpha D^p_\alpha, \vee_\alpha S^{p-1}_\alpha) \otimes h_q(F) & \xleftarrow{\{i_{\alpha*} \otimes 1\}} & \oplus_\alpha \pi^*_p(D^p, S^{p-1}) \otimes h_q(F) \\
\kappa \downarrow & & \kappa \downarrow & & \cong \downarrow \kappa \\
h_{p+q}(E|B^p, E|B^{p-1}) & \xleftarrow[\cong]{\bar{\chi}_*} & h_{p+q}(E_\chi, E_\chi|\vee_\alpha S^{p-1}_\alpha) & \xleftarrow[\cong]{\{\bar{i}_{\alpha*}\}} & \oplus_\alpha h_{p+q}(E_\chi|D^p_\alpha, E_\chi|S^{p-1}_\alpha).
\end{array}$$

The last κ on the right is an isomorphism by 15.25. The Hurewicz homomorphism $h: \pi^*_p(D^p, S^{p-1}) \to H_p(D^p, S^{p-1})$ is an isomorphism since (D^p, S^{p-1}) is $(p-1)$-connected. We are of course assuming h_* satisfies the wedge axiom.

For the case $p = 1$ we instead take $\chi: \vee_\alpha S^1_\alpha \to B^1$; χ and $\bar{\chi}$ are then homeomorphisms. ▯

15.27. Theorem (Leray–Serre). *If h_* is a homology theory with products satisfying the wedge axiom for CW-complexes and the WHE axiom, then for every fibration $p: E \to B$ orientable with respect to h_* and with B 0-connected there is a spectral sequence $\{E^r_{pq}, d^r\}$ converging to $h_*(E)$ and having*

$$E^2_{pq} \cong H_p(B; h_q(F)), \text{ all } p, q.$$

The spectral sequence is natural with respect to fibre maps.

Proof: First assume B is a CW-complex in $\mathscr{PT}$ with $B^0 = \{b_0\}$. We take two copies of the isomorphism κ, one for the fibration $p: E \to B$ and the theory h_*, one for the fibration $1: B \to B$ and the theory $H_*(-;\mathbb{Z})$. Thus we have an isomorphism

$$\begin{array}{ccccc}
h_{p+q}(E|B^p, E|B^{p-1}) & \xleftarrow[\cong]{\kappa} & \pi^*_p(B^p, B^{p-1}) \otimes h_q(F) & \xrightarrow[\cong]{h \otimes 1} & H_p(B^p, B^{p-1}) \otimes h_q(F) \\
\| & & & & \wr\| \\
E^1_{pq} & & & & C_p(B) \otimes h_q(F).
\end{array}$$

We take two copies of the diagram of 15.24 and stick them back to back, getting a commutative diagram

$$\begin{array}{ccccc}
h_{p+q}(E|B^p, E|B^{p-1}) & \xleftarrow{\kappa} & \pi_p^*(B^p, B^{p-1}) \otimes h_q(F) & \xrightarrow{h \otimes 1} & H_p(B^p, B^{p-1}) \otimes h_q(F) \\
\Big\downarrow \Delta & & \Big\uparrow k \otimes 1 & & \Big\downarrow d \otimes 1 \\
 & & \pi_p(B^p, B^{p-1}, b_0) \otimes h_q(F) & & \\
 & & \Big\downarrow \Delta \otimes 1 & & \\
 & & \pi_{p-1}(B^{p-1}, B^{p-2}, b_0) \otimes h_q(F) & & \\
 & & \Big\downarrow k \otimes 1 & & \\
h_{p+q-1}(E|B^{p-1}, E|B^{p-2}) & \xleftarrow{\kappa} & \pi_{p-1}^*(B^{p-1}, B^{p-2}) \otimes h_q(F) & \xrightarrow{h \otimes 1} & H_{p-1}(B^{p-1}, B^{p-2}) \otimes h_q(F)
\end{array}$$

in which $k \otimes 1$ is an epimorphism. This shows that our isomorphism $E^1_{pq} \cong C_p(B) \otimes h_q(F)$ is a chain map. Even the case $p = 0$ works out right because $h_q(E|B^0, E|B^{-1}) \cong C_0(B) \otimes h_q(F)$ and both

$$\partial : h_{q+1}(E|B^1, E|B^0) \to h_q(E|B^0, \varnothing)$$

and $d : C_1(B) \to C_0(B)$ are 0 (in the former case $\partial\kappa([w] \otimes x) = \alpha([w])_*(x) - x = 0$ and $\kappa : \pi_1^*(B^1, b_0) \otimes h_q(F) \to h_{q+1}(E|B^1, E|B^0)$ is an isomorphism). Thus κ induces a natural isomorphism $\hat{\kappa} : E^2_{pq} \cong H_p(B; h_q(F))$ for all p,q. Therefore the spectral sequence arising from the skeletal filtration of B is the required one.

For any fibration $p : E \to B$ with B 0-connected we can find a CW-substitute $f : B' \to B$ for B with $B'^0 = \{b_0'\}$. We take $p' : E' \to B'$ to be the induced fibration E_f; from the exact homotopy sequences of the fibrations $p : E \to B$, $p' : E' \to B'$ and the 5-lemma it follows that $\bar{f} : E' \to E$ is a weak homotopy equivalence. Therefore $\bar{f}_* : h_*(E') \to h_*(E)$ is an isomorphism. Hence the spectral sequence just constructed for $p' : E' \to B'$ satisfies $E^2_{pq} \cong H_p(B'; h_q(F)) \cong H_p(B; h_q(F))$ and converges to $h_*(E') \cong h_*(E)$. It is still natural, since CW-substitutes are. □

Remark 1. The Atiyah–Hirzebruch–Whitehead spectral sequence 15.7 follows from 15.27 as a special case—even if X is only a 0-connected space and not necessarily a CW-complex. One has only to apply 15.27 to the fibration $1 : X \to X$, which is always orientable.

Remark 2. There are relative versions of 15.27. In particular, suppose that $\dot{E} \subset E$ is a subspace such that $(p|\dot{E}) : \dot{E} \to B$ is also a fibration. Let $\dot{F} = F \cap \dot{E}$. Then there is a spectral sequence with

$$E^2_{pq} \cong H_p(B; h_q(F, \dot{F}))$$

and converging to $h_*(E, \dot{E})$. One considers subgroups of

$$h_{p+q}(E|B^p, E|B^{p-1} \cup \dot{E}|B^p)$$

instead. Or again if $B_0 \subset B$ is a subspace, then there is a spectral sequence with $E^2_{pq} \simeq H_p(B, B_0; h_q(F))$ and converging to $h_*(E, E|B_0)$. In particular there is the reduced Atiyah–Hirzebruch–Whitehead spectral sequence $\tilde{H}_p(X; \tilde{h}_q(S^0)) \Rightarrow \tilde{h}_*(X)$.

Remark 3. There is a cohomology version of all this. The interested reader should work out some of the details for himself. Note, however, that the analog of 15.5 does not hold in all cases because of the existence of $\lim^1$.

Remark 4. These spectral sequences behave well with respect to products. In particular, if $p_1: E_1 \to B_1, p_2: E_2 \to B_2$ are two fibrations, then $p_1 \times p_2: E_1 \times E_2 \to B_1 \times B_2$ is also a fibration with fibre $F_1 \times F_2$. We denote the spectral sequences of p_1, p_2 and $p_1 \times p_2$ by $E^r_{pq}(1)$, $E^r_{pq}(2)$, $E^r_{pq}(3)$ respectively. Then the $\times$-product induces a natural pairing

$$\times^r: E^r_{pq}(1) \otimes E^r_{st}(2) \to E^r_{p+s,q+t}(3)$$

for all p, q, r, s, t. If $d^r_\otimes$ denotes the differential on $E^r_{**}(1) \otimes E^r_{**}(2)$ given by $d^r_\otimes(a \otimes b) = d^r(1)a \otimes b + (-1)^{\deg a}\, a \otimes d^r(2)b$, where if $a \in E^r_{pq}(1)$ then $\deg a = p + q$, then $\times^r$ is a chain map: that is $d^r(3) \circ \times^r = \times^r \circ d^r_\otimes$. Moreover the square

$$\begin{array}{ccc} H(E^r_{**}(1)) \otimes H(E^r_{**}(2)) & \xrightarrow{\times} & H(E^r_{**}(1) \otimes E^r_{**}(2)) \\ \| \wr & & \downarrow \times^r_* \\ E^{r+1}_{**}(1) \otimes E^{r+1}_{**}(2) & \xrightarrow{\times^{r+1}} & E^{r+1}_{**}(3) \cong H(E^r_{**}(3)) \end{array}$$

commutes.

$$\times: h_*(E_1) \otimes h_*(E_2) \to h_*(E_1 \times E_2)$$

also induces a product

$$\times: F_{pq}(1) \otimes F_{st}(2) \to F_{p+s,q+t}(3)$$

and the induced map

$$\times^\infty: E^\infty_{pq}(1) \otimes E^\infty_{st}(2) \to E^\infty_{p+s,q+t}(3)$$

agrees with that obtained by taking dir lim over the $\times^r$'s. Finally

$$\times^2: E^2_{**}(1) \otimes E^2_{**}(2) \to E^2_{**}(3)$$

corresponds under $\hat{\kappa}$ to the usual homology $\times$-product

$$\times: H_p(B_1; h_q(F_1)) \otimes H_s(B_2; h_t(F_2)) \to H_{p+s}(B_1 \times B_2; h_{q+t}(F_1 \times F_2))$$

corresponding to the pairing $h_q(F_1) \otimes h_t(F_2) \xrightarrow{\times} h_{q+t}(F_1 \times F_2)$ of coefficient groups.

Checking the details of all these statements is extremely tedious. The basic pairing is induced by

$$h_{p+q}(X^p, X^{p-1}) \otimes h_{s+t}(Y^s, Y^{s-1}) \xrightarrow{\times}$$

$$h_{p+s+q+t}(X^p \times Y^s, X^p \times Y^{s-1} \cup X^{p-1} \times Y^s) \xrightarrow{i_*}$$

$$h_{p+s+q+t}((X \times Y)^{p+s}, (X \times Y)^{p+s-1}).$$

One must check that this pairing sends

$$Z^r_{pq}(1) \otimes Z^r_{st}(2) \quad \text{into } Z^r_{p+s,q+t}(3),$$

$$Z^r_{pq}(1) \otimes B^r_{st}(2) + B^r_{pq}(1) \otimes Z^r_{st}(2) \quad \text{into } B^r_{p+s,q+t}(3),$$

etc.

In cohomology we can apply the diagonal map to get an internal (cup) product which makes $\{E_r^{pq}\}$ into a bigraded ring and d_r into a *derivation*: $d_r ab = d_r a \cdot b + (-1)^{\deg a}\, a \cdot d_r b$.

Remark 5. We can also fit the maps $i_*: h_*(F) \to h_*(E)$ and $p_*: H_*(E) \to H_*(B)$ into this framework as the so-called "edge homomorphisms". The y-axis always forms a free edge in the diagram of E^r_{pq}—i.e. $E^r_{pq} = 0$ if $p < 0$.

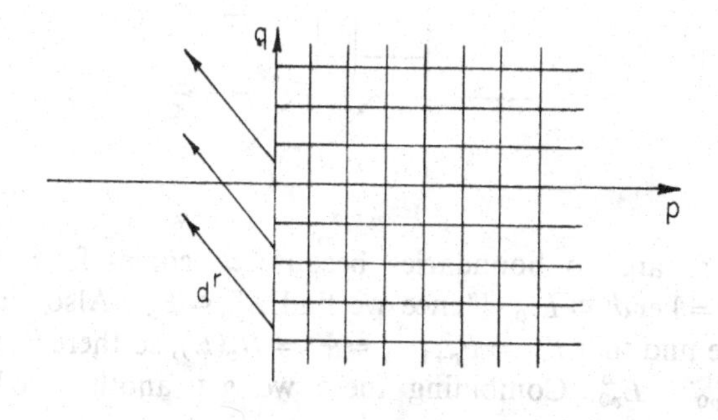

This means that d^r is always zero on the groups E^r_{0q}, or in other words every element of E^r_{0q} is always a cycle and hence E^{r+1}_{0q} is a quotient of E^r_{0q}.

Taking all the surjections $E^2_{0q} \to E^3_{0q} \to \cdots \to E^r_{0q} \to \cdots$ we get a surjection $E^2_{0q} \to E^\infty_{0q}$. But $E^\infty_{0q} = F_{0q} \subset h_q(E)$.

The composite

$$\textbf{15.28} \qquad h_q(F) \cong E^2_{0q} \to E^\infty_{0q} \to h_q(E)$$

is just i_*. To see this we consider the fibration $p': F \to \{b_0\}$ and the fibre map

$$\begin{array}{ccc} F & \xrightarrow{i} & E \\ {\scriptstyle p'}\downarrow & & \downarrow{\scriptstyle p} \\ \{b_0\} & \subset & B. \end{array}$$

Let $E^{r\prime}_{pq}$ denote the spectral sequence of p'; by naturality we have a commutative diagram

$$\begin{array}{ccccc} h_q(F) & \longrightarrow & E^\infty_{0q} & \longrightarrow & h_q(E) \\ \uparrow{\scriptstyle 1} & & \uparrow{\scriptstyle i^\infty_{0q}} & & \uparrow{\scriptstyle i_*} \\ h_q(F) & \longrightarrow & E^{\infty\prime}_{0q} & \longrightarrow & h_q(F). \end{array}$$

But examination of the definition of $E^{\infty\prime}_{0q}$ shows that $Z^{\infty\prime}_{0q} = h_q(F)$, $B^{\infty\prime}_{0q} = 0$ and the two lower horizontal arrows are both 1. Thus the edge homomorphism 15.28 is i_* as claimed.

If $h_* = H_*$, an ordinary homology theory, then $E^r_{pq} = 0$ if $q < 0$ also, so the x-axis becomes a free edge too.

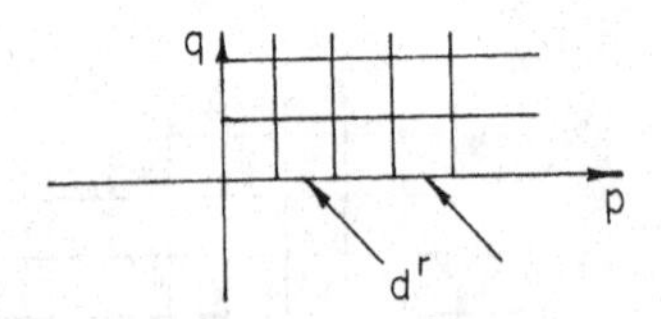

In E^r_{p0} there are no boundaries because d^r comes from a 0 group. Thus $E^{r+1}_{p0} = \ker d^r \subset E^r_{p0}$. Hence we find $E^\infty_{p0} \subset E^2_{p0}$. Also since $E^\infty_{pq} = 0$ if $q < 0$, we find that $F_{p0} = F_{p+1,-1} = \cdots = H_p(E)$, so there is a surjection $H_p(E) = F_{p0} \to E^\infty_{p0}$. Combining these we get another "edge homomorphism"

$$\textbf{15.29.} \qquad H_p(E) \to E^\infty_{p0} \to E^2_{p0} \cong H_p(B).$$

This homomorphism is in fact just p_*. The proof is similar to the one above; one considers the fibre map

$$\begin{array}{ccc} E & \xrightarrow{p} & B \\ {\scriptstyle p}\downarrow & & \downarrow{\scriptstyle 1} \\ B & \xrightarrow{1} & B \end{array}$$

and observes that

$$\begin{aligned} Z_{p0}^{\infty\,\prime} &= \operatorname{im}[H_p(B^p) \to H_p(B^p, B^{p-1})] \\ &\cong H_p(B^p) \end{aligned}$$

$$\begin{aligned} B_{p0}^{\infty\,\prime} &= \operatorname{im}[\Delta: H_{p+1}(B, B^p) \to H_p(B^p, B^{p-1})] \\ &\cong \operatorname{im}[\partial: H_{p+1}(B, B^p) \to H_p(B^p)] \end{aligned}$$

so that $E_{p,0}^{\infty\,\prime} \cong H_p(B^p)/\operatorname{im}\partial \cong H_p(B)$ and $H_p(B) \to E_{pq}^{\infty\,\prime} \to H_p(B)$ is in fact 1.

Let us now turn to some applications of 15.27. In order for the spectral sequence to be useful there must be some way to compute the differentials d^r. The easiest case is that in which there are so many gaps in the E^2-term that all the differentials must be zero. For example, if $E_{pq}^2 = 0$ if either p or q is odd, then since $d^2: E_{pq}^2 \to E_{p-2,q+1}^2$, we see that d^2 always has either zero domain or zero range and hence $d^2 = 0$. Thus $E_{**}^3 = E_{**}^2$. Again $d^3: E_{pq}^3 \to E_{p-3,q+2}^3$, so d^3 must always be zero. Continuing in this way we find $d^r = 0$ for all $r \geqslant 2$ and hence $E_{**}^\infty = E_{**}^2$.

15.30. Another case where there are large gaps in the E^2-term is the following: suppose $h^* = H^*(-;R)$ is ordinary cohomology with coefficients R, R a ring, and $p: E \to B$ is an orientable fibration with fibre an n-sphere, $n > 0$, – a so-called *spherical fibration*. Then

$$H^q(F;R) = \begin{cases} R & q = 0, n \\ 0 & q \neq 0, n. \end{cases}$$

Thus

$$E_2^{pq} = H^p(B; H^q(F;R)) = \begin{cases} H^p(B;R) & q = 0, n \\ 0 & q \neq 0, n. \end{cases}$$

There are only two non-zero rows in the E_r-diagram.

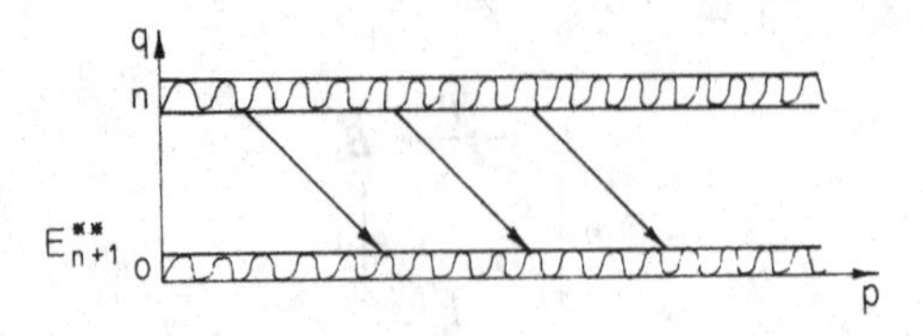

Moreover the differentials $d_r : E_r^{pq} \to E_r^{p+r,q-r+1}$ obviously must all vanish except for d_{n+1}. Clearly

$$E_{n+2}^{p,n} = \ker [d_{n+1} : E_{n+1}^{p,n} \to E_{n+1}^{p+n+1,0}]$$

$$E_{n+2}^{p,0} = \operatorname{coker} [d_{n+1} : E_{n+1}^{p-n-1,n} \to E_{n+1}^{p,0}].$$

Since there are no other non-zero differentials we have

$$E_{n+2}^{**} = E_{n+3}^{**} = \cdots = E_\infty^{**}.$$

Since $E_\infty^{pq} = 0$ if $q \neq 0, n$, we see that we have equalities

$$F^{m-n,n} = F^{p,q} = H^m(E;R) \quad \text{if } p+q=m, q \geqslant n,$$

$$F^{m-n+1,n-1} = F^{p,q} = F^{m,0} \quad \text{if } p+q=m, 0 \leqslant q \leqslant n-1,$$

and $F^{m,0} = E_\infty^{m,0}$. These facts yield the following exact sequences:

$$0 \to E_\infty^{p,n} \to E_2^{p,n} \xrightarrow{d_{n+1}} E_2^{p+n+1,0} \to E_\infty^{p+n+1,0} \to 0$$

$$0 \to E_\infty^{m,0} \to H^m(E;R) \to E_\infty^{m-n,n} \to 0.$$

Glueing these together and recalling that $E_2^{p0} \cong E_2^{pn} \cong H^p(B;R)$, we get a long exact sequence

15.31.

$$\cdots \longrightarrow H^m(E;R) \xrightarrow{\phi} H^{m-n}(B;R) \xrightarrow{d_{n+1}}$$

$$H^{m+1}(B;R) \longrightarrow H^{m+1}(E;R) \longrightarrow \cdots.$$

Closer examination shows that the map $H^{m+1}(B;R) \to H^{m+1}(E;R)$ is the edge homomorphism 15.29 and hence is just p^*. We can also describe d_{n+1} in a more useful way;

$$E_2^{pn} \cong H^p(B;H^n(S^n;R)), \quad E_2^{p0} \cong H^p(B;H^0(S^n;R)).$$

Let $u \in H^n(S^n, R)$ be a generator; we can also regard u as lying in $H^0(B; H^n(S^n; R)) = E_2^{0,n}$. Let $e = d_{n+1}u \in H^{n+1}(B; H^0(S^n; R)) = E_2^{n+1,0}$; e is called the *Euler class* of the spherical fibration. We may regard elements of $E_2^{p,n}$ as being of the form ux for $x \in H^p(B;R) = H^p(B; H^0(S^n; R)) = E_2^{p0}$. Thus by Remark 4 it follows $d_{n+1}(ux) = (d_{n+1}u) \cdot x + (-1)^{|u|} u \cdot d_{n+1}(x) = e \cdot x$. Thus

$$d_{n+1}: H^{m-n}(B;R) \to H^{m+1}(B;R)$$

is simply cup-product with the class $e: x \mapsto ex$. The sequence 15.31 is called the *Gysin sequence* of the spherical fibration $p: E \to B$.

15.32. Example. We have the Hopf fibration $S^{2n+1} \to CP^n$ with fibre S^1 for every $n \geqslant 1$. The fibration is always orientable because $\pi_1(CP^n, *) = 0$, $n \geqslant 1$. The Euler class e lies in $H^2(CP^n; R)$ and for $0 < m < 2n$ 15.31 becomes

$$0 \longrightarrow H^{m-1}(CP^n; R) \xrightarrow{e} H^{m+1}(CP^n; R) \longrightarrow 0.$$

Since we also have an exact sequence

$$0 \to H^1(CP^n; R) \to H^1(S^{2n+1}; R),$$

we see that $H^1(CP^n; R) = 0$ and hence

$$H^{2n}(CP^n; R) \cong H^{2n-2}(CP^n; R) \cong \cdots \cong H^2(CP^n; R) \cong H^0(CP^n; R) \cong R,$$

$$H^{2n-1}(CP^n; R) \cong H^{2n-3}(CP^n; R) \cong \cdots \cong H^1(CP^n; R) = 0,$$

and $H^{2r}(CP^n; R)$ is generated by e^r. Since CP^n is a $2n$-dimensional CW-complex, the higher cohomology groups must vanish. Thus we have proved the following.

15.33. Theorem. *As a ring $H^*(CP^n; R) \cong R[e]/(e^{n+1})$, a truncated polynomial algebra, where $e \in H^2(CP^n; R)$. Similarly $H^*(CP^\infty; R) \cong R[e]$.*

It would be tempting to apply 15.31 to the fibration $S^0 \to S^m \to RP^m$, but unfortunately we had to assume the fibre was an n-sphere with $n > 0$ in order that $H^n(F; R) \cong R$. We shall rederive the Gysin sequence later without the assumption that $n > 0$.

15.34. The spectral sequence collapses to a long exact sequence in a similar fashion in the case where the base B is a sphere S^n, $n > 0$; in this case we may take h^* to be any appropriate cohomology theory. The diagram of the E_2-term has only two non-zero columns, and clearly the only non-zero differential is $d_n: E_n^{0p} \to E_n^{n,p-n+1}$.

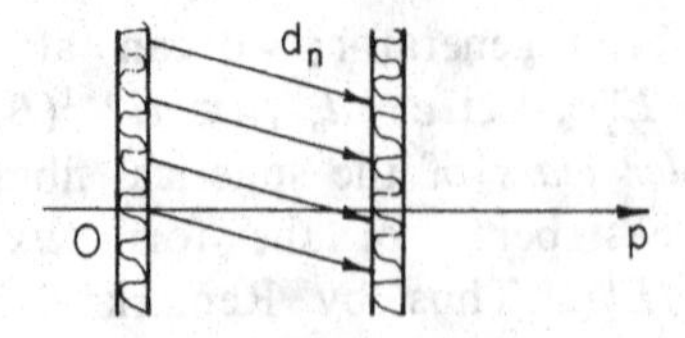

Just as before we end up with a long exact sequence

15.35.

$$\cdots \longrightarrow h^m(E) \xrightarrow{i^*} h^m(F) \xrightarrow{d_n} h^{m-n+1}(F) \xrightarrow{\psi} h^{m+1}(E) \longrightarrow \cdots,$$

where $i: F \to E$ is the inclusion. All we can say about d_n is that it is a derivation:

$$d_n(xy) = d_n x \cdot y + (-1)^{|x|} x \cdot d_n y, \quad x, y \in h^*(F).$$

$\psi: H^0(F) \to H^n(E)$ coincides with the edge homomorphism $H^0(F) \cong H^n(S^n; H^0(F)) \xrightarrow{p^*} H^n(E)$ when $h^* = H^*$. 15.35 is called the *Wang sequence.*

15.36. Exercise. Compute the cohomology ring $H^*(\Omega S^n; \mathbb{Z})$ being careful to distinguish the cases n even, n odd.

15.37. Example. Let us take an example which is important for computing the homotopy groups of spheres. We know $[S^3; H(\mathbb{Z}, 3)] \cong H^3(S^3; \mathbb{Z}) \cong \mathbb{Z}$; let $f: S^3 \to H(\mathbb{Z}, 3)$ represent a generator. Then from the path space fibration $P_3 \to H(\mathbb{Z}, 3)$ with fibre $\Omega H(\mathbb{Z}, 3) = H(\mathbb{Z}, 2)$ f induces a fibration $p: E \to S^3$ also with fibre $H(\mathbb{Z}, 2)$. From the exact sequence

$$\cdots \to \pi_r(H(\mathbb{Z}, 2), *) \to \pi_r(E, *) \to \pi_r(S^3, *) \xrightarrow{\partial} \pi_{r-1}(H(\mathbb{Z}, 2), *) \to \cdots$$

we find that

$$\pi_r(E, *) \cong \begin{cases} \pi_r(S^3) & r > 3 \\ 0 & r \leqslant 3. \end{cases}$$

The fact that $\partial: \pi_3(S^3, *) \to \pi_2(H(\mathbb{Z}, 2), *)$ is an isomorphism comes from the choice of f; we have a commutative diagram

$$\begin{array}{ccc} \pi_3(S^3, *) & \xrightarrow{\partial} & \pi_2(H(\mathbb{Z}, 2), *) \\ \cong \downarrow f_* & & \downarrow 1 \\ \pi_3(H(\mathbb{Z}, 3), *) & \underset{\cong}{\xrightarrow{\partial}} & \pi_2(H(\mathbb{Z}, 2), *). \end{array}$$

Thus anything we can deduce about $\pi_*(E,*)$ will tell us something about the higher homotopy groups of S^3.

Before we proceed let us identify $H(\mathbb{Z},2)$. From the exact homotopy sequence of the fibration $S^1 \to S^{2n+1} \to CP^n$ we see that

$$\pi_r(CP^n,*) \cong \begin{cases} \mathbb{Z} & r=2 \\ 0 & 2<r\leqslant 2n. \end{cases}$$

Therefore

$$\pi_r(CP^\infty,*) = \begin{cases} \mathbb{Z} & r=2 \\ 0 & \text{otherwise,} \end{cases}$$

so CP^∞ will do as $H(\mathbb{Z},2)$. From 15.33 we have $H^*(CP^\infty;\mathbb{Z}) \cong \mathbb{Z}[e]$, $e \in H^2(CP^\infty;\mathbb{Z})$.

Now the Wang sequence for $p: E \to S^3$ looks as follows:

$$\begin{array}{ccccccccccc} H^{2m-3}(CP^\infty) & \to & H^{2m}(E) & \to & H^{2m}(CP^\infty) & \xrightarrow{d^3} & H^{2m-2}(CP^\infty) & \to & H^{2m+1}(E) & \to & H^{2m+1}(CP^\infty). \\ \| & & & & \wr\| & & \wr\| & & & & \| \\ 0 & & & & \mathbb{Z} & & \mathbb{Z} & & & & 0 \end{array}$$

Since d_3 is a derivation, we have $d_3(e^m) = me^{m-1}$; thus $\ker d_3 = 0$, $\operatorname{coker} d_3 \cong \mathbb{Z}_m$, and we get

$$\left.\begin{array}{r} H^{2m}(E) = 0 \\ H^{2m+1}(E) \cong \mathbb{Z}_m \end{array}\right\} \quad m \geqslant 1.$$

Now the universal coefficient theorem for singular homology implies that

$$\left.\begin{array}{r} H_{2m}(E) \cong \mathbb{Z}_m \\ H_{2m-1}(E) = 0 \end{array}\right\} \quad m \geqslant 1.$$

If ${}_pG$ denotes the p-primary part of the group G, then we have

$${}_p[H_q(E)] = \begin{cases} \mathbb{Z}_p & q=2p \\ 0 & 0<q<2p, \end{cases}$$

for every prime p.

Now Serre [78] has proved a generalization of the Hurewicz isomorphism theorem which says, among other things, that if for a given p we have ${}_p[\pi_q(X,x_0)] = 0$ for $q<m$ then ${}_p[\tilde{H}_r(X)] = 0$, $0 \leqslant r < m$ and ${}_p[\pi_m(X,x_0)] \cong {}_p[\tilde{H}_m(X)]$.

Therefore in our case we have

$$_p[\pi_r(S^3,*)] \cong {}_p[\pi_r(E,*)] \cong {}_p[\tilde{\mathrm{H}}_r(E)] \cong \begin{cases} 0 & 3 < r < 2p \\ \mathbb{Z}_p & r = 2p. \end{cases}$$

For example $\pi_4(S^3,*) \cong \mathbb{Z}_2$, and the first 3-torsion which occurs is $\mathbb{Z}_3$ in $\pi_6(S^3,*)$.

Serre's generalized Hurewicz isomorphism theorem is proved by using spectral sequences. The reader who has followed this chapter so far should be quite capable of understanding the proof. See for example [78] or [80].

15.38. Let us look at one more important case where the spectral sequence can be reduced, at least partly, to an exact sequence. Suppose the fibration $p: E \to B$ is such that $H_p(B;\mathbb{Z}) = 0$, $0 < p < n$, $H_q(F;G) = 0$, $0 < q < m$, and F is 0-connected. Then the diagram of the E^2-term looks as follows.

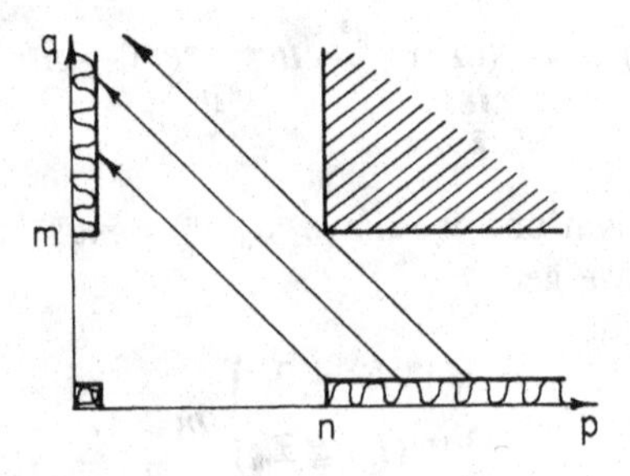

In the range indicated the only non-zero differentials are those which go all the way from the x-axis to the y-axis: $d^p\colon E^p_{p0} \to E^p_{0,p-1}$. Again we get exact sequences

$$0 \longrightarrow E^\infty_{p0} \longrightarrow E^2_{p0} \xrightarrow{d^p} E^2_{0,p-1} \longrightarrow E^\infty_{0,p-1} \longrightarrow 0,$$

$$0 \longrightarrow E^\infty_{0,p} \longrightarrow H_p(E;G) \longrightarrow E^\infty_{p,0} \longrightarrow 0,$$

which are valid in the range $p < n + m$. Putting them together we get a finite exact sequence

15.39.

$$H_{n+m-1}(E;G) \xrightarrow{p_*} H_{n+m-1}(B;G) \xrightarrow{\tau} H_{n+m-2}(F;G) \xrightarrow{i_*}$$

$$H_{n+m-2}(E;G) \cdots \longrightarrow H_2(B;G) \xrightarrow{\tau} H_1(F;G) \xrightarrow{i_*}$$

$$H_1(E;G) \xrightarrow{p_*} H_1(B;G),$$

where we use τ to denote d_p. τ is called the *transgression*. Thus in a certain range we get an exact sequence in homology which mimics the long exact sequence in homotopy. It is called the *Serre sequence*.

We can in fact define the transgression more generally as a map from a subgroup of $H_n(B;G)$ to a quotient of $H_{n-1}(F;G)$—namely

$$H_n(B;G) \cong E^2_{n0} \supset E^n_{n0} \xrightarrow{d^n} E^n_{0,n-1} \longleftarrow E^2_{0,n-1} \cong H_{n-1}(F;G).$$

We can give an alternative description of this map: the projection $p:E \to B$ defines a map

$$p_*: H_n(E,F;G) \to H_n(B,\{*\};G) = \tilde{H}_n(B;G).$$

Define $\bar{\tau}:\mathrm{im}\, p_* \to H_{n-1}(F;G)/\partial(\ker p_*)$ by taking $\bar{\tau}(x) = \{\partial(y)\}$ for $x \in \mathrm{im}\, p_*$ and $y \in H_n(E,F;G)$ such that $p_*(y) = x$. The elements of $\mathrm{im}\, p_* \subset \tilde{H}_n(B;G)$ (i.e. those on which $\bar{\tau}$ is defined) are called *transgressive*. We show now that $\bar{\tau} = \tau$.

15.40. Lemma. *For any filtration $\{X^n\}$ of X there is a commutative diagram*

$$\begin{array}{ccccccc}
h_p(X^{p-1}, X^0) & \xrightarrow{j_*} & h_p(X, X^0) & \longrightarrow & E^p_{p0} & \longrightarrow & 0 \\
\downarrow \partial & & \downarrow \partial & & \downarrow d^p & & \\
\partial h_p(X^{p-1}, X^0) & \subset & h_{p-1}(X^0) & \longrightarrow & E^p_{0,p-1} & \longrightarrow & 0
\end{array}$$

in which the rows are exact, all $p \geqslant 2$.

Proof: 15.40 expands into the monster pictured on page 362. Here we need the fact that $p \geqslant 2$ in order to have $2p-1 \geqslant p+1$ so that

$$\Delta h_{p+1}(X^{2p-1}, X^p) = \Delta h_{p+1}(X, X^p). \quad \square$$

15.41. Corollary. $\bar{\tau} = \tau$ *for* $p \geqslant 2$.

Proof: We assume $B^0 = \{b_0\}$ and E is filtered by the subspaces $X^n = p^{-1}B^n$. Then $h_n(X, X^0) = H_n(E,F;G)$, $h_{n-1}(X^0) = H_{n-1}(F;G)$,

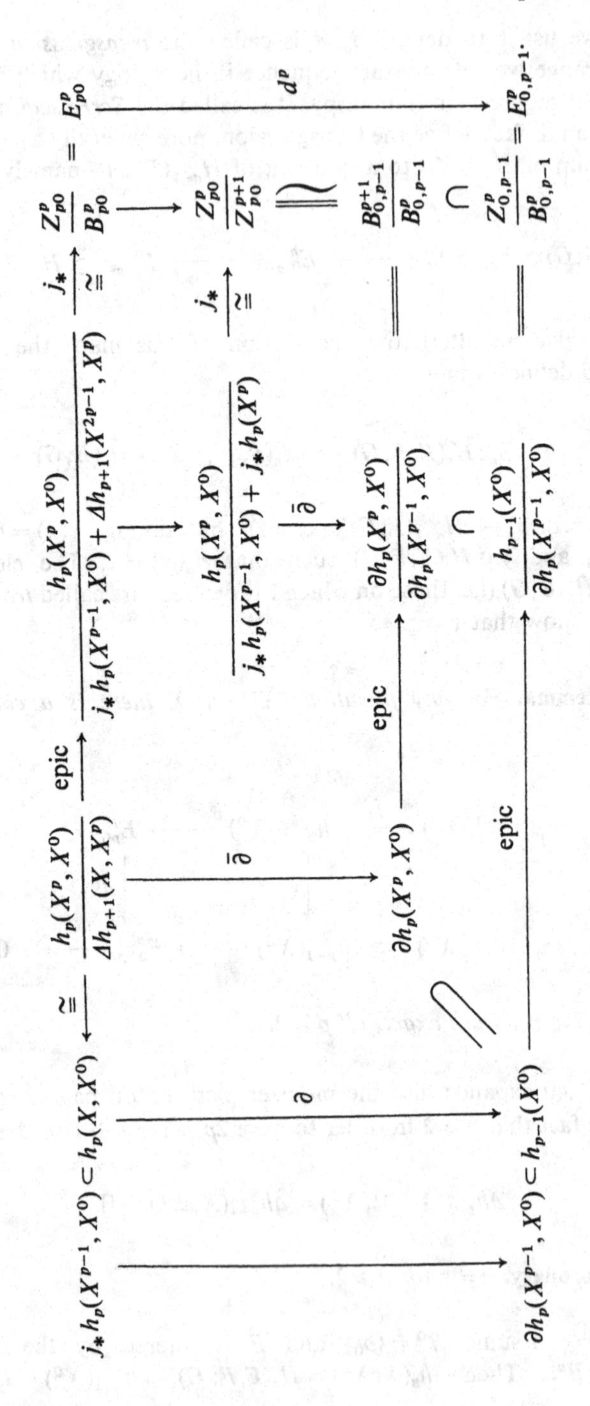

$h_n(X) = H_n(E; G)$, $E^2_{n0} = H_n(B; G)$. The result is contained in the commutative diagram

$$\begin{array}{ccccc}
\tilde{H}_n(B;G) & \cong & E^2_{n0} & & \\
\uparrow p_* & & \cup & & \\
H_n(E, F; G) & \longrightarrow & E^n_{n0} & \longrightarrow & 0 \\
\downarrow \partial & & \downarrow d^n & & \\
\tilde{H}_{n-1}(F;G) & \longrightarrow & E^n_{0,n-1} & \longrightarrow & 0, \quad n \geqslant 2. \quad \square
\end{array}$$

15.42. Of particular interest is the case of a fibration in which E is contractible. Then we can define a map $\sigma' : \tilde{H}_{n-1}(F; G) \to \tilde{H}_n(B; G)$ to be the composition

$$\tilde{H}_{n-1}(F;G) \underset{\cong}{\overset{\partial}{\longleftarrow}} H_n(E, F; G) \xrightarrow{p_*} H_n(B, \{b_0\}; G) = \tilde{H}_n(B;G).$$

From 15.41 we see that τ is the inverse of σ':

$$\tilde{H}_n(B;G) \supset \operatorname{im} \sigma' \to \tilde{H}_{n-1}(F;G)/\ker \sigma'.$$

Also in this case the Serre sequence 15.39 reduces to the isomorphisms

$$\sigma' : \tilde{H}_{p-1}(F;G) \cong \tilde{H}_p(B;G)$$

for $2 \leqslant p \leqslant 2n$ if B is n-connected, $n \geqslant 1$.

In the case of the fibration $\Omega B \to PB \to B$ the map σ' is often called the "homology suspension", which is a pity since that name would be better reserved for $\sigma : \tilde{H}_{n-1}(B; G) \to \tilde{H}_n(SB; G)$. The two are, however, closely related.

15.43. Proposition. *Suppose $f: X \to \Omega Y$ is any map and $f' : SX \to Y$ its adjoint. Then the diagram*

$$\begin{array}{ccc}
\tilde{H}_{n-1}(X;G) & \xrightarrow{f_*} & \tilde{H}_{n-1}(\Omega Y;G) \\
\downarrow \sigma & & \downarrow \sigma' \\
\tilde{H}_n(SX;G) & \xrightarrow{f'_*} & \tilde{H}_n(Y;G)
\end{array}$$

commutes for all n.

Proof: Define a map $\bar{f}: CX \to PY$ by $(\bar{f}[t,x])(s) = f(x)(st)$, $[t,x] \in CX$, $s \in I$. Then one checks immediately that the diagram

$$\begin{array}{ccccc} X & \xrightarrow{i_1} & CX & \xrightarrow{q} & SX \\ \downarrow{\scriptstyle f} & & \downarrow{\scriptstyle \bar{f}} & & \downarrow{\scriptstyle f'} \\ \Omega Y & \xrightarrow{i} & PY & \xrightarrow{p} & Y \end{array}$$

commutes, where $i_1(x) = [1,x]$, $x \in X$, and q is the obvious projection. Therefore in homology we have a commutative diagram

$$\begin{array}{ccccc} \tilde{H}_{n-1}(X;G) & \overset{\partial}{\underset{\cong}{\longleftarrow}} & H_n(CX,X;G) & \xrightarrow{q_*} & \tilde{H}_n(SX;G) \\ \downarrow{\scriptstyle f_*} & & \downarrow{\scriptstyle \bar{f}_*} & & \downarrow{\scriptstyle f'_*} \\ \tilde{H}_{n-1}(\Omega Y;G) & \overset{\partial}{\underset{\cong}{\longleftarrow}} & H_n(PY,\Omega Y;G) & \xrightarrow{p_*} & \tilde{H}_n(Y;G). \end{array}$$

(with $\sigma: \tilde{H}_{n-1}(X;G) \to \tilde{H}_n(SX;G)$ along the top and $\sigma': \tilde{H}_{n-1}(\Omega Y;G) \to \tilde{H}_n(Y;G)$ along the bottom) ∎

Now the adjoint of $1: SX \to SX$ is the map $i': X \to \Omega SX$ such that $i'(x)(s) = [s,x]$ for $x \in X$, $s \in I$.

15.44. Corollary. *For every n the diagram*

$$\begin{array}{ccc} \tilde{H}_n(X;G) & \xrightarrow{i'_*} & \tilde{H}_n(\Omega SX;G) \\ & {\scriptstyle \sigma}\searrow\cong \quad \swarrow{\scriptstyle \sigma'} & \\ & \tilde{H}_{n+1}(SX;G) & \end{array}$$

commutes.

15.45. Corollary. *If X is n-connected $(n \geqslant 1)$, then $i'_*: \tilde{H}_p(X;G) \to \tilde{H}_p(\Omega SX;G)$ is an isomorphism for $p \leqslant 2n+1$.*

Proof: This follows from 15.42 and 15.44, since SX is $(n+1)$-connected. ∎

Let $\Sigma: \pi_p(X,x_0) \to \pi_{p+1}(SX,*)$ be the homotopy suspension (6.23). It is easy to see that

$$\begin{array}{ccc} \pi_p(X,x_0) & \xrightarrow{i'_*} & \pi_p(\Omega SX,*) \\ & {\scriptstyle \Sigma}\searrow \quad /\!/ & \\ & \pi_{p+1}(SX,*) & \end{array}$$

commutes.

15.46. Theorem (Freudenthal). *If X is n-connected $(n \geqslant 1)$, then $\Sigma: \pi_p(X, x_0) \to \pi_{p+1}(SX, *)$ is an isomorphism for all $p \leqslant 2n$, epimorphism for $p = 2n + 1$.*

Proof: By 15.45 and the Whitehead Theorem (10.28) $i'_*: \pi_p(X, x_0) \to \pi_p(\Omega SX, *)$ is an isomorphism for $p \leqslant 2n$, an epimorphism for $p = 2n + 1$. ▯

Thus we have given a second proof of the Freudenthal suspension theorem which would be independent of the first proof if we had not used the Freudenthal suspension theorem to prove the Hurewicz isomorphism theorem and hence the Whitehead theorem. It is possible, however, to give an elementary proof of the Hurewicz isomorphism theorem (cf. [80]) which uses nothing except the definitions of homotopy and singular homology groups.

We turn to another important application of the Serre spectral sequence which will be valuable in the next chapter. In the following theorem we assume that H is any ring spectrum.

15.47. Theorem (Leray-Hirsch). *Let $p: (E, \dot{E}) \to B$ be a fibration pair with B 0-connected and fibres $(F, \dot{F})$. Suppose $e_1, e_2, \ldots, e_r \in H^*(E, \dot{E})$ are elements such that $i^* e_1, i^* e_2, \ldots, i^* e_r \in H^*(F, \dot{F})$ form a free basis for $H^*(F, \dot{F})$ as a module over $H^*(pt)$. $H^*(E, \dot{E})$ is a module over $H^*(B)$ with the module action given by $be = p^*(b) \cup e$ for $b \in H^*(B)$, $e \in H^*(E, \dot{E})$. As such $H^*(E, \dot{E})$ is a free $H^*(B)$-module with $\{e_1, e_2, \ldots, e_r\}$ as basis.*

Proof: First we observe that the fibration pair $p: (E, \dot{E}) \to B$ is orientable; for if $\omega: (I, \dot{I}) \to (B, b_0)$ is a loop in B, then the diagram

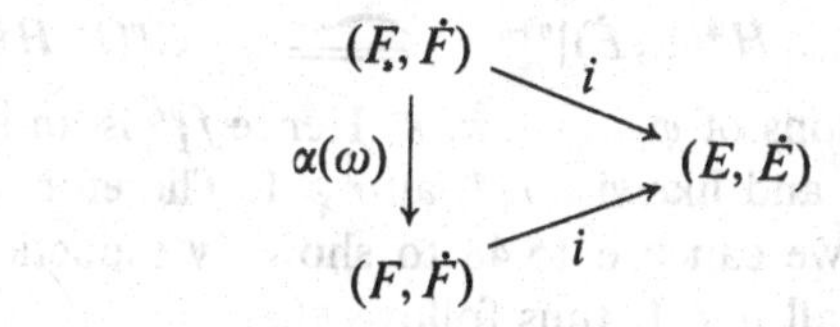

commutes up to homotopy, so $\alpha(\omega)^*(i^*(e_i)) = i^*(e_i)$ for all $i = 1, 2, \ldots, r$. But $i^*(e_1), \ldots, i^*(e_r)$ is a basis for $H^*(F, \dot{F})$; hence $\alpha(\omega)^* = 1$.

We define a cohomology theory h^* by

$$h^*(X, A) = H^*(X, A) \otimes_{H^*(pt)} H^*(F, \dot{F}).$$

h^* satisfies exactness because $H^*(F, \dot{F})$ is a free $H^*(pt)$-module. Let $f_i = i^*(e_i) \in H^*(F, \dot{F})$; we define homomorphisms

$$\phi_{X,Y}: h^*(X, Y) \to H^*(E|X, E|Y \cup \dot{E}|X)$$

for any $Y \subset X \subset B$ by taking $\phi_{X,Y}(x \otimes f_i) = p^* x \cup j^* e_i$, $x \in H^*(X, Y)$, where $j:(E|X,\dot{E}|X)\to(E,\dot{E})$ is the inclusion, and then extending linearly. $\phi_{X,Y}$ is natural with respect to inclusions and defines a *map of spectral sequences* (cf. 15.56) from the spectral sequence $\{\bar{E}_r^{pq},\bar{d}_r\}$ for h^* and the skeletal filtration of B, if B is a CW-complex, to the Serre spectral sequence $\{E_r^{pq},d_r\}$ of H on the fibration $(F,\dot{F})\to(E,\dot{E})\to B$: that is, the ϕ's define homomorphisms $f_r^{pq}:\bar{E}_r^{pq}\to E_r^{pq}$, $f^{pq}:\bar{F}^{pq}\to F^{pq}$ which commute with the differentials ($d_r\circ f_r^{pq}=f_r^{p+r,q-r+1}\circ\bar{d}_r$) and such that the diagrams

15.48.

$$\begin{array}{ccc} \bar{E}_{r+1}^{pq} & \xrightarrow{f_{r+1}^{pq}} & E_{r+1}^{pq} \\ \| \wr & & \| \wr \\ H(\bar{E}_r^{pq}) & \xrightarrow{H(f_r^{pq})} & H(E_r^{pq}) \end{array}$$

$$\begin{array}{ccccccccc} 0 & \to & \bar{F}^{p+1,q-1} & \to & \bar{F}^{pq} & \to & \bar{E}_\infty^{pq} & \to & 0 \\ & & \downarrow f^{p+1,q-1} & & \downarrow f^{pq} & & \downarrow f_\infty^{pq} & & \\ 0 & \to & F^{p+1,q-1} & \to & F^{pq} & \to & E_\infty^{pq} & \to & 0 \end{array}$$

commute. Moreover, the diagram

$$\begin{array}{ccc} \bar{E}_1^{pq} & \xrightarrow{f_1^{pq}} & E_1^{pq} \\ \| & & \| \\ [H^*(B^p,B^{p-1})\otimes_{H^*(pt)}H^*(F,\dot{F})]^{p+q} & \xrightarrow[\phi_{B^p,B^{p-1}}]{} & H^{p+q}(E|B^p,E|B^{p-1}\cup(\dot{E}|B^p)) \\ \bar{\kappa}\otimes 1 \,\| \wr & & \kappa \,\| \wr \\ [C^p(B;H^*(pt))\otimes_{H^*(pt)}H^*(F,\dot{F})]^{p+q} & \cong & C^p(B;H^q(F,\dot{F})) \end{array}$$

commutes by definitions of $\phi_{B^p,B^{p-1}}$, κ, $\bar{\kappa}$. Hence f_1^{pq} is an isomorphism. Therefore f_2^{pq} is also and likewise f_r^{pq}, all $r\geqslant 1$. Therefore f_∞^{pq} is also an isomorphism. Then we can use 15.48 to show by induction that f^{pq} is an isomorphism for all p,q. It thus follows that

$$\phi_{B,\phi}:H^*(B)\otimes_{H^*(pt)}H^*(F,\dot{F})\to H^*(E,\dot{E})$$

is an isomorphism. But this is synonymous with saying that $\{e_1,e_2,\ldots,e_r\}$ is a basis for $H^*(E,\dot{E})$ over $H^*(B)$. □

15.49. There is an analogous result in homology; we have an isomorphism

$$\psi:H_*(E,\dot{E})\to H_*(B)\otimes_{H_*(pt)}H_*(F,\dot{F})$$

given by $\psi(x)=\sum_{i=1}^r p_*(e_i\cap x)\otimes f_i^*$ for all $x\in H_*(E,\dot{E})$, where $(f_1^*,f_2^*,\ldots,f_r^*\}$ is the dual basis in $H_*(F,\dot{F})$ to $\{f_1,f_2,\ldots,f_r\}$. The proof is analogous to the one given above.

15.50. We can now give a new proof of the Thom isomorphism theorem 14.6. If $p:\dot{E} \to B$ is an $(n-1)$-sphere bundle with associated n-disk bundle $p:E \to B$, then a *Thom class* or *orientation class* for p is an element $t \in H^n(E, \dot{E})$ such that $j^*(t) \in H^n(D^n, S^{n-1})$ is a generator of $H^*(D^n, S^{n-1})$ as an $H^*(pt)$-module, where $j:(D^n, S^{n-1}) \to (E, \dot{E})$ is the inclusion of the fibre over b_0. Since we are tacitly assuming in this chapter that B is 0-connected, this definition of Thom class is equivalent to that in Chapter 14 for vector bundles (every $O(n)$-bundle has associated disk and sphere bundles).

15.51. Theorem (Thom Isomorphism Theorem). *If $p:\dot{E} \to B$ is an $(n-1)$-sphere bundle with associated disk bundle $p:E \to B$ and Thom class $t \in H^n(E, \dot{E})$, then the homomorphisms*

$$\Phi^*: H^q(B) \to H^{q+n}(E, \dot{E}) \qquad \Phi^*(z) = p^*(z) \cup t$$

$$\Phi_*: H_{q+n}(E, \dot{E}) \to H_q(B) \qquad \Phi_*(x) = p_*(t \cap x)$$

are isomorphisms for all $q \in \mathbb{Z}$.

Proof: These are just special cases of 15.47 and 15.49 with $r = 1$.

Remark. In particular it follows from the proof of 15.47 that the existence of a Thom class implies that the bundle pair $p:(E, \dot{E}) \to B$ is orientable in the sense of 15.14. Moreover, if $\xi = (B, p, E, \mathbb{R}^n)$ is a vector bundle regarded as having structure group $O(n)$, then a Thom class $t \in H^n(E, E - i(B))$ for the associated microbundle as in 14.5 defines a Thom class for the pair $(D(\xi), S(\xi))$:

$$H^n(E, E - i(B)) \cong H^n(D(\xi),\ D(\xi) - i(B)) \cong H^n(D(\xi), S(\xi)).$$

Thus "orientable" in the sense of 14.5 implies "orientable" in the sense of 15.14.

Exercise. To what extent is the converse true?

15.52. We can now reconstruct the Gysin sequence from the Thom theorem:

$$\begin{array}{ccccccc}
\cdots \longrightarrow & H^{q+n-1}(\dot{E}) & \xrightarrow{\ \delta\ } & H^{q+n}(E, \dot{E}) & \xrightarrow{\ j^*\ } & & \\
& \uparrow 1 & & \cong \uparrow \Phi^* & & & \\
\cdots \longrightarrow & H^{q+n-1}(\dot{E}) & \xrightarrow{\ \phi\ } & H^q(B) & \xrightarrow{\ \psi\ } & & \\
& & & & & & \\
& & & H^{q+n}(E) & \xrightarrow{\ i^*\ } & H^{q+n}(\dot{E}) & \xrightarrow{\ \delta\ } \cdots \\
& & & \cong \uparrow p^* & & \uparrow 1 & \\
& & & H^{q+n}(B) & \xrightarrow{\ p^*\ } & H^{q+n}(\dot{E}) & \longrightarrow \cdots
\end{array}$$

is commutative if we define ϕ to be $\Phi^{*-1} \circ \delta$, ψ to be $p^{*-1} \circ j^* \circ \Phi^*$. $p: E \to B$ is a homotopy equivalence since the disc bundle E contracts onto the 0-section, which is homeomorphic to B.

Exercise. Show that $\psi = d_{n+1}, \phi$ are as in 15.31 if $H = H(R)$.

Note that we have gained something; we now have the Gysin sequence for any ring spectrum H for which $p:(E, \dot{E}) \to B$ has a Thom class, and this includes even 0-sphere bundles. In particular we always have a Gysin sequence for mod 2 homology and cohomology.

15.53. Let us apply the Gysin sequence to compute $H^*(RP^n; \mathbb{Z}_2)$. The fibration $S^0 \to S^n \to RP^n$ gives an exact sequence

$$0 \longrightarrow H^0(RP^n; \mathbb{Z}_2) \xrightarrow{p^*} H^0(S^n; \mathbb{Z}_2) \longrightarrow$$

$$\cdots \longrightarrow H^q(S^n; \mathbb{Z}_2) \xrightarrow{\phi} H^q(RP^n; \mathbb{Z}_2) \xrightarrow{\psi} H^{q+1}(RP^n; \mathbb{Z}_2) \xrightarrow{p^*}$$

$$H^{q+1}(S^n; \mathbb{Z}_2) \longrightarrow \cdots \longrightarrow H^n(S^n; \mathbb{Z}_2) \xrightarrow{\phi} H^n(RP^n; \mathbb{Z}_2) \longrightarrow 0.$$

Since $H^q(S^n; \mathbb{Z}_2) = 0$ for $0 < q < n$, we see that $\psi: H^q(RP^n; \mathbb{Z}_2) \to H^{q+1}(RP^n; \mathbb{Z}_2)$ is an isomorphism for $0 < q < n-1$ and a monomorphism for $q = n-1$. Assuming $n \geqslant 1$, we have $p^*: H^0(RP^n; \mathbb{Z}_2) \cong H^0(S^n; \mathbb{Z}_2)$, so ψ is a monomorphism even for $q = 0$. Thus $\mathbb{Z}_2 = H^0(RP^n; \mathbb{Z}_2) \cong H^1(RP^n; \mathbb{Z}_2) \cong \cdots \cong H^{n-1}(RP^n; \mathbb{Z}_2)$ and $H^n(RP^n; \mathbb{Z}_2) \neq 0$. Since $H^n(S^n; \mathbb{Z}_2) = \mathbb{Z}_2$, $\phi: H^n(S^n; \mathbb{Z}_2) \to H^n(RP^n; \mathbb{Z}_2)$ must be an isomorphism, so even $\psi: H^{n-1}(RP^n; \mathbb{Z}_2) \to H^n(RP^n; \mathbb{Z}_2)$ is an isomorphism. Let $w = \psi(1)$ in $H^1(RP^n; \mathbb{Z}_2)$; this is the Euler class and ψ is multiplication by w. Hence the non-zero element of $H^q(RP^n; \mathbb{Z}_2)$ is w^q.

15.54. Theorem *As a ring* $H^*(RP^n; \mathbb{Z}_2) \cong \mathbb{Z}_2[w]/(w^{n+1})$. *Similarly* $H^*(RP^\infty; \mathbb{Z}_2) \cong \mathbb{Z}_2[w]$.

Exercise. Ponder the relation between the Euler class $e \in H^n(B; R)$ and the Thom class $t \in H^n(E, \dot{E}; R)$ for an $(n-1)$-sphere bundle $p: \dot{E} \to B$.

15.55. *Remark.* We have the covering $S^n \to RP^n$, so

$$\pi_q(RP^n, *) \cong \pi_q(S^n, *) = 0, \quad 1 < q < n.$$

Therefore $\pi_q(RP^\infty, *) = 0$, all $q > 1$. Since $\pi_1(RP^\infty, *) \cong \mathbb{Z}_2$, we may take RP^∞ as $H(\mathbb{Z}_2, 1)$; thus we have just computed $H^*(H(\mathbb{Z}_2, 1); \mathbb{Z}_2)$.

We conclude with two further general results about spectral sequences which have important applications.

15.56. Definition. A *map of spectral sequences* $\{f^r_{pq}\}$ is a collection of homomorphisms $f^r_{pq}: E^r_{pq} \to \bar{E}^r_{pq}$, $f_{pq}: F_{pq} \to \bar{F}_{pq}$ such that for each p, q, r

i) $$d^r \circ f^r_{pq} = f^r_{p-r,q+r-1} \circ d^r,$$

ii) the diagram

$$\begin{array}{ccc} E^{r+1}_{pq} & \xrightarrow{f^{r+1}_{pq}} & \bar{E}^{r+1}_{pq} \\ \| \wr & & \| \wr \\ H_*(E^r_{pq}) & \xrightarrow{f^r_{pq*}} & H_*(\bar{E}^r_{pq}) \end{array}$$ commutes

iii) the diagram

$$\begin{array}{ccc} E^{\infty}_{pq} & \xrightarrow{f^{\infty}_{pq}} & \bar{E}^{\infty}_{pq} \\ \| \wr & & \| \wr \\ \operatorname{dir\,lim}_r E^r_{pq} & \xrightarrow{\operatorname{dir\,lim} f^r_{pq}} & \operatorname{dir\,lim}_r \bar{E}^r_{pq} \end{array}$$ commutes, and

iv) the diagram

$$\begin{array}{ccccccccc} 0 & \longrightarrow & F_{p-1,q+1} & \longrightarrow & F_{pq} & \longrightarrow & E^{\infty}_{pq} & \longrightarrow & 0 \\ & & \downarrow f_{p-1,q+1} & & \downarrow f_{pq} & & \downarrow f^{\infty}_{pq} & & \\ 0 & \longrightarrow & \bar{F}_{p-1,q+1} & \longrightarrow & \bar{F}_{pq} & \longrightarrow & \bar{E}^{\infty}_{pq} & \longrightarrow & 0 \end{array}$$ commutes.

A fibre map

$$\begin{array}{ccc} E & \xrightarrow{f} & \bar{E} \\ \downarrow p & & \downarrow \bar{p} \\ B & \xrightarrow{f} & \bar{B} \end{array}$$ induces

a map of Serre spectral sequences; that is what we mean by saying the Serre spectral sequence is natural.

15.57. Theorem (Zeeman's Comparison Theorem). *Let* $\{f_{pq}\}:\{E_{pq}\} \to \{\bar{E}^r_{pq}\}$ *be a map of spectral sequences satisfying*

i) $E^r_{pq} = \bar{E}^r_{pq} = 0$ *if* $p < 0$ *or* $q < 0$;
ii) $E^2_{pq} = E^2_{p0} \otimes E^2_{0q}$, $\bar{E}^2_{pq} = \bar{E}^2_{p0} \otimes \bar{E}^2_{0q}$ *and* $f^2_{pq} = f^2_{p0} \otimes f^2_{0q}$;
iii) $E^{\infty}_{pq} = \bar{E}^{\infty}_{pq} = 0$ *for all* $(p,q) \neq (0,0)$;
iv) f^2_{p0} *is an isomorphism for all* $p \geqslant 0$.

Then f^2_{0q} *is an isomorphism for all* $q \geqslant 0$.

Proof: We must (regretfully) recall all the extra machinery inside a spectral sequence. We let $B^3_{pq} = \operatorname{im} d^2 \subset E^2_{pq}$, $Z^3_{pq} = \ker d^2 \subset E^2_{pq}$, so that we have

$$0 = B^2_{pq} \subset B^3_{pq} \subset Z^3_{pq} \subset Z^2_{pq} = E^2_{pq} \quad \text{and} \quad E^3_{pq} = Z^3_{pq}/B^3_{pq}.$$

Then $\operatorname{im} d^3 \subset E^3_{pq}$ has the form $\operatorname{im} d^3 = B^4_{pq}/B^3_{pq}$ and $\ker d^3 = Z^4_{pq}/B^3_{pq}$ for some subgroups $B^4_{pq} \subset Z^4_{pq}$ such that

$$0 = B^2_{pq} \subset B^3_{pq} \subset B^4_{pq} \subset Z^4_{pq} \subset Z^3_{pq} \subset Z^2_{pq} = E^2_{pq}, \; E^4_{pq} \cong Z^4_{pq}/B^4_{pq}.$$

Repeating this process, we get a filtration

$$0 = B^2_{pq} \subset B^3_{pq} \subset \cdots \subset B^r_{pq} \subset \cdots \subset B^\infty_{pq} \subset Z^\infty_{pq} \subset \cdots \subset Z^r_{pq} \subset \cdots \subset Z^2_{pq} = E^2_{pq}$$

with

$$\ker d^r = Z^{r+1}_{pq}/B^r_{pq}, \quad \operatorname{im} d^r = B^{r+1}_{pq}/B^r_{pq}, \quad E^r_{pq} = Z^r_{pq}/B^r_{pq},$$

$$B^\infty_{pq} = \bigcup_r B^r_{pq}, \quad Z^\infty_{pq} = \bigcap_r Z^r_{pq}, \quad E^\infty_{pq} \cong Z^\infty_{pq}/B^\infty_{pq}.$$

Finally the isomorphism $\operatorname{im} d^r \cong E^r_{pq}/\ker d^r$ becomes

$$Z^r_{pq}/Z^{r+1}_{pq} \cong B^{r+1}_{p-r,q+r-1}/B^r_{p-r,q+r-1}.$$

Since

$$d^r : E^r_{pq} \to E^r_{p-r,q+r-1} \quad \text{is } 0 \text{ if } r > p$$

we see that

$$Z^{r+1}_{pq}/B^r_{pq} = \ker d^r = E^r_{pq} \cong Z^r_{pq}/B^r_{pq}, \text{ so } Z^{r+1}_{pq} = Z^r_{pq} \quad \text{if } r > p.$$

Similarly

$$0 = \operatorname{im} d^r = B^{r+1}_{pq}/B^r_{pq} \quad \text{if } r > q + 1,$$

so that

$$B^{r+1}_{pq} = B^r_{pq} \quad \text{for } r > q + 1.$$

In particular, we get

$$Z^{p+1}_{pq} = Z^{p+2}_{pq} = \cdots = Z^\infty_{pq}, \quad B^{q+2}_{pq} = B^{q+3}_{pq} = \cdots = B^\infty_{pq}.$$

Since

$$Z^{p+1}_{pq}/B^{q+2}_{pq} = Z^\infty_{pq}/B^\infty_{pq} = E^\infty_{pq} = 0 \quad \text{if } (p,q) \neq (0,0),$$

it follows

$$Z^{p+1}_{pq} = B^{q+2}_{pq} \quad \text{if } (p,q) \neq (0,0).$$

The proof of the theorem really begins at this point. Suppose $f_{0q}^2: E_{0q}^2 \to \bar{E}_{0q}^2$ is an isomorphism for $q \leqslant Q$ (the case $q = 0$ follows from the assumption iv)). We wish to prove $f_{0,Q+1}^2$ is an isomorphism; in fact we shall show $f_{0,Q+1}^r$ is an isomorphism for all r by downward induction on r. For $r > Q+2$ we have $E_{0,Q+1}^r = E_{0,Q+1}^\infty = 0 = \bar{E}_{0,Q+1}^\infty = \bar{E}_{0,Q+1}^r$, so we can begin the induction. We have a commutative diagram with exact rows

$$\begin{array}{ccccccccc} 0 & \to & B_{0,Q+1}^{p+1}/B_{0,Q+1}^p & \to & E_{0,Q+1}^p & \to & E_{0,Q+1}^{p+1} & \to & 0 \\ & & \downarrow f & & \downarrow f_{0,Q+1}^p & & \downarrow f_{0,Q+1}^{p+1} & & \\ 0 & \to & \bar{B}_{0,Q+1}^{p+1}/\bar{B}_{0,Q+1}^p & \to & \bar{E}_{0,Q+1}^p & \to & \bar{E}_{0,Q+1}^{p+1} & \to & 0, \end{array}$$

so we can make the inductive step if we prove $f: B_{0,Q+1}^{p+1}/B_{0,Q+1}^p \to \bar{B}_{0,Q+1}^{p+1}/\bar{B}_{0,Q+1}^p$ is an isomorphism for all p, $2 \leqslant p \leqslant Q+2$. But we have

$$B_{0,Q+1}^{p+1}/B_{0,Q+1}^p \cong Z_{p,Q-p+2}^p/Z_{p,Q-p+2}^{p+1} \cong Z_{p,Q-p+2}^p/B_{p,Q-p+2}^{Q-p+4}$$

and similarly with bars. Thus it suffices to show that

$$f: Z_{p,Q-p+2}^p/B_{p,Q-p+2}^{Q-p+4} \to \bar{Z}_{p,Q-p+2}^p/\bar{B}_{p,Q-p+2}^{Q-p+4}$$

is an isomorphism for $2 \leqslant p \leqslant Q+2$.

What we shall show now is that $f: Z_{pq}^r \to \bar{Z}_{pq}^r$ is an isomorphism for $q + r \leqslant Q + 2$ and $f: B_{pq}^r \to \bar{B}_{pq}^r$ is an isomorphism for $q \leqslant Q$ (as above, we shall simply use f to denote any of the homomorphisms induced by the f_{pq}^r's). The result is clear for $r = 2$ because of ii), iv) and the inductive hypothesis on f_{0q}^2. Suppose the statement proved for some $r \geqslant 2$; then in the following commutative diagram f_1 and f_4 are isomorphisms if $q \leqslant Q$:

$$\begin{array}{ccccccc} Z_{p+r,q-r+1}^r & \xrightarrow{\text{epic}} & Z_{p+r,q-r+1}^r/Z_{p+r,q-r+1}^{r+1} \cong B_{pq}^{r+1}/B_{pq}^r & \xrightarrow{\text{monic}} & E_{pq}^2/B_{pq}^r \\ \downarrow f_1 & & \downarrow f_2 \qquad\qquad \downarrow f_3 & & \downarrow f_4 \\ \bar{Z}_{p+r,q-r+1}^r & \xrightarrow[\text{epic}]{} & \bar{Z}_{p+r,q-r+1}^r/\bar{Z}_{p+r,q-r+1}^{r+1} \cong \bar{B}_{pq}^{r+1}/\bar{B}_{pq}^r & \xrightarrow[\text{monic}]{} & \bar{E}_{pq}^2/\bar{B}_{pq}^r. \end{array}$$

Thus it follows that f_2 and f_3 are isomorphisms and therefore the inductive step for $r+1$ follows. This completes the induction over q also and hence the proof of the theorem. □

Remark. There is an analogous theorem in which one assumes instead that f_{0q}^2 is an isomorphism for $q \geqslant 0$ and then concludes that f_{p0}^2 must be an isomorphism for all $p \geqslant 0$. Both comparison theorems have their analogs in cohomology.

15.58. Definition. Let A be an algebra over a ring R. Then elements $x_1, x_2, \ldots, x_n, \ldots \in A$ are said to form a *simple system of generators* for A if the monomials $x_1^{\varepsilon_1} x_2^{\varepsilon_2} \ldots x_m^{\varepsilon_m}$ ($\varepsilon_i = 0$ or 1, $m \geqslant 0$) form a basis for A over R.

15.59. Examples. In $R[x]$ the monomials $x, x^2, x^4, \ldots, x^{2^n}, \ldots$ form a simple system of generators. In an exterior algebra $E(x_1, x_2, \ldots)$ the elements $x_1, x_2, \ldots$ form a simple system of generators.

15.60. Theorem (A. Borel). *Let $\Omega B \to PB \to B$ be the path fibration with B a simply connected H-group. Let $f_1, f_2, \ldots \in H_*(\Omega B; R)$ be elements satisfying:*

i) *for each n only finitely many f_i's lie in $H_n(\Omega B; R)$;*
ii) *$\sigma'(f_1), \sigma'(f_2), \ldots \in H_*(B; R)$ form a simple system of generators for the Pontrjagin ring.*

Then $H_*(\Omega B; R) \cong R[f_1, f_2, \ldots]$.

Proof: The idea of the proof is to construct an abstract spectral sequence $\{\bar{E}^r_{pq}\}$ which behaves as we conjecture the Serre spectral sequence $\{E^r_{pq}\}$ of this fibration ought to behave and a map

$$\{\phi^r_{pq}\}: \{\bar{E}^r_{pq}\} \to \{E^r_{pq}\}$$

of spectral sequences with ϕ^2_{p0} an isomorphism for all $p \geqslant 0$. Then we can apply 15.57. Certainly the Serre spectral sequence $\{E^r_{pq}\}$ satisfies the hypotheses of 15.57, for

$$E^2_{pq} \cong H_p(B; R) \otimes_R H_q(\Omega B; R) \cong E^2_{p,0} \otimes_R E^2_{0,q}$$

($H_*(B; R)$ is R-free) and $E^\infty_{pq} = 0$ if $(p, q) \neq (0, 0)$.

We construct a spectral sequence $\bar{E}^*_{**}(i)$ for each $i \geqslant 1$; let $f_i \in H_{n_i}(F; R)$ and give $\bar{E}^2_{**}(i)$ an R-basis consisting of elements

$$x(i, m) \in \bar{E}^2_{0, mn_i}(i), \quad y(i, m) \in \bar{E}^2_{n_i+1, mn_i}(i), \quad m \geqslant 0.$$

Define $d^r(i)$ by $d^r(i) = 0$ for $2 \leqslant r \leqslant n_i$ and $d^{n_i+1}(i)(y(i, m)) = x(i, m+1)$. Then only one element—$x(i, 0) = 1$—survives to $\bar{E}^{n_i+2}_{**}(i)$, so we may take $d^r(i) = 0$ for $r > n_i + 1$.

Now we define $\bar{E}^*_{**}$ by taking $\bar{E}^r_{**} = \bar{E}^r_{**}(1) \otimes \bar{E}^r_{**}(2) \otimes \ldots$ in the sense that we form $\bar{E}^r_{**}(1) \otimes \ldots \otimes \bar{E}^r_{**}(m)$ for each m and take the direct limit over m. d^r is defined by the usual formula for a differential on tensor products—i.e. d^r is a derivation.

Now by the Künneth theorem we have

$$\begin{aligned} H(\bar{E}^r_{**}) &= H(\bar{E}^r_{**}(1) \otimes \bar{E}^r_{**}(2) \otimes \cdots) \\ &\cong H(\bar{E}^r_{**}(1)) \otimes H(\bar{E}^r_{**}(2)) \otimes \cdots \\ &\cong \bar{E}^{r+1}_{**}(1) \otimes \bar{E}^{r+1}_{**}(2) \otimes \cdots \\ &= \bar{E}^{r+1}_{**}. \end{aligned}$$

Therefore $\{\bar{E}^r_{pq}\}$ is a spectral sequence.

Observe that $\bar{E}^2_{*0}$ has a basis consisting of the elements

$$y(1,0)^{\varepsilon_1} \otimes y(2,0)^{\varepsilon_2} \otimes \cdots \otimes y(m,0)^{\varepsilon_m}$$

($\varepsilon_i = 0$ or 1)—in other words $y(1,0)$, $y(2,0)$, ... form a simple system of generators for $\bar{E}^2_{*0}$. $\bar{E}^2_{0*}$ has a basis consisting of the elements

$$x(1, r_1) \otimes x(2, r_2) \otimes \ldots \otimes x(m, r_m)$$

for $r_1, \ldots, r_m \in \mathbb{N}$. Clearly $\bar{E}^2_{p,q} \cong \bar{E}^2_{p,0} \otimes \bar{E}^2_{0,q}$. Also assumption i) implies that $\bar{E}^r_{pq} = 0$ if r is sufficiently large, so $\bar{E}^\infty_{pq} = 0$ for $(p,q) \neq (0,0)$.

We now define ϕ^r_{pq}; let $b_i = \sigma'(f_i) \in H_{n_i+1}(B; R)$. Because $f_i \in \tau(b_i)$ we have

15.61.
$$d^r b_i = 0 \quad r < n_i + 1$$
$$d^{n_i+1} b_i = \{f_i\}.$$

We take

$$\begin{aligned} \phi^r(x(i,m)) &= \{f_i\}^m \\ \phi^r(y(i,m)) &= b_i\{f_i\}^m, \quad r \leqslant n_i + 1, \end{aligned}$$

where we use $x(i,m)$ to denote $1 \otimes \ldots \otimes x(i,m) \otimes 1 \ldots$ and $\{f_i\}$ denotes the class of f_i in E^r_{0,n_i}. We extend ϕ^r to be a map of algebras. By 15.61 and the definition of $\bar{d}^r$ it follows ϕ^r commutes with the differentials. We also have a commutative diagram

$$\begin{array}{ccc} \bar{E}^2_{p0} \otimes \bar{E}^2_{0q} & \xrightarrow{\cong} & \bar{E}^2_{pq} \\ \Big\downarrow \phi^2_{p0} \otimes \phi^2_{0q} & & \Big\downarrow \phi^2_{pq} \\ E^2_{p0} \otimes E^2_{0q} & \xrightarrow{\cong} & E^2_{pq}. \end{array}$$

Since $b_1, b_2, \ldots$ form a simple system of generators for $E^2_{*,0} \cong H_*(B; R)$, it follows ϕ^2_{*0} is an isomorphism. Thus all the hypotheses of 15.57 are satisfied, and we may conclude that $\phi^2_{0,*}: \bar{E}^2_{0,*} \to E^2_{0,*} \cong H_*(\Omega B; R)$ is an isomorphism. Therefore the monomials $f_1^{r_1} f_2^{r_2} \ldots f_m^{r_m}$ form a basis for $H_*(\Omega B; R)$—or in other words $H_*(\Omega B; R) \cong R[f_1, f_2, \ldots]$. ☐

We give the dual of 15.60 explicitly since it plays an important role in Chapter 18.

15.62. Theorem. *Let* $F \to E \to B$ *be a fibration with* E *contractible. Let* $b_1, b_2, \ldots \in H^*(B; R)$ *be elements satisfying:*

i) *for each* n *only finitely many* b_i*'s lie in* $H^n(B; R)$;

ii) $\sigma'(b_1)$, $\sigma'(b_2)$, $\ldots \in H^*(F; R)$ *form a simple system of generators.*

Then $H^*(B; R) \cong R[b_1, b_2, \ldots]$.

Comments

There can scarcely have been a student of mathematics who had to deal with spectral sequences and was not repelled or at least very confused by them on his first encounter. One needs much practice with spectral sequences before all those indices stop swimming before one's eyes and begin to take on some sensible meaning. The following chapters contain many important applications of spectral sequences, which should a) convince the reader that spectral sequences are very useful and b) give him some practice in using them. The reader should not, however, content himself with reading the applications given here, but should work out as many examples as he can find in order to get practice and a sense of familiarity with spectral sequences.

References

1. S.-T. Hu [48]
2. S. Mac Lane [56]
3. R. E. Mosher and M. C. Tangora [68]
4. J.-P. Serre [76, 78]
5. E. H. Spanier [80]

Chapter 16

Characteristic Classes

In Chapter 11 we saw that isomorphism classes of vector bundles with structure group $G(n)$ over a CW-complex X were classified by homotopy classes of maps $f:X \to BG(n)$ of X into the classifying space $BG(n)$ for $G(n)$-bundles. If ξ, η are two $G(n)$-bundles with classifying maps $f_\xi, f_\eta : X \to BG(n)$, then $\xi \simeq \eta$ if and only if $f_\xi \simeq f_\eta$. Suppose we wanted to prove ξ, η were *not* isomorphic. We might try to show that f_ξ and f_η were not homotopic. There are two disadvantages to this approach: i) given a vector bundle ξ as vector bundle it is usually very difficult to describe its classifying map f_ξ; ii) the problem of showing directly that two given functions are not homotopic is at least as difficult in general as showing two vector bundles are not isomorphic. However, we do have one standard trick for showing two functions are not homotopic: for any cohomology theory k^* if $f_\xi^* \neq f_\eta^* : k^*(BG(n)) \to k^*(X)$, then $f_\xi \not\simeq f_\eta$ and hence $\xi \not\simeq \eta$. Therefore we look for an appropriate k^* and some $x \in k^*(BG(n))$ such that $f_\xi^*(x) \neq f_\eta^*(x) \in k^*(X)$.

Note that since the homotopy class of f_ξ is completely determined by ξ, it follows the element $f_\xi^*(x) \in k^*(X)$ is completely determined by ξ, so we could equally well write $x(\xi)$ for $f_\xi^*(x)$. The elements $x(\xi) \in k^*(X)$ for various $x \in k^*(BG(n))$ are called *characteristic classes* of ξ. What we have just proved is the following.

16.1. Proposition. *Two $G(n)$-bundles ξ, η are isomorphic only if all their characteristic classes agree for all possible cohomology theories.*

Of course we do not claim that equality of characteristic classes for all theories implies $\xi \simeq \eta$.

The next theorem tells us that under favorable conditions on the cohomology theory h^* we can choose certain characteristic classes which behave well and are therefore readily computable. This theorem will turn out to yield as a bonus the possibility of computing $h^*(BG(n))$ in suitable cases.

16.2. Theorem. *Suppose h^* is a cohomology theory with products such that for each n there are elements $x_n \in h^2(CP^n)$ satisfying*

i) $h^*(CP^n) \cong h^*(pt)[x_n]/(x_n^{n+1})$;
ii) *the inclusion* $i: CP^n \to CP^{n+1}$ *gives* $i^* x_{n+1} = x_n$, $n \geqslant 1$.

Then for each $U(n)$-bundle ξ over a CW-complex X there are uniquely defined elements $c_0(\xi), c_1(\xi), \ldots, c_n(\xi)$ with $c_i(\xi) \in h^{2i}(X)$ depending only on the isomorphism class of ξ and satisfying

a) *if $\xi \to X$ is a bundle and $f: Y \to X$ a map, then $c_i(f^*\xi) = f^*(c_i(\xi))$, $0 \leqslant i \leqslant n$;*
b) $c_0(\xi) = 1$ *for all* ξ;
c) *if $\gamma \to CP^n$ is the Hopf $U(1)$-bundle over CP^n, then $c_1(\gamma) = x_n$;*
d) *if ξ is a $U(m)$-bundle and η is a $U(n)$-bundle, both over X, then $c_i(\xi \oplus \eta) = \sum_{j+k=i} c_j(\xi) c_k(\eta)$, $0 \leqslant i \leqslant n+m$, where we set $c_i(\xi) = 0$ if $i \geqslant m$.*

Proof: Before we begin the actual proof we make some auxiliary bundle constructions. CP^n is all complex lines l in $\mathbb{C}^{n+1}$ which pass through 0. The fibre of γ over l is the set of all points of $\mathbb{C}^{n+1}$ belonging to l. In other words $E(\gamma) = \{(l, y) \in CP^n \times \mathbb{C}^{n+1} : y \in l\}$. Now there is a standard trick for taking any $U(n)$-bundle $\xi \to X$ and "splitting off a line bundle": let $P(\xi)$ be the space of lines through 0 in all fibres of ξ; there is still a map $p': P(\xi) \to X$ which is a fibre bundle with fibre CP^{n-1}. Another way of describing $P(\xi)$ is this: let $S(\xi)$ denote the $(2n-1)$-sphere bundle of ξ; $U(1)$ acts on $S(\xi)$ freely and we take $P(\xi) = S(\xi)/U(1)$. This gives $P(\xi)$ a topology. If X is a CW-complex, then we may give $P(\xi)$ the structure of a CW-complex (cf. 12.26).

Now consider the induced bundle $p'^*\xi$ over $P(\xi)$. There is a monomorphism $\lambda_\xi \to p'^*\xi$ of bundles, where λ_ξ is the following line bundle over $P(\xi)$: $E(\lambda_\xi) = \{(l, y) \in P(\xi) \times E(\xi) : y \in l\}$. (Note that if $j: CP^{n-1} \to P(\xi)$ is the inclusion of a fibre, then $j^*\lambda_\xi = \gamma$). We let μ_ξ be the orthogonal complement of λ_ξ in $p'^*\xi$, so that $p'^*\xi \simeq \lambda_\xi \oplus \mu_\xi$. μ_ξ is a $U(n-1)$-bundle.

From the assumptions i), ii) on h^* it follows $\lim^1 h^*(CP^n) = 0$ and hence by 7.66 $h^*(CP^\infty) \cong \lim^0 h^*(CP^n) \cong h^*(pt)[[x_\infty]]$, where $x_\infty \in h^2(CP^\infty)$ is the unique element such that $i_n^* x_\infty = x_n$, $i_n: CP^n \to CP^\infty$ the inclusion. Let $f: P(\xi) \to CP^\infty$ classify the line bundle λ_ξ (we showed in 11.36 that we can take CP^∞ as $BU(1)$). If $j: CP^{n-1} \to P(\xi)$ is the inclusion of a fibre, then $f \circ j \simeq i_{n-1}$, so $j^* f^* x_\infty = x_{n-1}$. Let $y = f^* x_\infty \in h^2(P(\xi))$. Then $1, y, y^2, \ldots, y^{n-1}$ are elements in $h^*(P(\xi))$ such that $\{j^*1, j^*y, \ldots, j^*y^{n-1}\} = \{1, x_{n-1}, \ldots, x_{n-1}^{n-1}\}$ form a basis for $h^*(CP^{n-1})$ over $h^*(pt)$. The Leray–Hirsch theorem 15.47 then applied to the fibration $CP^{n-1} \to P(\xi) \to X$ says that $h^*(P(\xi))$ is a free $h^*(X)$-module with a basis $\{1, y, y^2, \ldots, y^{n-1}\}$.

The proof of 16.2 can now begin. We show uniqueness first. Note that $i_n^* \gamma = \gamma$ on CP^n (we also use γ for the Hopf $U(1)$-bundle over CP^∞). Therefore by a) and c) we have

$$i_n^*(c_1(\gamma)) = c_1(i_n^* \gamma) = c_1(\gamma) = x_n, \quad n \geqslant 1.$$

But x_∞ is the unique element such that $i_n^*(x_\infty) = x_n$, $n \geqslant 1$, so we have $c_1(\gamma) = x_\infty \in h^2(CP^\infty)$.

Now any $U(1)$-bundle $\xi \to X$ is classified by a map $f_\xi : X \to CP^\infty$, so by a) $c_1(\xi) = c_1(f_\xi^* \gamma) = f_\xi^*(c_1(\gamma)) = f_\xi^*(x_\infty)$, which shows $c_1(\xi)$ is uniquely determined for all $U(1)$-bundles ξ.

Suppose uniqueness has been proved for all $U(n-1)$-bundles and let $\xi \to X$ be a $U(n)$-bundle. Then $p'^* \xi = \lambda_\xi \oplus \mu_\xi$ over $P(\xi)$ and $c_1(\lambda_\xi) = f^*(x_\infty) = y$. Thus we have

$$\begin{aligned} p'^*(c_i(\xi)) = c_i(p'^*(\xi)) &= c_i(\lambda_\xi \oplus \mu_\xi) = c_i(\mu_\xi) + c_1(\lambda_\xi)\, c_{i-1}(\mu_\xi) \\ &= c_i(\mu_\xi) + y c_{i-1}(\mu_\xi). \end{aligned}$$

But $c_i(\mu_\xi)$ and $c_{i-1}(\mu_\xi)$ are uniquely determined according to the induction hypothesis. Thus $p'^*(c_i(\xi))$ is uniquely determined. Now by the Leray–Hirsch theorem $p'^* : h^*(X) \to h^*(P(\xi))$ is a monomorphism ($p'^*(x) = p'^*(x) \cdot 1 = x \cdot 1$), so it follows $c_i(\xi)$ is also uniquely determined.

Now we prove existence. We have seen that $1, y, \ldots, y^{n-1}$ form a basis for $h^*(P(\xi))$ over $h^*(X)$. Therefore we can express y^n as a linear combination

$$y^n = (-1)^{n+1} c_n(\xi) \cdot 1 + (-1)^n c_{n-1}(\xi) \cdot y + \cdots + c_1(\xi) \cdot y^{n-1}$$

for appropriate coefficients $c_1(\xi), \ldots, c_n(\xi) \in h^*(X)$. That is, we have

$$\textstyle\sum_{i=0}^n (-1)^i c_i(\xi)\, y^{n-i} = 0,$$

where we take $c_0(\xi) = 1$. Then b) follows by definition.

To prove a) one needs to demonstrate that $P(f^* \xi) \simeq f^* P(\xi)$, $\lambda_{f^*(\xi)} \simeq f^* \lambda_\xi$, where $\bar{f} : f^* P(\xi) \to P(\xi)$ covers f, and $\bar{f}^*(y_\xi) = y_{f^* \xi}$, all of which are straightforward. Then we compute

$$\begin{aligned} 0 = \bar{f}^*(\textstyle\sum_{i=0}^n (-1)^i c_i(\xi)\, y_\xi^{n-i}) &= \textstyle\sum_{i=0}^n (-1)^i f^*(c_i(\xi))\, \bar{f}^*(y_\xi)^{n-i} \\ &= \textstyle\sum_{i=0}^n (-1)^i f^*(c_i(\xi))\, y_{f^*\xi}^{n-i}. \end{aligned}$$

But the elements $c_i(f^* \xi)$ are the unique elements such that

$$\textstyle\sum_{i=0}^n (-1)^i c_i(f^* \xi)\, y_{f^*\xi}^{n-i} = 0,$$

so it follows that $f^*(c_i(\xi)) = c_i(f^* \xi)$, $0 \leqslant i \leqslant n$.

To prove c) we simply observe that $P(\gamma) = CP^n$ and $\lambda_\gamma = \gamma$, $y_\gamma = x_n$, so $c_1(\gamma) = x_n$.

We turn now to d); suppose ξ,η are $U(m)$-, $U(n)$-bundles, respectively. Then $P(\xi)$, $P(\eta)$ are subspaces of $P(\xi\oplus\eta)$. Let $U=P(\xi\oplus\eta)-P(\eta)$, $V=P(\xi\oplus\eta)-P(\xi)$; then $P(\xi)$ is a strong deformation retract of U, $P(\eta)$ is a strong deformation retract of V and $U\cup V=P(\xi\oplus\eta)$. Also

$$x_1=\sum_{i=0}^{m}(-1)^i c_i(\xi)\,y^{m-i},\qquad x_2=\sum_{j=0}^{n}(-1)^j c_j(\eta)\,y^{n-j}$$

are elements of $h^*(P(\xi\oplus\eta))$ $(y=c_1(\lambda_{\xi\oplus\eta}))$ such that $x_1|P(\xi)=0$ and hence $x_1|U=0$, $x_2|P(\eta)=0$ and hence $x_2|V=0$. Thus there are elements $x_1'\in h^*(P(\xi\oplus\eta),U)$, $x_2'\in h^*(P(\xi\oplus\eta),V)$ with $j_U^*x_1'=x_1$, $j_V^*x_2'=x_2$, $j_U:(P(\xi\oplus\eta),\varnothing)\to(P(\xi\oplus\eta),U)$, $j_V:(P(\xi\oplus\eta),\varnothing)\to(P(\xi\oplus\eta),V)$ the inclusions. Then

$$x_1'\cdot x_2'\in h^*(P(\xi\oplus\eta),U\cup V)=0,$$

so it follows $x_1x_2=j_U^*x_1'\cdot j_V^*x_2'=j^*(x_1'\cdot x_2')=0$. That is,

$$\begin{aligned}0&=\left(\sum_{i=0}^{m}(-1)^i c_i(\xi)\,y^{m-i}\right)\left(\sum_{j=0}^{n}(-1)^j c_j(\eta)\,y^{n-j}\right)\\&=\sum_{k=0}^{m+n}(-1)^k\left(\sum_{i+j=k}c_i(\xi)\,c_j(\eta)\right)y^{n+m-k}.\end{aligned}$$

But by definition the elements $c_k(\xi\oplus\eta)$ are the unique elements such that

$$0=\sum_{k=0}^{m+n}(-1)^k c_k(\xi\oplus\eta)\,y^{n+m-k}.$$

Hence

$$c_k(\xi\oplus\eta)=\sum_{i+j=k}c_i(\xi)\,c_j(\eta),\quad 0\leqslant k\leqslant m+n.\ \square$$

There are similar results for real and symplectic bundles.

16.3. Theorem. *Suppose h^* is a cohomology theory with products such that for each $n\geqslant 1$ there are elements $x_n\in h^1(RP^n)$ satisfying*

i) $h^*(RP^n)=h^*(pt)[x_n]/(x_n^{n+1})$;
ii) *the inclusion $i:RP^n\to RP^{n+1}$ gives $i^*x_{n+1}=x_n$.*

Then for each $O(n)$-bundle ξ over a CW-complex X there are uniquely defined elements $w_0(\xi)$, $w_1(\xi)$, ..., $w_n(\xi)$ with $w_i(\xi)\in h^i(X)$ depending only on the equivalence class of ξ and satisfying

a) $w_i(f^*\xi)=f^*(w_i(\xi))$, *all $f:Y\to X$;*
b) $w_0(\xi)=1$;
c) *if $\omega\to RP^n$ is the Hopf $O(1)$-bundle over RP^n, then $w_1(\omega)=x_n$;*
d) $w_k(\xi\oplus\eta)=\sum_{i+j=k}w_i(\xi)w_j(\eta)$.

16.4. Similarly if $h^*(HP^n)=h^*(pt)[x_n]/(x_n^{n+1})$ for classes $x_n\in h^4(HP^n)$, then for every $Sp(n)$-bundle there are classes $p_i(\xi)\in h^{4i}(X)$ with $p_1(\rho)=x_n$, $\rho\to HP^n$ the Hopf bundle.

Remark. One often defines the total class $c(\xi) = 1 + c_1(\xi) + c_2(\xi) + \cdots$ formally; the *Cartan formula* 16.2d) can then be written

$$c(\xi \oplus \eta) = c(\xi)\, c(\eta).$$

Similarly for $w(\xi) = 1 + w_1(\xi) + w_2(\xi) + \cdots$ we have $w(\xi \oplus \eta) = w(\xi)\, w(\eta)$. This is a purely formal device for convenience in calculation.

16.5. Examples. In 15.33 we showed $H^*(CP^n;\mathbb{Z}) \cong \mathbb{Z}[e]/(e^{n+1})$; moreover the Euler classes on the various CP^n are compatible, for if $i: CP^n \to CP^{n+1}$ is the inclusion then $i^* e_{n+1} = i^* d_2(1) = d_2(i^* 1) = d_2(1) = e_n$. Similarly $H^*(HP^n;\mathbb{Z}) \cong \mathbb{Z}[e']/(e'^{n+1})$, $e' \in H^4(HP^n;\mathbb{Z})$. In 15.54 we showed $H^*(RP^n;\mathbb{Z}_2) \cong \mathbb{Z}_2[w]/(w^{n+1})$. Hence we get *Chern classes* $c_i(\xi) \in H^{2i}(X;\mathbb{Z})$ for $U(n)$-bundles, *Pontrjagin classes* $p_i(\xi) \in H^{4i}(X;\mathbb{Z})$ for $Sp(n)$-bundles and *Stiefel–Whitney classes* $w_i(\xi) \in H^i(X;\mathbb{Z}_2)$ for $O(n)$-bundles.

In particular if $\omega_n \to BO(n)$ is the universal $O(n)$-bundle, then we may consider the *universal Stiefel–Whitney classes* $w_i = w_i(\omega_n)$ in $H^i(BO(n);\mathbb{Z}_2)$, $0 \leqslant i \leqslant n$. Likewise there are the *universal Chern classes* $c_i = c_i(\gamma_n) \in H^{2i}(BU(n);\mathbb{Z})$ and the *universal Pontrjagin classes* $p_i = p_i(\rho_n) \in H^{4i}(BSp(n);\mathbb{Z})$.

16.6. The following application of the Stiefel–Whitney classes illustrates the usefulness of characteristic classes. Suppose a differentiable manifold M^n is immersed in Euclidean space $\mathbb{R}^{n+k}$. Then the immersion has a normal bundle ν^k such that $\tau^n \oplus \nu^k \simeq \varepsilon^{n+k}$. Since $w(\varepsilon^{n+k}) = 1$, the Cartan formula gives $w(\tau^n)\, w(\nu^k) = 1$, so $w(\nu^k)$ is the formal inverse of $w(\tau^n)$. One often writes $w(M^n) = w(\tau^n)$ and calls $w(M^n)$ the *Stiefel–Whitney class of* M^n; then $w(\nu^k)$ is called the *dual* Stiefel–Whitney class and denoted by $\tilde{w}(M^n)$. Note that $\tilde{w}_r(M^n) = 0$ for $r > k$.

If M^n can be embedded in $\mathbb{R}^{n+k}$, then so can a tubular neighborhood of M^n; in other words there is an embedding $D(\nu^k) \to \mathbb{R}^{n+k}$. The diligent reader will have verified that if $t \in H^k(D(\nu^k), S(\nu^k);\mathbb{Z}_2)$ is the Thom class and $s: M^n \to D(\nu^k)$ the O-section, then $s^* j^*(t) \in H^k(M^n;\mathbb{Z}_2)$ is the Euler class $e(\nu^k)$ of ν^k (cf. 15.54 ff). Thus $s^* j^*(t) = \tilde{w}_k(M^n)$ (see proof of 16.10). We have a commutative diagram

$$\begin{array}{ccccc}
H^k(\mathbb{R}^{n+k};\mathbb{Z}_2) & & \xleftarrow{\quad j^* \quad} & & H^k(\mathbb{R}^{n+k}, \mathbb{R}^{n+k} - f(M^n);\mathbb{Z}_2) \\
\downarrow f^* & \searrow \bar{f}^* & & & \bar{f}^* \downarrow \cong \\
H^k(M_n;\mathbb{Z}_2) & \xleftarrow{s^*} & H^k(D(\nu^k);\mathbb{Z}_2) & \xleftarrow{j^*} & H^k(D(\nu^k), S(\nu^k);\mathbb{Z}_2)
\end{array}$$

where $f: M^n \to \mathbb{R}^{n+k}$ is the embedding and $\bar{f}: D(\nu^k) \to \mathbb{R}^{n+k}$ is an extension to a tubular neighborhood. The $\bar{f}^*$ at the right is an isomorphism by excision. Thus there is a $t' \in H^k(\mathbb{R}^{n+k}, \mathbb{R}^{n+k} - f(M^n); \mathbb{Z}_2)$ such that $\bar{f}^*(t') = t$; thus $\tilde{w}_k(M^n) = s^* j^* t = s^* j^* \bar{f}^* t' = f^* j^* t' = 0$ since $j^* t' \in H^k(\mathbb{R}^{n+k}; \mathbb{Z}_2) = 0$. Thus if M^n can be embedded in $\mathbb{R}^{n+k}$, then $\tilde{w}_r(M^n) = 0$ for $r \geqslant k$.

16.7. Lemma. *If τ^n denotes the tangent bundle of RP^n and ω the Hopf line bundle, then $(n+1)\omega \simeq \tau^n \oplus \varepsilon^1$.*

Proof: $(n+1)\omega$ denotes $\omega \oplus \omega \oplus \cdots \oplus \omega$. A typical point of $(n+1)\omega$ is of the form $(\{x\}, y_0 x, y_1 x, \ldots, y_n x)$ where $x \in S^n$, $y_0, y_1, \ldots, y_n \in \mathbb{R}$ and $\{x\}$ denotes the image of x in RP^n. We define $\phi: (n+1)\omega \to \tau^n \oplus \varepsilon^1$ by

$$\phi(\{x\}, y_0 x, y_1 x, \ldots, y_n x) = (\{x\}, y - (x \cdot y) x, x \cdot y).$$

Here we have used y to denote $(y_0, y_1, \ldots, y_n) \in \mathbb{R}^{n+1}$. The reader can easily write down ϕ^{-1}. □

16.8. Corollary. $w(RP^n) = (1 + x)^{n+1}$, $x \in H^1(RP^n; \mathbb{Z}_2)$ *the generator.*

Proof:

$$w(RP^n) = w(\tau^n) = w(\tau^n \oplus \varepsilon^1) = w((n+1)\omega) = w(\omega)^{n+1} = (1+x)^{n+1}. \quad \square$$

16.9. Proposition. *RP^n can be embedded in $\mathbb{R}^{n+1}$ only if $n = 2^r - 1$ for some r and can be immersed in $\mathbb{R}^{n+1}$ only if $n = 2^r - 1$ or $n = 2^r - 2$. If $n = 2^r$ then there is no immersion of RP^n in $\mathbb{R}^{2n-2}$ and no embedding in $\mathbb{R}^{2n-1}$.*

Proof: We know that if RP^n can be immersed in $\mathbb{R}^{n+1}$ then $\tilde{w}(RP^n) = 1$ or $\tilde{w}(RP^n) = 1 + x$. In the first case we would have $(1 + x)^{n+1} = 1$, which implies $n + 1 = 2^r$, some r. In the latter case $(1 + x)^{n+2} = 1$, which implies $n + 2 = 2^r$. The former case must hold if RP^n can be embedded. If $n = 2^r$, then $\tilde{w}(RP^n) = (1+x)^{-(n+1)} = (1+x)^{-n}(1+x)^{-1} = (1+x^n) \cdot (1 + x + \cdots + x^n) = 1 + x + \cdots + x^{n-1}$. □

Remark. Whitney showed any differentiable n-manifold can be immersed in $\mathbb{R}^{2n-1}$ and embedded in $\mathbb{R}^{2n}$.

16.10. Theorem.

$$H^*(BO(n); \mathbb{Z}_2) \cong \mathbb{Z}_2[w_1, w_2, \ldots, w_n]$$

$$H^*(BU(n); \mathbb{Z}) \cong \mathbb{Z}[c_1, c_2, \ldots, c_n]$$

$$H^*(BSp(n); \mathbb{Z}) \cong \mathbb{Z}[p_1, p_2, \ldots, p_n].$$

Proof: The first step is to show that we can regard $B\imath: BU(n-1) \to BU(n)$ as being the sphere bundle of $\gamma_n \to BU(n)$. Let $EU(n)$ denote the total space of the universal principal $U(n)$-bundle; in fact, any contractible space on which $U(n)$ acts freely will do (11.35). Then we have $BU(n) \simeq EU(n)/U(n)$. The total space of the sphere bundle of γ_n can be written $E(S(\gamma_n)) = EU(n) \times_{U(n)} S^{2n-1}$. But we can construct a homeomorphism

$$h: EU(n)/U(n-1) \to EU(n) \times_{U(n)} S^{2n-1}:$$

the map $g: EU(n) \to EU(n) \times_{U(n)} S^{2n-1}$ given by $g(e) = (e, s_0)$, where $s_0 = (0, 0, \ldots, 0, 1) \in S^{2n-1}$, satisfies $g(eA) = (eA, s_0) = (e, As_0) = (e, s_0) = g(e)$ for all $A \in U(n-1)$. Thus g induces

$$h: EU(n)/U(n-1) \to EU(n) \times_{U(n)} S^{2n-1}.$$

h^{-1} is described as follows: given $(e,s) \in EU(n) \times_{U(n)} S^{2n-1}$ choose an $A \in U(n)$ with $As_0 = s$ and then let $h^{-1}(e,s) = \{eA\}$. This turns out to be independent of the choice of A and continuous. Now observe that $EU(n)$ is a contractible space on which $U(n-1)$ acts freely, so we can take $BU(n-1) = EU(n)/U(n-1) \simeq E(S(\gamma_n))$, and $B\iota$ is just the projection.

Now consider the Gysin sequence of the spherical fibration $S^{2n-1} \to BU(n-1) \to BU(n)$:

$$\cdots \longrightarrow H^{m-1}(BU(n-1);\mathbb{Z}) \xrightarrow{\phi} H^{m-2n}(BU(n);\mathbb{Z}) \xrightarrow{e\cdot}$$

$$H^m(BU(n);\mathbb{Z}) \xrightarrow{B\iota^*} H^m(BU(n-1);\mathbb{Z}) \longrightarrow$$

where $e \in H^{2n}(BU(n);\mathbb{Z})$ is the Euler class. Suppose the theorem proved for $n-1$; it is true for $n = 1$, since $BU(1) \simeq CP^\infty$ and we have 15.33. Since $B\iota^*(c_k) = B\iota^*(c_k(\gamma_n)) = c_k(B\iota^*\gamma_n) = c_k(\gamma_{n-1}) = c_k$ for each k, $0 \leqslant k \leqslant n-1$, and since $B\iota^*$ is a ring homomorphism, it follows $B\iota^*$ is surjective and thus $\phi = 0$ and $e\cdot$ is injective. We see that $\ker B\iota^*$ is the ideal generated by e. But we also have $B\iota^*(c_n) = B\iota^*(c_n(\gamma_n)) = c_n(B\iota^*\gamma_n) = c_n(\gamma_{n-1}) = 0$, so $c_n \in \ker B\iota^* = (e)$. Let us say $c_n = \alpha e$, where

$$\alpha \in H^0(BU(n);\mathbb{Z}).$$

We wish to show $\alpha = \pm 1$. Take $X = CP^\infty \times \cdots \times CP^\infty$ (n factors) with $\xi = \gamma \times \gamma \times \cdots \times \gamma$ and classifying map $f: X \to BU(n)$; then the total Chern class $c(\xi)$ is

$$c(\xi) = c(\gamma \times \gamma \cdots \times \gamma) = \textstyle\prod_{i=1}^n (1 + y_i) = \sum_{i=0}^n \sigma_i(y_1, y_2, \ldots, y_n),$$

where σ_i is the ith elementary symmetric polynomial in $y_1, \ldots, y_n$ and where $y_i \in H^2(X;\mathbb{Z})$ is $\pi_i^* y$, $\pi_i: X \to CP^\infty$ the projection on the ith factor. In particular we have $c_n(\xi) = \sigma_n(y_1, y_2, \ldots, y_n) = y_1 y_2 \ldots y_n$ But

$c_n(\xi) = f^*(c_n) = f^*(\alpha e) = \alpha f^*(e)$, which shows that α must be ± 1 $(H^*(X;\mathbb{Z}) \cong \mathbb{Z}[y_1, y_2, \ldots, y_n])$.

The proofs for $BO(n)$, $BSp(n)$ are entirely analogous. $\square$

16.11. Corollary.

$$H^*(BO;\mathbb{Z}_2) \cong \mathbb{Z}_2[w_1, w_2, \ldots]$$
$$H^*(BU;\mathbb{Z}) \cong \mathbb{Z}[c_1, c_2, \ldots]$$
$$H^*(BSp;\mathbb{Z}) \cong \mathbb{Z}[p_1, p_2, \ldots].$$

Proof: $\lim^1 H^*(BO(n);\mathbb{Z}_2) = 0$ since all $B\iota^*$ are surjective, so

$$H^*(BO;\mathbb{Z}_2) \cong \lim{}^0 H^*(BO(n);\mathbb{Z}_2) \cong \mathbb{Z}_2[w_1, w_2, \ldots].$$

The other two cases are identical. $\square$

Let $X = CP^\infty \times CP^\infty \times \cdots \times CP^\infty$ as in the proof of 16.10. In $H^*(X;\mathbb{Z}) \cong \mathbb{Z}[y_1, y_2, \ldots, y_n]$ we have the subalgebra $S(y_1, y_2, \ldots, y_n)$ of symmetric polynomials and we know $S(y_1, y_2, \ldots, y_n) \cong \mathbb{Z}[\sigma_1, \sigma_2, \ldots, \sigma_n]$. For any sequence $\alpha = (\alpha_1, \alpha_2, \ldots, \alpha_n)$ of integers let c^α denote the monomial $c_1^{\alpha_1} c_2^{\alpha_2} \cdots c_n^{\alpha_n}$. As a corollary of the proof of 16.10 we get the following.

16.12. Proposition. $c^\alpha(\xi) = \sigma_1^{\alpha_1} \sigma_2^{\alpha_2} \cdots \sigma_n^{\alpha_n}$.

Proof: We showed that $c_i(\xi) = \sigma_i$. Hence the formula, since

$$c^\alpha(\xi) = f^*(c^\alpha) = f^*(c_1^{\alpha_1} c_2^{\alpha_2} \ldots c_n^{\alpha_n}) = (f^* c_1)^{\alpha_1} (f^* c_2)^{\alpha_2} \ldots (f^* c_n)^{\alpha_n} = \sigma_1^{\alpha_1} \sigma_2^{\alpha_2} \ldots \sigma_n^{\alpha_n}. \quad \square$$

16.13. Corollary. *The map $f: X \to BU(n)$ classifying ξ induces an isomorphism*

$$f^*: H^*(BU(n);\mathbb{Z}) \to H^*(X;\mathbb{Z})$$

of $H^*(BU(n);\mathbb{Z})$ onto $S(y_1, y_2, \ldots, y_n)$.

There are analogous results for $BO(n)$, $BSp(n)$.

We shall also want to know the homology of BO, BU, BSp. These spaces are H-groups, the sums in K-theory being given by the H-group structures. Thus $H_*(BU;\mathbb{Z})$, for example, is a ring with a Pontrjagin product: that is, if $\phi: BU \times BU \to BU$ is the H-product of BU, then we define a product on $H_*(BU;\mathbb{Z})$ by

$$H_*(BU;\mathbb{Z}) \otimes H_*(BU;\mathbb{Z}) \xrightarrow{\times} H_*(BU \times BU;\mathbb{Z}) \xrightarrow{\phi_*} H_*(BU;\mathbb{Z}).$$

This product is evidently associative and graded commutative—i.e. $ab = (-1)^{|a|\,|b|} ba$.

Now we have the inclusion $i: O(1) \to O$ inducing $Bi: BO(1) \to BO$; also $BO(1) \simeq RP^\infty$, so $H_*(BO(1);\mathbb{Z}_2)$ has a $\mathbb{Z}_2$-basis $1, x_1, x_2, \ldots$, where x_n is dual to $w_1^n \in H^*(BO(1);\mathbb{Z}_2)$. Let us also denote by x_n the element $Bi_*(x_n) \in H_*(BO;\mathbb{Z}_2)$. Similarly we have a basis $1, y_1, y_2, \ldots$ in $H_*(BU(1);\mathbb{Z})$ dual to $\{c_1^n\}$, and we get elements $1, y_1, y_2, \ldots$ in $H_*(BU;\mathbb{Z})$. Likewise there are elements $1, z_1, z_2, \ldots$ in $H_*(BSp;\mathbb{Z})$.

In addition to the Pontrjagin product on $H_*(BO;\mathbb{Z}_2)$, ϕ also induces a coproduct

$$H^*(BO;\mathbb{Z}_2) \xrightarrow{\phi^*} H^*(BO \times BO;\mathbb{Z}_2) \cong H^*(BO;\mathbb{Z}_2) \otimes_{\mathbb{Z}_2} H^*(BO;\mathbb{Z}_2),$$

which we shall denote by ϕ^*.

16.14. Proposition. *For $z \in H^*(BO;\mathbb{Z}_2)$ if $\phi^*(z) = \sum_i z_i \otimes z_i'$, then $z(\xi \oplus \eta) = \sum_i z_i(\xi) z_i'(\eta)$ for any bundles ξ, η.*

Proof: Let $f: X \to BO(m)$, $g: X \to BO(n)$ classify ξ, η respectively. Then $\xi \oplus \eta$ is classified by $\mu_{mn} \circ (f \times g) \circ \Delta$, where

$$\mu_{mn}: BO(m) \times BO(n) \to BO(m+n)$$

classifies $\omega_m \times \omega_n$. Hence

$$\begin{aligned} z(\xi \oplus \eta) &= \Delta^* \circ (f \times g)^* \circ \mu_{mn}^* \circ Bi_{m+n}^*(z) \\ &= \Delta^* \circ (f \times g)^* \circ (Bi_m \times Bi_n)^* \circ \phi^*(z) \\ &= \Delta^* \circ (f \times g)^*(\textstyle\sum_i Bi_m^* z_i \times Bi_n^* z_i') \\ &= \Delta^*(\textstyle\sum_i z_i(\xi) \times z_i'(\eta)) = \sum_i z_i(\xi) z_i'(\eta). \end{aligned}$$

Here we have used the fact that

$$\begin{array}{ccc} BO(m) \times BO(n) & \xrightarrow{\mu_{mn}} & BO(m+n) \\ \downarrow{\scriptstyle Bi_m \times Bi_n} & & \downarrow{\scriptstyle Bi_{m+n}} \\ BO \times BO & \xrightarrow{\phi} & BO \end{array}$$

commutes up to homotopy. $\square$

For the Hopf bundle $\omega \to RP^\infty$ we have

$$w^\alpha(\omega) = \begin{cases} x^k & \alpha = (k) \\ 0 & \text{otherwise.} \end{cases}$$

Thus $\langle w^\alpha, x_i\rangle = \langle w^\alpha, Bi_*(x_i)\rangle = \langle Bi^*(w^\alpha), x_i\rangle = \langle w^\alpha(\omega), x_i\rangle$

$$= \begin{cases} \langle x^k, x_i\rangle & \alpha = (k) \\ 0 & \text{otherwise} \end{cases} = \begin{cases} 1 & \alpha = (i) \\ 0 & \text{otherwise.} \end{cases}$$

Hence x_i is dual to $w^{(i)} = w_1^i$ in $H_*(BO; \mathbb{Z}_2)$. We can write

$$w^\alpha(\omega) = \textstyle\sum_{i\geq 0} \langle w^\alpha, x_i\rangle x^i.$$

Since every $z \in H^*(BO; \mathbb{Z}_2)$ is a $\mathbb{Z}_2$-linear combination of the w^α's, we thus have proved

16.15. $$z(\omega) = \textstyle\sum_{i\geq 0} \langle z, x_i\rangle x^i, \quad z \in H^*(BO; \mathbb{Z}_2).$$

Let $X, \xi_n \to X$ be as before: $X = RP^\infty \times \cdots \times RP^\infty$, $\xi_n = \omega \times \cdots \times \omega$.

16.16. Proposition. *For every $z \in H^*(BO; \mathbb{Z}_2)$ we have*

$$z(\xi_n) = \textstyle\sum_{l(\alpha)=n} \langle z, x_\alpha\rangle u_1^{\alpha_1} u_2^{\alpha_2} \ldots u_n^{\alpha_n},$$

where $u_i = \pi_i^* x$, $\pi_i: X \to RP^\infty$ the ith projection, $x_\alpha = x_{\alpha_1} x_{\alpha_2} \ldots x_{\alpha_n}$ and $l(\alpha) = m$ if $\alpha = (\alpha_1, \alpha_2, \ldots, \alpha_m)$.

Proof: The proof is by induction on n. It is true for $n = 1$ by 16.15. Suppose it proved for $n-1$ and suppose $\phi^* z = \sum_i z_i \otimes z_i'$. Then

$$\begin{aligned} z(\xi_n) &= z(\xi_{n-1} \times \omega) = \textstyle\sum_i z_i(\xi_{n-1}) \times z_i'(\omega) \\ &= \textstyle\sum_i \left(\sum_{l(\alpha)=n-1} \langle z_i, x_\alpha\rangle u_1^{\alpha_1} u_2^{\alpha_2} \ldots u_{n-1}^{\alpha_{n-1}}\right) \times \left(\sum_{j\geq 0} \langle z_i', x_j\rangle u_n^j\right) \\ &= \textstyle\sum_{j,\, l(\alpha)=n-1} \langle \sum_i z_i \otimes z_i', x_\alpha \otimes x_j\rangle u_1^{\alpha_1} \ldots u_{n-1}^{\alpha_{n-1}} u_n^j \\ &= \textstyle\sum_{j,\, l(\alpha)=n-1} \langle \phi^* z, x_\alpha \otimes x_j\rangle u_1^{\alpha_1} \ldots u_{n-1}^{\alpha_{n-1}} u_n^j \\ &= \textstyle\sum_{j,\, l(\alpha)=n-1} \langle z, x_\alpha x_j\rangle u_1^{\alpha_1} \ldots u_{n-1}^{\alpha_{n-1}} u_n^j \\ &= \textstyle\sum_{l(\beta)=n} \langle z, x_\beta\rangle u_1^{\beta_1} \ldots u_n^{\beta_n}, \qquad \beta = (\alpha_1, \ldots, \alpha_{n-1}, j). \quad \square \end{aligned}$$

16.17. Theorem.

$$\begin{aligned} H_*(BO; \mathbb{Z}_2) &\cong \mathbb{Z}_2[x_1, x_2, \ldots] \\ H_*(BU; \mathbb{Z}) &\cong \mathbb{Z}[y_1, y_2, \ldots] \\ H_*(BSp; \mathbb{Z}) &\cong \mathbb{Z}[z_1, z_2, \ldots]. \end{aligned}$$

Proof: We consider first the Gysin sequence of the spherical fibration $S^{2n-1} \to BU(n-1) \to BU(n)$:

$$\cdots \longleftarrow H_{m-1}(BU(n-1); \mathbb{Z}) \xleftarrow{\ \phi\ } H_{m-2n}(BU(n); \mathbb{Z}) \xleftarrow{\ d^2\ }$$
$$H_m(BU(n); \mathbb{Z}) \xleftarrow{\ Bi_*\ } H_m(BU(n-1); \mathbb{Z}) \longleftarrow \cdots.$$

By double induction over n and m we see that $H_m(BU(n);\mathbb{Z})$ is finitely generated for all n,m (this is clearly true for $H_m(BU(1);\mathbb{Z}) = H_m(CP^\infty;\mathbb{Z})$). From the universal coefficient sequence

$$0 \to \operatorname{Ext}(H_{m-1}(BU(n);\mathbb{Z}),\mathbb{Z}) \to H^m(BU(n);\mathbb{Z}) \to \operatorname{Hom}(H_m(BU(n);\mathbb{Z}),\mathbb{Z}) \to 0$$

we see that $H_*(BU(n);\mathbb{Z})$ can have no torsion for all $n \geqslant 1$. Thus $H_m(BU(n);\mathbb{Z})$ is finitely generated and free abelian, from which follows

$$H^m(BU(n);\mathbb{Z}) \cong \operatorname{Hom}(H_m(BU(n);\mathbb{Z}),\ \mathbb{Z})$$

and

$$H_m(BU(n);\mathbb{Z}) \cong \operatorname{Hom}(H^m(BU(n);\mathbb{Z}),\mathbb{Z}) \cong \operatorname{Hom}(H^m(BU;\mathbb{Z}),\mathbb{Z})$$

for all $n > m/2$. In particular

$$H_m(BU;\mathbb{Z}) \cong \operatorname{dir\,lim}_n H_m(BU(n);\mathbb{Z}) \cong \operatorname{Hom}(H^m(BU;\mathbb{Z}),\mathbb{Z}).$$

Thus $H_m(BU;\mathbb{Z})$ is finitely generated and free abelian and $\operatorname{rk} H_m(BU;\mathbb{Z}) = \operatorname{rk} H^m(BU;\mathbb{Z})$. It will therefore suffice to show that the monomials $y^\alpha = y_1^{\alpha_1} \cdots y_n^{\alpha_n}$ span $H_*(BU;\mathbb{Z})$.

Suppose there is some $z \in H^*(BU;\mathbb{Z})$ with $\langle z, y^\alpha \rangle = 0$ for all y^α. Then

$$f^*\, Bi_n^*(z) = z(\xi_n) = \textstyle\sum_\alpha \langle z, y^\alpha \rangle u_1^{\alpha_1} \ldots u_n^{\alpha_n} = 0$$

for all n, where $f: CP^\infty \times \cdots \times CP^\infty \to BU(n)$ classifies $\xi_n = \gamma \times \cdots \times \gamma$. By 16.13 f is a monomorphism, so $Bi_n^*(z) = 0$ for all n. This implies $z = 0$. Now this would suffice to show that the y^α span $H_*(BU;\mathbb{Z})$ if the coefficient ring were not $\mathbb{Z}$ but rather a field—for example $\mathbb{Z}_p$ for any prime p. For coefficients $\mathbb{Z}$, however, we now only know that for any $x \in H_*(BU;\mathbb{Z})$ there is an integer n such that nx is a linear combination of the y^α's.

Let $N = \mathbb{Z}[x_1, x_2, \ldots]$, grade $x_i = 2i$, and define a homomorphism $h: N \to H_*(BU;\mathbb{Z})$ of algebras by $h(x_i) = y_i$, $i \geqslant 1$. We know that $h \otimes 1: N \otimes \mathbb{Z}_p \to H_*(BU;\mathbb{Z}) \otimes \mathbb{Z}_p$ is an isomorphism for all primes p. Suppose that for some $x \in N$ there were a prime p which divided $h(x)$. Then $h(x) \otimes 1 = 0 \in H_*(BU;\mathbb{Z}) \otimes \mathbb{Z}_p$, so $x \otimes 1 = 0 \in N \otimes \mathbb{Z}_p$. Thus $p|x$. This completes the proof that $H_*(BU;\mathbb{Z}) \cong \mathbb{Z}[y_1, y_2, \ldots]$. The proofs for BO and BSp are analogous. □

16.18. Now that we know $H_*(BU;\mathbb{Z})$ we can address ourselves to the calculation of $H_*(K;\mathbb{Z})$. From 13.49 we know that $H_n(K;\mathbb{Z}) \cong$

$\operatorname{dir}_q \lim \tilde{H}_{n+2q}(BU;\mathbb{Z})$, where the homomorphisms of the direct system are

$$\tilde{H}_{n+2q}(BU;\mathbb{Z}) \xrightarrow{\sigma^2} \tilde{H}_{n+2q+2}(S^2 \wedge BU;\mathbb{Z}) \xrightarrow{B_*} \tilde{H}_{n+2q+2}(BU;\mathbb{Z}).$$

We shall denote these morphisms by $\underline{B}$; we must work out the effect of $\underline{B}$ on $\tilde{H}_*(BU;\mathbb{Z})$. The first step is the result that $\underline{B}$ annihilates all products; thus it will be necessary to determine only $\underline{B}(y_k)$ for each k. By 15.43 the diagram

$$\begin{array}{ccc} \tilde{H}_{n+2q}(BU;\mathbb{Z}) & \xrightarrow{B'_*} & \tilde{H}_{n+2q}(\Omega_0^2 BU;\mathbb{Z}) \\ \downarrow \sigma^2 & & \downarrow \sigma'^2 \\ \tilde{H}_{n+2q+2}(S^2 \wedge BU;\mathbb{Z}) & \xrightarrow{B_*} & \tilde{H}_{n+2q+2}(BU;\mathbb{Z}) \end{array}$$

commutes, so it will suffice to show σ' annihilates products once we have remarked that under $B': BU \to \Omega_0^2 BU$ the product ϕ on BU corresponds up to homotopy to the loop-space product μ_Ω on $\Omega_0^2 BU$. This is so because the Bott isomorphism $\tilde{K}^0(X) \cong \tilde{K}^0(S^2 \wedge X)$ preserves sums, implying that ϕ corresponds to $\Omega^2\phi$; $\Omega^2\phi$ and μ_Ω agree up to homotopy by 2.24 (cf. proof of 2.26).

16.19. Proposition. *The homology suspension $\tilde{h}_n(\Omega Y) \xrightarrow{\sigma'} \tilde{h}_{n+1}(Y)$ annihilates Pontrjagin products whenever h_* is an unreduced homology theory with products.*

Proof: Let $a \in \tilde{h}_p(\Omega Y)$, $b \in \tilde{h}_q(\Omega Y)$. We have a diagram (see opposite). By 7.10 we can choose $a' \in h_p(\Omega Y)$, $b' \in h_q(\Omega Y)$ with $j_* a' = a$, $j_* b' = b$, $p_* a' = 0 = p_* b' \in h_*(\{y_0\})$. Then $ab = j_* \mu_*(a' \times b') \in h_{p+q}(\Omega Y, \{\omega_0\}) = \tilde{h}_{p+q}(\Omega Y)$. Since $\times \circ (1 \otimes \partial) = (-1)^p \partial \circ \times$ by 13.56, the diagram commutes up to sign and we have $\sigma'(ab) = p_* a' \times \sigma'(b) = 0$; here $\mu': \Omega Y \times PY \to PY$ is the obvious map: $\mu'(\omega, w) = \omega * w$. ☐

Now if A is an augmented algebra, then the *indecomposable quotient* $Q(A)$ is $A/\phi(\bar{A} \otimes \bar{A})$, where $\bar{A}$ denotes the *augmentation ideal* $\ker[\varepsilon: A \to R]$ and $\phi: A \otimes A \to A$ is the product. Intuitively speaking one divides out all non-trivial products in A. Since $\underline{B}: \tilde{H}_*(BU;\mathbb{Z}) \to \tilde{H}_*(BU;\mathbb{Z})$ annihilates all products (note that $\overline{H_*(BU;\mathbb{Z})} = \tilde{H}_*(BU;\mathbb{Z})$) we can regard $\underline{B}$ as a homomorphism

$$\underline{B}: Q_m(H_*(BU;\mathbb{Z})) \to Q_{m+2}(H_*(BU;\mathbb{Z})).$$

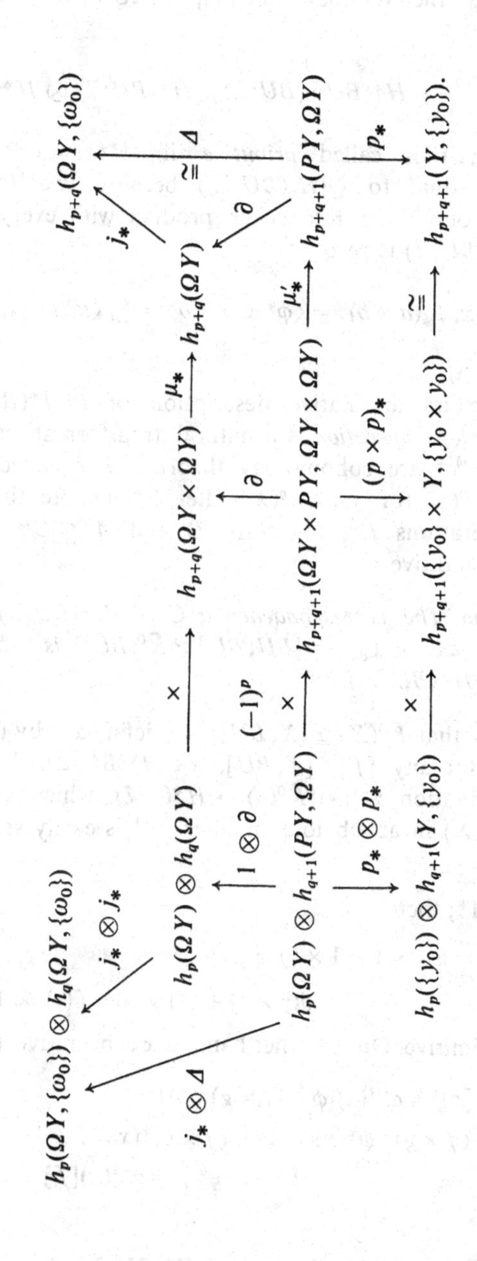
$h_p(\Omega Y, \{\omega_0\}) \otimes h_q(\Omega Y, \{\omega_0\})$
$h_{p+q}(\Omega Y, \{\omega_0\})$
$j_* \otimes j_*$
$j_* \otimes \Delta$
j_*
$\cong \Delta$
$h_p(\Omega Y) \otimes h_q(\Omega Y)$
$\times$
$h_{p+q}(\Omega Y \times \Omega Y)$
μ_*
$h_{p+q}(\Omega Y)$
$1 \otimes \partial$
$(-1)^p$
∂
∂
$h_p(\Omega Y) \otimes h_{q+1}(PY, \Omega Y)$
$\times$
$h_{p+q+1}(\Omega Y \times PY, \Omega Y \times \Omega Y)$
μ'_*
$h_{p+q+1}(PY, \Omega Y)$
$p_* \otimes p_*$
$(p \times p)_*$
p_*
$h_p(\{y_0\}) \otimes h_{q+1}(Y, \{y_0\})$
$\times$
$h_{p+q+1}(\{y_0\} \times Y, \{y_0 \times y_0\})$
$\cong$
$h_{p+q+1}(Y, \{y_0\})$.

Now the dual of $Q(H_*(BU;\mathbb{Z}))$ is the subgroup $P(H^*(BU;\mathbb{Z}))$ of *primitive elements*: the Whitney sum map $\phi: BU \times BU \to BU$ induces a coproduct

$$H^*(BU;\mathbb{Z}) \xrightarrow{\phi^*} H^*(BU \times BU;\mathbb{Z}) \cong H^*(BU;\mathbb{Z}) \otimes H^*(BU;\mathbb{Z}),$$

and $x \in H^*(BU;\mathbb{Z})$ is called *primitive* if $\phi^*(x) = 1 \otimes x + x \otimes 1$. $P(H^*(BU;\mathbb{Z}))$ is dual to $Q(H_*(BU;\mathbb{Z}))$ because $x \in H^*(BU;\mathbb{Z})$ is primitive if and only if its Kronecker product with every non-trivial product $ab \in H_*(BU;\mathbb{Z})$ is zero:

$$\langle x, ab\rangle = \langle x, \phi_*(a \times b)\rangle = \langle \phi^* x, a \times b\rangle = \textstyle\sum_i \langle x_i, a\rangle\langle x_i', b\rangle$$

if $\phi^* x = \sum_i x_i \otimes x_i'$.

We can give an alternative description of $P(H^*(BU;\mathbb{Z}))$. An (*unstable*) *cohomology operation* is a natural transformation $\theta: k^q(-) \to \bar{k}^r(-)$, where k^*, $\bar{k}^*$ are cohomology theories. θ is called *additive* if $\theta(x+y) = \theta(x) + \theta(y)$ for $x, y \in k^q(X)$. Let C^* denote the set of all cohomology operations $K^0(-) \to H^*(-;\mathbb{Z})$ and $A^* \subset C^*$ the set of those which are additive.

16.20. Proposition. *The correspondence $\alpha: C^* \to H^*(BU;\mathbb{Z})$ defined by $\theta \mapsto \theta(\gamma)$ (γ the class of 1_{BU} in $[BU;BU] \cong \tilde{K}^0(BU)$) is a bijection and maps A^* onto $P(H^*(BU;\mathbb{Z}))$.*

Proof: We know that $\tilde{K}^0(X) \cong [X;BU]$; we define α^{-1} by $(\alpha^{-1}(x))[f] = f^*x \in H^*(X;\mathbb{Z})$ for any $[f] \in [X;BU]$, $x \in H^*(BU;\mathbb{Z})$. This defines a natural transformation $\alpha^{-1}(x): \tilde{K}^0(-) \to H^*(-;\mathbb{Z})$, which can then be extended to $K^0(X)$ in an obvious manner. α^{-1} is easily seen to be the inverse of α.

Suppose $\theta \in A^*$; then

$$\phi^*\theta(\gamma) = \theta(\phi^*\gamma) = \theta(\gamma \times 1 + 1 \times \gamma) = \theta(\gamma \times 1) + \theta(1 \times \gamma) = \theta(\gamma) \otimes 1 + 1 \otimes \theta(\gamma);$$

that is, $\theta(\gamma)$ is primitive. On the other hand, if x is primitive, then

$$\begin{aligned}\alpha^{-1}(x)([f] + [g]) &= \alpha^{-1}(x)[\phi \circ (f \times g) \circ \Delta] = \\ \Delta^* \circ (f \times g)^* \phi^*(x) &= \Delta^* \circ (f \times g)^*(x \times 1 + 1 \times x) = \\ &f^* x + g^* x = \alpha^{-1}(x)[f] + \alpha^{-1}(x)[g].\end{aligned}$$

Thus $\alpha^{-1}(x) \in A^*$. □

We define $\bar{B}: P^n(H^*(BU;\mathbb{Z})) \to P^{n-2}(H^*(BU;\mathbb{Z}))$ by requiring the following square to commute for all $\theta \in A^*$:

$$\begin{array}{ccc}
\tilde{K}^0(X) & \xrightarrow{\cong} & \tilde{K}^0(S^2 \wedge X) \\
\bar{B}\theta \downarrow & & \downarrow \theta \\
\tilde{H}^{n-2}(X;\mathbb{Z}) & \underset{\cong}{\xrightarrow{\sigma^{-2}}} & \tilde{H}^n(S^2 \wedge X;\mathbb{Z}).
\end{array}$$

16.21. Proposition. *$\underline{B}$ and $\bar{B}$ are related by the formula*

$$\langle \theta, \underline{B}a \rangle = \langle \bar{B}\theta, a \rangle$$

for $\theta \in A^*$, $a \in Q_m(H_*(BU;\mathbb{Z}))$.

Proof:

$$\langle \bar{B}\theta(\gamma), a \rangle = \langle \sigma^2\, \theta(B^*\, \gamma), a \rangle = \langle \theta(B^*\, \gamma), \sigma^2\, a \rangle =$$
$$\langle B^*(\theta(\gamma)), \sigma^2\, a \rangle = \langle \theta(\gamma), B_*\, \sigma^2\, a \rangle = \langle \theta(\gamma), \underline{B}a \rangle = \langle \theta, \underline{B}a \rangle. \qquad \square$$

By naturality any $\theta \in A^*$ is completely determined by its value on the universal line bundle γ_1 over $BU(1)$: as in the proof of 16.2 given any bundle $\xi \to X$ we can find a map $f: Y \to X$ such that $f^*\xi$ is a sum of line bundles (we repeatedly split off line bundles) and $f^*: \tilde{H}^*(X;\mathbb{Z}) \to \tilde{H}^*(Y;\mathbb{Z})$ is a monomorphism. But for a line bundle $\lambda \to Y$ we can find a classifying map $g: Y \to BU(1)$ such that $\lambda = g^*\gamma_1$ and hence $\theta(\lambda) = \theta(g^*\gamma_1) = g^*\theta(\gamma_1)$. If $\theta \in A^n$, then $\theta(\gamma_1)$ can be any element of $H^n(BU(1);\mathbb{Z})$.

Let ϕ^r be the unique operation such that $\phi^r(\gamma_1) = y^r \in H^{2r}(BU(1);\mathbb{Z})$. Then we have $\langle \phi^r, y_q \rangle = \delta^r_q$.

16.22. Proposition. *$\bar{B}\phi^r = r\phi^{r-1}$, $r > 1$, and $\bar{B}\phi^1 = 0$.*

Proof: Consider the following commutative diagram

$$\begin{array}{ccccc}
\tilde{K}^0(X) & \xrightarrow{\cong} & \tilde{K}^0(S^2 \wedge X) & \longrightarrow & \tilde{K}^0(S^2 \times X) \\
\bar{B}\phi^r \downarrow & & \downarrow \phi^r & & \downarrow \phi^r \\
\tilde{H}^{2r-2}(X;\mathbb{Z}) & \xrightarrow{\sigma^{-2}} & \tilde{H}^{2r}(S^2 \wedge X;\mathbb{Z}) & \longrightarrow & \tilde{H}^{2r}(S^2 \times X;\mathbb{Z}).
\end{array}$$

Let η be the Hopf bundle over $S^2(\cong CP^1)$ and $w = \eta - 1 \in \tilde{K}^0(S^2)$; the top horizontal map is multiplication by w. The lower horizontal map is multiplication by $g \in \tilde{H}^2(S^2;\mathbb{Z})$, the generator. We take $X = BU(1)$ and apply the diagram to $\gamma_1 - 1$. We find

$$\begin{aligned}
g \times \bar{B}\phi^r(\gamma_1 - 1) &= \phi^r((\eta - 1) \times (\gamma_1 - 1)) \\
&= \phi^r(\eta \times \gamma_1 - 1 \times \gamma_1 - \eta \times 1 + 1 \times 1) \\
&= \phi^r(\eta \times \gamma_1) - \phi^r(1 \times \gamma_1) - \phi^r(\eta \times 1) + \phi^r(1 \times 1).
\end{aligned}$$

If $\mu_1 : BU(1) \times BU(1) \to BU(1)$ classifies $\gamma_1 \times \gamma_1$ and $h : S^2 \to BU(1)$ classifies η, then $\eta \times \gamma_1$ is classified by

$$S^2 \times BU(1) \xrightarrow{h \times 1} BU(1) \times BU(1) \xrightarrow{\mu_1} BU(1).$$

As $\mu_1^* y = y \times 1 + 1 \times y$ we calculate

$$\begin{aligned}\phi^r(\eta \times \gamma_1) &= \phi^r((h \times 1)^* \mu_1^* \gamma_1) = (h \times 1)^* \mu_1^* \phi^r(\gamma_1) = (h \times 1)^* \mu_1^*(y^r) \\ &= (h \times 1)^*(y \times 1 + 1 \times y)^r = (h \times 1)^* \left(\sum_{j=0}^{r} \binom{r}{j} y^j \times y^{r-j} \right) \\ &= \sum_{j=0}^{r} \binom{r}{j} g^j \times y^{r-j}.\end{aligned}$$

But $g^2 = 0$, so in fact $\phi^r(\eta \times \gamma_1) = 1 \times y^r + rg \times y^{r-1}$. Similarly

$$\phi^r(\eta \times 1) = \begin{cases} g \times 1 & r = 1 \\ 0 & r > 1. \end{cases}$$

Hence altogether

$$g \times \bar{B}\phi^r(\gamma_1 - 1) = \begin{cases} rg \times y^{r-1} & r > 1 \\ 0 & r = 1. \quad \square \end{cases}$$

16.23. Corollary. *$\beta y_k = (k+1) y_{k+1}$ mod decomposable elements for $k \geqslant 1$.*

Let $y_j^{j-k} \in H_{2j-2k}(K;\mathbb{Z})$ denote the image of $y_j \in \tilde{H}_{2j}(BU;\mathbb{Z})$, where BU is regarded as the $2k$th term in the spectrum K, $k \in \mathbb{Z}$, $j \geqslant 1$. 16.23 implies that $y_j^{j-k} = (j+1) y_{j+1}^{j-k}$ or in other words $y_j^k = (1/j!) y_1^k$ for all $k \in \mathbb{Z}, j \geqslant 1$.

16.24. Corollary. *$H_{2n}(K;\mathbb{Z}) \cong \mathbb{Q}$ with generator y_1^n, $n \in \mathbb{Z}$, and*

$$H_{2n+1}(K;\mathbb{Z}) = 0, \quad n \in \mathbb{Z}.$$

If $\hat{B} : \pi_q(BU, *) \to \pi_{q+2}(BU, *)$ is the periodicity map

$$\pi_q(BU, *) \xrightarrow{\Sigma^2} \pi_{q+2}(S^2 \wedge BU, *) \xrightarrow{B_*} \pi_{q+2}(BU, *)$$

and $t \in \pi_2(BU, *)$ the generator (corresponding to $\eta - 1 \in \tilde{K}^0(S^2)$), then $\hat{B}(t^q) = t^{q+1}$, $q \in \mathbb{Z}$. The diagram

$$\begin{array}{ccc} \pi_{2q}(BU, *) & \xrightarrow{\hat{B}} & \pi_{2q+2}(BU, *) \\ \downarrow h & & \downarrow h \\ \tilde{H}_{2q}(BU;\mathbb{Z}) & \xrightarrow{B} & \tilde{H}_{2q+2}(BU;\mathbb{Z}) \end{array}$$

commutes, so since $h(t) = y_1$ clearly it follows by induction that

$$h(t^q) = h(\hat{B}t^{q-1}) = \underline{B}h(t^{q-1}) = \underline{B}((q-1)!y_{q-1}) = q!y_q$$

mod decomposable elements. The diagram

$$\begin{array}{ccc} \pi_{2q}(BU,*) & \xrightarrow{\cong} & \pi_{2q}(K) \\ \downarrow h & & \downarrow h \\ \tilde{H}_{2q}(BU;\mathbb{Z}) & \longrightarrow & H_{2q}(K;\mathbb{Z}) \end{array}$$

also commutes, so we find that $h(t^q) = q!y_q^q = y_1^q$. But h is a ring homomorphism (the ring structures being given by the tensor product map $K \wedge K \to K$), so we also have $h(t^q) = h(t)^q = (y_1^1)^q$. Thus $y_1^q = (y_1^1)^q$; this holds for all $q \in \mathbb{Z}$.

16.25. Theorem. *$H_*(K;\mathbb{Z}) \cong \mathbb{Q}[u,u^{-1}]$, the ring of finite Laurent series over $\mathbb{Q}$. $u \in H_2(K;\mathbb{Z})$ is y_1^1.*

We could carry out a similar analysis to calculate $H_*(KO;\mathbb{Z})$ using the fact that every $(8n+4)$th term of KO is BSp.

In Chapter 13 we defined the Boardman map $B: F^*(X) \to (E \wedge F)^*(X)$ for ring spectra E and F. In particular we may take $E = H(\mathbb{Z})$ and $F = K$. Now $k^* = \operatorname{Hom}^*(H_*(-;\mathbb{Z}), H_*(K;\mathbb{Z}))$ is a cofunctor ($\operatorname{Hom}^r(\,,\,)$ means the group of homomorphisms of degree r between graded groups). k^* is in fact a cohomology theory; exactness follows from 16.25 and the fact that $\operatorname{Hom}(-,\mathbb{Q})$ is an exact cofunctor (cf. [56], p. 93). k^* even fulfills the wedge and WHE axioms. Moreover $\tau: (H(\mathbb{Z}) \wedge K)^* \to k^*$ defined by $(\tau[f])[g] = [\mu_H \circ (1 \wedge f) \circ g]$ for $f: X \to H(\mathbb{Z}) \wedge K$, $g: S^0 \to H(\mathbb{Z}) \wedge X$ and $\mu_H: H(\mathbb{Z}) \wedge H(\mathbb{Z}) \to H(\mathbb{Z})$ the product, is a natural transformation of cohomology theories which is clearly an isomorphism for $X = S^0$; thus τ is a natural equivalence. Furthermore since $\operatorname{Hom}(-,\mathbb{Q})$ is exact, $\operatorname{Ext}(A,\mathbb{Q}) = 0$ for all abelian groups A. From the universal coefficient theorem it thus follows that

$$\operatorname{Hom}^r(H_*(X;\mathbb{Z}), H_*(K;\mathbb{Z})) \cong \begin{cases} \prod_{q \geq 0} H^{2q}(X;\mathbb{Q}) & r \text{ even} \\ \prod_{q \geq 0} H^{2q+1}(X;\mathbb{Q}) & r \text{ odd.} \end{cases}$$

One writes $H^{\mathrm{ev}}(X;\mathbb{Q})$, respectively $H^{\mathrm{od}}(X;\mathbb{Q})$, for these products; the composite

$$K^*(X) \xrightarrow{B} (H(\mathbb{Z}) \wedge K)^*(X) \xrightarrow{\tau} \operatorname{Hom}^*(H_*(X;\mathbb{Z}), H_*(K;\mathbb{Z})) \cong H^{**}(X;\mathbb{Q})$$

is denoted by ch and called the *Chern character*. It is a ring homomorphism for every space X.

16.26. Proposition. *If γ denotes the Hopf bundle over CP^∞ then*

$$ch(\gamma) = 1 + y + \frac{y^2}{2!} + \frac{y^3}{3!} + \cdots = e^y.$$

Proof: If $H^*(X;\mathbb{Z})$ is free and of finite type, then

$$\mathrm{Hom}^*(H_*(X;\mathbb{Z}), H_*(K;\mathbb{Z})) \simeq H^*(X;\mathbb{Z}) \mathbin{\hat{\otimes}} H_*(K;\mathbb{Z}),$$

the isomorphism being (uncanonically) defined as follows: if $\{z_\alpha\}$ is a basis for $H^*(X;\mathbb{Z})$ with dual basis $\{x_\alpha\}$ in $H_*(X;\mathbb{Z})$, then

$$\phi \mapsto \textstyle\sum_\alpha z_\alpha \otimes \phi(x_\alpha) \quad \text{for} \quad \phi \in \mathrm{Hom}^*(H_*(X;\mathbb{Z}), H_*(K;\mathbb{Z})).$$

Now for $X = CP^\infty$ we have $\gamma - 1 \in \tilde{K}^0(BU(1))$ represented by the map $BU(1) \subset BU \to K$ including $BU(1)$ in BU and regarding BU as term 0 of K. Thus $\tau B(\gamma - 1)(y_i) = y_i^! \in H_{2i}(K)$. But $y_i^! = (1/i!)u^i$. Therefore

$$ch(\gamma) - 1 = ch(\gamma - 1) = \textstyle\sum_{i>0} y^i \otimes \tau B(\gamma - 1)(y_i) = \sum_{i>0} \frac{1}{i!} y^i \otimes u^i.$$

Identifying $[H^*(BU(1);\mathbb{Z}) \otimes H_*(K;\mathbb{Z})]_2$ with $\prod_{i\geq 0} H^{2i}(BU(1);\mathbb{Q})$, we get the desired result. □

For a general $U(n)$-bundle $\xi \to X$ we can compute $ch(\xi)$ as follows. If ξ is a $U(1)$-bundle, let $f: X \to BU(1)$ classify ξ. Then $ch(\xi) = ch(f^*\gamma) = f^*ch(\gamma) = f^*(e^y) = e^{f^*y} = e^{c_1(\xi)}$. For a $U(n)$-bundle we find a map $f: Y \to X$ such that $f^*\xi$ splits as a sum of line bundles $\lambda_1 \oplus \lambda_2 \oplus \cdots \oplus \lambda_n$ and $f^*: H^*(X;\mathbb{Z}) \to H^*(Y;\mathbb{Z})$ is a monomorphism as in the paragraph after 16.21. Then

$$f^*\,ch(\xi) = ch(f^*\xi) = ch(\lambda_1 \oplus \cdots \oplus \lambda_n) =$$
$$ch(\lambda_1) + \cdots + ch(\lambda_n) = e^{c_1(\lambda_1)} + \cdots + e^{c_1(\lambda_n)}.$$

We turn now to the theory of Chern classes in other cohomology theories. The following assumption will turn out to be sufficient to guarantee the hypotheses of 16.2.

16.27. Suppose E is a ring spectrum and suppose $y \in \tilde{E}^*(CP^\infty)$ is such that the inclusion $i: CP^1 \to CP^\infty$ sends y to a generator $i^*y \in \tilde{E}^*(CP^1) \cong \tilde{E}^*(S^2)$ over $\tilde{E}^*(S^0)$.

16.28. Examples. i) $y \in \tilde{H}^2(CP^\infty;\mathbb{Z})$ as in 16.10 is such an element.

ii) If $\gamma \to CP^\infty$ is the Hopf line bundle, then $Y = \gamma - 1 \in \tilde{K}^0(CP^\infty)$ is an element as in 16.27.

iii) We have a homotopy equivalence $BU(1) \simeq MU_1$: for the sphere bundle of the universal bundle $\gamma_1 \to BU(1)$ is the universal $U(1)$-bundle ($U(1) = S^1$) and hence is contractible. Therefore $D(\gamma_1) \to D(\gamma_1)/S(\gamma_1) = MU_1$ is a homotopy equivalence. But of course $D(\gamma_1) \simeq BU(1)$. Thus we have a map $f: CP^\infty \simeq MU_1 \xrightarrow{M(i)} \Sigma^2 MU$, where $M(i)$ is the map of spectra induced by the inclusion of MU_1 as the second term of the spectrum MU. The element $y = [f] \in \tilde{MU}^2(CP^\infty)$ has the property 16.27.

16.29. Proposition. *If E is a ring spectrum and $y^E \in \tilde{E}^q(CP^\infty)$ is as in 16.27, then $E^*(CP^n) \cong E^*(pt)[y^E]/((y^E)^{n+1})$ and $E^*(CP^\infty) \cong E^*(pt)[[y^E]]$, the ring of formal power series in y^E.*

(We have used y^E to denote the restriction of y^E to CP^n also.)

Proof: We show that the spectral sequence

$$\tilde{H}^*(CP^n; \tilde{E}^*(S^0)) \Rightarrow \tilde{E}^*(CP^n)$$

is trivial for each n. This is certainly true for $n = 1$, for then there is only one non-zero column in E_2^{**}. Observe that

$$F^{2,q-2}(\tilde{E}^*(CP^n)) = \ker[\tilde{E}^q(CP^n) \to \tilde{E}^q((CP^n)^1)] = \ker[\tilde{E}^q(CP^n) \to \tilde{E}^q(pt)] = \tilde{E}^q(CP^n)$$

because the 1-skeleton $(CP^n)^1$ is just a single point. Hence $y^E \in F^{2,q-2}$; let $\bar{y}$ be its image in $E_\infty^{2,q-2} \subset E_2^{2,q-2}$ (note that $E_\infty^{2,q-2} \subset E_2^{2,q-2}$ since $E_2^{p,*} = 0$ if $p \leqslant 1$). Since $\bar{y} \in E_\infty^{2,q-2}$, it follows $d_r\bar{y} = 0$, $r \geqslant 2$. But d_r is a derivation, so $d_r(\bar{y})^k = 0$, $k \geqslant 1$, $r \geqslant 2$. Now

$$\bar{y} \in E_2^{2,q-2} = H^2(CP^n; \tilde{E}^{q-2}(S^0))$$

has the property that its restriction to $\tilde{H}^2(CP^1; \tilde{E}^{q-2}(S^0))$ is a generator, so it must have the form uy for some invertible element $u \in \tilde{E}^{q-2}(S^0)$, $y \in \tilde{H}^2(CP^n; \mathbb{Z})$ as in 16.28i). It thus follows that $\bar{y}, (\bar{y})^2, \ldots, (\bar{y})^n$ form an $\tilde{E}^*(S^0)$-basis for E_2^{**}, and as d_r is $\tilde{E}^*(S^0)$-linear, it follows $d_r = 0$.

Since all differentials are 0 it follows $E_\infty^{**} \cong \tilde{H}^*(CP^n; \tilde{E}^*(S^0))$ has an $\tilde{E}^*(S^0)$-basis $\{\bar{y}, (\bar{y})^2, \ldots, (\bar{y})^n\}$. E_∞^{**} is free as a module over $\tilde{E}^*(S^0)$, so the extension problems

$$0 \to F^{p+1,*} \to F^{p,*} \to E_\infty^{p,*} \to 0$$

are trivial, and it thus follows that $\tilde{E}^*(CP^n)$ has an $\tilde{E}^*(S^0)$-basis $\{y^E, (y^E)^2, \ldots, (y^E)^n\}$. This is equivalent to the statement of the proposition.

We then have $\lim^1 E^*(CP^n) = 0$ and hence

$$E^*(CP^\infty) \cong \lim{}^0 E^*(CP^n) \cong E^*(pt)[[y^E]]. \quad \square$$

Thus the existence of the single element $y^E \in \tilde{E}^*(CP^\infty)$ satisfying 16.27 is enough to guarantee the existence of the generalized Chern classes c_i as in 16.2.

16.30. Proposition. *If E and y^E are as above then there are unique elements $\beta_i \in \tilde{E}_*(CP^n)$, $1 \leqslant i \leqslant n$, such that $\langle (y^E)^k, \beta_i \rangle = \delta_i^k$, $1 \leqslant i \leqslant n$, $1 \leqslant k \leqslant n$. Moreover $\{\beta_1, \beta_2, \ldots, \beta_n\}$ form an $\tilde{E}_*(S^0)$-basis for $\tilde{E}_*(CP^n)$. The coproduct $\psi: E_*(CP^n) \to E_*(CP^n) \otimes_{E_*(pt)} E_*(CP^n)$ induced by the diagonal $\Delta: CP^n \to CP^n \times CP^n$ is given by*

$$\psi\beta_i = \textstyle\sum_{j+k=i} \beta_j \otimes \beta_k.$$

These statements remain true even if $n = \infty$.

Proof: Just as there are external products in spectral sequences, so there is a Kronecker product pairing between the spectral sequences

$$\tilde{H}^*(X; \tilde{E}^*(S^0)) \Rightarrow \tilde{E}^*(X) \quad \text{and} \quad \tilde{H}_*(X; \tilde{E}_*(S^0)) \Rightarrow \tilde{E}_*(X).$$

On the E_2-level it agrees with the ordinary Kronecker product

$$\langle\, , \rangle: \tilde{H}^*(X; \tilde{E}^*(S^0)) \otimes \tilde{H}_*(X; \tilde{E}_*(S^0)) \to \tilde{E}_*(S^0).$$

We have seen that $\tilde{H}^*(CP^n; \tilde{E}^*(S^0))$ has an $\tilde{E}^*(S^0)$-basis

$$\{\bar{y}, (\bar{y})^2, \ldots, (\bar{y})^n\},$$

and hence $\tilde{H}_*(CP^n; \tilde{E}_*(S^0))$ must have a dual basis $\{\beta_1, \beta_2, \ldots, \beta_n\}$. We then have

$$\langle z, d^r \beta_i \rangle = \langle d_r z, \beta_i \rangle = 0$$

for all $z \in E_r^{**}$. Therefore $d_r \beta_i = 0$, $1 \leqslant i \leqslant n$, $r \geqslant 2$. Again the spectral sequence is trivial and we see that $\{\beta_1, \beta_2, \ldots, \beta_n\}$ form an $\tilde{E}_*(S^0)$-basis for $\tilde{E}_*(CP^n)$. For $n = \infty$ we have $\tilde{E}_*(CP^\infty) \cong \operatorname{dir lim} \tilde{E}_*(CP^n) =$ free $\tilde{E}_*(S^0)$-module with basis $\{\beta_1, \beta_2, \ldots\}$. ψ is dual to the cup product in $E^*(CP^\infty)$, so

$$\langle (y^E)^j \otimes (y^E)^k, \psi\beta_i \rangle = \langle (y^E)^{j+k}, \beta_i \rangle = \begin{cases} 1 & j+k=i \\ 0 & \text{otherwise.} \end{cases} \quad \square$$

We can also consider the images of the β_i's in $E_*(BU)$; we call them β_i again, and we let $\beta_0 = 1$.

16.31. Theorem. $E_*(BU) \cong E_*(pt)[\beta_1, \beta_2, \ldots]$ *and the coproduct ψ is given by*

$$\psi\beta_i = \textstyle\sum_{j+k=i} \beta_j \otimes \beta_k.$$

Proof: In the proof of 16.29 we saw that $\bar{y} \in \tilde{H}^2(CP^\infty; \tilde{E}^{q-2}(S^0))$ was of the form uy for an invertible $u \in \tilde{E}^{q-2}(S^0)$. Hence the β_i are of the form $u^{-i}y_i$ in $\tilde{H}_{2i}(CP^\infty; \tilde{E}_{i(q-2)}(S^0))$. Hence by 16.17 and the universal coefficient theorem it follows that the monomials in the β_i form an $E_*(pt)$-basis for $H_*(BU; E_*(pt))$. As $d_r\beta_i = 0$ in the spectral sequence for CP^∞, it follows the same must be true in the spectral sequence for BU. d_r is a derivation, so $d_r = 0$ on all monomials in the β_i. Since d_r is $E_*(pt)$-linear, it follows $d_r = 0$. Hence the spectral sequence is trivial and the proposition follows. □

If $i: CP^1 \to CP^\infty$ is the inclusion, then $i^*(y^E) \in \tilde{E}^*(CP^1) = \tilde{E}^*(S^2)$ may not be the canonical generator $\iota_2 \in \tilde{E}^2(S^2)$, but it must be of the form $i^*y^E = u\iota_2$ for some invertible element $u \in \tilde{E}^*(S^0)$.

Let $X = CP^\infty \times CP^\infty \times \cdots \times CP^\infty$ (n factors) and $\xi_n = \gamma \times \gamma \times \cdots \times \gamma$ as in 16.16. Exactly as in 16.16 we deduce the formula

$$z(\xi_n) = \sum_\alpha \langle z, \beta_\alpha \rangle u_1^{\alpha_1} u_2^{\alpha_2} \ldots u_n^{\alpha_n}$$

for all $z \in E^*(BU)$, where $\beta_\alpha = \beta_{\alpha_1}\beta_{\alpha_2}\cdots\beta_{\alpha_n}$ and $u_i = \pi_i^* y^E$. From 16.2c) and d) we have

$$c_n(\xi_n) = c_1(\gamma) \times c_1(\gamma) \times \cdots \times c_1(\gamma) = u^{-n} u_1 u_2 \ldots u_n.$$

Thus it follows that

$$\langle c_n, \beta_\alpha \rangle = \begin{cases} u^{-n} & \alpha = (1, 1, \ldots, 1) \\ 0 & \text{otherwise.} \end{cases}$$

In other words c_n is dual to $u^n \beta_1^n$, $n \geqslant 1$.

16.32. Theorem. *$E^*(BU(k)) \cong E^*(pt)[[c_1, c_2, \ldots, c_k]]$ and $E^*(BU) \cong E^*(pt)[[c_1, c_2, \ldots]]$, the ring of formal power series in c_1, c_2, ...*

Proof: The argument is similar to that in 16.30, except we are going in the other direction. Let $E_{pq}^r(0)$, $E_r^{pq}(1)$, $E_r^{pq}(2)$ denote the spectral sequences

$$H_*(BU; E_*(pt)) \Rightarrow E_*(BU)$$
$$H^*(BU; E^*(pt)) \Rightarrow E^*(BU)$$

and

$$E_r^{pq}(2) = \mathrm{Hom}_{E_*(pt)}^{-q}(E_{p*}^r(0), E_*(pt)).$$

Kronecker product gives a map of spectral sequences

$$\kappa: E_r^{pq}(1) \to E_r^{pq}(2).$$

κ is an isomorphism on E_2-terms by the universal coefficient theorem.

It thus follows that κ is an isomorphism on the E_r-term for all r and hence on the E_∞-term. We then have commutative diagrams with exact rows

$$\begin{array}{ccccc} 0 \to E_\infty^{pq}(1) & \longrightarrow & G^{pq}(1) & \longrightarrow & \\ \downarrow \kappa & & \downarrow \kappa & & \\ 0 \to \operatorname{Hom}_{E_*(pt)}^{-q}(E_{p,*}^\infty(0), E_*(pt)) & \to & \operatorname{Hom}_{E_*(pt)}^{-q}(F_{p,*}(0), E_*(pt)) & \to & \\ & & G^{p-1,q+1}(1) & \longrightarrow & 0 \\ & & \downarrow \kappa & & \\ & & \operatorname{Hom}_{E_*(pt)}^{-(q+1)}(F_{p-1,*}(0), E_*(pt)) & \to & 0 \end{array}$$

where

$$G^{pq} = \operatorname{coim}\,[E^{p+q}(X) \to E^{p+q}(X^p)] = E^{p+q}(X)/\ker[E^{p+q}(X) \to E^{p+q}(X^p)] = E^{p+q}(X)/F^{p+1,q-1}.$$

By induction it follows

$$\kappa : G^{pq}(1) \to \operatorname{Hom}_{E_*(pt)}^{-q}(F_{p,*}(0), E_*(pt))$$

is an isomorphism for all p,q. But $E^n(BU) = \lim_p^0 G^{p,n-p}(1)$ and

$$\operatorname{Hom}_{E_*(pt)}^n(E_*(BU), E_*(pt)) \cong \lim_p{}^0 \operatorname{Hom}_{E_*(pt)}^{n-1}(F_{p,*}(0), E_*(pt)).$$

Thus we have proved

$$E^*(BU) \cong \operatorname{Hom}_{E_*(pt)}^*(E_*(BU), E_*(pt)).$$

(Note: the proof just given actually works for any X such that $H_*(X; E_*(pt)) \Rightarrow E_*(X)$ is trivial and $H_*(X; E_*(pt))$ free over $E_*(pt)$.)

Let $\Delta^n : BU \to BU \times BU \times \cdots \times BU$ be the n-fold diagonal and m some monomial in the β_i. Suppose

$$(\Delta^n)_*(m) = \textstyle\sum_\alpha m_{\alpha_1} \otimes m_{\alpha_2} \otimes \cdots \otimes m_{\alpha_n}.$$

Then

$$\begin{aligned} \langle c_{i_1} c_{i_2} \ldots c_{i_n}, m \rangle &= \langle \Delta^{n*}(c_{i_1} \otimes c_{i_2} \cdots \otimes c_{i_n}), m \rangle \\ &= \langle c_{i_1} \otimes \cdots \otimes c_{i_n}, \textstyle\sum_\alpha m_{\alpha_1} \otimes m_{\alpha_2} \cdots \otimes m_{\alpha_n} \rangle \\ &= \textstyle\sum_\alpha \langle c_{i_1}, m_{\alpha_1} \rangle \langle c_{i_2}, m_{\alpha_2} \rangle \cdots \langle c_{i_n}, m_{\alpha_n} \rangle. \end{aligned}$$

This expression does not depend on the spectrum E and in particular is exactly the same as for $E = H(\mathbb{Z})$. Hence the E_2-term has an $E^*(pt)$-basis consisting of the monomials in the c_i, and therefore the result follows. The spectral sequence $H^*(BU(k); E^*(pt)) \Rightarrow E^*(BU(k))$ is also trivial. $\square$

16.33. In particular we have

$$\begin{aligned} K_*(BU) &\cong K_*(pt)[Y_1, Y_2, \ldots] \\ K^*(BU) &\cong K^*(pt)[[C_1, C_2, \ldots]] \\ MU_*(BU) &\cong MU_*(pt)[\beta_1, \beta_2, \ldots] \\ MU^*(BU) &\cong MU^*(pt)[[cf_1, cf_2, \ldots]]. \end{aligned}$$

The Chern classes cf_i were first described by Conner and Floyd [31], hence the notation.

16.34. There are corresponding results for BO and BSp. In particular the elements $(\gamma - 1) \otimes_H (\gamma - 1) \in \widetilde{KO}^0(S^4 \wedge HP^\infty) = \widetilde{KO}^{-4}(HP^\infty)$, $\gamma \to S^4 = HP^1$ and $\gamma \to HP^\infty$ the symplectic Hopf bundles, and

$$z = [HP^\infty \simeq MSp_1 \to \Sigma^4 MSp] \in \widetilde{MSp}^4(HP^\infty)$$

are elements satisfying the analog of 16.27. Hence we get

$$\begin{aligned} KO_*(BSp) &\cong KO_*(pt)[Z_1, Z_2, \ldots] \\ KO^*(BSp) &\cong KO^*(pt)[[P_1, P_2, \ldots]] \\ MSp_*(BSp) &\cong MSp_*(pt)[\sigma_1, \sigma_2, \ldots] \\ MSp^*(BSp) &\cong MSp^*(pt)[[pf_1, pf_2, \ldots]]. \end{aligned}$$

16.35. The usefulness of the element y^E is not yet exhausted; we shall also be able to compute $E_*(MU)$. First, however, we determine $H_*(MU; \mathbb{Z})$.

Since $\pi_1(BU(n), *) = 0$, $n \geqslant 1$, the universal bundle γ_n over $BU(n)$ is orientable with respect to any cohomology theory. In particular there is a unique Thom class $t_n \in \tilde{H}^{2n}(MU_n; \mathbb{Z})$ with $\iota_n^*(t_n) = \iota_{2n} \in \tilde{H}^{2n}(S^{2n}; \mathbb{Z})$ for every $n \geqslant 1$; in particular $t_1 = y \in \tilde{H}^2(MU_1; \mathbb{Z})$. For any $U(n)$-bundle ξ with classifying map $f: X \to BU(n)$ we get a Thom class $t(\xi)$ by taking $t(\xi) = M(f)^*(t_n) \in \tilde{H}^{2n}(M(\xi); \mathbb{Z})$. $t(\xi)$ satisfies $t(f^*\xi) = M(f)^*(t(\xi))$ for all $f: Y \to X$ and also $t(\xi \times \eta) = t(\xi) \wedge t(\eta)$ (why?).

We have Thom isomorphisms

$$\Phi_*: \tilde{H}_{q+2n}(MU_n; \mathbb{Z}) \to H_q(BU(n); \mathbb{Z}),$$

and indeed the diagrams

$$\begin{array}{ccccc} \tilde{H}_{q+2n}(MU_n;\mathbb{Z}) & \xrightarrow{\sigma^2} & \tilde{H}_{q+2n+2}(S^2MU_n;\mathbb{Z}) & \xrightarrow{M(i_n)_*} & \tilde{H}_{q+2n+2}(MU_{n+1};\mathbb{Z}) \\ \cong\downarrow \Phi_* & \swarrow \Phi_* & & & \cong\downarrow \Phi_* \\ H_q(BU(n);\mathbb{Z}) & & \xrightarrow{i_{n*}} & & H_q(BU(n+1);\mathbb{Z}) \end{array}$$

commute; for if $a \in \tilde{H}_*(MU_n;\mathbb{Z})$, then

$$\begin{aligned} \Phi_* \circ M(i_n)_* \circ \sigma^2(a) &= p_{n+1*}(t_{n+1} \cap M(i_n)_* \sigma^2(a)) \\ &= p_{n+1*} \circ D(i_n)_* \, [M(i_n)^*(t_{n+1}) \cap \sigma^2(a)] \\ &= i_{n*} \circ p_{n*}[t(\varepsilon^1 \oplus \gamma_n) \cap \sigma^2(a)] \\ &= i_{n*} \circ p_{n*} \, [\sigma^{-2}(t_n) \cap \sigma^2(a)] \\ &= i_{n*} \circ p_{n*}(t_n \cap a) = i_{n*} \circ \Phi_*(a). \end{aligned}$$

Therefore we may take direct limits (recall 13.49), obtaining a "stable Thom isomorphism"

$$\Phi_*: H_q(MU;\mathbb{Z}) \to H_q(BU;\mathbb{Z}), \quad q \in \mathbb{Z}.$$

16.36. Lemma. *$H_*(MU;\mathbb{Z})$ and $H_*(BU;\mathbb{Z})$ both have Pontrjagin products and Φ_* is a ring homomorphism.*

Proof: We wish to show that the diagram

$$\begin{array}{ccc} \tilde{H}_{q+2n}(MU_n) \otimes \tilde{H}_{p+2m}(MU_m) & \xrightarrow{\wedge} & \\ \downarrow \Phi_* \otimes \Phi_* & & \\ H_q(BU(n)) \otimes H_p(BU(m)) & \xrightarrow{\times} & \end{array}$$

$$\begin{array}{ccc} \tilde{H}_{q+p+2n+2m}(MU_n \wedge MU_m) & \xrightarrow{M(\mu_{n,m})_*} & \tilde{H}_{q+p+2n+2m}(MU_{n+m}) \\ \downarrow \Phi_* & & \downarrow \Phi_* \\ H_{q+p}(BU(n) \times BU(m)) & \xrightarrow{\mu_{n,m*}} & H_{q+p}(BU(n+m)) \end{array}$$

commutes. But if $a \in \tilde{H}_*(MU_n)$ and $b \in \tilde{H}_*(MU_m)$, then

$$\begin{aligned} \Phi_* \circ M(\mu_{nm})_*(a \wedge b) &= p_{n+m*}(t_{n+m} \cap M(\mu_{nm})_*(a \wedge b)) \\ &= p_{n+m*} \circ D(\mu_{nm})_*[M(\mu_{nm})^*(t_{n+m}) \cap (a \wedge b)] \\ &= \mu_{nm*} \circ (p_n \times p_m)_* \, [t(\gamma_n \times \gamma_m) \cap (a \wedge b)] \end{aligned}$$

$$= \mu_{nm*} \circ (p_n \times p_m)_*[(t_n \wedge t_m) \cap (a \wedge b)]$$
$$= \mu_{nm*} \circ (p_n \times p_m)_*[(t_n \cap a) \times (t_m \cap b)]$$
$$= \mu_{nm*}[p_{n*}(t_n \cap a) \times p_{m*}(t_m \cap b)]$$
$$= \mu_{nm*}(\Phi_*(a) \times \Phi_*(b)). \quad \square$$

16.37. Theorem. *$H_*(MU;\mathbb{Z}) \cong \mathbb{Z}[b_1, b_2, \ldots]$, where $\Phi_*(b_i) = y_i$.*

Now consider the map $f: CP^\infty \simeq MU(1) \xrightarrow{M(i)} \Sigma^2 MU$ of 16.28iii).

16.38. Proposition. $f_*(y_i) = b_{i-1}$, $i \geqslant 1$.

Proof: Consider the commutative diagram

$$\begin{array}{ccccc} & & H_{2i-2}(BU(1);\mathbb{Z}) & \xrightarrow{i_*} & H_{2i-2}(BU;\mathbb{Z}) \\ & & \uparrow \Phi_* & & \uparrow \Phi_* \\ \tilde{H}_{2i}(CP^\infty;\mathbb{Z}) & \xrightarrow{\cong} & \tilde{H}_{2i}(MU_1;\mathbb{Z}) & \xrightarrow{M(i)_*} & H_{2i-2}(MU;\mathbb{Z}). \end{array}$$

Recalling the identification $BU(1) \overset{p}{\underset{\simeq}{\leftarrow}} D(\gamma_1) \underset{\simeq}{\rightarrow} MU_1$, we compute $\Phi_*(y_i) = p_*(y \cap y_i) = y \cap y_i$. But $\langle y^k, y \cap y_i \rangle = \langle y^{k+1}, y_i \rangle = \delta_i^{k+1}$, so $y \cap y_i = y_{i-1}$, $i \geqslant 1$. The result follows. $\square$

There are analogous results for MO, MSp.

Now let us return to $y^E \in \tilde{E}^*(CP^\infty)$. Recall the invertible element $u = u^E \in \tilde{E}^*(S^0)$ such that $i^* y^E = u^E \iota_2 \in \tilde{E}^*(CP^1) = \tilde{E}^*(S^2)$. Taking 16.38 as our guide we define elements $b_i^E \in E_{2i}(MU)$ by the formula

$$b_i^E = f_*(u^E \beta_{i+1}), \quad i \geqslant 0.$$

The factor u^E is introduced so that $b_0^E = 1$ in $E_0(MU)$.

16.39. Theorem. $E_*(MU) \cong \tilde{E}_*(S^0)[b_1^E, b_2^E, \ldots]$.

Proof: Exactly as in the proof of 16.31 we find that the monomials in the b_i^E induce an $\tilde{E}_*(S^0)$-basis for the E^2-term of the spectral sequence $H_*(MU; \tilde{E}_*(S^0)) \Rightarrow E_*(MU)$. Again

$$d_r b_i^E = d_r f_*(u^E \beta_{i+1}) = f_*(d_r u^E \beta_{i+1}) = 0,$$

so it follows $d_r = 0$ for all $r \geqslant 2$. Thus the spectral sequence is trivial, and the theorem follows. $\square$

16.40. We can also show that all $U(n)$-bundles are orientable with respect to E. An argument identical to that in the proof of 16.32 shows that

$$E^*(MU) \cong \mathrm{Hom}^*_{\tilde{E}_*(S^0)}(E_*(MU), \tilde{E}_*(S^0)).$$

In particular there is a unique element $t \in E^0(MU)$ such that $t(1) = 1$ and

$t(m) = 0$ for all other monomials m in the b_i^E. For each n we let $M(i_n): MU_n \to \Sigma^{2n} MU$ be the "inclusion" of MU_n and take

$$t_n = M(i_n)^* \sigma^{-2n}(t) \in \tilde{E}^{2n}(MU_n).$$

For any $U(n)$-bundle $\xi \to X$ with classifying map $f: X \to BU(n)$ we define $t(\xi) = M(f)^*(t_n) \in \tilde{E}^{2n}(M(\xi))$ as in 16.35. In order to show $t(\xi)$ is a Thom class for ξ it suffices to show that t_n is a Thom class for γ_n, because we have a commutative diagram

$$\begin{array}{ccc} & S^{2n} & \\ {\scriptstyle j_b} \swarrow & & \searrow {\scriptstyle j_{f(b)}} \\ M(\xi) & \xrightarrow{M(f)} & MU_n \end{array}$$

for every $b \in X$. It is equivalent to show $\iota_n^*(t_n) \in \tilde{E}^{2n}(S^{2n})$ is a generator of $\tilde{E}^*(S^{2n})$. We have another commutative diagram

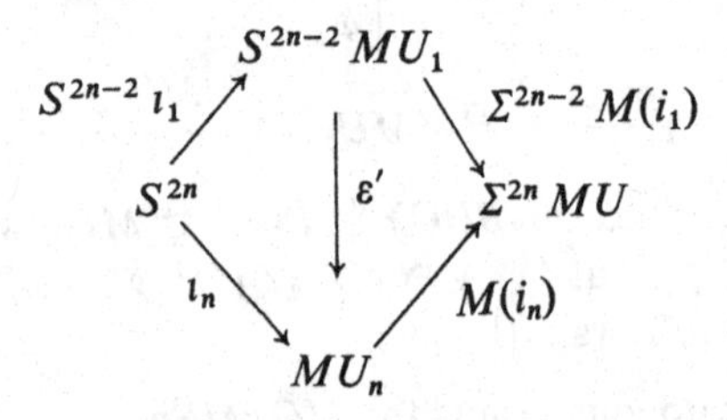

where $\varepsilon' = \varepsilon_{n-1} \circ S^2 \varepsilon_{n-2} \circ \cdots \circ S^{2n-4} \varepsilon_1$. Hence $\iota_n^*(t_n) = \iota_n^* M(i_n)^* \sigma^{-2n}(t) = \sigma^{-2n+2} \iota_1^* M(i_1)^*(\sigma^{-2} t)$. If $g_{2n} \in \tilde{E}_{2n}(S^{2n})$ denotes the canonical generator (represented by $\iota^E \wedge 1: S^0 \wedge S^{2n} \to E \wedge S^{2n}$), then

$$\begin{aligned} \langle \iota_n^* t_n, g_{2n} \rangle &= \langle \sigma^{-2n+2} \iota_1^* M(i_1)^* (\sigma^{-2} t), g_{2n} \rangle = \langle t, \sigma^{-2} M(i_1)_* \iota_{1*} g_2 \rangle \\ &= \langle t, \sigma^{-2} M(i_1)_* u^E \beta_1 \rangle = \langle t, b_0^E \rangle = 1. \end{aligned}$$

Thus $\iota_n^* t_n$ is a generator of $\tilde{E}^*(S^{2n})$.

Therefore we have Thom isomorphisms

$$\Phi_*: \tilde{E}_{q+2n}(M(\xi)) \to E_q(X)$$

$$\Phi^*: E^q(X) \to \tilde{E}^{q+2n}(M(\xi)).$$

By running the argument of 16.38 in reverse we can show that

$$\Phi_*(b_i^E) = u^E \beta_i \in E_{2i}(BU), \quad i \geqslant 1.$$

Warning: $E^*(MU)$ does not have a natural ring structure, so do not be tempted to say that $E^*(MU) \cong \tilde{E}^*(S^0)[[c_1', c_2', \ldots]]$ for $c_i' = \Phi^*(c_i)$. Of course, in the case $E = MU$ we do have a product on $MU^*(MU)$ given by composition of maps (see Chapter 17), but Φ^* is not a ring homomorphism: composition is not even commutative!

16.41. There are analogous results for BO and BSp. We give two specific examples:

$$KO_*(MSp) \cong \widetilde{KO}_*(S^0)[Z'_1, Z'_2, \ldots]$$

$$MSp_*(MSp) \cong \widetilde{MSp}_*(S^0)[\sigma'_1, \sigma'_2, \ldots].$$

16.42. One of the most important applications of characteristic classes is the theory of characteristic numbers of cobordism classes of manifolds. Suppose we are considering a class of manifolds M^n which are orientable—have a fundamental class—with respect to a ring spectrum. For example we might be considering U-manifolds and a spectrum E satisfying 16.27; let us take this specific case to illustrate the general procedure. Each U-manifold M^n can be embedded in S^{n+2k} for some large k, and the normal bundle ν^k of the embedding can be given a $U(k)$-structure. Let $f: M^n \to BU(k)$ classify ν^k. Then we can consider the characteristic classes $z(\nu^k) = f^*(z) \in E^*(M^n)$, $z \in E^*(BU(k))$, and also the elements $z[M^n] = \langle z(\nu^k), \sigma_{M^n} \rangle \in E_*(pt)$, σ_{M^n} the fundamental class. The elements $z[M^n]$ are called the *characteristic numbers* of M^n, and we shall see that they depend only on the cobordism class $[M^n]$ of M^n. They are called "numbers" because they were first defined for the case $E = H(\mathbb{Z})$, in which case they were elements of $H_*(pt; \mathbb{Z})$.

In particular we have the generalized Chern classes $c_i \in E^*(BU(k))$, $1 \leqslant i \leqslant k$, and so for each monomial $c^\alpha = c_1^{\alpha_1} c_2^{\alpha_2} \ldots c_k^{\alpha_k}$ we can form the *Chern number* $c^\alpha[M^n]$. Certain other polynomials $s_\alpha(c_1, c_2, \ldots, c_k)$ are also frequently considered.

There is another way of regarding characteristic numbers. The nth unitary cobordism group Ω_n^U is isomorphic to $\pi_n(MU)$, and we have the Hurewicz homomorphism

$$h: \pi_n(MU) \to E_n(MU).$$

By 16.39 $E_*(MU) \cong \tilde{E}_*(S^0)[b_1, b_2, \ldots]$, so for a cobordism class

$$[M^n] \in \Omega_n^U = \pi_n(MU)$$

we have $h[M^n] = \sum_\alpha n_\alpha b^\alpha$ for suitable coefficients $n_\alpha \in \tilde{E}_*(S^0) = E_*(pt)$. Here b^α denotes the monomial $b_1^{\alpha_1} b_2^{\alpha_2} \ldots b_r^{\alpha_r}$ if $\alpha = (\alpha_1, \alpha_2, \ldots, \alpha_r)$. What are the coefficients n_α? We are going to show that $n_\alpha = s_\alpha[M^n]$, where $s_\alpha \in E^*(BU)$ and $\{s_\alpha\}$ is the basis dual to the basis $\{u^{\|\alpha\|} \beta^\alpha\}$ in $E_*(BU)$, where $\|\alpha\| = \alpha_1 + \alpha_2 + \cdots + \alpha_n$.

16.43. Proposition. *For a U-manifold M^n and $z \in E^*(BU)$ we have $z[M^n] = \langle z, \Phi_* h[M^n] \rangle$ and hence $h[M^n] = \sum_\alpha s_\alpha[M^n] b^\alpha \in E_n(M^n)$.*

Proof: For this we need to recall the Pontrjagin–Thom construction 12.30 which gives us our isomorphism $\Omega_*^U \cong \pi_*(MU)$. Suppose M^n embedded in

S^{n+2k} with tubular neighborhood N, which we can identify with $D(\nu)$. Let $f: M^n \to BU(k)$ be the classifying map for the normal bundle ν. Then we have a commutative diagram

$$\begin{array}{ccccc} \tilde{E}_{n+2k}(S^{n+2k}) & \xrightarrow{p_*} & \tilde{E}_{n+2k}(M(\nu)) & \xrightarrow{M(f)_*} & \tilde{E}_{n+2k}(MU_k) \\ & & \cong \downarrow \Phi_* & & \cong \downarrow \Phi_* \\ & & E_n(M^n) & \xrightarrow{f_*} & E_n(BU(k)), \end{array}$$

where $p: S^{n+2k} \to M(\nu)$ is the map which on $N = D(\nu)$ is the projection $D(\nu) \to M(\nu)$ and outside N sends the rest of S^{n+2k} to $*$ in $M(\nu)$. From 14.40 we know that $\Phi_* p_*(\iota_{n+2k}) = \sigma_{M^n}$, where ι_{n+2k} is the generator of $\tilde{E}_{n+2k}(S^{n+2k})$.

Now $[M^n]$ is represented in $\pi_n(MU)$ by $\{[M(f) \circ p]\}$, so $h[M^n] = M(f)_* \circ p_*(\iota_{n+2k})$. For any $z \in E^*(BU)$ then we have

$$\begin{aligned} z[M^n] &= \langle f^*(z), \sigma_{M^n} \rangle = \langle f^*(z), \Phi_* p_*(\iota_{n+2k}) \rangle = \langle z, f_* \circ \Phi_* \circ p_* (\iota_{n+2k}) \rangle \\ &= \langle z, \Phi_* \circ M(f)_* \circ p_*(\iota_{n+2k}) \rangle = \langle z, \Phi_* \circ h[M^n] \rangle. \end{aligned}$$

Thus if $h[M^n] = \sum_\alpha n_\alpha b^\alpha$, then

$$z[M^n] = \langle z, \Phi_*(\textstyle\sum_\alpha n_\alpha b^\alpha) \rangle = \langle z, \sum_\alpha n_\alpha u^{\|\alpha\|} \beta^\alpha \rangle = \sum_\alpha n_\alpha \langle z, u^{\|\alpha\|} \beta^\alpha \rangle.$$

Taking $z = s_\gamma$ gives

$$s_\gamma[M^n] = \textstyle\sum_\alpha n_\alpha \langle s_\gamma, u^{\|\alpha\|} \beta^\alpha \rangle = n_\gamma. \quad \square$$

16.44. Corollary. *The characteristic numbers of a U-manifold M^n are cobordism invariants.*

One may also prove 16.44 directly using only the definition of "characteristic number" and "cobordism".

Note. The characteristic *classes* $z(\nu)$ certainly are not cobordism invariants; we get something invariant only after taking Kronecker product with σ_M. From the proof of 16.43 it is clear we have a specific choice of σ_M in mind: $\Phi_*(p_*(\iota_{n+2k}))$. Φ_* in turn depends on the choice of Thom class t, but this is fixed once we have fixed the universal Thom class $t \in E^0(MU)$. Otherwise the $z[M^n]$ are determined only up to a unit in $\tilde{E}^0(S^0)$.

Note. By algebraic manipulation one can show that the polynomials s_α are described as follows: we express the symmetric polynomial

$$\sum x_1 x_2 \ldots x_{\alpha_1} x_{\alpha_1+1}^2 \ldots x_{\alpha_2}^2 \ldots x_{\alpha_r}^r$$

in the form $s_\alpha(\sigma_1, \sigma_2, \ldots, \sigma_n)$. Then $s_\alpha = s_\alpha(c_1, c_2, \ldots, c_n)$. (See [55].)

16.43 gives us a means of computing $h[M^n]$ for specific manifolds M.

16.45. Proposition. *For all* $n \geqslant 0$

$$h[CP^n] = (b^{-n-1})_{2n} \in H_{2n}(MU;\mathbb{Z}),$$

where $b = 1 + b_1 + b_2 + \cdots$.

Proof: The tangent bundle τ of CP^n satisfies $\tau \oplus \varepsilon^1 \simeq (n+1)\gamma$, where $\gamma =$ the Hopf line bundle over CP^n (cf. 16.7). The stable normal bundle ν is the negative of τ, so if $\chi: BU \to BU$ is the operation of taking stable inverse ($\xi \mapsto -\xi$), then

$$\begin{aligned} s_y(\nu) &= (\chi^* s_y)(\tau) = (\chi^* s_y)((n+1)\gamma) = \textstyle\sum_\alpha \langle \chi^* s_y, y_\alpha \rangle y^{\alpha_1+\alpha_2+\cdots+\alpha_{n+1}} \\ &= \textstyle\sum_\alpha \langle s_y, \chi_* y_\alpha \rangle y^{\alpha_1+\cdots+\alpha_{n+1}}. \end{aligned}$$

To find $s_y[CP^n]$ we must take $\langle s_y(\nu), \sigma_{CP^n} \rangle$; but $\sigma_{CP^n} = y_n \in H_{2n}(CP^n;\mathbb{Z})$. Thus we get

$$\begin{aligned} s_y[CP^n] &= \textstyle\sum_{\alpha_1+\cdots+\alpha_{n+1}=n} \langle s_y, \chi_*(y_\alpha) \rangle \\ &= \langle s_y, \chi_*(\textstyle\sum_{\alpha_1+\cdots+\alpha_{n+1}=n} y_\alpha) \rangle \\ &= \langle s_y, \chi_*((1 + y_1 + y_2 + \cdots)^{n+1})_{2n} \rangle \\ &= \langle s_y, (\chi_*(1 + y_1 + y_2 + \cdots))_{2n}^{n+1} \rangle. \end{aligned}$$

It is left as an exercise for the reader to show that $\chi_*(1 + y_1 + y_2 + \cdots) = (1 + y_1 + y_2 + \cdots)^{-1}$. Thus

$$s_y[CP^n] = \langle s_y, (1 + y_1 + y_2 + \cdots)_{2n}^{-n-1} \rangle = \langle \Phi^* s_y, (b^{-n-1})_{2n} \rangle,$$

and therefore

$$h[CP^n] = \textstyle\sum_\alpha \langle \Phi^* s_\alpha, (b^{-n-1})_{2n} \rangle b^\alpha = (b^{-n-1})_{2n},$$

since $\{\Phi^* s_\alpha\}$ is the basis dual to $\{b^\alpha\}$. □

16.46. Examples.

$$b^{-2} = 1 - 2b_1 + \cdots, \quad \text{so } h[CP^1] = -2b_1.$$

$$b^{-3} = 1 - 3b_1 + 6b_1^2 - 3b_2 + \cdots, \quad \text{so } h[CP^2] = 6b_1^2 - 3b_2.$$

In general $(b^{-n-1})_{2n} = -(n+1)b_n +$ decomposables.

There are analogous results for other classical groups G. For example we get

$$h[RP^n] = (a^{-n-1})_n \in H_n(MO;\mathbb{Z}_2),$$

where $a = 1 + a_1 + a_2 + \cdots$ $(H_*(MO;\mathbb{Z}_2) \cong \mathbb{Z}_2[a_1, a_2, \ldots])$.

16.47. We may now use our knowledge of $H_*(BU;\mathbb{Z})$ and the Borel theorem 15.60 to prove the Bott periodicity theorem 11.60 for the unitary group. Since $\Omega U \simeq \mathbb{Z} \times \Omega SU$, it will suffice if we show $BU \simeq \Omega SU$. We begin with some geometric constructions.

In Chapter 11 we defined BU to be $BU = \bigcup_{k \geqslant 1} BU(k)$, where

$$BU(k) = \bigcup_{n \geqslant k} G_{k,n} = \bigcup_{n \geqslant k} \frac{U(n)}{U(n-k) \times U(k)}.$$

In fact, one can easily see that

$$\bigcup_{n \geqslant 1} \frac{U(2n)}{U(n) \times U(n)}$$

has the same homotopy type as $\bigcup_{k \geqslant 1} BU(k)$, so we may take

$$BU = \bigcup_{n \geqslant 1} \frac{U(2n)}{U(n) \times U(n)}$$

for our present purposes. Here we must think of $U(2n) \subset U(2n+2)$ as being all matrices of the form

$$\begin{pmatrix} 1 & 0 & 0 \\ 0 & A & 0 \\ 0 & 0 & 1 \end{pmatrix}, \quad A \in U(2n).$$

We regard CP^n as

$$\frac{U(n+1)}{U(n) \times U(1)};$$

if $U(n+1) \subset U(2n)$ is the subgroup of matrices

$$\begin{pmatrix} A & 0 \\ 0 & I_{n-1} \end{pmatrix}, \quad A \in U(n+1),$$

then the inclusion induces a map

$$\frac{U(n+1)}{U(n) \times U(1)} \xrightarrow{f_n} \frac{U(2n)}{U(n) \times U(n)}$$

which classifies the Hopf bundle γ over CP^n.

We now define maps

$$g_{n,m}: S^1 \wedge \frac{U(n+m)}{U(n) \times U(m)} \to SU(n+m)$$

by

$$g_{n,m}[t, A] = \begin{pmatrix} e^{\pi it} I_n & 0 \\ 0 & e^{-\pi it} I_m \end{pmatrix} A \begin{pmatrix} e^{-\pi it} I_n & 0 \\ 0 & e^{\pi it} I_m \end{pmatrix} A^{-1}.$$

The diagrams

$$\begin{array}{ccc} S^1 \wedge CP^n & \xrightarrow{g_{n,1}} & SU(n+1) \\ \big\downarrow Sf_n & & \cap \\ S^1 \wedge \dfrac{U(2n)}{U(n)\times U(n)} & \xrightarrow{g_{n,n}} & SU(2n) \end{array} \qquad \begin{array}{ccc} S^1 \wedge \dfrac{U(2n)}{U(n)\times U(n)} & \xrightarrow{g_{n,n}} & SU(2n) \\ \cap & & \cap \\ S^1 \wedge \dfrac{U(2n+2)}{U(n+1)\times U(n+1)} & \xrightarrow{g_{n+1,n+1}} & SU(2n+2) \end{array}$$

commute, and hence we get maps $h: S^1 \wedge CP^\infty \to SU$, $g: S^1 \wedge BU \to SU$ such that the diagram

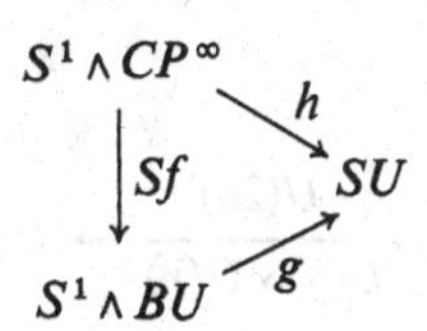

commutes.

Now BU is an H-space, the product being associated to the Whitney sum of complex vector bundles. We can describe a product for BU as follows: we think of U as acting on

$$\mathbb{C}^\infty = \{(x_1, x_2, \ldots): x_i \in \mathbb{C},\ i \geqslant 1,\ x_i \neq 0 \text{ only finitely many } i\}.$$

$U(2n)$ acts only on the first $2n$ coordinates. We can define

$$\bar{\Delta}: \mathbb{C}^\infty \oplus \mathbb{C}^\infty \to \mathbb{C}^\infty$$

by $\bar{\Delta}(e_i, 0) = e_{2i-1}$, $\bar{\Delta}(0, e_i) = e_{2i}$, $i \geqslant 1$. Then we can define

$$\mu_{2n}: U(2n) \times U(2n) \to U(4n)$$

by $\mu_{2n}(A, B) = \bar{\Delta} \circ (A \oplus B) \circ \bar{\Delta}^{-1}$. For example

$$\mu_2\left(\begin{pmatrix} a & b \\ c & d \end{pmatrix}, \begin{pmatrix} x & y \\ u & v \end{pmatrix}\right) = \begin{pmatrix} a & 0 & b & 0 \\ 0 & x & 0 & y \\ c & 0 & d & 0 \\ 0 & u & 0 & v \end{pmatrix}.$$

μ_{2n} induces

$$\mu_{2n}: \frac{U(2n)}{U(n)\times U(n)} \times \frac{U(2n)}{U(n)\times U(n)} \to \frac{U(4n)}{U(2n)\times U(2n)},$$

and the diagrams

$$\begin{array}{ccc} \dfrac{U(2n)}{U(n)\times U(n)} \times \dfrac{U(2n)}{U(n)\times U(n)} & \xrightarrow{\mu_{2n}} & \dfrac{U(4n)}{U(2n)\times U(2n)} \\ \Big\downarrow g'_{n,n}\times g'_{n,n} & & \Big\downarrow g'_{2n,2n} \\ \Omega SU(2n)\times \Omega SU(2n) & \xrightarrow{\Omega\mu_{2n}} & \Omega SU(4n) \end{array}$$

$$\begin{array}{ccc} \dfrac{U(2n)}{U(n)\times U(n)} \times \dfrac{U(2n)}{U(n)\times U(n)} & \xrightarrow{\mu_{2n}} & \dfrac{U(4n)}{U(2n)\times U(2n)} \\ \cap & & \cap \\ \dfrac{U(2n+2)}{U(n+1)\times U(n+1)} \times \dfrac{U(2n+2)}{U(n+1)\times U(n+1)} & \xrightarrow{\mu_{2n+2}} & \dfrac{U(4n+4)}{U(2n+2)\times U(2n+2)} \end{array}$$

commute, where

$$g'_{n,n}\colon \frac{U(2n)}{U(n)\times U(n)} \to \Omega SU(2n)$$

is the adjoint of $g_{n,n}$. Therefore the μ_{2n}'s define a map $\mu\colon BU\times BU\to BU$ which we can take as our Whitney sum map ($\mu_{2n}\colon U(2n)\times U(2n)\to U(4n)$ is homotopic in $U(4n+4)$ to the inclusion $h\colon U(2n)\times U(2n)\to U(4n)$ which we used in 11.46). Moreover the $g'_{n,n}$'s define a map $g'\colon BU\to \Omega SU$ which is a map of H-spaces, ΩSU being given the product $\Omega\mu$.

If $h'\colon CP^\infty\to\Omega SU$ is the adjoint of h, then the diagram

$$\begin{array}{ccc} & H_*(CP^\infty;\mathbb{Z}) & \\ {}^{f_*}\swarrow & & \searrow^{h'_*} \\ H_*(BU;\mathbb{Z}) & \xrightarrow{g'_*} & H_*(\Omega SU;\mathbb{Z}) \end{array}$$

commutes. We wish to show that g' is a homotopy equivalence. SU can be given the structure of a CW-complex, and so by the theorem of Milnor quoted in Chapter 9 ΩSU has the homotopy type of a CW-complex. Since $\pi_0(BU,*)=\pi_1(BU,*)=\pi_0(\Omega SU,\omega_0)=\pi_1(\Omega SU,\omega_0)=0$, it suffices to show that $g'_*\colon H_*(BU;\mathbb{Z})\to H_*(\Omega SU;\mathbb{Z})$ is an isomorphism. g'_* is a homomorphism of Pontrjagin rings and $H_*(BU;\mathbb{Z})\cong\mathbb{Z}[y_1,y_2,\ldots]$, where $y_i=f_*(y_i)$, $i\geqslant 1$, and $\{y_1,y_2,\ldots\}$ is the basis in $H_*(CP^\infty;\mathbb{Z})$. Thus it suffices to show that $H_*(\Omega SU;\mathbb{Z})\cong\mathbb{Z}[y'_1,y'_2,\ldots]$ with $y'_i=h'_*(y_i)$. Here the Pontrjagin ring structure in $H_*(\Omega SU;\mathbb{Z})$ is that induced by $\Omega\mu$; but from 2.25 we know that this product agrees with that induced by the loop space product.

16.48. Lemma. *The map*

$$(S^1 \wedge CP^n, S^1 \wedge CP^{n-1}) \xrightarrow{g_{n,1}} (SU(n+1), SU(n)) \xrightarrow{\pi} (S^{2n+1}, s_0)$$

induces an isomorphism of homology groups. Here π is the map $\pi(A) = Ae_1 \in S^{2n+1}$ for $A \in SU(n+1)$.

Proof: It suffices to show that the induced map

$$\frac{S^1 \wedge CP^n}{S^1 \wedge CP^{n-1}} \to S^{2n+1}$$

is bijective, since

$$\frac{S^1 \wedge CP^n}{S^1 \wedge CP^{n-1}}$$

and S^{2n+1} are Hausdorff and compact. Thus we are required to show that

a) $g_{n,1}(t, A)(e_1) = e_1 \Leftrightarrow t \in \dot{I}$ or

$$A(U(n) \times U(1)) \in \frac{\{1\} \times U(n)}{\{1\} \times U(n-1) \times U(1)} = CP^{n-1};$$

b) for all unit vectors $u \in \mathbb{C}^{n+1}$ with $u \neq e_1$ we can write $u = g_{n,1}(t, A)e_1$ for some $t \in I$, $A \in U(n+1)$.

For any $A \in U(n+1)$ we can write

$$e_1 = \alpha A(e_{n+1}) + (e_1 - \alpha A(e_{n+1})), \quad \alpha = \langle e_1, A(e_{n+1})\rangle.$$

Then $A(e_{n+1}) \perp e_1 - \alpha A(e_{n+1})$, and hence

$$\begin{pmatrix} e^{-\pi i t} I_n & 0 \\ 0 & e^{\pi i t} \end{pmatrix} A^{-1} e_1 = e^{\pi i t} \alpha e_{n+1} + e^{-\pi i t} A^{-1}(e_1 - \alpha A(e_{n+1})).$$

Therefore

$$g_{n,1}(t, A)\, e_1 = \begin{pmatrix} e^{\pi i t} I_n & 0 \\ 0 & e^{-\pi i t} \end{pmatrix} (e^{\pi i t} \alpha A(e_{n+1}) + e^{-\pi i t}(e_1 - \alpha A(e_{n+1})))$$

16.49. $$= \begin{pmatrix} I_n & 0 \\ 0 & e^{-2\pi i t} \end{pmatrix} (e_1 + (e^{2\pi i t} - 1)\, \alpha A(e_{n+1})).$$

From 16.49 it follows that $g_{n,1}(t,A)e_1 = e_1$ if and only if $t \in \dot{I}$ or $A(e_{n+1}) \perp e_1$—i.e.

$$A(U(n) \times U(1)) \in \frac{\{1\} \times U(n)}{\{1\} \times U(n-1) \times U(1)}.$$

Suppose u is a unit vector and $u \neq e_1$; then

$$u = e_1 + (u - e_1) = e_1 + \frac{\|u - e_1\|^2}{\langle e_1, u\rangle - 1} \cdot \frac{\langle e_1, u\rangle - 1}{\|u - e_1\|} \cdot \frac{u - e_1}{\|u - e_1\|}$$

$$= e_1 + (e^{2\pi i t} - 1)\langle e_1, v\rangle v$$

if we set

$$v = \frac{u - e_1}{\|u - e_1\|} \quad \text{and} \quad e^{2\pi i t} = -\frac{\overline{\langle e_1, u\rangle - 1}}{\langle e_1, u\rangle - 1}.$$

Now if $u \neq e_1$, then also

$$u' = \begin{pmatrix} I_n & 0 \\ 0 & e^{2\pi i t} \end{pmatrix} u \neq e_1,$$

so we can choose an $A \in U(n+1)$ with

$$A(e_{n+1}) = \begin{pmatrix} I_n & 0 \\ 0 & e^{2\pi i t} \end{pmatrix} v = \frac{u' - e_1}{\|u' - e_1\|}.$$

Then we have

$$u' = e_1 + (e^{2\pi i t} - 1)\langle e_1, A(e_{n+1})\rangle A(e_{n+1})$$

and thus by 16.49 $g_{n,1}(t,A)e_1 = u$. □

Let $z_{i+1} = \sigma(y_i) \in \tilde{H}_{2i+1}(SCP^{n-1};\mathbb{Z})$, $1 \leqslant i \leqslant n-1$.

16.50. Theorem. *$H_*(SU(n);\mathbb{Z}) \cong E(\lambda_2, \ldots, \lambda_n)$, where*

$$\lambda_i = g_{n-1,1*}(z_i) \in H_{2i-1}(SU(n);\mathbb{Z}), \quad 2 \leqslant i \leqslant n,\ n \geqslant 1.$$

Proof: The proof is by induction on n; the case $n = 2$ is clear since $SU(2) \cong S^3$. We have the fibration $SU(n) \to SU(n+1) \xrightarrow{\pi} S^{2n+1}$ and the resulting Wang sequence

$$\cdots \longrightarrow H_m(SU(n);\mathbb{Z}) \xrightarrow{i_*} H_m(SU(n+1);\mathbb{Z}) \xrightarrow{\psi}$$

$$H_{m-(2n+1)}(SU(n);\mathbb{Z}) \xrightarrow{d^{2n+1}} H_{m-1}(SU(n);\mathbb{Z}) \longrightarrow \cdots.$$

Now in the case of a fibration $F \xrightarrow{i} E \to S^n$ where E, F are H-groups and i an H-group map, we see from the definition of ψ

$$(H_m(E) \to E^\infty_{n,m-n} \subset E^2_{n,m-n} \cong H_n(S^n; H_{m-n}(F)) \cong H_{m-n}(F))$$

that with respect to the Pontrjagin product ψ satisfies $\psi(x \cdot i_*(y)) = \psi(x) \cdot y$ for $x \in H_*(E)$, $y \in H_*(F)$. Moreover we know $\psi: H_{2n+1}(SU(n+1); \mathbb{Z}) \to H_0(SU(n); \mathbb{Z}) \cong \mathbb{Z}$ coincides with the edge homomorphism

$$\pi_*: H_{2n+1}(SU(n+1); \mathbb{Z}) \to H_{2n+1}(S^{2n+1}; H_0(SU(n); \mathbb{Z})).$$

Since $\pi_*(\lambda_{n+1}) = \pi_* \circ g_{n,1*}(z_{n+1})$, it follows from 16.48 that $\pi_*(\lambda_{n+1}) = \pm \iota_{2n+1} \in H_{2n+1}(S^{2n+1}; \mathbb{Z})$, and thus $\psi(\lambda_{n+1}) = \pm 1$. Hence for any $x \in H_*(SU(n); \mathbb{Z})$ we have $\psi(\lambda_{n+1} \cdot i_*(x)) = \pm x$, which shows that ψ is surjective. We therefore have short exact sequences

$$0 \longrightarrow H_m(SU(n); \mathbb{Z}) \xrightarrow{i_*} H_m(SU(n+1); \mathbb{Z}) \xrightarrow{\psi} H_{m-2n-1}(SU(n); \mathbb{Z}) \longrightarrow 0.$$

We now show that $\lambda_2, \ldots, \lambda_{n+1}$ form a simple system of generators for $H_*(SU(n+1); \mathbb{Z})$. For any $u \in H_*(SU(n+1); \mathbb{Z})$ let $w = \psi(u)$; then $\psi(\lambda_{n+1} \cdot w) = \psi(\lambda_{n+1}) \cdot w = \pm 1 \cdot w = \pm \psi(u)$. Therefore $u \mp \lambda_{n+1} \cdot w \in \operatorname{im} i_*$; thus w and $u \mp \lambda_{n+1} \cdot w$ are linear combinations of the monomials $\lambda_2^{\varepsilon_2} \lambda_3^{\varepsilon_3} \ldots \lambda_n^{\varepsilon_n}$. That these monomials are linearly independent also follows easily. $\square$

We now consider the elements $y_i' = g'_{n,1*}(y_i) \in H_{2i}(\Omega SU(n+1); \mathbb{Z})$. Since $\sigma'(y_i') = \sigma'(g'_{n,1*}(y_i)) = g_{n,1*}(\sigma(y_i)) = g_{n,1*}(z_{i+1}) = \lambda_{i+1}$ for $i \geqslant 1$, we see that the hypotheses of the Borel theorem 15.60 are satisfied, and we obtain the following theorem.

16.51. Theorem. *$H_*(\Omega SU(n+1); \mathbb{Z}) \cong \mathbb{Z}[y_1', \ldots, y_n']$, where*

$$y_i' = g'_{n,1*}(y_i) \in H_{2i}(\Omega SU(n+1); \mathbb{Z}).$$

As a corollary it follows $g'_*: H_*(BU; \mathbb{Z}) \to H_*(\Omega SU; \mathbb{Z})$ is an isomorphism and hence $g': BU \to \Omega SU$ is a homotopy equivalence.

The homotopy equivalences $\mathbb{Z} \times BO \simeq \Omega^4 BSp$ and $\mathbb{Z} \times BSp \simeq \Omega^4 BO$ are constructed in three steps each:

$$\mathbb{Z} \times BO \simeq \Omega(U/O) \quad U/O \simeq \Omega(Sp/U) \quad Sp/U \simeq \Omega Sp$$

$$\mathbb{Z} \times BSp \simeq \Omega(U/Sp) \quad U/Sp \simeq \Omega(O/U) \quad O/U \simeq \Omega SO.$$

The proof that each of the six maps is a homotopy equivalence goes in principle as above but is more complicated because one must show that each map induces an isomorphism on $\mathbb{Z}_p$ homology for each prime p.

The proof that $\mathbb{Z} \times BU \simeq \Omega^2 BU$ given above is due to Cartan and Moore [29] and Dyer and Lashof [39]. It has the advantage for us that it uses techniques and results already at our disposal. It has the disadvantage that it does not really give the reader any geometric feeling for the reason why $\tilde{K}^0(S^n) \cong \tilde{K}^0(S^{n+2})$. For example, we have not shown that this isomorphism is just multiplication by $\{\gamma - 1\} \in \tilde{K}^0(S^2) = \tilde{K}^0(CP^1)$. The proof of Atiyah does give considerable geometric insight but has the disadvantage for our purposes that it uses methods which do not fit in well here. The reader should nevertheless make the acquaintance of Atiyah's proof—in [16], for example.

16.52. Exercise. i) if $\omega^+ \to BSO(n)$ denotes the universal $SO(n)$-bundle, then we let $p_i = (-1)^i c_{2i}(\omega^+ \otimes \mathbb{C}) \in H^{4i}(BSO(n); R)$ for any ring R. Show that if 2 is invertible in R then

$$H^*(BSO(2r); R) \cong R[p_1, p_2, \ldots, p_{r-1}, e]$$

$$H^*(BSO(2r+1); R) \cong R[p_1, p_2, \ldots, p_r],$$

where e is the Euler class $e(\omega^+)$. (Note that $e^2(\omega^+) = 0$ if n is odd and $e^2(\omega^+) = p_r$ if $n = 2r$.)

ii) $SO(2) \cong U(1) \cong S^1$, so $H_*(BSO(2); R)$ is a free R-module with basis $\{1, y_1, y_2, \ldots\}$, $y_i \in H_{2i}(BSO(2); R)$. Show that if 2 is invertible in R and $Bi: BSO(2) \to BSO$ is the map induced by the inclusion, then $Bi_*(y_{2k+1}) = 0$ for $k \geqslant 0$ and $H_*(BSO; R) \cong R[q_1, q_2, \ldots]$, $q_i = Bi_*(y_{2i}) \in H_{4i}(BSO; R)$.

16.53. Exercise. Compute $H_*(U(n); \mathbb{Z})$.

References

1. J. F. Adams [7]
2. M. F. Atiyah [16]
3. A. Borel and F. Hirzebruch [23]
4. R. Bott [24, 25]
5. P. E. Conner and E. E. Floyd [31]
6. E. Dyer and R. K. Lashof [39]
7. D. Husemoller [49]
8. A. Liulevicius [55]
9. J. W. Milnor [62]
10. N. Steenrod [81]

Chapter 17

Cohomology Operations and Homology Cooperations

In Chapter 13 we constructed the Hurewicz map

$$H\colon [X,Y] \to \mathrm{Hom}_{\tilde{E}_*(S^0)}(\tilde{E}_*(X), E_*(Y))$$

and the Boardman map

$$B\colon [X,Y] \to \mathrm{Hom}_{\tilde{E}^*(S^0)}(\tilde{E}^*(Y), E^*(X))$$

for any ring spectrum E. These maps can be regarded as translating essentially geometric problems (existence or non-existence of maps $f\colon X \to Y$) into algebraic problems (existence or non-existence of $\tilde{E}_*(S^0)$-module homomorphisms with given properties). If one is trying to show the non-existence of a map $f\colon X \to Y$ with certain properties, then one wants to show that no homomorphism $f_*\colon E_*(X) \to E_*(Y)$ with corresponding properties lies in $\mathrm{im}\, H$. Thus it is desirable to obtain good limits on the extent of $\mathrm{im}\, H$.

The following simple observation has proved extremely fruitful in algebraic topology. In any case the functors $E_*(-)$ take values in the category $\mathscr{GM}_E$ of graded $\tilde{E}_*(S^0)$-modules; suppose that in fact $E_*(-)$ always takes values in a smaller category $\mathscr{B} \subset \mathscr{GM}_E$. Then $\mathrm{im}\, H$ lies in $\mathrm{Hom}_{\mathscr{B}}(E_*(X), E_*(Y)) \subset \mathrm{Hom}_{\tilde{E}_*(S^0)}(E_*(X), E_*(Y))$, where $\mathrm{Hom}_{\mathscr{B}}(-,-)$ denotes the morphisms in the category $\mathscr{B}$. The cup product gives a good example: if E is a ring spectrum, then on $\mathscr{PT}'$ the cofunctor $E^*(-)$ takes values in the smaller category $\mathscr{GA}_E \subset \mathscr{GM}_E$ of graded *algebras* over $\tilde{E}^*(S^0)$. In other words, if $f^*\colon E^*(Y) \to E^*(X)$ is to qualify as induced by a map $f\colon X \to Y$ of spaces, then f^* had better be a ring homomorphism. For example, $H^*(S^2 \times S^3; \mathbb{Z})$ and $H^*(S^2 \vee S^3 \vee S^6; \mathbb{Z})$ are isomorphic as graded groups but not as rings, so there can be no homotopy equivalence $S^2 \times S^3 \simeq S^2 \vee S^3 \vee S^6$.

Steenrod discovered an algebra $A(p)^*$ over $\mathbb{Z}_p$ such that $H^*(X;\mathbb{Z}_p)$ is an $A(p)^*$-module for every spectrum X, and for every map $f\colon X \to Y$ the homomorphism $f^*\colon H^*(Y;\mathbb{Z}_p) \to H^*(X;\mathbb{Z}_p)$ is an $A(p)^*$-homomorphism. Thus $\mathrm{im}\, B \subset \mathrm{Hom}_{A(p)^*}(H^*(Y;\mathbb{Z}_p), H^*(X;\mathbb{Z}_p))$. The algebra structure of the *Steenrod algebra* $A(p)^*$ is rather complicated, as we shall see, and this complexity makes it somewhat difficult to work with; but from another point of view this very complexity is a great advantage: it is difficult for a homomorphism $f^*: H^*(Y;\mathbb{Z}_p) \to H^*(X;\mathbb{Z}_p)$ to be $A(p)^*$-linear, so $\mathrm{Hom}_{A(p)^*}(H^*(Y;\mathbb{Z}_p), H^*(X;\mathbb{Z}_p))$ is often very small. Many non-existence proofs became possible after Steenrod's discovery which had not been possible before.

If $\theta \in A(p)^q$ is any element then we can think of θ as a homomorphism

$$\theta: H^n(X;\mathbb{Z}_p) \to H^{n+q}(X;\mathbb{Z}_p), \quad \text{all } X, n,$$

by letting $\theta(x) = \theta \cdot x$ for $x \in H^n(X;\mathbb{Z}_p)$. When thought of in this way θ is called a *cohomology operation of degree q*. The Steenrod operations have the additional property that they commute with suspension:

$$\theta \circ \sigma = \sigma \circ \theta: H^{n+1}(\Sigma X;\mathbb{Z}_p) \to H^{n+q}(X;\mathbb{Z}_p).$$

17.1. Definition. For any cohomology theory k^* a *cohomology operation of type* (p,q) is a natural transformation $\theta: k^p(-) \to k^q(-)$ between set-valued cofunctors. θ is *additive* if θ is a natural transformation of group-valued cofunctors. A *stable* cohomology operation of degree q is a sequence of cohomology operations $\theta^n: k^n(-) \to k^{n+q}(-)$, $n \in \mathbb{Z}$, such that $\theta^n \circ \sigma = \sigma \circ \theta^{n+1}$, all $n \in \mathbb{Z}$. One usually denotes such a sequence $\{\theta^n\}$ simply by a single letter θ.

17.2. Let $A(E)^q$ denote the set of all stable cohomology operations of degree q for the cohomology theory $E^*(-)$, E a spectrum. We can define an addition $+$ on $A(E)^q$ by $(\theta + \phi)(x) = \theta(x) + \phi(x)$ for all $x \in E^*(X)$, all X. Then $A(E)^q$ becomes an abelian group. Composition of operations defines a pairing

$$A(E)^q \times A(E)^r \xrightarrow{\circ} A(E)^{q+r}.$$

17.3. Lemma. *If $\theta = \{\theta^n\}$ is a stable cohomology operation, then each θ^n is additive*

Proof: We have a commutative diagram

$$\begin{array}{ccccccc}
E^{n+1}(\Sigma X) \times E^{n+1}(\Sigma X) & \xleftarrow[\cong]{\{i_1^*, i_2^*\}} & E^{n+1}(\Sigma X \vee \Sigma X) & \xrightarrow{\mu'^*} & E^{n+1}(\Sigma X) & \xrightarrow{\sigma} & E^n(X) \\
\downarrow{\theta \times \theta} & & \downarrow{\theta} & & \downarrow{\theta} & & \downarrow{\theta} \\
E^{n+q+1}(\Sigma X) \times E^{n+q+1}(\Sigma X) & \xleftarrow[\cong]{\{i_1^*, i_2^*\}} & E^{n+q+1}(\Sigma X \vee \Sigma X) & \xrightarrow{\mu'^*} & E^{n+q+1}(\Sigma X) & \xrightarrow{\sigma} & E^{n+q}(X),
\end{array}$$

where $\mu': \Sigma X \to \Sigma X \vee \Sigma X$ is the coproduct on ΣX. Given $x, y \in E^n(X)$, let $x = \sigma(x')$, $y = \sigma(y')$. By the cohomology version of 7.37

$$\mu'^* \circ \{i_1^*, i_2^*\}^{-1}(x', y') = x' + y',$$

so we get

$$\begin{aligned}\theta(x+y) &= \theta(\sigma x' + \sigma y') = \theta \circ \sigma \circ \mu'^* \circ \{i_1^*, i_2^*\}^{-1}(x', y') \\ &= \sigma \circ \mu'^* \circ \{i_1^*, i_2^*\}^{-1} \circ (\theta \times \theta)(x', y') \\ &= \sigma(\theta x' + \theta y') = \theta\sigma x' + \theta\sigma y' = \theta x + \theta y. \quad \square\end{aligned}$$

17.4. Corollary. *Composition is bilinear and hence defines a homomorphism*

$$A(E)^p \otimes A(E)^q \to A(E)^{p+q}.$$

Proof: $[\theta \circ (\phi + \psi)](x) = \theta(\phi(x) + \psi(x)) = \theta\phi(x) + \theta\psi(x)$ by 17.3

$$[(\theta + \phi) \circ \psi](x) = \theta\psi x + \phi\psi(x). \quad \square$$

Since composition is clearly associative, it follows that composition makes $A(E)^* = \bigoplus_q A(E)^q$ into a graded ring. In fact it is a graded algebra over $\tilde{E}^*(S^0)$, for if $r \in \tilde{E}^*(S^0)$ and $\theta \in A(E)^*$, then we can let $r\theta$ be the stable operation such that $(\theta r)(x) = (\theta(x))r$ for all x.

17.5. Theorem. *There is an isomorphism $A(E)^* \to E^*(E)$ given by $\theta \mapsto \theta([1_E])$, $[1_E] \in E^0(E)$. The product in $E^*(E)$ is given by composition of maps: $[f][g] = [\Sigma^m f \circ g]$ if $f: E \to \Sigma^n E$, $g: E \to \Sigma^m E$.*

Proof: Let α denote the map defined by $\alpha(\theta) = \theta[1_E]$. Define $\beta: E^*(E) \to A(E)^*$ by $(\beta[f])[g] = [\Sigma^m f \circ g]$ for any $f: E \to \Sigma^n E$, $g: X \to \Sigma^m E$. Then $\alpha \circ \beta[f] = (\beta[f])[1_E] = [f \circ 1_E] = [f]$ and $(\beta \circ \alpha(\theta))[g] = [\Sigma^m \theta[1_E] \circ g] = g^* \theta[1_E] = \theta g^*[1_E] = \theta[g]$, so $\beta \circ \alpha(\theta) = \theta$. The last statement is evident. $\square$

17.6. Examples. i) The Steenrod algebra $A(p)^*$ is defined to be $A(H(\mathbb{Z}_p))^*$. Thus $A(p)^* \cong H^*(H(\mathbb{Z}_p); \mathbb{Z}_p)$ for each prime p. We shall compute $A(2)^*$ later by using the Serre spectral sequence to work out $\tilde{H}^*(H(\mathbb{Z}_2, q); \mathbb{Z}_2)$ for all $q \geqslant 1$.

ii) In the algebra $A(K)^* \cong K^*(K)$ we can easily think of the two operations $1 = \Psi^1$ and Ψ^{-1}, the operation carrying each virtual bundle to its complex conjugate. These two generate a subgroup $\mathbb{Z} \oplus \mathbb{Z}$. The author does not know whether there is more to $K^*(K)$; if not $A(K)^*$ will not be very useful.

We have just seen that for every X we have that $E^*(X)$ is a module over the algebra $A(E)^* = E^*(E)$. One might hope there would be a dual notion for homology, and indeed in 1967 J. F. Adams showed there is under two assumptions on E. Not too surprisingly, the object dual to $E^*(E)$ is $E_*(E)$. We are going to provide $E_*(E)$ with structure maps which make it into a Hopf algebra (the product is the Pontrjagin product coming from the product $E \wedge E \to E$ of the ring spectrum E). Then for every X $E_*(X)$ will be a comodule over $E_*(E)$.

17.7. Definition. If we describe modules in terms of commutative diagrams, then the definition looks as follows: let R be a ring, A an algebra over R with unit η. An *A-module* is an R-module M with an A-action—i.e. an R-homomorphism $\phi_M: A \otimes M \to M$ such that the diagrams

$$\begin{array}{ccc} R \otimes M & \xrightarrow{\eta \otimes 1} & A \otimes M \\ & \searrow^{\cong} & \downarrow{\scriptstyle \phi_M} \\ & & M \end{array} \qquad \begin{array}{ccc} A \otimes A \otimes M & \xrightarrow{\phi_A \otimes 1} & A \otimes M \\ \downarrow{\scriptstyle 1 \otimes \phi_M} & & \downarrow{\scriptstyle \phi_M} \\ A \otimes M & \xrightarrow[\phi_M]{} & M \end{array}$$

commute, where ϕ_A is the product of A. (Note that A is an A-module with A-action ϕ_A.)

If we now dualize by reversing arrows, we obtain the notion of a comodule; let C be a coalgebra over R with augmentation $\varepsilon: C \to R$. A *C-comodule* is an R-module L with a C-coaction—i.e. an R-homomorphism $\psi_L: L \to C \otimes L$ such that the diagrams

$$\begin{array}{ccc} R \otimes L & \xleftarrow{\varepsilon \otimes 1} & C \otimes L \\ & \nwarrow^{\cong} & \uparrow{\scriptstyle \psi_L} \\ & & L \end{array} \qquad \begin{array}{ccc} C \otimes C \otimes L & \xleftarrow{\psi_C \otimes 1} & C \otimes L \\ \uparrow{\scriptstyle 1 \otimes \psi_L} & & \uparrow{\scriptstyle \psi_L} \\ C \otimes L & \xleftarrow[\psi_L]{} & L \end{array}$$

commute, where ψ_C is the coproduct of C. (Note that ψ_C is a C-coaction for C.) There is a corresponding graded version of these definitions.

In what follows the ground ring R is $\tilde{E}_*(S^0)$ and C is $E_*(E)$. There is one subtle point to be observed, however: $E_*(E)$ is of course a left $\tilde{E}_*(S^0)$-module by the usual action:

$$\pi_*(E) \otimes \pi_*(E \wedge E) \longrightarrow \pi_*(E \wedge E \wedge E) \xrightarrow{(\mu \wedge 1)_*} \pi_*(E \wedge E).$$

But there is also a natural right action of $\tilde{E}_*(S^0)$ on $E_*(E)$:

$$\pi_*(E \wedge E) \otimes \pi_*(E) \longrightarrow \pi_*(E \wedge E \wedge E) \xrightarrow{(1 \wedge \mu)_*} \pi_*(E \wedge E).$$

Thus $E_*(E)$ is also a right $\tilde{E}_*(S^0)$-module, and *these two module structures are in general different*. Thus $E_*(E)$ is what we call a *bimodule* over $\tilde{E}_*(S^0)$. The notions of coalgebra, Hopf algebra, etc., look formally the same but we must be careful to distinguish the two module structures.

17.8. Theorem. *Suppose E is a commutative ring spectrum such that $E_*(E) = C$ is flat as a right module over $\tilde{E}_*(S^0) = R$. Then there are homomorphisms*

$$\phi: C \otimes C \to C \qquad \varepsilon: C \to R$$
$$\eta_L: R \to C \qquad c: C \to C$$
$$\eta_R: R \to C \qquad \psi_E: C \to C \otimes_R C$$

and $\psi_X: E_*(X) \to C \otimes_R E_*(X)$ *for all spectra* X, *with the following properties:*

i) *C is a Hopf algebra with commutative product ϕ having left and right units η_L, η_R and associative coproduct ψ_E having augmentation ε;*

ii) *if $\lambda \in R$ and $x \in C$, then*
$$\lambda x = \phi(\eta_L(\lambda) \otimes x), \qquad x\lambda = \phi(x \otimes \eta_R(\lambda));$$

iii) $\varepsilon\eta_L = 1 = \varepsilon\eta_R$, $c\eta_L = \eta_R$, $c\eta_R = \eta_L$, $\varepsilon c = \varepsilon$, $c^2 = 1$;

iv) $\psi_E(1) = 1 \otimes 1$ *and hence* $\psi_E\eta_L(\lambda) = \eta_L(\lambda) \otimes 1$ *and* $\psi_E\eta_R(\lambda) = 1 \otimes \eta_R(\lambda)$ *for all* $\lambda \in R$;

v) *ψ_X is natural with respect to maps of X;*

vi) *ψ_X is a coaction map;*

vii) *if* $\psi_X(x) = \sum_i e' \otimes x_i$ *and* $\psi_Y(y) = \sum_j e'' \otimes y_j$ *for* $x, x_i \in E_*(X)$, $y, y_j \in E_*(Y)$, $e'_j, e''_j \in C$, *then*
$$\psi_{X \wedge Y}(x \wedge y) = \textstyle\sum_{i,j} (-1)^{|x_i||e''_j|}\, e'_i e''_j \otimes (x_i \wedge y_j);$$

viii) *the diagram*
$$\begin{array}{ccc} C & \xrightarrow{c} & C \\ \downarrow{\scriptstyle \psi_E} & & \downarrow{\scriptstyle \psi_E} \\ C \otimes_R C & \xrightarrow{(c \otimes c) \circ T} & C \otimes_R C \end{array}$$
commutes;

ix) *the diagrams*
$$\begin{array}{ccc} C & \xrightarrow{\varepsilon} & R \\ \downarrow{\scriptstyle \psi_E} & & \downarrow{\scriptstyle \eta_L} \\ C \otimes_R C & \xrightarrow[\phi(1 \otimes c)]{} & C \end{array} \qquad \begin{array}{ccc} C & \xrightarrow{\varepsilon} & R \\ \downarrow{\scriptstyle \psi_E} & & \downarrow{\scriptstyle \eta_R} \\ C \otimes_R C & \xrightarrow[\phi(c \otimes 1)]{} & C \end{array}$$
commute;

x) $\psi_{S^0}: \tilde{E}_*(S^0) \to C \otimes_R \tilde{E}_*(S^0)$ *is just* η_L.

Proof: As we have already indicated, ϕ is to be the Pontrjagin product on $E_*(E)$. Let $\iota: S^0 \to E$ be the unit of E and μ its product; we define η_L, η_R, ε and c as follows:

$$\eta_L: \pi_*(E) \cong \pi_*(E \wedge S^0) \xrightarrow{(1 \wedge \iota)_*} \pi_*(E \wedge E)$$

$$\eta_R : \pi_*(E) \cong \pi_*(S^0 \wedge E) \xrightarrow{(\iota \wedge 1)_*} \pi_*(E \wedge E)$$

$$\varepsilon : \pi_*(E \wedge E) \xrightarrow{\mu_*} \pi_*(E)$$

$$c : \pi_*(E \wedge E) \xrightarrow{\tau_*} \pi_*(E \wedge E).$$

For any spectra X, Y we can define

$$m : \pi_*(X \wedge E) \otimes_{\pi_*(E)} \pi_*(E \wedge Y) \longrightarrow \pi_*(X \wedge E \wedge E \wedge Y) \xrightarrow{(1 \wedge \mu \wedge 1)_*} \pi_*(X \wedge E \wedge Y),$$

and if $\pi_*(X \wedge E)$ is flat as a right $\pi_*(E)$-module, then m is an isomorphism just as in 13.75. Then under the assumption that C is a flat right R-module we can define ψ_X to be

$$\pi_*(E \wedge X) \xrightarrow{\cong} \pi_*(E \wedge S^0 \wedge X) \xrightarrow{(1 \wedge \iota \wedge 1)_*} \pi_*(E \wedge E \wedge X) \xleftarrow{\cong} C \otimes_R \pi_*(E \wedge X).$$

We obtain ψ_E by simply taking $X = E$. The proofs of i)–x) then involve writing down numerous diagrams and checking that each commutes. For example, one of the diagrams involved in proving vi) looks as shown on facing page. The isomorphism $\pi_*(E \wedge X) \cong \pi_*(E \wedge S^0 \wedge X)$ is always understood. Square ① obviously commutes; ② and ③ commute by naturality of m and ④ commutes because of an easily verified associativity property of m. Thus $(1 \otimes \psi_X) \circ \psi_X = (\psi_E \otimes 1) \circ \psi_X$. Taking $X = E$ gives the associativity of ψ_E claimed in i). □

Note. c is an automorphism of C which takes the right action of R into the left action and conversely. Hence C is flat as a right R-module if and only if it is flat as a left R-module.

17.9. Examples. i) Let us take $E = H(\mathbb{Z}_p)$. Since $\mathbb{Z}_p = R$ is a field and $H(\mathbb{Z}_p)_*(H(\mathbb{Z}_p))$ turns out to be of finite type (i.e. $H_q(H(\mathbb{Z}_p); \mathbb{Z}_p)$ is finitely generated for each q), it follows from the universal coefficient theorem that $A(p)_* = H(\mathbb{Z}_p)_*(H(\mathbb{Z}_p))$ is the dual (in the sense of $\mathbb{Z}_p$-vector spaces) of the Steenrod algebra $A(p)^*$. Milnor studied this dual $A(p)_*$ and found that it was a very nice algebra; $A(p)_*$ is the tensor product of an exterior algebra and a polynomial algebra. Naturally the coproduct (which is dual to the product in $A(p)^*$) is somewhat hard to handle. We shall derive the structures of $A(2)^*$ and $A(2)_*$ in Chapter 18.

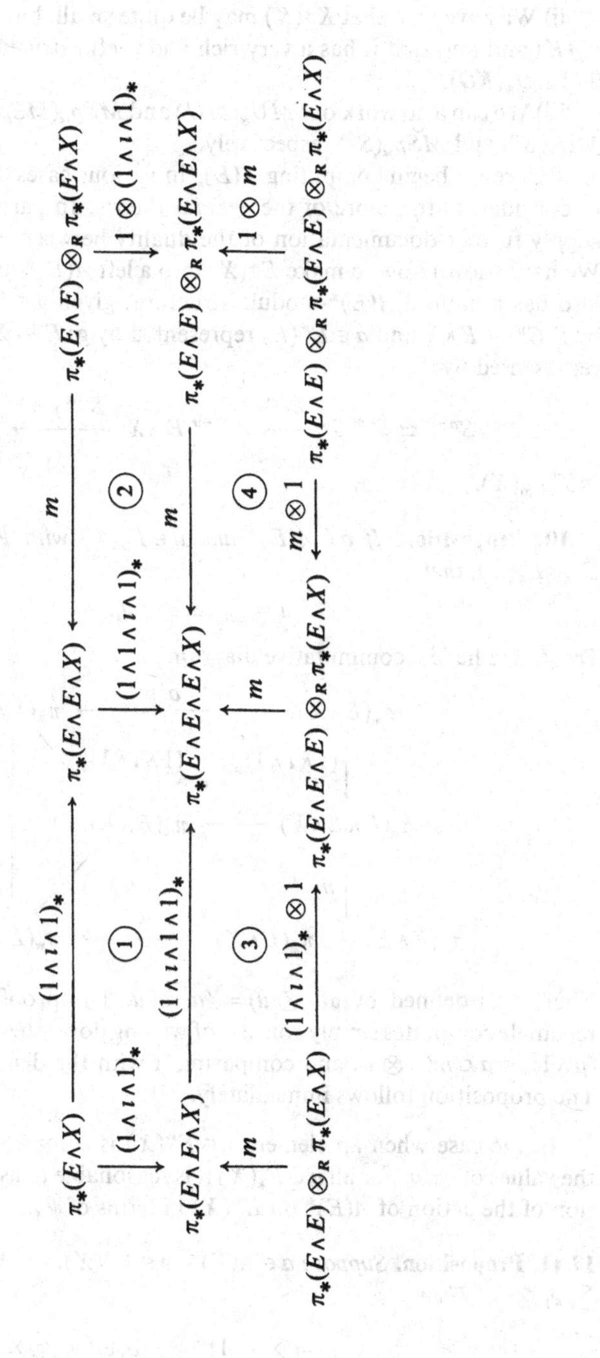

π*(E∧X)
(1∧ι∧1)*
π*(E∧E∧X)
m
π*(E∧E) ⊗R π*(E∧X)
(1∧ι∧1)*
1
(1∧1∧ι∧1)*
2
1 ⊗ (1∧ι∧1)*
π*(E∧E∧X)
(1∧ι∧1∧1)*
π*(E∧E∧E∧X)
m
π*(E∧E) ⊗R π*(E∧E∧X)
m
m
1 ⊗ m
π*(E∧E) ⊗R π*(E∧X)
(1∧ι∧1)* ⊗ 1
3
π*(E∧E∧E) ⊗R π*(E∧X)
m ⊗ 1
4
π*(E∧E) ⊗R π*(E∧E) ⊗R π*(E∧X)

ii) We have seen that $K^*(K)$ may be quite small, but we shall compute $K_*(K)$ and find that it has a very rich and useful structure. We shall also find $KO_*(KO)$.

iii) We can also work out $MU_*(MU)$ and $MSp_*(MSp)$ as modules over $\widetilde{MU}_*(S^0)$ and $\widetilde{MSp}_*(S^0)$ respectively.

Before we begin computing $A(E)_*$ in various cases, however, we wish to consider a little more of the general theory. In particular we wish to supply further documentation of the duality between $A(E)^*$ and $A(E)_*$. We have shown how to make $E^*(X)$ into a left $A(E)^*$-module, but $E_*(X)$ also has a natural $A(E)^*$-module structure: given $u \in E_n(X)$ represented by $f: S^n \to E \wedge X$ and $a \in E^m(E)$ represented by $g: E \to \Sigma^m E$ we let $a \cdot u$ be represented by

$$S^{n-m} \simeq \Sigma^{-m} S^n \xrightarrow{\Sigma^{-m} f} \Sigma^{-m} E \wedge X \xrightarrow{\Sigma^{-m} g \wedge 1} E \wedge X$$

in $E_{n-m}(X)$.

17.10. Proposition. *If $a \in A(E)^*$ and $u \in E_*(X)$ with $\psi u = \sum_i e_i \otimes u_i$ in $C \otimes E_*(X)$, then*

$$a \cdot u = \textstyle\sum_i \langle a, ce_i \rangle u_i.$$

Proof: We have a commutative diagram

$$\begin{array}{ccc}
\pi_*(E \wedge X) & \xrightarrow{\quad a \quad} & \pi_*(E \wedge X) \\
\downarrow (1 \wedge \iota \wedge 1)_* & \swarrow (1 \wedge \iota \wedge 1)_* & \downarrow 1 \\
\pi_*(E \wedge E \wedge X) & \xrightarrow{\quad a \quad} & \pi_*(E \wedge E \wedge X) \\
\uparrow m & \searrow (\mu \wedge 1)_* & \\
\pi_*(E \wedge E) \otimes_R \pi_*(E \wedge X) & \xrightarrow{\quad \alpha \quad} & \pi_*(E \wedge X),
\end{array}$$

where α is defined by $\alpha(e \otimes u) = \langle a, ce \rangle u$. The proof that the bottom rectangle commutes simply consists of writing down the map which defines $(\mu \wedge 1)_* \circ a \circ m(e \otimes u)$ and comparing it with the definition of $\langle a, ce \rangle u$. The proposition follows immediately. □

In the case when an element $z \in E^*(X)$ is completely determined by the values of $\langle z, u \rangle$ for all $u \in E_*(X)$ it is reasonable to ask for a determination of the action of $A(E)^*$ on $E^*(X)$ in terms of ψ_X.

17.11. Proposition. *Suppose $a \in A(E)^*$, $y \in E^*(X)$, $u \in E_*(X)$ and $\psi_X(u) = \sum_i e_i \otimes u_i$. Then*

$$\langle ay, u \rangle = \textstyle\sum_i (-1)^{|y||e_i|} \langle a, e_i \langle y, u_i \rangle \rangle.$$

The expression on the right makes sense because $e_l \in E_*(E)$ and $\langle y, u_l \rangle \in \tilde{E}_*(S^0)$, which acts on the right of $E_*(E)$.

Proof: Define $y_*: \pi_*(F \wedge X) \to \pi_*(F \wedge E)$ for any spectrum F as follows: suppose y is represented by $g: X \to \Sigma^p E$ and $f: S^r \to F \wedge X$ represents an element of $\pi_*(F \wedge X)$. Let $y_*([f])$ be the class of

$$S^{r-p} \simeq \Sigma^{-p} S^r \xrightarrow{\Sigma^{-p} f} \Sigma^{-p} F \wedge X \simeq F \wedge \Sigma^{-p} X \xrightarrow{1 \wedge \Sigma^{-p} g} F \wedge E.$$

Then one easily checks that $\langle ay, u \rangle = \langle a, y_* u \rangle$ for $a \in E^*(E)$, $u \in E_*(X)$. Now we set up another commutative diagram

$$\begin{array}{ccc}
\pi_*(E \wedge X) & \xrightarrow{\quad y_* \quad} & \pi_*(E \wedge E) \\
\Big\downarrow (1 \wedge \iota \wedge 1)_* & \swarrow (1 \wedge \iota \wedge 1)_* & \Big| \\
\pi_*(E \wedge E \wedge X) & \xrightarrow{\quad y_* \quad} \pi_*(E \wedge E \wedge E) & \Big| \, 1 \\
\Big\uparrow m & \searrow (1 \wedge \mu)_* & \Big\downarrow \\
\pi_*(E \wedge E) \otimes_R \pi_*(E \wedge X) & \xrightarrow{\quad \beta \quad} & \pi_*(E \wedge E),
\end{array}$$

where β is defined by $\beta(e \otimes u) = (-1)^{|y||e|} e\langle y, u \rangle$ for $e \in E_*(E)$, $u \in E_*(X)$. The proposition follows. $\square$

Finally, before we begin specific calculations, let us make explicit what these structures do for the Hurewicz homomorphism. We have

$$h: [X, Y] \to \mathrm{Hom}_{E_*(E)}(E_*(X), E_*(Y)),$$

where $\mathrm{Hom}_C(M, L)$ means homomorphisms $\phi: M \to L$ as comodules over the coalgebra C. In particular for $X = E(S^n)$ we have

$$h: \pi_n(Y) = [S^n, Y] \to \mathrm{Hom}_{E_*(E)}(\tilde{E}_*(S^n), E_*(Y)).$$

Now a homomorphism $\phi: M \to L$ of graded groups, modules, etc. such that $\phi(M_k) \subset L_{k+d}$ for all $k \in \mathbb{Z}$ is said to be of *degree d*. When we write $\mathrm{Hom}_A(M, L)$, we mean homomorphisms of degree 0. Let us write $\mathrm{Hom}_A^d(M, L)$ for the group of homomorphisms of degree d. Then composing with the iterated suspension $\sigma^n: \tilde{E}_r(S^0) \to \tilde{E}_{r+n}(S^n)$, we can regard h as a homomorphism

$$h: \pi_n(Y) \to \mathrm{Hom}^n_{E_*(E)}(\tilde{E}_*(S^0), E_*(Y)).$$

17.12. Definition. Let C be a Hopf algebra over a ring R with a 1 and let M be a comodule over C with coaction $\phi_M: M \to C \otimes_R M$. Then an element $m \in M$ is called *primitive* if $\phi_M(m) = 1 \otimes m$. The set $P(M)$ of all primitive elements is a subgroup of M (but not an R-submodule if C is a bimodule with distinct left and right R-actions).

17.13. Lemma. *There is an isomorphism* $\kappa: \mathrm{Hom}_C^n(R, M) \to P_n(M)$ *given by* $\kappa(\theta) = \theta(1)$, $\theta \in \mathrm{Hom}_C^n(R, M)$.

Proof: Here R is given the structure of a C-comodule by the map $\eta_L: R \to C \cong C \otimes_R R$, sending r to $r \cdot 1$ $(= r \cdot 1 \otimes 1)$. To say that $\theta: R \to M$ is a C-homomorphism means

$$\begin{array}{ccc} R & \xrightarrow{\eta_L} & C \otimes_R R \\ \Big\downarrow{\scriptstyle \theta} & & \Big\downarrow{\scriptstyle 1 \otimes \theta} \\ M & \xrightarrow{\phi_M} & C \otimes_R M \end{array}$$

commutes. Thus $\phi_M(\kappa(\theta)) = \phi_M(\theta(1)) = (1 \otimes \theta)(\eta_L(1)) = 1 \otimes \theta(1) = 1 \otimes \kappa(\theta)$. In other words $\kappa(\theta) \in P_n(M)$. Clearly κ is a homomorphism.

Suppose $p \in P_n(M)$; let $\theta_p: R \to M$ be defined by $\theta_p(r) = rp$, $r \in R$. Then for all $r \in R$ we have

$$\begin{aligned} \phi_M \circ \theta_p(r) &= \phi_M(rp) = r\phi_M(p) \quad (\phi_M \text{ is an } R\text{-module map}) \\ &= r(1 \otimes p) = r \cdot 1 \otimes p = r \cdot 1 \otimes \theta_p(1) = (1 \otimes \theta_p)(r \cdot 1 \otimes 1) \\ &= (1 \otimes \theta_p)(\eta_L(r)). \end{aligned}$$

Hence $\theta_p \in \mathrm{Hom}_C^n(R, M)$ and clearly $\kappa(\theta_p) = p$. On the other hand if $\theta, \theta' \in \mathrm{Hom}_C^n(R, M)$ are such that $\kappa(\theta) = \kappa(\theta')$, then $\theta(1) = \theta'(1)$, so for all $r \in R$ we have $\theta(r) = \theta(r \cdot 1) = r\theta(1) = r\theta'(1) = \theta'(r \cdot 1) = \theta'(r)$—i.e. $\theta = \theta'$. Hence κ is an isomorphism. ▯

17.14. Corollary. *For any ring spectrum E such that* $E_*(E)$ *is flat over* $\tilde{E}_*(S^0)$ *the Hurewicz homomorphism* $h: \pi_*(X) \to E_*(X)$ *has* $\mathrm{im}\, h \subset P(E_*(X)) =$ *the group of primitive elements of* $E_*(X)$ *under the coaction of* $E_*(E)$.

In many important cases one can determine $PE_*(X)$ without too much difficulty and hence obtain a bound on $\mathrm{im}\, h$.

17.15. Now the easiest of the Hopf algebras $E_*(E)$ to compute at this stage is $MU_*(MU)$.

17.16. Theorem. $MU_*(MU) \cong \tilde{MU}_*(S^0)[\beta_1', \beta_2', \ldots]$, *where* $\Phi_*(\beta_i') = \beta_i \in MU_*(BU)$ *and the standard map* $f: CP^\infty \simeq MU_1 \to \Sigma^2 MU$ *sends* β_i *to* β_{i-1}', $i \geqslant 1$. *If* $\beta' = 1 + \beta_1' + \beta_2' + \cdots$, *then the coproduct is given by*

$$\psi_{MU}(\beta_k') = \textstyle\sum_{i+j=k} (\beta')_{2i}^{j+1} \otimes \beta_j', \quad k \geqslant 0,$$

and the antiautomorphism c is given by $c(\beta_k') = m_k$, *where the inverse of the formal power series* $u = \sum_{k \geqslant 0} \beta'_k x^{k+1}$ is $x = \sum_{j \geqslant 0} m_j u^{j+1}$. *The augmentation* ε *is given by*

$$\varepsilon(1) = 1, \quad \varepsilon(\beta'_i) = 0, \quad i \geqslant 1.$$

η_L *is given by* $\eta_L(a) = a \cdot 1$ *for all* $a \in \widetilde{MU}_*(S^0)$.

Proof: The first two statements are just 16.39. The maps

$$\eta_R : MU \simeq S^0 \wedge MU \xrightarrow{\iota \wedge 1} MU \wedge MU$$

and

$$\eta_L : MU \simeq MU \wedge S^0 \xrightarrow{1 \wedge \iota} MU \wedge MU$$

induce natural transformations

$$\eta_R, \eta_L : MU_*(-) \to (MU \wedge MU)_*(-) \quad \text{and}$$

$$\bar{\eta}_R, \bar{\eta}_L : MU^*(-) \to (MU \wedge MU)^*(-).$$

In fact $\bar{\eta}_R$ is just the Boardman map, so we have a commutative diagram

$$\begin{array}{ccc} MU^*(X) & \xrightarrow{\bar{\eta}_R} & (MU \wedge MU)^*(X) \\ & \alpha \searrow \quad \swarrow p & \\ & \mathrm{Hom}_{\widetilde{MU}_*(S^0)}(MU_*(X), MU_*(MU)) & \end{array}$$

as in 13.77ff. Let $X = CP^\infty$ and let

$$y^R = \bar{\eta}_R(y), \quad y^L = \bar{\eta}_L(y) \in (MU \wedge MU)^*(CP^\infty),$$

$$\beta_i^R = \eta_R(\beta_i), \quad \beta_i^L = \eta_L(\beta_i) \in (MU \wedge MU)_*(CP^\infty).$$

The statement $f_*(\beta_i) = \beta'_{i-1}$, $i \geqslant 1$, can then be interpreted as follows:

17.17. $$y^R = \bar{\eta}_R[f] = \textstyle\sum_{i \geqslant 0} \beta'_i (y^L)^{i+1}.$$

Taking kth powers gives $(y^R)^k = \sum_{i \geqslant k} (\beta')^k_{2(i-k)} \cdot (y^L)^i$.

Now ψ_{CP^∞} is just

$$\widetilde{MU}_*(CP^\infty) \xrightarrow{\eta_L}$$

$$(\widetilde{MU \wedge MU})_*(CP^\infty) \cong MU_*(MU) \otimes_{\widetilde{MU}_*(S^0)} \widetilde{MU}_*(CP^\infty).$$

Suppose $\psi_{CP^\infty}(\beta_i) = \sum_{j=0}^{i} B^i_{i-j} \otimes \beta_j$, $i \geqslant 1$, for some coefficients

$$B^i_{i-j} \in MU_{2(i-j)}(MU).$$

This is the same as saying

$$\beta_i^L = \textstyle\sum_{j=0}^{i} B^i_{i-j} \beta_j^R.$$

Taking Kronecker product we find

$$B^i_{i-k} = \langle (y^R)^k, \textstyle\sum_{j=0}^{i} B^i_{i-j}\beta^R_j\rangle = \langle \sum_{i\geqslant k} (\beta')^k_{2(i-k)}(y^L)^i, \beta^L_i\rangle = (\beta')^k_{2(i-k)}.$$

Thus $\psi_{CP^\infty}(\beta_i) = \sum_{j=0}^{i} (\beta')^j_{2(i-j)} \otimes \beta_j$. Applying $1 \otimes f_*$ and using the naturality of the coaction ψ_X, we find that

$$\psi_{MU}(\beta'_{i-1}) = \textstyle\sum_{j=1}^{i} (\beta')^j_{2(i-j)} \otimes \beta'_{j-1}, \quad i \geqslant 1,$$

which is equivalent to the formula for ψ_{MU} in the statement of the theorem.

Since $\tau \circ \eta_R = \eta_L$, we see that applying c to the formula 17.17 yields

$$y^L = \textstyle\sum_{i\geqslant 0} c\beta'_i(y^R)^{i+1};$$

but by definition of the m_i's, we have $y^L = \sum_{i\geqslant 0} m_i(y^R)^{i+1}$. Therefore $c(\beta'_i) = m_i$ for $i \geqslant 0$.

For any $x \in MU_*(MU)$ it is easy to see that $\varepsilon(x) = \langle 1, x\rangle$. Now

$$\langle 1, \beta'_i\rangle = \langle 1, f_*(\beta_{i+1})\rangle = \langle f^*(1), \beta_{i+1}\rangle = \langle y, \beta_{i+1}\rangle = 0, \quad i \geqslant 1.$$

Thus $\varepsilon(\beta'_i) = 0$, $i \geqslant 1$. The statement about η_L is trivial. □

Remark 1. The formula for η_R is quite complicated; the interested reader is referred to [7].

Remark 2. Since the definition of the elements m_i is purely formal, their determination should be an algebraic manipulation. It is left as an exercise for the reader to show (by playing about with formal power series) that

$$m_i = \frac{1}{i+1}(\beta')^{-i-1}_{2i}.$$

For example

$$m_1 = -\beta'_1,$$
$$m_2 = 2\beta'^2_1 - \beta'_2,$$
$$m_3 = -5\beta'^3_1 + 5\beta'_1\beta'_2 - \beta'_3.$$

The algebra of cohomology operations $A(MU)^* = MU^*(MU)$ was determined by Novikov [70] and by Landweber [51]. For a clear account of these results see also [5].

There are corresponding results for MSp.

17.18. Theorem. *$MSp_*(MSp) \cong \widetilde{MSp}_*(S^0)[\sigma'_1, \sigma'_2, \ldots]$, where $\Phi_*(\sigma'_i) = \sigma_i$. The standard map g defining z satisfies $g_*(\sigma_i) = \sigma'_{i-1}$, $i \geqslant 1$. If $\sigma' = 1 + \sigma'_1 + \sigma'_2 + \cdots$, then*

$$\psi_{MSp}(\sigma'_k) = \textstyle\sum_{i+j=k} (\sigma')^{j+1}_{4i} \otimes \sigma'_j, \quad k \geqslant 0,$$

and the antiautomorphism c is given by $c(\sigma'_k) = n_k$, where n_k is the same *polynomial expression in the σ_k''s that m_k is in the β_k''s. The augmentation ε is given by $\varepsilon(1)=1$, $\varepsilon(\sigma_k')=0$, $k\geqslant 1$. η_L is given by $\eta_L(a)=a\cdot 1$ for all $a\in \widetilde{MSp}_*(S^0)$.*

Remark. There is a corresponding result for MO, but as it turns out that $MO\simeq \bigvee_\alpha \Sigma^{n_\alpha}H(\mathbb{Z}_2)$, cooperations for MO do not tell us anything we could not learn from $H(\mathbb{Z}_2)$.

Next we turn to the calculation of $K_*(K)$ and $KO_*(KO)$. By 13.49 we know that $K_q(K)\simeq \operatorname{dir\,lim}_n \tilde{K}_{q+2n}(BU)$, where the morphisms of the direct system are

$$\tilde{K}_{q+2n}(BU) \xrightarrow{\sigma^2} \tilde{K}_{q+2n+2}(S^2\wedge BU) \xrightarrow{B_*} \tilde{K}_{q+2n+2}(BU).$$

Similarly $KO_q(KO)\simeq \operatorname{dir\,lim}_n \widetilde{KO}_{q+8n+4}(BSp)$, where the morphisms of the direct system are

$$\widetilde{KO}_{q+8n+4}(BSp) \xrightarrow{\sigma^8} \widetilde{KO}_{q+8n+12}(S^8\wedge BSp) \xrightarrow{B_*} \widetilde{KO}_{q+8n+12}(BSp).$$

In fact we are going to describe $K_*(K)$ by giving its image in $K_*(K)\otimes\mathbb{Q}$, the latter being easy to describe.

17.19. Lemma. *For any two spectra E,F the map*

$$\wedge\otimes 1:\pi_*(E)\otimes\pi_*(F)\otimes\mathbb{Q}\to\pi_*(E\wedge F)\otimes\mathbb{Q}$$

is an isomorphism.

Proof: Think of E as fixed and F as variable. Tensoring with $\mathbb{Q}$ and $\pi_*(E)\otimes\mathbb{Q}$ preserves exactness, so $\pi_*(E)\otimes\pi_*(-)\otimes\mathbb{Q}$ and $\pi_*(E\wedge -)\otimes\mathbb{Q}$ are homology theories and $\wedge\otimes 1$ is a natural transformation which is an isomorphism for $F=S^0$ because $\pi_q(S^0)$ is a torsion group for all $q>0$ by a theorem of Serre (cf. [80], p. 515 e.g.) and $\pi_0(S^0)=\mathbb{Z}$. Therefore $\wedge\otimes 1$ is a natural equivalence. □

Now recall from Chapter 13 that $\pi_*(K)\simeq\mathbb{Z}[t,t^{-1}]$ and the complexification map $c:KO\to K$ maps $\bigoplus_n \pi_{4n}(KO)$ into $\mathbb{Z}[t^2,t^{-2}]$ in $\pi_*(K)$.

17.20. Corollary. *$K_*(K)\otimes\mathbb{Q}\simeq\mathbb{Q}[u,v,u^{-1},v^{-1}]$ and*

$$(c\wedge c)_*\otimes 1:\ KO_*(KO)\otimes\mathbb{Q}\to K_*(K)\otimes\mathbb{Q}$$

has image $\mathbb{Q}[u^2,v^2,u^{-2},v^{-2}]$.

17.21. Lemma. *$K_q(K)$ and $KO_{4q}(KO)$ are torsion-free for all $q\in\mathbb{Z}$ and hence $K_q(K)\to K_q(K)\otimes\mathbb{Q}$ and $KO_{4q}(KO)\to KO_{4q}(KO)\otimes\mathbb{Q}$ are monomorphisms for $q\in\mathbb{Z}$.*

Proof: $K_{q+2n}(BU)$ and $KO_{4q+8n+4}(BSp)$ are torsion-free by 16.33 and 16.34; therefore the same is true of the direct limits. □

Thus we can describe $K_*(K)$ and $\oplus_q KO_{4q}(KO)$ by describing which finite Laurent series $f(u,v)$ lie in their respective images.

Our analysis of the action of the Bott map

$$\underline{B}:\tilde{K}_{q+2n}(BU) \to \tilde{K}_{q+2n+2}(BU)$$

will be modelled on that for $H_*(K)$ in 16.18 through 16.25, although of course it will be somewhat more complicated. By 16.19 it follows that $\underline{B}$ defines a homomorphism

$$\underline{B}: Q_q(K_*(BU)) \to Q_{q+2}(K_*(BU))$$

between the indecomposable quotients, since $\underline{B}$ annihilates products. Again the dual groups are the groups $PK^*(BU)$ of primitive elements:

$$PK^*(BU) \cong \mathrm{Hom}_{\tilde{K}_*(S^0)}(QK_*(BU), \tilde{K}_*(S^0))$$

$$PKO^*(BSp) \cong \mathrm{Hom}_{\tilde{KO}_*(S^0)}(QKO_*(BSp), \widetilde{KO}_*(S^0)).$$

As before we can identify $PK^*(BU)$ with the group A^* of additive unstable cohomology operations

$$\phi: K^0(X) \to K^*(X).$$

We can then decompose A^* in the form

$$A^* = \tilde{K}_*(S^0) \oplus \tilde{A}^*,$$

where the first summand is generated by the operation Ψ^0 which carries every $U(n)$-bundle into the trivial bundle ε^n of the same dimension, and the second summand consists of operations which act as 0 when $X = pt$. We may also identify $\tilde{A}^*$ with the group of additive operations

$$\tilde{\phi}: \tilde{K}^0(X) \to \tilde{K}^*(X).$$

An operation $\phi \in A^*$ is completely determined by its value on the universal line bundle $\gamma \to BU(1)$, and this value may be any element of $K^*(BU(1)) \cong K^*(pt)[[Y]]$.

We now define particular operations which will be especially useful: let ϕ^r be the operation such that $\phi^r(\gamma) = Y^r$. The operation Ψ^{-1} which sends every virtual bundle to its complex conjugate also lies in A^*. $\Psi^{-1}\gamma = \bar{\gamma} = \gamma^{-1} = (1+Y)^{-1}$.

17.22. Lemma. Ψ^{-1} *commutes with all operations in* A^0.

Proof: Suppose $\phi \in A^0$ and $\phi(\gamma) = f(Y)$ for some power series f. Then

$$\phi(\Psi^{-1}(\gamma)) = f(\bar{\gamma} - 1) \quad \text{(since } \bar{\gamma} \text{ is a line bundle)} \; = f(\Psi^{-1}(Y));$$

but we also have $\Psi^{-1}(\phi(\gamma)) = \Psi^{-1}(f(Y)) = f(\Psi^{-1}(Y))$, since Ψ^{-1} commutes with sums, products and limits. □

Since $\Psi^{-1} \circ \Psi^{-1} = 1$, it follows that Ψ^{-1} splits A^0 into its +1 and −1 eigenspaces. We take some basic operations in each: let S^r be the operation with

$$S^r(\gamma) = (Y + \bar{Y})^r,$$

where $\bar{Y}$ denotes $\Psi^{-1} Y = (1+Y)^{-1} - 1$. Let A^r be the operation such that

$$A^r(\gamma) = (Y - \bar{Y})(Y + \bar{Y})^{r-1}.$$

17.23. Lemma. *If $x \in QK_*(BU)$ and either*
i) $\langle \phi^r, x\rangle = 0$, all $r \geqslant 1$, *or*
ii) $\langle S^r, x\rangle = 0 = \langle A^r, x\rangle$, *all* $r \geqslant 1$,
then $x = 0$.

Proof: i) is obvious and for ii) we note that

$$S^r(\gamma) = Y^{2r} + \text{higher powers of } Y$$

$$A^r(\gamma) = 2\,Y^{2r-1} + \text{higher powers of } Y. \quad □$$

Now we have complexification maps $c: KO \to K$ and $c': BSp \to BU$ which give

$$(c \wedge c')_*: \widetilde{KO}_{4n}(BSp) \to \tilde{K}_{4n}(BU).$$

The class $Z_n \in KO_{4n}(BSp)$ gives us an element in $Q_{4n}(K_*(BU))$ which we also denote by Z_n.

17.24. Lemma.
i) $\langle \phi^r, Y_n\rangle = \delta^r_n$
ii) $\langle S^r, Z_n\rangle = 2t^{2n}\,\delta^r_n$, $\quad \langle A^r, Z_n\rangle = 0$,
where $t \in \tilde{K}_2(S^0)$ is the generator.

Proof: i) is obvious. Since Z_n is in the image of $c: KO_*(BU) \to K_*(BU)$ we have

$$\langle A^r, Z_n\rangle = \langle A^r, \Psi^{-1} Z_n\rangle = \langle \Psi^{-1} A^r, Z_n\rangle = -\langle A^r, Z_n\rangle,$$

thus proving $\langle A^r, Z_n\rangle = 0$. Now consider the commutative diagram

$$\begin{array}{ccc} & BU & \\ {\scriptstyle c'}\nearrow & & \searrow{\scriptstyle s} \\ BSp & \xrightarrow{\;2\;} & BSp, \end{array}$$

where s is symplectification of virtual bundles and 2 denotes 1 + 1 in the sense of Whitney sum. Let $P^r \in P^{4r}(KO^*(BSp))$ be the unique element whose restriction to $BSp(1)$ is $Z^r \in KO^{4r}(BSp(1))$. The image of Z in $\tilde{K}^4(BU(1))$ is

$$t^{-2}(\gamma + \bar{\gamma} - 2) = t^{-2}(Y + \bar{Y}),$$

so the image of Z^r in $\tilde{K}^{4r}(BU(1))$ is $t^{-2r}(Y+\bar{Y})^r$. The map $s:BU \to BSp$ is an H-map, so it carries primitives to primitives, and therefore the image of P^r in $P^{4r}(K^*(BU))$ is $t^{-2r}S^r$. Then we calculate

$$\begin{aligned}\langle S^r, (c \wedge c')_* Z_n\rangle &= \langle t^{2r} cs^* P^r, cc'_* Z_n\rangle = t^{2r} c\langle 2^* P^r, Z_n\rangle \\ &= 2t^{2r} c\,\langle Z^r, Z_n\rangle = 2t^{2r}\delta^r_n. \quad \square\end{aligned}$$

Now we define $\bar{B}:\tilde{A}^* \to \tilde{A}^*$ by requiring the following diagram to commute for all $\phi \in \tilde{A}^*$ and all X:

$$\begin{array}{ccc} \tilde{K}^0(X) & \xrightarrow{\cong} & \tilde{K}^0(S^2 \wedge X) \\ {\scriptstyle \bar{B}\phi}\downarrow & & \downarrow{\scriptstyle \phi} \\ \tilde{K}^*(X) & \xrightarrow{\cong} & \tilde{K}^*(S^2 \wedge X). \end{array}$$

The horizontal arrows are the periodicity isomorphism. Then we have the following relation between $\bar{B}$ and $\underline{B}$:

$$\langle \phi, \underline{B}x\rangle = t\langle \bar{B}\phi, x\rangle, \quad x \in Q_*(K_*(BU)).$$

The proof is similar to that of 16.21.

Replacing 16.22 is the following calculation of $\bar{B}$.

17.25. Lemma. *If $\phi \in \tilde{A}^*$ satisfies $\phi(\gamma) = f(Y)$ for some formal power series f, then*

$$(\bar{B}\phi)(\gamma) = (1+Y)f'(Y) - f'(0).$$

(f' denotes the formal derivative of the power series f.)

Proof: Again we consider the commutative diagram

$$\begin{array}{ccccc} \tilde{K}^0(X) & \xrightarrow{\cong} & \tilde{K}^0(S^2 \wedge X) & \longrightarrow & \tilde{K}^0(S^2 \times X) \\ {\scriptstyle \bar{B}\phi}\downarrow & & \downarrow{\scriptstyle \phi} & & \downarrow{\scriptstyle \phi} \\ \tilde{K}^*(X) & \xrightarrow{\cong} & \tilde{K}^*(S^2 \wedge X) & \longrightarrow & \tilde{K}^*(S^2 \times X). \end{array}$$

Each horizontal arrow is multiplication by $w = \eta - 1$. Again we take $X = BU(1)$ and apply the diagram to $Y = \gamma - 1$, getting

$$\begin{aligned}
w \times (\bar{B}\phi)(Y) &= \phi(w \times Y) = \phi((\eta - 1) \times (\gamma - 1)) \\
&= \phi(\eta \times \gamma - 1 \times \gamma - \eta \times 1 + 1 \times 1) \\
&= \phi(\eta \times \gamma) - \phi(1 \times \gamma) - \phi(\eta \times 1) + \phi(1).
\end{aligned}$$

Thus

$$\begin{aligned}
w \times (\bar{B}\phi)(\gamma - 1) &= f(wY + w + Y) - f(Y) - f(w) + f(0) \\
&= w(1 + Y) f'(Y) - wf'(0).
\end{aligned}$$

Here we have used the fact that $w^2 = 0$ in $\tilde{K}^0(S^2)$. Therefore $(\bar{B}\phi)(\gamma - 1) = (1 + Y)f'(Y) - f'(0)$. □

Since $\bar{B}\Psi^{-1} = -\Psi^{-1}$, we see that $\bar{B}$ interchanges the $+1$ and -1 eigenspaces of Ψ^{-1}. In fact 17.25 gives the following explicit formulae.

17.26. Lemma.

$$\begin{aligned}
&\bar{B}\phi^r = r\phi^r + r\phi^{r-1} \\
&\bar{B}S^r = rA^r \\
&\bar{B}A^r = rS^r + (4r - 2)\, S^{r-1},
\end{aligned}$$

where we interpret ϕ^0 and S^0 as 0.

Proof: We regard $\tilde{A}^*$ as a quotient of A^*, and then we can neglect the term $f'(0)$ in 17.25. For ϕ^r we have $f(Y) = Y^r$ so

$$(1 + Y)f'(Y) = (1 + Y)\, r\, Y^{r-1} = r\, Y^r + r\, Y^{r-1}.$$

Thus $\bar{B}\phi^r = r\phi^r + r\phi^{r-1}$.

Next we take the relation $(1 + Y)(1 + \bar{Y}) = 1$ and differentiate:

$$(1 + \bar{Y}) + (1 + Y)\,\frac{d}{dY}\,\bar{Y} = 0.$$

Therefore we have

$$(1 + Y)\,\frac{d}{dY}\,(Y + \bar{Y}) = Y - \bar{Y},$$

$$(1 + Y)\,\frac{d}{dY}\,(Y - \bar{Y}) = Y + \bar{Y} + 2.$$

Moreover $(Y - \bar{Y})^2 = (Y + \bar{Y})^2 - 4Y\bar{Y} = (Y + \bar{Y})^2 + 4(Y + \bar{Y})$. Since for S^r we have $f(Y) = (Y + \bar{Y})^r$, we get

$$(1 + Y)f'(Y) = r(Y + \bar{Y})^{r-1}(Y - \bar{Y}),$$

which gives $\bar{B}S^r = rA^r$.

For A^r we have $f(Y)=(Y-\bar{Y})(Y+\bar{Y})^{r-1}$ and hence

$$(1+Y)f'(Y)=(Y+\bar{Y}+2)(Y+\bar{Y})^{r-1}+(r-1)(Y-\bar{Y})^2(Y+\bar{Y})^{r-2}$$
$$=r(Y+\bar{Y})^r+(4r-2)(Y+\bar{Y})^{r-1}.$$

Thus $\bar{B}A^r=rS^r+(4r-2)S^{r-1}$. ▯

17.27. Corollary. *In $QK_*(BU)$ we have*

$$BY_n=t(nY_n+(n+1)Y_{n+1})$$
$$(B)^2Z_n=n^2t^2Z_n+(2n+1)(2n+2)Z_{n+1}.$$

Proof: The first formula follows immediately from 17.26 and 17.24i). From 17.26 we get $(\bar{B})^2S^r=r^2S^r+2r(2r-1)S^{r-1}$ and $(\bar{B})^2A^r$ is a linear combination of A^r, A^{r-1}. Then the second formula above follows from 17.24ii). ▯

17.28. Corollary. *In $QK_*(BU)\otimes\mathbb{Q}$ we have*

$$t^nY_n=\frac{1}{n!}(B-t)(B-2t)\dots(B-(n-1)t)tY_1$$

$$Z_n=\frac{2}{(2n)!}(B^2-t^2)(B^2-2^2t^2)\dots(B^2-(n-1)^2t^2)Z_1.$$

Now if we consider BU as the $2n$th term of K and BSp as the $(8n+4)$th term of KO then we have homomorphisms

$$\iota_n:\tilde{K}_{2q}(BU)\to K_{2q-2n}(K)$$
$$\kappa_n:\widetilde{KO}_{4q+4}(BSp)\to KO_{4q-8n}(KO)$$

and $(c\wedge c)_*\circ\kappa_n=\iota_{4n+2}\circ(c\wedge c')_*:KO_q(BSp)\to K_{q-8n-4}(K)$.

17.29. Lemma. *The diagram*

$$\begin{array}{ccc}
\tilde{K}_{2q}(BU) & \xrightarrow{\iota_n} & K_{2q-2n}(K)\\
\downarrow B & & \downarrow v\cdot\\
\tilde{K}_{2q+2}(BU) & \xrightarrow{\iota_n} & K_{2q-2n+2}(K)
\end{array}$$

commutes, where $v\in K_2(K)$ is $\eta_R(t)$.

Proof: One simply takes a class $[f]\in\tilde{K}_{2q}(BU)$ and writes down representatives for $v\cdot\iota_n[f]$ and $\iota_nB[f]$ to see that they are the same. ▯

Let us denote $\iota_n(Y_q)$ by Y_q^n and $\iota_n(Z_q)$ by Z_q^n. Then 17.29 implies that $Y_q^n=v^{-n}Y_q^0$, $Z_q^{4n+2}=v^{-4n}Z_q^2$.

17.30. Proposition. *In* $K_*(K) \otimes \mathbb{Q}$

$$u^n Y_n^0 = p_n'(u,v) = \frac{1}{n!}\, v(v-u)(v-2u)\ldots(v-(n-1)u), \quad n \geqslant 1,$$

$$Z_{n+1}^2 = q_n'(u,v) = \frac{2}{(2n+2)!}\,(v^2-u^2)(v^2-2^2u^2)\ldots(v^2-n^2u^2), \quad n \geqslant 0.$$

Proof: This follows immediately from 17.28 and 17.29 once one remarks that multiplication by t in $\tilde{K}_*(BU)$ goes into multiplication by $u = \eta_L(t)$ in $K_*(K)$. $tY_1 \in \tilde{K}_2(CP^1)$ is the generator, so $\iota_0(tY_1) = v$. Similarly $Z_1 \in \widetilde{KO}_4(HP^1)$ is the generator, so $\iota_2(Z_1) = 1 \in K_0(K)$. □

17.31. Now $Q\tilde{K}_*(BU)$ is a free $\mathbb{Z}[t,t^{-1}]$-module with basis $\{Y_1, Y_2, \ldots\}$, so we see that the Y_q^n span $K_*(K)$ over $\mathbb{Z}[u,u^{-1}]$. Since $Y_q^n = v^{-n}Y_q^0$, it follows the polynomials $p_n'(u,v)$ span $K_*(K)$ over $\mathbb{Z}[u,u^{-1},v^{-1}]$. We have thus identified the image of $K_*(K)$ in $K_*(K) \otimes \mathbb{Q} \cong \mathbb{Q}[u,v,u^{-1},v^{-1}]$. Similarly the image of $\oplus_n Q_{4n}\widetilde{KO}_*(BSp)$ in $Q\tilde{K}_*(BU)$ is a free $\mathbb{Z}[t^4, 2t^2, t^{-4}]$-module with basis $\{Z_1, Z_2, \ldots\}$, so we see that the image of $\oplus_n KO_{4n}(KO)$ in $K_*(K) \otimes \mathbb{Q}$ is spanned over $\mathbb{Z}[u^4, 2u^2, u^{-4}, v^{-4}]$ by the polynomials $q_n'(u,v)$, $n \geqslant 0$.

17.31 gives a complete description of $K_*(K)$ and $KO_{4n}(KO)$, but we can give a handier description of the elements in these images by doing a little algebra. Theorem 17.34 gives a complete characterization of the finite Laurent series which lie in $K_*(K)$ or $KO_{4n}(KO)$; the next few results lead up to that theorem.

17.32. Lemma. *Let* $f(t) \in \mathbb{Q}[t]$ *be a polynomial such that* $f(k) \in \mathbb{Z}$ *for each integer* k. *Then* f *can be written as a* $\mathbb{Z}$*-linear combination of the binomial polynomials*

$$p_n(t) = \frac{1}{n!}\, t(t-1)(t-2)\ldots(t-(n-1)), \quad n \geqslant 0.$$

(p_0 *is interpreted as* 1.) *If in addition* f *satisfies* $f(-t) = f(t)$, *then* f *can be written as a* $\mathbb{Z}$*-linear combination of* 1 *and the polynomials*

$$q_n(t) = \frac{2}{(2n)!}\, t^2(t^2-1^2)(t^2-2^2)\ldots(t^2-(n-1)^2), \quad n \geqslant 1.$$

(*Note that* q_0 *would have to be interpreted as* 2, *but we do not consider* q_0.)

Proof: Clearly the polynomials p_n do satisfy the condition $p_n(k) \in \mathbb{Z}$ for all $k \in \mathbb{Z}$. Since $q_n(t) = 2p_{2n}(t+n) - p_{2n-1}(t+n-1)$ the q_n do also. Clearly $q_n(-t) = q_n(t)$, $n \geqslant 1$.

We now prove the second statement of the lemma, as the first is similar and easier. We proceed by induction on the degree of the polynomial f. If $\deg f = 0$, the result is obvious. Suppose the lemma true for polynomials of degree $\leqslant 2(d-1)$. Let $f(t)$ be a polynomial of degree $\leqslant 2d$ satisfying $f(k) \in \mathbb{Z}$ for all $k \in \mathbb{Z}$ and $f(-t) = f(t)$. Define $\delta^2 f$ to be the polynomial

$$(\delta^2 f)(t) = f(t+1) - 2f(t) + f(t-1).$$

Then $\delta^2 f$ also satisfies the conditions of 17.32 and is easily seen to have degree $\leqslant 2(d-1)$, so by the inductive hypothesis

$$(\delta^2 f)(t) = n_0 + \textstyle\sum_{r \geqslant 1} n_r q_r(t)$$

for some integer coefficients $n_0, n_1, \ldots$. Since $n_0 = (\delta^2 f)(0) = 2(f(1) - f(0))$, we see that n_0 is even. Since one readily checks that

$$\delta^2 q_n = q_{n-1}, n \geqslant 2, \quad \delta^2 q_1 = 2,$$

we see that if we take

$$g(t) = (n_0/2) q_1 + \textstyle\sum_{r \geqslant 1} n_r q_{r+1}(t),$$

then $g(-t) = g(t)$ and $\delta^2 g = \delta^2 f$. Thus we must have $f(t) = a \cdot 1 + bt + g(t)$ for some $a, b \in \mathbb{Q}$. But $f(t) - g(t)$ is even, so $b = 0$. Also $a = f(0) \in \mathbb{Z}$. This proves that f is of the required form. □

17.33. Lemma. i) *Suppose $f(u,v)$ is a finite Laurent series in $\mathbb{Q}[u, u^{-1}, v, v^{-1}]$ which satisfies the condition*

$$f(ht, kt) \in \mathbb{Z}\left[t, t^{-1}, \frac{1}{hk}\right]$$

for all non-zero integers h, k. Then $f(u,v)$ can be written as a $\mathbb{Z}[u, u^{-1}, v^{-1}]$-linear combination of the polynomials

$$p'_n(u,v) = \frac{1}{n!} v(v-u)(v-2u)\ldots(v-(n-1)u), \quad n \geqslant 1.$$

ii) *Suppose $f(u,v)$ is a finite Laurent series in $\mathbb{Q}[u, u^{-1}, v, v^{-1}]$ satisfying the conditions*

$$f(-u, v) = f(u, v) = f(u, -v)$$

$$f(ht, kt) \in \mathbb{Z}\left[t^4, 2t^2, t^{-4}, \frac{1}{hk}\right], \quad h, k \in \mathbb{Z} - \{0\}.$$

Then $f(u,v)$ can be written as a $\mathbb{Z}[u^4, 2u^2, u^{-4}, v^{-4}]$-linear combination of the polynomials

$$q_n'(u,v) = \frac{2}{(2n+2)!}(v^2-u^2)(v^2-2^2u^2)\dots(v^2-n^2u^2), \quad n \geqslant 0.$$

Proof: We prove ii); i) is similar but easier. $f(-u,v)=f(u,v)=f(u,-v)$ implies that f is a Laurent series in u^2 and v^2. By separating $f(u,v)$ into its homogeneous components, we see it suffices to consider the case where f is homogeneous. Suppose f is homogeneous of degree $2r$; if r is even we let $f'(u,v)=u^{-2r}f(u,v)$, $f'(u,v)=(u^{-2r}/2)f(u,v)$ if r is odd. Then f' is homogeneous of degree 0, and we have

$$f'(u,v)=g(u^{-1}v) \quad \text{for some} \quad g(w) \in \mathbb{Q}[w^2, w^{-2}].$$

The finite Laurent series g has only finitely many coefficients, so we let m be the highest power to which any prime occurs in the denominator of any of these coefficients. Multiplying g by a suitable positive power of $w^4=u^{-4}v^4$, we see that it is sufficient to consider the case in which g is a polynomial divisible by w and w^m. In particular $g(0)=0$. If $k \in \mathbb{Z}$ is an integer with $k \neq 0$, then $g(k)=f'(t,kt)$. If p is a prime dividing k, then p does not occur in the denominator of $g(k)$ by construction. If p is a prime not dividing k, then $g(k)=f'(t,kt)$, which by assumption lies in $\mathbb{Z}[1/k]$, and so p does not occur in the denominator of $g(k)$. Thus we must have $g(k) \in \mathbb{Z}$. Therefore g satisfies the hypotheses of 17.32, so $f'(u,v)=g(u^{-1}v)$ can be written as a $\mathbb{Z}$-linear combination of the polynomials $1=q_0'$ and

$$q_n(u^{-1}v) = \frac{2}{(2n)!}u^{-2n}v^2(v^2-u^2)(v^2-2^2u^2)\dots(v^2-(n-1)^2u^2).$$

But we can write

$$q_{2n}(u^{-1}v) = u^{-4n}\{(4n+2)(4n+1)q_{2n}'(u,v)+4n^2u^2q_{2n-1}'(u,v)\}$$
$$q_{2n-1}(u^{-1}v) = u^{-4n}\{4n(4n-1)u^2q_{2n-1}'(u,v)+(2n-1)^2u^4q_{2n-2}'(u,v)\}.$$

Thus 17.33 follows. □

Combining 17.31 with 17.33 we get our main theorem.

17.34. Theorem. i) *The image of $K_*(K)$ in $K_*(K) \otimes \mathbb{Q} \cong \mathbb{Q}[u,v,u^{-1},v^{-1}]$ consists precisely of those finite Laurent series $f(u,v)$ such that*

$$f(ht,kt) \in \mathbb{Z}\left[t, t^{-1}, \frac{1}{hk}\right]$$

for all $h,k \in \mathbb{Z}-\{0\}$.

ii) *The image of* $\bigoplus_n KO_{4n}(KO)$ *in* $K_*(K) \otimes \mathbb{Q}$ *consists precisely of those finite Laurent series* $f(u,v)$ *such that*

$$f(-u,v) = f(u,v) = f(u,-v)$$

and

$$f(ht,kt) \in \mathbb{Z}\left[t^4, 2t^2, t^{-4}, \frac{1}{hk}\right]$$

for all $h,k \in \mathbb{Z} - \{0\}$.

Now $KO_*(KO)$ is a module over $\widetilde{KO}_*(S^0)$, so we have an action map

$$\rho: \widetilde{KO}_q(S^0) \otimes KO_0(KO) \to KO_q(KO), \quad q \in \mathbb{Z}.$$

17.35. Theorem. *The map* ρ *is an isomorphism for all* q. *Explicitly*

$$KO_q(KO) \cong \begin{cases} \mathbb{Z}_2 \otimes KO_0(KO) & q \equiv 1, 2 \bmod 8 \\ 0 & q \equiv 3, 5, 6, 7 \bmod 8. \end{cases}$$

Proof: If $q \equiv 0 \bmod 4$, then the statement follows immediately from 17.34. We show that $\widetilde{KO}_q(S^0) \otimes KO_0(KO) \xrightarrow{\rho} KO_q(KO)$ is an epimorphism. Every element of $\widetilde{KO}_q(BSp)$ can be written in the form $ax + by$ for some $a,b \in \widetilde{KO}_*(S^0)$ and $x \in \widetilde{KO}_0(BSp)$, $y \in \widetilde{KO}_4(BSp)$. Passing to the direct limit we find the same is true in $KO_q(KO)$. By the remark at the beginning of this proof we can write y in the form cz for $c \in \widetilde{KO}_4(S^0)$, $z \in KO_0(KO)$. This shows that $\widetilde{KO}_q(S^0) \otimes KO_0(KO) \to KO_q(KO)$ is an epimorphism. That completes the proof for all cases except $q \equiv 1, 2 \bmod 8$.

To show ρ is a monomorphism when $q \equiv 1, 2 \bmod 8$ we must examine the maps

$$\kappa_n: \widetilde{KO}_q(BSp) \to KO_{q-8n-4}(KO)$$

more closely. The diagram

$$\begin{array}{ccc} \widetilde{KO}_q(BSp) & \xrightarrow{B_0} & \widetilde{KO}_{q+8}(BSp) \\ \Big\downarrow (c \wedge c')_* & & \Big\downarrow (c \wedge c')_* \\ \widetilde{K}_q(BU) & \xrightarrow{B^4} & \widetilde{K}_{q+8}(BU) \end{array}$$

commutes, so from 17.27 we find

$$B_0 Z_n = n^4 y Z_n + \{n^2 + (n+1)^2\}\binom{2n+2}{2} x Z_{n+1} + 4!\binom{2n+4}{4} Z_{n+2}.$$

Here we have used x to denote the generator of $\pi_4(KO)$ (which maps to $2t^2$ in $\pi_4(K)$) and y to denote the generator of $\pi_8(KO)$ (which maps to t^4

in $\pi_8(K)$). Hence if α denotes the non-zero element of $\widetilde{KO}_q(S^0) \cong \mathbb{Z}_2$ ($q \equiv 1$ or $2 \bmod 8$) then

$$B_0(\alpha Z_n) = n^4 \, \alpha y Z_n = \begin{cases} \alpha y Z_n & n \text{ odd} \\ 0 & n \text{ even,} \end{cases}$$

since $\alpha x = 0 = 2\alpha$. Therefore $B_0(\sum_{k \geq 0} n_k \alpha y^{n-k} Z_{2k+1}) = 0$ for integer coefficients n_k if and only if $n_k \equiv 0 \bmod 2$, $k \geq 0$. Passing to the limit, we find

$$\kappa_n(\textstyle\sum_{k \geq 0} n_k \, \alpha y^{n-k} Z_{2k+1}) = 0$$

if and only if $n_k \equiv 0 \bmod 2$, $k \geq 0$.

Now suppose $\rho(\alpha \otimes f) = 0$ for some $f \in KO_0(KO)$. Then we can write f in the form

$$f = \kappa_n(\textstyle\sum_{k \geq 0} n_k \, y^{n-k} Z_{2k+1} + \sum_{k \geq 1} n'_k \, x y^{n-k} Z_{2k}).$$

Since the diagram

$$\begin{array}{ccc} \widetilde{KO}_q(S^0) \otimes Q_{8n+4} \widetilde{KO}_*(BSp) & \xrightarrow{\rho} & Q_{q+8n+4} \widetilde{KO}_*(BSp) \\ \downarrow 1 \otimes \kappa_n & & \downarrow \kappa_n \\ \widetilde{KO}_q(S^0) \otimes KO_0(KO) & \xrightarrow{\rho} & KO_q(KO) \end{array}$$

commutes, we get

$$\begin{aligned} 0 &= \rho(\alpha \otimes f) \\ &= \rho \circ (1 \otimes \kappa_n)[\alpha \otimes (\textstyle\sum_{k \geq 0} n_k \, y^{n-k} Z_{2k+1} + \sum_{k \geq 1} n'_k \, x y^{n-k} Z_{2k})] \\ &= \kappa_n(\textstyle\sum_{k \geq 0} n_k \, \alpha y^{n-k} Z_{2k+1}), \end{aligned}$$

from which it follows by the previous paragraph that $n_k \equiv 0 \bmod 2$, $k \geq 0$. But in $K_*(K) \otimes \mathbb{Q}$ the element $\kappa_n(\sum_{k \geq 1} n'_k x y^{n-k} Z_{2k})$ corresponds to

$$\textstyle\sum_{k \geq 1} 2n'_k u^{4n-4k+2} v^{-4n} q'_{2k-1}(u, v) = \sum_{k \geq 1} 2n'_k (u^{-1} v)^{-4n-2} q_{2k}(u^{-1} v),$$

which shows f is divisible by 2 in $KO_0(KO)$ and hence $\alpha \otimes f = 0$. Thus ρ is a monomorphism. □

17.36. Exercise. Show that $f(u, v)$ gives 0 in $\mathbb{Z}_2 \otimes KO_0(KO)$ if and only if $f(h, k) \in 2\mathbb{Z}[1/hk]$ for all odd integers h, k.

17.37. Theorem. *The structure maps of $K_*(K)$ and $KO_*(KO)$ are given by the following formulae:*

i) $\psi(u) = u \otimes 1$, $\psi(v) = 1 \otimes v$;

ii) $\varepsilon(u) = t = \varepsilon(v)$;

iii) $c(u) = v$, $c(v) = u$;

iv) *if $\alpha \in \widetilde{KO}_1(S^0)$ is the non-zero element, then $\eta_R(\alpha) = \eta_L(\alpha)$.*

Proof: Let $\iota: S^0 \to K$ be the unit; we know that $\pi_1(S^0) \cong \mathbb{Z}_2$ is generated by the Hopf map η, and $\iota_*(\eta) = \alpha$, or in other words

$$\alpha \in \operatorname{im}[h: \pi_1(S^0) \to \widetilde{KO}_1(S^0)] \subset P\widetilde{KO}_1(S^0)$$

by 17.14. 17.8x) says ψ_{S^0} is just

$$\eta_L: \widetilde{KO}_*(S^0) \to KO_*(KO) \cong KO_*(KO) \otimes_{\widetilde{KO}_*(S^0)} \widetilde{KO}_*(S^0),$$

so since α is primitive, we have $\eta_L(\alpha) = 1 \otimes \alpha = 1 \cdot \alpha \otimes 1 = \eta_R(\alpha)$, thus proving iv).

By definition $u = \eta_L(t)$, $v = \eta_R(t)$ and hence by 17.8iv) $\psi u = \psi \eta_L(t) = \eta_L(t) \otimes 1 = u \otimes 1$ and $\psi v = \psi \eta_R(t) = 1 \otimes \eta_R(t) = 1 \otimes v$. By 17.8 $c \circ \eta_R = \eta_L$ and $c \circ \eta_L = \eta_R$, so it follows $c(u) = v$ and $c(v) = u$. Similarly $\varepsilon \circ \eta_R = 1 = \varepsilon \circ \eta_L$, so $\varepsilon(u) = t = \varepsilon(v)$. ▯

We now have complete information on the Hopf algebras $K_*(K)$ and $KO_*(KO)$. Let us make some applications of this information.

In 17.14 we saw that $\operatorname{im}[h: \pi_*(X) \to E_*(X)] \subset P(E_*(X))$; for example $h: \pi_*(MU) \to P(K_*(MU))$. We now know $K_*(K)$ and $K_*(MU) \cong \tilde{K}_*(S^0)[Y_1', Y_2', \ldots]$, so we may try to compute $P(K_*(MU))$. The map $f: CP^\infty \simeq MU_1 \to \Sigma^2 MU$ sends $t^i Y_i \in \tilde{K}_{2i}(CP^\infty)$ to $Y_{i-1}' \in K_{2i-2}(MU)$. On the other hand the map $CP^\infty \simeq BU(1) \subset BU \xrightarrow{i_0} K$ sends $t^i Y_i$ to p_i' in $K_*(K)$.

From Chapter 16 we know that $U(n)$-bundles are orientable with respect to K, and in particular we have the universal Thom class $t \in K^0(MU)$, which we can represent by a map $\mu: MU \to K$ of spectra. This map also defines a natural transformation

$$\mu: MU_* \to K_*$$

of homology theories and also of cohomology theories. From the definition of μ it follows that

$$\underline{\mu}: MU_*(CP^\infty) \to K_*(CP^\infty)$$

sends β_i to $t^i Y_i$ and hence

$$\underline{\mu}: MU_*(MU) \to K_*(MU)$$

sends β_i' to Y_i'. We have a homotopy-commutative diagram

$$\begin{array}{ccc} \Sigma^2 MU & \xrightarrow{\Sigma^2 \mu} & \Sigma^2 K \\ \uparrow{\scriptstyle i} & & \uparrow{\scriptstyle i_1} \\ MU_1 \simeq & BU(1) & \subset BU, \end{array}$$

where we regard BU as K_2, and hence it follows that $\mu_*: K_*(MU) \to K_*(K)$ sends Y_i' to

$$i_{1*}(t^{i+1} Y_{i+1}) = \frac{(v-u)(v-2u)\dots(v-iu)}{(i+1)!}.$$

We set

$$p_n = \frac{(v-u)(v-2u)\dots(v-nu)}{(n+1)!} \in K_{2n}(K).$$

Then we see that

$$(\mu \wedge \mu)_* = \mu_* \circ \underline{\mu}: MU_*(MU) \to K_*(K)$$

sends β_i' to p_i, $i \geqslant 1$.

Now it is easy to see that $\underline{\mu}$ commutes with the cooperation maps ψ in the sense that the following diagram commutes

$$\begin{array}{ccc} MU_*(X) & \xrightarrow{{}_{MU}\psi_X} & MU_*(MU) \otimes_{\widetilde{MU}_*(S^0)} MU_*(X) \\ \downarrow \underline{\mu} & & \downarrow (\mu \wedge \mu)_* \otimes \underline{\mu} \\ K_*(X) & \xrightarrow{{}_K\psi_X} & K_*(K) \otimes_{\tilde{K}_*(S^0)} K_*(X). \end{array}$$

Taking $X = MU$ and applying 17.16 we get the following.

17.38. Proposition. *On $K_*(MU)$ the coaction ψ_{MU} is given by*

$$\psi_{MU}(Y_k') = \textstyle\sum_{i+j=k} (P^{j+1})_{2i} \otimes Y_j', \quad k \geqslant 0,$$

where $P = 1 + p_1 + p_2 + \cdots$. Hence on $K_*(CP^\infty)$ the coaction ψ_{CP^∞} is given by

$$\psi_{CP^\infty}(t^k Y_k) = \textstyle\sum_{i+j=k} (P^j)_{2i} \otimes t^j Y_j, \quad k \geqslant 0.$$

17.39. Corollary. *On the elements $p_i \in K_*(K)$ the coproduct ψ_K is given by*

$$\psi_K(p_k) = \textstyle\sum_{i+j=k} (P^{j+1})_{2i} \otimes p_j, \quad k \geqslant 0.$$

Analogous results hold for KO and MSp.

17.40. Proposition. *On $KO_*(MSp)$ the coaction ψ_{MSp} is given by*

$$\psi_{MSp}(Z_k') = \textstyle\sum_{i+j=k} (Q^{j+1})_{4i} \otimes Z_j', \quad k \geqslant 0,$$

where $Q = 1 + q_1 + q_2 + \cdots, q_n = q_n'(u,v)$. Hence on $KO_*(HP^\infty)$ the coaction ψ_{HP^∞} is given by

$$\psi_{HP^\infty}(Z_k) = \textstyle\sum_{i+j=k} (Q^j)_{4i} \otimes Z_j, \quad k \geqslant 0.$$

17.41. Corollary. *On the elements $q_i \in KO_*(KO)$ the coproduct ψ_{KO} is given by*

$$\psi_{KO}(q_k) = \sum_{i+j=k} (Q^{j+1})_{4i} \otimes q_j, \quad k \geqslant 0.$$

Let $G_n \subset K_{2n}(MU)$ be the intersection with the submodule generated by monomials $Y'^{\alpha} = Y_1'^{\alpha_1} Y_2'^{\alpha_2} \dots Y_r'^{\alpha_r}$ for $|\alpha| = \alpha_1 + 2\alpha_2 + \cdots + r\alpha_r \leqslant n$. Then it follows that $h(\pi_{2n}(MU)) \subset P(K_*(MU)) \cap G_n$, because $G_n = F_{2n,0} \subset K_{2n}(MU)$. Let $G_n \cap P(K_*(MU)) = P_{2n}$. We calculate P_{2n} for small values of n.

dim 2: A typical element of G_1 is of the form $A Y_1' + Bt$, $A, B \in \mathbb{Z}$; then

$$\begin{aligned}\psi_{MU}(A Y_1' + Bt) &= 1 \otimes A Y_1' + Ap_1 \otimes 1 + Bu \otimes 1 \\ &= 1 \otimes (A Y_1' + Bt) + (A - 2B) p_1 \otimes 1.\end{aligned}$$

This element is primitive if and only if $A - 2B = 0$, so the general element of P_2 is of the form

$$B(2 Y_1' + t), \quad B \in \mathbb{Z}.$$

If one computes $h[CP^1] \in K_2(MU)$ in a fashion analogous to 16.46 (cf. [8]), one finds $h[CP^1] = -2 Y_1' - t$, so we see that $\operatorname{im} h = P_2$ in dim 2.

dim 4: The typical element of G_2 is of the form $A Y_2' + B Y_1'^2 + Ct Y_1' + Dt^2$, for $A, B, C, D \in \mathbb{Z}$.

$$\begin{aligned}\psi_{MU}(A Y_2' + B Y_1'^2 + Ct Y_1' + Dt^2) &= A[1 \otimes Y_2' + 2p_1 \otimes Y_1' + p_2 \otimes 1] \\ &\quad + B[1 \otimes Y_1'^2 + 2p_1 \otimes Y_1' + p_1^2 \otimes 1] \\ &\quad + C[u \otimes Y_1' + up_1 \otimes 1] + Du^2 \otimes 1 \\ &= 1 \otimes [A Y_2' + B Y_1'^2 + Ct Y_1' + Dt^2] \\ &\quad + 2(A + B - C) p_1 \otimes Y_1' \\ &\quad + \tfrac{1}{12}(v - u)\{(2A + 3B - 12D) v \\ &\quad - (4A + 3B - 6C + 12D) u\} \otimes 1.\end{aligned}$$

This element is primitive if and only if

$$\begin{aligned}A + B - C &= 0 \\ 2A + 3B - 12D &= 0 \\ 4A + 3B - 6C + 12D &= 0,\end{aligned}$$

or in other words $A = 3C - 12D$, $B = -2C + 12D$. Hence the general element of P_4 is of the form

$$C(3 Y_2' - 2 Y_1'^2 + t Y_1') + D(-12 Y_2' + 12 Y_1'^2 + t^2), \quad C, D \in \mathbb{Z}.$$

We see therefore that P_4 is generated by

$$3Y_2' - 2Y_1'^2 + tY_1'$$
$$4Y_1'^2 + 4tY_1' + t^2 = (2Y_1' + t)^2$$

or also by

$$3Y_2' - 6Y_1'^2 - 3tY_1' - t^2 = -h[CP^2]$$
$$(2Y_1' + t)^2 \qquad = h[CP^1]^2.$$

At this point one might conjecture that

i) $[CP^i]$ could be chosen as a polynomial generator for Ω_*^U, $i \geqslant 1$, and

ii) $h:\pi_*(MU) \to P(K_*(MU))$ is an isomorphism.

Further calculation, however, shows that these conjectures are incompatible. In P_6 one finds the new polynomial generator is of the form $2Y_3'+$ decomposables, whereas we know from 16.46 that $h[CP^3] = -4Y_3'+$ decomposables.

Let us calculate $PKO_*(MSp)$ in a few low dimensions. Again we let $G_n \subset KO_{4n}(MSp)$ be the intersection with the submodule generated by monomials $Z'^{\alpha} = Z_1'^{\alpha_1} Z_2'^{\alpha_2} \ldots Z_r'^{\alpha_r}$ with $|\alpha| = \alpha_1 + 2\alpha_2 + \cdots + r\alpha_r \leqslant n$ and $P_k = P(KO_k(MSp)) \cap G_n$ for $4n \leqslant k \leqslant 4n+3$.

dim 1: By 17.37iv) $\eta_R(\alpha) = \eta_L(\alpha)$, where $\alpha \in \widetilde{KO}_1(S^0) \cong \mathbb{Z}_2$ is the non-zero element. Now $G_0 \cap KO_1(MSp)$ is $\mathbb{Z}_2$ generated by $\alpha \cdot 1$, and we have

$$\psi_{MSp}(\alpha \cdot 1) = \alpha(1 \otimes 1) = \eta_L(\alpha) \otimes 1 = \eta_R(\alpha) \otimes 1 = 1 \otimes \alpha \cdot 1.$$

Hence $\alpha \cdot 1$ is primitive, so $P_1 \cong \mathbb{Z}_2$ generated by α.

Now we have $\pi_1^S \cong \Omega_1^{fr} \to \Omega_1^{Sp}$, and the image of the non-zero element $[\eta] \in \pi_1^S$ in $\widetilde{KO}_1(S^0)$ is α, so we see that in dim 1 we have $\operatorname{im} h = P_1$. A representative for the non-zero element in Ω_1^{Sp} can be taken to be S^1 with $\omega \otimes \mathbb{H}$ as stable normal bundle ($\omega \to S^1 \cong RP^1$ the Hopf bundle).

dim 2: $G_0 \cap KO_2(MSp) = \mathbb{Z}_2$ generated by $\alpha^2 \cdot 1$ and $\psi_{MSp}(\alpha^2 \cdot 1) = 1 \otimes \alpha^2 \cdot 1$, so $P_2 \cong \mathbb{Z}_2$ generated by $\alpha^2 = h[S^1]^2$.

dim 3: $KO_3(MSp) = 0$, so $P_3 = 0$.

dim 4: For the generator $x \in \widetilde{KO}_4(S^0)$ we have

$$\psi_{MSp}(x \cdot 1) = x \cdot (1 \otimes 1) = x \cdot 1 \otimes 1 = 2u^2 \otimes 1$$

(x maps to $2u^2$ in $KO_*(KO) \subset \mathbb{Q}[u, v, u^{-1}, v^{-1}]$). A typical element of $G_1 \cap KO_4(MSp)$ is of the form $AZ_1' + Bx$, $A, B \in \mathbb{Z}$, and we have

$$\begin{aligned}\psi_{MSp}(AZ_1' + Bx) &= 1 \otimes AZ_1' + Aq_1 \otimes 1 + 2Bu^2 \otimes 1 \\ &= 1 \otimes (AZ_1' + Bx) + (A - 24B)\, q_1 \otimes 1.\end{aligned}$$

This element is primitive if and only if $A - 24B = 0$, so $P_4 \cong \mathbb{Z}$ generated by $24Z_1' + x$.

dim 5: $G_4 \cap KO_5(MSp) \cong \mathbb{Z}_2$ generated by $\alpha Z_1'$.

$$\psi(\alpha Z_1') = \alpha(1 \otimes Z_1' + q_1 \otimes 1) = 1 \otimes \alpha Z_1' + \alpha q_1 \otimes 1.$$

By 17.35 and 17.36 it follows $\alpha q_1 = 0$, for $q_1'(h,k) \in 2\mathbb{Z}[1/hk]$ for all odd h, k. Thus $\alpha Z_1'$ is primitive, and $P_5 \cong \mathbb{Z}_2$ generated by $\alpha Z_1'$.

dim 6: Similarly $P_6 \cong \mathbb{Z}_2$ generated by $\alpha^2 Z_1'$.

dim 7: $KO_7(MSp) = 0$, so $P_7 = 0$.

dim 8:
$$\begin{aligned}\psi_{MSp}(AZ_2' + BZ_1'^2 + CxZ_1' + Dy) = 1 \otimes AZ_2' &+ 2Aq_1 \otimes Z_1' \\ &+ Aq_2 \otimes 1 + 1 \otimes BZ_1'^2 \\ &+ 2Bq_1 \otimes Z_1' + Bq_1^2 \otimes 1 \\ &+ 2Cu^2 \otimes Z_1' + 2Cu^2 q_1 \otimes 1 \\ &+ Du^4 \otimes 1.\end{aligned}$$

To simplify this expression we use the relation $2u^2 q_1 = 8q_1^2 - 20q_2$, getting

$$\begin{aligned}\psi_{MSp}(AZ_2' + BZ_1'^2 + CxZ_1' + Dy) &= 1 \otimes (AZ_2' + BZ_1'^2 + CxZ_1' + Dy) \\ &- 2Cv^2 \otimes Z_1' - Dv^4 \otimes 1 + 2Aq_1 \otimes Z_1' + Aq_2 \otimes 1 \\ &+ 2Bq_1 \otimes Z_1' + Bq_1^2 \otimes 1 + 2Cu^2 \otimes Z_1' + 8Cq_1^2 \otimes 1 - 20Cq_2 \otimes 1 \\ &+ Du^4 \otimes 1 = 1 \otimes (AZ_2' + BZ_1'^2 + CxZ_1' + Dy) \\ &+ (2A + 2B - 24C)q_1 \otimes Z_1' \\ &+ \{(A - 20C + 240D)q_2 + (B + 8C - 240D)q_1^2\} \otimes 1.\end{aligned}$$

This element is primitive if and only if

$$2A + 2B - 24C = A - 20C + 240D = B + 8C - 240D = 0,$$

or in other words $A = 20C - 240D$, $B = -8C + 240D$. Thus $P_8 \cong \mathbb{Z} \oplus \mathbb{Z}$ generated by

$$\begin{gathered}20Z_2' - 8Z_1'^2 + xZ_1' \\ -240Z_2' + 240Z_1'^2 + y,\end{gathered}$$

or equivalently by

$$\begin{gathered}20Z_2' - 8Z_1'^2 + xZ_1' \\ 144Z_1'^2 + 12xZ_1' + y = \tfrac{1}{4}(24Z_1' + x)^2.\end{gathered}$$

dim 9: $G_2 \cap KO_9(MSp)$ is generated by $\alpha Z_2'$, $\alpha Z_1'^2$ and $\alpha y \cdot 1$. We find

$$\begin{aligned}&\psi_{MSp}(\alpha Z_2') = 1 \otimes \alpha Z_2' + \alpha q_2 \otimes 1 \\ &\psi_{MSp}(\alpha Z_1'^2) = 1 \otimes \alpha Z_1'^2 \\ &\psi_{MSp}(\alpha y \cdot 1) = \alpha u^4 \otimes 1 = 1 \otimes \alpha y - \alpha(v^4 - u^4) \otimes 1 = 1 \otimes \alpha y\end{aligned}$$

since
$$v^4 - u^4 = 240(q_1^2 - q_2).$$

Now $\alpha q_2 \neq 0$ since

$$q_2'(1,3) = \frac{1}{360}(9-1)(9-4) = \frac{1}{9} \notin 2\mathbb{Z}\left[\frac{1}{3}\right].$$

Hence $P_9 \cong \mathbb{Z}_2 \oplus \mathbb{Z}_2$ generated by $\alpha Z_1'^2$, αy.

dim 10: Similarly $P_{10} \cong \mathbb{Z}_2 \oplus \mathbb{Z}_2$ generated by $\alpha^2 Z_1'^2$, $\alpha^2 y$.

dim 11: $KO_{11}(MSp) = 0$, so $P_{11} = 0$.

We can make a table of values:

n	0	1	2	3	4	5	6	7	8	9	10	11
$P_n(KO_*(MSp))$	$\mathbb{Z}$	$\mathbb{Z}_2$	$\mathbb{Z}_2$	0	$\mathbb{Z}$	$\mathbb{Z}_2$	$\mathbb{Z}_2$	0	$\mathbb{Z} \oplus \mathbb{Z}$	$\mathbb{Z}_2 \oplus \mathbb{Z}_2$	$\mathbb{Z}_2 \oplus \mathbb{Z}_2$	0.

From 17.14 we know $h(\pi_n(MSp)) \subset P_n(KO_*(MSp))$ for every n. For $n \leqslant 8$ one can construct explicit Sp-manifolds M^n with $h[M^n]$ = the generator calculated above (cf. [75]). Thus $h(\pi_n(MSp)) = P_n$, $n \leqslant 8$.

17.42. Exercise. By the remarks following 17.21 we can define unique additive (unstable) cohomology operations $\Psi^k: K^0(X) \to K^0(X)$ for $k \in \mathbb{Z}$ by taking $\Psi^k(\gamma) = \gamma^k = (1+Y)^k$.

a) By using the splitting method (cf. proof of 16.21) show that each Ψ^k is multiplicative: $\Psi^k(xy) = \Psi^k(x)\Psi^k(y)$ for $x, y \in K^0(X)$.

b) Show that $\Psi^k \circ \Psi^l = \Psi^{kl}$ for all $k, l \in \mathbb{Z}$.

c) Determine Ψ^k on $K^0(S^{2n})$ for each $k \in \mathbb{Z}$, $n \geqslant 0$.

Comments

In Chapter 18 we shall determine $E_*(E)$ for $E = H(\mathbb{Z}_2)$ and describe $E_*(E)$ for $E = H(\mathbb{Z}_p)$, p odd. Then in Chapters 19 and 20 come the really important applications of these Hopf algebras. For applications of cohomology operations see 18.24 and [68].

References

1. J. F. Adams [5, 6, 7, 8]
2. J. F. Adams, A. S. Harris and R. M. Switzer [11]
3. P. S. Landweber [51]
4. R. E. Mosher and M. C. Tangora [68]
5. S. P. Novikov [70]

Chapter 18

The Steenrod Algebra and its Dual

Having determined $E_*(E)$ for $E = MU$, MSp, K and KO we turn now to a determination of $E_*(E)$ and $E^*(E)$ for $E = H(\mathbb{Z}_2)$. Although historically $H^*(H(\mathbb{Z}_2);\mathbb{Z}_2)$ and $H_*(H(\mathbb{Z}_2);\mathbb{Z}_2)$ were known as much as fifteen years before $KO_*(KO)$, for example, we shall see that the calculation of these Hopf algebras is at least as difficult as the calculation we just did for KO. In theory we could proceed as follows: for each n we have a fibration $H(\mathbb{Z}_2,n-1) \to PH(\mathbb{Z}_2,n) \to H(\mathbb{Z}_2,n)$, and the total space $PH(\mathbb{Z}_2,n)$ is contractible (the path space PY is contractible for any Y). Thus in the Serre spectral sequence for this fibration $E^\infty_{pq} = 0$ if $(p,q) \neq (0,0)$. With a little luck this should enable us to determine $\tilde{H}_*(H(\mathbb{Z}_2,n);\mathbb{Z}_2)$ from a knowledge of $\tilde{H}_*(H(\mathbb{Z}_2,n-1);\mathbb{Z}_2)$. The process could begin at $n=1$ since $\tilde{H}_*(H(\mathbb{Z}_2,1);\mathbb{Z}_2) \cong \tilde{H}_*(RP^\infty;\mathbb{Z}_2)$, which is a $\mathbb{Z}_2$-vector space with basis $\{x_i : i \geqslant 1\}$, $x_i \in \tilde{H}_i(RP^\infty;\mathbb{Z}_2)$. Then we would have $H_q(H(\mathbb{Z}_2);\mathbb{Z}_2) \cong \operatorname{dir\,lim}_n \tilde{H}_{q+n}(H(\mathbb{Z}_2,n);\mathbb{Z}_2)$. This is precisely how Cartan did determine this algebra (see [28]) using some heavy guns from homological algebra. We shall take a different approach, however; we shall construct some specific cohomology operations—the Steenrod squares Sq^i—and show that they generate the algebra $A^* = H^*(H(\mathbb{Z}_2);\mathbb{Z}_2)$. It will then not be too difficult to determine the dual $A_* = H_*(H(\mathbb{Z}_2);\mathbb{Z}_2)$. We shall also indicate the results for $H(\mathbb{Z}_p)$, p an odd prime.

For now we shall consider only the prime 2, so for simplicity we shall write H instead of $H(\mathbb{Z}_2)$ and $H^*(X,A)$ instead of $H^*(X,A;\mathbb{Z}_2)$. At one point we shall have to consider integral cohomology, and then we shall write $H^*(X,A;\mathbb{Z})$ while continuing to use $H^*(X,A)$ for $\mathbb{Z}_2$-cohomology.

The idea for the Steenrod squares is the following: the diagonal map $\Delta: X \to X \times X$ satisfies $\tau \circ \Delta = \Delta$, where $\tau: X \times X \to X \times X$ is the "switch map". Any chain map $D_0: S(X) \to S(X) \otimes S(X)$ inducing the diagonal $H_0(X) \to H_0(X) \otimes H_0(X)$ is called a *diagonal approximation*; e.g. if $\rho: S(X \times X) \to S(X) \otimes S(X)$ is a chain equivalence as in the Eilenberg–Zilber theorem (13.30), then $\rho \circ \Delta_\# : S(X) \to S(X) \otimes S(X)$ is a diagonal approximation. But a diagonal approximation need not satisfy

$T \circ D_0 = D_0$, where $T(x \otimes y) = (-1)^{|x||y|}(y \otimes x)$. However, there is a chain homotopy $D_1 : S(X) \to S(X) \otimes S(X)$ of degree 1 between D_0 and $T \circ D_0 : \partial D_1 + D_1 \partial = T \circ D_0 - D_0$. Similarly $T \circ D_1$ may not be D_1, but there will be a D_2 of degree 2 .... The construction relies heavily on the acyclic models theorem 13.26.

18.1. Proposition. *There are natural homomorphisms* $D_j : S(X) \to S(X) \otimes S(X)$ *of degree* $j, j \geqslant 0$, *such that*

i) D_0 *is a chain map inducing the diagonal* $H_0(X) \to H_0(X) \otimes H_0(X)$;

ii) $dD_j + (-1)^{j+1} D_j d = D_{j-1} + (-1)^j TD_{j-1}, j > 0$.

If $\{D_j\}$ *and* $\{D'_j\}$ *are two such sequences, then there are natural homomorphisms* $E_j : S(X) \to S(X) \otimes S(X)$ *of degree* $j, j \geqslant 0$, *such that*

iii) $E_0 = 0$;

iv) $dE_{j+1} + (-1)^j E_{j+1} d = -E_j + (-1)^{j+1} TE_j + D_j - D'_j, j \geqslant 0$.

Proof: We apply 13.26 to the category $\mathscr{C}$ of chain complexes over a ring R, where in our case R is to be $\mathbb{Z}[t]/(t^2 - 1)$, the group ring of $\mathbb{Z}_2$. Let W be the R-free chain complex

$$\cdots \longrightarrow R \xrightarrow{1+t} R \xrightarrow{1-t} R \longrightarrow \cdots \xrightarrow{1-t} R \xrightarrow{\varepsilon} \mathbb{Z};$$

in other words, W is the free graded R-module with one generator $w_k \in W_k$ for each $k \geqslant 0$ and equipped with differential d given by

$$d(w_k) = [1 + (-1)^k t]\, w_{k-1}, \quad k \geqslant 1,$$

and augmentation ε with $\varepsilon(1) = \varepsilon(t) = 1$.

Consider the functors $F, G : \mathscr{T} \to \mathscr{C}$ given by

$$F(X) = W \otimes S(X) \qquad (t \cdot (w \otimes u) = tw \otimes u)$$
$$G(X) = S(X) \otimes S(X) \quad (t(u \otimes u') = T(u \otimes u')).$$

Then F is free on models $M = \{\Delta^n : n \geqslant 0\}$ and G is acyclic on M (by the Künneth theorem 13.13). 13.26 says there is a chain map $\phi : F \to G$ inducing the diagonal. We define the D_j by taking

18.2. $$D_j(a) = \phi(w_j \otimes a), \quad a \in S(X).$$

Writing down the condition that ϕ be a chain map yields i) and ii) for the collection $\{D_j\}$. On the other hand given $\{D_j\}$ satisfying i) and ii), we can define a diagonal-inducing chain map ϕ by using 18.2 and extending linearly. Chain homotopies correspond to collections $\{E_j\}$ satisfying iii) and iv). ▯

Now the D_j give us cochain maps

$$D_j^* : \mathrm{Hom}(S(X) \otimes S(X), \mathbb{Z}) \to \mathrm{Hom}(S(X), \mathbb{Z})$$

of degree $-j$. Given cochains $c \in \operatorname{Hom}(S_n(X), \mathbb{Z})$, $d \in \operatorname{Hom}(S_m(X), \mathbb{Z})$ we have their tensor product $c \otimes d \in \operatorname{Hom}((S(X) \otimes S(X))_{n+m}, \mathbb{Z})$, and we define

18.3. $$c \cup_i d = D_i^*(c \otimes d) \in \operatorname{Hom}(S_{n+m-i}(X), \mathbb{Z}).$$

The cochain $\cup_i$-product enjoys the following properties:

18.4. $\cup_i$ is natural: $f^*(c \cup_i d) = f^* c \cup_i f^* d$ for any map $f: Y \to X$, $c, d \in S^*(X; \mathbb{Z})$.

18.5.
$$\begin{aligned} \delta(c \cup_i c') = (-1)^i \delta c \cup_i c' + (-1)^{i+n} c \cup_i \delta c' \\ - (-1)^i c \cup_{i-1} c' - (-1)^{nm} c' \cup_{i-1} c \end{aligned}$$

for $c \in S^n(X; \mathbb{Z})$, $c' \in S^m(X; \mathbb{Z})$.

Proof: For every $a \in S(X)$ we have

$$\begin{aligned} \delta(c \cup_i c')(a) &= D_i^*(c \otimes c')(da) = (c \otimes c')(D_i\, da) \\ &= (c \otimes c')((-1)^i dD_i a - (-1)^i D_{i-1} a - TD_{i-1} a) \\ &= (-1)^i \delta(c \otimes c')(D_i a) - (-1)^i (c \otimes c')(D_{i-1} a) \\ &\quad - (-1)^{nm}(c' \otimes c)(D_{i-1} a) \\ &= ((-1)^i \delta c \otimes c' + (-1)^{i+n} c \otimes \delta c')(D_i a) \\ &\quad - [(-1)^i c \otimes c' + (-1)^{nm} c' \otimes c](D_{i-1} a) \\ &= [(-1)^i \delta c \cup_i c' + (-1)^{i+n} c \cup_i \delta c' - (-1)^i c \cup_{i-1} c' \\ &\quad - (-1)^{nm} c' \cup_{i-1} c](a). \quad \square \end{aligned}$$

18.6.
$$(c_1 + c_2) \cup_i (d_1 + d_2) = c_1 \cup_i d_1 + c_1 \cup_i d_2 + c_2 \cup_i d_1 + c_2 \cup_i d_2.$$

This follows from the linearity of D_i^*.

We now define $Sq^i c$ to be $c \cup_{n-i} c$ if $0 \leqslant i \leqslant n$, 0 if $i > n$ for
$$c \in \operatorname{Hom}(S_n(X), \mathbb{Z}).$$

Sq^i satisfies the following:

18.7. Sq^i is natural: $f^* Sq^i c = Sq^i f^* c$.

18.8. If c vanishes mod 2 on $S(A)$, $A \subset X$, then so does $Sq^i c$ (this is a particular case of naturality).

18.9. If $\delta c = 0 \bmod 2$, then $\delta Sq^i c = 0 \bmod 2$.

Proof:
$$\delta Sq^i c = \delta(c \cup_{n-i} c) = \delta c \cup_{n-i} c + c \cup_{n-i} \delta c + 2c \cup_{n-i-1} c = 0 \bmod 2. \quad \square$$

18.10. If $c = \delta d \bmod 2$, then

$$Sq^i c = \delta[d \cup_{n-i} c + d \cup_{n-i-1} d] \bmod 2, \quad 0 \leqslant i \leqslant n.$$

Proof:

$$\begin{aligned}
&\delta[d \cup_{n-i} c + d \cup_{n-i-1} d] \\
&\quad = \delta d \cup_{n-i} c + d \cup_{n-i} \delta c + d \cup_{n-i-1} c + c \cup_{n-i-1} d + \delta d \cup_{n-i-1} d \\
&\qquad + d \cup_{n-i-1} \delta d + 2d \cup_{n-i-2} d \\
&\quad = c \cup_{n-i} c = Sq^i c \bmod 2. \quad \square
\end{aligned}$$

18.11. If c_1, c_2 are cocycles mod 2, then

$$Sq^i(c_1 + c_2) = Sq^i c_1 + Sq^i c_2 + \delta(c_1 \cup_{n-i+1} c_2) \bmod 2.$$

Proof: This is obvious if $i > n$; otherwise

$$\begin{aligned}
\delta(c_1 \cup_{n-i+1} c_2) &= \delta c_1 \cup_{n-i+1} c_2 + c_1 \cup_{n-i+1} \delta c_2 + c_1 \cup_{n-i} c_2 + c_2 \cup_{n-i} c_1 \\
&= c_1 \cup_{n-i} c_2 + c_2 \cup_{n-i} c_1 \bmod 2.
\end{aligned}$$

Thus

$$\begin{aligned}
Sq^i(c_1 + c_2) &= (c_1 + c_2) \cup_{n-i} (c_1 + c_2) \\
&= c_1 \cup_{n-i} c_1 + c_1 \cup_{n-i} c_2 + c_2 \cup_{n-i} c_1 + c_2 \cup_{n-i} c_2 \\
&= Sq^i c_1 + Sq^i c_2 + \delta(c_1 \cup_{n-i+1} c_2) \bmod 2. \quad \square
\end{aligned}$$

18.7–18.11 imply that we can define a natural homomorphism $Sq^i : H^n(X, A; \mathbb{Z}_2) \to H^{n+i}(X, A; \mathbb{Z}_2)$ for each n, all (X, A), by taking

$$Sq^i\{c\}_2 = \{Sq^i c\}_2.$$

Sq^i does not depend on the choice of the sequence $\{D_j\}$, for if $\{D'_j\}$ is another such sequence, then by 18.1 there is a sequence $\{E_j\}$ satisfying 18.1iii) and iv), and we have for any cocycle c

$$\begin{aligned}
0 &= (c \otimes c)[dE_{n-i+1} + E_{n-i+1} d + E_{n-i} + TE_{n-i} + D_{n-i} + D'_{n-i}](a) \\
&= (c \otimes c)[E_{n-i+1} d + D_{n-i} + D'_{n-i}](a) \\
&= [Sq^i c + Sq^{i\prime} c + \delta E^*_{n-i+1}(c \otimes c)](a) \bmod 2
\end{aligned}$$

for all $a \in S(X)$. Thus $\{Sq^i c\}_2 = \{Sq^{i\prime} c\}_2$.

By reducing mod 2 we can also regard Sq^i as a homomorphism

$$Sq^i : H^n(X, A; \mathbb{Z}) \to H^{n+i}(X, A; \mathbb{Z}_2).$$

18.12. Proposition. *The operations Sq^i satisfy*

a) $Sq^0 = 1$;

b) *Sq^1 is the Bockstein (connecting homomorphism) associated with the coefficient sequence*

$$0 \to \mathbb{Z}_2 \to \mathbb{Z}_4 \to \mathbb{Z}_2 \to 0;$$

c) *if* $x \in H^n(X,A;\mathbb{Z}_2)$, *then* $Sq^n x = x^2$;
d) *if* $x \in H^n(X,A;\mathbb{Z}_2)$, *then* $Sq^i x = 0$ *for* $i > n$;
e) $\delta Sq^i = Sq^i \delta$, $\delta : H^{n-1}(A;\mathbb{Z}_2) \to H^n(X,A;\mathbb{Z}_2)$;
f) $Sq^1 Sq^{2j} = Sq^{2j+1}$, $Sq^1 Sq^{2j+1} = 0, j \geqslant 0$;
g) $Sq^i xy = \sum_{j+k=i} Sq^j x Sq^k y$.

Proof: d) is obvious from the definitions. Let $\rho : S(X \times Y) \to S(X) \otimes S(Y)$ be a chain equivalence as in 13.30. If $\Delta : X \to X \times X$ is the diagonal, then D_0 and $\rho \circ \Delta_\# : S(X) \to S(X) \otimes S(X)$ are both diagonal-inducing chain maps and hence chain homotopic. Therefore we may use D_0 to define $\cup$-products; in other words

$$\{c \cup_0 d\} = \{D_0^*(c \otimes d)\} = \{\Delta^\# \circ \rho^*(c \otimes d)\} = \{c\} \cup \{d\}.$$

Thus $Sq^n x = x \cup_0 x = x \cup x = x^2$ for $x \in H^n(X,A;\mathbb{Z}_2)$.

We next prove e). Recall how δ is defined; if c is a cocycle in $S^{n-1}(A;\mathbb{Z}_2)$, then we choose a cochain $d \in S^{n-1}(X;\mathbb{Z}_2)$ with $i^\# d = c$. Then $i^\# \delta d = \delta i^\# d = \delta c = 0$, so δd may be regarded as lying in $S^n(X,A;\mathbb{Z}_2)$. We then let $\delta\{c\} = \{\delta d\}$. Thus $Sq^i\{c\}$ and $Sq^i \delta\{c\}$ are represented by $c \cup_{n-i-1} c$ and $\delta d \cup_{n-i} \delta d$ respectively. Let $d' = d \cup_{n-i} \delta d + d \cup_{n-i-1} d$. Then $i^\# d' = i^\# d \cup_{n-i} i^\#(\delta d) + i^\# d \cup_{n-i-1} i^\# d = c \cup_{n-i-1} c \bmod 2$, while

$$\delta d' = \delta d \cup_{n-i} \delta d \bmod 2,$$

which shows that $\delta Sq^i\{c\} = Sq^i \delta\{c\}$.

Let $\beta : H^{n-1}(-;\mathbb{Z}_2) \to H^n(-;\mathbb{Z}_2)$ be the Bockstein associated to the coefficient sequence $0 \to \mathbb{Z}_2 \xrightarrow{\alpha} \mathbb{Z}_4 \xrightarrow{r} \mathbb{Z}_2 \to 0$ and let

$$S(j) = \begin{cases} 1 & j \text{ even} \\ 0 & j \text{ odd}. \end{cases}$$

We show next that $\beta Sq^j = S(j) Sq^{j+1}$; given this result both b) and f) will follow from a). If $\{c\} \in H^{n-1}(X,A;\mathbb{Z}_2)$, then we choose a cochain $d \in S^{n-1}(X,A;\mathbb{Z}_4)$ with $r_\#(d) = c$. Then $r_\#(\delta d) = \delta r_\#(d) = \delta c = 0$, so $\delta d = \alpha_\# c'$ for some $c' \in S^n(X,A;\mathbb{Z}_2)$. $\beta\{c\} = \{c'\}$. $Sq^j\{c\}$ is represented by $c \cup_{n-j-1} c$. $\delta c = 2b$ for some integral class $b \in S^n(X,A;\mathbb{Z})$. We have

$$\begin{aligned} \delta(c \cup_{n-j-1} c) = (-1)^{n-j-1} 2b \cup_{n-j-1} c + (-1)^{j-1} c \cup_{n-j-1} 2b \\ - (-1)^{n-j-1} c \cup_{n-j-2} c - (-1)^{n-1} c \cup_{n-j-2} c. \end{aligned}$$

Hence $\beta Sq^j\{c\}$ is represented by the mod 2 cocycle

$$b \cup_{n-j-1} c + c \cup_{n-j-1} b + S(j) c \cup_{n-j-2} c.$$

But $b \cup_{n-j-1} c + c \cup_{n-j-1} b = \delta(c \cup_{n-j} b) \bmod 2$, and hence $\beta Sq^j\{c\} = S(j) Sq^{j+1}\{c\}$, as claimed.

We turn now to a proof of a). We know from 15.54 that $H^*(RP^2) \cong \mathbb{Z}_2[x]/(x^3)$. We have $\beta Sq^0 x = Sq^1 x = x^2 \neq 0$ by c). Therefore $Sq^0 x \neq 0$ and hence $Sq^0 x = x$, x being the only non-zero element of $H^1(RP^2)$. Let $f: S^1 \to RP^2$ be the inclusion of the 1-cell; then $f^* x = g_1$, the generator of $\tilde{H}^1(S^1)$. Hence $Sq^0 g_1 = Sq^0 f^* x = f^* Sq^0 x = f^* x = g_1$. From e) it follows that $\sigma Sq^n = Sq^n \sigma$ for all n, so if $g_n \in \tilde{H}^n(S^n)$ is the generator, then $Sq^0 g_n = Sq^0 \sigma^{-n+1} g_1 = \sigma^{-n+1} Sq^0 g_1 = \sigma^{-n+1} g_1 = g_n$. By the Hopf isomorphism theorem 10.32 $\psi: [X, S^n] \to \tilde{H}^n(X; \mathbb{Z})$ is an isomorphism if $\dim X \leqslant n$. Also the Bockstein sequence associated with the coefficient sequence

$$0 \longrightarrow \mathbb{Z} \xrightarrow{2} \mathbb{Z} \xrightarrow{r} \mathbb{Z}_2 \longrightarrow 0$$

ends with $\cdots \to H^n(X;\mathbb{Z}) \xrightarrow{2} H^n(X;\mathbb{Z}) \xrightarrow{r_*} H^n(X;\mathbb{Z}_2) \to 0$ if X is n-dimensional. Therefore given any $x \in \tilde{H}^n(X;\mathbb{Z}_2)$, we can choose an $x' \in \tilde{H}^n(X;\mathbb{Z})$ with $r_*(x') = x$ and a map $h: X \to S^n$ with $\psi[h] = x'$. Then

$$x = r_* x' = r_* h^*(g_n) = h^*(r_* g_n) = h^*(g_n),$$

so

$$Sq^0 x = Sq^0 h^*(g_n) = h^*(Sq^0 g_n) = x.$$

For any arbitrary CW-complex X with n-skeleton X^n, the inclusion $i: X^n \to X$ induces a monomorphism $i^*: \tilde{H}^n(X) \to \tilde{H}^n(X^n)$, so it follows $Sq^0 = 1$ on X. For arbitrary topological spaces we prove $Sq^0 = 1$ by using CW-substitutes. Finally, for any pair (X, A) we have

$$H^n(X, A) \xleftarrow{\cong} H^n(X \cup CA, CA) \xrightarrow{\cong} H^n(H \cup CA),$$

so the relative case follows from the absolute case.

This leaves only g) still to be proved. Let

$$\rho: S(X \times Y) \to S(X) \otimes S(Y), \quad \sigma: S(X) \otimes S(Y) \to S(X \times Y)$$

be chain homotopy inverses given by 13.30. If $r: W \to W \otimes W$ is defined by

$$r(w_i) = \textstyle\sum_{j+k=i} (-1)^{jk} w_j \otimes t^j w_k$$

and extended to be R-linear, then the composite

$$W \otimes S(X \times Y) \xrightarrow{r \otimes \rho} W \otimes W \otimes S(X) \otimes S(Y) \xrightarrow{1 \otimes T \otimes 1}$$

$$W \otimes S(X) \otimes W \otimes S(Y) \xrightarrow{\phi_X \otimes \phi_Y}$$

$$S(X) \otimes S(X) \otimes S(Y) \otimes S(Y) \xrightarrow{1 \otimes T \otimes 1}$$

$$S(X) \otimes S(Y) \otimes S(X) \otimes S(Y) \xrightarrow{\sigma \otimes \sigma} S(X \times Y) \otimes S(X \times Y)$$

is a diagonal-inducing chain map, and hence will do as $\phi_{X\times Y}$. Note that $\sigma \circ \rho$ is chain homotopic to 1, so for any cocycle c and cycle a we have $c(\sigma \circ \rho(a)) = c(a)$. Similarly for $\rho \circ \sigma$. Thus for any cocycles $c \in S^n(X;\mathbb{Z}_2)$, $d \in S^m(Y;\mathbb{Z}_2)$ and cycles $a \in S(X;\mathbb{Z}_2)$, $b \in S(Y;\mathbb{Z}_2)$ we find (calculating mod 2)

$$\begin{aligned}
[Sq^i(c \times d)](a \times b) &= [Sq^i \rho^*(c \otimes d)]\,\sigma(a \otimes b) \\
&= [\rho^*(c \otimes d) \otimes \rho^*(c \otimes d)]\,\phi_{X\times Y}(w_{n+m-i} \otimes \sigma(a \otimes b)) \\
&= [c \otimes d \otimes c \otimes d](\rho \otimes \rho) \circ (\sigma \otimes \sigma) \circ (1 \otimes T \otimes 1) \circ \\
&\quad (\phi_X \otimes \phi_Y) \circ (1 \otimes T \otimes 1) \circ \\
&\quad (r \otimes \rho)(w_{n+m-i} \otimes \sigma(a \otimes b)) \\
&= [c \otimes d \otimes c \otimes d](1 \otimes T \otimes 1) \circ (\phi_X \otimes \phi_Y) \circ \\
&\quad (1 \otimes T \otimes 1)\big(\textstyle\sum_{j+k=n+m-i} w_j \otimes t^j w_k \otimes a \otimes b\big) \\
&= \textstyle\sum_{j+k=n+m-i} (c \otimes c \otimes d \otimes d) \circ \\
&\quad (\phi_X \otimes \phi_Y)(w_j \otimes a \otimes t^j w_k \otimes b) \\
&= \textstyle\sum_{j+k=n+m-i} [(c \otimes c)(\phi_X(w_j \otimes a)] \\
&\quad [(d \otimes d)(T^j \phi_Y(w_k \otimes b))] \\
&= \textstyle\sum_{j+k=n+m-i} (Sq^{n-j}(c))(a) \cdot (Sq^{m-k}(d))(b) \\
&= \textstyle\sum_{r+s=i} (Sq^r c \times Sq^s d)(a \times b).
\end{aligned}$$

Since all elements of $H_*(X \times Y)$ are of the form $\sum_i \{a_i \times b_i\}$ for $\{a_i\} \in H_*(X)$, $\{b_i\} \in H_*(Y)$, we see that

$$Sq^i(x \times y) = \sum_{j+k=i} Sq^j x \times Sq^k y$$

for all $x \in H^*(X)$, $y \in H^*(Y)$. We then have only to apply the diagonal map to obtain the formula for $\cup$-products. □

Of course property e) implies that $\sigma \circ Sq^n = Sq^n \circ \sigma$ for all n and hence that the Sq^i are stable cohomology operations—that is $Sq^i \in A^*$, $i \geqslant 0$.

We now begin the computation of $A^* = H^*(H)$. From 8.37 we have an exact sequence

$$0 \to \lim_n{}^1 \tilde{H}^{q+n-1}(H_n) \to H^q(H) \to \lim_n{}^0 \tilde{H}^{q+n}(H_n) \to 0,$$

where the homomorphisms of the inverse system are

$$\tilde{H}^{q+n}(H_n) \xrightarrow{\varepsilon^*_{n-1}} \tilde{H}^{q+n}(SH_{n-1}) \underset{\cong}{\xrightarrow{\sigma}} \tilde{H}^{q+n-1}(H_{n-1}),$$

or equivalently (15.43)

$$\tilde{H}^{q+n}(H_n) \xrightarrow{\sigma'} \tilde{H}^{q+n-1}(\Omega H_n) \xrightarrow[\cong]{\varepsilon'^*_{n-1}} \tilde{H}^{q+n-1}(H_{n-1}).$$

By 15.42 σ' is an isomorphism for $n > q+1$, since H_n is $(n-1)$-connected. Therefore $\varprojlim^1_n \tilde{H}^{q+n-1}(H_n)$ vanishes for all q, so we can compute $H^q(H)$ by taking $\varprojlim^0_n \tilde{H}^{q+n}(H_n)$.

We know $H_1 \simeq RP^\infty$, so we have $H^*(H_1) \cong \mathbb{Z}_2[x]$. We are going to compute $H^*(H_n)$ inductively using the Serre spectral sequence of the fibration

$$H_{n-1} \simeq \Omega H_n \to PH_n \to H_n.$$

Let $\iota_n \in H^n(H_n)$ be the class corresponding to $1 \in [H_n, H_n]$; for $n = 1$ we have $\iota_1 = x$. In $H^*(H_2)$ we have the elements

$$\iota_2, Sq^1 \iota_2, Sq^2 Sq^1 \iota_2, Sq^4 Sq^2 Sq^1 \iota_2, \ldots.$$

From the definition of σ' and the fact that the Sq^i commute with δ it follows that the Sq^i commute with σ'. Moreover $\sigma' \iota_n = \iota_{n-1}$ for all $n \geq 2$. Therefore we have

$$\sigma'(Sq^{2^{s-1}} Sq^{2^{s-2}} \ldots Sq^1 \iota_2) = Sq^{2^{s-1}} \ldots Sq^1 \iota_1 = x^{2^s}.$$

The E_2-term of the Serre spectral sequence then looks as follows

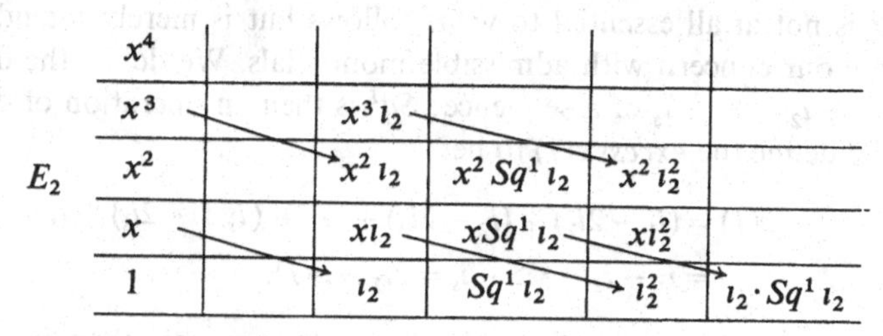

The E_3-term looks like this

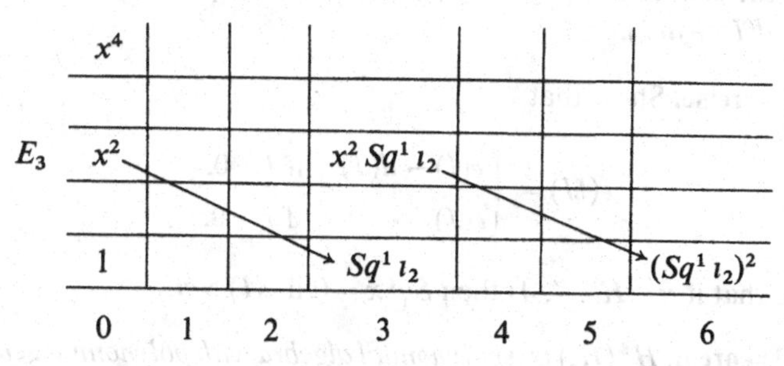

The pattern suggests that $H^*(H_2)$ is a polynomial algebra on the generators

$\iota_2, Sq^1\iota_2, Sq^2 Sq^1 \iota_2, \ldots$, and indeed this is true. To prove this we can apply the Borel Theorem 15.62. We have

$$\sigma'(Sq^{2^{n-1}} Sq^{2^{n-2}} \ldots Sq^2 Sq^1 \iota_2) = x^{2^n}$$

for each $n \geqslant 1$. But from the Remark 15.59 we know that the elements $x, x^2, x^4, \ldots x^{2^n}, \ldots$ form a simple system of generators for

$$H^*(H(\mathbb{Z}_2, 1); \mathbb{Z}_2) \cong \mathbb{Z}_2[x].$$

Thus the Borel theorem applies.

In order to deal with the case $n > 2$, however, we shall need some notation. Let $I = (i_1, i_2, \ldots, i_s)$ be any sequence of integers; $I = 0$ will mean I is the empty sequence. We define

$$Sq^I = Sq^{i_1} Sq^{i_2} \ldots Sq^{i_s} \quad (Sq^0 = 1, \text{ of course}).$$

We call a sequence I *admissible* if $I = 0$ or $i_1 \geqslant 2i_2$, $i_2 \geqslant 2i_3, \ldots, i_{s-1} \geqslant 2i_s$, $i_s \geqslant 1$; the monomials Sq^I for admissible I are called *admissible monomials*. As the Sq^i's satisfy the *Adem relations*

$$Sq^a Sq^b = \sum_{c=0}^{[a/2]} \binom{b-c-1}{a-2c}_2 Sq^{a+b-c} Sq^c \quad \text{for } a < 2b,$$

of which 18.12f) is a special case, we see that every monomial Sq^I is expressible as a linear combination of the admissible monomials. This remark is not at all essential to what follows but is merely intended to motivate our concern with admissible monomials. We define the degree $d(I) = i_1 + i_2 + \cdots + i_s$ of a sequence; Sq^I is then an operation of degree $d(I)$. We define the *excess* $e(I)$ to be

$$\begin{aligned} e(I) &= (i_1 - 2i_2) + (i_2 - 2i_3) + \cdots + (i_{s-1} - 2i_s) + i_s \\ &= i_1 - i_2 - \cdots - i_s = 2i_1 - d(I). \end{aligned}$$

If $I = (i_1, i_2, \ldots, i_s)$, $J = (j_1, j_2, \ldots, j_t)$ are two sequences, then we let IJ denote the sequence $(i_1, i_2, \ldots, i_s, j_1, \ldots, j_t)$. Thus $Sq^{IJ} = Sq^I \circ Sq^J$. Clearly $d(IJ) = d(I) + d(J)$.

18.13. Exercise. Show that

$$e(IJ) = \begin{cases} e(I) - d(J) & \text{if } I \neq 0. \\ e(J) & \text{if } I = 0. \end{cases}$$

Deduce that if $x \in H^n(X, A)$ then $Sq^I x = 0$ if $e(I) > n$.

18.14. Theorem. *$H^*(H_n)$ is a polynomial algebra with polynomial generators all $Sq^I \iota_n$ such that I is admissible and $e(I) < n$.*

Proof: The assertion is clearly true for $n=1$, and we have just proved it for $n=2$ also, since $e(I)<2$ means I is of the form 0 or $(2^{s-1},2^{s-2},\ldots,2,1)$ for some s. We proceed by induction on n.

Assume the theorem true for n and let $x_1,x_2,\ldots$ be some ordering of the elements $Sq^I\iota_n$, I admissible and $e(I)<n$. Let $x_i\in H^{q_i}(H_n)$; then $q_i=d(I)+n$ if $x_i=Sq^I\iota_n$. We introduce certain admissible sequences

$$J(q,r)=\begin{cases}(2^{r-1}q,2^{r-2}q,\ldots,2q,q) & r\geqslant 1\\ 0 & r=0.\end{cases}$$

Note that

$$e(J(q,r))=\begin{cases}q & r\geqslant 1\\ 0 & r=0\end{cases}$$

and $d(J(q,r))=q(2^r-1)$. Now the elements $\{x_i^{2^r}:i\geqslant 1,r\geqslant 0\}$ form a simple system of generators for $H^*(H_n)$. $x_i^{2^r}$ can also be written

$$Sq^{J(q_i,r)}x_i=Sq^{J(q_i,r)}Sq^I\iota_n=Sq^{J(q_i,r)I}\iota_n \text{ if } x_i=Sq^I\iota_n.$$

Since we have

$$\sigma'(Sq^{J(q_i,r)I}\iota_{n+1})=Sq^{J(q_i,r)I}\sigma'(\iota_{n+1})=Sq^{J(q_i,r)I}\iota_n=x_i^{2^r},$$

we see that Borel's theorem applied to the fibration $H_n\to PH_{n+1}\to H_{n+1}$ implies that $H^*(H_{n+1})$ is a polynomial algebra on generators

$$Sq^{J(n+d(I),r)I}\iota_{n+1}$$

as I runs through all admissible sequences with $e(I)<n$, $r\geqslant 0$. $J(n+d(I),r)I$ is clearly admissible; the only point which needs checking is the "joint", but $n+d(I)\geqslant 2i_1$ because $n>e(I)=2i_1-d(I)$. By 18.13 we have

$$e(J(n+d(I),r)I)=\begin{cases}e(I) & r=0\\ n+d(I)-d(I) & r\geqslant 1.\end{cases}$$

Therefore $e(J(n+d(I),r)I)<n+1$ for all admissible I with $e(I)<n$ and $r\geqslant 0$. Hence we see that the theorem will be proved if we show that every admissible sequence I with $e(I)=n$ can be written in the form

$$I=J(n+d(I'),r)I'$$

for some $r>0$ and some admissible I' with $e(I')<n$.

Given I let r be the first integer such that $i_r>2i_{r+1}$; then we can write

$$I=J(i_r,r)I',$$

where $I' = (i_{r+1}, i_{r+2}, \ldots, i_s)$. Certainly $r \geqslant 1$, so

$$n = e(I) = e(J(i_r, r)) - d(I') = i_r - d(I') > 2i_{r+1} - d(I') = e(I').$$

Also $i_r = n + d(I')$. ☐

Now the homomorphisms of our inverse system $\{\tilde{H}^{q+n}(H_n)\}$ are

$$\tilde{H}^{q+n}(H_n) \xrightarrow{\varepsilon_{n-1}^*} \tilde{H}^{q+n}(SH_n) \xrightarrow{\sigma} \tilde{H}^{q+n-1}(H_{n-1}),$$

and in $H^*(SH_n)$ all products are trivial by 13.66. Thus the homomorphisms of the inverse system annihilate all products in $H^*(H_n)$ and only indecomposable elements—that is, linear combinations of the $Sq^I \iota_n$, $e(I) < n$—survive.

18.15. Theorem. *$A^* = H^*(H)$ has a $\mathbb{Z}_2$-basis consisting of the admissible monomials Sq^I. Therefore as an algebra A^* is generated by the elements $1 = Sq^0, Sq^1, Sq^2 \ldots$.*

Proof: In the paragraph following the proof of 18.12 we saw that $A^* \cong \lim^0 \tilde{H}^*(H_n)$, where the morphisms of the inverse system are given by

$$\tilde{H}^{q+n}(H_n) \xrightarrow{\sigma'} \tilde{H}^{q+n-1}(\Omega H_n) \cong \tilde{H}^{q+n-1}(H_{n-1}),$$

and σ' is an isomorphism if $n > q + 1$. Thus we have

$$A^q \cong \lim_n{}^0 \tilde{H}^{n+q}(H_n) = \tilde{H}^{n+q}(H_n)$$

for any $n > q$. But from 18.14 we know $\tilde{H}^{n+q}(H_n)$ has a $\mathbb{Z}_2$-basis consisting of the elements $Sq^I \iota_n$ with I admissible, $d(I) = q$ and $e(I) < n$. Now $d(I) = q < n$ automatically implies $e(I) < n$, so we can say $\tilde{H}^{n+q}(H_n)$ has a $\mathbb{Z}_2$-basis $\{Sq^I \iota_n : I \text{ admissible}, d(I) = q\}$. Since the map $A^q \to \lim^0 \tilde{H}^{n+q}(H_n)$ is given by $\theta \mapsto \{\theta(\iota_n)\}$, the theorem follows. ☐

Since $A_* = H_*(H)$ is of finite type over $\mathbb{Z}_2$ (this follows from Serre's $\mathscr{C}$-theory, cf. [78], [80]), the universal coefficient theorem implies that $A_* = H_*(H)$ is the vector space dual $\mathrm{Hom}_{\mathbb{Z}_2}(H^*(H); \mathbb{Z}_2)$ of A^*. We now address ourselves to the determination of A_*.

We know that $H_1 \simeq RP^\infty$ and $H_*(H_1)$ has a $\mathbb{Z}_2$-basis $x_1, x_2, \ldots$ dual to the monomials $\{x^i\}$ in $H^*(H_1)$. Moreover H_1 is an H-space, so there is a Pontrjagin product

$$\mu_* : H_*(RP^\infty) \otimes H_*(RP^\infty) \to H_*(RP^\infty)$$

and a coproduct

$$\mu^* : H^*(RP^\infty) \to H^*(RP^\infty) \otimes H^*(RP^\infty).$$

$\mu^*(x)$ must have the form $\mu^*(x) = x \otimes 1 + 1 \otimes x$ since $\mu \circ (*, 1) \simeq 1 \simeq$

$\mu \circ (1, *)$—that is, the constant map $*$ is a homotopy unit. μ^* is a homomorphism of rings, so

$$\mu^*(x^k) = (x \otimes 1 + 1 \otimes x)^k = \sum_{i+j=k} \binom{k}{i}_2 x^i \otimes x^j.$$

Thus we calculate

$$\langle x^k, x_i x_j \rangle = \langle x^k, \mu_*(x_i \otimes x_j) \rangle = \langle \mu^* x^k, x_i \otimes x_j \rangle$$

$$= \sum_{n+m=k} \binom{k}{n}_2 \langle x^n \otimes x^m, x_i \otimes x_j \rangle = \begin{cases} \binom{k}{i}_2 & i+j=k \\ 0 & \text{otherwise.} \end{cases}$$

Thus we have

$$x_i x_j = \left(\frac{(i+j)!}{i!j!}\right)_2 x_{i+j};$$

in other words $H_*(RP^\infty)$ is what is called a *divided polynomial algebra* over $\mathbb{Z}_2$. One finds that x_k is indecomposable only if $k = 2^s$ for some s.

Now we know that decomposable elements are annihilated by the homomorphisms of the direct system $\{\tilde{H}_{q+n}(H_n)\}$, so only the elements x_{2^s} in $\tilde{H}_{2^s}(H_1)$ have a chance of surviving into $H_*(H)$. Let us denote the image of x_{2^s} in $H_{2^s-1}(H)$ by ξ_s, $s \geqslant 1$. That is, if $i: H_1 \to \Sigma H$ is the natural map, then $i_* x_{2^s} = \xi_s$. On the other hand

$$i^* Sq^I = Sq^I x = \begin{cases} x^{2^s} & \text{if } I = (2^{s-1}, 2^{s-2}, \ldots, 2, 1) = J(s, 1) \\ 0 & \text{otherwise.} \end{cases}$$

Hence we have

$$\langle Sq^I, \xi_k \rangle = \langle Sq^I, i_*(x_{2^k}) \rangle = \langle i^* Sq^I, x_{2^k} \rangle = \langle Sq^I x, x_{2^k} \rangle$$

$$= \begin{cases} \langle x^{2^s}, x_{2^k} \rangle & I = J(s, 1) \\ 0 & \text{otherwise} \end{cases} = \begin{cases} 1 & I = J(k, 1) \\ 0 & I \neq J(k, 1). \end{cases}$$

Thus ξ_k is the element dual to $Sq^{J(k,1)}$.

We can write the above result in the form

$$Sq^I x = \sum_{s \geqslant 0} \langle Sq^I, \xi_s \rangle x^{2^s},$$

which seems a little silly until we observe that every $a \in A^*$ is a $\mathbb{Z}_2$-linear combination of the Sq^I's, so we have

18.16. $$ax = \sum_{s \geqslant 0} \langle a, \xi_s \rangle x^{2^s}, \text{ all } a \in A^*.$$

Next we observe that the product $\mu: H \wedge H \to H$ induces a coproduct

$$\phi^*: H^*(H) \to H^*(H \wedge H) \cong H^*(H) \otimes H^*(H).$$

18.17. Proposition. *For any $a \in A^* = H^*(H)$ and $x, y \in H^*(X)$, if $\phi^* a = \sum_i a_i \otimes a_i'$ then $a(xy) = \sum_i (a_i x)(a_i' y)$.*

Proof: For any $z \in H^*(Z) = [Z, H]^*$ represented by $f: Z \to \Sigma^p H$, define $z^*: H^*(H) \to H^*(Z)$ by $z^*(a) = a(z)$—in other words, if $a = [g: H \to \Sigma^q H]$, then $z^*(a) = [\Sigma^p g \circ f: Z \to \Sigma^{p+q} H]$. Then the following diagram commutes for all $x \in H^*(X)$, $y \in H^*(Y)$:

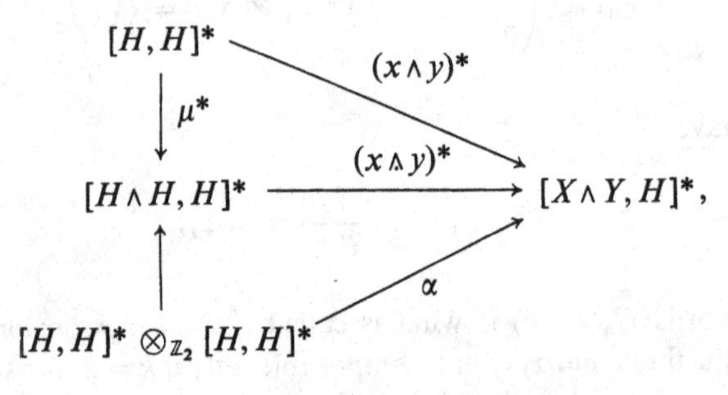

where $x \wedge y$ means $X \wedge Y \xrightarrow{x \wedge y} \Sigma^p H \wedge \Sigma^q H \cong \Sigma^{p+q} H \wedge H$ and $\alpha(a \otimes a') = a(x) \wedge a'(y)$ for $a, a' \in [H, H]^*$. Since $a(x \wedge y) = (x \wedge y)^*(a)$, the proposition follows by applying the diagonal map Δ^*. □

Remark. The proposition applies more generally to any ring spectrum E such that $E^*(E) \otimes_{E^*(S^0)} E^*(E) \to E^*(E \wedge E)$ is an isomorphism.

Now let $X = RP^\infty \times RP^\infty \times \cdots \times RP^\infty$ (n factors) and $p_i: X \to RP^\infty$ be the projection on the ith factor. We let $y_i = p_i^*(x) \in H^1(X)$.

18.18. Proposition. *For any $a \in A^*$*

$$a(y_1 y_2 \ldots y_n) = \textstyle\sum_\alpha \langle a, \xi_\alpha \rangle y_1^{2^{\alpha_1}} y_2^{2^{\alpha_2}} \ldots y_n^{2^{\alpha_n}},$$

where α runs over all sequences $\alpha = (\alpha_1, \alpha_2, \ldots, \alpha_n)$ of integers and ξ_α denotes $\xi_{\alpha_1} \xi_{\alpha_2} \ldots \xi_{\alpha_n}$.

Proof: We proceed by induction on n; the case $n = 1$ is just 18.16. Suppose the result proved for $n - 1$; then for n and any $a \in A^*$ with $\phi^* a = \sum_i a_i \otimes a_i'$ we have

$$\begin{aligned} a(y_1 y_2 \ldots y_n) &= \textstyle\sum_i a_i(y_1 \ldots y_{n-1}) a_i'(y_n) \\ &= \textstyle\sum_i (\sum_\beta \langle a_i, \xi_\beta \rangle y_1^{2^{\beta_1}} \ldots y_{n-1}^{2^{\beta_{n-1}}})(\sum_j \langle a_i', \xi_j \rangle y_n^{2^j}) \\ &= \textstyle\sum_{\beta, j} \langle \sum_i a_i \otimes a_i', \xi_\beta \otimes \xi_j \rangle y_1^{2^{\beta_1}} \ldots y_{n-1}^{2^{\beta_{n-1}}} y_n^{2^j} \\ &= \textstyle\sum_{\beta, j} \langle \phi^* a, \xi_\beta \otimes \xi_j \rangle y_1^{2^{\beta_1}} \ldots y_{n-1}^{2^{\beta_{n-1}}} y_n^{2^j} \end{aligned}$$

$$= \sum_{\beta, j} \langle a, \xi_\beta \xi_j \rangle y_1^{2\beta_1} \ldots y_{n-1}^{2\beta_{n-1}} y_n^{2^j}$$
$$= \sum_\alpha \langle a, \xi_\alpha \rangle y_1^{2\alpha_1} \ldots y_n^{2\alpha_n}$$

(where $\alpha = (\beta_1, \beta_2, \ldots, \beta_{n-1}, j)$). □

Now in $H^*(X) \cong \mathbb{Z}_2[y_1, y_2, \ldots, y_n]$ we have the subalgebra S of symmetric polynomials. The fundamental theorem on symmetric polynomials asserts that $S = \mathbb{Z}_2[\sigma_1, \sigma_2, \ldots, \sigma_n]$, where $\sigma_1, \sigma_2, \ldots, \sigma_n$ are the elementary symmetric polynomials defined by

$$\prod_{i=1}^n (t + y_i) = \sum_{j=0}^n \sigma_j(y_1, \ldots, y_n) t^{n-j}.$$

In particular $\sigma_n = y_1 y_2 \ldots y_n$. We order the monomials $\sigma_1^{\alpha_1} \sigma_2^{\alpha_2} \ldots \sigma_n^{\alpha_n}$ in S by ordering the sequences $(\alpha_1, \alpha_2, \ldots, \alpha_n)$ lexicographically from the right: that is $\sigma_1^{\alpha_1} \sigma_2^{\alpha_2} \ldots \sigma_n^{\alpha_n} > \sigma_1^{\beta_1} \ldots \sigma_n^{\beta_n}$ if $\alpha_n > \beta_n$ or $\alpha_n = \beta_n$ and $\alpha_{n-1} > \beta_{n-1}$ or if....

18.19. Proposition. *If I is an admissible sequence with $d(I) \leqslant n$, then $Sq^I \sigma_n = \sigma_n \cdot P_I$, where P_I is a symmetric polynomial of the form $P_I = \sigma_{i_1} \sigma_{i_2} \cdots \sigma_{i_s} +$ monomials of lower order.*

Proof: We proceed by induction on $s =$ length of I. If we formally write $Sq = Sq^0 + Sq^1 + Sq^2 + \cdots$, then the Cartan formula 18.12g) can be interpreted as saying that Sq is a ring homomorphism:

$$Sq(xy) = Sq(x)\, Sq(y).$$

Thus

$$Sq(\sigma_n) = Sq(\prod_{i=1}^n y_i) = \prod_{i=1}^n (Sq y_i) = \prod_{i=1}^n (y_i + y_i^2) = \sigma_n \cdot \prod_{i=1}^n (1 + y_i)$$
$$= \sigma_n (\sum_{j=0}^n \sigma_j) = \sum_{j=0}^n \sigma_n \sigma_j.$$

Thus $Sq^i \sigma_n = \sigma_n \sigma_i$, proving the proposition for the case $s = 1$.

Now assume the result for admissible I' of length less than s and suppose $I = (i_1, i_2, \ldots, i_s)$ is admissible. Let $I' = (i_2, i_3, \ldots, i_s)$. Then the Cartan formula gives

$$Sq^I(\sigma_n) = Sq^{i_1} Sq^{I'}(\sigma_n) = Sq^{i_1}(\sigma_n P_{I'}) = \sum_{j=0}^{i_1} Sq^j(\sigma_n) Sq^{i_1 - j}(P_{I'})$$
$$= \sigma_n \sigma_{i_1} P_{I'} + \sum_{j=0}^{i_1 - 1} \sigma_n \sigma_j Sq^{i_1 - j}(P_{I'})$$
$$= \sigma_n \sigma_{i_1} \sigma_{i_2} \ldots \sigma_{i_s} + \sigma_n \sigma_{i_1}(\text{lower terms of } P_{I'})$$
$$+ \sum_{j=0}^{i_1 - 1} \sigma_n \sigma_j Sq^{i_1 - j}(P_{I'}).$$

One easily sees that $\sigma_n \sigma_{i_1} \sigma_{i_2} \cdots \sigma_{i_s}$ is the highest term in this expression. □

18.20. Theorem. *$A_* \cong \mathbb{Z}_2[\xi_1, \xi_2, \ldots]$. The coproduct ψ is given on the generators by the formula*

$$\psi \xi_k = \sum_{i+j=k} \xi_j^{2^i} \otimes \xi_i.$$

Proof: Since the monomials $\sigma_n \sigma_{i_1} \sigma_{i_2} \cdots \sigma_{i_s}$ are linearly independent in

$S \subset H^*(X)$, it follows the elements $Sq^I \sigma_n$ for $d(I) \leqslant n$ are linearly independent. Suppose there were some $a \in A^*$ such that $\langle a, \xi_\alpha \rangle = 0$ for all monomials in the ξ_i's; then by 18.18 we would have

$$a(\sigma_n) = \textstyle\sum_\alpha \langle a, \xi_\alpha \rangle y_1^{2\alpha_1} \ldots y_n^{2\alpha_n} = 0$$

for all n. But we know we can write a in the form $a = \sum_I c_I Sq^I$ for suitable coefficients $c_I \in \mathbb{Z}_2$. Thus we have $0 = a(\sigma_n) = \sum_I c_I Sq^I \sigma_n$ for all n. This is impossible unless $c_I = 0$ for all I—in other words $a = 0$. Thus the monomials in the ξ_i's span A_*.

We can set up a bijection $\alpha : \mathscr{I} \to \mathscr{A}$ between the set $\mathscr{I}$ of admissible sequences I and the set $\mathscr{A}$ of all sequences $\alpha = (\alpha_1, \ldots, \alpha_n)$ by taking

$$\alpha(i_1, i_2, \ldots, i_s) = (i_1 - 2i_2, i_2 - 2i_3, \ldots, i_s).$$

Let ξ^α be the monomial $\xi_1^{\alpha_1} \xi_2^{\alpha_2} \ldots \xi_n^{\alpha_n}$; then

$$\deg \xi^\alpha = \alpha_1 + (2^2 - 1)\alpha_2 + \cdots + (2^n - 1)\alpha_n.$$

Hence for an admissible sequence I we have $\deg Sq^I = \deg \xi^{\alpha(I)}$. Therefore there are precisely as many monomials ξ^α of a given degree as there are admissible monomials Sq^I of the same degree, so the ξ^α must be linearly independent. This proves that $A_* \cong \mathbb{Z}_2[\xi_1, \xi_2, \ldots]$.

To prove the formula for ψ we first compute $(ab)x$ for $a, b \in A^*$. We find

$$(ab)\,x = a(bx) = a(\textstyle\sum_k \langle b, \xi_k \rangle x^{2^k}) = \sum_k \langle b, \xi_k \rangle a(x^{2^k}).$$

Let $\Delta^{2^k} : RP^\infty \to RP^\infty \times \cdots \times RP^\infty$ be the 2^k-fold diagonal; then $x^{2^k} = \Delta^{2^k *}(y_1 y_2 \ldots y_{2^k})$. Thus

$$\begin{aligned} ax^{2^k} &= \Delta^{2^k *}(a(y_1 \ldots y_{2^k})) = \Delta^{2^k *}(\textstyle\sum_\alpha \langle a, \xi_\alpha \rangle y_1^{2\alpha_1} \ldots y_n^{2\alpha_n}) \quad (n = 2^k) \\ &= \textstyle\sum_\alpha \langle a, \xi_\alpha \rangle x^{2\alpha_1 + 2\alpha_2 + \cdots + 2\alpha_n}. \end{aligned}$$

All the terms here vanish mod 2 except when α is of the form $(j, j, \ldots, j)$ for some j, because if $\alpha_m \neq \alpha_n$ for some $m < n$, then the term for

$$\alpha' = (\alpha_1 \ldots, \alpha_n, \ldots, \alpha_m, \ldots)$$

will cancel the term for α. Thus

$$ax^{2^k} = \textstyle\sum_j \langle a, \xi_j^{2^k} \rangle x^{2^{j+k}}.$$

Therefore

$$\begin{aligned} (ab)(x) &= \textstyle\sum_k \langle b, \xi_k \rangle (\sum_j \langle a, \xi_j^{2^k} \rangle x^{2^{j+k}}) \\ &= \textstyle\sum_m \langle a \otimes b, \sum_{j+k=m} \xi_j^{2^k} \otimes \xi_k \rangle x^{2^m}. \end{aligned}$$

But also

$$(ab)(x) = \textstyle\sum_m \langle ab, \xi_m \rangle x^{2^m}.$$

Now 17.11 says

$$\langle ay,u\rangle = \sum_i \langle a, e_i\langle y,u_i\rangle\rangle = \sum_i \langle a,e_i\rangle\langle y,u_i\rangle = \sum_i \langle a\otimes y, e_i\otimes u_i\rangle$$
$$= \langle a\otimes y, \psi u\rangle$$

for $a\in A^*$, $y\in H^*(X)$, $u\in H_*(X)$. If we apply this result with $y=b\in H^*(H)$, $u=\xi_m\in H_*(H)$, then we get

$$(ab)(x) = \sum_m \langle a\otimes b, \psi\xi_m\rangle x^{2^m}.$$

Since this holds for all $a,b\in A^*$, it follows that

$$\psi\xi_m = \sum_{j+k=m} \xi_j^{2^k}\otimes\xi_k. \quad \square$$

Thus although the multiplication on A^* is not commutative and is rather complicated, the multiplication on A_* is as nice as possible. Of course the coproduct on A_* (which is dual to the messy product on A^*) is a bit complicated, but nevertheless easier to remember than the Adem relations

$$Sq^a\,Sq^b = \sum_{c=0}^{[a/2]} \binom{b-c-1}{a-2c}_2 Sq^{a+b-c}\,Sq^c \quad \text{for } a<2b.$$

In theory of course the product ψ^* on A^* is completely determined by ψ and hence one ought to be able to derive the Adem relations from what we now know about ψ. We will not attempt to do so, however.

Now let us indicate what the corresponding results are for $H(\mathbb{Z}_p)$, p an odd prime. The analog of 18.12 is the following.

18.21. Theorem. *For each i there is a natural homomorphism*

$$P^i: H^q(X,A;\mathbb{Z}_p) \to H^{q+2i(p-1)}(X,A;\mathbb{Z}_p)$$

for all $q\in\mathbb{Z}$, $(X,A)\in\mathcal{T}^2$, *satisfying*

a) $P^0=1$;
b) *if* $x\in H^{2n}(X,A;\mathbb{Z}_p)$, *then* $P^n x = x^p$;
c) *if* $x\in H^n(X,A;\mathbb{Z}_p)$, *then* $P^i x = 0$ *for* $2i>n$;
d) $\delta P^i = P^i\delta$;
e) $P^i(xy) = \sum_{j+k=i} P^j x P^k y$.

There is also the Bockstein β associated to the coefficient sequence

$$0\to\mathbb{Z}_p\to\mathbb{Z}_{p^2}\to\mathbb{Z}_p\to 0.$$

S^∞ is contractible and it is possible to give a properly discontinuous action of $\mathbb{Z}_p$ on S^∞; therefore $L=S^\infty/\mathbb{Z}_p$ is an $H(\mathbb{Z}_p,1)$. One finds that $H^*(L;\mathbb{Z}_p)\cong E(x)\otimes\mathbb{Z}_p[y]$, $x\in H^1(L;\mathbb{Z}_p)$, $y\in H^2(L;\mathbb{Z}_p)$, where $\beta x=-y$ and $\beta y=0$, and where $E(x)$ means the exterior algebra.

One considers sequences $I = (\varepsilon_0, s_1, \varepsilon_1, s_2, \ldots, s_k, \varepsilon_k)$ with $s_i \in \mathbb{N}$, $\varepsilon_i = 0$ or 1 and monomials

$$P^I = \beta^{\varepsilon_0} P^{s_1} \beta^{\varepsilon_1} P^{s_2} \ldots P^{s_k} \beta^{\varepsilon_k}.$$

I and P^I are *admissible* if $s_i \geqslant p s_{i+1} + \varepsilon_i$, $i \geqslant 1$. The analog of 18.15 is the following.

18.22. Theorem. *$A^* = H(\mathbb{Z}_p)^*(H(\mathbb{Z}_p))$ has a $\mathbb{Z}_p$-basis consisting of the admissible monomials P^I.*

One then takes $\xi_k \in A_* = H(\mathbb{Z}_p)_*(H(\mathbb{Z}_p))$ to be the dual of P^{J_k}, where $J_k = (0, p^{k-1}, 0, p^{k-2}, 0, \ldots, 0, p, 0, 1, 0)$ and $\tau_k \in A_*$ to be the dual of $P^{J'_k}$, where $J'_k = (0, p^{k-1}, 0, \ldots, p, 0, 1, 1)$. Then

$$\deg \xi_k = 2(p^k - 1)$$

$$\deg \tau_k = 2p^k - 1.$$

Since τ_k has odd degree $\tau_k^2 = 0$, all $k \geqslant 0$. τ_k, ξ_k can also be described as elements in the image of $\tilde{H}_*(L; \mathbb{Z}_p) \to H_*(H(\mathbb{Z}_p); \mathbb{Z}_p)$.

18.23. Theorem. *$A_* = H(\mathbb{Z}_p)_*(H(\mathbb{Z}_p)) \cong E(\tau_0, \tau_1, \ldots) \otimes \mathbb{Z}_p[\xi_1, \xi_2, \ldots]$. The coproduct ψ is given by*

$$\psi \xi_k = \sum_{i+j=k} \xi_j^{p^i} \otimes \xi_i$$

$$\psi \tau_k = \tau_k \otimes 1 + \sum_{i+j=k} \xi_j^{p^i} \otimes \tau_i.$$

There is no use trying to pretend the proofs of 18.21–18.23 are completely analogous to those of 18.12, etc. This is one of those cases so frequent in mathematics where the prime 2 is exceptional. Nevertheless the construction of the P^i's is sufficiently similar to that of the Sq^i's that Steenrod does the two simultaneously in [82].

18.24. The following example illustrates the usefulness of cohomology operations in homotopy theory. Suppose a map $f: S^{n-1} \to S^m$ is null-homotopic, i.e. homotopic to the constant map f_0. Then $X = S^m \cup_f e^n$ is homotopy equivalent to $S^m \cup_{f_0} e^n = S^m \vee S^n$, as one can easily verify. Now $\tilde{H}^*(X; \mathbb{Z}_2)$ has two generators over $\mathbb{Z}_2$: $x \in \tilde{H}^m(X; \mathbb{Z}_2)$ and $y \in \tilde{H}^n(X; \mathbb{Z}_2)$, and since $X \simeq S^m \vee S^n$

$$\tilde{H}^*(X; \mathbb{Z}_2) \cong \tilde{H}^*(S^m; \mathbb{Z}_2) \oplus \tilde{H}^*(S^n; \mathbb{Z}_2).$$

Indeed the projection $S^m \vee S^n \xrightarrow{p} S^m$ sends the generator $g_m \in \tilde{H}^m(S^m; \mathbb{Z}_2)$ to $x \in \tilde{H}^m(X; \mathbb{Z}_2)$. Thus for $a \in A^q, q > 0$, we have $ax = ap^*(g_m) = p^*(ag_m) = p^*(0) = 0$. Hence for any map $f: S^{n-1} \to S^m$ if there is an $a \in A^*$ with $ax = y$, it follows $f \not\simeq f_0$.

As an illustration let us show that $S\eta: S^4 \to S^3$ is not null homotopic, where $\eta: S^3 \to S^2$ is the Hopf map. This result is not implied by the

Freudenthal suspension theorem, for $\Sigma:\pi_3(S^2,*)\to\pi_4(S^3,*)$ is not in the range where Σ is a monomorphism. Now $X=S^2\cup_\eta e^4$ is just CP^2 and so $H^*(X;\mathbb{Z}_2)\cong\mathbb{Z}_2[y]/(y^3)$. In particular $y^2=Sq^2y$. Now $SX=S^3\cup_{S\eta}e^5$, and there are generators $x=\sigma^{-1}(y)$ in $\tilde{H}^3(SX;\mathbb{Z}_2)$, $z=\sigma^{-1}(y^2)\in\tilde{H}^5(SX;\mathbb{Z}_2)$. By stability of Sq^2 it follows

$$Sq^2x=Sq^2\sigma^{-1}(y)=\sigma^{-1}Sq^2(y)=\sigma^{-1}(y^2)=z.$$

Therefore $S\eta$ is not null homotopic. In fact, one can show by other methods that $\pi_4(S^3,*)\cong\mathbb{Z}_2$, so it is generated by $[S\eta]$.

Comments

I confess to a sense of having shirked my duty in this chapter. In keeping with the spirit of Chapter 17 I really should have computed $H_*(H(\mathbb{Z}_p,n);\mathbb{Z}_p)$ for $n\geqslant 1$ and then taken $H_q(H(\mathbb{Z}_p);\mathbb{Z}_p)=\operatorname{dir}\lim_n\tilde{H}_{q+n}(H(\mathbb{Z}_p,n);\mathbb{Z}_p)$; that is, I should have employed the method of Cartan [28]. I failed to do so because I could not bring the necessary homological algebra into a sufficiently palatable form. The approach given here does have the virtue that the Steenrod squares are actually constructed rather than being introduced as certain elements in a dual basis for A^*. On the other hand a proof of 18.23 does not seem to be possible along the lines given for the prime 2—an unsatisfactory state of affairs. The good student is urged to have a look at Cartan.

References

1. H. Cartan [28]
2. J. W. Milnor [61]
3. R. Mosher and M. Tangora [68]
4. J.-P. Serre [77]
5. E. H. Spanier [80]
6. N. Steenrod and D. B. A. Epstein [82]

Chapter 19

The Adams Spectral Sequence and the e-Invariant

In Chapter 17 we saw that the Hopf algebras $A(E)_* = E_*(E)$, $A(E)^* = E^*(E)$ can be very useful in proving the non-existence of maps $f:X \to Y$ (if f exists, then $f_*: E_*(X) \to E_*(Y)$ must be $A(E)_*$-linear) and in giving information about the images of the Hurewicz homomorphism

$$H:[X,Y] \to \operatorname{Hom}(E_*(X), E_*(Y))$$

and the Boardman homomorphism

$$B:[X,Y] \to \operatorname{Hom}(E^*(Y), E^*(X)).$$

Their usefulness goes further, however: Adams has shown that with the aid of $A(E)_*$ and $A(E)^*$ many existence proofs become possible and indeed that there is a spectral sequence

$$\operatorname{Ext}^{s,t}_{E_*(E)}(E_*(X), E_*(Y)) \Rightarrow [X,Y]_*$$

whose E_2-term is the purely algebraic construct $\operatorname{Ext}^{s,t}_{E_*(E)}(E_*(X), E_*(Y))$ (derived from $\operatorname{Hom}^t_{E_*(E)}(E_*(X), E_*(Y))$) and converging to the semi-geometric object $[X,Y]_n = [\Sigma^n X, Y]$. The edge homomorphism of this spectral sequence is just the Hurewicz homomorphism

$$H:[X,Y]_n \to \operatorname{Hom}^n_{E_*(E)}(E_*(X), E_*(Y)).$$

In other words, knowing the comodule structures of $E_*(X)$, $E_*(Y)$ *almost* (remember the differentials!) gives us enough information to be able to determine $[X,Y]_*$. Another way of regarding the Adams spectral sequence, as it is called, is this: it measures the failure of the Hurewicz homomorphism to be an isomorphism.

Just as the Künneth theorem 13.31 was a mixture of algebra and geometry, so the Adams spectral sequence is clearly a mixture of algebra ($\operatorname{Ext}^{**}_C(M,N)$) and geometry ($[X,Y]_*$). We do the necessary algebraic preparations first.

In Chapter 13 we saw how the fact that $\operatorname{Hom}(-,-)$ was not exact led to a derived functor $\operatorname{Ext}(A,B)$ defined for all abelian groups A, B. We can carry out an analogous construction for left modules M, N over a ring R, except that we will get a whole string of functors $\mathrm{Ext}_R^p(-,-)$ for all $p \geqslant 0$.

19.1. Definition. An R-module P is *projective* if and only if for any diagram of modules

$$\begin{array}{ccccc} & & P & & \\ & & \downarrow{\scriptstyle \pi} & & \\ B & \xrightarrow{\gamma} & C & \longrightarrow & 0 \end{array}$$

(γ an epimorphism) there is a module homomorphism $\pi': P \to B$ with $\gamma \circ \pi' = \pi$.

19.2. Example. Every free R-module is projective.

Now we can construct $\mathrm{Ext}_R^p(M, N)$ as follows. We take a *projective resolution* of M; that is, we take an exact sequence

$$0 \longleftarrow M \xleftarrow{\varepsilon} X_0 \xleftarrow{d_0} X_1 \xleftarrow{d_1} X_2 \longleftarrow \cdots$$

of R-modules with X_k projective for $k \geqslant 0$. We can always do this, and in fact we can even assume the X_k are free; for example, we could take X_0 to be the free R-module with generators all elements of M and ε the obvious homomorphism. We can then take $X_1 \xrightarrow{d_0} \ker \varepsilon$ to be any epimorphism with X_1 free, and so on. Then for any R-module N we have the cochain complex

$$\mathrm{Hom}_R(X_0, N) \xrightarrow{d_0^\#} \mathrm{Hom}_R(X_1, N) \xrightarrow{d_1^\#} \mathrm{Hom}_R(X_2, N) \longrightarrow \cdots.$$

We take the cohomology of this cochain complex and set

$$\mathrm{Ext}_R^p(M, N) = H^p(\mathrm{Hom}_R(X_*, N)), \quad p \geqslant 0.$$

It is not clear that this definition is independent of the choices made; this we shall show in the next paragraph.

Suppose $f: M \to M'$ is an R-homomorphism and

$$0 \longleftarrow M \xleftarrow{\varepsilon} X_0 \xleftarrow{d_0} X_1 \xleftarrow{d_1} X_2 \longleftarrow \cdots$$

$$0 \longleftarrow M' \xleftarrow{\varepsilon'} X_0' \xleftarrow{d_0'} X_1' \xleftarrow{d_1'} X_2' \longleftarrow \cdots$$

are projective resolutions of M, M' respectively. If we think of $\{X_i, d_i\}$, $\{X_i', d_i'\}$ as chain complexes, then we can find a chain map $\{f_i: X_i \to X_i'\}$ such that $\varepsilon' \circ f_0 = f \circ \varepsilon$, and any two such chain maps are chain homotopic. The proof of this fact goes just as in the proof of the acyclic models theorem 13.26. The fact that the X_i are not free is unimportant here; "projective" is enough. Taking $f = 1_M$, we see that any two projective resolutions of M are canonically chain homotopy equivalent, and hence $\mathrm{Ext}_R^p(M, N)$ is well defined. For general f the chain map $\{f_i\}$ gives us a homomorphism

$$\mathrm{Ext}_R^p(f, 1): \mathrm{Ext}_R^p(M', N) \to \mathrm{Ext}_R^p(M, N), \quad p \geqslant 0.$$

If $g: M' \to M''$ is another R-homomorphism, then we easily see that $\mathrm{Ext}_R^p(f, 1) \circ \mathrm{Ext}_R^p(g, 1) = \mathrm{Ext}_R^p(g \circ f, 1)$. Also $\mathrm{Ext}_R^p(1, 1) = 1$, so $\mathrm{Ext}_R^p(-,-)$ is a cofunctor of its first variable. It is simple to see that it is a functor of its second variable.

19.3. Proposition. *The functors* $\mathrm{Ext}_R^p(-,-)$ *have the following properties:*

i) $\mathrm{Ext}_R^0(M, N) = \mathrm{Hom}_R(M, N)$;

ii) $\mathrm{Ext}_R^p(M, N) = 0$ *for* $p > 0$ *if* M *is projective;*

iii) *if* $0 \to N_1 \to N_2 \to N_3 \to 0$ *is an exact sequence of* R*-modules, then there is a long exact sequence*

$$0 \to \mathrm{Hom}_R(M, N_1) \to \mathrm{Hom}_R(M, N_2) \to \mathrm{Hom}_R(M, N_3) \to \mathrm{Ext}_R^1(M, N_1) \to \cdots$$

$$\cdots \to \mathrm{Ext}_R^p(M, N_2) \to \mathrm{Ext}_R^p(M, N_3) \to \mathrm{Ext}_R^{p+1}(M, N_1) \to \cdots.$$

Proof: i) Since $0 \leftarrow M \leftarrow X_0 \leftarrow X_1$ is exact in a projective resolution for M, it follows

$$\mathrm{Hom}_R(M, N) = \ker \mathrm{Hom}\,(d_0, 1) = H^0(\mathrm{Hom}_R(X_*, N)) = \mathrm{Ext}_R^0(M, N).$$

ii) If M is projective, then $0 \leftarrow M \overset{1}{\leftarrow} M \leftarrow 0$ is a projective resolution of M, so we have $\mathrm{Ext}_R^p(M, N) = 0$ for $p > 0$.

iii) We get a short exact sequence of cochain complexes

$$0 \to \mathrm{Hom}_R(X_*, N_1) \to \mathrm{Hom}_R(X_*, N_2) \to \mathrm{Hom}_R(X_*, N_3) \to 0$$

because the X_i are projective. 19.3iii) is just the resulting cohomology exact sequence. □

Now we can generalize all of this somewhat. Suppose R is a graded commutative ring with 1, A is a graded algebra over R and M, N are graded (left) A-modules. Then we have the groups $\mathrm{Hom}_A^q(M, N)$ of A-homomorphisms $f: M \to N$ of degree q—i.e. $f(M_n) \subset N_{n+q}$, $n \in \mathbb{Z}$. We can take a resolution

$$0 \longleftarrow M \overset{\varepsilon}{\longleftarrow} X_0 \overset{d_0}{\longleftarrow} X_1 \overset{d_1}{\longleftarrow} X_2 \longleftarrow \cdots$$

of M by projective graded A-modules X_i (the definition of "projective" is formally the same) and obtain a cochain complex

$$\mathrm{Hom}_A^q(X_0, N) \to \mathrm{Hom}_A^q(X_1, N) \to \cdots$$

for each $q \in \mathbb{Z}$. We then define

$$\mathrm{Ext}_A^{p,q}(M,N) = H^p(\mathrm{Hom}_A^q(X_*,N)), \quad p \geqslant 0, q \in \mathbb{Z}.$$

19.4. Definition. If $\phi: A \otimes_R A \to A$ is the product on A and V is an R-module, then we can give $A \otimes_R V$ the structure of an A-module by taking

$$A \otimes_R (A \otimes_R V) \cong (A \otimes_R A) \otimes_R V \xrightarrow{\phi \otimes 1} A \otimes_R V$$

as the action map. Any A-module of this form is called an *extended* A-module.

Extended A-modules are handy for the following reason.

19.5. Proposition. *For any R-module V and A-module N there is a natural isomorphism $\Phi: \mathrm{Hom}_A^q(A \otimes_R V, N) \cong \mathrm{Hom}_R^q(V,N)$, $q \in \mathbb{Z}$. If A is flat over R, then there is a natural isomorphism*

$$\mathrm{Ext}_A^{p,q}(A \otimes_R V, N) \cong \mathrm{Ext}_R^{p,q}(V,N), \quad p \geqslant 0, q \in \mathbb{Z}.$$

Proof: We define Φ and $\Psi: \mathrm{Hom}_R^q(V,N) \to \mathrm{Hom}_A^q(A \otimes_R V, N)$ by

$$\Phi(f)(v) = f(1 \otimes v), \quad f \in \mathrm{Hom}_A^q(A \otimes_R V, N), \quad v \in V,$$

$$\Psi(g)(a \otimes v) = ag(v), \quad g \in \mathrm{Hom}_R^q(V,N), \quad a \in A, v \in V.$$

These are clearly inverses of one another. In somewhat fancier terms (which permit dualization) we could describe Φ, Ψ as follows: let $\eta: R \to A$ be the unit of A. Then $\Phi(f)$ is the composite

$$V \cong R \otimes_R V \xrightarrow{\eta \otimes 1} A \otimes_R V \xrightarrow{f} N.$$

Let $\phi_N: A \otimes_R N \to N$ be the A-action map for N. Then $\Psi(g)$ is the composite

$$A \otimes_R V \xrightarrow{1 \otimes g} A \otimes_R N \xrightarrow{\phi_N} N.$$

Now suppose A is flat over R and let

$$0 \leftarrow V \leftarrow U_0 \leftarrow U_1 \leftarrow \cdots$$

be a resolution of V by projective R-modules. It follows that

$$0 \leftarrow A \otimes_R V \leftarrow A \otimes_R U_0 \leftarrow A \otimes_R U_1 \leftarrow \cdots$$

is a projective resolution of $A \otimes_R V$ over A (why is $A \otimes_R U_i$ projective?). Thus we can compute $\mathrm{Ext}_A^{p,q}(A \otimes_R V, N)$ by using the cochain complex $\mathrm{Hom}_A^q(A \otimes_R U_i, N) \cong \mathrm{Hom}_R^q(U_i, N)$, which just gives $\mathrm{Ext}_R^{p,q}(V,N)$. $\square$

Now we can dualize everything above and consider a coalgebra C over R and C-comodules K,L. We can take a resolution

$$0 \leftarrow K \leftarrow X_0 \leftarrow X_1 \leftarrow X_2 \leftarrow \cdots$$

of K by projective C-comodules X_i (the definition of "projective" is still formally the same). We then let $\mathrm{Ext}_C^{p,q}(K,L) = H^p(\mathrm{Hom}_C^q(X_*,L))$ for $p \geqslant 0, q \in \mathbb{Z}$.

19.6. Definition. If $\psi: C \to C \otimes_R C$ is the coproduct on C and V is an R-module, then we can give $C \otimes_R V$ the structure of a C-comodule by taking

$$C \otimes_R V \xrightarrow{\psi \otimes 1} (C \otimes_R C) \otimes_R V \cong C \otimes_R (C \otimes_R V)$$

as the C-coaction map. Any C-comodule of this form is called an *extended* C-comodule.

19.7. Proposition. *For every R-module V and C-comodule K there is a natural isomorphism* $\Phi: \mathrm{Hom}_C^q(K, C \otimes_R V) \cong \mathrm{Hom}_R^q(K, V)$, $q \in \mathbb{Z}$. *Hence there are natural isomorphisms*

$$\mathrm{Ext}_C^{p,q}(K, C \otimes_R V) \cong \mathrm{Ext}_R^{p,q}(K,V), \quad p \geqslant 0, q \in \mathbb{Z}.$$

Proof: The inverse isomorphisms Φ, Ψ are constructed in a manner dual to those in 19.5. If

$$0 \leftarrow K \leftarrow X_0 \leftarrow X_1 \leftarrow X_2 \leftarrow \cdots$$

is a projective resolution of K over C, then

$$\mathrm{Ext}_C^{p,q}(K, C \otimes_R V) = H^p(\mathrm{Hom}_C^q(X_*, C \otimes_R V)) \cong H^p(\mathrm{Hom}_R^q(X_*, V)).$$

Suppose we have an epimorphism $\gamma: A \to B$ of R-modules A,B. Then the square

$$\begin{array}{ccc} \mathrm{Hom}_R^q(X_i, A) & \longrightarrow & \mathrm{Hom}_R^q(X_i, B) \\ \wr\| & & \wr\| \\ \mathrm{Hom}_C^q(X_i, C \otimes_R A) & \longrightarrow & \mathrm{Hom}_C^q(X_i, C \otimes_R B). \end{array}$$

commutes, and the bottom horizontal arrow is an epimorphism since X_i is projective over C and $C \otimes_R A \to C \otimes_R B \to 0$ is exact. Thus the upper horizontal arrow is an epimorphism too, showing that the X_i are projective over R. Hence $H^p(\mathrm{Hom}_R^q(X_*, V)) \cong \mathrm{Ext}_R^{p,q}(K, V)$. □

Now in special cases there is still another way to construct $\mathrm{Ext}_C^{p,q}(K,L)$ —one which we shall employ in deriving the Adams spectral sequence.

19.8. Proposition. *If K is a C-comodule which is projective as an R-module, then we may construct* $\mathrm{Ext}_C^{p,q}(K,L)$ *as follows: suppose*

$$0 \to L \to C \otimes_R V_0 \to C \otimes_R V_1 \to \cdots$$

is an exact sequence of C-comodules (the V_i may be arbitrary R-modules). Then $\mathrm{Ext}_C^{p,q}(K,L) \cong H^p(\mathrm{Hom}_C^q(K, C \otimes_R V_*))$, $p \geq 0$, $q \in \mathbb{Z}$.

Proof: Let

$$Q_i = \mathrm{im}\,[C \otimes_R V_{i-1} \to C \otimes_R V_i] = \ker\,[C \otimes_R V_i \to C \otimes_R V_{i+1}].$$

Then we get short exact sequences

$$0 \to L \to C \otimes_R V_0 \to Q_1 \to 0$$

$$0 \to Q_1 \to C \otimes_R V_1 \to Q_2 \to 0$$

$$\cdots$$

These short exact sequences give long exact sequences

$$\cdots \to \mathrm{Ext}_C^{p-1,q}(K, C \otimes_R V_i) \to \mathrm{Ext}_C^{p-1,q}(K, Q_{i+1}) \to$$

$$\mathrm{Ext}_C^{p,q}(K, Q_i) \to \mathrm{Ext}_C^{p,q}(K, C \otimes_R V_i) \to \cdots.$$

But $\mathrm{Ext}_C^{p,q}(K, C \otimes_R V_i) \cong \mathrm{Ext}_R^{p,q}(K, V_i) = 0$ for $p \geq 1$, since K is projective as an R-module. Thus we get an isomorphism

$$\mathrm{Ext}_C^{p,q}(K, Q_i) \cong \mathrm{Ext}_C^{p-1,q}(K, Q_{i+1}), \quad p \geq 2.$$

In particular

$$\mathrm{Ext}_C^{p,q}(K,L) \cong \mathrm{Ext}_C^{p-1,q}(K, Q_1) \cong \cdots \cong \mathrm{Ext}_C^{1,q}(K, Q_{p-1}).$$

We also have exact sequences

$$\mathrm{Hom}_C^q(K, C \otimes_R V_{p-1}) \to \mathrm{Hom}_C^q(K, Q_p) \to \mathrm{Ext}_C^{1,q}(K, Q_{p-1}) \to 0$$

and

$$0 \to \mathrm{Hom}_C^q(K, Q_p) \to \mathrm{Hom}_C^q(K, C \otimes_R V_p) \to \mathrm{Hom}_C^q(K, C \otimes_R V_{p+1}),$$

so we see that

$$\begin{aligned}\mathrm{Ext}_C^{1,q}(K, Q_{p-1}) &= \mathrm{Hom}_C^q(K, Q_p)/\mathrm{im}\,\mathrm{Hom}_C^q(K, C \otimes_R V_{p-1}) \\ &\cong \frac{\ker\,[\mathrm{Hom}_C^q(K, C \otimes_R V_p) \to \mathrm{Hom}_C^q(K, C \otimes_R V_{p+1})]}{\mathrm{im}\,[\mathrm{Hom}_C^q(K, C \otimes_R V_{p-1}) \to \mathrm{Hom}_C^q(K, C \otimes_R V_p)]} \\ &= H^p(\mathrm{Hom}_C^q(K, C \otimes_R V_*)). \quad \square\end{aligned}$$

That completes the algebra. We can now turn to the geometry. We began the chapter by describing the most general form of the Adams spectral sequence: its E_2-term is $E_2^{s,t} = \mathrm{Ext}^{s,t}_{E_*(E)}(E_*(X), E_*(Y))$ and it converges to $[X, Y]_*$ (with appropriate assumptions on E). Now in practice this full-scale Adams spectral sequence is almost never used; one usually wants to take $X = S^0$, getting a spectral sequence converging to $\pi_*^S(Y)$. If we therefore assume $X = S^0$ from the beginning, we shall spare ourselves some complicated details and still have the essential ideas.

19.9. Theorem. *Let E be a ring spectrum such that $E_*(E)$ is flat over $\tilde{E}_*(S^0)$. For every spectrum Y there is a natural spectral sequence $\{E_r^{s,t}, d_r\}$ with*

i) $d_r: E_r^{s,t} \to E_r^{s+r,t+r-1}$ *for all r, s, t;*

ii) $E_2^{s,t} \cong \mathrm{Ext}^{s,t}_{E_*(E)}(\tilde{E}_*(S^0), E_*(Y))$;

iii) $E_{r+1}^{s,t} \subset E_r^{s,t}$ *for* $r > s+1$ *and* $\bigcap_{r>s+1} E_r^{s,t} = E_\infty^{s,t}$.

There is a filtration

$$\pi_n(Y) = [\Sigma^n S^0, Y] = F^{0,n} \supset F^{1,n+1} \supset F^{2,n+2} \supset \cdots \supset F^{s,n+s} \supset \cdots$$

and

iv) *a monomorphism* $0 \to F^{s,t}/F^{s+1,t+1} \to E_\infty^{s,t}$, $s \geqslant 0$.

The edge homomorphism $\pi_n(Y) \to E_\infty^{0,n} \to E_2^{0,n} \cong \mathrm{Hom}^n_{E_*(E)}(\tilde{E}_*(S^0), E_*(Y))$ is the Hurewicz homomorphism.

19.10. *Remark.* Observe that this theorem does not say the spectral sequence converges; for this to be true we would need to know that the following groups vanished:

a) $A^{s,t} = \mathrm{coker}[F^{s,t}/F^{s+1,t+1} \to E_\infty^{s,t}]$, $s \geqslant 0$,

b) $D^n = \bigcap_{s \geqslant 0} F^{s,n+s}$.

If the $A^{s,t} = 0$ for all s, t, then we could define a new filtration

$$\pi_n(Y)/D^n = \hat{F}^{0,n} \supset \hat{F}^{1,n+1} \supset \cdots \supset \hat{F}^{s,n+s} \supset \cdots$$

with $\hat{F}^{s,n+s} = F^{s,n+s}/D^n$, all $s \geqslant 0$. Then we would still have

$$\hat{F}^{s,t}/\hat{F}^{s+1,t+1} \cong F^{s,t}/F^{s+1,t+1} \cong E_\infty^{s,t}, \quad \text{all } s, t,$$

but in addition $\bigcap \hat{F}^{s,t} = 0$. Thus if the $A^{s,t}$ vanish, we can say that the spectral sequence converges to the groups $\pi_*(Y)/D^*$.

We shall show that for $E = H(\mathbb{Z}_p)$ the groups $A^{s,t}$ vanish and $D^n \subset \pi_n(Y)$ is the subgroup of all elements of finite order prime to p. In general one must put additional assumptions on E and Y to guarantee that $A^{s,t} = 0$ for all s, t.

Proof of 19.9: $\tilde{E}_*(S^0)$ is certainly $\tilde{E}_*(S^0)$-projective—even free. Therefore we may use 19.8 to compute $\mathrm{Ext}_C^{s,t}(\tilde{E}_*(S^0), E_*(Y))$: we only need to find a resolution

$$0 \to E_*(Y) \to C \otimes_R V_0 \to C \otimes_R V_1 \to \cdots$$

of $E_*(Y)$ by extended C-comodules, where we use the abbreviations $R = \tilde{E}_*(S^0)$, $C = E_*(E)$. This motivates the following construction.

Let $Y_0 = Y$ and take a cofibre sequence

$$Y_0 \xrightarrow{f_0} E \wedge Y_0 \longrightarrow Y_1 \longrightarrow \Sigma Y_0 \longrightarrow \cdots,$$

where f_0 is the map $Y_0 \simeq S^0 \wedge Y_0 \xrightarrow{\imath \wedge 1} E \wedge Y_0$. Repeat this process: if Y_n has been defined, take a cofibre sequence

$$Y_n \xrightarrow{f_n} E \wedge Y_n \longrightarrow Y_{n+1} \longrightarrow \Sigma Y_n \longrightarrow \cdots,$$

thus defining Y_{n+1}.

Note that the diagram

$$\begin{array}{ccc} \pi_*(E \wedge Y_n) & \xrightarrow{\quad 1 \quad} & \pi_*(E \wedge Y_n) \\ \| \wr & \searrow^{f_{n*}} & \uparrow (\mu \wedge 1)_* \\ \pi_*(E \wedge S^0 \wedge Y_n) & \xrightarrow{(1 \wedge \imath \wedge 1)_*} & \pi_*(E \wedge E \wedge Y_n) \end{array}$$

commutes (where μ is the product on E), which shows that

$$E_*(Y_n) \xrightarrow{f_{n*}} E_*(E \wedge Y_n)$$

is a monomorphism. Thus the exact sequence

$$\cdots \to E_q(Y_n) \to E_q(E \wedge Y_n) \to E_q(Y_{n+1}) \to E_{q-1}(Y_n) \to \cdots$$

splits up into short exact sequences

$$0 \to E_q(Y_n) \to E_q(E \wedge Y_n) \to E_q(Y_{n+1}) \to 0.$$

Glueing these together, we get a long exact sequence

$$0 \to E_*(Y) \to E_*(E \wedge Y_0) \to E_*(E \wedge Y_1) \to \cdots.$$

Moreover $E_*(E \wedge Y_n) \cong E_*(E) \otimes_{\tilde{E}_*(S^0)} E_*(Y_n)$ by the weak Künneth theorem 13.75, so the comodules $E_*(E \wedge Y_n)$ are indeed extended C-comodules. Thus we have a resolution of the sort needed to construct $\mathrm{Ext}_C^{s,t}$; we can take

$$\mathrm{Ext}_C^{s,t}(R, E_*(Y)) = H^s(\mathrm{Hom}_C^t(\tilde{E}_*(S^0), E_*(E \wedge Y_n))).$$

Now the spectra Y_n give us a sequence

$$Y = Y_0 \leftarrow \Sigma^{-1} Y_1 \leftarrow \Sigma^{-2} Y_2 \leftarrow \cdots \leftarrow \Sigma^{-n} Y_n \leftarrow \cdots$$

called the *Adams filtration* of Y, which can be made into a filtration in the usual sense if we replace Y by the infinite mapping telescope. Let us use the symbol $Y_s/Y_{s'}$ for the cofibre of $\Sigma^{s-s'}Y_{s'} \to Y_s$ for $s \leqslant s'$. In particular we have $Y_s/Y_{s+1} = E \wedge Y_s$. With this notation we can now define a spectral sequence associated to the Adams filtration of Y in a manner similar to the way in which we constructed a spectral sequence for an ordinary filtration.

We let

$$Z_r^{s,t} = \mathrm{im}[\pi_t(Y_s/Y_{s+r}) \to \pi_t(Y_s/Y_{s+1})];$$

$$B_r^{s,t} = \mathrm{im}[\Delta: \pi_{t-r+2}(Y_{s-r+1}/Y_s) \to \pi_t(Y_s/Y_{s+1})], \quad r \leqslant s+1,$$

where Δ is the composite

$$\pi_{t-r+2}(Y_{s-r+1}/Y_s) \to \pi_t(Y_s) \to \pi_t(Y_s/Y_{s+1});$$

$$Z_\infty^{s,t} = \mathrm{im}[\pi_t(Y_s) \to \pi_t(Y_s/Y_{s+1})];$$

$$B_r^{s,t} = B_\infty^{s,t} = \mathrm{im}[\Delta: \pi_{t-s+1}(Y/Y_s) \to \pi_t(Y_s/Y_{s+1})], \quad r > s+1;$$

$$F^{s,t} = \mathrm{im}[\pi_t(Y_s) \to \pi_{t-s}(Y)];$$

$$A^{s,t} = \mathrm{im}[\pi_t(Y_s/Y_{s+1}) \to \pi_t(Y_{s+1})] \cap \textstyle\bigcap_{r\geqslant 1} \mathrm{im}[\pi_{t+r}(Y_{s+r+1}) \to \pi_t(Y_{s+1})].$$

We then have inclusions

$$0 = B_1^{s,t} \subset B_2^{s,t} \subset \cdots \subset B_r^{s,t} \subset B_{r+1}^{s,t} \subset \cdots$$
$$\subset B_\infty^{s,t} \subset Z_\infty^{s,t} \subset \cdots \subset Z_{r+1}^{s,t} \subset Z_r^{s,t} \subset \cdots \subset Z_1^{s,t}.$$

For example, the inclusion $B_r^{s,t} \subset B_{r+1}^{s,t}$ follows from the commutative diagram

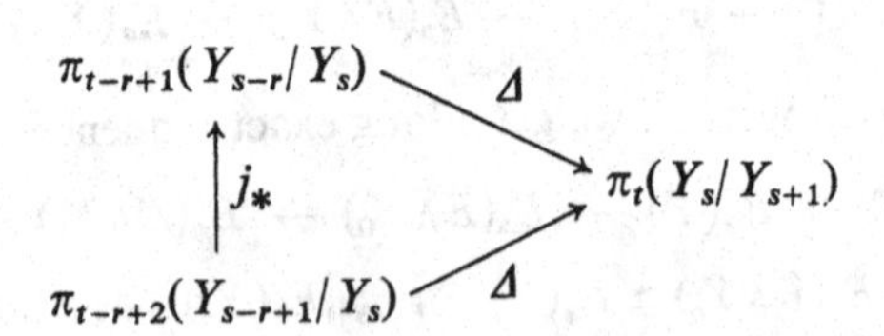

where the map $j: Y_{s-r+1}/Y_s \to Y_{s-r}/Y_s$ is induced by $1: Y_s \to Y_s$, $\Sigma^{-1}Y_{s-r+1} \to Y_{s-r}$. We take $Z_{\infty}^{s,t} = \bigcap_{r\geqslant 1} Z_r^{s,t}$; then we have

$$Z_\infty^{s,t} \subset Z_\infty^{s,t} \subset Z_r^{s,t}, \quad \text{all } r.$$

We let

$$E_r^{s,t} = Z_r^{s,t}/B_r^{s,t}, \quad E_\infty^{s,t} = Z_\infty^{s,t}/B_\infty^{s,t}.$$

Since $B_\infty^{s,t} = B_r^{s,t}$ for $r > s+1$, we see that there is a monomorphism

$$E_{r+1}^{s,t} \to E_r^{s,t}, \quad r > s+1,$$

and

$$E_\infty^{s,t} = \bigcap_{r>s+1} E_r^{s,t} = \operatorname{inv\,lim}_r E_r^{s,t}.$$

From the commutative diagram

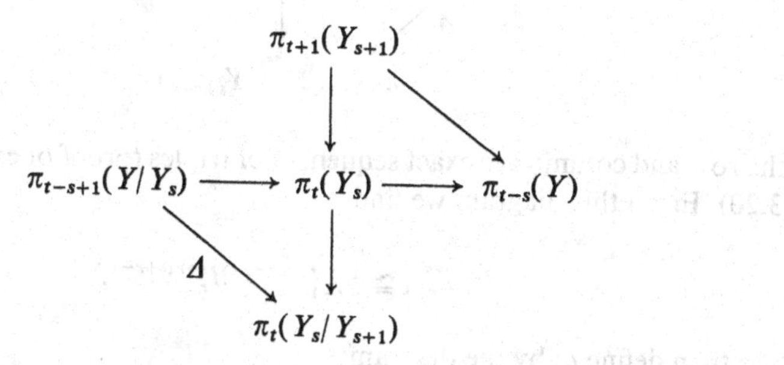

we see that $F^{s+t}/F^{s+1,t+1} \cong Z_\infty^{s,t}/B_\infty^{s,t}$. Thus we have an exact sequence

$$0 \to Z_\infty^{s,t}/B_\infty^{s,t} \to Z_\infty^{s,t}/B_\infty^{s,t} \to Z_\infty^{s,t}/Z_\infty^{s,t} \to 0$$

which becomes

$$0 \to F^{s,t}/F^{s+1,t+1} \to E_\infty^{s,t} \to A^{s,t} \to 0$$

once we have shown that $Z_\infty^{s,t}/Z_\infty^{s,t} \cong A^{s,t}$. To this end we have the commutative diagram

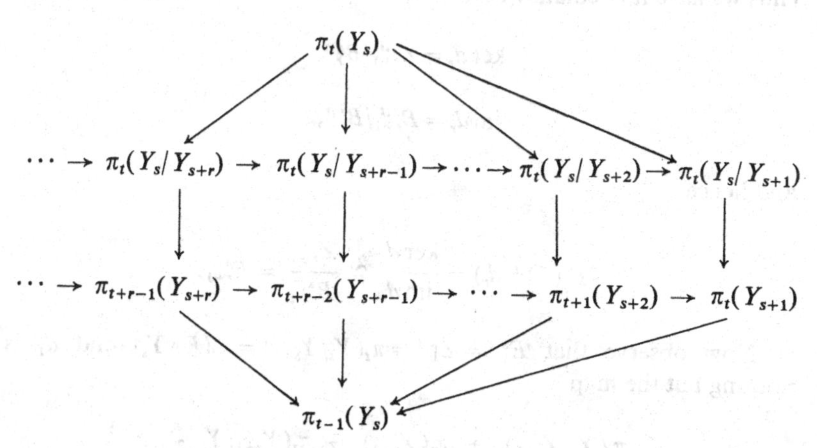

in which "vertical" sequences are exact. From this follows the exact sequence $0 \to Z_\infty^{s,t} \to Z_\infty^{s,t} \to A^{s,t} \to 0$.

Now in the commutative diagram

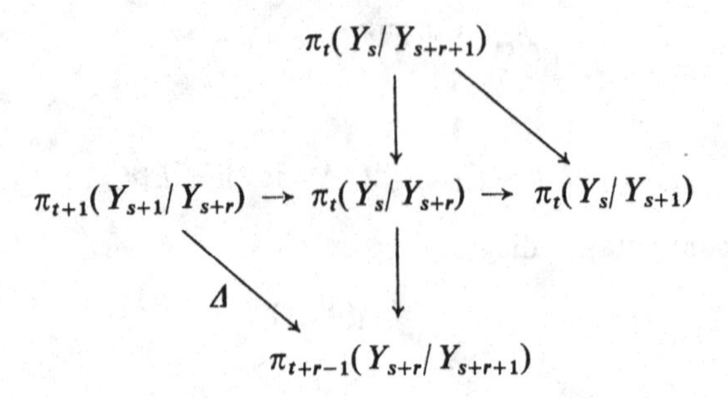

the row and column are exact sequences of triples (proof of exactness as in 3.20). From this diagram we find

$$Z_r^{s,t}/Z_{r+1}^{s,t} \cong B_{r+1}^{s+r,t+r-1}/B_r^{s+r,t+r-1}.$$

We then define d_r by the diagram

$$\begin{array}{ccc} Z_r^{s,t}/B_r^{s,t} \xrightarrow{\text{epic}} Z_r^{s,t}/Z_{r+1}^{s,t} \cong B_{r+1}^{s+r,t+r-1}/B_r^{s+r,t+r-1} & \xrightarrow{\text{monic}} & Z_r^{s+r,t+r-1}/B_r^{s+r,t+r-1} \\ \| \wr & & \| \wr \\ E_r^{s,t} & \xrightarrow{d_r} & E_r^{s+r,t+r-1}. \end{array}$$

Thus we have immediately that

$$\ker d_r = Z_{r+1}^{s,t}/B_r^{s,t}$$

$$\operatorname{im} d_r = B_{r+1}^{s,t}/B_r^{s,t},$$

and hence

$$H^*(E_r^{s,t}, d_r) = \frac{\ker d_r}{\operatorname{im} d_r} \cong \frac{Z_{r+1}^{s,t}}{B_{r+1}^{s,t}} = E_{r+1}^{s,t}.$$

Now observe that $E_1^{s,t} = Z_1^{s,t} = \pi_t(Y_s/Y_{s+1}) = \pi_t(E \wedge Y_s)$ and d_1 is nothing but the map

$$\pi_t(Y_s/Y_{s+1}) \to \pi_t(Y_{s+1}) \to \pi_t(Y_{s+1}/Y_{s+2}).$$

We also have a commutative diagram

$$\begin{array}{ccc}
\pi_t(E\wedge Y_s) & \longrightarrow & \pi_t(Y_{s+1}) \longrightarrow \\
\downarrow h & & \downarrow h \\
\mathrm{Hom}^t_C(\tilde{E}_*(S^0), E_*(E\wedge Y_s)) & \rightarrow & \mathrm{Hom}^t_C(\tilde{E}_*(S^0), E_*(Y_{s+1})) \rightarrow
\end{array}$$

$$\begin{array}{c}
\pi_t(E\wedge Y_{s+1}) \\
\downarrow h \\
\mathrm{Hom}^t_C(\tilde{E}_*(S^0), E_*(E\wedge Y_{s+1})).
\end{array}$$

Now for any spectrum Z the Hurewicz homomorphism

$$h: \pi_t(E\wedge Z) \rightarrow \mathrm{Hom}^t_C(\tilde{E}_*(S^0), E_*(E\wedge Z))$$

is an isomorphism, for we have a commutative square

$$\begin{array}{ccc}
\pi_t(E\wedge Z) & \xrightarrow{\cong} & \mathrm{Hom}^t_{\tilde{E}_*(S^0)}(\tilde{E}_*(S^0), E_*(Z)) \\
\downarrow h & & \cong \downarrow \Psi \\
\mathrm{Hom}^t_C(\tilde{E}_*(S^0), E_*(E\wedge Z)) & \xleftarrow[\cong]{\mathrm{Hom}(1, m_*)} & \mathrm{Hom}^t_C(\tilde{E}_*(S^0), E_*(E)\otimes_{\tilde{E}_*(S^0)} E_*(Z)).
\end{array}$$

Hence we see that

$$\cdots \longrightarrow E_1^{s-1,t} \xrightarrow{d_1} E_1^{s,t} \xrightarrow{d_1} E_1^{s+1,t} \longrightarrow \cdots$$

is chain isomorphic to the cochain complex $\{\mathrm{Hom}^t_C(\tilde{E}_*(S^0), E_*(E\wedge Y_s))\}$ which we use to compute $\mathrm{Ext}^{s,t}_C(\tilde{E}_*(S^0), E_*(Y))$. Therefore we have

$$E_2^{s,t} \cong \mathrm{Ext}^{s,t}_C(\tilde{E}_*(S^0), E_*(Y)).$$

The statement about the edge homomorphism should now be practically obvious.

Our next task must be to show that this construction is natural; in the process it will come out that the spectral sequence is independent of the choice of Adams filtration.

Suppose $f: Y \rightarrow Y'$ is a map; then we have a diagram

$$\begin{array}{ccccccccc}
Y_0 & \xrightarrow{\iota\wedge 1} & E\wedge Y_0 & \longrightarrow & Y_1 & \longrightarrow & \Sigma Y_0 & \longrightarrow & \cdots \\
\downarrow f_0 & & \downarrow 1\wedge f_0 & & \downarrow f_1 & & \downarrow \Sigma f_0 & & \\
Y'_0 & \longrightarrow & E\wedge Y'_0 & \longrightarrow & Y'_1 & \longrightarrow & \Sigma Y'_0 & \longrightarrow & \cdots
\end{array}$$

in which we can fill in a map $f_1: Y \to Y_1'$ to make the diagram commute (up to homotopy) because of 8.31. By induction we construct $f_i: Y_i \to Y_i'$ for $i \geqslant 0$. The f_i then define a map of spectral sequences. Now it is not true that the f_i are uniquely determined (even up to homotopy) but two different choices lead to cochain homotopic maps $\pi_t(E \wedge Y_i) \to \pi_t(E \wedge Y_i')$ and hence to the same map of E_2-terms. It thus follows the two maps of spectral sequences agree on E_r for $r \geqslant 2$. Also if $f = 1: Y \to Y$ then we get cochain homotopy equivalences between the E_1-terms of the spectral sequences arising from two different Adams filtrations. Hence the E_r-terms are isomorphic for $r \geqslant 2$. ▯

We now consider some conditions under which convergence is guaranteed.

19.11. Proposition. *If $\pi_q(S^0) \xrightarrow{\iota_*} \pi_q(E)$ is an isomorphism for $q \leqslant 0$ and an epimorphism for $q = 1$ and if $\pi_q(Y) = 0$ for $q < N$ for some $N \in \mathbb{Z}$, then $A^{s,t} = D^s = 0$ for all s, t, and hence the Adams spectral sequence converges to $\pi_*(Y)$* (cf. 19.10).

Remark. The first hypothesis says that

$$\pi_q(E) = \begin{cases} 0 & q < 0 \\ \mathbb{Z} & q = 0 \\ \mathbb{Z}_2 \text{ or } 0 & q = 1. \end{cases}$$

Both $E = MU$ and $E = MSp$ satisfy this hypothesis.

Proof: We wish to prove first that $\pi_r(Y_n) = 0$ for $r < N + 2n$; we proceed by induction on n, noting that the statement is true for $n = 0$. Suppose $\pi_r(Y_n) = 0$ for $r < N + 2n$; then we have an exact sequence

$$\pi_{2n+N+1}(Y_n) \xrightarrow{(\iota \wedge 1)_*} \pi_{2n+N+1}(E \wedge Y_n) \longrightarrow \pi_{2n+N+1}(Y_{n+1}) \longrightarrow$$
$$\pi_{2n+N}(Y_n) \xrightarrow{(\iota \wedge 1)_*} \pi_{2n+N}(E \wedge Y_n) \longrightarrow \pi_{2n+N}(Y_{n+1}) \longrightarrow$$
$$\pi_{2n+N-1}(Y_n) = 0.$$

By 7.55 the Hurewicz homomorphism

$$(\iota \wedge 1)_*: \pi_q(Y_n) \to \pi_q(E \wedge Y_n) = E_q(Y_n)$$

is an isomorphism for $q \leqslant N + 2n$ and an epimorphism for $q = N + 2n + 1$. Hence $\pi_{2n+N+1}(Y_{n+1}) = \pi_{2n+N}(Y_{n+1}) = 0$, completing the induction.

Thus $\pi_{t+r}(Y_{s+r+1}) = 0$ for $r > t - 2s - N - 2$. Since

$$A^{s,t} = \operatorname{im}[\pi_t(Y_s/Y_{s+1}) \to \pi_t(Y_{s+1})] \cap \bigcap_{r \geqslant 1} \operatorname{im}[\pi_{t+r}(Y_{s+r+1}) \to \pi_t(Y_{s+1})]$$

we have $A^{s,t} = 0$ for all s, t. Similarly $\pi_{n+r}(Y_r) = 0$ for $r > n - N$, so $D^n = \bigcap_{r \geq 0} \text{im}\,[\pi_{n+r}(Y_r) \to \pi_n(Y)] = 0$. □

Now the other important case we have to consider is $E = H(\mathbb{Z}_p)$, which does not satisfy the hypothesis of 19.11. We can show in this case also that $A^{s,t} = 0$ and can compute D^*.

19.12. Proposition. *If $E = H(\mathbb{Z}_p)$ and $\pi_q(Y) = 0$ if $q < N$, some $N \in \mathbb{Z}$, then an element is in $D^n = \bigcap_{r \geq 0} F^{r,n+r} \subset \pi_n(Y)$ if and only if it is divisible by p^s for every $s \geq 0$. If further $H_*(Y)$ is of finite type, then D^* is the subgroup of all torsion elements whose order is prime to p and $A^{s,t} = 0$ for all s, t. Therefore the Adams spectral sequence converges to the p-primary part ${}_p\pi_*(Y) = \pi_*(Y)/D^*$ of $\pi_*(Y)$.*

Proof: We showed earlier that $Y_n \xrightarrow{\iota \wedge 1} E \wedge Y_n$ induces a monomorphism $E_*(Y_n) \xrightarrow{(\iota \wedge 1)_*} E_*(E \wedge Y_n)$, from which it follows that the homomorphisms $E_*(\Sigma^{-1} Y_{n+1}) \to E_*(Y_n)$ are zero for all n.

As an inductive hypothesis suppose that we have proved that if $x \in \pi_n(Y)$ is divisible by p^{s-1} then $x \in F^{s-1,n+s-1}$, all n, all Y. The statement is trivial for $s = 1$. Suppose that for some Y an element $x \in \pi_n(Y)$ is divisible by p^s, say $x = p^s y$ for some $y \in \pi_n(Y)$. Then $(\iota \wedge 1)_*(py) = p(\iota \wedge 1)_*(y) \in H_n(Y; \mathbb{Z}_p)$, so $(\iota \wedge 1)_*(py) = 0$. Hence since

$$\Sigma^{-1} Y_1 \xrightarrow{i_1} Y_0 \xrightarrow{\iota \wedge 1} E \wedge Y_0$$

is a cofibre sequence, there is a $z \in \pi_n(\Sigma^{-1} Y_1) = \pi_{n+1}(Y_1)$ with $i_{1*} z = py$. Therefore $i_{1*}(p^{s-1} z) = p^{s-1} i_{1*}(z) = p^s y = x$. Now

$$Y_1 \leftarrow \Sigma^{-1} Y_2 \leftarrow \Sigma^{-2} Y_3 \leftarrow \cdots$$

is an Adams filtration of Y_1, so we may apply the inductive hypothesis to $p^{s-1} z \in \pi_{n+1}(Y_1)$ to conclude there is a $w \in \pi_{n+s}(Y_s)$ with

$$i_{2*} \circ i_{3*} \circ \cdots \circ i_{s*}(w) = p^{s-1} z.$$

Hence $i_{1*} \circ i_{2*} \circ \cdots \circ i_{s*}(w) = i_{1*}(p^{s-1} z) = x$—i.e. $x \in F^{s,n+s}$. This completes the induction step.

The converse is a little harder. Suppose some $x \in \pi_n(Y)$ is not divisible by p^s for some s; we shall show that $x \notin F^{t,n+t}$ for some t. We can add maps of spectra, so we consider the map $Y \xrightarrow{p^s} Y$, by which we mean p^s copies of 1_Y added together. The induced map on homotopy $\pi_*(Y) \to \pi_*(Y)$ is multiplication by p^s. We let W be the cofibre of $p^s : Y \to Y$; from the homotopy sequence of $Y \xrightarrow{p^s} Y \xrightarrow{f} W$ we find that $\pi_q(W) = 0$ if $q < N$. Moreover, we have a short exact sequence

$$0 \to \text{coker}\, p^s \to \pi_*(W) \to \ker p^s \to 0,$$

and since $p^s(\text{coker}\, p^s) = p^s(\ker p^s) = 0$, it follows that $p^{2s} \pi_*(W) = 0$.

Now we construct a filtration of W as follows. Let π_m be the first non-zero homotopy group of W (we know $m \geqslant N$). The map $\pi_m(W) \to \pi_m(W) \otimes \mathbb{Z}_p$ can be realized by a map $W \to \Sigma^m H(\pi_m \otimes \mathbb{Z}_p)$ by a spectrum analog of 6.39. Let W_1 be the cofibre: that is,

$$W_0 \to \Sigma^m H \to W_1 \to \Sigma W_0$$

is a cofibre sequence. The exact homotopy sequence becomes

$$0 \to \pi_{m+1}(W_1) \to \pi_m(W) \to \pi_m(W) \otimes \mathbb{Z}_p,$$

which shows that $\pi_{m+1}(W_1) \cong p\pi_m(W_0)$. Thus $p^{2s-1}\pi_{m+1}(W_1) = 0$; moreover $p^{2s}\pi_r(W_1) = 0$ for $r > m+1$ and $\pi_r(W_1) = 0$ for $r < m+1$. Thus we may repeat this process until we have $\pi_{m+2s}(W_{2s}) = 0$, $p^{2s}\pi_r(W_{2s}) = 0$, $r > m+2s$, $\pi_r(W_{2s}) = 0$ for $r < m+2s$. Then we set to work again, killing off $\pi_{m+2s+1}(W_{2s})$, and eventually we find a t with $\pi_{n+t}(W_t) = 0$.

Now we show that we can find maps $f_s: Y_s \to W_s$ so that $f_0 = f$ and

$$\begin{array}{ccc} \Sigma^{-1} Y_{s+1} & \xrightarrow{\Sigma^{-1} f_{s+1}} & \Sigma^{-1} W_{s+1} \\ \downarrow{\scriptstyle i_{s+1}} & & \downarrow{\scriptstyle i'_{s+1}} \\ Y_s & \xrightarrow{f_s} & W_s \end{array}$$

commutes up to homotopy for all $s \geqslant 0$. We have the cofibre sequences

$$\Sigma^{-1} W_{s+1} \xrightarrow{i'_{s+1}} W_s \xrightarrow{j_s} \Sigma^k H(\pi_k(W_s) \otimes \mathbb{Z}_p)$$

$$\Sigma^{-1} Y_{s+1} \xrightarrow{i_{s+1}} Y_s \xrightarrow{\iota \wedge 1} H(\mathbb{Z}_p) \wedge Y_s$$

which define W_{s+1}, Y_{s+1} respectively. At the beginning of this proof we remarked that the maps $(i_{s+1})_*: H_*(\Sigma^{-1} Y_{s+1}; \mathbb{Z}_p) \to H_*(Y_s; \mathbb{Z}_p)$ were all zero. The universal coefficient theorem

$$H^*(Y_s; \pi_k(W_s) \otimes \mathbb{Z}_p) \cong \operatorname{Hom}_{\mathbb{Z}_p}(H_*(Y_s; \mathbb{Z}_p), \pi_k(W_s) \otimes \mathbb{Z}_p)$$

shows that also $i^*_{s+1} = 0$ on $H^*(Y_s; \pi_k(W_s) \otimes \mathbb{Z}_p)$. If we regard

$$[j_s] \in [W_s, \Sigma^k H(\pi_k(W_s) \otimes \mathbb{Z}_p)] = H^k(W_s; \pi_k(W_s) \otimes \mathbb{Z}_p)$$

as a cohomology element, then we have

$$[j_s \circ f_s \circ i_{s+1}] = (f_s \circ i_{s+1})^*[j_s] = i^*_{s+1} \circ f^*_s[j_s] = 0.$$

But $[j_s \circ f_s \circ i_{s+1}]$ can also be written $j_{s*}[f_s \circ i_{s+1}]$ and

$$[\Sigma^{-1} Y_{s+1}, \Sigma^{-1} W_{s+1}] \xrightarrow{(i'_{s+1})_*} [\Sigma^{-1} Y_{s+1}, W_s] \xrightarrow{j_{s*}} [\Sigma^{-1} Y_{s+1}, \Sigma^k H]$$

is exact, so $j_{s*}[f_s \circ i_{s+1}] = 0$ implies

$$[f_s \circ i_{s+1}] = (i'_{s+1})_*[\Sigma^{-1} f_{s+1}]$$

for some element $[\Sigma^{-1} f_{s+1}] \in [\Sigma^{-1} Y_{s+1}, \Sigma^{-1} W_{s+1}]$. Thus we may construct the maps f_s for all s by induction.

Now suppose x were in $F^{t,n+t}$ for the particular t found above. We have the commutative diagram

$$\begin{array}{ccccc} & & \pi_{n+t}(Y_t) & \xrightarrow{f_{t*}} & \pi_{n+t}(W_t) = 0 \\ & & \downarrow i_* & & \downarrow i'_* \\ \pi_n(Y) & \xrightarrow{p^s} & \pi_n(Y) & \xrightarrow{f_*} & \pi_n(W) \end{array}$$

in which the bottom line is exact. $x \in F^{t,n+t}$ means there is a $y \in \pi_{n+t}(Y_t)$ with $i_* y = x$. Thus $f_* x = i'_* \circ f_{t*}(y) = 0$, so by exactness $x = p^s z$ for some $z \in \pi_n(Y)$. This contradicts the original assumption that x was not divisible by p^s. Hence we must have $x \notin F^{t,n+t}$.

Now

$$A^{s,t} = \operatorname{im}[\pi_t(E \wedge Y_s) \to \pi_t(Y_{s+1})] \cap \bigcap_{r \geqslant 1} \operatorname{im}[\pi_{t+r}(Y_{s+r+1}) \to \pi_t(Y_{s+1})]$$

and $Y_{s+1} \leftarrow \Sigma^{-1} Y_{s+2} \leftarrow \Sigma^{-2} Y_{s+3} \leftarrow \cdots$ is an Adams filtration of Y_{s+1}. Hence by what we just proved above every element of

$$\bigcap_{r \geqslant 1} \operatorname{im}[\pi_{t+r}(Y_{s+r+1}) \to \pi_t(Y_{s+1})]$$

is divisible by p^n for all $n \geqslant 0$. On the other hand $\pi_t(E \wedge Y_s) = H_t(Y_s; \mathbb{Z}_p)$, so every element of $\operatorname{im}[\pi_t(E \wedge Y_s) \to \pi_t(Y_{s+1})]$ is of order p. Now if $H_*(Y)$ is of finite type, then an inductive argument shows that $H_*(Y_{s+1})$ is also of finite type. It follows then from Serre's $\mathscr{C}$-theory (cf. [78] or [80]) that $\pi_*(Y_{s+1})$ is of finite type. In a group of finite type, however, the only element of order p which is divisible by p^n for all n is 0. Thus $A^{s,t} = 0$. The statement that $D^* =$ the subgroup of torsion elements of order prime to p is also clear. □

The reader will have noticed that the indexing of the Adams spectral sequence is set up somewhat differently from that in the spectral sequences encountered heretofore. For this reason it turns out to be convenient to graph it with s along the y-axis and $t - s$ along the x-axis. Then the differentials go as in the picture below.

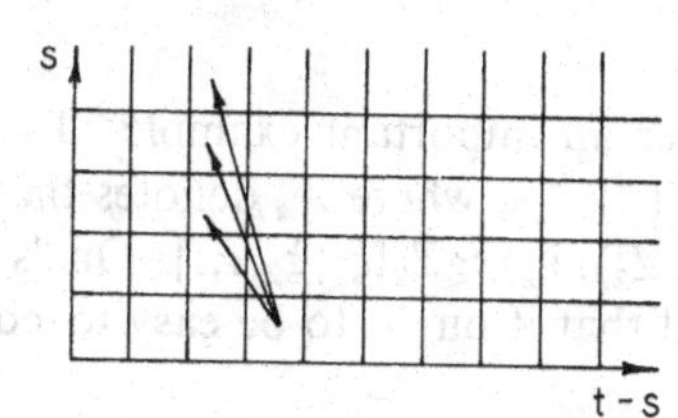

That is, d_r goes one column to the left and r rows up. Also all the groups in a given vertical column, say $t-s=n$, have to do with one particular homotopy group—namely $\pi_n(Y)$.

19.13. Before we turn to specific examples we must indicate how the Adams spectral sequence behaves with respect to products, for the product structure is often valuable for computing differentials d_r and solving the extension problems

$$0 \to F^{s+1,t+1} \to F^{s,t} \to E_\infty^{s,t} \to 0.$$

Let us denote the Adams spectral sequence of Y by $E_r^{s,t}(Y)$. If Y, Z are two spectra, then there is a natural pairing

$$E_r^{s,t}(Y) \otimes E_r^{s',t'}(Z) \to E_r^{s+s',t+t'}(Y \wedge Z)$$

such that the differential d_r is a derivation:

$$d_r(y \wedge z) = d_r y \wedge z + (-1)^{s+t} y \wedge d_r z$$

for $y \in E_r^{s,t}(Y)$, $z \in E_r^{s',t'}(Z)$. There is also a pairing of the filtration groups

$$F^{s,t}(Y) \otimes F^{s',t'}(Z) \to F^{s+s',t+t'}(Y \wedge Z)$$

which agrees with the inverse limit pairing

$$E_\infty^{s,t}(Y) \otimes E_\infty^{s',t'}(Z) \to E_\infty^{s+s',t+t'}(Y \wedge Z).$$

Moreover, the exterior product $E_*(Y) \otimes E_*(Z) \to E_*(Y \wedge Z)$ gives a pairing

$$\mathrm{Hom}_C^t(R, E_*(Y)) \otimes \mathrm{Hom}_C^{t'}(R, E_*(Z)) \to \mathrm{Hom}_C^{t+t'}(R, E_*(Y \wedge Z)),$$

which induces a pairing

$$\mathrm{Ext}_C^{s,t}(R, E_*(Y)) \otimes \mathrm{Ext}_C^{s',t'}(R, E_*(Z)) \to \mathrm{Ext}_C^{s+s',t+t'}(R, E_*(Y \wedge Z)),$$

and this pairing agrees with the one

$$E_2^{s,t}(Y) \otimes E_2^{s',t'}(Z) \to E_2^{s+s',t+t'}(Y \wedge Z)$$

on E_2-terms.

19.14. Let us consider an important example : $Y = S^0$, $E = H(\mathbb{Z}_2)$. The E_2-term then is $\mathrm{Ext}_{A_*}^{s,t}(\mathbb{Z}_2, \mathbb{Z}_2)$, where A_* denotes the dual of the Steenrod algebra: $A_* = H_*(H(\mathbb{Z}_2); \mathbb{Z}_2) \cong \mathbb{Z}_2[\xi_1, \xi_2, \ldots]$. One's first reaction might be that $\mathbb{Z}_2$ is so small that it ought to be easy to compute $\mathrm{Ext}_{A_*}^{s,t}(\mathbb{Z}_2, \mathbb{Z}_2)$.

But remember that we need a resolution

$$0 \leftarrow \mathbb{Z}_2 \leftarrow X_0 \leftarrow X_1 \leftarrow X_2 \leftarrow \cdots$$

of $\mathbb{Z}_2$ by A_*-comodules X_i which are A_*-free—in other words the X_i are huge. In fact $\mathrm{Ext}_{A_*}^{s,t}(\mathbb{Z}_2, \mathbb{Z}_2)$ is only known for a finite range of s. The first few terms of the E_2-term look as follows (cf. [1] or [68]):

s \ $t-s$	0	1	2	3	4	5	6	7	8
4	h_0^4							$h_0^3 h_3$	
3	h_0^3			$h_0^2 h_2 = h_1^3$				$h_0^2 h_3$	α
2	h_0^2		h_1^2	$h_0 h_2$			h_2^2	$h_0 h_3$	$h_1 h_3$
1	h_0	h_1		h_2				h_3	
0	1								

where blank squares denote zero groups and elements written into squares denote generators of $\mathbb{Z}_2$-summands. There is an $h_i \in E_2^{1,2^i}$ for every $i \geqslant 0$ (these correspond to $\xi_1^{2^i} \in A_{2^i}$; see 19.17 below). We have employed the multiplicative structure of the spectral sequence to write various generators as products of others. All groups above the dashed line are zero.

Now it turns out that in this part of the spectral sequence all differentials vanish. In fact, one can see immediately that the only element on which d_r has a chance of being non-zero is h_1. Suppose we had $d_r h_1 = h_0^{r+1}$ for some r. Now $h_0 h_1 = 0$, so we would have $0 = d_r(h_0 h_1) = h_0(d_r h_1) = h_0^{r+2}$. But all powers of h_0 are non-zero, so $d_r h_1 = h_0^{r+1}$ is impossible.

Hence in the part of the spectral sequence drawn we have $E_\infty^{s,t} = E_2^{s,t}$. There is still the extension problem to be solved. Let us consider the case $t - s = 0$ first. It is easy to see that $h_0 \in E_2^{1,1}$ corresponds to a map $f: S^n \to S^n$ of degree 2 in π_0^S. As an inductive hypothesis suppose we have shown that $[f]^s \in F^{s,s}$ maps to h_0^s in $E_\infty^{s,s}$. We have a commutative square

$$\begin{array}{ccc} F^{1,1} \otimes F^{s,s} & \longrightarrow & F^{s+1,s+1} \\ \downarrow & & \downarrow \\ E_\infty^{1,1} \otimes E_\infty^{s,s} & \longrightarrow & E_\infty^{s+1,s+1}, \end{array}$$

from which it follows that $[f]^{s+1}$ goes to h_0^{s+1}. In particular $[f]^s$ is non-zero for all s. This is the same as saying that the class $[1] \in {}_2\pi_0^S$ is of infinite order. Now $h: \pi_0^S \to \tilde{H}_0(S^0; \mathbb{Z})$ sends $[1]$ to 1, so $[1]$ is not divisible by 2. Therefore $[1]$ generates a $\mathbb{Z}$-summand in ${}_2\pi_0^S$. Suppose that ${}_2\pi_0^S \simeq \mathbb{Z} \oplus A$;

then $F^{s,s} \cong 2^s\mathbb{Z} \oplus A_s$ for some subgroup A_s of A. Since $F^{s,s}/F^{s+1,s+1} \cong \mathbb{Z}_2$, it follows $A_s = A_{s+1}$ for all $s \geqslant 0$. But $\bigcap_{s\geqslant 0} F^{s,s} = 0$, so we must have $A = 0$. That is, ${}_2\pi_0^S \cong \mathbb{Z}$ generated by [1], which we already knew, of course.

A similar (but easier) argument shows that the tower h_2, h_0h_2, $h_0^2h_2$ for $t-s=3$ gives ${}_2\pi_3^S \cong \mathbb{Z}_8$ rather than $\mathbb{Z}_2 \oplus \mathbb{Z}_4$ or $\mathbb{Z}_2 \oplus \mathbb{Z}_2 \oplus \mathbb{Z}_2$. This argument works whenever we have a tower of the form $x, h_0x, h_0^2x, \ldots$.

Thus we see that the 2-primary part of π_n^S in the range $n \leqslant 7$ is as follows:

n	0	1	2	3	4	5	6	7
${}_2\pi_n^S$	$\mathbb{Z}$	$\mathbb{Z}_2$	$\mathbb{Z}_2$	$\mathbb{Z}_8$	0	0	$\mathbb{Z}_2$	$\mathbb{Z}_{16}$

A look in [87], where the stable stems π_n^S are computed by a different method, gives

n	0	1	2	3	4	5	6	7
π_n^S	$\mathbb{Z}$	$\mathbb{Z}_2$	$\mathbb{Z}_2$	$\mathbb{Z}_{24}$	0	0	$\mathbb{Z}_2$	$\mathbb{Z}_{240}$

(Compare with 15.37).

19.15. Exercise. There is yet another way to compute Ext_C^{**}.

a) An R-module J is called *injective* if for all exact sequences $0 \to A \xrightarrow{\alpha} B$ of R-modules and for all R-homomorphisms $\rho: A \to J$ there is an R-homomorphism $\rho': B \to J$ with $\rho' \circ \alpha = \rho$. The definition is the same for A-modules (A an R-algebra) and C-comodules (C an R-coalgebra). Show that for every C-comodule K we can find an injection $K \to J$ of K into an injective C-comodule J. Hence deduce that every C-comodule K has an *injective resolution*

$$0 \longrightarrow K \longrightarrow Y_0 \xrightarrow{d^0} Y_1 \xrightarrow{d^1} Y_2 \longrightarrow \cdots$$

(i.e. the sequence is exact and each Y_i is injective over C).

b) Show that if J is injective over C then $\mathrm{Ext}_C^{p,q}(K,J) = 0$ for all C-comodules K, $p \geqslant 1$.

c) Show that if $0 \to L \to Y_0 \to Y_1 \to Y_2 \to \cdots$ is an injective resolution of the C-comodule L, then $\mathrm{Ext}_C^{p,q}(K,L) \cong H^p(\mathrm{Hom}_C^q(K, Y_*))$, $p \geqslant 0$, $q \in \mathbb{Z}$, K any C-comodule.

19.16. Exercise. Show that 13.9 can be generalized to the case of graded C-comodules, C a coalgebra over a graded ring R: that is $\mathrm{Ext}_C^{1,q}(M,N)$ can be identified with the group of equivalence classes of extensions $0 \to N \xrightarrow{\alpha} Q \xrightarrow{\beta} M \to 0$ of M by N with $\deg\alpha + \deg\beta = -q$.

19.17. Proposition. *If C is a coalgebra over the graded ring R and $P_q(C)$ the group of primitive elements of grading q, then*

$$\mathrm{Ext}_C^{1,q}(R,R) \cong P_q(C)/(\eta_L - \eta_R)(R_q).$$

Proof: Suppose $0 \to R \xrightarrow{\alpha} M \xrightarrow{\beta} R \to 0$ is an extension defining an element of $\mathrm{Ext}_C^{1,q}(R,R)$. As an R-module $M \cong R\cdot x + R\cdot y$ for $x = \alpha(1) \in M_p$, some $y \in M_{p+q}$ with $\beta(y) = 1$. Then

19.18. $$\psi x = 1 \otimes x,$$

$$\psi y = 1 \otimes y + h \otimes x, \quad \text{some } h \in C_q.$$

From $(\psi \otimes 1) \circ \psi = (1 \otimes \psi) \circ \psi$ and $(\varepsilon \otimes 1) \circ \psi = 1$ it follows that h must be primitive. Now y is not uniquely defined; we can replace y by $y + kx$ for any $k \in R_q$. Then h is replaced by $h + (\eta_L - \eta_R)(k)$. Thus only the residue class $\{h\} \in P_q(C)/(\eta_L - \eta_R)(R_q)$ is determined.

On the other hand given $h \in P_q(C)$, we can define $\{M_h\} \in \mathrm{Ext}_C^{1,q}(R,R)$ by setting $M = R\cdot x + R\cdot y$ with $x \in M_0$, $y \in M_q$ and defining ψ by 19.18. Clearly $\{M_{h+(\eta_L-\eta_R)(k)}\} = \{M_h\}$ for any $k \in R_q$. □

Since for A_* we have $\eta_L = \eta_R$ and $P_q(A_*) = \mathbb{Z}_2$ generated by $\xi_1^{2^i}$ if $q = 2^i$, $P_q(C) = 0$ if q is not of the form 2^i, we see that $\mathrm{Ext}_{A_*}^{1,*}(\mathbb{Z}_2, \mathbb{Z}_2)$ is as given in 19.14.

19.19. In 19.11 we showed that the Adams spectral sequence converges under certain conditions which included $\pi_q(E) = 0$ for $q < 0$. Unfortunately this condition is not fulfilled by $E = K$ or $E = KO$. Nevertheless Theorem 19.9 still holds for $E = K$ or KO, and in particular the edge homomorphism

$$\pi_n(Y) \to E_\infty^{0,n} \to E_2^{0,n} \cong \mathrm{Hom}^n_{K_*(K)}(\tilde{K}_*(S^0), K_*(Y))$$

is the Hurewicz homomorphism h_K. For $Y = S^0$, however,

$$\mathrm{Hom}^n_{K_*(K)}(\tilde{K}_*(S^0), \tilde{K}_*(Y)) \cong \mathbb{Z} \text{ or } 0,$$

whereas $\pi_n^S = \pi_n(S^0)$ is a finite group for $n > 0$. Thus all elements of π_n^S, $n > 0$, lie in $\ker h_K$, so that h_K is not useful for investigating π_*^S. However, we know that $\ker h_K = F^{1,n+1} \subset \pi_n(Y)$, so we have the map

$$\ker h_K = F^{1,n+1} \to E_\infty^{1,n+1} \subset E_2^{1,n+1}$$

$$\cong \mathrm{Ext}^{1,n+1}_{K_*(K)}(\tilde{K}_*(S^0), \tilde{K}_*(S^0)) \cong \frac{P_{n+1}(K_*(K))}{(\eta_L - \eta_R)(\tilde{K}_{n+1}(S^0))}.$$

This homomorphism was introduced by Adams and called the e_C-invariant. There is the analogous invariant

$$e_R:\ker h_{KO} \to \frac{P_{n+1}(KO_*(KO))}{(\eta_L - \eta_R)(\widetilde{KO}_{n+1}(S^0))}.$$

It thus becomes of interest to determine $P_*(K_*(K))$ and $P_*(KO_*(KO))$.

19.20. Proposition. *If we make $C = \mathbb{Q}[u, u^{-1}, v, v^{-1}]$ into a graded coalgebra over the ring $R = \mathbb{Z}[t, t^{-1}]$ by setting $t \cdot f = uf$, $f \cdot t = fv$ for $f \in C$ and $\psi u = u \otimes 1$, $\psi v = 1 \otimes v$,* grade u = grade $v = 2$, *then*

$$P_{2n}(C) = \mathbb{Q} \cdot (u^n - v^n) \quad \textit{for all } n \in \mathbb{Z}.$$

Proof: Suppose $f = \sum a_i u^i v^{n-i} \in C_{2n}$, $a_i \in \mathbb{Q}$. Then $\psi f = \sum a_i u^i \otimes v^{n-i}$, so

$$\begin{aligned}\psi f - 1 \otimes f - f \otimes 1 &= \sum a_i u^i \otimes v^{n-i} - \sum a_i v^i \otimes v^{n-i} - f \otimes 1 \\ &= \sum_{i \neq 0, n} a_i p_i \otimes v^{n-i} + [a_n p_n - f] \otimes 1\end{aligned}$$

if $p_i = u^i - v^i$, $i \geqslant 1$. This expression is 0 if and only if $a_n p_n - f = 0$—that is $f = a_n(u^n - v^n)$. ▯

19.21. Corollary. *$P_{2n}(K_*(K)) \cong \mathbb{Z}$ with generator $(u^n - v^n)/m(|n|)$ for all $n \in \mathbb{Z} - \{0\}$, where $m(n)$ denotes the largest positive integer such that $(x^n - 1)/m(n) \in \mathbb{Z}[1/x]$ for all $x \in \mathbb{Z} - \{0\}$, $n > 0$.*

$$P_0(K_*(K)) = P_{2n+1}(K_*(K)) = 0 \quad \textit{for all } n \in \mathbb{Z}.$$

19.22. Proposition.

$$\mathrm{Ext}^{1,q}_{K_*(K)}(\tilde{K}_*(S^0), \tilde{K}_*(S^0)) \cong \begin{cases} \mathbb{Z}_{m(|n|)} & q = 2n,\ n \neq 0 \\ 0 & q = 0 \textit{ or } q \text{ odd.} \end{cases}$$

Proof: Clearly $(\eta_L - \eta_R)(\tilde{K}_{2n}(S^0)) \cong \mathbb{Z}$ generated by $(\eta_L - \eta_R)(t^n) = u^n - v^n$. ▯

Thus $\mathrm{Ext}^{1,q}_{K_*(K)}(\tilde{K}_*(S^0), \tilde{K}_*(S^0))$ is a finite group for all q, and one might hope that $e_C: \pi^S_{2n-1} \to \mathbb{Z}_{m(|n|)}$ would be non-zero, thus providing some information about π^S_{2n-1}, $n \geqslant 1$. We shall see that this is indeed the case for n even.

19.23. Lemma.

$$P_{4n}(KO_*(KO)) \cong \mathbb{Z} \textit{ generated by } \begin{cases} \dfrac{u^{2n} - v^{2n}}{m(|2n|)} & n \text{ even} \\ 2\dfrac{u^{2n} - v^{2n}}{m(|2n|)} & n \text{ odd.} \end{cases}$$

Proof: This follows from 19.20 and 17.34. ▯

19.24. Corollary.

$$\mathrm{Ext}^{1,4n}_{\widetilde{KO}_*(KO)}(\widetilde{KO}_*(S^0), \widetilde{KO}_*(S^0)) \cong \begin{cases} \mathbb{Z}_{m(|2n|)} & n \neq 0 \\ 0 & n = 0. \end{cases}$$

Proof: $(\eta_L - \eta_R)(y^n) = u^{4n} - v^{4n}$, $(\eta_L - \eta_R)(xy^n) = 2(u^{4n+2} - v^{4n+2})$. □

It would clearly be desirable to have more information about the integers $m(n)$. We shall see that they are related to the Bernoulli numbers. If we expand $x/(e^x - 1)$ in a power series, then we get a series of the form

$$\frac{x}{e^x - 1} = \Sigma_{t=0}^{\infty} \beta_t \frac{x^t}{t!},$$

where $\beta_{2s+1} = 0$ for $s > 0$. The coefficients β_t are related to the Bernoulli numbers B_s by the formula

$$\beta_{2s} = (-1)^{s-1} B_s, \quad s > 0.$$

Let $\hat{m}(t)$ be the number-theoretic function defined as follows. We write $v_p(n)$ for the exponent to which the prime p occurs in the decomposition of n into prime powers:

$$n = 2^{v_2(n)} 3^{v_3(n)} 5^{v_5(n)} \ldots.$$

Then

$$v_p(\hat{m}(t)) = \begin{cases} 0 & t \not\equiv 0(p-1) \\ 1 + v_p(t) & t \equiv 0(p-1), \ p \text{ odd}, \end{cases}$$

$$v_2(\hat{m}(t)) = \begin{cases} 1 & t \not\equiv 0\ (2) \\ 2 + v_2(t) & t \equiv 0\ (2). \end{cases}$$

In particular $\hat{m}(2t+1) = 2$ for all $t \geqslant 0$. In [3] Adams does the necessary number theory to prove the following facts:

i) $\hat{m}(2t)$ is the denominator of $B_t/4t$ when the latter is expressed in lowest terms;

ii) let $f(k)$ take non-negative integer values for $k \in \mathbb{Z}$ and let $m(f,t)$ be the greatest common divisor of the integers $x^{f(x)}(x^t - 1)$, $x \in \mathbb{Z}$. Then $m(f,t)$ divides $\hat{m}(t)$ for all t and f; and for each t there is a function f such that $m(f,t) = \hat{m}(t)$.

19.25. Proposition. $\hat{m}(t) = m(t)$.

Proof: For each fixed t there is a function $f_t(x)$ (non-negative) such that

$$\frac{x^{f_t(x)}(x^t - 1)}{m(t)} \in \mathbb{Z}$$

for all $x \in \mathbb{Z}$ (definition of $m(t)$). Thus $m(t)|m(f_t, t)$ and hence $m(t)|\hat{m}(t)$.

On the other hand, for each fixed t there is a function $f_t(x)$ (non-negative) such that

$$\frac{x^{f_t(x)}(x^t - 1)}{\hat{m}(t)} \in \mathbb{Z}$$

for all $x \in \mathbb{Z}$ (by ii)). That is

$$\frac{x^t - 1}{\hat{m}(t)} \in \mathbb{Z}[1/x] \quad \text{for } x \in \mathbb{Z} - \{0\}.$$

Therefore $\hat{m}(t) \leqslant m(t)$. □

Clearly

$$\begin{aligned} \operatorname{Ext}^{1,q}_{KO_*(KO)}(\widetilde{KO}_*(S^0), \widetilde{KO}_*(S^0)) &\cong P_q(\widetilde{KO}_*(KO))/(\eta_L - \eta_R)(\widetilde{KO}_*(S^0)) \\ &= 0 \quad \text{for } q \equiv 3, 5, 6 \text{ or } 7 \bmod 8. \end{aligned}$$

We can also compute these groups for $q \equiv 1, 2 \bmod 8$. From 17.35 we know $\widetilde{KO}_q(S^0) \otimes KO_0(KO) \to KO_q(KO)$ is an isomorphism, so we ask which $f(u^2, v^2) \in KO_0(KO)$ have the property that $\alpha u^{4n} f(u^2, v^2)$ is primitive. The argument is like that of 19.20; if $f = \sum a_i u^{-2i} v^{2i}$, then

$$\begin{aligned} \psi f - u^{-4n} v^{4n} \otimes f - f \otimes 1 &= \sum a_i u^{-2i} \otimes v^{2i} \\ &\qquad - \sum a_i u^{-4n} v^{4n-2i} \otimes v^{2i} - f \otimes 1 \\ &= \textstyle\sum_{i \neq 0, 2n} a_i u^{-4n} (u^{4n-2i} - v^{4n-2i}) \otimes v^{2i} \\ &\qquad + [a_0(1 - u^{-4n} v^{4n}) - f] \otimes 1. \end{aligned}$$

Thus if $\alpha u^{4n} f$ is to be primitive we must have $\alpha u^{4n} f = a_0 \alpha (u^{4n} - v^{4n})$.

19.26. Proposition. $\operatorname{Ext}^{1,q}_{KO_*(KO)}(\widetilde{KO}_*(S^0), \widetilde{KO}_*(S^0)) \cong \mathbb{Z}_2$ *for* $q \equiv 1, 2 \bmod 8$.

Proof:

$$P_q(KO_*(KO)) = \mathbb{Z}_2 \cdot \left(\alpha \cdot \frac{u^{4q} - v^{4q}}{m(|4q|)} \right). \quad □$$

Our next task is to show that $e_R: \pi_r^S \to \operatorname{Ext}^{1,r+1}_{KO_*(KO)}(\widetilde{KO}_*(S^0), \widetilde{KO}_*(S^0))$ is surjective for $r \equiv 0, 1, 3, 7 \bmod 8$.

G. W. Whitehead has defined a homomorphism $J: \pi_r(SO(q)) \to \pi_{r+q}(S^q)$ as follows: we regard S^{r+q} as the boundary $\partial(D^{r+1} \times D^q) = S^r \times D^q \cup D^{r+1} \times S^{q-1}$ with $S^r \times D^q \cap D^{r+1} \times S^{q-1} = S^r \times S^{q-1}$. Given $f: S^r \to SO(q)$ we define $g: S^r \times S^{q-1} \to S^{q-1}$ by $g(x, y) = f(x) \cdot y$ for $(x, y) \in S^r \times S^{q-1}$. We extend g to a map $S^r \times D^q \to D^q = H^q_+ \subset S^q$ by again taking $g(x, y) = f(x) \cdot y$ for $x \in S^r$, $y \in D^q$. We extend g to a map

$g: D^{r+1} \times S^{q-1} \cong CS^r \times S^{q-1} \to CS^{q-1} = H^q_- \subset S^q$ by setting $g([t,x],y) = [t,g(x,y)]$. The result is a map $g: S^{r+q} \to S^q$, and we set $J[f] = [g]$. For a proof that J is a homomorphism see [49]. Stabilizing we get the stable J-homomorphism

$$J: \pi_r(SO) \to \pi_r^S.$$

We shall show that

$$\pi_r(SO) \xrightarrow{J} \pi_r^S \xrightarrow{e_R} \mathrm{Ext}^{1,r+1}_{KO_*(KO)}(\widetilde{KO}_*(S^0), \widetilde{KO}_*(S^0))$$

is surjective.

We also define a map $J': \pi_{r+1}(BSO) \to \pi_r^S$ by requiring

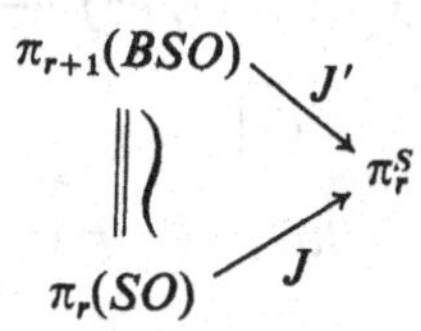

to commute.

19.27. Lemma. *Let $f': S^{r+1} \to BSO(q)$ be a map and $\xi = f'^* \omega_q^+$ the induced bundle over S^{r+1}. By 12.26 $M(\xi)$ is of the form $S^q \cup_g e^{q+r+1}$, and in fact we can find a homotopy equivalence $M(\xi) \simeq S^q \cup_{J'f'} e^{q+r+1}$ which is of degree 1 on the two cells.*

Proof: Let $f: S^r \to SO(q)$ be any map; we shall show that $X = S^q \cup_{Jf} e^{q+r+1}$ is homotopy equivalent to $M(\xi)$, where $f': S^{r+1} \to BSO(q)$ is the adjoint of f and $\xi = f'^*(\omega_q^+)$. We have described $Jf: S^{q+r} \to S^q$ above; we do not change the homotopy type of X if we identify $H^q_- \subset S^q$ to a point.

The disk bundle of $f'^*(\xi)$ can be described as follows: we regard S^{r+1} as D^{r+1} with S^r identified to a point. Then $D(\xi)$ is obtained from $(D^{r+1} \times D^q) \,\dot\cup\, D^q$ by identifying $(x,y) \in S^r \times D^q$ with $f(x) \cdot y \in D^q$. Then $M(\xi)$ is obtained by identifying $(D^{r+1} \times S^{q-1}) \cup S^{q-1}$ to a point. But then we have precisely X/H^q_-. □

By composition we also obtain maps $J'_C: \pi_{r+1}(BU) \to \pi_r^S$ and $J'_H: \pi_{r+1}(BSp) \to \pi_r^S$.

We now describe a way of computing $e_R[f]$ for any map

$$f: S^{4r+q-1} \to S^q.$$

Since $0 = h_{KO}[f] \in \widetilde{KO}_*(S^0)$, we know that f factors through the first filtration spectrum Y_1:

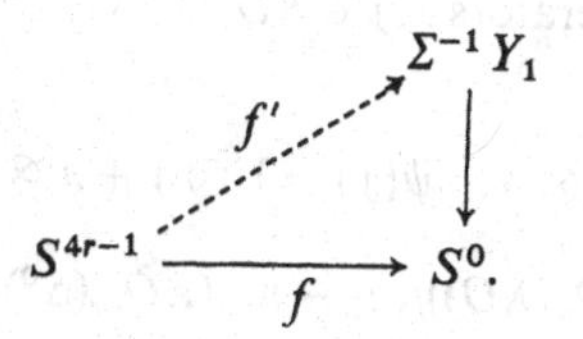

In fact we have a homotopy commutative diagram

$$\begin{array}{ccccccc}
S^{4r-1} & \xrightarrow{f} & S^0 & \xrightarrow{j} & S^0 \cup_f e^{4r} & \xrightarrow{k'} & S^{4r} \\
\downarrow f' & & \| & & \downarrow k & & \downarrow \Sigma f' \\
\Sigma^{-1} Y_1 & \longrightarrow & S^0 & \longrightarrow & Y_1/Y_0 & \longrightarrow & Y_1
\end{array}$$

for some map $k: S^0 \cup_f e^{4r} \to Y_1/Y_0 = KO \wedge S^0 \simeq KO$. Thus the diagram

$$\begin{array}{ccccccccc}
0 & \longrightarrow & \widetilde{KO}_*(S^0) & \xrightarrow{j_*} & KO_*(S^0 \cup_f e^{4r}) & \xrightarrow{k'_*} & \widetilde{KO}_*(S^0) & \longrightarrow & 0 \\
 & & \| & & \downarrow k_* & & \downarrow (\Sigma f')_* & & \\
0 & \longrightarrow & \widetilde{KO}_*(S^0) & \longrightarrow & KO_*(KO) & \longrightarrow & KO_*(Y_1) & \longrightarrow & KO_*(Y_1/Y_2)
\end{array}$$

commutes. Checking the definition of $e_R[f]$ shows that it is precisely the class

$$\{\Sigma f'_*\} \in \mathrm{Ext}^{1,4r}_{KO_*(KO)}(\widetilde{KO}_*(S^0), \widetilde{KO}_*(S^0)) = \frac{\ker[\mathrm{Hom}^{4r}_{KO_*(KO)}(\widetilde{KO}_*(S^0), KO_*(Y_1/Y_2)) \to \mathrm{Hom}^{4r}_{KO_*(KO)}(\widetilde{KO}_*(S^0), KO_*(Y_2/Y_3))]}{\mathrm{im}[\mathrm{Hom}^{4r}_{KO_*(KO)}(\widetilde{KO}_*(S^0), KO_*(Y_0/Y_1)) \to \mathrm{Hom}^{4r}_{KO_*(KO)}(\widetilde{KO}_*(S^0), KO_*(Y_1/Y_2))]}.$$

On the other hand

$$0 \longrightarrow \widetilde{KO}_*(S^0) \xrightarrow{j_*} KO_*(S^0 \cup_f e^{4r}) \xrightarrow{k'_*} \widetilde{KO}_*(S^0) \longrightarrow 0$$

is an extension—that is an element of $\mathrm{Ext}^{1,4r}_{KO_*(KO)}(\widetilde{KO}_*(S^0), \widetilde{KO}_*(S^0))$ (cf. 19.16). If we follow through the construction of the correspondence between the various definitions of Ext, then we see that this extension corresponds exactly to the class $\{\Sigma f'_*\} = e_R[f]$.

If we now choose generators $x, y \in KO_*(S^0 \cup_f e^{4r})$ such that $j_*(1) = x$, $k'_*(y) = 1$, then

$$\psi(x) = 1 \otimes x, \quad \psi(y) = 1 \otimes y + h \otimes x$$

and $e_R[f] = \{h\} \in P_{4r}(KO_*(KO))/(\eta_L - \eta_R)(\widetilde{KO}_{4r}(S^0))$ (19.17).

Let us write X_f for $S^0 \cup_f e^{4r}$. Then we also have

$$KO^*(X_f) \cong \widetilde{KO}^*(S^0)\cdot\alpha \oplus \widetilde{KO}^*(S^0)\cdot\beta$$

for suitable $\alpha, \beta \in KO^*(X_f)$, and we can even choose α, β so that $\langle\alpha, x\rangle = \langle\beta, y\rangle = 1$, $\langle\alpha, y\rangle = \langle\beta, x\rangle = 0$. Since $KO^*(X_f)$ is a free $\widetilde{KO}^*(S^0)$-module, we may regard the Boardman map as a map

$$B: KO^*(X_f) \to KO^*(X_f) \otimes_{\widetilde{KO}_*(S^0)} KO_*(KO)$$

and the Hurewicz map H as a map

$$H: KO_*(X_f) \to KO_*(KO) \otimes_{\widetilde{KO}_*(S^0)} KO_*(X_f).$$

At first one might think H is identical with ψ, but examination of the definitions shows that the diagram

$$\begin{array}{ccc} & \overset{H}{\nearrow} & KO_*(KO) \otimes_{\widetilde{KO}_*(S^0)} KO_*(Y) \\ KO_*(Y) & & \big\downarrow c \otimes 1 \\ & \underset{\psi}{\searrow} & KO_*(KO) \otimes_{\widetilde{KO}_*(S^0)} KO_*(Y) \end{array}$$

commutes for all spectra Y. Thus $H(x) = 1 \otimes x$ and $H(y) = 1 \otimes y + c(h) \otimes x$. Since h is of the form $a(u^{2r} - v^{2r})$ for some $a \in \mathbb{Q}$ and $c(u) = v$, $c(v) = u$, it follows $H(y) = 1 \otimes y - h \otimes x$.

Suppose $\alpha \in KO^*(X_f)$ is represented by $g: X_f \to \Sigma^q KO$; by definition of B we have

$$B(\alpha) = \alpha \otimes g_*(x) + \beta \otimes g_*(y).$$

On the other hand B and H are dual (13.78), so we compute

$$\begin{aligned} 1 = \langle\alpha, x\rangle = \kappa(B(\alpha) \otimes H(x)) &= \kappa([\alpha \otimes g_*(x) + \beta \otimes g_*(y)] \otimes [1 \otimes x]) \\ &= g_*(x) \end{aligned}$$

and

$$\begin{aligned} 0 = \langle\alpha, y\rangle = \kappa(B(\alpha) \otimes H(y)) &= \kappa([\alpha \otimes 1 + \beta \otimes g_*(y)] \\ &\quad \otimes [1 \otimes y - h \otimes x]) = g_*(y) - h. \end{aligned}$$

Thus

$$e_R[f] = \{g_*(y)\} \in P_{4r}(KO_*(KO))/(\eta_L - \eta_R)(\widetilde{KO}_{4r}(S^0)).$$

Now consider $f': S^{4r} \to BSp(q)$. We wish to determine $e_R J'_H[f']$. For $X_{J'_H f'}$ we may take $M(f'^*\xi_q)$, and for $\alpha \in \widetilde{KO}^*(X_{J'_H f'})$ we may take the Thom class $t(f'^*\xi_q) = M(f')^*(t_q)$. Now the KO-theory Thom class is represented by a map $\mu_R: MSp \to KO$ (cf. Chapter 17), so α is represented by the composite

$$X_{J'_H f'} \xrightarrow{M(f')} MSp_q \xrightarrow{i_q} \Sigma^{4q} MSp \xrightarrow{\Sigma^{4q}\mu_R} \Sigma^{4q} KO.$$

For $y \in KO_*(M(f'^*\xi_q))$ we may take $\Phi_*^{-1}(\iota_{4r})$, $\iota_{4r} \in \widetilde{KO}_{4r}(S^{4r})$ the generator. Thus we have

$$e_R J'_H[f'] = \{\mu_{R_*} M(f')_*(\Phi_*^{-1}(\iota_{4r}))\} = \{\mu_{R_*} \Phi_*^{-1} f'_*(\iota_{4r})\} = \{\mu_{R_*} \Phi_*^{-1} h_{KO}[f']\}.$$

Hence we have proved the following.

19.28. Proposition. *The following diagram commutes for all r:*

$$\begin{array}{ccccc}
 & KO_{4r}(BSp) & \xleftarrow[\cong]{\Phi_*} KO_{4r}(MSp) & \xrightarrow{\mu_R} & KO_{4r}(KO) \\
 {}^{h_{KO}}\nearrow & & & & \downarrow q \\
\pi_{4r}(BSp) & \xrightarrow{J'_H} & \pi_{4r-1}^S & \xrightarrow{e_R} & \dfrac{KO_{4r}(KO)}{(\eta_L - \eta_R)(\widetilde{KO}_{4r}(S^0))},
\end{array}$$

where q is the natural projection.

If we denote the generator of $\pi_4(BSp)$ by z, then we can write the generator of $\pi_{8q+4}(BSp)$ as $y^q \cdot z$, the generator of $\pi_{8q+8}(BSp)$ as $xy^q \cdot z$, where we use the product $\mu_2 : \pi_*(BO) \otimes \pi_*(BSp) \to \pi_*(BSp)$ to give $\pi_*(BSp)$ the structure of a $\pi_*(BO)$-module and $x \in \pi_4(BO)$, $y \in \pi_8(BO)$ are the standard generators. We must determine $h_{KO}(y^q \cdot z)$ and $h_{KO}(xy^q \cdot z)$.

For any $f: S^r \to X$, X a CW-complex, we have $h_{KO}[f] = f_*(\iota_r)$ and hence for $a, b \in \widetilde{KO}^*(X)$

$$\langle ab, h_{KO}[f]\rangle = \langle ab, f_*(\iota_r)\rangle = \langle f^*(a) f^*(b), \iota_r\rangle = 0,$$

since all products in $\widetilde{KO}^*(S^r)$ vanish. Thus $h_{KO}[f]$ annihilates all nontrivial products in $\widetilde{KO}^*(X)$ and therefore $h_{KO}[f]$ is primitive with respect to the coproduct

$$KO_*(X) \xrightarrow{\Delta_*} KO_*(X \times X) \cong KO_*(X) \otimes_{KO_*(pt)} KO_*(X)$$

if $KO_*(X)$ is flat over $KO_*(pt)$. $Q(KO^*(BSp))$ is generated over $KO^*(pt)$ by the Pontrjagin classes $P_1, P_2, \ldots, P_k, \ldots$, so $P(KO_*(BSp))$ has a dual basis over $KO_*(pt)$ consisting of the *Newton polynomials* $N_1, N_2, \ldots, N_k, \ldots$ (cf. note following 16.44).

Now $h_{KO}(z) = Z_1 \in KO_4(BSp)$ and

$$\begin{array}{ccc} \pi_{4r}(BSp) & \xrightarrow{x\cdot} & \pi_{4r+4}(BSp) \\ \downarrow c'_* & & \downarrow c'_* \\ \pi_{4r}(BU) & & \pi_{4r+4}(BU) \\ \downarrow h_K & & \downarrow h_K \\ K_{4r}(BU) & \xrightarrow{2B^2} & K_{4r+4}(BU) \end{array} \qquad \begin{array}{ccc} \pi_{4r}(BSp) & \xrightarrow{y\cdot} & \pi_{4r+8}(BSp) \\ \downarrow c'_* & & \downarrow c'_* \\ \pi_{4r}(BU) & & \pi_{4r+8}(BU) \\ \downarrow h_K & & \downarrow h_K \\ K_{4r}(BU) & \xrightarrow{B^4} & K_{4r+8}(BU) \end{array}$$

commute, so from 17.27 it follows

$$h_{KO}(y^q \cdot z) = \frac{(4q+2)!}{2} Z_{2q+1}$$
$$+ \text{ decomposables and terms in } Z_1, Z_2, \ldots, Z_{2q}$$

$$h_{KO}(xy^{q-1} \cdot z) = \frac{(4q)!}{2} Z_{2q} + \text{decomposables and terms in } Z_1, Z_2, \ldots, Z_{2q-1}.$$

But $h_{KO}(y^q \cdot z)$ and $h_{KO}(xy^{q-1} \cdot z)$ are $\mathbb{Z}[t^4, 2t^2, t^{-4}]$-linear combinations of the N_k's and

$$N_k = (-1)^{k-1} k Z_k \text{ mod decomposables,}$$

so we have

$$h_{KO}(y^q \cdot z) = (4q+1)!\, N_{2q+1} + \text{terms of form } a_r y^r N_{2q-2r+1}$$
$$\text{and } b_r xy^r N_{2q-2r}$$

$$h_{KO}(xy^{q-1} \cdot z) = -\frac{(4q-1)!}{2} N_{2q} + \text{ terms of form } a_r y^r N_{2q-2r}$$
$$\text{and } b_r xy^r N_{2q-2r-1}.$$

Φ_* and μ_{R*} are both ring homomorphisms, and as in Chapter 17 we have

$$\Phi_*^{-1}(Z_i) = Z_i', \quad \mu_{R_*}(Z_i') = q_i'(u, v) = \frac{2}{(2i+2)!}(v^2 - u^2) \ldots (v^2 - i^2 u^2).$$

Hence

$$\mu_{R_*} \Phi_*^{-1} h_{KO}(y^q \cdot z) = (4q+1)!\, N_{2q+1}(q_1', \ldots, q_{2q+1}')$$
$$+ \text{ terms of form } a_r u^{2r} N_{2q-r+1}, \quad r > 0,$$

$$\mu_{R*} \Phi_*^{-1} h_{KO}(xy^{q-1} \cdot z) = -\frac{(4q-1)!}{2} N_{2q}(q_1', \ldots, q_{2q}')$$
$$+ \text{ terms of form } a_r u^{2r} N_{2q-r}, \quad r > 0.$$

$e_R J'_H(y^q \cdot z) = \{a_{2q+1}(v^{4q+2} - u^{4q+2})\}$ for some $a_{2q+1} \in \mathbb{Q}$; clearly a_{2q+1} must be the coefficient of v^{4q+2} in $(4q+1)!\,N_{2q+1}(q'_1, \ldots, q'_{2q+1})$. Similarly $e_R J'_H(xy^{q-1} \cdot z) = \{a_{2q}(v^{4q} - u^{4q})\}$, where a_{2q} is the coefficient of v^{4q} in

$$-\frac{(4q-1)!}{2} N_{2q}(q'_1, \ldots, q'_{2q}).$$

The Newton polynomials N_k satisfy the recurrence relation

$$N_k - N_{k-1}q'_1 + N_{k-2}q'_2 - \cdots + (-1)^{k-1} N_1 q'_{k-1} + (-1)^k k q'_k = 0$$

(cf. [55]). We therefore have

$$\sum_{s=1}^k (-1)^{k+1} \frac{2\varepsilon_s a_s}{(2s-1)!\,(2k-2s+2)!} + (-1)^k k \frac{2}{(2k+2)!} = 0$$

where

$$\varepsilon_s = \begin{cases} 1 & s \text{ odd} \\ 2 & s \text{ even.} \end{cases}$$

In other words

$$\sum_{s=1}^k \binom{2k+2}{2s} (2s\varepsilon_s a_s) - k = 0.$$

Let $\beta'_{2s} = 2s\varepsilon_s a_s$, $s \geqslant 1$; then we have

19.29. $$\sum_{s=1}^k \binom{2k+2}{2s} \beta'_{2s} - k = 0 \quad \text{for } k \geqslant 1.$$

Following 19.24 we defined numbers β_s by

$$\frac{x}{e^x - 1} = \sum_{s=0}^\infty \beta_s \frac{x^s}{s!}.$$

If we compare coefficients of x^{k+1}, $k \geqslant 1$, in

$$x = \frac{x}{e^x - 1}(e^x - 1) = \left(\sum_{s=0}^\infty \beta_s \frac{x^s}{s!}\right)\left(\sum_{t=1}^\infty \frac{x^t}{t!}\right)$$

$$= \sum_{k=0}^\infty \left(\sum_{s=0}^k \binom{k+1}{s} \beta_s\right) \frac{x^{k+1}}{(k+1)!},$$

we see that the β_s satisfy

$$\sum_{s=0}^k \binom{k+1}{s} \beta_s = 0, \quad k \geqslant 1.$$

One sees immediately that $\beta_0 = 1$, $\beta_1 = -1/2$. The function

$$f(x) = \frac{x}{e^x - 1} - 1 + \frac{x}{2}$$

is readily seen to be an even function, so it follows $\beta_{2r+1} = 0$ for $r \geqslant 1$. Hence the β_s satisfy

$$\sum_{s=1}^{k} \binom{2k+2}{2s} \beta_{2s} - k = 0.$$

Therefore the unique solution of 19.29 is

$$a_s = \frac{\beta'_{2s}}{2s\varepsilon_s} = \frac{\beta_{2s}}{2s\varepsilon_s} = \frac{(-1)^{s-1} B_s}{2s\varepsilon_s}, \quad s \geqslant 1.$$

Thus we have proved the following.

19.30. Proposition.

i)

$$e_R J'_H(y^q \cdot z) = \left\{ \frac{B_{2q+1}}{4q+2} (v^{4q+2} - u^{4q+2}) \right\} = \left\{ \frac{2z}{m(4q+2)} (v^{4q+2} - u^{4q+2}) \right\}$$

for some $z \in \mathbb{Z}$ relatively prime to $m(4q+2)$.

ii) $$e_R J'_H(xy^{q-1} \cdot z) = \left\{ \frac{-B_{2q}}{8q} (v^{4q} - u^{4q}) \right\} = \left\{ \frac{z}{m(4q)} (v^{4q} - u^{4q}) \right\}$$

for some $z \in \mathbb{Z}$ relatively prime to $m(4q)$.

Comparing with 19.23 we see that we have proved the following.

19.31. Theorem.

$$\pi_{4q-1}(SO) \xrightarrow{J} \pi_{4q-1}^S \xrightarrow{e_R} \frac{P_{4q}(KO_*(KO))}{(\eta_L - \eta_R)(\widetilde{KO}_{4q}(S^0))} \cong \mathbb{Z}_{m(2q)}$$

is surjective for all $q \geqslant 1$.

19.32. Corollary. *$\pi_q(SO) \xrightarrow{J} \pi_q \xrightarrow{e_R} \mathbb{Z}_2$ is an isomorphism for all $q \equiv 0, 1 \bmod 8$, $q > 1$.*

Proof: If $z \in \pi_{8n-1}(SO)$ is a generator, then $e_R J(z)$ is a generator of $\mathbb{Z}_{m(4n)} \cong \mathrm{Ext}^{1,8n}_{KO_*(KO)}(\widetilde{KO}_*(S^0), \widetilde{KO}_*(S^0))$. Let $\eta \in \pi_1^S$ be the non-zero element; then $h_{KO}(\eta) \in \widetilde{KO}_1(S^0)$ is the element we have denoted by α. From the product structure of the Adams spectral sequence one readily deduces the following: if $e_R(x)$ is defined, then $e_R(xy)$ is defined and

$$e_R(xy) = e_R(x)\, h_{KO}(y) \quad \text{for all } x, y \in \pi_*^S.$$

Composition of maps defines a map $\pi_q(SO) \times \pi_{q+s}(S^q) \to \pi_{q+s}(SO)$, and one readily sees that $J[f \circ g] = J[f] \cdot [g]$, where $\cdot$ denotes the product in π_*^S. Thus

$$e_R J(z \circ \eta) = e_R J(z) \cdot h_{KO}(\eta) = e_R J(z)\alpha \neq 0 \in \mathrm{Ext}^{1,8n+1}_{KO_*(KO)}(\widetilde{KO}_*(S^0), \widetilde{KO}_*(S^0)).$$

Similarly $e_R J(z \circ \eta^2) \neq 0$. $\square$

In [3] Adams shows that for the generator $g \in \pi_{4n-1}(SO)$ and for every $k \in \mathbb{Z}$ there is a positive integer $f(k)$ such that

$$2k^{f(k)}(k^{2n} - 1)g \in \ker J.$$

Since $m(f, 2n)$ was defined to be the greatest common divisor of all integers $k^{f(k)}(k^{2n} - 1)$, it follows $2m(f, 2n)g \in \ker J$. In particular the order of $\mathrm{im} J$ divides $2m(f, 2n)$. But we also had $m(f, 2n) | m(2n)$, so we have

$$o(\mathrm{im} J) | 2m(2n).$$

In fact Adams showed $o(\mathrm{im}J) | m(2n)$ if $4n - 1 \equiv 3 \bmod 8$ and conjectured that $o(\mathrm{im}J) | m(2n)$ when $4n - 1 \equiv 7 \bmod 8$. This conjecture was proved by Quillen in [72]. Putting these facts together with 19.31 gives

$$\mathrm{im}[J : \pi_{4n-1}(SO) \to \pi_{4n-1}^S] \cong \mathbb{Z}_{m(2n)}.$$

Moreover, since the diagram

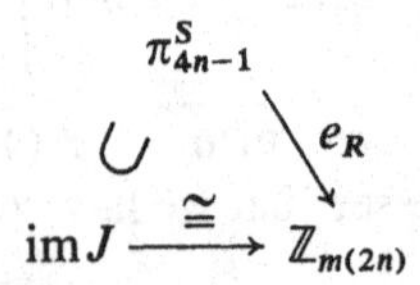

commutes, we see that $\mathrm{im}J$ is a direct summand in π_{4n-1}^S for all $n \geq 1$. Thus we have at least isolated a non-trivial summand in π_n^S for all $n \equiv 0, 1, 3, 7 \bmod 8$.

Comments

The reader will find a discussion of the full Adams spectral sequence as well as a more detailed discussion of the question of convergence in [8]. In the next chapter we give a major application of the Adams spectral sequence—namely to the calculation of the cobordism groups. The reader interested in seeing more calculations of the stable stems should have a look at [6].

In [3] Adams defined certain Ext groups by an ad hoc method. The reader will find it a worthwhile exercise showing that those Ext groups are isomorphic with our $\mathrm{Ext}^{**}_{K_*(K)}(-,-)$ and that the invariant e_C defined by Adams is the same as ours on the stable groups π^S_*. Adams' e_C was defined on the unstable groups and gives information about unstable phenomena, which ours does not. [3] contains many additional interesting applications of the *e*-invariant to the determination of homotopy elements.

References

1. J. F. Adams [1, 3, 4, 8]
2. D. Husemoller [49]
3. A. Liulevicius [55]
4. R. Mosher and M. Tangora [68]
5. D. Quillen [72]

Chapter 20

Calculation of the Cobordism Groups

In Chapter 12 we gave the structure of various cobordism groups $\Omega_*^G \cong \widetilde{MG}_*(S^0) = \pi_*(MG)$ without giving any proofs. As it turns out, the determination of the homotopy groups $\pi_*(MG)$ is a problem to which the Adams spectral sequence is particularly well adapted for the reason that the homology $H_*(MG;\mathbb{Z}_p)$ turns out to have a relatively simple structure as A_*-comodule for $G = O$, U, SO and SU. In this chapter we shall use the Adams spectral sequence to compute Ω^O, Ω^U and Ω^{SO}. We shall also prove the theorem of Stong and Hattori which says that $\pi_*(MU) \xrightarrow{h} P(K_*(MU))$ is an isomorphism.

In Chapter 16 we computed the Pontrjagin rings $H_*(MO;\mathbb{Z}_2) \cong \mathbb{Z}_2[a_1, a_2, \ldots]$, $H_*(MU;\mathbb{Z}) \cong \mathbb{Z}[b_1, b_2, \ldots]$ and $H_*(MSp;\mathbb{Z}) \cong \mathbb{Z}[q_1, q_2, \ldots]$. For calculation with the Adams spectral sequence for $E = H(\mathbb{Z}_2)$, for example, we need to know also the comodule structures of these over $A_* = A(H(\mathbb{Z}_2))_*$.

Let $M = H_*(MO;\mathbb{Z}_2)$; we wish to find the coaction map $\psi: M \to A_* \otimes_{\mathbb{Z}_2} M$. What we shall show is that there is an isomorphism

$$f: M \to A_* \otimes_{\mathbb{Z}_2} H$$

of A_*-comodules, where

$$H = \mathbb{Z}_2[u_2, u_4, u_5, \ldots]$$

is a polynomial algebra over $\mathbb{Z}_2$ with generators $u_k \in H_k$ for all k not of the form $2^s - 1$. In fact we shall show that f is an isomorphism of algebras.

20.1. Definition. Let C be a Hopf algebra and L a C-comodule with coaction map $\psi: L \to C \otimes L$. L is called an *algebra over* C if it has an associative product $\mu: L \otimes L \to L$ which is compatible with the comodule structure—i.e.

$$\begin{array}{ccccc}
L \otimes L & & \xrightarrow{\quad\mu\quad} & & L \\
\downarrow{\scriptstyle \psi\otimes\psi} & & & & \downarrow{\scriptstyle \psi} \\
C \otimes L \otimes C \otimes L & \xrightarrow{1\otimes T\otimes 1} & C \otimes C \otimes L \otimes L & \xrightarrow{\phi\otimes\mu} & C \otimes L
\end{array}$$

commutes, where ϕ is the product on C.

20.2. Examples. i) $H_*(MO;\mathbb{Z}_2)$, $H_*(MU;\mathbb{Z}_2)$ etc. are all algebras over A_*, because their products are defined by maps of spectra and the A_*-coaction is natural and satisfies 17.8vii).

ii) If B is any algebra over $\mathbb{Z}_2$ with product $\phi_B : B \otimes B \to B$, then $A_* \otimes B$ is an algebra over A_* with product

$$(A_* \otimes B) \otimes (A_* \otimes B) \xrightarrow{1\otimes T\otimes 1} A_* \otimes A_* \otimes B \otimes B \xrightarrow{\phi\otimes\phi_B} A_* \otimes B.$$

In particular the $A_* \otimes H$ which appears above is an algebra over A_*.

Note. Whenever we are working with modules, algebras, etc., over $\mathbb{Z}_p$—as we shall be throughout this chapter—we shall simply write $\otimes$ rather than $\otimes_{\mathbb{Z}_p}$, because all tensor products will be over the ground field.

Recall our theorem $\mathrm{Hom}_C^t(K, C \otimes_R V) \cong \mathrm{Hom}_R^t(K, V)$; given an R-homomorphism $g: K \to V$ we constructed a C-homomorphism $\bar{g}: K \to C \otimes_R V$ by taking $K \xrightarrow{\psi_K} C \otimes_R K \xrightarrow{1\otimes g} C \otimes_R V$.

20.3. Lemma. *If K is an algebra over C and V is an algebra over R and $g: K \to V$ is a homomorphism of R-algebras, then $\bar{g}$ is a homomorphism of C-algebras.*

Proof: We have a commutative diagram

$$\begin{array}{ccccc}
K \otimes K & & \xrightarrow{\quad\phi_K\quad} & & K \\
\downarrow{\scriptstyle \psi_K\otimes\psi_K} & & (1) & & \downarrow{\scriptstyle \psi_K} \\
C \otimes K \otimes C \otimes K & \xrightarrow{1\otimes T\otimes 1} & C \otimes C \otimes K \otimes K & \xrightarrow{\phi_C\otimes\phi_K} & C \otimes K \\
\downarrow{\scriptstyle 1\otimes g\otimes 1\otimes g} & (2) & \downarrow{\scriptstyle 1\otimes 1\otimes g\otimes g} & (3) & \downarrow{\scriptstyle 1\otimes g} \\
C \otimes V \otimes C \otimes V & \xrightarrow{1\otimes T\otimes 1} & C \otimes C \otimes V \otimes V & \xrightarrow{\phi_C\otimes\phi_V} & C \otimes V,
\end{array}$$

in which square (1) commutes because K is an algebra over C, square (2) obviously commutes and square (3) commutes because g is a morphism of algebras over R. But this diagram says $\bar{g} = (1 \otimes g) \circ \psi_K$ is a morphism of algebras over C. □

Since $\psi_M: M \to A_* \otimes M$ is compatible with the product on M, it suffices to determine $\psi_M(a_k)$ for each k. By the naturality of ψ_M, it will be enough to find $\psi_{RP^\infty}(x_{k+1})$ for each k. But we recall the map $\alpha: RP^\infty \simeq H(\mathbb{Z}_2, 1) \to \Sigma^{-1} H(\mathbb{Z}_2)$ which sends x_{2^k} to $\xi_k \in A_{2^k-1}$, and we know that $\psi\xi_k = \sum_{s=0}^k \xi_{k-s}^{2^s} \otimes \xi_s$ in $A_* \otimes A_*$. Therefore $\psi x_{2^t} = \sum_{s=0}^t \xi_{t-s}^{2^s} \otimes x_{2^s}$ mod $A_* \otimes$ (decomposables). In fact we can give ψx_k exactly for all k.

20.4. Lemma. $\psi x_k = \sum_{i+j=k} (X^j)_i \otimes x_j$, $k \geqslant 1$, *where* $X = 1 + \xi_1 + \xi_2 + \cdots$.
Proof: The proof is similar to that of 18.20. We recall the formula 17.11

$$\langle \theta y, u\rangle = \textstyle\sum_i \langle \theta, e_i \langle y, u_i\rangle\rangle \quad \text{for} \quad \theta \in A^*, \quad y \in H^*(X; \mathbb{Z}_2),$$

and

$$u \in H_*(X; \mathbb{Z}_2) \quad \text{with} \quad \psi u = \textstyle\sum_i e_i \otimes u_i.$$

We apply this to $X = RP^\infty$ with $y = x^i$, $u = x_k$, getting

$$\langle \theta x^i, x_k\rangle = \textstyle\sum_j \langle \theta, e_j \langle x^i, x_j\rangle\rangle = \langle \theta, e_i\rangle \quad \text{if} \quad \psi x_k = \sum_j e_j \otimes x_j.$$

We now compute θx^i; let $\Delta^i: RP^\infty \to RP^\infty \times \cdots \times RP^\infty$ be the i-fold diagonal ($\Delta^i(z) = (z, z, \ldots, z)$). Let $u_j \in H^1(RP^\infty \times \cdots \times RP^\infty; \mathbb{Z}_2)$ be $\pi_j^* x$, where $\pi_j: RP^\infty \times \cdots \times RP^\infty \to RP^\infty$ is the jth projection. Then $x^i = \Delta^{i*}(u_1 u_2 \ldots u_i)$. Hence

$$\begin{aligned} \theta x^i &= \theta \Delta^{i*}(u_1 u_2 \ldots u_i) = \Delta^{i*}\theta(u_1 u_2 \ldots u_i) \\ &= \Delta^{i*}\left(\textstyle\sum_\alpha \langle \theta, \xi_\alpha\rangle u_1^{2^{\alpha_1}} u_2^{2^{\alpha_2}} \ldots u_i^{2^{\alpha_i}}\right) \\ &= \textstyle\sum_\alpha \langle \theta, \xi_\alpha\rangle x^{2^{\alpha_1} + 2^{\alpha_2} + \cdots + 2^{\alpha_i}} \\ &= \textstyle\sum_{j \geqslant i} \langle \theta, (X^i)_{j-i}\rangle x^j. \end{aligned} \tag{18.18}$$

Thus

$$\langle \theta, e_i\rangle = \langle \theta x^i, x_k\rangle = \textstyle\sum_{j \geqslant i} \langle \theta, (X^i)_{j-i}\rangle \langle x^j, x_k\rangle = \langle \theta, (X^i)_{k-i}\rangle, \theta \in A^*.$$

That is, $e_i = X_{k-i}^i$ for all i. □

20.5. Corollary. $\psi a_k = \sum_{i+j=k} (X^{j+1})_i \otimes a_j$, *all* $k \geqslant 0$.

We can now define a map of algebras $g: M \to H$ by

$$g(a_k) = \begin{cases} u_k & \text{if } k \text{ not of form } 2^t - 1, t > 0, \\ 0 & \text{if } k = 2^t - 1, \text{ some } t > 0. \end{cases}$$

We let $f = \bar{g}: M \to A_* \otimes H$; by 20.3 f is a morphism of algebras over A_*.

Let $A_*^t \subset A_*$ be the subalgebra generated by $1, \xi_1, \ldots, \xi_t$ and $H^{(m)} \subset H$ the subalgebra of H generated by $1, g(a_1), g(a_2), \ldots, g(a_m)$.

20.6. Lemma. i) $f(a_{2^t-1}) \equiv \xi_t \otimes 1 \bmod \bar{A}_*^{t-1} \otimes H^{(2^t-2)}$,

ii) $f(a_k) \equiv 1 \otimes u_k \bmod \bar{A}_*^{t-1} \otimes H^{(k-1)}$

if $2^{t-1} - 1 < k < 2^t - 1$.

($\bar{A}_* = \ker(\varepsilon: A_* \to \mathbb{Z}_2) =$ elements of degree > 0.)

Proof:

$$f(a_{2^t-1}) = (1 \otimes g)\psi(a_{2^t-1}) \equiv (1 \otimes g)(\xi_t \otimes 1 + 1 \otimes a_{2^t-1})$$
$$\equiv \xi_t \otimes 1 \bmod \bar{A}_*^{t-1} \otimes H^{(2^t-2)}.$$

For any k we have $\psi(a_k) \equiv 1 \otimes a_k \bmod \bar{A}_*^{t-1} \otimes M^{(k-1)}$, where $M^{(k-1)}$ is the subalgebra generated by $1, a_1, a_2, \ldots, a_{k-1}$, $2^{t-1} - 1 \leqslant k < 2^t - 1$. Thus if $k \neq 2^{t-1} - 1$ we have

$$f(a_k) = (1 \otimes g)\psi(a_k) \equiv 1 \otimes g(a_k) \bmod \bar{A}_*^{t-1} \otimes H^{(k-1)}$$
$$\equiv 1 \otimes u_k \bmod \bar{A}_*^{t-1} \otimes H^{(k-1)}. \quad \square$$

20.7. Theorem. *$f: M \to A_* \otimes H$ is an isomorphism.*

Proof: One can easily check that M_n and $(A_* \otimes H)_n$ have the same dimension as vector spaces over $\mathbb{Z}_2$ for every n. Hence it will suffice to show f is an epimorphism. We show by double induction on t, m that $\operatorname{im} f$ contains $A_*^t \otimes H^{(m)}$ for every t, m.

Suppose $A_*^{t-1} \otimes H^{(2^t-2)} \subset \operatorname{im} f$. Then from 20.6i) it follows that $\xi_t \otimes 1 \in \operatorname{im} f$ and thus also $A_*^t \otimes H^{(0)} \subset \operatorname{im} f$. Suppose we have proved $A_*^t \otimes H^{(m-1)} \subset \operatorname{im} f$ for some m, $1 \leqslant m < 2^{t+1} - 1$. If $m = 2^s - 1$ for some $s, 0 < s \leqslant t$, then $H^{(m)} = H^{(m-1)}$. Otherwise it follows from 20.6ii) that $1 \otimes u_m \in \operatorname{im} f$. Because f is a map of A_*-algebras, it follows that $A_*^t \otimes H^{(m)} \subset \operatorname{im} f$. Hence by induction we have $A_*^t \otimes H^{(2^{t+1}-2)} \subset \operatorname{im} f$. This completes the induction step. $\square$

Thus $H_*(MO; \mathbb{Z}_2) \cong A_* \otimes H$—the best possible outcome, for by 19.7 we now have

$$E_2^{s,t} = \operatorname{Ext}_{A_*}^{s,t}(\mathbb{Z}_2, A_* \otimes H) \cong \operatorname{Ext}_{\mathbb{Z}_2}^{s,t}(\mathbb{Z}_2, H) = \begin{cases} \operatorname{Hom}_{\mathbb{Z}_2}^t(\mathbb{Z}_2, H) & s = 0 \\ 0 & s > 0 \end{cases}$$

$$= \begin{cases} H_t & s = 0 \\ 0 & s > 0. \end{cases}$$

Thus the spectral sequence collapses, and we get the following theorem.

20.8. Theorem. *The Hurewicz homomorphism*

$$h: {}_2\pi_*(MO) \to P(H_*(MO; \mathbb{Z}_2)) \cong P(A_* \otimes H) \cong H$$

is an isomorphism, and hence ${}_2\pi_*(MO) \cong \mathbb{Z}_2[u_2, u_4, u_5, \ldots]$.

By 12.24 every non-zero element of Ω_*^O is of order 2, so $\pi_*(MO) = {}_2\pi_*(MO)$. Thus we have computed the unoriented cobordism ring Ω_*^O:

$$\Omega_*^O \simeq \mathbb{Z}_2[N_2, N_4, N_5, \ldots], \quad N_k \in \Omega_k^O \text{ for all } k \neq 2^t - 1 \text{ for some } t > 0.$$

Moreover, we see that we can choose as polynomial generators $N_k \in \Omega_k^O$ any cobordism class such that

$$f(h(N_k)) = 1 \otimes u_k$$

i.e. any N_k such that $h(N_k) = a_k \bmod M^{(k-1)}$. Expressed in terms of characteristic numbers, this says that we may take any N_k with

$$s_{(0,0,\ldots,1)}(N_k) = 1.$$

Next we consider MU and MSp. Let us write $L = H_*(MU;\mathbb{Z}_2)$, $K = H_*(MSp;\mathbb{Z}_2)$.

20.9. Proposition. *There are isomorphisms $f: M \to L$, $g: L \to K$ of (ungraded) algebras over $\mathbb{Z}_2$ such that the diagram*

$$\begin{array}{ccc} M & \xrightarrow{\psi_M} & A_* \otimes M \\ {\scriptstyle f}\downarrow & & \downarrow{\scriptstyle \alpha \otimes f} \\ L & \xrightarrow{\psi_L} & A_* \otimes L \\ {\scriptstyle g}\downarrow & & \downarrow{\scriptstyle \alpha \otimes g} \\ K & \xrightarrow{\psi_K} & A_* \otimes K \end{array}$$

commutes, where $\alpha: A_* \to A_*$ is the squaring map $\alpha(x) = x^2$, $x \in A_*$ (α is an algebra homomorphism).

Proof: For any complex bundle ξ let $r\xi$ be the underlying real bundle and let $c: RP^\infty \to CP^\infty$ classify $\omega \otimes \mathbb{C}$, ω the real Hopf bundle. If $\eta \to CP^\infty$ denotes the complex Hopf bundle, then we compute

$$c^*(w_2 r\eta) = w_2(c^* r\eta) = w_2(rc^*\eta) = w_2(\omega \oplus \omega) = x^2.$$

In particular $w_2 r\eta \neq 0$, so we must have $w_2 r\eta = y \in H^2(CP^\infty;\mathbb{Z}_2)$. Moreover it follows $c^*(y) = x^2$ and therefore

$$c_*(x_k) = \begin{cases} y_i & k = 2i \\ 0 & k \text{ odd.} \end{cases}$$

From 20.4 we have $\psi_{RP^\infty}(x_k) = \sum_{i+j=k} (X^j)_i \otimes x_j$. Thus

$$\begin{aligned} \psi_{CP^\infty}(y_k) &= \psi_{CP^\infty} c_*(x_{2k}) = (1 \otimes c_*)\psi_{RP^\infty}(x_{2k}) \\ &= (1 \otimes c_*)(\textstyle\sum_{i+j=2k} (X^j)_i \otimes x_j) \\ &= \textstyle\sum_{i+2j=2k} (X^{2j})_i \otimes y_j = \sum_{i+j=k} \alpha(X_i^j) \otimes y_j. \end{aligned}$$

It follows that

$$\psi_{MU}(b_k) = \sum_{i+j=k} \alpha(X_i^{j+1}) \otimes b_j \quad \text{for } k \geqslant 1.$$

Hence the upper half of 20.9 will commute if we define f by $f(a_k) = b_k$ for $k \geqslant 1$ and require f to be an algebra homomorphism. Similarly the lower half will commute if we define g by $g(b_k) = q_k$, $k \geqslant 1$. ▯

Let $H' = \mathbb{Z}_2[v_2, v_4, v_5, \ldots]$ with $v_k \in H'_{2k}$ for all $k \neq 2^t - 1$, all $t > 0$, and let $H'' = \mathbb{Z}_2[w_2, w_4, w_5, \ldots]$ with $w_k \in H''_{4k}$, $k \neq 2^t - 1$, all $t > 0$, (these w_k are not to be confused with the Stiefel–Whitney classes).

20.10. Theorem. *There are isomorphisms of graded algebras over A_**

$$\kappa' : H_*(MU; \mathbb{Z}_2) \to \alpha A_* \otimes H'$$
$$\kappa'' : H_*(MSp; \mathbb{Z}_2) \to \alpha^2 A_* \otimes H''.$$

Proof: We have a diagram

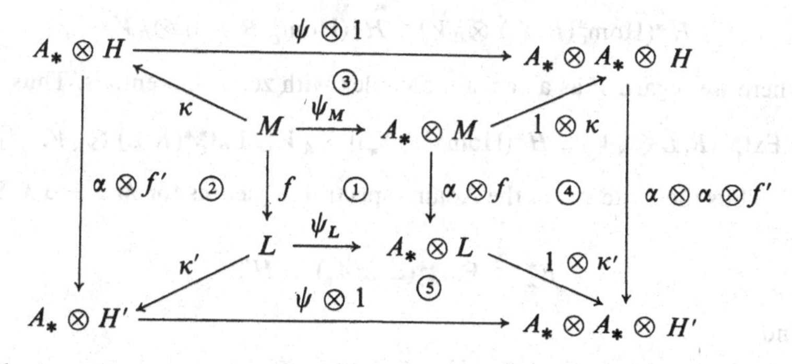

where $\kappa : M \to A_* \otimes H$ is the isomorphism of 20.7 and κ' is defined to make square ② commute if $f' : H \to H'$ is the map of algebras defined by $f(u_i) = v_i$, $i \geqslant 2$. Then the square ④ commutes also. Square ① commutes by construction of f, square ③ by construction of κ, and the large outer square commutes because α is a map of coalgebras (why?). Thus it follows that square ⑤ commutes—which just says that κ' is a map of comodules. Note that κ' preserves grading. Now we repeat the argument to construct κ''. ▯

If L is a C-comodule and V is an R-module, the $L \otimes_R V$ is also a C-comodule with coaction map

$$L \otimes_R V \xrightarrow{\psi_L \otimes 1} (C \otimes_R L) \otimes_R V \cong C \otimes_R (L \otimes_R V).$$

20.11. Lemma. *Suppose R is a field, V a graded vector space over R, C a coalgebra over R and L a C-comodule; then there is an isomorphism (not natural)*

$$\mathrm{Ext}_C^{**}(R, L \otimes_R V) \cong \mathrm{Ext}_C^{**}(R, L) \otimes_R V.$$

Proof: Choose a homogeneous R-basis $\{v_\alpha\}$ for V; that is, each v_α lies in $V_{i(\alpha)}$ for some $i(\alpha)$. Then we define

$$\Phi: \mathrm{Hom}_C^*(R, L \otimes_R V) \to \mathrm{Hom}_C^*(R, L) \otimes_R V$$

by letting $\Phi(\phi) = \sum_\alpha \phi_\alpha \otimes v_\alpha$ for each $\phi \in \mathrm{Hom}_C^*(R, L \otimes_R V)$, where $\phi(1) = \sum_\alpha \phi_\alpha(1) \otimes v_\alpha$ (an R-homomorphism ϕ_α is uniquely determined by giving $\phi_\alpha(1)$). Φ is an isomorphism (this is evident if one thinks of $\mathrm{Hom}_C^*(R, -)$ as $P(-)$).

Now let $0 \to L \to Y_0 \to Y_1 \to Y_2 \to \cdots$ be an injective resolution of L over C (cf. 19.15). Then each $Y_i \otimes_R V$ is also injective (why?), so $0 \to L \otimes_R V \to Y_0 \otimes_R V \to Y_1 \otimes_R V \to \cdots$ is an injective resolution of $L \otimes_R V$ over C. Hence

$$\mathrm{Ext}_C^{**}(R, L \otimes_R V) \cong H^*(\mathrm{Hom}_C^*(R, Y_* \otimes_R V)) \cong H^*(\mathrm{Hom}_C^*(R, Y_*) \otimes_R V).$$

Now by the Künneth theorem we have

$$H^*(\mathrm{Hom}_C^*(R, Y_*) \otimes_R V) \cong H^*(\mathrm{Hom}_C^*(R, Y_*)) \otimes_R V,$$

where we regard V as a cochain complex with zero differentials. Thus

$$\mathrm{Ext}_C^{**}(R, L \otimes_R V) \cong H^*(\mathrm{Hom}_C^*(R, Y_*)) \otimes_R V \cong \mathrm{Ext}_C^{**}(R, L) \otimes_R V. \quad \square$$

Thus the E_2-terms of the Adams spectral sequences for MU and MSp are

$$E_2^{**} \cong \mathrm{Ext}_{A_*}^{**}(\mathbb{Z}_2, \alpha A_*) \otimes H',$$

and

$$E_2^{**} \cong \mathrm{Ext}_{A_*}^{**}(\mathbb{Z}_2, \alpha^2 A_*) \otimes H''$$

respectively. Therefore we turn now to the problem of computing $\mathrm{Ext}_C^{**}(K, D)$ for D an appropriate subcoalgebra of C. This will, of course, involve some more homological algebra.

20.12. Suppose that A is an algebra over R and M is a right A-module, N a left A-module. Then we construct $M \otimes_A N$ by taking $M \otimes_R N$ and introducing further relations

$$m \otimes an = ma \otimes n \quad \text{for all} \quad m \in M, n \in N, a \in A.$$

In other words we have an exact sequence

$$M \otimes_R A \otimes_R N \xrightarrow{\phi_M \otimes 1 - 1 \otimes \phi_N} M \otimes_R N \longrightarrow M \otimes_A N \longrightarrow 0,$$

where ϕ_M, ϕ_N are the A-action maps of M, N. Now we can dualize this definition, obtaining the *cotensor product* $K \square_C L$ of a right C-comodule

K and a left C-comodule L, where C is a coalgebra over R. Explicitly, $K \square_C L$ is defined by the exact sequence

$$0 \longrightarrow K \square_C L \longrightarrow K \otimes_R L \xrightarrow{\psi_K \otimes 1 - 1 \otimes \psi_L} K \otimes_R C \otimes_R L,$$

where ψ_K, ψ_L are the coaction maps of K,L. $\square_C$ is a covariant functor of two variables. One can easily prove associativity relations like

$$M \otimes_R (N \square_C Q) \cong (M \otimes_R N) \square_C Q$$

if R is a field.

Suppose E is a connected coalgebra (i.e. $E_0 = R$) over a field R with a commutative coproduct, $F \subset E$ a subcoalgebra. (The example we have in mind is $A \cdot (\overline{\alpha(A)}) \subset A$; why is $A \cdot \overline{\alpha(A)}$ a subcoalgebra?) Let $B = E/\bar{F}$, where $\bar{F}$ denotes as usual the elements of positive grading in F. We shall write $E//F$ for $E/E\bar{F}$ if E is a Hopf algebra, F a Hopf subalgebra and $E\bar{F} = \bar{F}E$.

20.13. Exercise. The coproduct $\psi_E : E \to E \otimes_R E$ induces a homomorphism $\psi_B : B \to B \otimes_R B$; show that B is a coalgebra with coproduct ψ_B and that the homomorphism

$$E \xrightarrow{\psi_E} E \otimes_R E \xrightarrow{1 \otimes p} E \otimes_R B,$$

where $p : E \to B$ is the projection, is a coaction map making E into a right comodule over B. We denote $(1 \otimes p) \circ \psi_E$ by ψ_E^B.

20.14. Exercise. Let $\sigma_l : E \to E \otimes_R B$ be defined by $\sigma_l(e) = e \otimes 1$ for all $e \in E$. Show that F is the kernel of

$$E \xrightarrow[\psi_E^B - \sigma_l]{} E \otimes_R B.$$

20.15. Proposition. *There is a natural isomorphism $\phi : E \square_B R \to F$ of left E-comodules.*

Proof: The coaction on R as a B-comodule is given by $\psi_R^B(r) = r \cdot 1 \otimes 1$, all $r \in R$. We have a commutative diagram

$$\begin{array}{ccccccc}
0 \longrightarrow & E \square_B R & \longrightarrow & E \otimes_R R & \xrightarrow{\psi_E^B \otimes 1 - 1 \otimes \psi_R^B} & E \otimes_R B \otimes_R R \\
& \downarrow \phi & & \| \wr & & \wr \| \\
0 \longrightarrow & F & \longrightarrow & E & \xrightarrow{\psi_E^B - \sigma_l} & E \otimes_R B
\end{array}$$

with exact rows, which defines $\phi: E \square_B R \to F$. To show that ϕ is a map of E-comodules we have a diagram

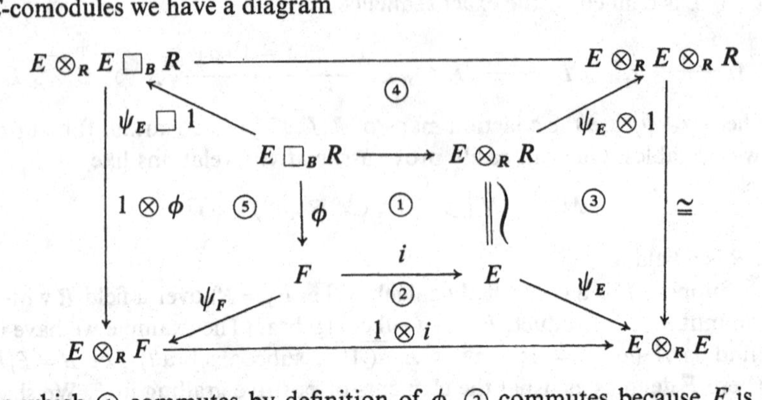

in which ① commutes by definition of ϕ, ② commutes because F is a subcoalgebra of E, ③ commutes because ψ_E is an R-module morphism, ④ commutes by definition of $\psi_E \square 1$ and the outer square commutes because ① does. Hence ⑤ commutes, which says ϕ is an E-comodule morphism. $\square$

20.16. Proposition. *If E, F and B are as above, then there is a natural isomorphism*

$$\mathrm{Hom}_E^t(L, E \square_B M) \cong \mathrm{Hom}_B^t(L, M)$$

for every left E-comodule L and left B-comodule M. Hence there is an isomorphism

$$\mathrm{Ext}_E^{s,t}(L, E \square_B M) \cong \mathrm{Ext}_B^{s,t}(L, M), \quad s \geqslant 0, t \in \mathbb{Z}.$$

Proof: This proposition generalizes our earlier result

$$\mathrm{Ext}_E^{s,t}(L, E \otimes_R M) \cong \mathrm{Ext}_R^{s,t}(L, M);$$

we simply take $F = E$ in 20.16. Indeed we have a diagram with exact rows

$$\begin{array}{ccccc} 0 \to \mathrm{Hom}_E^t(L, E \square_B M) & \to & \mathrm{Hom}_E^t(L, E \otimes_R M) & \to & \mathrm{Hom}_E^t(L, E \otimes_R B \otimes_R M) \\ & & \Phi \downarrow \uparrow \Psi & & \\ 0 \to \mathrm{Hom}_B^t(L, M) & \to & \mathrm{Hom}_R^t(L, M), & & \end{array}$$

where the top line comes from applying $\mathrm{Hom}_E^t(L, -)$ to the exact sequence defining $E \square_B M$. We have only to check that

$$\Phi(\mathrm{Hom}_E^t(L, E \square_B M)) \subset \mathrm{Hom}_B^t(L, M)$$

and

$$\Psi(\mathrm{Hom}_B^t(L, M)) \subset \mathrm{Hom}_E^t(L, E \square_B M).$$

a) Suppose $g: L \to M$ is a B-morphism; then $\Psi(g)$ is the composite

$$L \xrightarrow{\psi_L^E} E \otimes_R L \xrightarrow{1 \otimes g} E \otimes_R M.$$

We shall show that $(\psi_E^B \otimes 1 - 1 \otimes \psi_M^B) \circ \Psi(g) = 0$, which implies that $\Psi(g) \in \mathrm{Hom}_E^t(L, E \,\square_B\, M)$.

$$\begin{aligned}
&(\psi_E^B \otimes 1) \circ (1 \otimes g) \circ (\psi_L^E)\\
&= (1 \otimes 1 \otimes g) \circ (\psi_E^B \otimes 1) \circ \psi_L^E\\
&= (1 \otimes 1 \otimes g) \circ (1 \otimes p \otimes 1) \circ (\psi_E^E \otimes 1) \circ \psi_L^E \quad \text{(by definition of } \psi_E^B)\\
&= (1 \otimes 1 \otimes g) \circ (1 \otimes p \otimes 1) \circ (1 \otimes \psi_L^E) \circ \psi_L^E\\
&= (1 \otimes 1 \otimes g) \circ (1 \otimes \psi_L^B) \circ \psi_L^E \quad \text{(by definition of } \psi_L^B)\\
&= (1 \otimes \psi_M^B) \circ (1 \otimes g) \circ \psi_L^E.
\end{aligned}$$

b) Suppose $f: L \to E \otimes_R M$ is an E-morphism; then $\Phi(f)$ is the composite

$$L \xrightarrow{f} E \otimes_R M \xrightarrow{\varepsilon \otimes 1} R \otimes_R M \underset{\kappa}{\cong} M.$$

We shall show that if $f \in \mathrm{Hom}_E^t(L, E \,\square_B\, M)$—i.e. $(\psi_E^B \otimes 1) \circ f = (1 \otimes \psi_M^B) \circ f$—then $\Phi(f) \in \mathrm{Hom}_B^t(L, M)$—i.e.

$$\psi_M^B \circ \Phi(f) = [1 \otimes \Phi(f)] \circ \psi_L^B.$$

$$\begin{aligned}
[1 \otimes \Phi(f)] \circ \psi_L^B &= (1 \otimes \kappa) \circ (1 \otimes \varepsilon \otimes 1) \circ (1 \otimes f) \circ (p \otimes 1) \circ \psi_L^E\\
&= (1 \otimes \kappa) \circ (1 \otimes \varepsilon \otimes 1) \circ (p \otimes 1 \otimes 1) \circ (1 \otimes f) \circ \psi_L^E\\
&= (1 \otimes \kappa) \circ (1 \otimes \varepsilon \otimes 1) \circ (p \otimes 1 \otimes 1) \circ (\psi_E^E \otimes 1) \circ f\\
&= (p \otimes 1) \circ (1 \otimes \kappa) \circ (1 \otimes \varepsilon \otimes 1) \circ (\psi_E^E \otimes 1) \circ f\\
&= (p \otimes 1) \circ \kappa \circ (\varepsilon \otimes 1 \otimes 1) \circ (\psi_E^E \otimes 1) \circ f\\
&= \kappa \circ (\varepsilon \otimes 1 \otimes 1) \circ (1 \otimes p \otimes 1) \circ (\psi_E^E \otimes 1) \circ f\\
&= \kappa \circ (\varepsilon \otimes 1 \otimes 1) \circ (\psi_E^B \otimes 1) \circ f\\
&= \kappa \circ (\varepsilon \otimes 1 \otimes 1) \circ (1 \otimes \psi_M^B) \circ f \quad \text{(by assumption on } f)\\
&= \psi_M^B \circ \kappa \circ (\varepsilon \otimes 1) \circ f = \psi_M^B \circ \Phi(f).
\end{aligned}$$

At this point we must quote without proof a theorem of Milnor and Moore [67]: under the given assumptions on E, F and B there is an isomorphism $E \cong F \otimes_R B$ as right B-comodules. Since R is a field (and therefore F is a vector space over R) it follows that E is free as a B-comodule. Therefore the functor $E \,\square_B -$ preserves exactness. The proof of the Milnor–Moore theorem should be readily comprehensible to any reader who has understood what we have said so far about comodules and cotensor products.

Now choose a projective resolution $0 \leftarrow L \leftarrow X_0 \leftarrow X_1 \leftarrow \cdots$ of L over E. The X_i are also projective as B-comodules, for given a diagram

$$\begin{array}{ccccc} & & X & & \\ & & \downarrow{\scriptstyle \pi} & & \\ C & \xrightarrow{\gamma} & D & \longrightarrow & 0 \end{array}$$

of B-comodules, where X is projective over E, we have a commutative diagram

$$\begin{array}{ccc} \mathrm{Hom}_B^*(X, C) & \longrightarrow & \mathrm{Hom}_B^*(X, D) \\ \wr\| & & \wr\| \\ \mathrm{Hom}_E^*(X, E \,\square_B\, C) & \rightarrow & \mathrm{Hom}_E^*(X, E \,\square_B\, D) \end{array}$$

in which the bottom horizontal arrow is an epimorphism, and hence the upper one is also. Therefore

$$\begin{aligned} \mathrm{Ext}_E^{s,t}(L, E \,\square_B\, M) &= H^s(\mathrm{Hom}_E^t(X_*, E \,\square_B\, M)) \cong H^s(\mathrm{Hom}_B^t(X_*, M)) \\ &= \mathrm{Ext}_B^{s,t}(L, M), \quad s \geqslant 0,\, t \in \mathbb{Z}. \quad \square \end{aligned}$$

20.17. Corollary. $\mathrm{Ext}_E^{s,t}(L, F) \cong \mathrm{Ext}_B^{s,t}(L, R)$ *for all E-comodules L.*

Hence in the Adams spectral sequences for MU and MSp we have

$$E_2^{**} \cong \mathrm{Ext}_{B_1}^{**}(\mathbb{Z}_2, \mathbb{Z}_2) \otimes H'$$

and

$$E_2^{**} \cong \mathrm{Ext}_{B_2}^{**}(\mathbb{Z}_2, \mathbb{Z}_2) \otimes H''$$

respectively, where

$$B_1 = A_*/A_* \cdot \overline{\alpha(A_*)}, \quad B_2 = A_*/A_* \cdot \overline{\alpha^2(A_*)}.$$

B_1 is easy to describe; all squares x^2 become zero in B_1, so

$$B_1 = E(\xi_1, \xi_2, \ldots),$$

an exterior algebra over $\mathbb{Z}_2$ on generators $\xi_i \in (B_1)_{2^i-1}$, $i \geqslant 1$. B_2 is more complicated. Note that $\psi_{B_1}(\xi_i) = \xi_i \otimes 1 + 1 \otimes \xi_i$.

It is also not difficult to write down a resolution of $\mathbb{Z}_2$ by extended comodules over B_1. Let us look first at an exterior algebra $E(x)$ on one generator with $\psi x = x \otimes 1 + 1 \otimes x$. We take $Y_* = E(x) \otimes \mathbb{Z}_2[y]$ with $Y_s = E(x) \otimes y^s$, with grade y = grade x and with differential $\delta: Y_s \rightarrow Y_{s+1}$ given by requiring that it be a derivation and $\delta x = y$, $\delta y = 0$. Then $\delta y^s = 0$ and $\delta x y^s = y^{s+1}$ for all $s \geqslant 1$, so it is clear that (Y_*, δ) is acyclic, and one easily checks that δ is a comodule map over $E(x)$. Now for an exterior algebra $E(\xi_1, \xi_2, \ldots)$ with only finitely many generators ξ_i in each grading we can take the tensor product of the above resolutions for each $E(\xi_i)$ (after all, $E(\xi_1, \xi_2, \ldots) \cong E(\xi_1) \otimes E(\xi_2) \otimes \cdots$). Thus we get

$$Y_* = E(\xi_1, \xi_2, \ldots) \otimes \mathbb{Z}_2[y_1, y_2, \ldots],$$

grade y_i = grade ξ_i and differential δ given by $\delta\xi_i = y_i$, $\delta y_i = 0$ and the requirement that δ be a derivation. We then have $Y_n = E(\xi_1, \xi_2, \ldots)$-comodule generated by all the monomials $y^\alpha = y_1^{\alpha_1} y_2^{\alpha_2} \ldots y_r^{\alpha_r}$ with $\alpha_1 + \alpha_2 + \cdots + \alpha_r = n$. By the Künneth theorem it follows Y_* is acyclic.

Thus if $B_1 = E(\xi_1, \xi_2, \ldots)$, then

$$\begin{aligned} \mathrm{Hom}^*_{B_1}(\mathbb{Z}_2, Y_*) &= \mathrm{Hom}^*_{B_1}(\mathbb{Z}_2, B_1 \otimes \mathbb{Z}_2[y_1, y_2, \ldots]) \\ &\cong \mathrm{Hom}^*_{\mathbb{Z}_2}(\mathbb{Z}_2, \mathbb{Z}_2[y_1, y_2, \ldots]) \\ &\cong \mathbb{Z}_2[y_1, y_2, \ldots], \end{aligned}$$

and under this isomorphism the differential $\mathrm{Hom}(1, \delta)$ becomes 0. Thus we see that

$$\mathrm{Ext}^{**}_{B_1}(\mathbb{Z}_2, \mathbb{Z}_2) \cong \mathbb{Z}_2[q_0, q_1, \ldots] \quad q_k \in \mathrm{Ext}^{1, 2^{k+1}-1}_{B_1}(\mathbb{Z}_2, \mathbb{Z}_2),$$

where $q_k = y_{k+1}$, $k \geqslant 0$.

20.18. Exercise. Check that multiplication by $q_0 \in \mathrm{Ext}^{1,1}_{B_1}(\mathbb{Z}_2, \mathbb{Z}_2)$ agrees with multiplication by h_0 from the Adams spectral sequence for π^S_*.

Combining 20.10, 20.11, 20.17 and the above calculation, we find that in the Adams spectral sequence for MU we have

$$E_2^{**} \cong \mathbb{Z}_2[q_0, q_1, \ldots] \otimes H' = \mathbb{Z}_2[q_0] \otimes \mathbb{Z}_2[m_1, m_2, \ldots],$$

where we take

$$m_k = \begin{cases} v_k \in E_2^{0,2k} & \text{for } k \text{ not of the form } 2^t - 1, \\ q_t \in E_2^{1,2k+1} & \text{for } k = 2^t - 1. \end{cases}$$

In particular $E_2^{s,t} = 0$ if $t - s$ is odd, so since d_r lowers $t - s$ by 1 for all r, it follows all differentials d_r vanish. The E_∞-term then looks as follows.

$s \uparrow$	0	1	2	3	4	
	$\vdots$		$\vdots$		$\vdots$ $\vdots$	
4	q_0^4		$q_0^3 m_1$		$q_0^2 m_1^2$ $q_0^4 m_2$	$\cdots$
3	q_0^3		$q_0^2 m_1$		$q_0 m_1^2$ $q_0^3 m_2$	$\cdots$
2	q_0^2		$q_0 m_1$		m_1^2 $q_0^2 m_2$	$\cdots$
1	q_0		m_1		$q_0 m_2$	$\cdots$
0	1				m_2	$\cdots$

$t - s \rightarrow$.

Each tower $m^\alpha, q_0 m^\alpha, q_0^2 m^\alpha, \ldots$ for a given monomial m^α in the m_k gives a $\mathbb{Z}$ summand in ${}_2\pi_*(MU)$.

20.19. Theorem. *There is an isomorphism of groups*

$$ {}_2\pi_*(MU) \cong \mathbb{Z}[m_1, m_2, \ldots] \quad (\text{grade } m_k = 2k)$$

and the Hurewicz homomorphism ${}_2\pi_t(MU) \to PH_t(MU;\mathbb{Z}_2) \cong H_t'$ *is a surjection. In particular* $\pi_*(MU)$ *has no 2-torsion.*

20.20. Exercise. Show that a closed manifold M^{2n} is unoriented cobordant to a U-manifold if and only if it is unoriented cobordant to $V^n \times V^n$ for some closed manifold V^n.

[Hint: What is $\operatorname{im}[H_*(MU;\mathbb{Z}_2) \to H_*(MO;\mathbb{Z}_2)]$?]

Now it also turns out that $\pi_*(MU)$has no p-torsion for any odd prime p. To show this we must investigate the mod p Adams spectral sequence for MU. The odd prime calculations for MU turn out to be remarkably similar to the mod 2 calculations for MO. We shall set up these calculations in such a way as to emphasize the similarity.

As before we shall write $\tilde{H}_*(CP^\infty;\mathbb{Z}_p) =$ free $\mathbb{Z}_p$-vector space with basis $y_1, y_2, \ldots$ and $H_*(MU;\mathbb{Z}_p) \cong \mathbb{Z}_p[b_1, b_2, \ldots]$, where $f: CP^\infty \to \Sigma^2 MU$ sends y_{i+1} to b_i, $i \geqslant 0$.

20.21. Proposition. $\alpha y = \sum_i \langle \alpha, \xi_i \rangle y^{p^i}$ *for all* $\alpha \in A^*$, *and hence*

$$\psi y_k = \textstyle\sum_{i+j=k} (X^j)_{2i} \otimes y_j, \quad k \geqslant 0.$$

Here A_* *denotes* $H_*(H(\mathbb{Z}_p);\mathbb{Z}_p)$ *and* $X = 1 + \xi_1 + \xi_2 + \cdots$.

Proof: First we recall that $A_* \cong E(\tau_0, \tau_1, \ldots) \otimes \mathbb{Z}_p[\xi_1, \xi_2, \ldots]$. The dual of τ_i is the admissible monomial

$$P^{p^{i-1}} P^{p^{i-2}} \ldots P^p P^1 \beta$$

and the dual of ξ_i is the admissible monomial

$$P^{p^{i-1}} P^{p^{i-2}} \ldots P^p P^1.$$

Since $H^{2k+1}(CP^\infty;\mathbb{Z}_p) = 0$, all k, it follows any admissible monomial containing a β must be 0 on $H^*(CP^\infty;\mathbb{Z}_p)$. $y \in H^2(CP^\infty;\mathbb{Z}_p)$, so any monomial of excess greater than 1 will annihilate y. Thus we get

$$P^I y = \begin{cases} y^{p^s} & I = (0, p^{s-1}, 0, p^{s-2}, 0, \ldots, p, 0, 1, 0) \\ 0 & \text{otherwise.} \end{cases}$$

We can also express this by writing

$$P^I y = \textstyle\sum_i \langle P^I, \xi_i \rangle y^{p^i}, \quad \text{all admissible } P^I.$$

But the P^I's form a $\mathbb{Z}_p$-basis for A^*. Hence $\alpha y = \sum_i \langle \alpha, \xi_i \rangle y^{p^i}$, all $\alpha \in A^*$. Just as we calculated αx^k for all α in the mod 2 Steenrod algebra, so we find

$$\alpha y^k = \textstyle\sum_i \langle \alpha, (X^k)_{2(k-i)} \rangle y^i.$$

Therefore from the formula $\langle \alpha z, u \rangle = \sum_j \langle \alpha, e_j \langle z, u_j \rangle \rangle$ follows the formula 20.21. □

20.22. Corollary. $\psi b_k = \sum_{i+j=k} (X^{j+1})_{2i} \otimes b_j, \quad k \geqslant 0.$

We now let $H' = \mathbb{Z}_p[v_1, v_2, \ldots]$ with $v_k \in H'_{2k}$ for all k not of the form $k = p^t - 1$, and we define $g: H_*(MU; \mathbb{Z}_p) \to H'$ by

$$g(b_k) = \begin{cases} v_k & \text{if } k \neq p^t - 1 \text{ for all } t > 0, \\ 0 & \text{if } k = p^t - 1 \text{ for some } t > 0. \end{cases}$$

Then g induces a homomorphism $f: H_*(MU; \mathbb{Z}_p) \to C \otimes H'$ of algebras over A_*, where $C \subset A_*$ is the Hopf subalgebra $\mathbb{Z}_p[\xi_1, \xi_2, \ldots]$.

20.23. Theorem. $f: H_*(MU; \mathbb{Z}_p) \to C \otimes H'$ is an isomorphism of comodules over A_*.

The proof is virtually identical to that of 20.7.

20.24. Corollary. In the mod p Adams spectral sequence for MU we have

$$\begin{aligned} E_2^{**} &\cong \operatorname{Ext}_{A_*}^{**}(\mathbb{Z}_p, C \otimes H') \cong \operatorname{Ext}_{A_*}^{**}(\mathbb{Z}_p, C) \otimes H' \\ &\cong \operatorname{Ext}_B^{**}(\mathbb{Z}_p, \mathbb{Z}_p) \otimes H'. \end{aligned}$$

where $B = A_*/A_* \cdot \overline{C} \cong E(\tau_0, \tau_1, \ldots)$ with $\psi_B(\tau_i) = \tau_i \otimes 1 + 1 \otimes \tau_i$.

Here we have used 20.11 and 20.17 again.

But now a resolution of $\mathbb{Z}_p$ over the exterior algebra $E(\tau_0, \tau_1, \ldots)$ is constructed in exactly the same way as for $E(\xi_1, \xi_2, \ldots)$ earlier; we merely replace $\mathbb{Z}_2$ by $\mathbb{Z}_p$ everywhere.

Hence we get

$$\operatorname{Ext}_B^{**}(\mathbb{Z}_p, \mathbb{Z}_p) \cong \mathbb{Z}_p[q_0, q_1, \ldots] \quad q_k \in \operatorname{Ext}_B^{1, 2p^k - 1}(\mathbb{Z}_p, \mathbb{Z}_p).$$

Therefore in the Adams spectral sequence for MU we have

$$E_2^{**} \cong \mathbb{Z}_p[q_0, q_1, \ldots] \otimes H' \cong \mathbb{Z}_p[q_0] \otimes \mathbb{Z}_p[m_1, m_2, \ldots],$$

where we take

$$m_k = \begin{cases} v_k \in E_2^{0,2k} & \text{if } k \neq p^t - 1, \text{ all } t > 0, \\ q_t \in E_2^{1,2k+1} & \text{if } k = p^t - 1, \text{ some } t > 0. \end{cases}$$

Again $E_2^{s,t} = 0$ if $t - s$ is odd, so all $d_r = 0$. Multiplication by q_0 in E_∞^{**} corresponds to multiplication by p in $\pi_*(MU)$, so we get the following theorem.

20.25. Theorem. *${}_p\pi_*(MU) \cong \mathbb{Z}[m_1, m_2, \ldots]$ as groups, and the Hurewicz homomorphism*

$$h: {}_p\pi_t(MU) \to PH_t(MU; \mathbb{Z}_p) \cong H'$$

is a surjection. In particular $\pi_*(MU)$ has no p-torsion for any prime p and hence is torsion free.

20.26. Corollary. *$h: \pi_*(MU) \to H_*(MU; \mathbb{Z})$ is a monomorphism.*

Proof: $(h \otimes 1): \pi_*^S(-) \otimes \mathbb{Q} \to H_*(-; \mathbb{Z}) \otimes \mathbb{Q}$ is a natural transformation of homology theories which is an isomorphism for $X = S^0$ and hence is a natural equivalence. Hence we have a commutative diagram

$$\begin{array}{ccc} \pi_*(MU) & \longrightarrow & \pi_*(MU) \otimes \mathbb{Q} \\ h \downarrow & & \cong \downarrow h \otimes 1 \\ H_*(MU; \mathbb{Z}) & \longrightarrow & H_*(MU; \mathbb{Z}) \otimes \mathbb{Q} \end{array}$$

in which $\pi_*(MU) \to \pi_*(MU) \otimes \mathbb{Q}$ is a monomorphism, since $\pi_*(MU)$ is torsion free. Thus the corollary follows. □

20.27. Theorem. *There is a ring isomorphism*

$$\pi_*(MU) \cong \mathbb{Z}[M_1, M_2, \ldots], \quad M_i \in \pi_{2i}(MU).$$

Moreover the polynomial generators M_i can be chosen so that

$$h(M_i) = \begin{cases} b_i \bmod \textit{decomposables if } i \textit{ not of form } p^t - 1, \textit{ all primes } p \\ p^{f(i)} b_i \bmod \textit{decomposables if } i = p^t - 1, \textit{ some prime } p, t > 0 \end{cases}$$

for some function $f(i)$ such that $f(i) = 0$ if $i + 1$ is not a power of a prime.

Proof: First we show that $\pi_*(MU) \otimes \mathbb{Z}_p \cong \mathbb{Z}_p[P_1, P_2, \ldots]$ with $P_i \in \pi_{2i}(MU)$ for every prime p. We choose $P_i \in \pi_{2i}(MU)$ to correspond to $m_i \in E_\infty^{**}$. That the monomials $P^\alpha = P_1^{\alpha_1} P_2^{\alpha_2} \ldots P_r^{\alpha_r}$ in the P_i's are linearly independent in $\pi_*(MU) \otimes \mathbb{Z}_p$ is easy to see. To show that these monomials span $\pi_*(MU) \otimes \mathbb{Z}_p$ we need the following lemma.

20.28. Lemma. *Suppose $\pi_*(Y)$ is of finite type, $q \in \mathbb{Z}$, p a prime and $s > 0$. If $G \subset \pi_q(Y)$ denotes the subgroup of elements divisible by p^s, then there is a t with $F^{t,q+t} \subset G$.*

Proof: Since $\pi_q(Y)$ is finitely generated and every element $x \in \pi_q(Y)/G$ satisfies $p^s \cdot x = 0$, $\pi_q(Y)/G$ is finite. Now

$$\frac{\pi_q(Y)}{G} = \frac{F^{0,q} + G}{G} \supset \frac{F^{1,q+1} + G}{G} \supset \cdots \supset \frac{F^{t,q+t} + G}{G} \supset \cdots$$

is a filtration of $\pi_q(Y)/G$ and from 19.12 it follows

$$\bigcap_{t\geqslant 0}\frac{F^{t,q+t}+G}{G}=\frac{\bigcap_{t\geqslant 0}F^{t,q+t}+G}{G}=0.$$

Thus we must have $\dfrac{F^{t,q+t}+G}{G}=0$ for some finite t—that is $F^{t,q+t}\subset G$. ▯

Now we can show that the monomials P^α span $\pi_*(MU)\otimes\mathbb{Z}_p$. If we take $Y=MU$ and $s=1$ in 20.28 we find a t so that $F^{t,q+t}\otimes\mathbb{Z}_p=0$. We show that the monomials P^α of grading q and filtration at least s span $F^{s,q+s}\otimes\mathbb{Z}_p$ by downward induction over s starting at $s=t$. If

$$m^{\alpha_1},m^{\alpha_2},\ldots,m^{\alpha_r}$$

are the monomials in the m's in $E_\infty^{s-1,q+s-1}$, then any $x\in F^{s-1,q+s-1}\otimes\mathbb{Z}_p$ can be written in the form

$$x=\textstyle\sum_{i=1}^r n_{\alpha_i}P^{\alpha_i}\bmod F^{s,q+s}\otimes\mathbb{Z}_p$$

because the projection $F^{s-1,q+s-1}\to E_\infty^{s-1,s+q-1}$ sends P^{α_i} to m^{α_i}, $1\leqslant i\leqslant r$. But by the inductive assumption every element of $F^{s,q+s}\otimes\mathbb{Z}_p$ is a linear combination of monomials P^α.

The first part of 20.27 will follow from the next lemma.

20.29. Lemma. *Let R be a commutative graded ring of finite type with $R\otimes\mathbb{Z}_p\cong\mathbb{Z}_p[P_1,P_2,\ldots]$, $P_i=P_i(p)\in R_{2i}$ for each $i\geqslant 1$ and each prime p. Then $R\cong\mathbb{Z}[M_1,M_2,\ldots]$ for some $M_i\in R_{2i}$, $i\geqslant 1$.*

Proof: For the indecomposable quotients we have $Q_{2i}(R)\otimes\mathbb{Z}_p\cong Q_{2i}(R\otimes\mathbb{Z}_p)\cong\mathbb{Z}_p$ (with generator P_i) for $i\geqslant 1$ and all primes p. Thus we must have $Q_{2i}(R)\cong\mathbb{Z}$. Similarly $Q_{2i+1}(R)=0$, $i\geqslant 0$. Let $M_i\in R_{2i}$ be an element whose class in $Q_{2i}(R)$ generates $Q_{2i}(R)$. We let $R'=\mathbb{Z}[x_1,x_2,\ldots]$ with grade $x_i=2i$ and define $f\colon R'\to R$ to be the ring homomorphism with $f(x_i)=M_i$, $i\geqslant 1$. Then

$$f\otimes 1\colon R'\otimes\mathbb{Z}_p\to R\otimes\mathbb{Z}_p$$

satisfies $f(x_i)\otimes 1=P_i\otimes\alpha_i$ mod decomposables, some $\alpha_i\neq 0\in\mathbb{Z}_p$, so $f\otimes 1$ is an isomorphism for all primes p.

Suppose $f(r')=0$ for some $r'\in R'$; then $f\otimes 1(r'\otimes 1)=0$ and hence $r'\otimes 1=0\in R'\otimes\mathbb{Z}_p$ for all primes p. That implies that $p|r'$ for all primes. Therefore $r'=0$, so f is injective. On the other hand if $f(r')$ is divisible by a prime p, then $f(r')\otimes 1=0\in R\otimes\mathbb{Z}_p$ and hence $r'\otimes 1=0\in R'\otimes\mathbb{Z}_p$—i.e. r' is divisible by p. Since R_{2q} and R'_{2q} are free abelian groups of the same rank, it follows f is an isomorphism. ▯

The second part of 20.27 follows easily from the choice of the M_i's.

Remark. Milnor showed $f(i)=1$ for all i with $i+1$ a power of a prime.

To discuss $\operatorname{im} h$ in greater detail we need K-theory; $H_*(H(\mathbb{Z});\mathbb{Z})$ is not well behaved, so we cannot talk about $P(H_*(MU;\mathbb{Z}))$. We shall show that for the K-theory Hurewicz homomorphism $h:\pi_*(MU) \to K_*(MU)$ we have $\operatorname{im} h = P(K_*(MU))$.

20.30. Definition. For any abelian group G we have the homomorphism $i_G: G \to G \otimes \mathbb{Q}$ given by $i_G(g) = g \otimes 1$. We say an element $y \in G \otimes \mathbb{Q}$ is *integral* if $y = i_G(g)$ for some $g \in G$. We say a homomorphism $\theta: G \to G'$ is *integrality preserving* if for all $y \in G \otimes \mathbb{Q}$ we have y is integral if and only if $(\theta \otimes 1)(y)$ is integral.

Remarks. i) If y is integral clearly $(\theta \otimes 1)(y)$ is also.

ii) θ is integrality preserving if and only if $\theta(g) = ng'$ for some $g \in G$, $g' \in G'$, $n \in \mathbb{Z}$ implies $g = nh$ for some $h \in G$.

20.31. Exercise. Show that if $\{\theta_n\}:\{G_n\} \to \{G_n'\}$ is a morphism of direct systems and each θ_n is integrality preserving, then so is $\operatorname{dir\,lim} \theta_n$.

Now in general $h:\pi_*^S(X) \to K_*(X)$ is induced by the unit $\iota: S^0 \to K$ of the spectrum K. Suppose ι is represented by a function $\{i_{2n}: S^{2n} \to BU\}$ (remember $K_{2n} = BU$). We observe that $K_{2q}(MU) = \pi_{2q}(K \wedge MU) \cong \pi_{2q}(MU \wedge K) = MU_{2q}(K) \cong \operatorname{dir\,lim} \widetilde{MU}_{2q+2n}(BU)$. Hence we have a commutative diagram

$$\begin{array}{ccccc}
\operatorname{dir\,lim}_n \widetilde{MU}_{2p+2n}(S^{2n}) & \cong & \widetilde{MU}_{2q}(S^0) & \cong & \pi_{2q}(MU) \\
\Big\downarrow {\scriptstyle \operatorname{dir\,lim}(i_{2n})_*} & & \Big\downarrow {\scriptstyle \iota_*} & & \Big\downarrow {\scriptstyle h} \\
\operatorname{dir\,lim}_n \widetilde{MU}_{2q+2n}(BU) & \cong & MU_{2q}(K) & \cong & K_{2q}(MU).
\end{array}$$

We wish to prove that h preserves integrality, so it will clearly suffice to prove that i_{2n*} does for every $n \geqslant 0$.

20.32. Lemma. *$h:\pi_0(MU) \to K_0(MU)$ is integrality preserving.*

Proof: By 20.31 it suffices to show $h:\pi_{2n}(MU(n)) \to \tilde{K}_{2n}(MU(n))$ is integrality preserving for each n. $MU(n)$ is $(2n-1)$-connected, so

$$h:\pi_{2n}(MU(n)) \to \tilde{H}_{2n}(MU(n)) \cong \mathbb{Z}$$

is an isomorphism. Consider the morphism

$$\pi_{2n}(MU(n)) \longrightarrow \tilde{K}_{2n}(MU(n)) \xrightarrow{\langle \tau_n, - \rangle} \tilde{K}_0(S^0),$$

where $\tau_n \in \tilde{K}^{2n}(MU(n))$ is the Thom class as in 16.40; if $j_{2n}: S^{2n} \to MU(n)$ is the inclusion of the fibre (and therefore a term of the identity $\iota: S^0 \to MU$), then we have

$$\langle \tau_n, h(1) \rangle = \langle \tau_n, h[j_{2n}] \rangle = \langle \tau_n, j_{2n*}(g_{2n}) \rangle = \langle j_{2n}^* \tau_n, g_{2n} \rangle = 1$$

($g_{2n} \in \tilde{K}_{2n}(S^{2n})$ the generator). This shows that h is integrality preserving—in other words that $h(1)$ is divisible only by ± 1. $\square$

20.33. Proposition. (Stong, Hattori) *$h: \pi_*(MU) \to K_*(MU)$ is integrality preserving.*

Proof: We shall prove that $i_{2n*}: \widetilde{MU}_{2n+2q}(S^{2n}) \to MU_{2n+2q}(BU)$ is integrality preserving for all n, q.

Case $q=0$: we have a commutative diagram

$$\begin{array}{ccc} \widetilde{MU}_{2n}(S^{2n}) & \xrightarrow{\cong} & \operatorname{dir\,lim} \widetilde{MU}_{2m}(S^{2m}) \cong \pi_0(MU) \\ \downarrow i_{2n*} & & \operatorname{dir\,lim}(i_{2m*}) \downarrow \qquad \downarrow h \\ \widetilde{MU}_{2n}(BU) & \longrightarrow & \operatorname{dir\,lim} \widetilde{MU}_{2m}(BU) \cong K_0(MU), \end{array}$$

and by 20.32 h is integrality preserving. It then follows i_{2n*} is also.

Case $q>0$: $\widetilde{MU}_*(S^{2n})$ is a free $\widetilde{MU}_*(S^0)$-module with one generator $g_{2n} \in \widetilde{MU}_{2n}(S^{2n})$. Hence a typical element of $\widetilde{MU}_{2n+2q}(S^{2n}) \otimes \mathbb{Q}$ is of the form $\omega g_{2n} \otimes 1/d$ for some $\omega \in \widetilde{MU}_{2q}(S^0)$ and integer d. Suppose $i_{2n*}(g_{2n}) = \sum_\alpha \omega_\alpha \beta^\alpha$, where $\beta^\alpha = \beta_1^{\alpha_1}\beta_2^{\alpha_2}\ldots\beta_r^{\alpha_r}$ is a monomial in the polynomial generators $\beta_i \in MU_{2i}(BU)$ and $\omega_\alpha \in \widetilde{MU}_*(S^0)$. Then

$$(i_{2n*} \otimes 1)(\omega g_{2n} \otimes 1/d) = \textstyle\sum_\alpha \omega\omega_\alpha \beta^\alpha \otimes 1/d.$$

If this element is integral, then d must divide each of the coefficients $\omega\omega_\alpha$. Since we know $\widetilde{MU}_*(S^0) = \pi_*(MU)$ is a polynomial ring, this means that for each α we have $d = e_\alpha f_\alpha$ with $e_\alpha | \omega$, $f_\alpha | \omega_\alpha$. Let $e = \text{l.c.m.}\{e_\alpha\}$ and $f = \text{g.c.d.}\{f_\alpha\}$. Then $d = ef$, $e|\omega$ and $f|\omega_\alpha$ for all α. Hence

$$(i_{2n*} \otimes 1)(g_{2n} \otimes 1/f) = \textstyle\sum_\alpha \omega_\alpha \beta^\alpha \otimes 1/f$$

is integral; but this implies that $g_{2n} \otimes 1/f$ is integral by the case $q=0$ above. This can only be so if $f = \pm 1$, which implies $d|\omega$. Thus $\omega g_{2n} \otimes 1/d$ is indeed integral. $\square$

20.34. Theorem. *$h: \pi_*(MU) \to P(K_*(MU))$ is an isomorphism.*

Proof: $K_*(-) \otimes \mathbb{Q}$ is a homology theory and $K^*(-) \otimes \mathbb{Q}$ is a cohomology theory on the category of finite spectra (the wedge axiom is not satisfied on $\mathcal{SP}$). Hence by Brown's theorem there is a spectrum $K\mathbb{Q}$ representing $K^*(-) \otimes \mathbb{Q}$ on the category of finite spectra. The natural transformation $K^*(-) \to K^*(-) \otimes \mathbb{Q}$ (sending x to $x \otimes 1$) defines a map $\alpha: K \to K\mathbb{Q}$ of representing spectra. Then $K\mathbb{Q}_*(X) \cong K_*(X) \otimes \mathbb{Q}$ for all finite spectra X, and taking direct limits we find the same is true for all spectra. We then have a commutative diagram

$$\begin{array}{ccccc}
\pi_*(MU) & \xrightarrow[\text{mono}]{j} & \pi_*(MU) \otimes \mathbb{Q} & & \\
\downarrow h & & \downarrow h_{\mathbb{Q}} & \searrow h \otimes 1 & \\
P(K_*(MU)) & \xrightarrow[\text{mono}]{\alpha} & P(K\mathbb{Q}_*(MU)) & & \\
\cap & & \cap & & \\
K_*(MU) & \xrightarrow[\text{mono}]{\alpha} & K\mathbb{Q}_*(MU) & \cong & K_*(MU) \otimes \mathbb{Q}.
\end{array}$$

Here α is monomorphic since $K_*(MU)$ is torsion free. We are going to show that $h_{\mathbb{Q}}$ is an isomorphism. From this fact it follows immediately that h is a monomorphism. Suppose $y \in P(K_*(MU))$; if $h_{\mathbb{Q}}$ is surjective then $\alpha(y) = h_{\mathbb{Q}}(z)$ for some $z \in \pi_*(MU) \otimes \mathbb{Q}$. But h is integrality preserving by 20.33, so z is integral. That is, there is a $w \in \pi_*(MU)$ with $j(w) = z$, or in other words $\alpha \circ h(w) = h_{\mathbb{Q}} \circ j(w) = \alpha(y)$, and therefore $h(w) = y$. This shows that h is an epimorphism.

Now in fact $h_{\mathbb{Q}}: \pi_*(X) \otimes \mathbb{Q} \to P(E\mathbb{Q}_*(X))$ is an isomorphism for all spectra E, X with $\pi_0(E)$ torsion-free. From 17.19 we have the isomorphism

$$\wedge \otimes 1 : \pi_*(E) \otimes \pi_*(X) \otimes \mathbb{Q} \to \pi_*(E \wedge X) \otimes \mathbb{Q} \cong E\mathbb{Q}_*(X).$$

With this identification $h_{\mathbb{Q}}$ becomes the map

$$\pi_*(X) \otimes \mathbb{Q} \to \pi_*(E) \otimes \pi_*(X) \otimes \mathbb{Q}$$

given by $y \mapsto 1 \otimes y$. Moreover the diagram

$$\begin{array}{ccc}
E\mathbb{Q}_*(X) & \xrightarrow{\psi_X} & E\mathbb{Q}_*(E\mathbb{Q}) \otimes_{E\tilde{\mathbb{Q}}_*(S^0)} E\mathbb{Q}_*(X) \\
\wr\| & & \wr\| \\
\pi_*(E) \otimes \pi_*(X) \otimes \mathbb{Q} & \xrightarrow{\Psi} & \pi_*(E) \otimes \pi_*(E) \otimes \pi_*(X) \otimes \mathbb{Q}
\end{array}$$

commutes if $\Psi(x \otimes y) = x \otimes 1 \otimes y$ for $x \in \pi_*(E)$, $y \in \pi_*(X) \otimes \mathbb{Q}$. Thus the subgroup of primitive elements is precisely the subgroup of elements of the form $1 \otimes y$ in $\pi_*(E) \otimes \pi_*(X) \otimes \mathbb{Q} \cong E\mathbb{Q}_*(X)$—in other words precisely the image of $h_{\mathbb{Q}}$. $h_{\mathbb{Q}}$ is clearly a monomorphism. □

20.35. At the end of Chapter 17 we computed $P(K_*(MU))$ in low dimensions. We now see that we may take those calculations as a determination of the K-theory Hurewicz images of polynomial generators $M_1, M_2, \ldots$ for $\pi_*(MU) \cong \Omega_*^U$. In particular $[CP^3]$ will not serve as the polynomial generator in dimension 6.

20.36. *Remark.* In [31] Conner and Floyd show that there is a close connection between MU and K on the one hand and MSp and KO on the other. One might well conjecture that the following analog to 20.34 holds:

$$h:\pi_*(MSp) \to P(KO_*(MSp)) \quad \text{is an isomorphism.}$$

However, $\pi_*(MSp)$ is known in low dimensions, and we may compare the known results with our calculations at the end of Chapter 17. It turns out that the conjecture is true in dimensions less than 9. However, $\pi_9(MSp) \cong \mathbb{Z}_2$, whereas we found $P_9(KO_*(MSp)) \cong \mathbb{Z}_2 \oplus \mathbb{Z}_2$. [It is not known whether h is a monomorphism or whether the weaker conjecture "$h:\pi_n(MSp) \to P_n(KO_*(MSp))$ is an isomorphism for $n \equiv 0 \bmod 4$" is true or false.]

With Ω_*^U firmly in hand we turn next to the determination of Ω_*^{SO}. The mod p Adams spectral sequence for $\pi_*(MSO)$ for odd primes p is simple, so we consider the case p odd first. We have a Thom isomorphism $\Phi_*:H_*(MSO;\mathbb{Z}_p) \to H_*(BSO;\mathbb{Z}_p)$. According to exercise 16.52 we have $H_*(BSO;\mathbb{Z}_p) \cong \mathbb{Z}_p[q_1,q_2,\ldots]$, where $Bi_*:H_*(BSO(2);\mathbb{Z}_p) \to H_*(BSO;\mathbb{Z}_p)$ satisfies $Bi_*(y_{2i+1}) = 0$, $i \geqslant 0$, and $Bi_*(y_{2i}) = q_i$, $i \geqslant 1$. Now we have a commutative diagram

$$\begin{array}{ccc} BU(1) & \cong & BSO(2) \\ \downarrow{\scriptstyle Bi} & & \downarrow{\scriptstyle Bi} \\ BU & \xrightarrow{\ r\ } & BSO, \end{array}$$

where r is the "forgetful" map which regards a unitary bundle as an oriented real bundle. From it we see that $r_*(y_{2i}) = q_i$ and $r_*(y_{2i+1}) = 0$, all $i \geqslant 0$. We let $q_i' \in H_{4i}(MSO;\mathbb{Z}_p)$ be such that $\Phi_*(q_i') = q_i$, $i \geqslant 1$. If $Mr:MU \to MSO$ is the map induced by $r:BU \to BSO$, then we have

$$Mr_*(b_i) = \begin{cases} q_k' & i = 2k, \\ 0 & i \text{ odd.} \end{cases}$$

Hence from 20.22 we get the following lemma.

20.37. Lemma. $\psi q_k' = \sum_{i+j=k} (X^{2j+1})_{4i} \otimes q_j'$, $k \geqslant 0$.

Thus if we let $H'' = \mathbb{Z}_p[z_1,z_2,\ldots]$, with $z_k \in H_{4k}''$ for every k not of the form $(p^t - 1)/2$ for some $t \geqslant 1$, then we have the following.

20.38. Lemma. $H_*(MSO;\mathbb{Z}_p) \cong C \otimes H''$ *as algebras over* A_*.

Hence in the Adams spectral sequence for MSO we have

$$E_2^{**} \cong \mathbb{Z}_p[q_0,q_1,\ldots] \otimes H'' \cong \mathbb{Z}_p[q_0] \otimes \mathbb{Z}_p[n_1,n_2,\ldots]$$

with

$$n_k = \begin{cases} z_k \in E_2^{0,4k} & \text{if } k \text{ not of the form } (p^t-1)/2, \\ q_t \in E_2^{1,4k+1} & \text{if } k = (p^t-1)/2 \text{ for some } t > 0. \end{cases}$$

20.39. Theorem. *There is an isomorphism of groups ${}_p\pi_*(MSO) \cong \mathbb{Z}[n_1, n_2, \ldots]$ and the Hurewicz homomorphism*

$$h: {}_p\pi_*(MSO) \to P(H_*(MSO; \mathbb{Z}_p))$$

is an epimorphism. In particular $\pi_*(MSO)$ has no odd torsion.

20.40. Exercise. Show that if p is an odd prime, then ${}_p\pi_*(MSp) \cong \mathbb{Z}[p_1, p_2, \ldots]$ as groups, grade $p_i = 4i$.

Hint: Show that $r': BSp \to BSO$ induces an isomorphism

$$H_*(BSp; \mathbb{Z}_p) \cong H_*(BSO; \mathbb{Z}_p).$$

The discussion of the mod 2 Adams spectral sequence for MSO is more interesting. What we shall prove is that

$$H^*(MSO; \mathbb{Z}_2) \cong A^*/A^* Sq^1 \otimes H^* \oplus A^* \otimes Z^*,$$

where $H^* = \mathbb{Z}_2[p_1, p_2, \ldots]$ and Z^* is some $\mathbb{Z}_2$-vector space. It will follow that

$$H_*(MSO; \mathbb{Z}_2) \cong (A^*/A^* Sq^1)^* \otimes H \oplus A_* \otimes Z.$$

We first get a grip on $(A^*/A^* Sq^1)^*$.

A^* acts on A_* on both the left and the right as follows:

$$\langle b, a\eta \rangle = \langle ba, \eta \rangle \quad a, b \in A^*$$
$$\langle b, \eta a \rangle = \langle ab, \eta \rangle \quad \eta \in A_*.$$

We also have $c(\eta a) = c^*(a) c(\eta)$, $c(a\eta) = c(\eta) c^*(a)$, $a \in A^*$, $\eta \in A_*$, where $c: A_* \to A_*$ is the canonical anti-automorphism and $c^*: A^* \to A^*$ is its dual.

20.41. Lemma. $Sq^1 \xi_i = \xi_{i-1}^2$, $i \geqslant 1$, *and* $\xi_i Sq^1 = \begin{cases} \xi_0 & i = 1 \\ 0 & i > 1. \end{cases}$

Sq^1 is a derivation, both from the left and from the right.

Proof: Recall the formula 17.11:

$\langle ay, u \rangle = \sum_i \langle a, e_i \langle y, u_i \rangle \rangle$ if $a \in A^*$, $y \in H^*(X; \mathbb{Z}_2)$, $u \in H_*(X; \mathbb{Z}_2)$ and $\psi u = \sum_i e_i \otimes u_i$.

Applying this to the case $X = H(\mathbb{Z}_2)$ gives

$\langle a, Sq^1 \xi_i \rangle = \langle a Sq^1, \xi_i \rangle = \sum_j \langle a, \xi_{i-j}^{2^j} \langle Sq^1, \xi_j \rangle \rangle = \langle a, \xi_{i-1}^2 \rangle$ for all $a \in A^*$.

Thus $Sq^1 \xi_i = \xi_{i-1}^2$. Similarly

$$\langle a, \xi_i Sq^1 \rangle = \langle Sq^1 a, \xi_i \rangle = \textstyle\sum_j \langle Sq^1, \xi_{i-j}^{2^j} \langle a, \xi_j \rangle\rangle$$

$$= \begin{cases} \langle a, \xi_0 \rangle & i = 1 \\ 0 & i \neq 1 \end{cases} \quad \text{all } a \in A^*.$$

Thus

$$\xi_i Sq^1 = \begin{cases} \xi_0 & i = 1 \\ 0 & i > 1. \end{cases}$$

Let $\phi: A_* \otimes A_* \to A_*$ be the product and $\phi^*: A^* \to A^* \otimes A^*$ the coproduct which is dual to it. Then we must have $\phi^* Sq^1 = Sq^1 \otimes 1 + 1 \otimes Sq^1$. Let $a \in A^*$, $n, \eta' \in A_*$ and suppose $\phi^* a = \sum_i a_i \otimes a_i'$. Then

$$\begin{aligned}
\langle a, Sq^1(\eta\eta') \rangle &= \langle aSq^1, \phi(\eta \otimes \eta') \rangle = \langle \phi^*(aSq^1), \eta \otimes \eta' \rangle \\
&= \langle (\textstyle\sum_i a_i \otimes a_i')(Sq^1 \otimes 1 + 1 \otimes Sq^1), \eta \otimes \eta' \rangle \\
&= \textstyle\sum_i \{\langle a_i Sq^1, \eta \rangle \langle a_i', \eta' \rangle + \langle a_i, \eta \rangle \langle a_i' Sq^1, \eta' \rangle\} \\
&= \textstyle\sum_i \{\langle a_i, Sq^1 \eta \rangle \langle a_i', \eta' \rangle + \langle a_i, \eta \rangle \langle a_i', Sq^1 \eta' \rangle\} \\
&= \langle \textstyle\sum_i a_i \otimes a_i', Sq^1 \eta \otimes \eta' + \eta \otimes Sq^1 \eta' \rangle \\
&= \langle \phi^* a, Sq^1 \eta \otimes \eta' + \eta \otimes Sq^1 \eta' \rangle \\
&= \langle a, (Sq^1 \eta)\eta' + \eta(Sq^1 \eta') \rangle.
\end{aligned}$$

Hence $Sq^1(\eta\eta') = (Sq^1\eta)\eta' + \eta(Sq^1\eta')$. The proof that

$$(\eta\eta') Sq^1 = \eta(\eta' Sq^1) + (\eta Sq^1)\eta'$$

is similar. □

Now we have an exact sequence

$$A^* \xrightarrow{RSq^1} A^* \longrightarrow A^*/A^* Sq^1 \longrightarrow 0,$$

where RSq^1 means action on the right by Sq^1. Taking duals gives

$$0 \longrightarrow (A^*/A^* Sq^1)^* \longrightarrow A_* \xrightarrow{LSq^1} A_*$$

and applying c gives

$$0 \longrightarrow c(A^*/A^* Sq^1)^* \longrightarrow A_* \xrightarrow{RSq^1} A_*$$

since $c^* Sq^1 = Sq^1$.

20.42. Corollary. $c(A^*/A^* Sq^1)^* \simeq \mathbb{Z}_2[\xi_1^2, \xi_2, \xi_3, \ldots]$.

Now in $H^0(MSO)$ (we drop the coefficient group $\mathbb{Z}_2$ for brevity) we have the element t corresponding to the Thom classes $t_n \in \tilde{H}^n(MSO_n)$.

Define $\kappa^*: A^* \to H^*(MSO)$ by $\kappa^*(a) = a \cdot t$ for $a \in A^*$. This is the same as the composite

$$A^* \cong A^* \otimes \mathbb{Z}_2 \xrightarrow{1 \otimes \eta} A^* \otimes H^*(MSO) \xrightarrow{\psi^*} H^*(MSO),$$

where $\eta(1) = t$ and all tensor products are over $\mathbb{Z}_2$. Dualizing gives $\kappa: H_*(MSO) \to A_*$, which is

$$H_*(MSO) \xrightarrow{\psi} A_* \otimes H_*(MSO) \xrightarrow{1 \otimes \varepsilon} A_* \otimes \mathbb{Z}_2 \cong A_*,$$

ε being the augmentation of $H_*(MSO)$.

Now in the exact sequence

$$0 \to H^n(MSO_n) \to H^n(BSO(n)) \to H^n(BSO(n-1)) \to \cdots$$

(where we regard MSO_n as the mapping cone of the sphere bundle $BSO(n-1) \to BSO(n)$ (cf. 16.11)) the Thom class t_n maps to $w_n \in H^n(BSO(n))$. (Note: $H^*(BSO(n)) \cong \mathbb{Z}_2[w_2, w_3, \ldots, w_n]$; see [49], for example). Now $Sq^1 w_n = w_n w_1 = 0$, since $w_1 = 0$ in $H^*(BSO(n))$; so $Sq^1 t_n = 0$, all n, and hence $Sq^1 t = 0$. Therefore κ^* induces $\bar{\kappa}^*: A^*/A^* Sq^1 \to H^*(MSO)$ and $\bar{\kappa}: H_*(MSO) \to (A^*/A^* Sq^1)^*$.

20.43. Proposition. *$\bar{\kappa}^*$ is a monomorphism, or equivalently $\bar{\kappa}$ is an epimorphism.*

Proof: First we introduce some notation: let $\xi^\alpha = \xi_1^{\alpha_1} \ldots \xi_r^{\alpha_r}$ denote a typical monomial in A_*; then $\{\xi^\alpha\}$ is a $\mathbb{Z}_2$-basis for A_*. Let $\{Sq^\alpha\}$ denote the dual basis in A^*. (Note: In this notation it happens that $Sq^{(k)}$ (the dual of $\xi^{(k)} = \xi_1^k$) is Sq^k, but in general $Sq^{(\alpha_1, \alpha_2, \ldots, \alpha_r)}$ is *not* $Sq^{\alpha_1} Sq^{\alpha_2} \ldots Sq^{\alpha_r}$).

Now we know that $(A^*/A^* Sq^1)^*$ is generated as an algebra by $c(\xi_1^2)$, $c(\xi_2)$, $c(\xi_3)$, ... Thus if we can show that the subgroup $P_* \subset A_*$ generated by $c(\xi_1^2)$, $c(\xi_2)$, ... is in the image of $\bar{\kappa}$, then since $\bar{\kappa}$ is a ring homomorphism, it will follow that $\bar{\kappa}$ is surjective. Dually, it would suffice to show that κ^* was a monomorphism on $P =$ subgroup of A^* generated by $c^*(Sq^{(2)}) = c^*(Sq^2)$, $c^*(Sq^{(0,1)}) = Sq^{(0,1)}$, ... $c^*(Sq^{(0,\ldots,0,1)}) = Sq^{(0,\ldots,0,1)}$, ... Let us abbreviate $(0,\ldots,0,1)$ by Δ_i if the 1 is in the ith place. The Sq^{Δ_i} satisfy the relation

$$Sq^{\Delta_{i+1}} = Sq^1 Sq^{2\Delta_i} + Sq^{2\Delta_i} Sq^1, \quad i \geqslant 1,$$

(cf. [61]), so we compute

$$\kappa^*(c^*(Sq^2)) = c^*(Sq^2) \cdot t = \Phi^*(w_2)$$

$$\begin{aligned}
\kappa^*(Sq^{\Delta_{i+1}}) &= (Sq^1 Sq^{2\Delta_i} + Sq^{2\Delta_i} Sq^1) \cdot t = Sq^1(\Phi^*(w_{2^{i+1}-2} + \text{lower terms})) \\
&= \Phi^*(Sq^1 w_{2^{i+1}-2} + \text{lower terms})) \\
&= \Phi^*(w_{2^{i+1}-1} + \text{lower terms}), \quad i \geqslant 1.
\end{aligned}$$

Here we have used the fact that Sq^1 commutes with Φ^* because $Sq^1 t_n = 0$. Thus we see that $\bar{\kappa}^*$ is injective and hence $\bar{\kappa}$ is surjective. $\square$

Since $Sq^1 \circ Sq^1 = 0$, we can think of Sq^1 as a differential on any A^*-module M and compute the cohomology $H^*(M; Sq^1)$. In particular we find:

i) $H^*(H^*(MSO); Sq^1) \cong \mathbb{Z}_2[w_2^2, w_4^2, \ldots]$. In fact, since Sq^1 commutes with Φ^*, it suffices to show that $H^*(H^*(BSO); Sq^1) \cong \mathbb{Z}_2[w_2^2, w_4^2, \ldots]$. But $Sq^1 w_{2n} = w_{2n+1}$, so if we write $H^*(BSO) = \mathbb{Z}_2[w_2, w_3] \otimes \mathbb{Z}_2[w_4, w_5] \otimes \ldots$ and use the Künneth theorem the result follows.

ii) $H^*(A^*/A^* Sq^1; Sq^1) \cong \mathbb{Z}_2$; for $A^*/A^* Sq^1$ has a $\mathbb{Z}_2$-basis consisting of the classes of the admissible monomials Sq^I with $I = 0$ or $i_r \geqslant 2$.

Also

$$Sq^1 Sq^I = \begin{cases} Sq^{I'} & i_1 \text{ even}, I \neq 0 \\ 0 & i_1 \text{ odd or } I = 0, \end{cases}$$

where $I' = (i_1 + 1, i_2, \ldots i_r)$ if $I = (i_1, i_2, \ldots, i_r)$. Thus the result follows.

Hence if we let $H^* = \mathbb{Z}_2[p_1, p_2, \ldots]$, $p_i \in H^*_{4i}$, and define

$$\theta : A^*/A^* Sq^1 \otimes H^* \to H^*(MSO)$$

by $\theta(\{a\} \otimes p_\alpha) = a\Phi^*(w_2^{2\alpha_1} w_4^{2\alpha_2} \ldots w_{2r}^{2\alpha_r})$ for $a \in A^*$, then θ is a homomorphism of A^*-modules and

$$\theta^* : H^*(A^*/A^* Sq^1 \otimes H^*; Sq^1) \cong H^*(H^*(MSO); Sq^1).$$

Now let $M = H^*(MSO)$, $\bar{M} = M/\bar{A}^* M$, where $\bar{A}^*$ denotes elements of A^* of positive degree. Let $\rho : M \to \bar{M}$ be the projection and choose a $\mathbb{Z}_2$-subspace $Z^* \subset M$ such that $\rho | Z^*$ is a monomorphism and

$$\bar{M} \cong \rho(\theta(A^*/A^* Sq^1 \otimes H^*)) \oplus \rho(Z^*);$$

in other words, we split off $\rho\theta(A^*/A^* Sq^1 \otimes H^*)$ as a direct summand of $\bar{M}$, choose a basis for the complement $\rho(Z^*)$ and pull the basis elements back to M to form Z^*. Let $N = A^*/A^* Sq^1 \otimes H^* \oplus A^* \otimes Z^*$ and extend θ to N by $\theta(a \otimes z) = az$ for all $a \in A^*$, $z \in Z^*$.

20.44. Theorem. *$\theta : N \to H^*(MSO)$ is an isomorphism.*

Proof: Let $N^{(n)}$, $M^{(n)}$ denote the A^*-submodules of N, M generated by $N_i, M_i, i \leqslant n$, respectively. We show by induction on n that $\theta : N^{(n)} \to M^{(n)}$ is an isomorphism. From the choice of Z^* it follows that $\theta(N^{(n)}) = M^{(n)}$.

Suppose $n = 0$; $N^{(0)} = A^*/A^* Sq^1 \otimes \{1\}$ and $M^{(0)} = A^* \cdot t$. Moreover, $\theta | N^{(0)}$ is just $\bar{\kappa}^*$, so by 20.43 $\theta | N^{(0)}$ is a monomorphism.

Suppose we have proved $\theta: N^{(n-1)} \to M^{(n-1)}$ is an isomorphism and let $\lambda: N/N^{(n-1)} \to M/M^{(n-1)}$ be the map induced by θ. We wish to prove $\lambda|(N^{(n)}/N^{(n-1)})$ is a monomorphism. We let $P \subset N$ be the subspace generated by elements of the form $h, z, Sq^1 z$ for $h \in H_n^*$, $z \in Z_n^*$. Note that P can be regarded as subspace of $N/N^{(n-1)}$.

20.45. Lemma. *$\lambda|P$ is a monomorphism.*

Proof: We first note that since $H^*(A^*; Sq^1) = 0$, $\theta^*: H^*(N; Sq^1) \to H^*(M; Sq^1)$ is still an isomorphism. As $\theta: N^{(n-1)} \to M^{(n-1)}$ is an isomorphism, it follows that $\lambda^*: H^*(N/N^{(n-1)}; Sq^1) \to H^*(M/M^{(n-1)}; Sq^1)$ is also an isomorphism.

Suppose $v \in P$ and $\lambda(v) = 0$. Observe that $\dim v = n$ or $n+1$; we consider the two cases separately.

$\dim v = n$: then $v = h + z$, $h \in H_n^*$, $z \in Z_n^*$.

$$\lambda(v) = 0 \text{ means } \theta(h+z) \in M_n^{(n-1)};$$

thus by the choice of Z^*, $z = 0$. Then h represents a non-bounding Sq^1-cycle in N such that $\lambda(h) = 0$; it follows $h = 0$.

$\dim v = n+1$: then $v = Sq^1 z$, some $z \in Z_n^*$. If $\lambda(v) = 0$, then $\lambda(z)$ is a Sq^1-cycle in $M/M^{(n-1)}$, so since λ^* is surjective, it follows $\lambda(z) = \lambda(h) + Sq^1 z'$ for some $h \in H_n^*$, $z' \in (M/M^{(n-1)})_{n-1}$. But this means $z' = 0$, so $\lambda(z) = \lambda(h)$. Thus $\lambda(h+z) = 0$. By the previous case it follows $h = z = 0$. □

Now we return to the induction step in the proof of 20.44. The multiplication $\mu: MSO \wedge MSO \to MSO$ gives a coproduct $\mu^*: M \to M \otimes M$ on $M = H^*(MSO)$, and the diagram

$$\begin{array}{ccc} A^* \otimes M & \xrightarrow{\psi_M^*} & M \\ \downarrow{\scriptstyle \psi^* \otimes \mu^*} & & \downarrow{\scriptstyle \mu^*} \\ A^* \otimes A^* \otimes M \otimes M & \xrightarrow[(\psi_M^* \otimes \psi_M^*) \circ T]{} & M \otimes M \end{array}$$

commutes because the A^*-action ψ_M^* is natural.

Let $p: M \to M/M^{(n-1)}$ be the projection. If $u \in H_n^* \oplus Z_n^*$, then $\mu^* \theta(u) = 1 \otimes \theta u \bmod M \otimes M^{(n-1)}$, hence for any $v \in P$ we have

$$(1 \otimes p)\mu^* \theta(v) = 1 \otimes \lambda(v).$$

Now choose bases $h_1, h_2, \ldots, h_r$ for H_n^*, $z_1, z_2, \ldots, z_s$ for Z_n^*. Then P has a basis $\{v_i\} = \{h_1, \ldots, h_r, z_1, \ldots, z_s, Sq^1 z_1, \ldots, Sq^1 z_s\}$, and any $v \in N^{(n)}/N^{(n-1)}$ has an expression in the form $v = \sum_i a_i v_i$, where $a_i \notin A^* Sq^1$: initially we write v as

$$v = \sum c_{I,j} Sq^I h_j + \sum d_{J,k} Sq^J z_k,$$

where $Sq^I \notin A^* Sq^1$, and then for each Sq^J for which $j_r = 1$ we write $Sq^J z_k = Sq^{J'} Sq^1 z_k$ with $J' = (j_1, j_2, \ldots, j_{r-1})$. Let $m = \max\{\dim a_i\}$ and let $a_{i_1}, a_{i_2}, \ldots, a_{i_l}$ be the a_i's of dimension m. If $\lambda(v) = 0$ then $\theta(v) \in M^{(n-1)}$ and hence $0 = (1 \otimes p)\mu^* \theta(v) = \sum a_{i_j} \cdot t \otimes \lambda(v_{i_j}) + \sum b_k \cdot t \otimes m_k$ for some $m_k \in M$, $b_k \in A$ with $\dim b_k < m$. But we showed that $\lambda | P$ was a monomorphism, so the $\lambda(v_{i_j})$ are linearly independent, and hence $a_{i_j} \cdot t = 0$, all j. But $a_{i_j} \cdot t = 0$ implies $a_{i_j} \in A^* Sq^1$ by 20.43. This is a contradiction, so $\lambda(v) = 0$ implies $v = 0$. This completes the induction step. □

20.46. Corollary. *$H_*(MSO) \cong (A^*/A^* Sq^1)^* \otimes H \oplus A_* \otimes Z$, where*

$$H \cong \mathbb{Z}_2[q_1, q_2, \ldots],\ q_i \in H_{4i},$$

and Z is some $\mathbb{Z}_2$ vector space.

Now by 20.17 we have

$$\operatorname{Ext}^{**}_{A_*}(\mathbb{Z}_2, (A^*/A^* Sq^1)^*) \cong \operatorname{Ext}^{**}_{A_*//(A^*/A^* Sq^1)^*}(\mathbb{Z}_2, \mathbb{Z}_2).$$

But $A_*//(A^*/A^* Sq^1)^* \cong A_*//c(A^*/A^* Sq^1)^* \cong E(\zeta_1)$, the exterior algebra generated by ξ_1. We know how to give a resolution of $\mathbb{Z}_2$ over an exterior algebra; we get

$$\operatorname{Ext}^{**}_{A_*}(\mathbb{Z}_2, (A^*/A^* Sq^1)^*) \cong \mathbb{Z}_2[q_0].$$

20.47. Corollary. $\operatorname{Ext}^{**}_{A_*}(\mathbb{Z}_2, H_*(MSO)) \cong \mathbb{Z}_2[q_0] \otimes H \oplus Z$.

Now the q_0-towers occur only in every fourth dimension, so no d_r can go from one tower to another. Suppose $d_r(z) = q_0^r h$ for some $r \geq 2$, $z \in Z$, $h \in H$. Then $0 = d_r(q_0 z) = q_0 d_r(z) = q_0^{r+1} h$, which is a contradiction. Hence we see that all differentials are zero—for a slightly less trivial reason than before.

20.48. Theorem. ${}_2\pi_*(MSO) \cong \mathbb{Z}[n_1, n_2, \ldots] \oplus Z$ *as groups*, grade $n_i = 4i$. *The Hurewicz homomorphism $h: {}_2\pi_*(MSO) \to P(H_*(MSO; \mathbb{Z}_2))$ is a surjection.*

In fact it turns out that $\Omega^{SO}_*/\text{Tors}$ is a polynomial algebra on generators $N_i \in \Omega^{SO}_{4i}/\text{Tors}$; this can be proved by an argument similar to that in 20.27.

To calculate Ω^{SU}_* one has a commutative diagram

$$\begin{array}{ccc} H_*(MSO; \mathbb{Z}_2) & \xrightarrow{\psi} & A_* \otimes H_*(MSO; \mathbb{Z}_2) \\ \downarrow h & & \downarrow \alpha \otimes h \\ H_*(MSU; \mathbb{Z}_2) & \xrightarrow{\psi} & A_* \otimes H_*(MSU; \mathbb{Z}_2), \end{array}$$

where $\alpha: A_* \to A_*$ is the squaring map and h is an isomorphism of ungraded algebras obtained by factoring the isomorphism $f: H_*(MO;\mathbb{Z}_2) \to H_*(MU;\mathbb{Z}_2)$ composed with $H_*(MSO;\mathbb{Z}_2) \to H_*(MO;\mathbb{Z}_2)$ through $H_*(MSU;\mathbb{Z}_2) \to H_*(MU;\mathbb{Z}_2)$. It thus follows that

$$H_*(MSU;\mathbb{Z}_2) \cong \alpha(A^*/A^*Sq^1)^* \otimes H' \oplus \alpha(A_*) \otimes Z'.$$

We already know $\mathrm{Ext}_{A_*}^{**}(\mathbb{Z}_2, \alpha(A_*))$. In [54] Liulevicius constructs a resolution for $\alpha(A^*/A^*Sq^1)^*$ over A_* and hence determines

$$\mathrm{Ext}_{A_*}^{**}(\mathbb{Z}_2, H_*(MSU;\mathbb{Z}_2)).$$

Anderson, Brown and Peterson in [13] take this result, show that the only non-zero differential in the Adams spectral sequence is d_2 and then determine ${}_2\pi_*(MSU)$. The calculation of ${}_p\pi_*(MSU)$ for odd p is not difficult. It turns out, of course, that $\pi_*(MSU)$ has no odd torsion.

The next most difficult calculation was that of $\pi_*(M\mathrm{Spin})$. In [14] Anderson, Brown and Peterson show that

$$H^*(M\mathrm{Spin};\mathbb{Z}_2) \cong A^*/A^*(Sq^1, Sq^2) \otimes X \oplus A^*/A^*Sq^3 \otimes Y \oplus A^* \otimes Z$$

using an argument similar to, but much longer than, the one we just gave to show that

$$H^*(MSO;\mathbb{Z}_2) \cong A^*/A^*Sq^1 \otimes H^* \oplus A^* \otimes Z^*.$$

They then determine Ω_*^{Spin} by using KO-theory characteristic numbers rather than by using the Adams spectral sequence. Again it is not difficult to show Ω_*^{Spin} has no odd torsion.

About Ω_*^{Sp} we at least know that it only has 2-torsion and $\Omega_*^{Sp}/\mathrm{Tors} \cong \mathbb{Z}[x_1, x_2, \ldots]$, grade $x_k = 4k$—as groups but not as rings. N. Ray, D. Segal and others have computed Ω_n^{Sp} for small n.

Comments

The calculation of $H_*(MO;\mathbb{Z}_2)$ as A_*-comodule given here is based on that of Liulevicius in [53]. The proof of 20.27 follows closely that of Adams in [10]; for a different proof see [7]. The proofs of 20.32, 20.33 were given by Adams in a lecture in Manchester in 1967.

There is one feature of this chapter which strikes me as rather unsatisfactory. On the one hand we compute $\pi_*(MU)$ by using the Adams spectral sequences for $E = H(\mathbb{Z}_p)$, all primes p. On the other hand we remark that the "right" Hurewicz homomorphism to use for studying $\pi_*(MU)$ is $h: \pi_*(MU) \to K_*(MU)$ and proceed to justify this claim by proving 20.34. Clearly it would be much more satisfying to compute $\pi_*(MU)$ and prove 20.34 simultaneously by using the Adams spectral sequence for K. True, we do not know that this spectral sequence converges in general, but in this case it might well turn out that the spectral sequence was trivial or "almost trivial" in some appropriate sense which would guarantee convergence and make 20.34 obvious. Unfortunately I do not see how to make this approach work just now. Adams and Liulevicius in [12] have proved 20.34 by using the Adams spectral sequence of another spectrum BP.

References

1. J. F. Adams [7, 10, 12]
2. D. W. Anderson, E. H. Brown and F. P. Peterson [13, 14]
3. P. E. Conner and E. E. Floyd [31, 32]
4. A. Dold [33]
5. D. Husemoller [49]
6. A. Liulevicius [53, 54]
7. J. W. Milnor [61, 64]
8. J. W. Milnor and J. C. Moore [67]
9. R. E. Stong [83]
10. R. Thom [84, 85]
11. C. T. C. Wall [89, 90, 92]

Bibliography

1. Adams, J. F.: On the structure and applications of the Steenrod algebra. Comm. Math. Helv. **32**, 180–214 (1958).
2. Adams, J. F.: Vector fields on spheres. Ann. Math. **75**, 603–632 (1962).
3. Adams, J. F.: On the groups $J(X)$, I–IV, Topology **2**, 181–195; **3**, 137–171; 193–222; and **5**, 21–71 (1963, 1965 and 1966).
4. Adams, J. F.: Stable homotopy theory. Lecture Notes in Mathematics, Vol. 3. Berlin–Heidelberg–New York: Springer 1964.
5. Adams, J. F.: S. P. Novikov's work on operations on complex cobordism. Chicago: University of Chicago Mathematics Lecture Notes 1967.
6. Adams, J. F.: Lectures on generalised cohomology. In: Lecture Notes in Mathematics, Vol. 99, pp. 1–138. Berlin–Heidelberg–New York: Springer 1969.
7. Adams, J. F.: Quillen's work on formal groups and complex cobordism. Chicago: University of Chicago Mathematics Lecture Notes 1970.
8. Adams, J. F.: Stable homotopy and generalised homology. Chicago: University of Chicago Mathematics Lecture Notes 1971.
9. Adams, J. F.: A variant of E. H. Brown's representability theorem. Topology **10**, 185–198 (1971).
10. Adams, J. F.: Algebraic topology: a student's guide, London Mathematical Society Lecture Note Series 4. Cambridge: Cambridge University Press 1972.
11. Adams, J. F., Harris, A. S., Switzer, R. M.: Hopf algebras of cooperations for real and complex K-theory, Proc. London Math. Soc. (3) **23**, 385–408 (1971).
12. Adams, J. F., Liulevicius, A.: The Hurewicz homomorphism for MU and BP. J. London Math. Soc. **5**, 539–545 (1972).
13. Anderson, D. W., Brown, E. H., Peterson, F. P.: SU cobordism, KO-characteristic numbers and the Kervaire invariant. Ann. Math. **83**, 54–67 (1966).
14. Anderson, D. W., Brown, E. H., Peterson, F. P.: The structure of the spin cobordism ring. Ann. Math. **86**, 271–298 (1967).
15. Atiyah, M. F.: Thom complexes, Proc. London Math. Soc. (3) **11**, 291–310 (1961).
16. Atiyah, M. F.: K-theory. New York: Benjamin 1967.
17. Atiyah, M. F., Bott, R., Shapiro, A.: Clifford modules. Topology **3**, suppl. 1, 3–38 (1964).
18. Atiyah, M. F., Hirzebruch, F.: Vector bundles and homogeneous spaces. Proc. of Symposia in Pure Mathematics, Vol. 3, Differential Geometry. Amer. Math. Soc., 7–38 (1961).
19. Atiyah, M. F., Singer, I.M.: The index of elliptic operators on compact manifolds. Bull. Amer. Math. Soc. **69**, 422–433 (1963).
20. Barratt, M. G.: Track groups I, II. Proc. London Math. Soc. **5**, 71–106, 285–329 (1955).
21. Boardman, J. M.: Stable homotopy theory, mimeographed notes. Warwick, 1966.
22. Boardman, J. M., Vogt, R. M.: Homotopy-everything H-spaces. Bull. Amer. Math. Soc. **74**, 1117–1122 (1968).
23. Borel, A., Hirzebruch, F.: Characteristic classes and homogeneous spaces I, II, III. Amer. J. Math. **80**, 458–538; **81**, 315–382; and **82**, 491–504 (1958, 1959 and 1960).

24. Bott, R.: The stable homotopy of the classical groups. Ann. Math. **70**, 313–337 (1959).
25. Bott, R.: Lectures on $K(X)$. New York: Benjamin 1969.
26. Browder, W.: The Kervaire invariant of framed manifolds and its generalization. Ann. of Math. **90**, 157–186 (1969).
27. Brown, E. H.: Cohomology theories. Ann. Math. **75**, 467–484 (1962); with a correction, Ann. Math. **78**, 201 (1963).
28. Cartan, H.: Séminaire H. Cartan 1954–1955, Paris.
29. Cartan, H.: Séminaire H. Cartan 1959–1960, Paris.
30. Conner, P. E., Floyd, E. E.: Differentiable periodic maps. Berlin–Heidelberg–New York: Springer 1964.
31. Conner, P. E., Floyd, E. E.: The relation of cobordism to K-theories. Lecture Notes in Mathematics, Vol. 28. Berlin–Heidelberg–New York: Springer 1966.
32. Conner, P. E., Floyd, E. E.: Torsion in SU-bordism. Memoirs Amer. Math. Soc. **60**, 1966.
33. Dold, A.: Erzeugende der Thomschen Algebra $\mathscr{N}$. Math. Z. **65**, 25–35 (1956).
34. Dold, A.: Partitions of unity in the theory of fibrations. Ann. Math. **78**, 223–255 (1963).
35. Dold, A.: Halbexakte Homotopiefunktoren. Lecture Notes in Mathematics, Vol. 12. Berlin–Heidelberg–New York: Springer 1966.
36. Dold, A.: Lectures on algebraic topology. Berlin–Heidelberg–New York: Springer 1972.
37. Dyer, E.: The functors of algebraic topology. In: MAA Studies in Mathematics, Vol. 5, Studies in modern topology (P. J. Hilton, edit.). Englewood Cliffs, N.J.: Prentice Hall 1968.
38. Dyer, E.: Cohomology theories. New York: Benjamin 1969.
39. Dyer, E., Lashof, R. K.: A topological proof of the Bott periodicity theorems. Annali di Math. **54**, 231–254 (1961).
40. Eilenberg, S., Steenrod, N.: Foundations of algebraic topology. Princeton Mathematical Series 15. Princeton: Princeton University Press 1952.
41. Greenberg, M.: Lectures on algebraic topology. New York: Benjamin 1967.
42. Hilton, P. J.: An introduction to homotopy theory. Cambridge Tracts in Mathematics and Mathematical Physics 43. Cambridge: Cambridge University Press 1953.
43. Hilton, P. J.: Homotopy theory and duality. New York: Gordon and Breach 1965.
44. Hilton, P. J., Stammbach, U.: A course in homological algebra. Berlin–Heidelberg–New York: Springer 1971.
45. Hilton, P. J., Wylie, S.: Homology theory. Cambridge: Cambridge University Press 1960.
46. Hirzebruch, F.: Topological methods in algebraic geometry (3rd edit., translated). Berlin–Heidelberg–New York: Springer 1966.
47. Hopf, H.: Über die Abbildungen der dreidimensionalen Sphäre auf due Kugelfläche. Math. Ann. **104**, 637–665 (1931).
48. Hu, S.-T.: Homotopy theory. Pure and Applied Mathematics VIII. New York: Academic Press 1959.
49. Husemoller, D.: Fibre bundles. New York: McGraw-Hill 1966.
50. Kervaire, M. A., Milnor, J. W.: Groups of homotopy spheres I. Ann. Math. 77, 504–537 (1963).
51. Landweber, P. S.: Cobordism operations and Hopf algebras. Trans. Amer. Math. Soc. **129**, 94–110 (1967).
52. Lang, S.: Introduction to differentiable manifolds. New York: Interscience Publishers 1962.
53. Liulevicius, A.: A proof of Thom's theorem. Comm. Math. Helv. **37**, 121–131 (1962).
54. Liulevicius, A.: Notes on homotopy of Thom spectra. Amer. J. Math. **86**, 1–16 (1964).
55. Liulevicius, A.: On characteristic classes, mimeographed lecture notes from the Nordic Summer School in Mathematics. Aarhus, 1968.

56. Mac Lane, S.: Homology. Berlin–Heidelberg–New York: Springer 1963.
57. Mac Lane, S.: Natural associativity and commutativity. Rice University Studies **49**, No. 4, Houston, 1963.
58. Massey, W. S.: Algebraic topology: an introduction. New York: Harcourt, Brace and World 1967.
59. Maunder, C. R. F.: Algebraic topology. London: Van Nostrand 1970.
60. Milnor, J. W.: Differential topology, mimeographed notes. Princeton: Princeton University 1958.
61. Milnor, J. W.: The Steenrod algebra and its dual. Ann. Math. **67**, 150–171 (1958).
62. Milnor, J. W.: Lectures on characteristic classes, mimeographed notes. Princeton: Princeton University 1957.
63. Milnor, J. W.: On spaces having the homotopy type of a *CW*-complex. Trans. Amer. Math. Soc. **90**, 272–280 (1959).
64. Milnor, J. W.: On the cobordism ring Ω^* and a complex analogue, Part I. Amer. J. Math. **82**, 505–521 (1960).
65. Milnor, J. W.: On axiomatic homology theory. Pacific J. Math. **12**, 337–341 (1962).
66. Milnor, J. W.: Microbundles I. Topology **3**, suppl. 1, 53–81 (1964).
67. Milnor, J. W., Moore, J. C.: On the structure of Hopf algebras. Ann. Math. **81**, 211–264 (1965).
68. Mosher, R. E., Tangora, M. C.: Cohomology operations and applications in homotopy theory. New York: Harper and Row 1968.
69. Munkres, J. R.: Elementary differential topology. Princeton: Princeton University Press 1966.
70. Novikov, S. P.: The methods of algebraic topology from the viewpoint of cobordism theories. Izv. Akad. Nauk S.S.S.R., Seria Matematicheskia **31**, 855–951 (1967).
71. Puppe, D.: Homotopiemengen und ihre induzierten Abbildungen I. Math. Z. **69**, 299–344 (1958).
72. Quillen, D.: The Adams conjecture. Topology **10**, 67–80 (1970).
73. Ray, Nigel: Some results in generalized homology, *K*-theory and bordism. Proc. Camb. Phil. Soc. **71**, 283–300 (1972).
74. Ray, Nigel: The symplectic bordism ring. Proc. Camb. Phil. Soc. **71**, 271–282 (1972).
75. Ray, Nigel: Realizing symplectic bordism classes. Proc. Camb. Phil. Soc. **71**, 301–305 (1972).
76. Serre, J.-P.: Homologie singulière des espaces fibrés. Ann. Math. **54**, 425–505 (1951).
77. Serre, J.-P.: Cohomologie modulo 2 des complexes d'Eilenberg-MacLane. Comm. Math. Helv. **27**, 198–231 (1953).
78. Serre, J.-P.: Groupes d'homotopie et classes de groupes abéliens. Ann. Math. **58**, 258–294 (1953).
79. Spanier, E. H.: Function spaces and duality. Ann. Math. **70**, 338–378 (1959).
80. Spanier, E. H.: Algebraic topology. New York: McGraw-Hill 1966.
81. Steenrod, N.: The topology of fibre bundles. Princeton Mathematical Series 14. Princeton: Princeton University Press 1951.
82. Steenrod, N., Epstein, D. B. A.: Cohomology operations. Annals of Mathematics Studies 50. Princeton: Princeton University Press 1962.
83. Stong, R. E.: Notes on cobordism theory. Mathematical Notes. Princeton: Princeton University Press 1968.
84. Thom, R.: Quelques propriétés globales des variétés differentiables. Comm. Math. Helv. **28**, 17–86 (1954).
85. Thom, R.: Travaux de Milnor sur le cobordisme. Paris: Séminaire Bourbaki 1958/59.
86. Thomas, E.: Seminar on fibre spaces. Lecture Notes in Mathematics, Vol. 13. Berlin–Heidelberg–New York: Springer 1966.
87. Toda, H.: Composition methods in homotopy groups of spheres. Annals of Mathematics Studies 49. Princeton: Princeton University Press 1962.
88. Vick, James W.: Homology theory. New York and London: Academic Press 1973.

89. Wall, C. T. C.: Determination of the cobordism ring. Ann. Math. **72**, 292–311 (1960).
90. Wall, C. T. C.: A characterization of simple modules over the Steenrod algebra mod 2. Topology **1**, 249–254 (1962).
91. Wall, C. T. C.: Finiteness conditions for *CW*-complexes, Ann. Math. **81**, 56–69 (1965).
92. Wall, C. T. C.: Addendum to a paper of Conner and Floyd. Proc. Camb. Phil. Soc. **62**, 171–175 (1966).
93. Whitehead, G. W.: Generalized homology theories. Trans. Amer. Math. Soc. **102**, 227–283 (1962).
94. Whitehead, G. W.: Homotopy theory. Cambridge, Mass.: The M.I.T. Press 1966.
95. Whitehead, J. H. C.: Combinatorial homotopy I, II. Bull. Amer. Math. Soc. **55**, 213–245, 453–496 (1949).
96. Whitehead, J. H. C.: The mathematical works of J. H. C. Whitehead. London–New York: Pergamon Press 1962.

Index